POWER PLANTS AND POWER SYSTEMS CONTROL 2003

A Proceedings volume from the 5th IFAC Symposium,
Seoul, South Korea, 15 – 19 September 2003

Edited by

K.Y. LEE
Department of Electrical Engineering,
The Pennsylvania State University, PA, USA

and

M.C. SHIN
Information and Computer Engineering,
Sungkyunkwan University, Suwon-city, Korea

(In three volumes)

Volume 1

Published for the

INTERNATIONAL FEDERATION OF AUTOMATIC CONTROL

by

ELSEVIER LTD

ELSEVIER Ltd
The Boulevard, Langford Lane
Kidlington, Oxford OX5 1GB, UK

Elsevier Internet Homepage
http://www.elsevier.com

Consult the Elsevier Homepage for full catalogue information on all books, journals and electronic products and services.

IFAC Publications Internet Homepage
http://www.elsevier.com/locate/ifac

Consult the IFAC Publications Homepage for full details on the preparation of IFAC meeting papers, published/forthcoming IFAC books, and information about the IFAC Journals and affiliated journals.

First edition 2004

Library of Congress Cataloging in Publication Data

A catalogue record for this book is available from the Library of Congress

British Library Cataloguing in Publication Data

A catalogue record for this book is available from the British Library

ISBN 0-08-044210 2
ISSN 1474-6670

Printed in Great Britain

To Contact the Publisher

Elsevier welcomes enquiries concerning publishing proposals: books, journal special issues, conference proceedings, etc. All formats and media can be considered. Should you have a publishing proposal you wish to discuss, please contact, without obligation, the publisher responsible for Elsevier's industrial and control engineering publishing programme:

Christopher Greenwell
Publishing Editor
Elsevier Ltd
The Boulevard, Langford Lane
Kidlington, Oxford
OX5 1GB, UK

Phone: +44 1865 843230
Fax: +44 1865 843920
E.mail: c.greenwell@elsevier.com

General enquiries, including placing orders, should be directed to Elsevier's Regional Sales Offices – please access the Elsevier homepage for full contact details (homepage details at the top of this page).

5th IFAC SYMPOSIUM ON POWER PLANTS AND POWER SYSTEMS CONTROL 2003

Sponsored by
International Federation of Automatic Control (IFAC)
Technical Committee on Power Plants and Power Systems Control

Organized by
KIEE (The Korean Institute of Electrical Engineers)
on behalf of ICASE, NMO of IFAC

Co-sponsored by
CIGRE
IEEE Power Engineering Society

Supported by
KEPCO (Korea Electric Power Corporation)
Korea Southern Power Co., Ltd
Korea Western Power Co., Ltd
Korea Hydro & Nuclear Power Co., Ltd
Korea East-West Power Co., Ltd
Korea Midland Power Co., Ltd
Korea South-East Power Co., Ltd
Korea Power Exchange
KOPEC (Korea Power Engineering Corporation, Inc.)
KERI (Korea Electrotechnology Research Institute)
KEPRI (Korea Electric Power Research Institute)
EESRI (Electrical Engineering & Science Research Institute)
Korea Electrical Contractors Association

International Programme Committee (IPC)

National Organizing Committee (NOC)

Chair: Dr. Rimtaig Lee
Vice-Chair: Prof. Young-Hyun Moon
Editors: Prof. Kwang Y. Lee
Prof. Myung-Chul Shin
Secretary: Prof. Kyung-Bin Song

Jae-Seok Choi
Gilsoo Jang
Jae Won Jang
Heui Tai Kang
Baik Kim
Chul-Hwan Kim
Jae-Chul Kim
Jae-Eon Kim
Jin-O Kim
Jung-Hoon Kim
Kwang-Ho Kim
Byung-Ha Lee
Kwang-Ho Lee
Seung-Jae Lee
Young-Hwan Moon
Yong-Taek Oh
June Ho Park
Dae-Seok Rho
Kyoung-Soo Ro
Hyo-Seong Song
Kyung-Bin Song

Steering Committee

Chair: Prof. Hee-Ro Kwak
Co-Chairs: Dr. Tay-Ho Jung, Mr. Kap-Koo Yoon

Jeong-Shik Ahn
Young-Sik Baek
Jong-Woong Choe
Jin-Boo Choo
Yong-Chul Chung
KabJu Hwang
Hoyong Kim
Il-Nam Kim
In-Sup Kim
Kern-Joong Kim
Tae-Won Kwon
InKyu Lee
Jin-Moon Lee
Seung-Chul Lee
Tae-Kyoo Oh
Dong-Wook Park
Jin Bae Park
Sang Doug Park
Jang Sun Seo
Hee Seung Shin
Joong-Rin Shin

Sponsoring Committee:
Chair: Mr. In-Seok Koh
Co-Chairs: Mr. In Soon An, Mr. Young-Ki Baek,
Members:
Mr. Yang Uoo Choe
Dr. Rimtaig Lee
Mr. Yong-Oh Lee
Mr. Young Cheol Lee
Dr. Pyung-Sam Ji
Dr. Tay-Ho Jung
Mr. Chang-Joon Kim
Mr. Chun Chae Kim
Mr. Young-Joon Kim
Mr. Yung-Chul Kim
Dr. Young-Han Kwun
Mr. Haeng-Soon Yoon
Dr. Meng-Hyun Yoon

Secretary: Prof. Jung-Hoon Kim

Local Technical Program Committee:

Chair: Prof. Jong-Keun Park
Co-Chairs: Dr. Young-Chang Kim, Prof. Sae-Hyuk Kwon, Dr. Hong-Woo Rhew

Members
Sang-Hoon Chang
Bong-Hee Kim
Kyu-Ho Kim
Moon-Duk Kim
Se-Ho Kim
Seog-Joo Kim
Byong-Jun Lee
Hahk Sung Lee
Heung-Jae Lee
Jong-Beom Lee
Ick-Hun Lim
Seung-Il Moon

PREFACE

This Proceedings volume of the IFAC Symposium on Power Plants and Power Systems Control 2003 held in Seoul, South Korea contains 203 papers presented in 32 oral sessions and a poster session. In addition to these contributed papers, plenary speeches, papers in three panel sessions and a summary of a tutorial session were presented. All of the papers are original and were provided by specialists from 26 different countries.

The papers duly presented at the Symposium are archived in this Proceedings volume published by Elsevier Ltd., Oxford, UK. Upon recommendation, some of the papers will also be published in the IFAC Journals after going through a formal review process.

All this is made possible by the efforts of the contributors and we would like to take this opportunity to thank all delegates for participating. We would also like to thank the organizers, sponsors, co-sponsors and the industrial partners who supported the meeting financially.

The Editors:
Kwang Y. Lee, The Pennsylvania State University, USA
Myong-Chul Shin, Sungkyunkwan University, Korea

CONTENTS

VOLUME 1

PLENARY PAPER

TUTORIAL PAPER

ORAL PAPERS

Power Plant Operation and Control I

Power System Modeling I

Power System Monitoring and Diagnosis

Distribution System Operation

Electricity Pricing/Bidding Schemes

Power System Control I

Power System Protection

VOLUME 2

ORAL PAPERS (continued)

Operation of Competitive Electricity Markets I

Power Plant Modeling I

Transmission/Distribution Apparatus

Power System Stability I

Unit Commitment

Simulators/Simulation Methods

Power System Restoration

Voltage and Reactive Power Control

Power System Stabilizer

Power Plant Operation and Control VI

Power System Analysis I

Power System Modeling II

Operation of Competitive Electricity Markets II

Power Plant Modeling II

VOLUME 3

POSTER PAPERS (continued)

ELSEVIER
IFAC
PUBLICATIONS
www.elsevier.com/locate/ifac

AMALGAMATION OF ADAPTIVE CONTROL AND AI TECHNIQUES: APPLICATIONS TO GENERATOR EXCITATION CONTROL

O. P. Malik
University of Calgary
Calgary, Canada

Abstract: Adaptive control techniques can provide excellent performance and improve the dynamic performance of plants with non-linear characteristics and variable parameters by allowing the controller parameters to be adjusted as the operating conditions change. The robustness of such controllers can be improved by employing artificial intelligence (AI) techniques. It is possible to either implement the entire algorithm using AI techniques or integrate the analytical and AI techniques such that some functions are performed using the analytical approach while the rest are performed using AI techniques. Examples of successful implementation of various approaches as applied to generator excitation control are given. Copyright© IFAC 2003

Keywords: Generator Excitation Control; PSS; Adaptive Control; AI Control

1. Introduction

Adaptive control can be described as the changing of controller parameters on-line based on the changes in system operating conditions. Whenever an adaptive controller detects changes in system operating conditions, it responds by determining a new set of control parameters.

An adaptive controller based on analytical techniques can provide excellent performance and improve the dynamic performance of the plant by allowing the parameters of the controller to adjust as the operating conditions change. However, proper care needs to be taken in the design of the analytical algorithms to make them robust, especially under large disturbances.

The controller robustness can be improved by employing artificial intelligence (AI) techniques such as fuzzy logic and neural networks (NNs). It is possible that either the entire algorithm may be implemented using AI techniques or the analytical and AI techniques be integrated such that some functions are performed using analytical approach while the rest are performed using AI techniques. Examples of successful implementation of all three approaches, i.e. purely analytical, purely AI and integrated are given in this presentation.

2. Adaptive Controllers

The adaptive control theory provides a possible way to solve many of the problems associated with the control of non-linear time-varying systems, such as power systems. Two distinct approaches – direct adaptive control and indirect adaptive control – can be used to control a plant adaptively. In the direct control, the parameters of the controller are directly adjusted to reduce some norm of the output error. In the indirect control, the parameters of the plant are estimated as the elements of a vector at any instant k, and the parameters vector of the controller is adapted based on the estimated plant vector.

A general configuration of the indirect adaptive control as a self-tuning controller is shown in Figure 1. At each sampling instant, the input and output of the generating unit are sampled and a plant model is obtained by some on-line identification algorithm to represent the dynamic behaviour of the generating unit at that instant in time. It is expected that the model obtained at each sampling instant can track the system operating conditions.

The required control signal is computed based on the identified model. Various control techniques can be used to compute the control. All control algorithms assume that the identified model is the true mathematical description of the controlled system. Various approaches are illustrated by application to the development of an adaptive power system stabilizer (PSS) to improve damping and stability of an electric generating unit.

Power systems are non-linear and operate over a wide range. For example, the gain of the plant increases with generator load. Also, the phase lag of the plant increases as the a.c. system becomes stronger. Due to the non-linear characteristics, wide operating conditions and unpredictability of perturbations in a power system, the conventional PSS (CPSS), a linear controller, generally cannot maintain the same quality of performance under all conditions of operation.

To further improve the performance and stability of the power system, various other approaches using the linear quadratic optimal control, H-infinity, variable structure, rule based, and AI techniques [1-9] have been proposed in the literature to design a fixed parameter PSS. One common feature of all fixed parameter controllers is that the design is done off-line.

3. Analytical Approach Based Adaptive PSS

In the analytical approach to the design of an adaptive PSS, sampled data design techniques are used to compute control in the following way:

Select a sampling frequency, f_s, about ten times the normal frequency of oscillation to be damped.

Each sampling interval $T(= 1/f_s)$, update the system model parameters. A number of identification algorithms, in recursive form, have been developed to determine the transfer function of the controlled plant in the discrete domain.

Use the updated estimates of the parameters to compute the control based on the control strategy chosen. Various control strategies, among them optimal, minimum variance, pole-zero assignment, pole assignment, pole shift, have been proposed.

Extensive amount of work has been done to develop and implement an adaptive PSS based on the above strategy. Such a PSS can adjust its parameters on-line according to the environment in which it works and can provide good damping over a wide range of operating conditions of the power system.

3.1 Self-adjusting Pole-shift Control Strategy

The common procedure in process control is to compare the actual measured values of the output with the desired values and the difference, the control error, is fed as input to the process via a regulator and an actuator. Various criteria are available for the computation of the control.

In the self-adjusting pole-shift control strategy, the closed-loop poles of the controlled system are shifted towards the center in the z-plane by a factor less than one. This factor is varied on-line to always produce maximum damping contribution without exceeding the control limits. Using this technique, the desired control law is obtained as:

$$u(t) = f[\theta_s(t), y(t), u(t-T)] \tag{1}$$

where: $\theta_s(t)$ is the system parameter vector

$y(t)$ is the output vector $[y(t), y(t-T) \ldots]'$

$U(t-T)$ is the control vector $[u(t-T)\ u(t-2T) \ldots]'$

$'$ denotes the transpose, and

T is the sampling period.

If the parameter vector is known, control to meet specific performance criterion can be computed directly. However, the dynamics of a complex non-linear system vary with time depending upon the operating conditions, disturbances, etc. To determine the desired control, such a system may be modeled by a linear low order discrete model with time-varying parameters. An on-line real-time adaptive controller then consists of two stages: (i) identification stage for on-line model; and (ii) control stage for the computation of control using the identified model parameters.

3.2 System Model

The generating unit is described by a discrete ARMA model of the form:

$$A(z^{-1})\,y(t) = B(z^{-1})\,u(t) + e(t) \tag{2}$$

where $A(z^{-1})$ and $B(z^{-1})$, polynomials in the delay operator z^{-1}, are of the form:

$$A(z^{-1}) = 1 + a_1 z^{-1} + \Lambda + a_i z^{-i} + \Lambda + a_{n_a} z^{-n_a} \tag{3}$$

$$B(z^{-1}) = b_1 z^{-1} + \Lambda + b_i z^{-i} + \Lambda + b_{n_b} z^{-n_b} \tag{4}$$

$$n_a \geq n_b$$

the variables y(t) and u(t) are the system output and system input, respectively, and e(t) is assumed to be a sequence of independent random variables with zero mean.

3.3 System Parameter Estimation

The control is computed based on the identified model parameters, a_i and b_i. Thus to compute the control appropriate to the varying conditions the system parameters have to be estimated on-line. The correctness of the identification determines the preciseness of the identified model that tries to reflect the true system. For a time-varying system the tracking ability of the identification method is very desirable.

An on-line estimate of the system parameters is obtained by providing in the regulator a mathematical model having a desired structure describing the actual process. Such a model may be expressed as:

$$\hat{y}(t) = g[\theta_m, \xi(t)] \quad (5)$$

where: $\hat{y}(t)$ is the predicted (estimated) value of the system output.

θ_m is the model parameter vector, and

$\xi(t)$ is the information known at the time of prediction.

The model parameter vector may either be constant, θ_m, or be a function of time, $\theta_m(t)$. For the model to track the system dynamics, i.e. tune itself to the system, its parameters must be updated continuously at an interval that is consistent with the time constants of the system.

Several methods can be used to obtain an estimate for the model parameter vector, $\theta_m(t)$ [10]. A commonly used technique of achieving a continuous tracking of the system behaviour is the recursive least-squares (RLS) parameter estimation technique. It minimizes the square of the error between the actual system output and the model output, and the estimated parameter vector, $\hat{\theta}_m(t)$, is given by:

$$\hat{\theta}_m(t) = h[\hat{\theta}_m(t-T), P(t), \xi(t)] \quad (6)$$

where P(t) is the covariance matrix of the error of estimates. In general terms it contains the entire history of the process.

To enhance the ability of the identifier to track the operating conditions of the actual system, a forgetting factor is used to discount the importance of the older data. It can be chosen as a constant or a variable. A variable forgetting factor, employed to improve the tracking ability especially under large disturbances, is calculated on-line every sampling interval [11].

3.4 Self-tuning Regulator

The parameters of the model of a given structure, estimated as above, are used in the control algorithm to compute the updated control. Description of the complete algorithm, called the 'self-tuning regulator' is given in [11] and a block-diagram of the regulator is shown in Fig. 1. Because the control is based on the estimated model parameter vector, $\hat{\theta}_m(t)$, eqn. (1) now becomes:

$$u(t) = f[\hat{\theta}_m(t), y(t), U(t-T)] \quad (7)$$

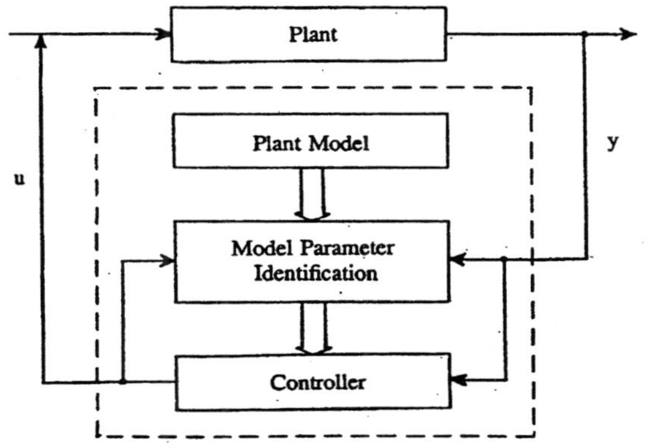

Figure 1 Block diagram of a Self-Tuning Controller

3.5 Performance Studies with Pole-shifting Control PSS

Performance of the self-tuning adaptive controller based on the pole-shifting control algorithm has been investigated by conducting simulation studies on single machine [11, 12] and multi-machine system [13], on a single machine [14] and on a multi-machine physical model [15] in the laboratory and on a 400MW thermal machine under fully loaded conditions connected to the system [16].

Results of a simulation study to demonstrate the effect of the adaptive PSS (APSS) on the transient stability margin are shown in Table 1. With the single machine infinite bus system initially operating at 0.95 pu power, 0.9 pf lag, a three phase to ground fault was applied near the sending end of one transmission line. It can be seen from Table 1 that the APSS provides the largest maximum clearance time.

Table 1 Transient stability margin results

	Without PSS	With CPSS	With APSS
Maximum Clearing Time	120 ms	150 ms	165 ms

This APSS was implemented on a microprocessor and tested in real-time on a physical model of a single-machine infinite bus system. With the system operating at a stable operating point, the APSS was applied and the torque reference increased gradually to the level, P = 1.307 pu, pf = 0.95 lead, v_t

= 0.950 pu. At this load, the system was still stable with the APSS.

At 5s, Fig. 2, the APSS was replaced by the CPSS. After the switch over, the system began to oscillate and diverge, which means that the CPSS is unable to keep the system stable at this load level. At about 25s, the APSS was switched back to control the unstable system and the system came under control very quickly as shown in Fig. 2. This test demonstrates that the ASPSS can provide a larger dynamic stability margin than the CPSS. Also, more power can be transmitted with the help of the APSS if an overload operation is necessary under certain circumstances.

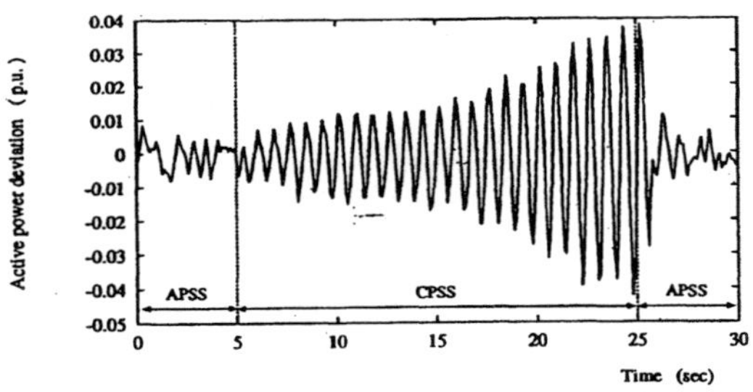

Figure 2 Dynamic stability improvement by the APSS

4. Adaptive PSS with ANN Predictor and ANN Controller

Identification of the power plant model using an on-line recursive identification technique is a computationally extensive task. Neural networks offer the alternative of a model-free method. An adaptive neural network based controller using indirect adaptive control method has been developed. It combines the advantages of neural networks with the good performance of the adaptive control. This controller employs the learning ability of the neural networks in the adaptation process and is trained each sampling period.

The controller consists of two sub-networks as shown in Fig. 3. One network is an adaptive neuro-identifier (ANI) that identifies the power plant in terms of its internal weights and predicts the dynamic characteristics of the plant. It is based on the inputs and outputs of the plant and does not need the states of the plant. The second sub-network is an adaptive neuro-controller (ANC) that provides the necessary control action to damp the oscillations of the power plant.

The success of the control algorithm depends upon the accuracy of the identifier in predicting the dynamic behaviour of the plant. The ANI and ANC are initially trained off-line over a wide range of operating conditions and a wide spectrum of possible disturbances. After the off-line training stage, the controller is hooked up in the system. Further training of the ANI and ANC is done on-line every sampling period. On-line training enables the controller to track the plant variations as they occur and to provide control signal accordingly.

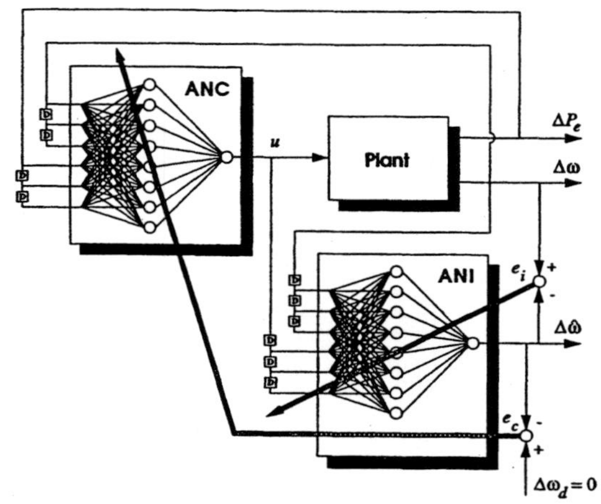

Figure 3 Controller structure for single-machine study

Employing a feedforward multilayer network in each of the two sub-networks, an adaptive neural network based PSS has been built [17]. The two networks are trained further in each sampling period using an on-line version of the back-propagation algorithm. The errors used to train the ANI and ANC are both scalar and the learning is done only once in each sampling period for each of the two sub-networks. This simplifies the training algorithm in terms of the computation time.

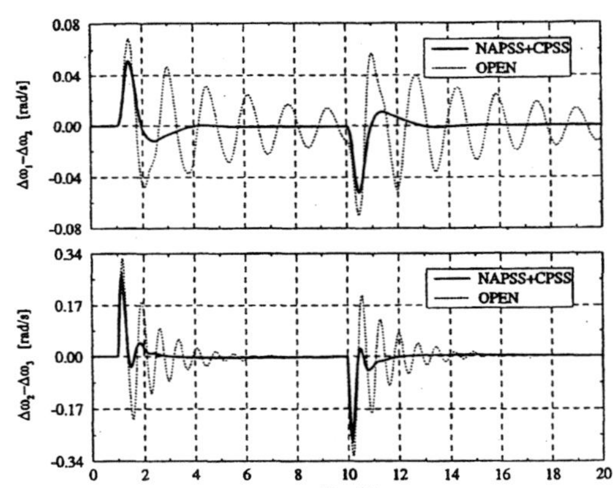

Figure 4 System response with NAPS installed on generators G1 and G3 and CPSS on G2 , G4 ,G5

System performance in response to a three-phase to ground fault on one circuit of a double-circuit transmission line in a five machine interconnected network is shown in Fig. 4. The adaptive neural network based PSSs were installed on two generators and CPSSs were installed on the other three generators. It can be seen that both the local mode and the interarea mode oscillations are damped effectively.

5. Adaptive Network Based Fuzzy Logic Controller

The characteristics of fuzzy logic and neural networks complement each other in respect of their pros and cons. That offers the possibility of using a hybrid neuro-fuzzy approach in the form of an adaptive network based Fuzzy Logic controller (FLC) whereby it is possible to take advantage of the positive features of both Fuzzy Logic and neural networks. Such a system can automatically find an appropriate set of rules and membership functions [18].

5.1 Architecture

In the neuro-fuzzy controller, the system is implemented in the framework of a network architecture. Considering the functional form of the fuzzy logic controller, Fig. 5, it becomes apparent that the FLC can be represented as a five layer feedforward network, in which each layer corresponds to one specific function with the node functions in each layer being of the same type. With this network representation of the fuzzy logic system, it is straightforward to apply the back-propagation or a similar method to adjust the parameters of the membership functions and inference rules.

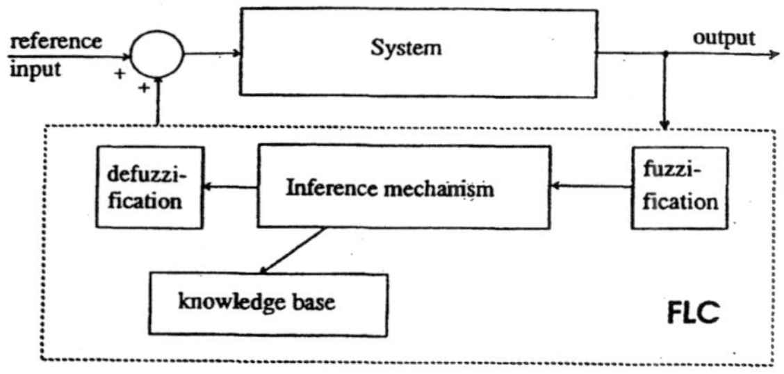

Figure 5 Basic structure of fuzzy logic controller

In this network, the links between the nodes from one layer to the next layer only indicate the direction of flow of signals and part or all of the nodes contain the adjustable parameters. These parameters are specified by the learning algorithm and should be updated according to the given training data and a gradient-based learning procedure to achieve a desired input/output mapping. It can be used as an identifier for non-linear dynamic systems or as a non-linear controller with adjustable parameters.

5.2 Training and Performance

Because the neuro-fuzzy controller has the property of learning, fuzzy rules and membership functions of the controller can be tuned automatically by the learning algorithm. Learning is based on the error in the controller output. Thus it is necessary to know the error that can be evaluated by comparing the output of the neuro-fuzzy controller and a desired controller.

To train this controller as an adaptive network based fuzzy PSS (ANF PSS), training data was obtained from a self-optimizing pole-shifting APSS. Training was performed over a wide range of operating conditions of the generating unit including various types of disturbances. Based on earlier experience, seven linguistic variables for each input variable were used to get the desired performance.

Extensive simulation [19] and experimental studies with the ANF PSSS show that it can provide good performance over a wide operating range and can significantly improve the dynamic performance of the system.

5.3 Self-learning ANF PSS

In the above case the ANF PSS was trained by data obtained from a desired controller. However, in a general situation, the desired controller may not be available. In that case, the neuro-fuzzy controller can be trained using a self-learning approach [20].

In the self-learning approach two neuro-fuzzy systems are used in a manner similar to Figure 10, one acting as the controller and the other acting as the predictor. The plant identifier can compute the derivative of the plant's output with respect to the plant's input by means of the back-propagation process illustrated by the line passing through the forward identifier and continuing back through the neuro-fuzzy controller that uses it to learn the control rule.

The self-learning ANF PSS was initially trained off-line on a power system simulation model over a wide range of operating conditions and disturbances. Electric power deviation and its integral were used as the input to the stabilizer. The ANF PSS with the parameters, membership functions and inference rules obtained from the off-line training procedure, was implemented on a DSP mounted on a PC and its performance was evaluated on a physical model of a power system in the laboratory. A digital CPSS was also implemented in the same environment on the DSP board for comparative studies.

Out of the various tests, results for a 0.25 pu step decrease in the input torque reference applied at 1s and removed at 9s with the generator operating at 0.9 pu power, 0.85 pf lag and 1.10 pu V_t are shown in Fig.

6. The ANF PSS provides a consistently good performance for either of the two disturbances.

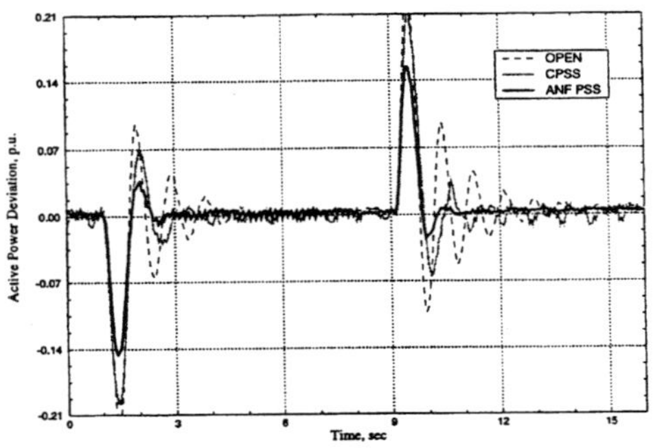

Figure 6 Comparison of ANF PSS and CPSS responses to a 0.25 p.u. step torque disturbance (P= 0.9 *p.u.*, 0.85 power pf lag)

Simulation studies on a single machine connected to a constant voltage bus and on a multi-machine power system [21] and experimental studies on a physical model of a power system have demonstrated the effectiveness of the ANN PSS in improving the performance of a power system over a wide operating range and a broad spectrum of disturbances.

5.4 Neuro-fuzzy Controller Architecture Optimization

Adaptive fuzzy systems offer a potential solution to the knowledge elicitation problem. The controller structure, expressed in terms of the number of membership functions and the number of inference rules, is usually derived by trial and error. The number of inference rules has to be determined from the standpoint of overall learning capability and generalization capability.

The above problem can be resolved by employing a genetic algorithm to determine the structure of the adaptive fuzzy controller. By employing both genetic algorithm and adaptive fuzzy controller, the inference rules parameters can be tuned and the number of membership functions can be optimized at the same time.

6. Adaptive PSS with ANN Identifier and Pole-shift Control

A self-tuning adaptive PSS described above can improve the dynamic performance of the synchronous generator by allowing the parameters of the PSS to adjust as the operating conditions change. However, proper care needs to be taken in the design of the RLS algorithm for identification to make it stable, especially under large disturbances.

It is possible to overcome the problems of RLS identification by using an ANN for identifying the system model parameters. An analytical technique, such as the pole-shift control, can be retained to compute the control signal. One approach, using a radial basis function (RBF) network for model parameter identification, is described below. The APSS shown in Figure 6 now consists of an ANN identifier and the pole-shifting control algorithm described above.

The RBF network, Fig. 7, is used to identify the system model parameters, a_i, b_i, eqns. (3) and (4). The network consists of three layers: the input, hidden and output layers. The input vector is:

$$V(t) = [\Delta P_e(t-T),\ \Delta P_e(t-2T),\ \Delta P_e(t-3T),\ u(t-T),\ u(t-2T),\ u(t-3T)] \tag{8}$$

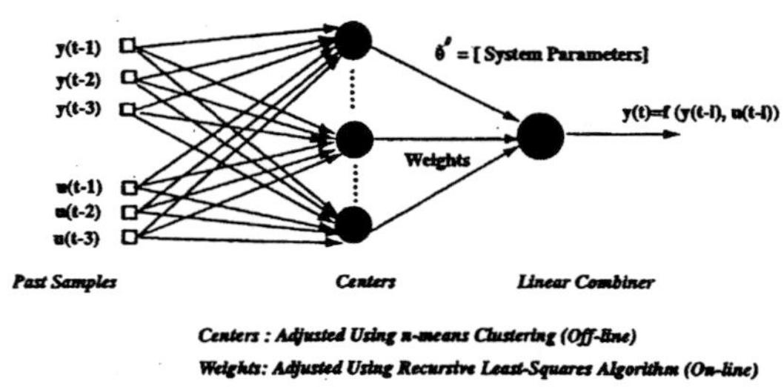

Figure 7 Radial basis function network model

Each of the six input variables is assigned to an individual node in the input layer and passes directly to the hidden layer without weights. The hidden nodes, called the RBF centers, calculate the Euclidean distance between the centers and the network input vector. The result is passed through a widely used Gaussian function characterized by a response which has a maximum value of 1 when the distance between the input vector and the center is zero. Thus a radial basis neuron acts as a detector which produces 1 whenever the input vector is identical to the center (active neuron). The other neurons with centers quite different from the input vector will have outputs near zero (non-active neurons).

The connections between the hidden neurons and the output node are linear weighted sums as described by the equation:

$$y = \sum_{i=1}^{nh} \theta' \exp.\left(-\frac{\|p-c_i\|^2}{\sigma^2} \right) \tag{9}$$

where c_i, σ, θ' and nh are the centers, widths, weights and the number of hidden layer neurons, respectively.

To make the proposed RBF identifier faster for on-line applications, the hidden layer is created as a competitive layer wherein the center closest to the input vector becomes the winner and all the other non-active centers are deactivated. Also, the scalar weights were modified as a vector θ′ whose size equals the size of the input vector. The weight vector is given by:

$$\theta'(t) = \left[a_1^{'}\ a_2^{'}\ a_3^{'}\ b_1^{'}\ b_2^{'}\ b_3^{'}\right] \qquad (10)$$

Linearizing the output of the RBF, y(t) = f [y(t-1), u(t-1)], by Taylor series expansion at each sampling instant, a one-to-one relationship between the weight vector θ′ and the system model parameters θ'_m (eqn. 7) can be obtained. These parameters are then used in computing the control signal.

The RBF identifier was first trained off-line to choose appropriate centers using data collected at a number of operating points for various disturbances. The n-means clustering algorithm used for training yielded 15 centers for the RBF model. After the off-line training, the weights (system parameters) were updated on-line to obtain the appropriate control signal using the pole-shifting controller. A 100ms sampling period was chosen for digital implementation.

Results of an experimental study for a 0.10 pu decrease in torque reference applied at 10s and removed at 20s, with the generator operating at 0.6 pu power, 0.92 pf lead and V_t of 0.99 pu are shown in Fig. 8. It can be seen that the APSS can provide a well damped response.

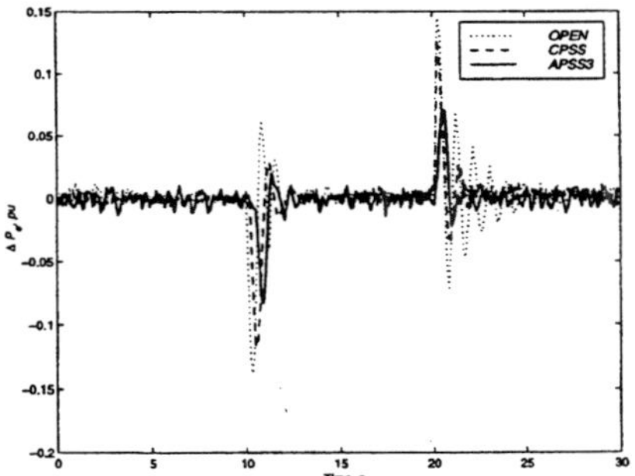

Figure 8 ΔP_e response for 0.1 p.u. input torque reference step change with APSS

7. Adaptive PSS with Fuzzy Logic Identifier and Pole-Shift controller

Takagi-Sugeno (TS) fuzzy systems have been successfully employed in the design of stabilization control of non-linear systems.

A non-linear plant can be represented by a set of linear models interpolated by membership functions of a TS fuzzy model. Although the TS system identifier is a NARMAX model, at each sample an average linear discrete ARMA model can be determined to identify the controlled plant according to the current active rules. This ARMA model can be used to determine the control signal by the pole-shifting control strategy. Using this approach, a self-tuning adaptive controller has been developed and applied as a PSS [22].

The proposed single-input single-output TS model used for the identification of dynamic systems is composed of fuzzy rules , the consequent part of which provides the rule output at time k based on the past inputs and past outputs with fuzzy sets designed in universe of discourse. The consequent part of the rule then identifies the parameters of a desired order discrete model of the plant. Two parallel on0line learning procedures, one each for the identification of premise and consequent parameters, are used to track the plant in real-time [23].

In the proposed TS system for generating unit identification, two input signals, the apst control input, u(k-1), and the past generator speed output, y(k-1), are used to identify a 3rd. order model of the plant. The output at sample k is the estimated generator speed output, $\hat{y}(k)$. The TS system is trained by using the steepest descent algorithm for the premise parameters and RLS algorithm for the consequent parameters using the error of the system output and the estimated TS output. Initially a set of three equally spaced membership functions, over the normalized universe, are used for the inputs of the system.

The response of the system with the TS system based identifier and pole-shift controller based adaptive PSS has been stucied for various disturbances at different operating conditions. One illustrative result for a three phase to ground fault is shown in Fig. 9.

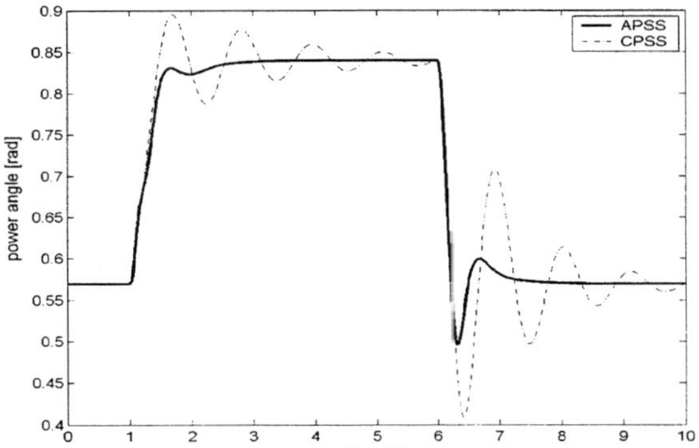

Figure 9. Three phase to ground fault at The middle of one transmission line and successful reclosure (P=0.95 pu, 0.9 pf lag)

8. Adaptive PSS with RLS Identifier and Fuzzy Logic Control

Fuzzy logic controllers (FLCs) have attracted considerable attention as candidates for novel computational systems because of the advantages they offer over the conventional computational systems. They have been successfully applied to the control of non-linear dynamic systems, especially in the field of adaptive control, by making use of on-line training.

A self-learning adaptive fuzzy logic controller has been developed. Only the inputs and outputs of the plant are measured and there is no need to determine the states of the plant. Using on-line training by the steepest descent method and the identified system model, the adaptive FLC is able to track the plant variations as they occur and compute the control.

In the proposed controller, a discrete model of the plant is first identified using the RLS parameter identification method. This allows a continuous tracking of the system behaviour.

The control learning is based on the prediction of the identified model. The identified model output is used as input to the Mamdani –type PD controller [24]. The center points of the controller inputs are updated [23] by treating them exactly the same as the weights of an NN and by using the steepest descent algorithm with chain rule.

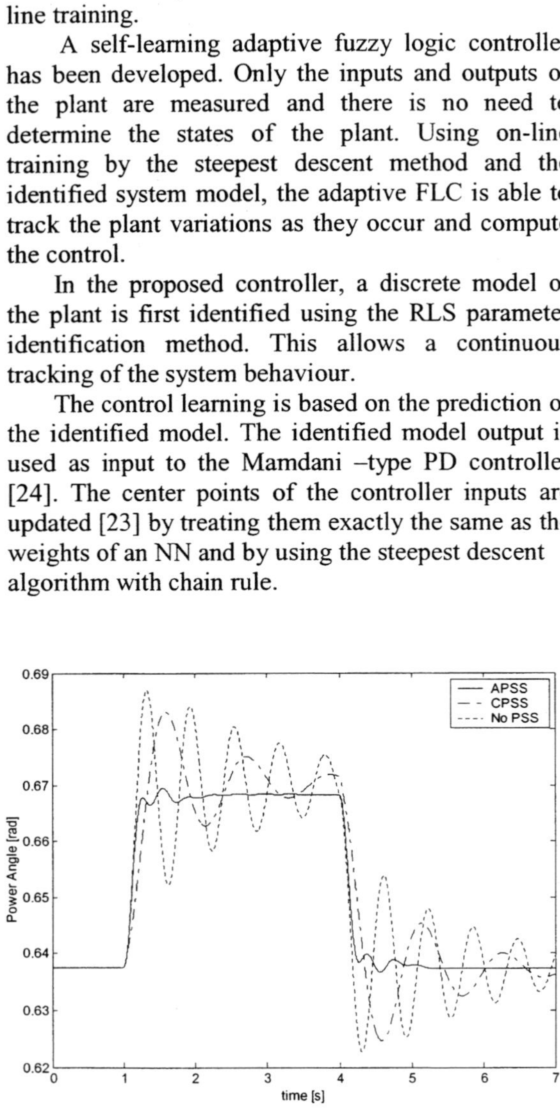

Figure 10. Response to a 0.05 *p.u.* step increase in torque and return to initial condition (P= 0.95 *p.u.*, 0.9 power pf lag)

The proposed adaptive FLC has been applied as an adaptive fuzzy PSS (AFPSS) [25]. For the AFPSS, the generating unit is identified as a 3rd. order model and two input signals, the generator speed deviation and its derivative, with an initial set of seven equally spaced membership functions over the normalized universe of discourse. The output, the supplementary control signal, also having seven membership functions, is added to the AVR summing junction. A number of simulation studies have been performed for various disturbances at different operating conditions. An illustrative result for a 0.05 pu increase in torque and return to initial condition, shown in Fig. 10, demonstrates the performance of this AFPSS.

9. Model Reference Adaptive Controller Using FLC

The objective of an MRAC system is to update the controller parameters such that the closed-loop system maintains a performance specified by a reference model. It requires a suitable model, an adaptive mechanism and a controller.

An adaptive PSS based on the MRAC principle (Fig. 11) has been developed [26]. In the adaptation mechanism, error between the reference model output and the actual plant output is used to update the controller parameters by a gradient method. An FLC with self-learning capability is used to adapt the system performance to track the reference model. Two inputs, generator speed deviation and its derivative, and the supplementary control output , each have seven membership functions. The FLC uses the Mamdani-type fuzzy PD rule base [24]. Updating the center points of the controller input membership functions, i.e. the weights of the fuzzy controller, using the backpropagation algorithm provides it with a self-learning capability. It can thus adapt the system performance to track the reference model.

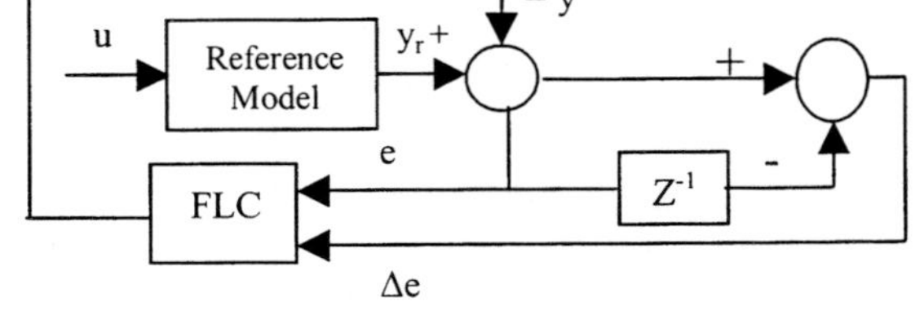

Figure 11. System configuration

Determination of a suitable reference model is an important requirement in a successful MRAC. If the plant characteristics and control constraints do not allow the plant to maintain the performance specified by the reference model, the whole system will have a poor performance. In this case, an off-line model identification has been used to obtain a dynamic model for the synchronous machine. In order to identify the reference model, a perturbation signal is applied to the generating unit. Viewing the generating unit as a second order linear system, a roughly estimated pole location variation is obtained through a variety of operating conditions and disturbances. An average pole location, determined from these studies, is then shifted by a certain amount with the consideration of the control constraints.

Results of the studies show that this APSS provides good damping over a wide operating range and improves performance of the system. An illustrative example showing the system response to a three phase to ground fault with the self-learning MRAC based FLC and a fixed center FLC is given in Fig. 12.

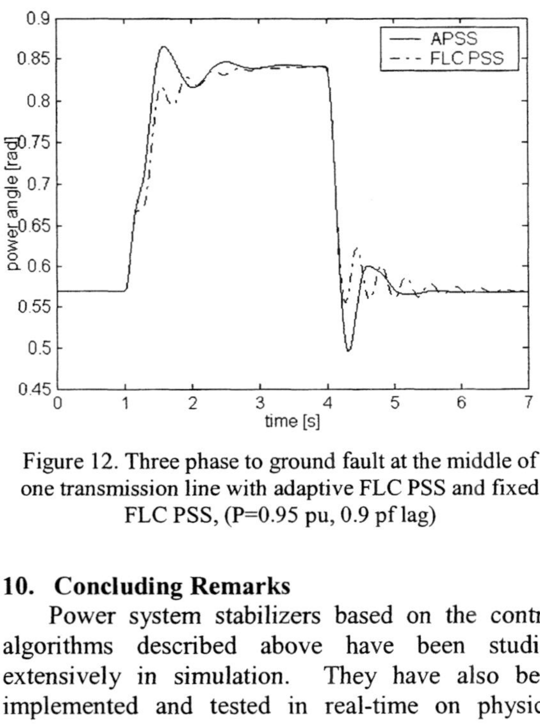

Figure 12. Three phase to ground fault at the middle of one transmission line with adaptive FLC PSS and fixed FLC PSS, (P=0.95 pu, 0.9 pf lag)

10. Concluding Remarks

Power system stabilizers based on the control algorithms described above have been studied extensively in simulation. They have also been implemented and tested in real-time on physical models in the laboratory with very encouraging results. The pole-shifting control algorithm based adaptive PSS has also been tested on a multi-machine physical model [15], on a 400 MW thermal machine under fully loaded conditions connected to the system [16], and is now in regular service in a hydro power station after extensive testing in the field [27]. These studies have shown clearly the advantages of the advanced control techniques and intelligent systems.

References

1. Larsen, E.W. and Swann, D.A., "Applying power system stabilizer: Parts 1-3", IEEE Trans. On Power Apparatus and Systems, Vol. PAS-100(6), 1981, pp. 3017-3046.
2. Kundur, P., Lee, D.C. and Zein el-Din, H.M., "Power system stabilizers for thermal units: analytical techniques and on-site validation", IEEE Trans. on Power Apparatus and Systems, Vol. PAS-100(1), 1981, pp. 81-95.
3. El-Metwally, M.M., Rao, N.D. and Malik, O.P., "Experimental results on the implementation of an optimal control of synchronous machines", IEEE Trans. on Power Apparatus and Systems, Vol. PAS-94(4), 1975, pp. 1192-1200.
4. Chen, S. and Malik, O.P., "An H_∞ optimization based power system stabilizer design", IEE Proceedings-Generation, Transmission and Distribution, Vol. 142(2), 1995, pp. 179-184.
5. Chan, W.C. and Hsu, Y.Y., "An optimal variable structure stabilizer for power system stabilizer", IEEE Trans. on Power Apparatus and Systems, Vol. PAS-102, 1983, pp. 1738-1746.
6. Hiyama, T., "Application of rule-based stabilizer controller to electric power system", IEE Proceedings C, Vol. 136(3), 1989, pp. 175-181.
7. Zadeh, L.A. et. al., "Calculus of fuzzy restriction in fuzzy sets and their application to cognitive and decision process", Academic Press, 1975, pp. 1-40.
8. El-Metwally, K.A., Hancock, G.C. and Malik, O.P., "Implementation of a fuzzy logic PSS using a micro-controller and experimental test results", IEEE Trans. on Energy Conversion, Vol. 11(1), 1996, pp. 91-96.
9. Zhang, Y., Malik, O.P. and Chen, G.P., "Artificial neural network power system stabilizer in multi-machine power system environment", IEEE Trans on Energy Conversion, Vol. 10(1), 1995, pp. 147-155.
10. Eykhoff, P., "System Identification, Wiley London, 1974.
11. Cheng, S.J., Chow, Y.S., Malik, O.P., and Hope, G.S., "An adaptive synchronous machine stabilizer", IEEE Trans. on Power Systems, Vol. PWRS-1(3), 1986, pp. 101-109.
12. Malik, O.P., Chen, G.P., Hope, G.S., Qin, Y.H. and Yu, G.Y., "Adaptive self-optimizing pole-shifting control algorithm", IEE Proceedings-D, Vol. 139(5), 1992, pp. 429-438.
13. Cheng, S.J., Malik, O.P. and Hope, G.S., "Damping of multi-modal oscillations in power systems using a dual-rate adaptive stabilizer", IEEE Trans. on Power Systems, Vol. PWRS-3(1), 1988, pp. 101-108.
14. Chen, G.P., Malik, O.P. and Hancock, G.C., "Implementation and experimental studies of an adaptive self-optimizing power system stabilizer", Control Engineering Practice, Vol. 2(6), 1994, pp. 969-977.
15. Malik, O.P., Stroev, V.A., Strobel, V.A., Hancock, G.C. and Beim, R.S., "Experimental studies with power system stabilizers on a physical model of a multi-machine power system", IEEE Trans. on Power Systems, Vol. 11(2), 1996, pp. 807-812.
16. Malik, O.P., Mao, C.X., Prakash, K.S., Hope, G.S., and Hancock, G.C., "Tests with a microcomputer based adaptive synchronous machine stabilizer on a

400 MW thermal unit", IEEE Trans. on Energy Conversion, Vol. 8(1), 1993, pp. 6-12.
17. Shamsollahi, P. and Malik, O.P., "An adaptive power system stabilizer using on-line trained neural networks", IEEE Trans. on Energy Conversion, Vol. 12(4), 1997, pp. 382-387.
18. Jang, J.G., "Adaptive-network-based fuzzy inference system", IEEE Trans. on Systems, Man and Cybernetics, Vol. 23(3), 1993, pp. 665-684.
19. Hariri, A. and Malik, O.P., "A fuzzy logic based power system stabilizer with learning ability", IEEE Trans. on Energy Conversion, Vol. 11(4), 1996, pp. 721-727.
20. Jang, J.G., "Self-learning fuzzy controllers based on temporal backpropagation", IEEE Trans on Neural Networks, Vol. 3(5), 1992, pp. 714-723.
21 Hariri, A. and Malik, O.P., "A self-learning adaptive-network-based fuzzy logic power system stabilizer in multi-machine power system", Engineering Intelligent Systems. In Print.
22. Abdelazim, Tamer and Malik, O.P., "Fuzzy logic based identifier and pole-shifting controller for PSS application", Proceedings, IEEE Power Engineering Society 2003 General Meeting, July 13-17, 2003, Toronto, Canada.
23. Adams, J.M. and Rattan, K.S., "Backpropagation learning for a fuzzy controller with partitioned membership functions", Proceedings, NAFIPS 2002 Annual Meeting of the North American Fuzzy Information Processing Society, pp. 172-177.
24. Mamdani, M., "Application of fuzzy algorithm for control of simple dynamic plant", Proc. IEE, Vol. 121(12), 1974, pp. 1585-1588.
25. Abdelazim, Tamer and Malik, O.P., "An adaptive power system stabilizer using on-line self-learning fuzzy system", Proceedings, IEEE Power Engineering Society 2003 General Meeting, July 13-17, 2003, Toronto, Canada.
26. Abdelazim, T.M. and Malik, O.P., "Gradient-based model reference adaptive control for PSS application", Procedings, Intelligent Systems Applications to Power systems, August 31-September 3, 2003, Limnos, Greece.
27. Eichmann, Armin, Kohler, Allessandro, Malik, O.P. and Taborda, Jose, "A prototype self-tuning adaptive power system stabilizer for damping active power swings", Conference Proceedings, Vol. 1, IEEE Power Engineering Society 2000 Summer Meeting, 16-20 July, Seattle, pp. 122-127.

ELSEVIER
IFAC
PUBLICATIONS
www.elsevier.com/locate/ifac

CONCEPTS AND TOOLS FOR MONITORING ELECTRICITY MARKETS

Dr. Lawrence Jones and Rick Schaal

ALSTOM-Energy Automation & Information, 11120 NE 33rd Place Bellevue, Washington 98004

Abstract: Electricity markets are being formed in many parts of world as part of industry restructuring. Tools and IT infrastructure for operating these markets have been developed and continue to evolve as market rules and their geographical footprints change. In contrast, tools for monitoring electricity markets, analysis of markets structure and the behavior of participants, e.g. GENCOs, and market power detection, have not received much attention. In North America the importance of Market Monitoring and Mitigation became very apparent as a result of the events in the California market and the ENRON trading strategies. As new electricity markets are being designed, it is important that systems and rules for market monitoring should be put in place before the markets begin operation. This module will provide an overview of market monitoring activities, analytical tools, and some of the best practices from different markets around the world. Perspectives on the evolution of tools and the role of emerging software technologies such as multi-agent technology, and advanced database systems, will be discussed. *Copyright © 2003 IFAC*

Keywords: Electricity Markets, GENCOS, Computer Interfaces, Databases, Implementation, and Software Tools.

1. INTRODUCTION

Market systems perform functions required to operate and settle markets in energy and ancillary services. The implementation of these functions and associated algorithms are driven by operational requirements that include performance of computational and data-handling speed and robustness. One consequence is that Market Operations (MO) may wish to eliminate non-essential data (from the view of real-time market operations) from databases used for current market clearing functions, particularly where this can improve performance.

The RTO/ISO Market Monitoring Unit (MMU) is tasked with the responsibility to ensure effective and efficient markets. To accomplish this, the MMU needs access to all relevant data from market operations as well as other data not directly available from the market. Differences between the MO and MMU requirements necessitate the development of separate MMU systems that serve the needs of the MMU while avoiding compromise of MO systems. To date, there has been limited development of separate MMU systems.

This paper attempts to identify major market monitoring activities, the analytical tools that complement MO systems, and permit extraction of data that currently is not available or that may be eliminated from the MO system due to performance considerations. The paper further explores some of the best practices from different markets around the world, which have been applied to an Open Market design concept.

2. MARKET MONITORING ACTIVITIES

An effective Market Monitoring System will enable a MMU to analyze market results and to focus on and quickly assess the impact of anomalous behavior on market outcomes in order to detect attempts at exercising market power. The type of services offered in a typical monitoring package can provide:

- Exactly reproduce market clearing results;
- Perform scenario analysis on market clearing results to explore the affects of system conditions or participant conduct;
- Analyze key aspects and variables affecting the market;
- Automatically screen the market to identify and report on anomalous bidding behavior; and
- Determine market impact of anomalous bidding behavior on the market relative to a competitive benchmark.

It should be emphasized that an effective Market Monitoring System will provide the capability to exactly reproduce market clearing results and provide the basis for direct and accurate assessment of the impacts of physical and economic withholding. The system will further support scenario development to assess the market impacts of bidding behavior and bid manipulation, bilateral schedules, outages, and maintenance schedules. In summary, a Market Monitoring System will allow an oversight entity to analyze the critical aspects of the real-time markets being cleared by the Market Operator (MO) today and, importantly, will support efficient and low-cost upgrades to monitor the MO's markets of the future.

3. ANALYTICAL TOOLS FOR MARKET MONITORING

Quantitative Analytical Tools are needed evaluate collected data. These functional tools will assist the RTO/ISO MMU in identifying and managing market practices to ensure effective and successful operation of energy markets and transmission grids

Specifically developed would be applied to:

- Performance metrics and indices
- Market-clearing reconstruction
- Market-clearing scenario simulation
- Transmission network security & adequacy analysis
- Security-constrained economic dispatch adapted for Unit Commitment and Production Costing evaluation

In addition to the Analytical Tools, data interface mechanisms will need to be constructed for collection and analysis of the data. These include data interfaces:

- From the MO's data-warehouse to the Market DB
- From other MO available sources
- Between the Market DB and SAS
- Between the Market DB and MCE
- From other data sources for Unit Commitment and Production Costing
- From other data sources as needed.

4. OPEN MARKET DESIGN (OMD)

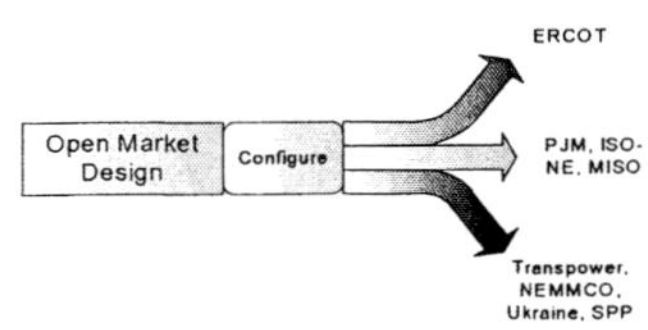

Figure 1: Open Market Design Concept

The OMD is refreshed with best practices, technology, new energy products and user enhancements to keep the assembly of OMD assets growing in value.

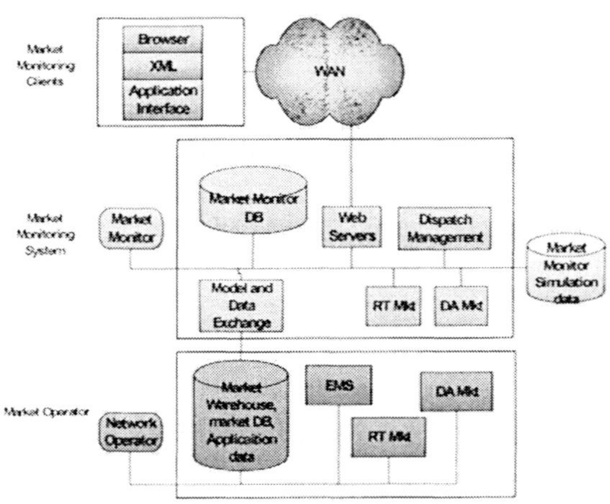

Figure 2: OMD Design

Typically the market monitoring systems are configured per the above diagram. In general, market system is functionally separate from the Market Monitoring System. This approach was first implemented by ALSTOM ESCA in 1996 and has proven to be the model of choice for ISO and RTO implementations. ALSTOM EAI is continually advancing the technology keeping the systems and the support environment from becoming out of date and becoming a technology risk. For market monitoring, the environment is updated to reflect changes in the production markets. Figure 3 shows an overview example of a Market Monitoring.

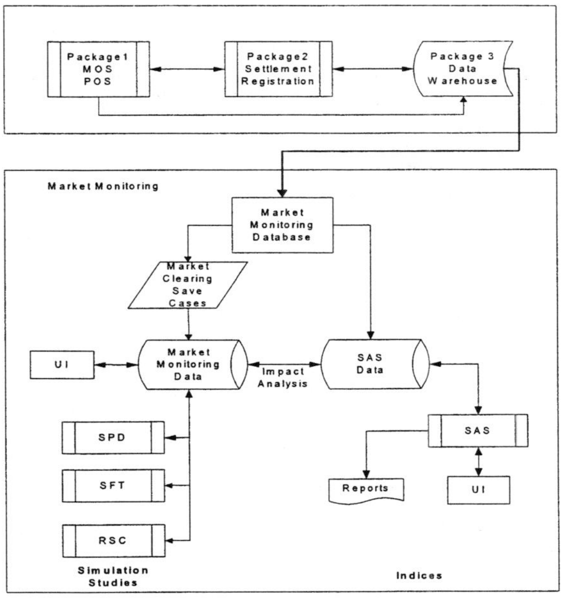

Figure 3: Market Monitoring Overview

5. CONCLUSION

Integration of a Market Monitoring system will provide the MMU the ability to better understand where the market is at any given moment while allowing for the future view of where it will follow. Monitoring the Market in this manner will allow for a more efficient and reliable Electricity Market.

ELSEVIER
IFAC
PUBLICATIONS
www.elsevier.com/locate/ifac

OPTIMAL BIDDING STRATEGIES FOR A SUPPLIER IN AN ELECTRICITY MARKET

Juhwan Jung*, Jae-Chul Kim*, Haili Song, and Chen-Ching Liu*****

**LG Industrial Systems, Co., Ltd., Korea*
*** ALSTOM EAI, Bellevue, WA, USA*
**** Electrical Engineering Department, University of Washington, Seattle, WA, USA*

Abstract: In this presentation, the panellist will review two state-of-the-art methodologies that can be used for spot and bilateral electricity markets. This article summarizes the basic principles of the methodologies from the suppliers' perspective. The methodologies can be used to determine optimal bidding strategies for a competitive spot market and a bilateral market to avoid risks. *Copyright © 2003 IFAC*

Keywords: Electricity Market, Markov Decision Process, Game Theory, Nash Equilibrium, Spot Market, Bilateral Market.

1. INTRODUCTION

Along the deregulated power system structures, the ownerships of generation, transmission, and distribution systems are being separated. In other words, the power industry is evolving into an open-access, competitive environment (R. J. Kaye *et al.*, 1990). The market participants bid / offer the volume and price of power that they are willing to buy / sell through the designed market structure (e.g., spot and / or bi-lateral markets) (J. W. M. Cheng *et al.*, 1998). The overall goal of these transactions is to maximize each firm's profit.

Generally speaking, in a deregulated environment, electricity and services can be sold or purchased through spot (short term) and bilateral (long term) markets. Market participants offer bids to the spot market which eventually determines the volume and clearing price of a transaction (H. Rudnick *et al.*, 1997). The market participants can also use bilateral electricity contracts in order to avoid and manage risks in the spot market (J. Ancona, 1997). As a result, market participants need to estimate the possible effects of various policies / strategies in order to maximize the expected return for one bidding period (J. W. Lamont *et al.*, 1997). Even though there have been a considerable amount of research to study the participants' bidding strategy and behaviour for the spot / bilateral markets, this paper summarizes two state-of-the-art methodologies as follows:

- Optimal electricity bidding strategy based on Markov Decision Process (MDP) (Haili Song *et al.*, 2000): a method that provides optimal bidding strategies for a spot market transaction,
- Nash Equilibrium (NE) bidding strategies in a bilateral electricity market (Haili Song *et al.*, 2002): a method that provides a strategy for a bilateral market and helps the market participants avoid possible risks.

These methodologies are studied and designed from the generation company's (GENCO's) perspective.

2. GENCO'S OPTIMAL BIDDING STRATEGY BY MDP

The purpose of the proposed method is to optimize the expected return over a planning period which can vary from a GENCO to another. The proposed method is able to provide a systematic way to solve multiple state probabilistic decision-making problems.

In an MDP, each state consists of state variables and the state transition from a state to another occurs based on the state transition probabilities (R. A. Howard, 1960). After one state transition, the decision-maker receives a reward value. Over a planning horizon, these reward values for state transitions are accumulated for the plan and one can decide an optimal plan which has the highest (or the lowest) an accumulated value among the plans. In the proposed method, each state can be decided by the values: production limit, spot price, today's load demand, and tomorrow's load forecast. The state transition probability for each transition can be computed as follows:

$$P_r(i,j,a) = P_r(SP(j,t)\ \&\ PL(j)\ \&\ LF(i,t)) * P_r(i, LF(j,t))$$

, where i: the current state, j: the next state, $SP(j,t)$: spot price at load period t in state j, $PL(j)$: the production limit in state j, $LF(i,t)$: the forecasted load at load period t in state i.

In order to forecast spot price in a state (i.e., $SP(j,t)$), the competitors' (i.e., other GENCOs) bidding behaviour should be modelled by bid prices and the volume of power that they will bid. This model can be built based on the decision maker's knowledge and information. The reward value for a state transition from i to j and the decision option a (one option can result in a couple of scenarios due to the stochastic nature of the problem) is the sum of the rewards for individual scenarios weighted by the conditional probabilities of these scenarios as follows:

$$r(i,j,a) = \sum_{t=1}^{T} \{r(i,s) * [\Pr(i,s) / \sum_{s\in\Omega(i,j)} \Pr(i,s)]\}$$

,where $r(i,j,a)$: reward corresponding to the transition from state i to state j, given the option a, $r(i,s)$: reward for a scenario s, $\Omega(i,j)$: subset of scenarios that result in the spot price, cleared quantity, and production limit of state j from those of state i, and $Pr(i,s)$: probability of a scenario s.

For each scenario, the reward value (i.e., $r(i,s)$) is a difference between the revenue and the cost of the decision-maker as a supplier. More detailed explanations can be found in (Haili Song *et al.*, 2000). As mentioned earlier, the reward value for every scenario is computed for the overall planning horizon and the optimal bidding decision can be made based on the accumulated reward values (i.e., $r(i,j,a)$). Fig. 1 briefly illustrates the proposed MDP. In Fig. 1, τ indicates the terminal stage of the overall planning horizon.

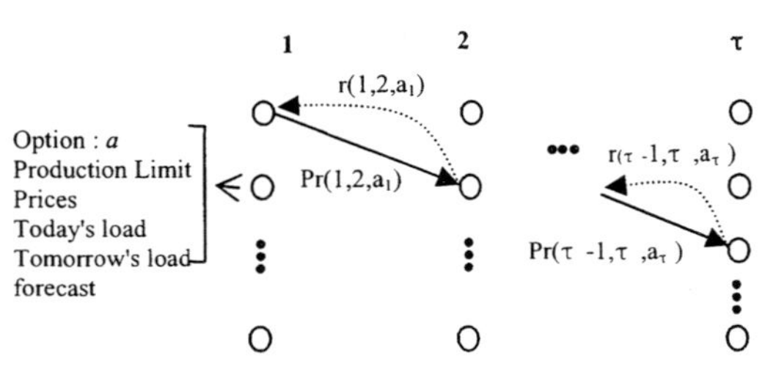

Fig. 1. Illustration of the MDP

3. NASH EQUILIBRIUM (NE) BIDDING STRATEGIES IN A BILATER ELECTRICITY MARKET

Game theory has been used to analyze problems when there are objectives conflicting among interacting decision-makers (R. W. Ferreo *et al.*, 1997). In a bilateral market, there are multiple sellers (GENCOs) and multiple buyers (loads) for a transaction. GENCOs submit bids to a load and the lowest bid, which is lower than the load's willingness to pay, is the winning bid. Therefore it is important for a GENCO to decide an optimal bidding strategy with respect to all the other players' / participants' strategies. Generally NE has been regarded as a universal solution concept in the game theory (X. Bai *et al.*, 1997). The whole purpose of the proposed method is to find the optimal bidding strategy (i.e., optimal bidding price) based on the game theory so that market equilibrium can be reached.

Suppose there are n loads and m generators that bid prices to the loads. Then the process to find an optimal bidding strategy is as follows:

- Step 1: Formation of the augmented cost matrix

Note that $c(g,l) = t(g,l)$ for g=1, …, m, and $c(m+l, l) = w(l)$. $w(l)$ indicates load l's willingness to pay and $t(g,l)$ indicates the total cost for generator g supplying load block l. The symbol ∞ indicates that the load does not participate in the bidding of another load (no resale could happen). The augmented matrix C is formulated as follows:

$$C = \begin{bmatrix} & \cdots & \\ \cdots & t(g,l) & \cdots \\ & \cdots & \\ w(1) & \infty & \infty \\ \infty & w(l) & \infty \\ \infty & \infty & w(n) \end{bmatrix} = [c(g,l)]_{(m+n)*n}$$

- Step 2: Finding a generator-load combination (GLC)

A GLC consists of n generator-load pairs (g_1, l_1), …, (g_n, l_n) such that g_1, …, g_n and l_1, …,l_n are independent each other. The generator set G_A can be represented as $G_A = \{g_A(1), g_A(2), \ldots, g_A(n)\}$.

- Step 3: Building an optimal bid price matrix

The total cost of a combination, i.e., G_A, can be calculated as follows:

$$CC(A) = \sum_{l=1,\cdots,n} c(g_A(l), l)$$

The proposed method is to find the GLC which cost is the lowest total cost. In other words,

$$CC(A^*) = CC^* = \min_{A\in S} CC(A)$$

,where $A^* = \{(g_{A^*}(l), l\}, l=1, \ldots, n)$ and G_{A^*} is the set of generators to form the combination A^*. As a result, the optimal bid price matrix can be determined based on the following procedure:

a) if $g \notin G_A^*$, then generator g bids at its cost for every load. This generator has zero profit in NE.

b) if $g \in G_A^*, (g = g_{A^*}(l))$, then remove row g in the augmented cost matrix C, and calculate

$$CC_{-g}^* = \min_{A\in S_{-g}} (CC(A))$$

$$\pi(g) = CC_{-g}^* - CC^*$$

If $\pi(g) > 0$, then $\pi(g)$ is the profit margin that generator g expects to make in NE. Consequently generator g's NE bid price for the load l and any other load l' should be:

$$b(g,l) = \pi(g) + c(g,l)$$

$$b(g,l') = \pi(g) + c(g,l')$$

Therefore the optimal bid price matrix can be built as follows:

$$B=\begin{bmatrix} b(g_1,1) & \cdots & b(g_1,n) \\ \vdots & \cdots & \vdots \\ b(g_m,1) & \cdots & b(g_m,n) \end{bmatrix}$$

There are a couple of necessary and sufficient conditions for the NE combination but more detailed explanations are provided in (Haili Song *et al.*, 2002). In this paper, it is assumed that one GENCO owns only one generator. However the proposed method can be extended to the case when one GENCO owns more than one generator so that a more sophisticated strategy, which considers cooperation of the generators belonging to the same GENCO, can be provided. This method is also presented in (Haili Song *et al.*, 2002).

4. CONCLUSION

Two methodologies that can be used for optimal bidding strategies for spot and bilateral markets are summarized in this paper. The proposed MDP method is expected to provide a systematic way to decide an optimal bidding strategy. There might be, however, a 'curse of dimensionality' problem when the number of states and / or stages is large. Therefore states and stages should be designed optimally in order to avoid such a problem. It is also clear that one needs to have complete information on the generators' cost and the loads' willingness to pay in order to apply the proposed NE method. This might not be realistic for a practical system. Since the proposed NE method can be used to avoid risks in the market, if the two proposed methods are well organized in a real software system, a GENCO is expected to maximize its profit and to be able to avoid possible market risks.

ACKNOWLEDGEMENT

The authors would like to acknowledge the contributions of the APT (Advanced Power Technologies) Consortium team members and the guidance from Professor Jacques Lawarrée and Mr. Robert W. Dahlgren at the University of Washington.

REFERENCES

H. Rudnick, R. Varela and W. Hogan (1997). Evaluation of alternatives for power system coordination and pooling in a competitive environment. *IEEE Trans. Power Systems*, vol. 12, no. 2. pp. 605-613

Haili Song, Chen-Ching Liu, Jacques Lawarrée and Robert W. Dahlgren (2000). Optimal Electricity Supply Bidding by Markov Decision Process. *IEEE Trans. Power Systems*, **vol. 15**, no. 2.

Haili Song, Chen-Ching Liu and Jacques Lawarrée (2002). Nash Equilibrium Bidding Strategies in a Bilateral Electricity Market. *IEEE Trans. Power Systems*, vol. **17**, no. 1.

J. Ancona (1997) A bids solicitation and selection method for developing a competitive spot priced electric market. *IEEE Trans. Power Systems*, **vol. 12**, no. 2, pp. 743-748.

J. W. Lamont and S. Rajan (1997). Strategic bidding in an energy brokerage. *IEEE Trans. Power Systems*, **vol. 12**, no, 4, pp. 1729-1733.

J. W. M. Cheng, F. D. Galiana and D. T. McGillis (1998). Studies of bilateral contracts with respect to steady-state security in a deregulated environment. *IEEE Trans. Power Systems*, **vol. 13**, pp. 1020-1025.

R. A. Howard (1960), *Dynamic programming and Markov process*, NewYork: The Technology Press of the MIT and John Wiley & Sons, Inc.

R. J. Kaye, H. R. Outhred and C. H. Bannister (1990). Forward contracts for the operation of an electricity industry under spot pricing. *IEEE Trans. Power Systems*, **vol. 5**, pp. 46-52.

R. W. Ferreo, S. M. Shahidehpour and V. C. Ramesh (1997). Transaction analysis in deregulated power systems using game theory. *IEEE Trans. Power Systems*, **vol. 12**, no. 3, pp. 1340-1347.

X. Bai, S. M. Shahidehpour, V. C. Ramesh and E. Yu (1997). Transmission analysis by Nash game method. *IEEE Trans. Power Systems*, **vol. 12**, pp. 1046-1052.

ELSEVIER
IFAC
PUBLICATIONS
www.elsevier.com/locate/ifac

OPTIMAL BIDDING STRATEGIES: AN EMPIRICAL CONJECTURAL APPROACH

Tengshun Peng and Kevin Tomsovic

School of EECS, Washington State University, Pullman, WA

Abstract: The optimal bidding strategy for a generation company is a complex decision making problem involving numerous uncertainties. Generally, all market participants attempt to refine their strategies to earn greater profit. Success in the market requires not only successful forecasting of demand and other market conditions but also anticipating rival behaviours. The main objective of this paper is to suggest one use of historical data for purposes of strategic bidding. A conjectural model based on the market clearing process is presented. In this model, participants estimate a mark-up function for their competitors in the market. Based on these estimates, an optimal bid is found. Numerical examples highlight the methodology. *Copyright © 2003 IFAC*

Keywords: Auctions, bidding strategies, conjectural model, risk.

1. INTRODUCTION

The optimal bidding strategy problem is a complex decision making problem involving numerous uncertainties. Generally, all market participants attempt to refine their strategies to earn greater profit. Success in the market requires not only successful forecasting of demand and other market conditions but also anticipating rival behaviours. As a result, many researchers have proposed a game theoretic model to address this problem (Ferrero, *et al.*, 1997; Hobbs and Kelley, 1992; Bai, *et al.*, 1997; Peng and Tomsovic, in press).

There is evidence to suggest that the energy market acts mostly like an oligopoly market (David and Wen, 2000). Game theory is often applied in oligopoly markets and certainly gaming has taken place in real markets. In an oligopoly market, a market participant's behaviour will affect the market clearing price (MCP). The participant's bidding strategy problem is to determine the bids to maximize one's own benefit.

In the typical electric power market, market participants submit bids and the intersection between the aggregate supply and aggregate demand curve is the market equilibrium point that determines the MCP and the winning bids. According to economics theory, there are several ways to consider strategic market interactions, including: pure competition, Cournot strategy, Stackelberg model, conjectural model, and so on. Particularly appropriate for the bidding problem here is the Cournot model (Day, *et al.*, 2002).

Due to the complexity of a real market with numerous participants, most of these theoretical models are too unwieldy to apply to a representative system model. Instead, numerous simplifications are needed to make application possible. For example, frequently there are no pure Nash equilibrium points and instead a mixed strategy is introduced (Mas-Colell, *et al.* 1995). Under mixed strategies, the optimal solution should be a probability density function. Unfortunately, this is difficult to apply as guidance to bidding activity. A deterministic decision is needed despite the underlying risks and various probabilistic outcomes.

A conjecture model can be used to consider market participants interactions. The conjecture model acts as a generalized Cournot model in that each market participant attempts to guess a rival's activity corresponding to the price change (Song, *et al.* 2003). Since many power markets have now been in operation for some time (albeit with frequent changes in the market rules), there exists significant historical data that can be analyzed to help participants understand likely competitor behaviour. At a minimum, market participants forecast energy prices based on historical prices. Beyond this, individual market bids are made publicly available in many markets, e.g., in California, all public bids are posted online by the ISO 6 months after the day of submission. Such information provides valuable information about a rival's likely behaviours. Thus, it is feasible to apply a conjecture model to address bidding strategies.

The main objective of this paper is to consider the best use of specific historical data. To begin, a model of the optimal bidding strategy problem is developed. Then, a statistical application method is introduced based on different price forecast technique.

2. PROBLEM DEFINITION

In economics, there are two types of conjecture models: the general conjectural variations (GCV) and the conjectured supply function (CSF). While both models require a "guess" of rival activities beforehand, the difference between these two conjectural models lies in the focus on a rival's price or the rival's supply function. The CSF model is adopted in this paper as the more appropriate for assessing bidding strategies in an electricity market.

2.1 CSF Model

Assume there are n market participants, let k_i denote market participant i's decision. Both the market clearing price and the quantity are a function of all market participants' decision variables denoted here by $MCP(k_1,k_2,\cdots k_n)$ and $p_1(k_1,k_2,\cdots k_n)$, respectively. Assume without loss of generality the bidding decision for individual market participant 1 is desired. The conjecture variation is defined as the belief of the ith market participant's response to its rivals. Here, we are interested in the bidding strategy k_i, and the conjecture variation can be represented by $k_i(k_1)$. This means that market participant 1 will guess all other decision variables k_i as the function of one's own activity, that is,

$$k_i(k_1) = f_i(k_1), \quad \forall i = 2,\cdots,m \qquad (1)$$

Since it is not possible to know a rival's actions precisely, it is necessary to represent the deviation of the ith firm's behaviour from this forecast. Here, the error is represented as:

$$k_i(k_1) = f_i(k_1)(1+e_i), \forall i = 2,\cdots,m \qquad (2)$$

where $f_i(k_1)$ are the forecast functions, which are static functions given the current market situation, and e_i are random variables representing the error in the forecast process. Since this function arises from historical data analysis, it represents the belief that these market participants will continue to bid in the same manner. Absent other information, the $k_i(k_1)$ are assumed to normally distributed since this is the most tractable mathematically. Then

$$E(k_i(k_1)) = f_i(k_1), \quad \forall i = 2,\cdots,m \qquad (3)$$

$$Var(k_i(k_1)) = f_i(k_1)^2 \operatorname{var}(e_i), \forall i = 2,\cdots,m \qquad (4)$$

It follows that $e_i \sim N(0,\sigma_i^2)$, since given a market situation $f_i(k_1)$ is determininistic. The decision problem is to maximize profit by choosing the decision variables k_1 such that:

$$\begin{aligned} \max_{k_1} \ & \pi = MCP_0 \cdot p_1 - C_1(p_1) \\ s.t \quad & k_2 = k_2(k_1) \\ & \vdots \\ & k_n = k_n(k_1) \\ & MCP_0 = MCP(k_1,k_2,\cdots,k_n), \\ & p_1 = p_1(k_1,k_2,\cdots,k_n) \end{aligned} \qquad (5)$$

where $C_1(p_1)$ is the generation cost function, p_1 represents the MW output awarded in the auction and MCP_0 is the market clear price in \$/MWhr for the time period of interest.

2.2 Decision Variable k_i

Let the generation cost be represented by a quadratic function as

$$C_i(p_i) = \frac{1}{2}c_i p_i^2 + b_i p_i + a_i \qquad (6)$$

Since the constant term a_i will not affect the maximization result (assuming a unit will be committed), let $a_i = 0$ with the marginal generation cost function of the form:

$$MC = c_i p_i + b_i \qquad (7)$$

In an electrical power market, the market participants submit bidding curves that provide the energy price for different levels of generation output. The market operator will determine the winning bids and the MCP according to the particular set of market rules. The decision problem of interest here is the optimal bidding curve that the market participant submits to the market. Normally, a staircase bidding curve is adopted, and for simplicity here, the linearization process as depicted in figure 1 is applied. This bidding curve is represented by the following linear function:

$$IC(p_i) = \alpha_i p_i + \beta_i \qquad (8)$$

Variables α_i and β_i represent the bidding coefficients participant i submits, which form the decision vector $k_i = [\alpha_i \quad \beta_i]$ for problem (5). We make a further assumption that both α_i and β_i are directly related to the MC by a constant level of mark-up. Thus, (5) simplifies to determination of a scalar, k_i. That is let:

$$k_i = {\alpha_i}/{c_i} = {\beta_i}/{b_i} \qquad (9)$$

Then it is easy to see that:

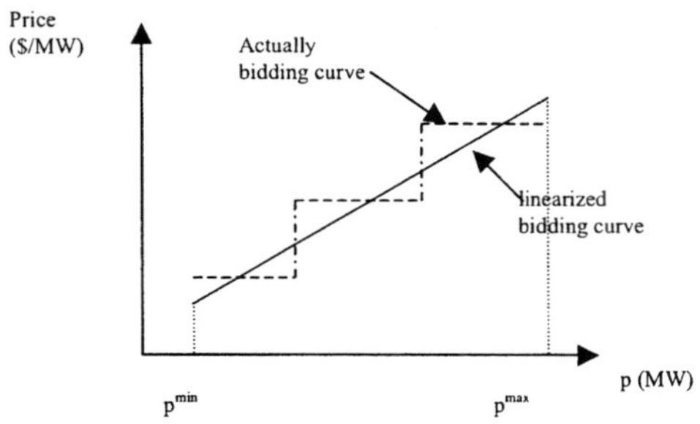

Fig. 1 Linearized Bidding Curve

$$IC(p_i) = \alpha_i p_i + \beta_i = k_i(c_i p_i + b_i) \qquad (10)$$

2.3 Conjecture Process

While each market participant may have more than one bidding strategy, in the development here we consider only one bidding strategy, and then subsequently address multiple bidding strategies in the numerical examples. Under a single strategy assumption, there are $m-1$ forecast functions needed by each decision maker to reach a decision with m market participants.

With the linearized bidding curve and market clearing process assumption, it follows that

$$p_1 = \frac{p(k_1, k_2(k_1), \cdots, k_m(k_1))}{k_1 c_1} - \frac{b_1}{c_1} \qquad (11)$$

Since the e_i are random variables, the $k_i(k_1)$ and $p(k_1, k_2(k_1), \cdots, k_m(k_1))$ are random variables as well. Thus, the profit maximization problem is a maximum expected profit problem.

2.4 Risk

Since the rival behaviour is not deterministic and knowledge of their behaviours is imperfect, there is risk in any bidding strategy. From a decision stand point, it is necessary to consider the consequence of competitors' actions that deviate significantly from the forecast function. In general, each market participant is willing to take a certain amount of risk to earn more in return. A related problem arises in investment where risk is often represented by the standard deviation in the expected profit. Then risk is addressed by the so-called Portfolio Selection Problem (Farrell, 1983), which is adopted widely in analysis of stock investments.

Portfolio management can be best characterized as obtaining the highest long-run return at the lowest risk (Farrell, 1983). There are two common formulations of portfolio selection: one, to minimize variance (risk) subject to achieving a specified level of return; and two, to maximize return subject to achieving a specified level of variance. Here, the former form is applied. Rewrite (5) as follows

$$\begin{aligned} &\max_{k_1} E(\pi) = E\big(MCP \cdot p_1 - C_1(p_1)\big) \\ &s.t \quad k_2 = k_2(k_1) \\ &\qquad \vdots \\ &\qquad k_m = k_m(k_1) \\ &\qquad MCP = MCP(k_1, k_2, \cdots k_m), \\ &\qquad p_1 = p_1(k_1, k_2, k_m) \\ &\qquad \sigma_\pi^2 \le \sigma_*^2 \end{aligned} \qquad (12)$$

where σ_*^2 represents the maximum acceptable risk determined a priori by the market participant. Those wishing to minimize the risk subject to achieving a specified level of return are essentially solving the dual of the above.

Now since the cost function is quadratic and noting that the third and forth moments of e_i are $E(e_i^3) = 0$ $E(e_i^4) = 3\sigma^4$, simple algebraic manipulation yields:

$$\begin{aligned} E(\pi) = &\left(\frac{1}{k_1 c_1} - \frac{1}{2k_1^2 c_1}\right)\left(\sigma_p^2 + \mu_p^2\right)\mu_p - \dots \\ &\frac{b_1}{c_1}\mu_p + \frac{b_1^2}{2c_1} \end{aligned} \qquad (13)$$

and

$$\begin{aligned} \sigma_\tau^2 = &\left(\frac{1}{k_1 c_1} - \frac{1}{2k_1^2 c_1}\right)^2 \sigma_{p^2}^2 - \dots \\ 2 * &\left(\frac{1}{k_1 c_1} - \frac{1}{2k_1^2 c_1}\right) \cdot \frac{b_1}{c_1}\left(E\left(p^3\right) - \left(\sigma_p^2 + \mu_p^2\right)\mu_p\right) + \dots \\ &\left(\frac{b_1}{c_1}\right)^2 \sigma_p^2 \end{aligned} \qquad (14)$$

where $\sigma_{mcp^2}^2 = E\left(MCP^4\right) - E\left(MCP^2\right)^2$, is the variance of MCP^2, σ_{mcp}^2 and μ_{mcp} are the variance and mean of MCP, respectively. The following section introduces a price forecasting approach to determine the necessary inputs for solving this problem. The expect profit and variance can be found by standard optimization routines given the statistics for errors in the prediction process.

3. PRICE ESTIMATION MODEL

This section places the preceding development in the context of an electricity market. Assume all other exogenous variables, including the load level, transmission limits, and fuel price, are known. The energy price is determined by the market participants' bidding strategies. Using a linear model forecast, the expected price can be represented by:

$$MCP_{\exp} = \left(\sum_{i=1}^{m} s_i k_i + s_0\right)(1 + e_{mcp}) \qquad (15)$$

Where s_i are the sensitivity coefficients between price and bidding strategies, s_0 is a constant term corresponding to the exogenous variables, and $e_{mcp} \sim N\left(0, \sigma_{mcp}^2\right)$ represents the error from the price estimation model. Rearranging

$$MCP_{\exp} = \left(s_1 k_1 + \sum_{i=2}^{m} s_i f_i(k_1)(1 + e_i) + s_0 \right)(1 + e_{mcp}) \quad (16)$$

with e_i an independent random variable. Now, separating the deterministic and the random variables, and defining the deterministic term

$$p_0 = s_1 k_1 + \sum_{i=2}^{m} s_i f_i(k_1) + s_0 \quad (17)$$

and the stochastic term,

$$MCP_{\exp} = \left(p_0 + \sum_{i=2}^{m} s_i f_i(k_1) e_i \right)(1 + e_{mcp}) \quad (18)$$

We have the following:

$$\mu_{mcp} = E(MCP_{\exp}) = p_0 \quad (19)$$

and

$$\sigma_{mcp}^2 = \left(\sum_{i=2}^{m+1} s_i^2 f_i(k_1)^2 \sigma_i^2 \right)(1 + \sigma_{mcp}^2) \quad (20)$$

4. NUMERICAL EXAMPLE

4.1 Numerical Example 1

First, consider a system with four market participants. The forecast error variances and coefficients are given by:

$$\sigma_2^2 = \sigma_3^2 = \sigma_3^2 = 0.25, \ \sigma_{mcp}^2 = 100$$

Assume that using statistical data analysis for a given market operation situation, participant 1 forecasts that the rivals have split strategies as given by the following bidding functions:

$$k_2(k_1) = \begin{cases} (1.0 + 0.1k_1)(1 + e_2), & k_1 \le 2.0 \\ (0.2 + 0.5k_1)(1 + e_2), & k_1 > 2.0 \end{cases},$$

$$k_3(k_1) = \begin{cases} (1.1 + 0.2k_1)(1 + e_3), & k_1 \le -0.5 \\ (1.2 + 0.3k_1)(1 + e_3), & k_1 > -0.5 \end{cases},$$

$$k_4(k_1) = \begin{cases} (2.0 - 0.1k_1)(1 + e_4), & k_1 \le 1.8 \\ (2.54 - 0.4k_1)(1 + e_4), & k_1 > 1.8 \end{cases}$$

Further assume the expected price function can be represented by the following linear function:

$$MCP_{\exp} \quad (7.973k_1 \quad 4.46k_2 + 82.68k_3 + 11.6k_4 + 4.014)(1 + e_{mcp})$$

ELSEVIER
IFAC
PUBLICATIONS
www.elsevier.com/locate/ifac

ANALYSIS OF ELECTRICITY MARKET BIDDING CHARACTERISTICS AND CLEARING PRICE FOR MODELING GENERATION CAPACITY GROWTH

Satoru Ihara*, Mark Sanford*, Andrew Kos*
Tetsuo Sasaki, Toshihisa Kadoya****

**GE Power Systems Schenectady, New York, USA*
***Kansai Electric Power Company Osaka, Japan*

Abstract: During the past few years U.S. electric generation capacity has grown at unprecedented levels. Now almost as quickly as it began, the growth era is drawing to a close. It is useful to examine market behaviors during this period to determine if they can provide any general insights about the effects of deregulation. In 2003 GE Power Systems Energy Consulting (GE PSEC) performed an analysis of the PJM system (Pennsylvania, New Jersey, Maryland) in the U.S. deregulated power market in order to establish a basis for a generation growth model. This paper discusses that study, its results and derived conclusions. *Copyright © 2003 IFAC*

Keywords: Generation Market, Deregulation, Boom Cycle, Bust Cycle, Capacity Growth Forecasting, Supply Margin

1. INTRODUCTION

During the past few years U.S. electric generation capacity has grown at unprecedented levels. Now almost as quickly as it began, the growth era is drawing to a close. It is useful to examine market behaviors during this period to determine if they can provide any general insights about the effects of deregulation.

Generation markets are regional. Inter-regional flows are constrained by network. Storage capability (e.g., pumped hydro and battery energy storage systems) is limited and expensive. In the U.S. deregulation was also regional with only some of the regulatory policy changes occurring at the national level.

The initial movement towards deregulation introduced considerable uncertainty for existing utilities and new power plant developers. New plant construction slowed and the reserve margin steadily decreased in many regions. During the summer of 1998 marginal energy prices in parts of the U.S. soared to levels previously unimaginable. A worldwide economic expansion and the eventual emergence of deregulated markets in the U.S. caused an explosion of plant construction and investment activity.

Production capacity often grows rapidly in commodity industries (boom cycles) in response to rising prices as shortages of supply increase earnings for market incumbents. The U.S. electricity generation industry appears to have followed this pattern. Adding generation capacity is a lengthy process with regulatory obstacles and delays. Many developers initiated construction of new plants simultaneously, all seeking the same economic gain. As a result, a large volume of new capacity entered the system in a very short period.

Before deregulation, utilities used some form of least-cost planning to make generation expansion decisions. The least-cost expansion process was designed to add only the amount of new generation needed to assure that consumer demands could be met reliably. Planners selected from among many generation options in order to provide service at the lowest total cost. As a result, generation additions tended to be more continuous. With deregulation, independent developers add new capacity in response to price signals. The resulting step-wise increase of generation capacity follows prices with substantial delays, causing reserve margin fluctuations.

After several years of a boom cycle, the generation market has entered a bust cycle. Competition, brought on by the abundance of new generation, has driven energy prices down in many regions. Market participants try to keep their plants running to cover their fixed costs. Ultimately, weaker participants will exit the market or will be acquired by stronger participants who try to benefit from economies of scale and from the market devaluation of distressed assets.

To some extent, boom-bust cycles appear to be inherent in the deregulated electricity market. The initial cycle in the U.S. appears to have been aggravated by the extended period of construction inactivity that preceded deregulation. It is challenging to predict the duration, frequency and severity of these cycles. Generation capacity growth forecasting is a new task in deregulated markets that replaces regulated market least-cost planning. For a study of the generation capacity growth, it is necessary to model the electricity market including the bidding agent characteristics and price formation mechanisms as well as the decision process of the power plant developers. Since no forecast is perfect, the ultimate objective of these models is to predict generation capacity growth under various scenarios.

2. THE ELECTRICITY PRICE DRIVERS TO GENERATION CAPACITY GROWTH

In 2003 GE Power Systems Energy Consulting (GE PSEC) performed an analysis of the PJM system (Pennsylvania, New Jersey, Maryland) in the U.S. deregulated power market in order to establish a basis for a generation growth model. A description of the market, its data, and the process used in the analysis is as follows:

- This study considers the PJM system specifically, although many of the observations may apply more broadly as well. The PJM system provides an opportunity to examine market behaviors and derive some potential relationships that could serve as the basis for a generation growth model. Although PJM did not experience as rapid a generation growth as some other parts of the U.S., it did have an operating competitive power market throughout the boom years. Data are available from PJM about clearing prices, energy demand, and even bid prices from individual generators. In the analysis process, an attempt was made to model the electricity market with the PJM market as an example. First, the PJM day-ahead market data available at web for 1999 April to 2001 October was studied. Identification of the relevant parameters characterizing the actions of the bidding agents and those influencing the actions was tried. Only the bidding data of the first 12 largest companies (in terms of total MW generation) was studied.

The studied data showed the following:

- While several price spikes occurred in July 1999, the total number of hours that the PJM price exceeded \$100/MWh was small. The PJM price was less than \$50/MWh for most of the time in July 1999. Even so, all the companies obtained a majority of their energy sales revenues for that month during the price spikes.
- The price spikes were strongly correlated to the scarcity of the supply margin.
- The bid price curves of most companies were flat, low and relatively invariant.
- Three companies, however, offered small amounts of generation at high prices. These high priced offers acted like antennae to detect the supply margin deficiencies. After a high price, these companies seemed to offer additional capacity at higher prices than normal. These observations could suggest a speculation strategy based on predicting tomorrow's price spike from today's clearing price and tomorrow's load forecast. The capacity that was always offered at high prices reduced the need to estimate the competitor status, the total generation capacity available, and the network congestion status in determining when to expect next-day price spikes.
- Variation of price curves gradually decreased after 1999 summer, fewer price spikes appeared, and offers in the mid-price range (\$50 to \$70/MWh) gradually increased.
- The gas price spiked in the winter of 2000/2001. The electricity price followed the gas price, went up quickly and down slowly. This behavior of the electricity price after a spike in the gas price seems consistent with characteristics of a Cournot equilibrium.

3. CONCLUDING REMARKS

The observed generator bid characteristics are conservative and practical. It does not appear that any generator attempted to create an initial price spike. This suggests at least one theory for bidding behavior during this period: The market sellers know that today's price is the best indicator for tomorrow's price (and that Friday's price is the best indicator for the following Monday's price). Today's price spike indicates that there is either a generation deficiency, successful speculation by others, or both. These observations suggest that the speculator model can be summarized as "Speculate tomorrow if there is a price spike today and if tomorrow's load is similar to today's load. Otherwise, do not speculate tomorrow."

To demonstrate the potential effect of this bid behavior model, the GE MAPS software program was used to simulate (Fig. 1) the 28,000-bus study system composed of the NY, PJM, NE, and Canadian markets. This approach enabled combined effects of the speculation and the network congestion into the New York City area as a purely fictitious example in which loads, generator availability, and the amount of capacity subject to speculative bidding behavior could be varied. These studies showed that the initial price spike typically requires very tight supply conditions. Once the initial spike occurs, it can be sustained in subsequent days if sufficient speculative bidding is assumed. In the absence of extreme levels of speculation, however, this model predicts prices to return to normal as a result of natural variations in load demand.

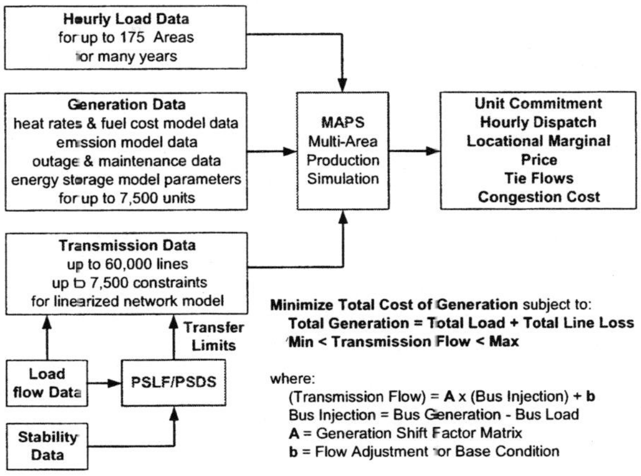

Fig. 1. GE MAPS Diagram

ELSEVIER
IFAC
PUBLICATIONS
www.elsevier.com/locate/ifac

REAL-TIME VOLTAGE STABILITY ASSESSMENT OF TRANSMISSION CORRIDORS

Mats Larsson *,[1] **Christian Rehtanz** *
Joachim Bertsch **

* *Corporate Research, ABB Switzerland Ltd.*
** *Utility Automation, ABB Switzerland Ltd.*

Abstract: Transmission corridors between interconnected power systems or within countrywide transmission systems are limiting network performance or even market activities. We describe a state-of-the-art wide-area measurement system which is commercially available, and a new method to assess the voltage stability of a transmission corridor. Using only topological information, several lines can be grouped into a so-called virtual transmission corridor, whose parameters can be calculated from wide-area measurements. A simple computational procedure for stability analysis, makes it possible to provide approximate stability margins with an update rate of less than a second. *Copyright © 2003 IFAC*

Keywords: Power Transmission, Voltage Stability, Real-time, Transmission Corridors, Phasor Measurement Units, PMU, Voltage Stability, Wide Area Control, Wide Area Protection, Wide Area Monitoring.

1. INTRODUCTION

Transmission corridors between interconnected power systems or within countrywide transmission systems are limiting network performance or even market activities. Transfer limits due to voltage stability limits are most often based on off-line calculation, and must therefore be conservative enough to accomodate all credible operating conditions. At the same time phasor measurements units are becoming an increasingly accepted and available technology. Using these phasor measurements it is possible to monitor the power system using a time resolution down to milliseconds. This gives the possibility of a wide range of stability monitoring and control applications. This paper focuses on the monitoring and operation of transmission corridors with the help of synchronized phasor information. We present a new and simple method to calculate the voltage stability margin for transmission corridors based on on-line phasor measurements. Since this method uses information about the actual operating conditions, the calculated voltage stability limit can be less conservative than those calculated using only off-line information. The method is described in detail and illustrated by simulation of an actual power system which contains a transmission corridor.

[1] Corresponding author.

2. SYSTEM SETUP

The wide area platform for dynamic monitoring of transmission systems comprises of hardware:

- Phasor Measurement Units (PMU)
- Communication Links
- Central Unit (Personal Computer)

and software:

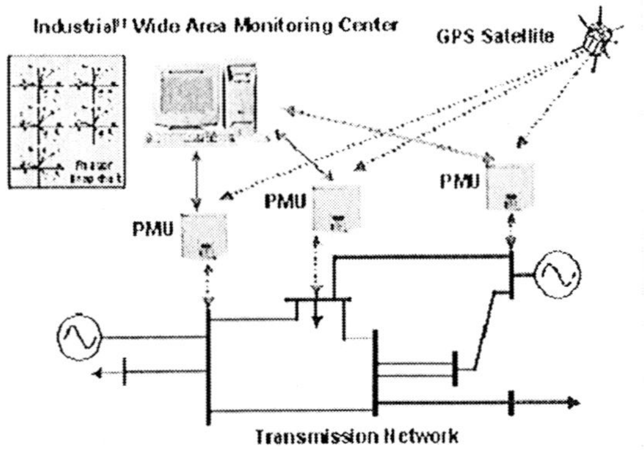

Fig. 1. Setup of wide area platform based on synchronized phasor measurements.

- Data preprocessing package
- Basic services
- Specific individual applications
- Graphical user interface (GUI)
- Package containing model/data of the supervised power system and coordination with other software packages

PMUs are placed in the substations to allow observation of a critical part of the supervised power system under any operation conditions (network islanding, outages of lines, generators etc.), taking into account a certain degree of redundancy to provide sufficient results also in a case of unavailability of some data (PMU outage, communication failure etc.). Measured data are sent via dedicated communication channels/links to a central unit, which is a central computational unit where the collected measurements are synchronized and sorted, yielding the snapshot of the power system state. This setup is shown in Fig. 1. The snapshot is then processed by the Basic Services package (BS), which is part of the central unit. Basic Services denote the set of algorithms included in all installations of the wide area platform for different applications and they are comprising the following capabilities:

- ability to provide needed data for any application
- fast execution - leaving sufficient time for running applications within the sampling interval
- robustness - resistance against poor quality of some input data (unavailability, out of range, synchronization problems etc.)

Applications, which are attached to the output of BS, address various phenomena occurring in power systems, such as frequency instability, voltage instability etc. They predict the state of the power system and trigger appropriate actions if an incipient instability is detected. Their output as well as the output of BS are displayed to the power system operator by an ergonomic GUI.

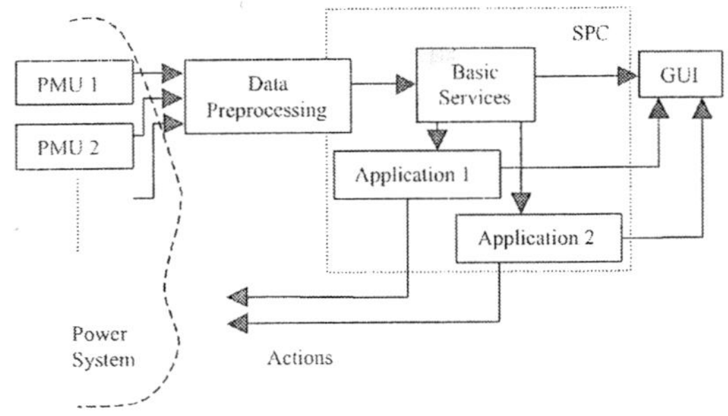

Fig. 2. Functional architecture of wide area measurement platform based on synchronized phasor measurements.

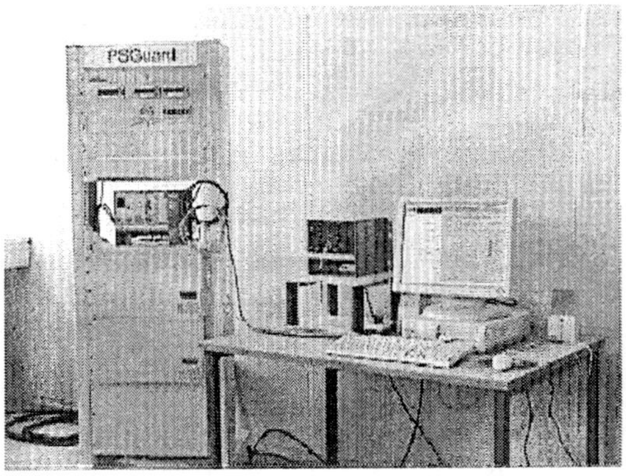

Fig. 3. Laboratory setup of the wide area monitoring platform.

The functional structure described above is illustrated in Fig. 2. The voltage stability assessment method described in Section 3 is one example of a method that can be applied in such an algorithm. Other examples have been previously published (Larsson and Karlsson, 2001; Larsson and Rehtanz, 2001; Rehtanz and Bertsch, 2001). The laboratory setup of the wide area platform for monitoring, protection and optimization comprises all the components of a field installation with the example of the supervision of a transmission line model. The setup is shown in Fig. 3.

3. VOLTAGE STABILITY MONITORING FOR TRANSMISSION CORRIDORS

Practical application of methods for detection of voltage instability are generally based solely on local measurements. The first methods were based on a local voltage measurement; if voltage becomes abnormally low (below a preset threshold value), voltage instability is assumed to be present and load is disconnected with a certain time-delay until voltage returns above the threshold value. This technique is normally referred to as Undervoltage Load Shedding (UFLS) and is still the most widely used technique to guard against voltage instability in high-voltage power networks

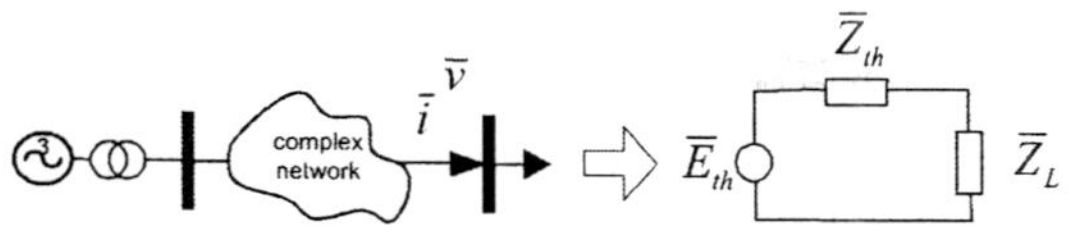

Fig. 4. Illustration Thevenin equivalent estimation in VIP-type approaches.

(CIGRE, 2000). The drawback of this method is that the voltage threshold must be set in advance, and thus the relay cannot adapt to changing operating conditions in the power system.

Other approaches are based on the estimation of a Thevenin equivalent of the network at a single bus. The current through a single line feeding that bus is used to estimate an equivalent Thevenin equivalent of a potentially complex network and the generation at the remote end. In this paper we refer to these as VIP-type approaches. Various variations on this theme have been presented by (Vu *et al.*, 1997; Balanathan *et al.*, 1998; Quantaince and Bertsch, 2001). The drawback of these methods is that a single set of (local) measurements do not contain enough information to directly compute the parameters of the Thevenin equivalent, however these can in principle be estimated using a least-squares method once two or more sets of measurements are available. A necessary condition for accurate estimation of these parameters is that sufficient change in the measurements has occurred due to load change. During this time the feeding network and generation is assumed to remain constant. Therefore the estimation is noise-sensitive and introduces a time-delay comparable to that of standard SCADA systems (several minutes). Recent developments on this type of scheme (Warland and Holen, 2002) include also measurements at remote buses to assist in the estimation of the Thevenin equivalents. Although this paper does not in detail describe the estimation procedure applied, the simulations in the paper show a similar time delay as those observed with the standard VIP-type approaches.

The main contribution in this paper is a method that uses measurements from both ends of the transmission corridor, enabling the splitting of the estimation part into two stages. Firstly, the parameters of a T-equivalent of the transmission corridor can be determined through direct calculation and therefore without the time delay seen in VIP-type approaches. Secondly a Thevenin equivalent of the feeding generators is computed. Once the parameters of the T- and Thevenin equivalents are known, stability analysis can be carried out analytically and various stability indicators calculated using the method described in Section 3.3. The main advantage over the present state of the art is that parameters of the equivalent network can be computed from a single set of phasor

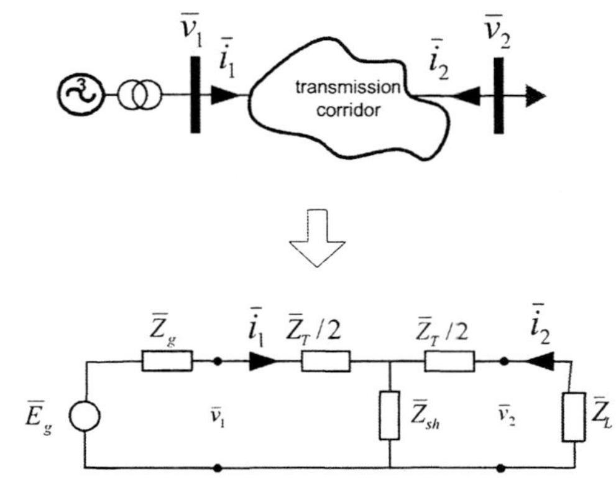

Fig. 5. Illustration of the T- and Thevenin equivalents of the transmission corridor and generation in our approach.

measurements, and thus this method does not have the time delay of the VIP-type approaches. This benefit is achieved through the placement of additional phasor measurements.

3.1 Calculation of T- and Thevenin equivalents

The calculation of the Thevenin equivalent is carried out in two stages in order to benefit from the fact that we have measurements at both ends of the transmission corridor. First we calculate the parameters of a T-equivalent of the actual transmission corridor. Any load or generation that may be present in the transmission corridor as shown in Fig. 5. This load or generation is then implicitly included in the shunt impedance.

Applying Ohm's and Kirchoff's laws, with the known complex quantities (measured phasors) $\bar{v}_1$, $\bar{i}_1$ and $\bar{v}_2$, $\bar{i}_2$, we can calculate the complex impedances $\bar{Z}_T$, $\bar{Z}_{sh}$ and $\bar{Z}_L$ as follows

$$\bar{Z}_T = 2\frac{\bar{v}_1 - \bar{v}_2}{\bar{i}_1 - \bar{i}_2} \tag{1}$$

$$\bar{Z}_{sh} = \frac{\bar{v}_1\bar{i}_2 - \bar{v}_2\bar{i}_1}{\bar{i}_2^2 - \bar{i}_1^2} \tag{2}$$

$$\bar{Z}_L = \frac{\bar{v}_2}{-\bar{i}_2} \tag{3}$$

The complex voltage $\bar{E}_g$ and impedance of the equivalent voltage source $\bar{Z}_g$ cannot be simultaneously calculated in the same straightforward way, so one of them must be assumed to be known to avoid the time-delay of an estimation procedure similar to that in the VIP-type approaches. If the generators have voltage controllers and can be assumed to stay within their capability limits, $\bar{E}_g$ can assumed to be constant and $\bar{Z}_g$ could then be calculated using:

$$\bar{Z}_g = \frac{\bar{E}_g - \bar{v}_1}{\bar{i}_1} \tag{4}$$

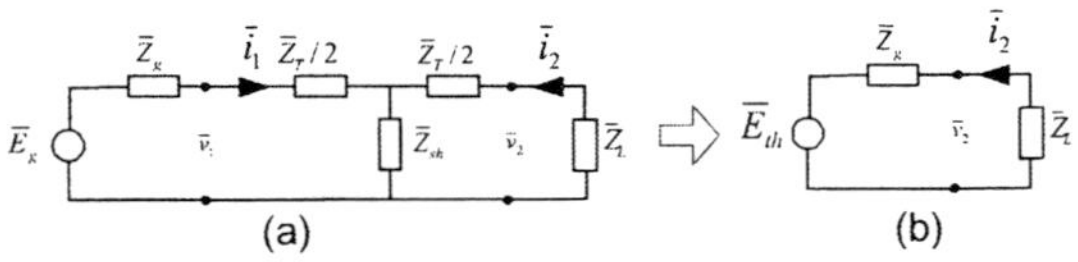

Fig. 6. Illustration of T- and Thevenin equivalents of transmission corridor and generation (a), and the reduction to a second Thevenin equivalent modeling the combined generation and transmission corridor (b).

However, in most practical cases it is more realistic to assume that $\bar{Z}_g$ is known since it typically comprises of the step-up transformers and short transmission line to the beginning of the transmission corridor. It is therefore preferential to calculate the equivalent complex voltage of the generators as follows:

$$\bar{E}_g = \bar{v}_1 + \bar{Z}_g \bar{i}_1 \tag{5}$$

Once we have calculated the parameters of the T- and Thevenin equivalents, a second Thevenin equivalent for the combined generation and transmission corridor can be calculated as follows:

$$\bar{Z}_{th} = \frac{\bar{Z}_T}{2} + \frac{1}{\frac{1}{\bar{Z}_{sh}} + \frac{1}{\bar{Z}_T/2+\bar{Z}_g}} \tag{6}$$

and

$$\bar{E}_{th} = \bar{v}_2 \frac{\bar{Z}_{th} + \bar{Z}_L}{\bar{Z}_L} \tag{7}$$

Based on the second Thevenin equivalent in Fig. 6b, stability analysis can be performed analytically in a straightforward way. This process is described in the application example below.

3.2 Network Reduction

Consider the example network diagram in Fig. 7. To apply the estimation procedure, first the main load and generation centers must be identified. In this case, a distinct generation center can be found in the area north of Cut 1, which contain three major generators and some shunt compensation but only a few minor loads. Between Cuts 1 and 2 is an area with no generation equipment and only a few minor loads. This is the transmission corridor, whose stability is of interest. In the equivalencing procedure described above, the loads within the corridor will be implicitly included in the shunt impedance.

South of Cut 2 is an area with predominantly load character. There are some minor generators, but in cases where the voltage stability is endangered, these generators would have exceeded their capability limits and thus no longer contribute to stabilization. It is therefore reasonable to include them in the shunt impedance modelling the load.

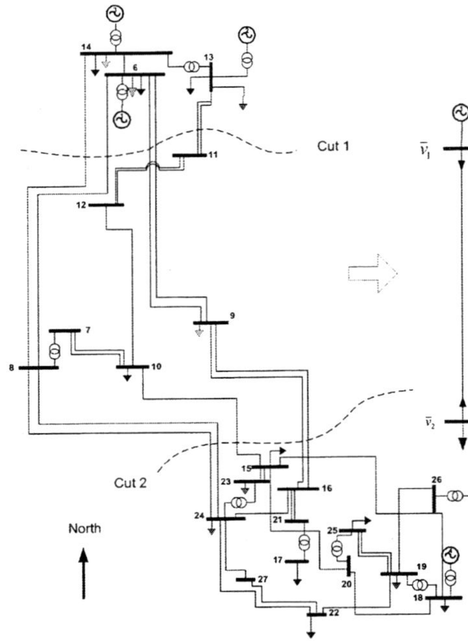

Fig. 7. Example network diagram, showing a part of an actual customer network.

After identifying the region boundaries, which are given by the two transfer cuts we can define two *virtual buses*, one for each end of the transmission corridor. These are the buses directly adjacent to a cut. Buses 6, 13 and 14 of the original system are grouped into virtual bus 1, and buses 24, 15 and 16 into virtual bus 2. The part of the system between cuts 1 and 2 becomes the *virtual transmission corridor*. At least one voltage in the area of each virtual bus and the currents on each line crossing a cut must be measured.

We can then compute the currents at either end of the virtual transmission corridor using

$$\bar{i}_i = \left(\frac{p_{cut-i} + jq_{cut-i}}{\bar{v}_i} \right)^* \quad i \in 1,2 \tag{8}$$

For example, here p_{cut-i} and q_{cut-i} refer to the sum of the power transfers through cut i, and $\bar{v}_i$ as the average of the voltages of the buses included in in virtual bus i.

3.3 Stability Analysis

This section outlines how the computed virtual transmission corridor model can be used to carry out computation of stability margins.

Based on the second Thevenin equivalent in Fig. 6, stability analysis can be performed analytically. The complex power delivered to the load impedance $\bar{Z}_L$ can be written

$$\bar{S}_L = \bar{Z}_L \left| \frac{\bar{E}_{th}}{\bar{Z}_{th} + \bar{Z}_L} \right|^2 \quad (9)$$

Assuming that load will evolve with constant power factor, we can set $\bar{Z}_L = k\bar{Z}_{L0}$, where k is a scale factor modeling change in the load impedance and $\bar{Z}_{L0}$ the present value of load impedance as calculated according to (3).

To find the point of maximum possible power transfer we need to compute the maximum of

$$p_L = \Re\left[\bar{S}_L\right] = \Re\left[k\bar{Z}_{L0}\left|\frac{\bar{E}_{th}}{\bar{Z}_{th} + k\bar{Z}_{L0}}\right|^2\right] \quad (10)$$

with respect to the load impedance scale factor k. Differentiating (10) and solving for the first and second order conditions we find that the critical load scale factor, where no further increase in p_L is possible, is given by

$$k_{crit} = \left|\frac{\bar{Z}_{th}}{\bar{Z}_{L0}}\right| \quad (11)$$

After insertion of $k = k_{crit}$, (10) becomes

$$p_{Lmax} = \Re\left[k_{crit}\bar{Z}_{L0}\left|\frac{\bar{E}_{th}}{\bar{Z}_{th} + (k_{crit}\bar{Z}_{L0})}\right|^2\right] \quad (12)$$

Extensive field measurements reported by Karlsson and Hill (1994) have shown that normally at least part of the load has constant power characteristics, and the point of maximum power transfer as given by (12) then also becomes a loadability limit. Past this limit there is a loss of equilibrium and voltage collapse will occur (Van Cutsem and Vournas, 1998, pp. 29). Eq. (12) therefore also becomes a stability limit.

The voltage $\bar{v}_2$ at the virtual load bus is given by

$$\bar{v}_2 = \bar{E}_{th}\frac{k\bar{Z}_{L0}}{\bar{Z}_{th} + k\bar{Z}_{L0}} \quad (13)$$

Based on (11)-(13), various stability margins can be defined as follows; In terms of load impedance (in percent)

$$MARGIN_Z = 100(1 - k_{crit}) \quad (14)$$

in terms of active power delivered to the load bus (in p.u.)

$$MARGIN_P = \begin{cases} p_{Lmax} - p_L & \text{if } \bar{Z}_L > \bar{Z}_{th} \\ 0 & \text{if } \bar{Z}_L < \bar{Z}_{th} \end{cases} \quad (15)$$

For example, if the value calculated by (15) is 4 p.u., voltage instability will occur if the active power load increase is larger than 4 p.u.

Fig. 8 shows simulation results where the the two lines between buses 9 and 16 are tripped at 20 s and at 30 s, which makes the system voltage unstable, and eventually there is a collapse at about 190 s. In the simulation, we have used a load model with an instantaneous impedance characteristic which recovers to constant power with a time constant of 60 s.

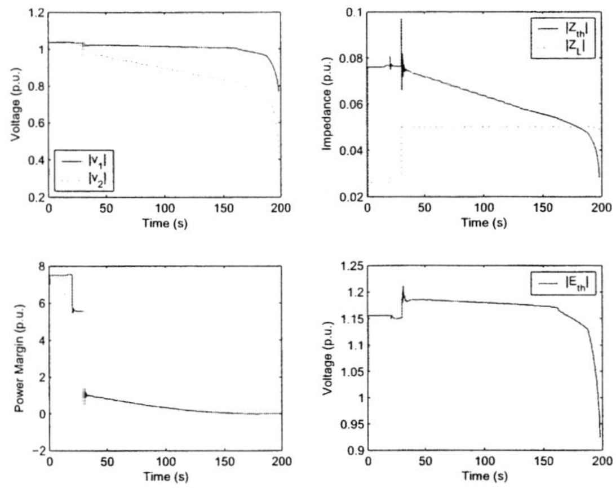

Fig. 8. The estimated Thevenin voltage and impedance, load impedance and power margins during the disturbance scenario.

The top-left figure shows the calculated voltages of the two virtual buses, and the top right figure shows the estimated load and Thevenin impedances. The two sharp step increases in the Thevenin impedance are due to the line trippings and the decrease in the load impedance is due to load recovery dynamics. In this case, the stability boundary is crossed at about 170 s, as indicated by the crossing of the two curves, and the collapse of the system progresses rapidly after this point. The bottom left figure shows the power margin. Note the two sharp steps in the power margin which are due to the two line trippings. The estimated Thevenin voltage is shown in the bottom right plot.

Fig. 9 shows the so-called PV-curves at four different instants. Before the disturbance, at 19.8 s, the power transfer is about 14 p.u. compared to the maximum 22 p.u., as given by the rightmost point of the PV-curve. At time 39.8 s, the fast transients following the two line trippings have settled and we can see that the maximum transfer is now about 13.5 p.u. Because of the initial load relief provided by the initial impedance characteristic the load power has dropped to about 12.6 p.u. however the constant power characteristic of the load drives the operating point towards the transfer limit and eventually the stability boundary is crossed at 170 s. Note that if we consider that the load in the long-term has constant power characteristics known, the instability could be detected directly following the line trippings (since the pre-fault loading was larger than the post-fault transfer limit).

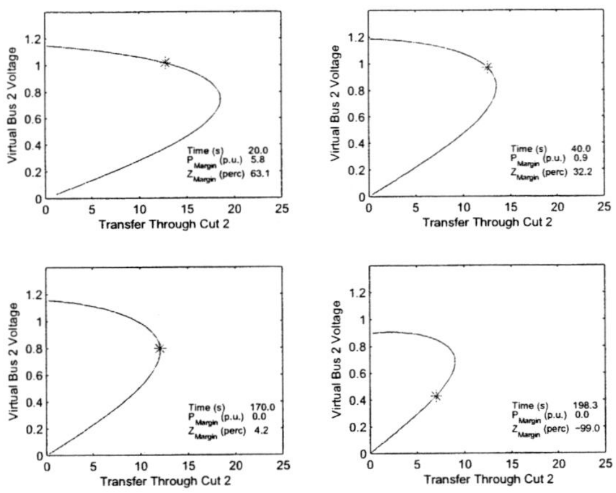

Fig. 9. PV-curves at four different instants during the collapse scenario. The distance between the current operating point (marked by the asterisk) and the rightmost point of each curve indicates the stability margin.

4. FUTURE WORK

Referring to the method described in Section 3, the estimated Thevenin equivalent of the network is now used only to assess the stability of the power system. The information in Fig. 9 will be presented online to the operator and thereby visualize the degree of stability to the operators. However, collapse scenarios may evolve too quickly for the operator to respond. The immediate continuation of theoretical work will be to extend the method with so it can calculate automatic corrective control actions when instability is detected. Another issue is the accuracy of the power margin that is calculated. Since the method internally uses a reduced representation of the power system, the power margin computed is an approximation. The initial results based on the power system in Fig. 7 indicate that the approximation is good and that the method is accurate, but future work is necessary to validate the method also on other network topologies.

A pilot installation of the wide-area platform in a European power system is now completed and the performance is continuously monitored. The implementation of the voltage stability assessment method is also part of the future work.

5. CONCLUSIONS

A commercially available platform for wide-area monitoring and control based on phasor measurements has been described. A new method for the voltage stability assessment of transmission corridors has been described. Using measurements at either end of the transmission corridor, a reduced equivalent of the network can be constructed and used for analytical stability assessment. The result of the stability is displayed using a dynamically updated PV-curve and numerical stability margins.

6. ACKNOWLEDGEMENT

This work was partly funded by a grant from the Bundesamt für Bildung und Wissenschaft, Bern, Schweiz in the Control and Computation project (IST-2001-33520).

REFERENCES

Balanathan, R, N C Pahalawaththa, U D Annakkage and P W Sharp (1998). Undervoltage load shedding to avoid voltage instability. *IEE Proceedings Generation, Transmission and Distribution* **145**(2), 175–81.

CIGRE (2000). System protection schemes in power networks. Technical report. CIGRE Task Force 38.02.19.

Karlsson, D and D J Hill (1994). Modelling and identification of nonlinear dynamic loads in power systems. *IEEE Transactions on Power Systems* **9**(1), 157–63.

Larsson, M and C Rehtanz (2001). Predictive frequency stability control based on wide-area phasor measurements. In: *Power Engineering Society Summer Meeting*. Vol. 1. IEEE. pp. 233–8.

Larsson, M and D Karlsson (2001). System protection scheme against voltage collapse based on heuristic search and predictive control. *IEEE Transactions of Power Systems, to appear.*

Quantaince, W.H. and J. et.al Bertsch (2001). Method and device for assessing the stability of an electric power transmission network. number 20020123849, 20020097055, 20010021896. US Patent Applications.

Rehtanz, C and J Bertsch (2001). A new wide area protection system. In: *Bulk Power System Dynamics and Control V, August 26-31, 2001, Onomichi, Japan.*

Van Cutsem, T and C Vournas (1998). *Voltage Stability of Electric Power Systems.* Power Electronics and Power Systems Series. Kluwer Academic Publishers.

Vu, K, M M Begovic, D Novosel and M M Saha (1997). Use of local measurements to estimate voltage-stability margin. In: *20th International Conference on Power Industry Computer Applications.* IEEE. pp. 318–323.

Warland, L and A T Holen (2002). Estimation of distance to voltage collapse: Testing an algorithm based on local measurements. In: *14th PSCC Sevilla.*

ELSEVIER
IFAC
PUBLICATIONS
www.elsevier.com/locate/ifac

STRATEGIES TO INCREASE TRANSFER LIMIT IN TERMS OF VOLTAGE STABILITY

Byongjun Lee[*], Hwachang Song[*], and Hae-Kon Nam[§]

** Advanced Power System Research Center, Korea University*
§ Dept. of Electrical Engr., Chonnam National University

Abstract: This paper presents strategies to enhance transfer limits in terms of voltage stability. First, this paper discusses definitions of transfer capability and transfer limits as well as their determination. In fact, adequate concept of transfer capability and transfer limit needs to be selected depending on power industry structure. Presently, Korea power market basically uses concept of transfer limits of interface lines for congestion management. Also, this paper describes strategies of enhancing transfer limits of systems using OPF formulations. In numerical simulation, an example of KEPCO (Korea Electric Power Corporation) system is shown. *Copyright © 2003 IFAC*

Keywords: Available transfer capability, transfer capability, voltage stability enhancement.

1. INTRODUCTION

Voltage stability has become one of the limiting factors to maintain system security in the recent power system planning and operation. Introduction of competition to power industry has been making power systems more vulnerable in terms of voltage security because power flow patterns are less regulated and existing transmission facilities are fully utilized by diverse power flow pattern. In this environment, independent system operators (ISO) need to establish the adequate procedure for determining transfer capability or transfer limits on key transmission lines, considering their own power industry structure. This paper first discusses transfer capability and transfer limits in terms of voltage stability.

Assuming that the systems are constrained by transfer limits in a specific condition, transmission systems need to be operated to enhance the limits, because, in the open access environment, transmission systems should support competitive generation and supply for efficient operation of power market. This paper also discusses methodologies for enhancing voltage stability limits of active power transfers. This concept of enhancing transfer limits of voltage stability is close to that of determining remedial action in voltage stability assessment. In the literature, there are various algorithms for remedial action in terms of voltage stability, but this paper focuses on OPF formulations for that purpose. Recently, OPF tools considering voltage security have been drawing attention of power system engineers and researchers because OPF has the function of dispatching active and reactive power generation that affect market significantly.

2. TRANSFER LIMITS

First, this section discusses transfer capability and transfer limits in terms of voltage stability to differentiate transfer limits from transfer capability.

As to transfer capability, ATC (Available Transfer Capability) definition is generally referred. In NERC report [1], definition of ATC is a measure of the transfer capability remaining in the physical transmission network for further commercial activity over and above already committed uses. To determine ATC, TTC (total transfer capability) needs to be obtained in advance. Transfer capability – this paper uses this notation to indicate ATC or TTC, is constrained by voltage limits, thermal limits and stability limits. From the words "further commercial activity" of the definition, transfer capability closely depends on direction of transaction.

On the other hand, transfer limits are closer to physical meaning. To determine transfer limit, a specific physical object is needed. This paper confines the object to a set of key transmission lines pre-defined by system engineers. Key transmission lines correspond to EHV interface lines connecting among areas or regions. Like transfer capability, transfer limits of interface flows are constrained by voltage limits, thermal limits and stability limits.

Presently, the Korean power industry is operated with a mandatory cost-based pool rule in the first reform process. Thus, ATC concept is not necessary because individual transaction is not allowed out of pool. To solve congestion problems, transfer limits of interface flow is used. Fig. 1 shows a map of KEPCO EHV transmission system. Metropolitan region in Fig. 1 has more than 40% of system total load, so Metropolitan interface lines are mainly congested, especially by voltage stability limits. To cope with the congestion problem, it is basically required determining voltage stability limit of the interface.

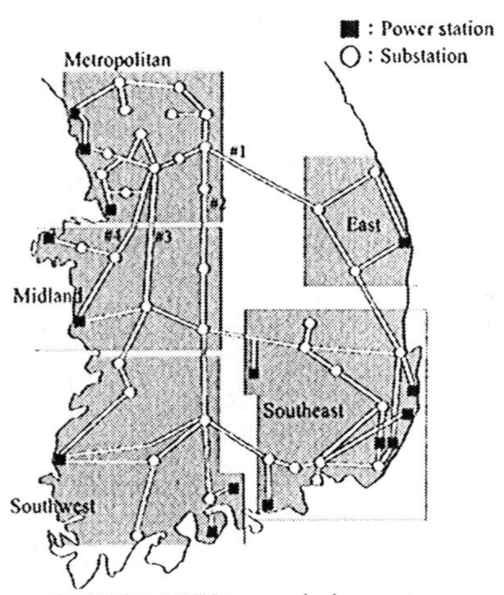

Fig. 1 A map of KEPCO EHV transmission system.

The procedure for determining secure limits of interface flow with respect to voltage stability is described as follows, in Fig. 2.

Step 1: *Choose the study region and interface lines connecting the study region and neighboring regions.*

Step 2: *Select severe contingencies to be applied.*

Step 3: *Determine a realistic direction of generation shift.*

Step 4: *Construct f-V curves of the normal case and contingent cases with generation shift direction..*

Step 5: *Determine the worst contingency with the least interface flow margin and find the quantity of generation shift at the maximum point p_1.*

Step 6: *Find the interface flow level (p_2) in the f-V curve of the normal case. At this level, the quantity of generation shift is the same as the one obtained in Step 5.*

Step 7: *Determine the interface flow level (p_3) with the α% (or MW) reserve margin as a secure limit in terms of steady-state voltage stability.*

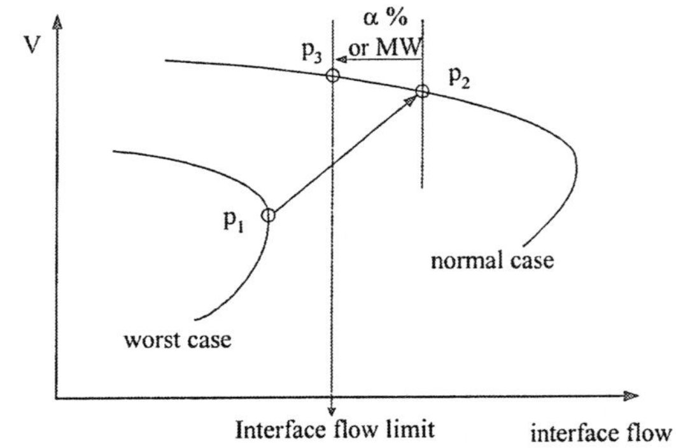

Fig 2. Illustration of determining interface flow limit

Difference between the interface flow limit and ATC is that generation shift is used to stress the system other than load increase of the study region. Because stressing the system with generation shift doesn't change system load, the limit using the procedure is adequate to apply to operational limit at the corresponding snap shot.

3. ENHANCEMENT OF TRANSFER LIMITS WITH OPF

This section describes strategies to enhance transfer limits or transfer capability perspective to voltage stability. Practically, the strategies are close to countermeasure determination in voltage security assessment (VSA). For countermeasure in terms of voltage stability, various methodologies can be considered, but this paper concentrates on application of OPF formulations.

In the literature [2-10], many researchers tried to incorporate voltage stability constraints into OPF formulations. The studies can be divided into two parts. First one is application of voltage stability constraint to one operating point, and the other is consideration of post-contingent voltage stability. Formulation of the former application can be presented as follows:

$$\begin{aligned} &\min\ f(x) \\ &s.t.\quad g(x)=0 \\ &\qquad h_{\min} \le h(x) \le h_{\max} \qquad (1) \\ &\qquad vs_{\min} \le vs(x) \end{aligned}$$

where $f(\cdot)$, $g(\cdot)$, $h(\cdot)$ and $vs(\cdot)$ represent the objective function, network equation functions, inequality constraint functions of operational limits in the normal state, and voltage stability constraint function of the current operating point respectively. Here, it is important what voltage stability index is utilized for voltage stability constraints. According to [10], there are two types of indices, given state based indices and large deviation based indices. In the case of applying given state based indices such as L-index, it is easy to incorporate those indices into OPF formulation, but adequate threshold value needs to be determined with experience based decision making. On the other hand, when applying large deviation based indices like MW and MVAr margin indices, more complicated formulations are needed to incorporate them into OPF.

Next, consider OPF formulations with post-contingent voltage stability constraints. A typical formulation is as follows:

$$\begin{aligned}
\min \quad & f(x_0) \\
s.t. \quad & g_0(x_0)=0 \\
& h_{\min,0} \le h_0(x_0) \le h_{\max,0} \\
& g_1(x_1)=0 \\
& vs_{\min} \le vs_1(x_1) \\
& g_2(x_2)=0 \qquad\qquad (2) \\
& vs_{\min} \le vs_2(x_2) \\
& \vdots \\
& g_{nc}(x_{nc})=0 \\
& vs_{\min} \le vs_{nc}(x_{nc})
\end{aligned}$$

where $f(\cdot)$, $g(\cdot)$, $h(\cdot)$ and $vs(\cdot)$ have the same definitions with (1), and the subscripts 0, 1, 2..., nc denote the normal state and contingent states respectively. The objective of the formulation is to determine preventive dispatch and/or control considering post-contingent voltage stability.

Assuming that there is a function that represents voltage stability of a post-contingent state in a closed form, it is easy to incorporate the function into the normal state OPF formulation in order for preventive control considering post-contingent voltage stability. However, there is no voltage stability index that can be evaluated without a power flow solution of the given state; that is, at least post-contingent power flow equations need to be incorporated into the normal OPF problem to accomplish the objective of the formulation. This type of problem is in the category of contingency constrained optimal power flow (CCOPF). To solve the problem, decomposition methods are usually used. Non-decomposed approaches are possible, but they need to handle very large dimension problems.

Fig. 3 shows the flow chart of one possible decomposed approach for preventive control considering post-contingent voltage stability. In the formulation, the most important thing is how to combine pre- and post-contingent states; that is, a kind of medium for linking pre- and post-contingent states is necessary. It can be sensitivity calculated in the evaluation of voltage stability in post-contingent states, or it can be other nonlinear parameters representing voltage stability of post-contingent states.

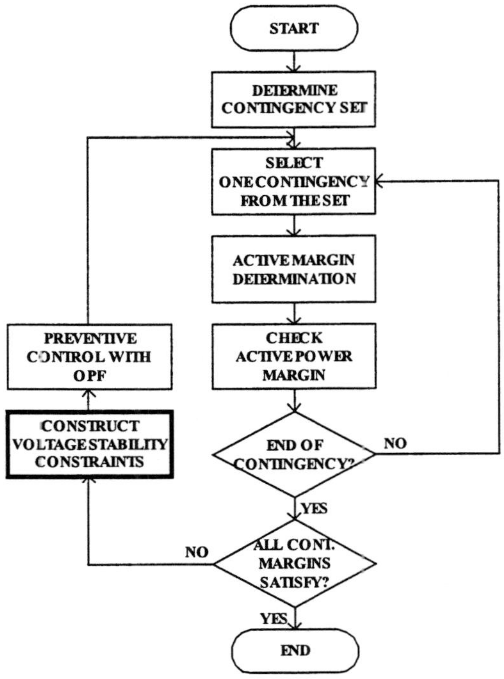

Fig. 3. Flow chart of a possible preventive control with OPF

4. Case Study

This section shows an example applying 2003 peak KEPCO system. First, using the procedure presented in section 2, voltage stability limits of active power flow on Metropolitan interface lines are determined. Then, with RCCOPF in [10], which uses effective reactive reserve as the medium to combine pre- and post-contingent states, preventive control is performed in order that post-contingent interface flow margins satisfy the given requirement. Table I shows specification of the KEPCO system.

Table I. Specification of the KEPCO system

Total demand	47783.2 MW
Total generation	48595.5 MW
Metropolitan demand	19897.2 MW
Metropolitan interface flow	8997.7 MW

Using modified CPF, which traces a path of power flow solution in the given direction of generation dispatch, interface flow vs. voltage (f-V) curves of contingencies are constructed. In this simulation, 74 contingencies containing outages of 345kV and 765kV routes (double circuit tower outages) are considered. Fig. 4 shows f-V curves of five severe contingencies. The worst contingency is #18, which has little interface flow margin; thus, the interface flow level with α % or MW is determined as the interface flow limit in terms of voltage stability as shown in Fig. 5.

Next, with RCCOPF, preventive control in the normal state is performed to enhance the interface flow limit. It is assumed that the required margin of post-contingency is 200 MW. In Fig. 6, f-V curves before and after RCCOPF control are shown.

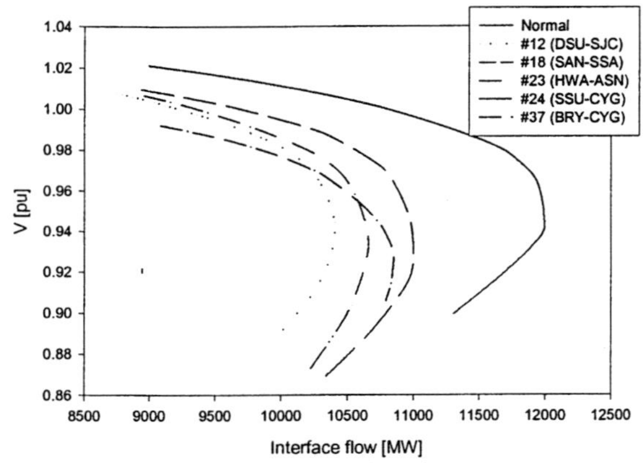

Fig. 4. f-V curves of five severe contingenies

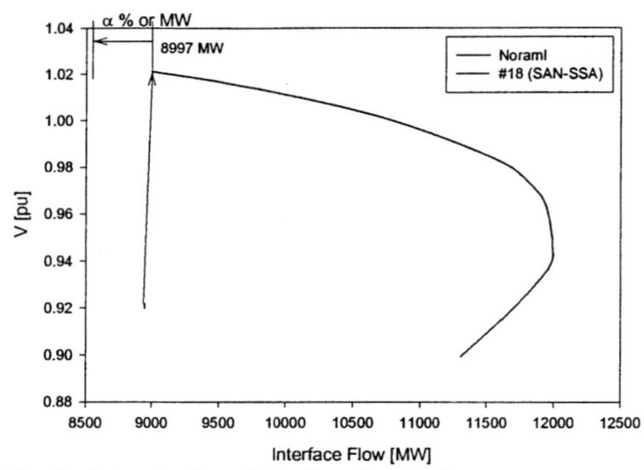

Fig. 5. Determination of interface flow limit

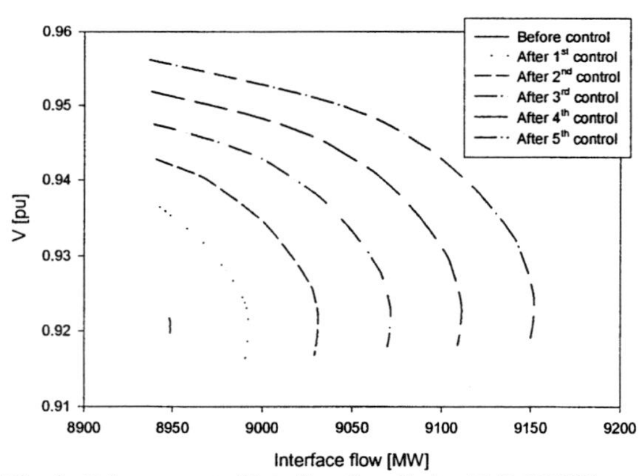

Fig. 6. Enhancement of interface flow limit with RCCOPF

After the 5th preventive control, the worst post-contingent state has the interface flow margin of 215 MW. Table II shows interface flow margins of five severe contingencies before and after RCCOPF control. Fig. 7 shows that the interface flow increases by about 200 MW after RCCOPF execution.

Table II. Interface flow margins of five severe contingencies before and after RCCOPF control

Outage	Interface flow margin [MW]					
	Before	1st	2nd	3rd	4th	5th
#12	1656.6	1688.2	1713.5	1740.9	1766.1	1794.0
#18	1.7	79.8	90.6	132.1	173.3	215.0
#23	1697.7	1728.2	1756.8	1787.4	1809.8	1844.7
#24	2055.7	2091.3	2120.6	2152.3	2182.0	2214.2
#37	1760.8	1795.3	1828.3	1859.5	1894.2	1928.3

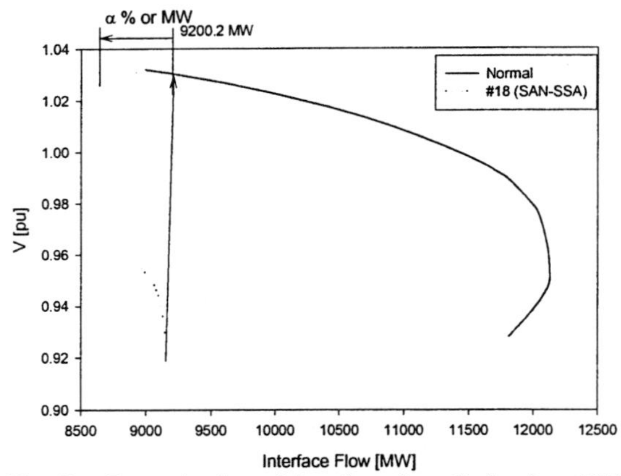

Fig. 7. Determination of interface flow limit after RCCOPF execution

5. CONCLUSIONS

In the open access environment, transfer capability and transfer limits of key transmission lines are important for efficient market operation in term of congestion management. For application of Korea power industry, transfer limit of interface flow, which is distinguished from transfer capability (ATC or TTC), is defined. This paper presents the procedure for determining voltage stability limit of interface flow.

Recently, OPF formulations considering voltage stability constraints have been drawing the attention of researchers, because of the need of dispatching tools for power markets. OPF formulations are categorized into two parts, application of one operating point and multi-operating points to take into account contingency. For OPF with post-contingent voltage stability constraints, a scheme of linking pre- and post- contingent states is necessary.

REFERENCES

[1] Available Transfer Capability Definitions and Determination: A Frame Work for Determining Available Transfer Capabilities of the Interconnected Networks for Commercially Viable Electricity Market, North American Electric Reliability Council (NERC), June 1996.

[2] B.H. Lee, K.Y. Lee, "Dynamic and static voltage stability enhancement of power systems," IEEE Trans. on Power Systems, Vol. 8, No. 8, Feb. 1993.

[3] X. Wang, G.C. Ejebe, J. Tong, J.G. Waight, "Preventive/Corrective Control for Voltage Stability Using Direct Interior Point Method," IEEE Trans. on Power Systems, Vol. 13, No. 3, Aug. 1998.

[4] W. Rosehart, C. Cañizares, V. Quintana, "Costs of Voltage Security in Electricity Markets," IEEE PES Summer Meeting 2000.

[5] D. Chattopadhyay, D. Gan, "Dispatch optimization incorporating transient and voltage stability constraints," IEEE PES Summer Meeting 2000.

[6] S. Kim, T.-Y. Song, M.-H. Jeong, B. Lee, Y.-H. Moon, J.-Y. Namkung, G. Jang, "Development of Voltage Stability Constrained Optimal Power Flow (VSCOPF), IEEE PES Summer Meeting 2001.

[7] J. Kubokawa, R. Inoue, H. Sasaki, "A solution of optimal power flow with voltage stability constraints," PowerCon 2000.

[8] N. A. Alemadi, R. A. Schlueter, "Dispatching for preventive and corrective control of voltage collapse in a deregulated power system," in Proc. Bulk Power System Dynamics and Control V.

[9] E. Vaahedi, Y. Mansour, C. Fuchs, S. Granville, M. L. Latore and H. Hamadanzadeh, "Dynamic security constrained optimal power flow/VAr planing," IEEE Trans. on Power Systems, vol. 16, no. 1, Feb. 2001, pp. 38-43.

[10] H. Song, B. Lee, S.-H. Kwon, V. Ajjarapu, "Reactive Reserve Based Contingency Constrained Optimal Power Flow (RCCOPF) for Enhancement of Voltage Stability Margins," will be published in IEEE Trans. Power Systems.

[11] CIGRÉ Task Force 38.02.11, *Indices Predicting Voltage Collapse Including Dynamic Phenomena*, 1994.

ELSEVIER
IFAC
PUBLICATIONS
www.elsevier.com/locate/ifac

Voltage stability and Preventive Control in China

Zhaoxu Liu
Electric Power Research Institute of China
Qinghe, Beijing 100085, P.R.China
Email: liuzhx@epri.ac.cn

Abstract: For the last two decades the rapid increase in power demands in China, stressed the transmission and generation close to the security limits. In addition, the national interconnection and power sector restructure are also presenting challenges to power system security and stability. Combined with economic development, extreme high load increasing rate, long distance power transmission and HVDC links in recent years, voltage instability problem have drawn much attention than ever before. In this paper, the revised power system stability criteria related with voltage stability are presented. Philosophies and preventive control are introduced. Difficulties and unfavorable trends are presented and discussed. *Copyright © 2003 IFAC*

Key words
Voltage stability, preventive security and stability control, security and stability criterion

INTRODUCTION

The power industry of main land China has developed very quickly in last two decades with the state annual average increasing rate of about 7-8%. The total installed generation has reached 353GW by 2002 standing at the second place in the world. Seven regional power systems have been formed with individual generation installation ranging from 21 to 56GW. The national interconnection and power sector restructure are developing actively. The north-east regional system has connected with north China regional system via 500kV AC line, east and middle China regional systems have been interconnected by two 500kV HVDC links. Other interconnections are in construction or planning (Fig.1). The engine of developing national interconnections is bulk power transmission caused by uneven distributed power resources and load centers. The 80% of the total coal reserve are located in three northern provinces while 67% of the hydraulic reserve is located in southwest parts of the country. The major load centers, however, are located along coastal provinces. The Three Gorge power plant will be in the center of the interconnections. The main stream of electric power flow will be generally from west to east and from north to south. By the end of 2020, the total power flow from west to east is estimated to be 120GW , composed of north pass 42GW, middle pass 40GW and south pass 30GW. Long distance power transmission is required. Remote generators will be operating synchronously over hundreds or even thousands of km. The national grid company and south China grid company have been set up and operating. There are 5 Power Generation companies that have been set up and in operation.

Large load centers such as Beijing, Shanghai and Guangzhou are importing much power from remote power plants and areas via long distance AC and/or HVDC transmission. The major concern is still the stability problems. Mostly are rotor angle stability problems while small signal stability problems increased obviously. They are related with long distance heavy power transmission and longitudinal weak interconnections. Voltage stability problems are related to receiving end load centers with large percent power injection. The main grid did not experience voltage instability or collapse incident. However some small or radial remote areas experienced frequency and voltage combined instability. Attentions have been paid to strengthening of the grid and reactive power generation.

In this paper, the incidents and grid performance will be presented first, followed by voltage stability related criteria of the recent revised state power system security and stability criterion. Then the emergency controls adopted in China are introduced. The difficulties and unfavorable trends in the country are presented and discussed. Finally, conclusions are given.

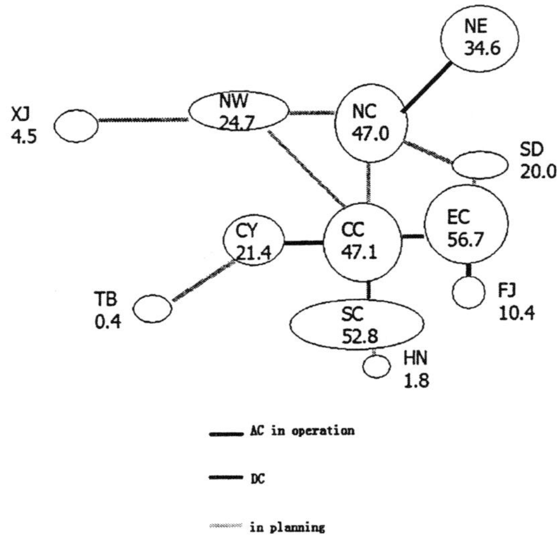

Fig. 1 The national interconnection (V: 500kV, P: GW)

INCIDENTS AND GRID PERFORMANCE

According to statistics, there are several features:

(1). The power system incidents from the year 1981 to 2002 decreased from 19.5 to 3. 86% of them are caused by line or bus multiple contingencies or developing multiple contingencies. Although incidents decreased but the cost of load curtailment, loss of energy and social influence increased.

(2). The initial trigger events are : human, natural, equipment, protection and uncertain effects. .

(3). The incidents caused by malfunction of protection increased from 48.1% to 68.4% while by primary equipments decreased from 75% to 36.8% during period of 1985-1995.

(4). The most incidents are rotor angle instability. Other forms of instability are listed as: frequency instability, tie line power oscillation caused by small signal instability and voltage instability.

(5). The system did not experienced long

term voltage instability. But in some cases, local system with n-1 contingency caused sever voltage drop. Radial or boundary area system islanding caused combined frequency and voltage collapse.

CRITERION

The national power system security and stability criterion was set up in 1981 and revised in 2001 to adapt to new requirements. To prevent voltage instability, the following items are involved:

(1). The defense plan of three-defense-line criterion against major failure of power system should be satisfied so that the reasonable stability levels can be ensured. Any single element outage should not result in angle instability, frequency or voltage collapse.

(2). Voltage stability should be studied and ensured in planning and operation. Reactive power generation redundancy should be able to maintain key bus voltage requirement for both normal and post fault operation.

(3). The steady bus voltage drop caused by single element outage including one set of reactive compensation device outage should not be severe such as exceed 5% of the normal voltage.

(4). In case of the dynamic voltage drops to under certain level such as 0.75 p.u. lasting for longer than 1 second in a stability simulation, the situation is judged not acceptable.

(5). The steady state voltage stability margin calculated by (V-Vcr)/V in normal operation should not be less than 10-15%, and the margin should not be less than 8% after single element outage.

(6). In voltage stability simulation, detail dynamic model should be considered including dynamic load model, generator and its control systems, rotor and armature current limiters, OLTC and reactive compensators.

(7). The receiving end system should be enhanced with sufficient voltage support from local generator and reactive compensation. The receiving end system should have sufficient dynamic reactive power reserve to guarantee voltage stability.

(8). Reactive compensation should be distributed to keep reactive power balance in each voltage level and in each local zone respectively to avoid reactive power cascade transfer between different voltage networks, from long distance or multi-stage transformers.

(9). Automatic or manual emergency load shedding or planned grid separation scheme should be available in load centers to prevent load center from voltage collapse.

PREVENTIVE CONTROL

(1). Emergency Control (EC) with system protection system (SPS) for fast load shedding.

(2). Automatic under voltage load shedding (UVLS).

(3). Adequate dynamic reactive power reserve.

(4). Automatic voltage control (AVC).

(5). Generation and grid enhancement, emphasize is given to load center system enhancement.

DIFFICULTIES AND UNFAVORABLE TREND

There are some trends that are not favorable for maintaining voltage stability:

(1). Air condition load increasing very quickly. In some large cities the air condition load accounts for 30-40% of the summer peak load.

(2). Large load center dependence on getting power from remote plants or area.

(3). Fast load increasing leads to

inadequate generation which in turn stresses the weak grid. Most of the power systems are facing power/ electricity supply shortage, spinning reserve and transmission redundancy reduction.

(4). Loosing of vertical integrated coordination and cooperation between generation plants and grid in planning, operation and dynamic performance. Uncertainty of unit commitment increased.

(5). Difficulty in local or customer owned reactive power compensation management.

(6). Multiple HVDC links integrated into the same load center such in Guangdong and east China system.

(7). Dynamic load modeling considering air condition load and motors in severe voltage drop is difficult but very important for identifying true dynamics process and instability.

(8). Long term power system planning is substituted by short-term planning or individual engineering in some places.

CONCLUSIONS

(1). Adequate generation, both active and reactive generation, and transmission network are fundamental for preventing voltage instability. Long-term generation and transmission planning, good operating planning and monitoring are important. The stability criterion has been proven effective to keep our system secure and stable. It is even important in the liberal electric power market.

(2). In the new environment how can many good criterion and experience be followed is important. New rules need to be set up and should be followed by all the market players.

(3). Simple criterion and steady state voltage stability assessment methods based on reactive power balance are helpful for operating monitoring. Simple, quick and robust index to give an estimation of the voltage stability margin of the present or near future operating condition are helpful and welcome in operating environment. The dynamic voltage stability assessment is necessary for on-line/real-time and fault analysis. Both methods need to be improved.

(4). Dynamic load modeling is critical for simulating voltage stability, especially in the case dynamic voltage dip.

(5). Adjustable reactive facility is important in modern large system where too complex grid operation and modern society consumption behavior exhibited modern "wealthy ill" that is unfavorable for voltage stability.

BIOGRAPHY

Zhaoxu Liu joined the Electric Power Research Institute of China in 1964. He is now the advisor of Power System department of EPRI of China. He is senior member of IEEE and CSEE. He is a member of revise group of the state security and stability criterion. His present interests are power system security and stability assessment, on line dynamic security assessment and emergency control, interaction and performance coordination between plants and grid in deregulated environment, planning of grid interconnections.

ELSEVIER
IFAC
PUBLICATIONS
www.elsevier.com/locate/ifac

SHUNT COMPENSATION FOR VOLTAGE STABILITY

Carson W. Taylor

Bonneville Power Administration
Vancouver, Washington USA

Abstract: I describe the many available options in applying shunt compensation. I describe situations where higher-cost power electronic based devices are justified, and discuss methods to reduce cost of shunt compensation. Several promising new concepts are described.

Keywords: power systems, voltage collapse, voltage stability, reactive power, shunt compensation, capacitor banks, static var compensators, STATCOM.

1. INTRODUCTION

Power system voltage stability is closely associated with reactive power, with reactive power shunt compensation both a contributor and countermeasure to voltage instability (Taylor, 1994; Van Cutsem and Vournas, 1998). Contributing to voltage instability, shunt capacitor banks have an unstable falling output as voltage sags.

Applying shunt compensation involves many considerations. There are many options both for mechanically switched devices and power electronic based devices. Synchronous condensers are also an option. The many available options are not always appreciated or understood by planners, resulting in inefficient, ineffective, or overly costly installations. Part of the problem is researchers and manufacturers who oversell the benefits of the more costly power electronic based devices.

Aspects of shunt compensation application include:

- Short-term (transient) versus long-term voltage stability;
- Coordination of shunt compensation with generator reactive power control;
- Coordination of mechanically switched and power electronic based compensation;
- Coordination with transformer tap changing.
- Large transmission-connected shunt compensation versus distributed compensation at distribution voltage;
- Enhanced mechanically switched compensation;
- Enhanced power electronic based compensation;.
- Restructured industry impacts.

2. SHUNT COMPENSATION APPLICATION

Principles and best practice in shunt compensation applications include:

1. Low-cost mechanically switched shunt capacitor/ reactor banks should be used for bulk reactive power needs (Nedwick, *et al.*, 1995). This allows reactive power reserve at generators, synchronous condensers, and power electronic based devices. Reactive power reserves are at equipment with automatic continuous control.

2. Reactive power output of generators, synchronous condensers and power electronic devices

can be used as a "pilot" to control shunt capacitor/reactor bank switching.

3. Reactive power compensation should be located close to the area of need. Reactive power flow through transformers should be minimized to avoid significant I^2X losses. Preference is to locate shunt capacitor banks on the regulated side of LTC transformers. This usually means that capacitor banks are first installed at distribution stations and on long distribution feeders. Some distribution capacitor banks may be fixed (unswitched) for lower cost (Hsu, *et al.*, 2000).

4. Economies of scale, however, may justify reactive power flow through transformers. Capacitor/reactor bank cost is best expressed as a fixed cost, plus incremental cost per kVar. The fixed cost includes many items, the largest being the switch or circuit breaker cost.

5. Voltage profile should be kept flat to the extent possible, with voltages held close to maximum during heavy load conditions. This minimizes reactive power losses, and supports line charging and shunt capacitor banks. Good practice is to control sending end transmission voltages with generators, and receiving end voltages with capacitor/reactor banks. Where lines are loaded well below or above surge impedance loading, shunt compensation near power plants may also be justified to increase generator reactive power reserves.

6. Mechanically switched capacitor banks should be used for long-term voltage stability where key instability mechanisms are generator field current limiting and load restoration by tap changing.

7. Power electronic devices, SVCs and STATCOMs, have application for short-term voltage stability problems, primarily voltage recovery following a short circuit with stalling and slow tripping of air conditioner compressor motors (Diaz de Leon and Taylor, 2000).

8. For low probability events such as multiple outages during extreme weather and heavy load, shunt compensation may not be justified. Undervoltage load shedding may then be used to ensure stability (Taylor, 1994).

9. Where possible, short time ratings of equipment should be exploited to reduce cost.

10. Life-cycle costs should be compared, as some options have significant operating losses or maintenance costs.

11. Innovative digital control for mechanical switching, similar to thyristor switched capacitors) should be considered.

We next discuss individual equipment used for shunt compensation

2.1 Generators and Synchronous Condensers

Similar to synchronous condensers, synchronous generators perform shunt compensation functions, and should be coordinated with shunt compensation. Control to enhance voltage stability is similar for generators and synchronous condensers.

Voltage stability is improved by tight control of the transmission voltage. Rather than terminal voltage control, line drop compensation or high side voltage control should be used (Taylor, 2000). Reactive droop compensation (reverse line drop compensation) should be avoided.

Line drop compensation of 50–80% of the step-up transformer reactance is beneficial. This is simply implemented with zero equipment cost within automatic voltage regulators (AVRs), resulting in high-speed control. Plant design should ensure that terminal voltage variations do not exceed limits for the generator, and that voltage at auxiliary loads also do not exceed limits.

Power plant supervisory control (SCADA) is often used to control high side voltage to a schedule provided by the transmission company (Brandt, *et al.* 1991; Davies and Midford, 2000). This is beneficial for long-term voltage stability but has no effect for short-term voltage stability.

High-speed high side voltage control has been developed (Kitamura, *et al.*, 2000) and demonstrated at a large hydroelectric power plant in the U.S. Pacific Northwest. This control uses terminal voltage and current measurements similar to line drop compensation, and can either be specified as part of digital AVRs or be added as standalone equipment for connection to existing AVRs. It includes cross-current compensation for use when two generators are connected to the same step-up transformer winding. Similar to SVCs, a high side droop function is included. Figure 1 shows field test results.

Synchronous condensers usually have higher capital and O&M costs than other shunt compensation. However, condensers can greatly improve both short-term and long-term voltage stability. Compared to power electronic devices, time-overload capability is substantial.

Hydroelectric generators are sometimes equipped with water depression compressors for synchronous condenser operation. Steam or gas turbine driven generators are sometimes converted for synchronous condenser operation, or equipped with a clutch for condenser operation (Cameron and Kay, 1999; Lefrancois. and McGough, 1990).

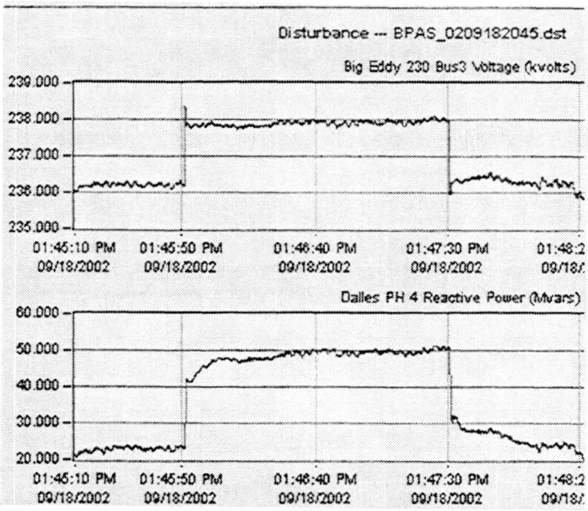

Fig. 1. Response of two 95-MVA generators with high side voltage control for switching a 168 MVAr, 241.5-kV capacitor bank. Units are connected to the same transformer winding. Response (with large high side droop of 6%) is about four times the response of normal AVR connection.

New technology generator and synchronous condensers are being developed using high voltage stator windings or superconducting field windings. Enhanced performance for voltage stability is likely.

With restructured industry and separate generation and transmission companies, generator voltage control and reactive power supply requirements must be specified by industry standards and contract agreements. For example, generating companies should be required by standards to operate in automatic voltage control mode at all times. Generating companies should provide reactive power for the step-up transformer and auxiliary equipment losses, and provide "free" reactive power to the network within a lagging and leading power factor range. Generator companies receive payments from transmission companies for reactive "energy" (MVAr-hours) outside the free range. Transmission companies thus have incentive to install transmission shunt compensation to minimize payments. Requirements for step-up transformer tap settings, line drop compensation, and high side voltage control are typically negotiated.

2.2 *Shunt Capacitor Banks*

Because of low cost including virtually zero losses, shunt capacitor banks are the first choice for transmission and distribution shunt compensation. (Shunt reactors are also used, particularly with long EHV lines.) Shunt capacitor/reactor banks typically provide the base reactive power needs, allowing reactive power reserves at generators and power electronic devices.

Although improvements in shunt capacitor banks are not as dramatic as in power electronics, technology has progressed. Notable are highly reliable SF6 and vacuum switches, and high power density polypropylene film capacitor units produced with advanced manufacturing methods. The development of simpler, less expensive fuseless banks is significant (Thiel, *et al.*, 1992).

Remote or wide-area signals for switching decisions can be effective. Specifically, high reactive power output of generator units can add intelligence for shunt capacitor bank switching (Taylor, 1994).

Advanced capacitor bank concepts are developed. One concept is termed *CAPS* for CAPacitor bank series Shorting (Taylor and Van Leuven, 1996; Bruns *et al.*, 2001). Figures 2 and 3 show the scheme. Using a lower voltage/lower cost switch than the main switch, a portion of a bank is shorted to reduce bank reactance, thereby boosting output. For emergency use, the temporary overvoltage capability of capacitors can be exploited for many minutes.

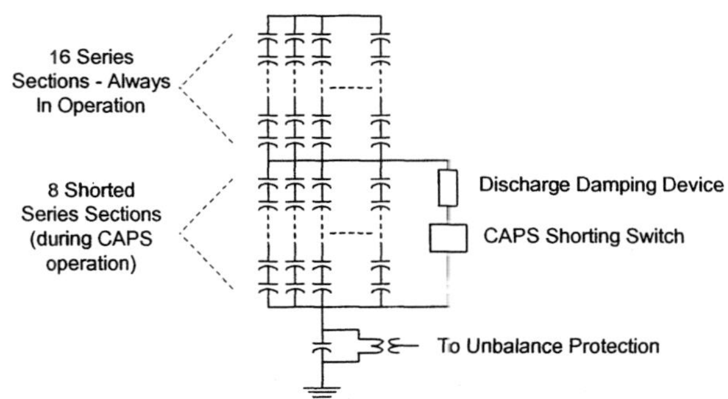

Fig. 2. One-line diagram of 550-kV, 460/691 MVAr CAPS bank.

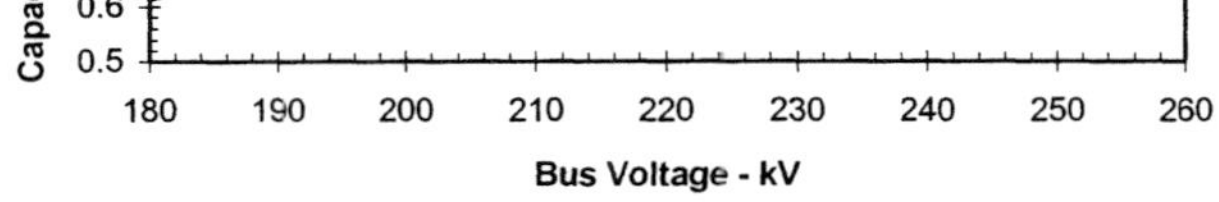

Fig. 3. CAPS bank output characteristic with 50% boost (230-kV bank). Bus voltage change for switching is ignored by showing vertical lines.

Reactive power output of capacitor banks is proportional to the voltage squared—contributing to voltage instability. CAPS compensates.

Banks can also be designed for continuous rating in the shorted condition. Softer switching results from a two-step bank. Compared to two conventional banks, CAPS may be around 20% less expensive—mainly because of reduced switchgear costs.

2.3 Static Var Compensators (SVCs)

Transmission-level static var compensators were introduced in the 1970s, and are now mature technology. Components include thyristor-controlled reactors (TCRs) providing continuous control, thyristor-switched capacitors (TSCs), thyristor-switched reactors (TSRs), harmonic filters, and coupling transformers.

SVCs are higher cost than mechanically switched shunt compensation, but provide faster and smoother control with unrestricted switching frequency.

SVCs typically include several TSCs and one TCR, with the TCR slightly larger that the largest TSC. The devices are normally continuously rated.

There are several ways to reduce SVC (and STATCOM) cost in some applications:

1. Connect SVCs to tertiary windings of autotransformers. This is especially attractive where design can be standardized at many substations (same type of autotransformer, similar SVCs—possibly relocatable).

2. Use several smaller distributed power electronic devices rather than a large transmission-connected device (Paserba *et al.*, 1994: Diaz de Leon and Taylor, 2000). Reactive power losses from the device location to critical motor loads are reduced.

3. Use SVC controls to command mechanical switching at the SVC site, or at nearby substations. A smaller SVC then can be used. Mechanical switching is most often used to restore reactive power reserve at the SVC. Small output of the SVC during normal operation minimizes losses. Fast mechanical switching can also be commanded following a disturbance, with the SVC preventing possible overvoltage from mechanical switching of a large capacitor or reactor bank.

4. Use discrete switching rather than continuous control (TSCs and TSRs only). TCRs are not used, avoiding the cost and losses of harmonic filters. TCR losses for continuous control are also avoided. Digital control facilitates discrete switching. Possible benefits of continuous control for power quality, however, must be evaluated.

5. Use short-time rated components. High output is generally needed for only short periods. SVC coupling transformers can be overloaded for many seconds. Time-overvoltage capability of capacitor banks can be exploited.

An example of the last three methods is the Forbes SVC in northern Minnesota, USA (Paulsson, *et al.*, 1994; Sybille *et al.*, 1996). The total short-time rating is 450 MVAr inductive and 1000 MVAr capacitive, including two 300 MVAr 500-kV mechanically switched capacitor banks. The SVC continuous rating is only 149 MVAr inductive and 110 MVAR capacitive, while the ten-second rating is 450 MVAr inductive and 400 MVAr capacitive. The coupling transformer continuous rating is only 168 MVAr with a ten-second rating of 535 MVAr. Two TSRs and three TSCs are configured for 19 discrete steps of 55 MVAr each.

2.4 STATCOMs

STATCOMs (STATic COMpensators) are similar to SVCs in many respects and many of the above comments apply. For example, STATCOMs can (and usually should) initiate mechanical switching of local and remote capacitor banks (Reed, *et al.*, 2001; Diaz de Leon and Taylor, 2000).

STATCOMs use voltage-sourced converters to generate an electronic voltage behind a reactance (such as a coupling transformer reactance). At device boost or buck limits, output is constant current down to a quite low voltage—somewhat similar to synchronous condensers. The constant current provides superior performance in a voltage collapse or voltage recovery situation compared to the constant reactance characteristic of SVCs at capacitive limit (Larsen, *et al.*, 1992). On the other hand, SVCs (TCRs or TSRs) have superior performance for overvoltages—reactive power output proportional to voltage squared.

STATCOMs are more stable and faster responding in weak system applications—where power electronic devices are most needed. (Stability of SVCs is ensured by gain supervisors/optimisers resulting in reduced gain and slower response.)

As with SVCs, there are many design possibilities. Technology is less mature with ongoing improvements in thyristor or transistor components, and in overall STATCOM topology. A key aspect is the method to reduce voltage harmonics. A high pulse number is realized with special transformers. Alternatively, pulse width modulation is used, with the switching frequency a key design factor. Gate-Turn Off thyristors (GTOs) or newer (Integrated) Gate Commutated Thyristors (IGCTs/GCTs) or Insulated-Gate Bipolar Transistors (IGBTs) are used.

As with SVCs, operating losses are important in the evaluation of STATCOMs. For example, continuously operating IGBT-based devices with high switching frequency have high losses even at small output.

When SVCs or STATCOMs are justified, performance-based specifications require vendors to compete based on application-specific criteria including life-cycle costs. Various types of SVCs or STATCOMs are allowed.

3. SIMULATION EXAMPLE

Figure 4 shows a reduced power system. For a three phase, four-cycle fault at the midpoint of one of the two lines, Figure 5 compares performance of mechanically switched compensation, a SVC, and a synchronous condenser, all rated 125 MVAr. See Taylor (1994) for more detail.

The application requires re-acceleration of motor load after fault clearing. Mechanically switched compensation is least expensive, but has performance disadvantages. The switching must be fast and the compensation must be large enough to avoid instability, but not so large as to cause overvoltage.

The SVC and synchronous condenser performance are similar; 230-kV voltage is regulated in each case. The synchronous condenser is more robust in that the fault voltage dip and motor deceleration is reduced, and temporary overload capability is available. SVC (and STATCOM) ratings are referred to the transmission side, while the synchronous condenser rating is the stator rating. At high output, the reactive power loss in the condenser transformer is substantial.

For the same system, comparisons between SVCs and STATCOMs are made (Noroozian, *et al.*, 2003).

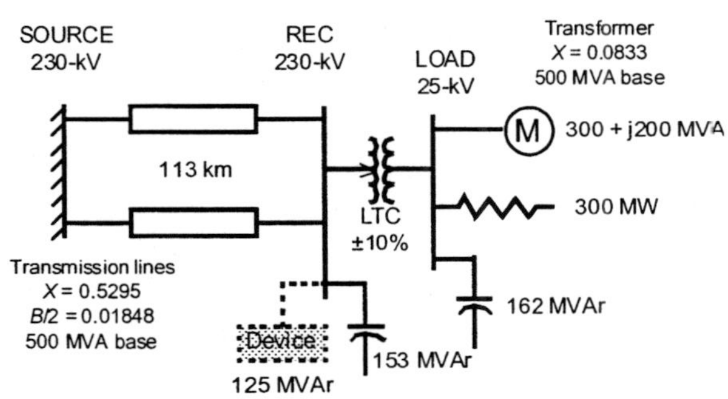

Fig. 4. Equivalent system (Taylor, 1994).

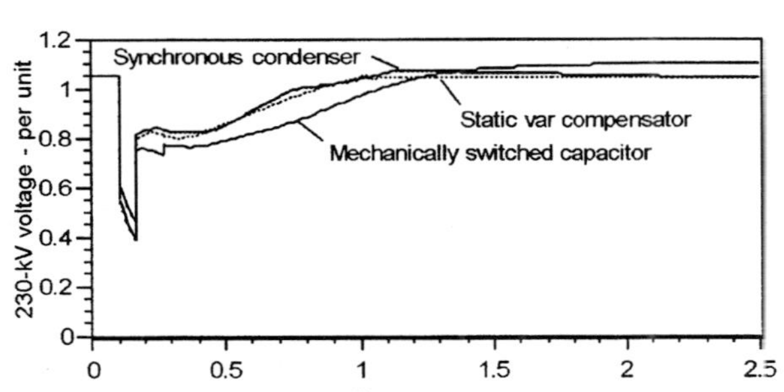

Fig. 5. Voltage recovery for three-phase midpoint fault with permanent line outage. Three reactive power compensation options.

4. MISCELLANEA

4.1 Limits to shunt compensation

What are the limits to shunt compensation? This is often asked. For example, the well-known *P-V* curves for a radial load (thevenin equivalent of source) show voltage at the nose point rising into the operating range as shunt compensation is increased. Also, the voltage change for capacitor bank switching increases as the *effective* short circuit capacity decreases. (Effective short circuit capacity is basic short circuit capacity minus the capacitor bank MVA.)

Design with nose voltage within operating voltage range should be allowed. Otherwise, static compensators—which improve performance and extend loadability but cause a flatter upper portion of a nose curve — could not be used. Monitoring of voltage security must be based primarily on reactive power reserves rather than on voltage magnitude.

Limits to shunt compensation limits should be evaluated by dynamic simulation.

4.2 Criteria for short-term voltage stability

Short-term voltage stability is associated with re-acceleration of motors following short circuits, and with fast voltage recovery to prevent loss of other load (e.g., electronically controlled load). Industry reliability criteria do not exist, although some have inappropriately used a voltage dip criterion developed for angle stability swings.

Voltage recovery criteria could be linked to the CBEMA/ITIC curve (http://www.itic.org/technical/iticurv.pdf). For example, voltage at the (equivalent) load should recover to 80% within 0.5 seconds after fault initiation. This assumes normal transmission network fault clearing time, and a greater than 80% voltage recovery at the low side of bulk power delivery transformers. The criterion would prevent massive loss of electronic equipment.

4.3 Shunt compensation cost range

As discussed above, equipment and life-cycle costs depend on many factors. Only rough guides can be given.

Installed equipment costs of \$5/\$20/\$10 per kVAr are suggested for unswitched distribution, switched distribution, and switched 115-kV capacitor banks, respectively (Hsu, et al., 2000). Bonneville Power Administration estimates of installed cost are around \$1.4M + \$5.25/kVAr for large 500-kV banks, and \$0.32M + \$5.4/kVAr for large 230-kV banks. Cost range for transmission level SVCs, STATCOMs, and synchronous condensers are in the \$30-55/kVAr range, with the STATCOMs and synchronous condensers at the upper end.

5. IN SUMMARY

In applying shunt compensation, first the speed requirements must be determined. For short-term voltage stability, SVCs or STATCOMs may be required.

Shunt compensation application should be coordinated with generator voltage control.

The many available options should be understood, and allowed in performance-based specifications for competitive procurements.

ACKNOWLEDGEMENTS

Discussions with many engineers as part of BPA, IEEE, CIGRE, and seminar activities helped shape this paper. Much of the paper is based on BPA practice. BPA has around 14,000 MVAr of shunt capacitor banks. BPA 500-kV capacitor banks are in the 200–460 MVAr range, with 230-kV banks up to 169 MVAr.

REFERENCES

Brandt, D., R. Wachal, R. Valiquette, and R. Wierckx (1991). Closed Loop Testing of a Joint Var Controller using a Digital Real-Time Simulator, *IEEE Tran. on Power Systems*, Vol. 6, No. 3, pp. 1140–1146, August 1991.

Bruns, D.P., G.R. Newcomb, S.A. Miske, C.W. Taylor, G.E. Lee, and A. Edris (2001). Shunt Capacitor Bank Series Group Shorting (CAPS) Design and Application, *IEEE Trans. on Power Delivery,* Vol. 16, No. 1, pp. 24–32, Jan. 2001. Discussion and closure in October 2001 issue.

Cameron, R. and T.W. Kay (1999). Conversion of Two Zion 1220 MVA Generators to Synchronous Condenser Operation, *Proceedings of American Power Conference*, Chicago.

Davies, J. B., and L. E. Midford (2000). High Side Voltage Control at Manitoba Hydro," panel session on Power Plant Secondary (High Side) Voltage Control, *Proc. of IEEE/PES 2000 Summer Meeting*, 16–20 July 2000.

Diaz de Leon II, J.A. and C. W. Taylor (2000) Understanding and Solving Short-Term Voltage Stability Problems, *Proc. of IEEE/PES 2000 Summer Meeting*.

Hsu, S.-M., H. J. Holley, W. M. Smith, and D. G. Piatt (2000). Voltage Profile Improvement Project at Alabama Power Company: A Case Study, *Proc. of the IEEE/PES 2000 Summer Meeting*.

Kitamura, H., M. Shimomura, and J. Paserba (2000). Improvement of Voltage Stability by the Advanced High Side Voltage Control Regulator, *Proc. of IEEE/PES 2000 Summer Meeting*.

Larsen, E., N. Miller, S. Nilsson, and S. Lindgren (1992). Benefits of GTO-Based Compensation Systems for Electric Utility Applications, *IEEE Trans. on Power Delivery*, Vol. 7, No. 4, pp. 2056–2064, October 1992.

Lefrancois, M.J. and J.C. McGough (1990). Highlights of the B.C. Hydro Burrard Generating Units Operating as Synchronous Condensers, *Proc. of American Power Conference*, Chicago.

Nedwick, P., A.F. Mistr, Jr., and E.B. Croasdale (1995). Reactive Management: A Key to Survival in the 1990s, *IEEE Trans. on Power Systems*, Vol. 10, pp. 1036–1043, May 1995.

Noroozian, M., Å.N. Petersson, B. Thorvaldson, B. A. Nilsson, and C.W. Taylor (2003). Benefits of SVC and STATCOM for Electric Utility Application, *Proc. of IEEE 2003 T&D Show*.

Paserba, J.J., D.J. Leonard, N.W. Miller, S.T. Naumann, M.G. Lauby, and F.P. Sener (1994). Coordination of a Distribution Level Continuously Controlled Compensation Device with Existing Substation Equipment for Long Term VAr Management, *IEEE Trans. on Power Delivery*, Vol. 9, pp. 1034–1040, April 1994.

Paulsson, L., A. Silva, B. Thorvaldsson, and R. Gonzalez (1994). Combined Short and Long Term Ratings of Static Var System Components for Steady State and Transient Voltage Support in a 500-kV Power Grid, invited paper 24, *IV Symposium of Specialists in Electric Operational and Expansion Planning*, Foz do Iguaçu, Brazil, 23–27 May 1994.

Reed, G., J. Paserba, T. Croasdaile, M. Takeda, Y. Hamaski, T. Aritsuka, N. Morishima, S. Jochi, I. Iyoda, M. Nambu, N. Toki, L. Thomas, G. Smith, D. LaForest, W. Allard, and D. Haas (2001). The VELCO STATCOM-Based Transmission System Project, *Proc. of the IEEE/PES 2001 Winter Meeting*.

Sybille, G, P. Giroux, S. Dellwo, R. Mazur, and G. Sweezy (1996), Simulator and Field Testing of Forbes SVC, *IEEE Trans. on Power Delivery*, Vol. 11, No. 3, pp. 1507–1514, July 1996.

Taylor, C.W. (1994). *Power System Voltage Stability*, McGraw-Hill, New York.

Taylor, C.W. and A. L. Van Leuven (1996). CAPS: Improving Power System Stability Using the Time-Overvoltage Capability of Large Shunt Capacitor Banks, *IEEE Trans. on Power Delivery*, Vol. 11, pp. 783–792, April 1996.

Taylor, C.W. (2000). Line Drop Compensation, High Side Voltage Control, Secondary Voltage Control—why not control a generator like a static var compensator? *Proc. of IEEE/PES 2000 Summer Meeting*.

Thiel, P.H., J. E. Harder, and G. E. Taylor (1992). Fuseless Capacitor Banks, *IEEE Trans. on Power Delivery*, vol. 7, pp. 1009-1015, April 1992.

Van Cutsem, T. and C. Vournas (1998). *Voltage Stability of Electric Power Systems*, Kluwer Academic Publishers.

ELSEVIER
IFAC
PUBLICATIONS
www.elsevier.com/locate/ifac

VOLTAGE STABILITY AND REACTIVE POWER ISSUES IN THE NATIONAL GRID SYSTEM

Y. H. Song

Brunel University, Uxbridge UB8 3PH, UK

Abstract – The reactive power requirement on the National Grid system have risen as a result of increased power flows and the shorter timescale on which power stations are commissioned and withdrawn from services. Power transfers across internal system boundaries are found in some cases to be determined by voltage limits and the need to avoid voltage collapse, rather than thermal capability of the transmission plant. In addition, a separate reactive power market is also running by the National Grid Transco. This paper reports on the voltage stability and reactive power issues in the UK National Grid System. A brief introduction to the National Grid System is given first. Then the paper presents voltage stability analysis methods developed for the system with detailed descriptions of an improved Singular Value Decomposition method for identifying weak busbars in terms of voltage stability. Numerical results are included on the power system with over 2600 busbars. *Copyright © 2003 IFAC*

Keywords: Voltage Stability; Reactive Power; National Grid

1. Introduction

1.1 The National Grid System

The National Grid Transco owns and operates the transmission system in England and Wales. This consists of a main 400kV network and subsidiary 275kV networks around major demand centers. The system is heavily meshed and consists largely of double-circuit transmission lines. It is connected, via supergrid transformers, to the distribution networks (operated at voltages of 132kV and below) which are owned by the Regional Electricity Companies. The transmission system is also connected to generators owned by separate generating companies. In addition, there are interconnections to the French and Scottish electricity transmission systems.

As well as maintaining the infrastructure of the transmission system, NGT has the role of running the balancing market under NETA to keep the demand and system balance. At the present time, majority of the active power generation is in the north of the country, while much of the demand is in the south, resulting in a predominant north to south transfer. Reactive compensation plants are therefore installed to ensure satisfactory voltage regulation [1].

The requirement of reactive support on the NGT system is heavily influenced by two effects. Firstly, the company has little control over the short and long generation pattern, except through its charges (to generators and consumers) for use of the transmission system. Secondly, for environmental reasons, it is desirable to maximize the use of existing wayleaves. As demand and power transfers have increased, line loadings have risen and may reach three or four times natural loading following circuit trips. With large power flows on circuits, the reactive power losses on the transmission are likely to increase. There must be sufficient reactive capability on the system, from generators and compensation plant, to ensure that an acceptable voltage profile can be maintained for normal states of operation and contingent states.

1.2 The voltage stability problem

Automatic control actions in a power system are designed to counter any drop in voltage which may result from an increase in demand or circuit failure. Heavily loaded systems can reach a state where sensitivities are reversed so that control actions actually make the situation worse. Control is then lost and the voltage falls progressively, s situation known as voltage collapse. Within the NGT, it has been identified that transfers across internal boundaries are limited not only by the thermal capability of the transmission plants [2], but also by the need to avoid transmission system constraints including the possibilities of system collapse due to voltage problems. The company installs reactive compensation to secure adequate voltage regulation and voltage collapse margins. Provision of adequate reactive reserves and maintenance of voltage collapses margins thus becomes a significant factor in economic operation of the power system. Thus reactive power market has been set up.

1.3 Voltage stability analysis

Analytical techniques have been developed for voltage stability analyses with the ultimate objective of providing better tools for operation planning and real-time operation. Reference [3] describes the development of a method for voltage stability analysis using improved sensitivity techniques to rapidly calculate the transfer limit of any defined boundary. This is extended to include the modeling of synchronous compensators and SVCs, as well as on-load tap changers. Reference [4, 5] applies a singular value decomposition method to voltage stability analysis in such as calculating voltage stability index, identifying the most weakest buses and areas, and ranking transmission lines. Reference [6] reports on the recent improvement and implementation of these two methods at the NGT system.

1.4 Reactive power market

Restructuring of the electricity industry in the UK began in 1990, when ownership and operation of the transmission system in England and Wales was vested to National Grid. Northern Ireland and Scotland own and operate their transmission systems separately for historical and geographical reasons. At that time it was decided that all generators of 50 MW capacity or more were required to provide mandatory reactive power support service. This was a stipulation of the Grid Code, which governs connection to the National Grid transmission system, as agreed with the regulator of the UK electricity industry.

Gradual restructuring of the electricity industry has continued in the UK over the last decade. It is possible to focus on the developments of reactive power markets in England and Wales over that period of time. Initially, an interim arrangement was introduced in 1990 with regard to compensating generators for supplying reactive power [7]. Compensation was based on a fixed payment reflecting the amount of reactive power produced in the year 1989 1990, prior to the onset of restructuring and privatisation. This arrangement was considered temporary and was to be reviewed after the completion of a new metering system that was capable of measuring reactive power flows [8]. When this was achieved an alternative scheme was proposed during the mid-1990 that was based on both MVAr utilisation and capability components. An essential feature of the proposed scheme was its potential for development to a full reactive power market mechanism [7].

The scheme was introduced in phases from 1997 2000. Again, all generators of 50 MW capacity or more were required to provide a mandatory reactive power support service. In order to receive payment for this service, generators must enter into either a default payment mechanism (DPM) or a tendering system. The tendering system consists of either an obligatory reactive power service (ORPS) or an enhanced reactive power service (ERPS). As an incentive for generators to enter the tendering system, the DPM ratio of capability to utilisation was introduced as 80:20 in 1997 and was them modified incrementally to 0:100 up to the year 2000 [8, 9]. Alternatively, the generators can tender specific prices for capability and utilisation with regard to ORPS or ERPS [8, 9]. The rounds for tenders have been held every six months since 1997 and the ninth tender round was for contracts effective from 1 April 2002. The tenders are for supplying a reactive service for a period of twelve months or more. If a tender is accepted the generator or provider of the reactive power service has a reasonably guaranteed level of income. The income a generation unit could receive from a reactive power service is dependent upon both the number of service providers and the demand within a particular area. On average approximately 50% of tenders have been accepted over the last five years and therefore approximately 50% of eligible generation units have reached bilateral market agreements with regard to reactive power services over that time. If a tender is rejected and the generation unit is eligible it enters the DPM. Tenders for ERPS can only be submitted by generation units with a reactive power capability in excess of that stipulated by the Grid Code. Up until now only a small number of generation units have tendered an ERPS and the majority of generation units have tendered an ORPS. A detailed account of submitted and accepted tenders for all tender rounds from 1997 2002 is available from National Grid [10]. It should be noted that the ORPS does not include reactive power services from synchronous compensators or SVCs. However, ERPS can include services from such compensators and indeed any other plant that can generate or absorb reactive power.

Finally, the cost of providing reactive power services is currently recovered by National Grid via use of system charges in the balancing mechanism [11]. The costs of reactive power contracts are recovered by NG via the daily Incentivised Balancing Cost (IBC). These are the costs that NG is incentivised to manage and are the basis upon which incentive payment to (or from) NG is calculated. The reactive power contract costs are included in the daily Balancing Services Contract Costs Allocation (BSCCA) that is a component of the IBC. Finally, it should also be noted that black start capability, frequency response and warming contracts are also included in the BSCCA.

2. Description Of Algorithm

Governing Theory

The revised SVD algorithm contains two parts. Firstly, it employs the continuation load flow to find the voltage collapse limit, and the calculated loading margin is used as voltage collapse relative indicator. Secondly, at the collapse limit, it applies singular value decomposition to Jacobian and performs weakest busbar ranking. The new algorithm follows the following steps:

Step1: Perform the first load flow calculation at the normal operation point.

Step2: Scale loads using Uniform Reactive Loading pattern,

$$Q_{L,i} = Q_{L0,i} * (1 + I_k) \quad i = 1,2,\Lambda ,n \tag{1}$$

where $Q_{L,i}$ is the reactive load at busbar i, $Q_{L0,i}$ is its initial value, and I_k is the increment factor which is calculated by the Binomial Searching Technique.

Step3: Perform load flow calculation and convergence check.

Repeat *Step2* and *Step 3* until the voltage collapse limit is reached. The percentage of load increment I_k will be used as the voltage collapse relative indicator. If it is less than the defined threshold (for example 5%), then a warning message will be print out to indicate this case is close to voltage stability limit. At the voltage collapse point, this value is equal to zero, which means the system reactive load cannot be increased at that point.

Step4: At the voltage collapse point, apply singular value decomposition to Jacobian and perform weakest busbar ranking in terms of the sensitivity of voltage, tap ratio, and power injection.

To improve the reliability, accuracy and efficiency of the algorithm, several techniques are introduced, which are detailed in the following sections.

Singular Value Decomposition and Busbar Ranking

Consider the following polar-form linear power flow equations

$$\begin{bmatrix} \Delta \mathbf{P} \\ \Delta \mathbf{Q} \end{bmatrix} = \mathbf{J} \cdot \begin{bmatrix} \Delta \boldsymbol{\theta} \\ \Delta \mathbf{V} \\ \Delta \mathbf{T} \end{bmatrix} \tag{2}$$

where $\mathbf{P}$ and $\mathbf{Q}$ are the active and reactive power injections, $\boldsymbol{\theta}$ and $\mathbf{V}$ are the nodal angles and voltage magnitudes, $\mathbf{T}$ are the tap ratios of variable transformers. Apply singular value decomposition to Jacobian $\mathbf{J}$

$$\mathbf{J} = \mathbf{R} \cdot [diag(w_j)] \cdot \mathbf{S}^T = \sum_{j=1}^{n} w_j \cdot \mathbf{r}_j \cdot \mathbf{s}_j^T \tag{3}$$

where w_j are the singular values, $\mathbf{r}_j$ and $\mathbf{s}_j$ are the left and right singular vectors.

If $\mathbf{J}$ is not singular, the effect on $[\Delta \boldsymbol{\theta} \quad \Delta \mathbf{V} \quad \Delta \mathbf{T}]^T$ of a small change in active and reactive power injections can be written as

$$\begin{bmatrix} \Delta \boldsymbol{\theta} \\ \Delta \mathbf{V} \\ \Delta \mathbf{T} \end{bmatrix} = \mathbf{J}^{-1} \cdot \begin{bmatrix} \Delta \mathbf{P} \\ \Delta \mathbf{Q} \end{bmatrix} = \sum_{j=1}^{n} (1/w_j) \cdot \mathbf{s}_j \cdot \mathbf{r}_j^T \cdot \begin{bmatrix} \Delta \mathbf{P} \\ \Delta \mathbf{Q} \end{bmatrix} \tag{4}$$

Close to the voltage collapse point, where a singular value is almost zero, the system response is entirely determined by the minimum singular value $w_{\min}$ and its corresponding left singular vector $\mathbf{r}_{\min}$ and right singular vector $\mathbf{s}_{\min}$, that is

$$\begin{bmatrix} \Delta \boldsymbol{\theta} \\ \Delta \mathbf{V} \\ \Delta \mathbf{T} \end{bmatrix} \approx (1/w_{\min}) \cdot \mathbf{s}_{\min} \cdot \mathbf{r}_{\min}^T \cdot \begin{bmatrix} \Delta \mathbf{P} \\ \Delta \mathbf{Q} \end{bmatrix} \tag{5}$$

where the vectors $\mathbf{r}_{\min}$ and $\mathbf{s}_{\min}$ have been normalized so that $\mathbf{r}_{\min}^T \mathbf{r}_{\min} = \mathbf{1}$, and $\mathbf{s}_{\min}^T \mathbf{s}_{\min} = \mathbf{1}$.

Let

$$\begin{bmatrix} \Delta \mathbf{P} \\ \Delta \mathbf{Q} \end{bmatrix} = \mathbf{r}_{\min} \tag{6}$$

then

$$\begin{bmatrix} \Delta \boldsymbol{\theta} \\ \Delta \mathbf{V} \\ \Delta \mathbf{T} \end{bmatrix} = \frac{\mathbf{s}_{\min}}{w_{\min}} \tag{7}$$

Thus, the left singular vector $\mathbf{r}_{\min}$ provides a prototype of power injection changes at busbars, and the right singular vector $\mathbf{s}_{\min}$ provides a prototype of busbar angle, voltage and tap variations. The minimum singular value $w_{\min}$ is a measure of how close to singularity the power flow Jacobian matrix is. If the minimum singular value is equal to zero, the studied matrix is singular and no power flow solution can be obtained. This can be interpreted as an infinite sensitivity of the power flow solution to small perturbations in the parameter values.

The weakest busbar ranking can be carried out based on the values of the right singular vector $\mathbf{s}_{\min}$. The maximum entry of the voltage part of $\mathbf{s}_{\min}$ indicates the most sensitive busbar voltage or tap ratio. That is, if node k is the weakest busbar, then

$$|\Delta V_k| = \max\{ \ |\Delta V_i| \ , \ i \in J_V \ \} \tag{8}$$

or

$$|\Delta T_k| = \max\{ \ |\Delta T_i| \ , \ i \in J_T \ \} \tag{9}$$

where J_V is a set of load buses which correspond to busbar voltage magnitudes, and J_T correspond to tap ratios.

Similarly, the ranking of the power injections with most effect on system can be obtained from the values of the reactive part of $\mathbf{r}_{\min}$. The maximum entry corresponds to the most effective power injection.

The Inverse Power Method can be used to calculate the minimum singular value and its corresponding left and right singular vectors.

Some references use the minimum singular value as an indicator on voltage collapse margin. However, in our studies, it is found that the minimum singular value is not an absolute index and its threshold value varies with different systems and operating conditions. It is not suggested to use this value to indicate the proximity to voltage collapse.

Uniform Reactive Loading

A uniform loading pattern concerning all reactive loads in system is used for calculation of continuation load flow. The function of Singular Value Decomposition is to provide effective indices to perform weakest busbar ranking for heavily loaded systems. As mentioned above, for a stable system at normal operation point without suffering voltage problem, busbar ranking is not reliable.

A suitable loading should be applied to push the system close to the collapse point and make the problematic busbars merge, and consequently make the SVD ranking more reliable.

Theoretically, the lack of adequate reactive support in the network could lead to voltage collapse. A system is prone to collapse if there are heavy power flows into an area with insufficient reactive reserves. The contingency, resulting from the reactive loading which would make the generators exhaust their reactive reserve one by one, represents the worst case scenario linked to the voltage instability. The test results suggested that, for weakest busbar ranking, the reactive loading is more reliable than the real power loading and the system loading (concerning both real and reactive load).

The calculated voltage collapse relative indicator is the margin of reactive loading. This margin does not translate system loading in any way. However, for system operation, the practical margin concerning both real and reactive loading is required and more useful. Unfortunately, it cannot be obtained from this algorithm.

Binomial Searching Technique

A form of binomial searching algorithm is applied to compute the increment factor for uniform reactive loading. The following general formula is used

$$I_k = \sum_{j=1}^{k} a \cdot r^{n-1} \tag{10}$$

where I_k (%) is the total increment factor at iteration k, a (%) is the initial increment factor, and $r = 2$.

The algorithm is applied in two steps.

1) The 'Limit Locating' Step

The increase is doubled at each iteration until a limit is hit. If a limit is hit then the forward step is re-based back to previous iteration and the 'Close In' step is carried out. A limit is such a point where the load flow fails to converge.

2) The 'Close In' Step

The increment calculation is re-based at the last iteration before the limit was hit. The increase is halved and the calculation is carried out forward using half as a ratio. A formula is similar to (10) but with $r = 1/2$ and a being half of the current increment.

The process will alternate and continue until the increase is less than a tolerance. In the program, the initial increase is set to 4% by default and the tolerance is 1%. An illustration is shown in Fig. 1.

Apparently, the binomial searching technique would allow accelerating the search for the collapse point. Furthermore, this method is more efficient and accurate than the fixed increment algorithm. It is because, given a finite number of iterations allowed, the fixed increment algorithm may either lead to the predicted limit not being close to the actual limit (because of a large incremental step size), or to the limit not being reached (because of a too small step size).

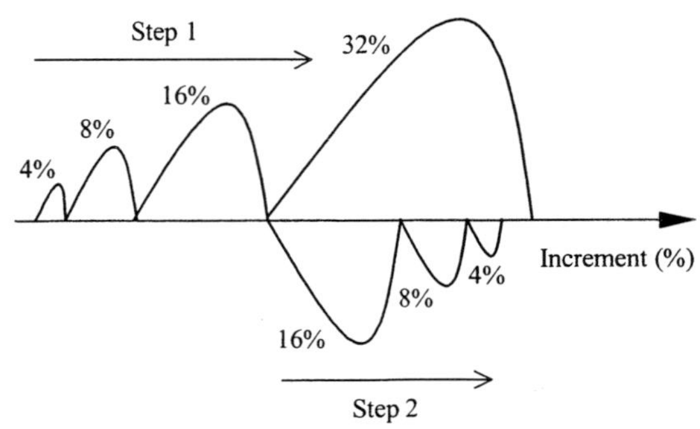

Fig. 1. Illustration of binomial searching algorithm

Polar-Form Optimal Multiplier

The polar-form Optimal Multiplier, being able to optimize the solution path of the AC load flow calculation, is used to improve the load flow convergence and make the calculation of voltage collapse indicator more accurate.

It is well known that, close to the voltage collapse point where the minimum singular value of Jacobian is almost zero, the non-linearity is beginning to have significant effect on system behavior. This will aggravate the ill condition of the system and make the load flow convergence worse. The Optimal Multiplier has been claimed to be able to deal with ill- conditioned network and improve the load flow convergence. However, due to the requirement of a quadratic formulation, the standard Optimal Multiplier is only based on the rectangular-form load flow equations. To apply Optimal Multiplier in polar-form load flow, some further work is done in this research. The technical issues pertaining to this implementation are the variable and function conversions between rectangular and polar coordinates.

Consider the following equation given in.

$$y_s - y(x^r) - \mu J^r \Delta x^r - \mu^2 y(\Delta x^r) = 0 \tag{11}$$

where the superscript r and p denote rectangular and polar quantities respectively, μ is the Optimal Multiplier. Since the same mismatch function yields the same value for the same complex vector, it yields

$$y_s - y(x^p) - \mu J^r \Delta x^r - \mu^2 y(\Delta x^p) = 0 \tag{12}$$

In polar-form Optimal Multiplier method, it will be arranged that the correction vector Δx is the same regardless of which coordinate systems its elements are expressed in. To calculate the second coefficient $J^r \Delta x^r$, the Taylor series expansion of the coordinate transform $\Phi(\cdot)$ is considered

$$\Delta x^r = \Phi(\Delta x^p) \approx W \Delta x^p + \tfrac{1}{2} \begin{bmatrix} [\Delta x^p]^T H_1 \Delta x^p \\ [\Delta x^p]^T H_2 \Delta x^p \\ \vdots \\ [\Delta x^p]^T H_n \Delta x^p \end{bmatrix} \tag{13}$$

where $W = \dfrac{\partial \Phi}{\partial x^p}$ is the Jacobian of $\Phi(\cdot)$ and $H_i = \dfrac{\partial^2 \Phi_i}{\partial x_j^p \partial x_k^p}$ is its Hessien matrix. Then it follows that

$$J^r \Delta x^r \approx J^p \Delta x^p + J^p \Delta \Delta x^p \tag{14}$$

where

$$\Delta\Delta x^p = \tfrac{1}{2} W^{-1}\left[\left[\Delta x^p\right]^T H_i \Delta x^p \right]_{n\times 1}$$

Consider the state vector has the following form

$$x^p = \begin{bmatrix} \theta_1 & v_1 & \theta_2 & v_2 & \Lambda & \theta_n & v_n \end{bmatrix}$$

then the Jacobian and Hessien of $\Phi(\cdot)$ are

$$W = diag\left\{ \Lambda \quad \begin{bmatrix} -v_i \sin\theta_i & \cos\theta_i \\ v_i \cos\theta_i & \sin\theta_i \end{bmatrix} \quad \Lambda \right\}$$

$$H_i = diag\left\{ O \quad \Lambda \quad \begin{bmatrix} -v_i \cos\theta_i & -\sin\theta_i \\ -\sin\theta_i & 0 \end{bmatrix} \quad \Lambda \quad O \right\}$$

and

$$H_{i+1} = diag\left\{ O \quad \Lambda \quad \begin{bmatrix} -v_i \sin\theta_i & \cos\theta_i \\ \cos\theta_i & 0 \end{bmatrix} \quad \Lambda \quad O \right\}$$

$$i = 1,3,5,\Lambda\ ,2n-1$$

So the $\Delta\Delta x^p$ can be determined as

$$\Delta\Delta x^p = \begin{bmatrix} \dfrac{\Delta\theta_1 \Delta v_1}{v_1} & -\dfrac{1}{2} v_1 \Delta\theta_1^2 & \dfrac{\Delta\theta_2 \Delta v_2}{v_2} & -\dfrac{1}{2} v_2 \Delta\theta_2^2 & \Lambda \end{bmatrix}^T$$

Let

$$a = y_s - y(x^p),\ b = -J^p \Delta x^p - J^p \Delta\Delta x^p,\ c = -y(\Delta x^p) \quad (15)$$

To determine μ, the following cost function is considered

$$Min\ F(\mu) = 1/2 \sum_{i=1}^{n} (a_i + \mu b_i + \mu^2 c_i)^2 \quad (16)$$

Then μ can be obtained by solving the equation

$$\frac{dF}{d\mu} = g_0 + g_1 \mu + g_2 \mu^2 + g_3 \mu^3 = 0 \quad (17)$$

where $g_0 = \sum_{i=1}^{n} (a_i b_i)$, $g_1 = \sum_{i=1}^{n} (b_i^2 + 2a_i c_i)$, $g_2 = 3\sum_{i=1}^{n} (b_i c_i)$, and $g_3 = 2\sum_{i=1}^{n} (c_i^2)$.

Since the multiplier μ would slow down the convergence slightly during the first few iterations due to the approximation of the polar-form formulation, this functionality is invoked after a defined number of iterations. The number currently is set to 4.

4. Test Results

Several large-scale realistic networks have been tested to verify the reliability of the new developed algorithm. The voltage and load ranking results of one typical network with 2681 busbars are given in Table I. The standard SVD results at different load levels (normal operation point, midway to collapse limit and collapse limit) are also given.

TABLE I

RANKING OF WEAKEST BUSBARS IN LARGE SCALE NETWORK

Rank	*SVD-std*			*SVD-new*	
	100%	106%	112%	Voltage	Load
1	DUNOC8	BUSHC1	WIENC1	WIENC1	CELLC2
2	DUNOC9	CELLC2	CELLC2	BUSHC1	WIENC1
3	TONGC1	BISWB1	BUSHC1	CELLC2	BUSHC1
4	NETSC1	PENNB1	IRONC1	IRONC1	IRONC1
5	WHTBCS	KITWB1	COVEB1	COVEB1	KITWC1
6	WHTBC5	OLDBB1	OCKHB1	HAMHB1	HAMHC1
7	WHTLCS	FECKB1	HAMHB1	OCKHB1	OLDBC1
8	WHTLC5	PENNA9	BUSHB1	BUSHB1	COVEC1
9	GLLEC1	PENNA8	OLDBB1	KITWB1	OCKHC1
10	DOUNB1	OCKHB1	KITWB1	OLDBB1	BUSTC2
:	:	:	:	:	:

It can be seen that, in large-scale network, the standard SVD method gives different ranking results for different load levels. The ranking only becomes reliable when the system is close to the collapse point. The new developed SVD algorithm can always give the reliable ranking results since it performs singular value decomposition and busbar ranking at the voltage collapse limit. The ranking of voltage and reactive power injection produces similarly results in ordering weak busbars, which indicate the area with the most sensitive voltage and relatively heavy load.

To verify the load ranking results, the voltage collapse margins with regard to different group loading are calculated. At the normal operation point, the total system load is 47237MW. According to load ranking results, define the top 20 loads as Group A (4569MW) and the second 20 as Group B (6429MW). Keep increasing consumer demand of each group and run load flow until the collapse is reached. The busbars listed first should indicate the loads with most effect on system voltage stability. The results are shown in Table II.

TABLE II

VOLTAGE COLLAPSE MARGINS WITH REGARD TO DIFFERENT GROUP LOADING

Loading Pattern	System Load (MW)		Load Increment	
	Operation Point	Collapse Point	MW	Percentage of system load
Group A (Top 20)	47237	47788	551	1.2%
Group B (2nd 20)		49903	2666	5.6%

The smaller load increment indicates the most dangerous loads. It can be seen that the ranking of reactive power injection produces the correct results, that is, the top loads of ranking list are more important or dangerous than other loads. The loads with most effect on the system voltage stability can be identified and listed.

Acknowledgment

The author gratefully acknowledges the support of the National Grid Transco Company plc. of England and Wales.

References

[1] The national Grid Transco Seven Year Statement 2002

[2] J.H. Grundy, H.P. Johnson, C. Proudfoot, Transmission constraint management on the National Grid system and the effect upon the commercial market place, Proceedings of the IEE International Conference on Power System Control and Management, London, 1996

[3] A.O. Ekwue, H.B. Wan, D.T.Y. Cheng, Y.H. Song, Voltage stability analysis on the NGC system, Electric Power Systems Research, 47 (1998) 173-180

[4] A.O. Ekwue, H.B. Wan, D.T.Y. Cheng, Y.H. Song,, Singular value decomposition method for voltage stability analysis on the National Grid system (NGC), Electrical Power and Energy Systems, 21 (1999) 425-432
[5] A. O. Ekwue, D. T. Y. Cheng, Y. H. Song and H. B. Wan, "Integrated Methods for Voltage Stability Analysis of the NGC System," *IEE Colloquium on Voltage Collapse (Digest No: 1997/101)*, pp.3/1-3/5, London, UK, April 1997
[6] H. Li, Y.H. Song, Identification of weak busbars in large scale power system, IEEE Powercon, Kuming, 2003
[7] N.H. Dandachi, M.J. Rawlins, O. Alsac, M. Prais and B. Stott, "OPF for reactive pricing studies on the NGC System," IEEE Trans. Power Systems, vol. 11, no. 1, pp. 226-232, 1996.
[8] National Grid, 'Obligatory and Enhanced Reactive Power Services: Methodology Document for the Aggregation of Reactive Power Metering,' June, 2000.
[9] S. Ahmed and G. Strbac, 'A method for simulation and analysis of reactive power market,' In Power Industry Computer Applications, pp. 337-341, 1999.
[10] National Grid, 'National Grid Reactive Market Report: Ninth Tender Round for Obligatory and Enhanced Reactive Power Services,' May, 2002.
[11] National Grid, 'A Guide to Balancing Services Use of System (BSUoS),' March, 2001.
[12] Y. H. Song, H. B. Wan and A. T. Johns, "Kohonen Neural Network Based Approach to Voltage Weak Buses/Areas Identification," *IEE Proc.-Generation, Transmission and Distribution*, vol. 144, No. 3, pp.340-344, May 1997

ELSEVIER
IFAC
PUBLICATIONS
www.elsevier.com/locate/ifac

New Concept of Energy Delivery Network for Practical Use of Distributed Generations and Customer Services - Committee Report of IEE Japan -

Ryuichi Yokoyama

Graduate School of Engineering
Tokyo Metropolitan University

Koichi Nara

Dept. of Systems Engineering
Ibaraki University

I. INTRODUCTION

The electric utilities are moving toward deregulation. With the degree of future demand for electricity being unclear, power market players are reluctant to commit to long-term capital investment, with the result that construction of new large-scale power plants and also transmission and distribution infrastructure is typically being avoided. At the same time power facilities are being upgraded both in responses to business strategies in the power business and for supply reliability purposes.

Given these seemingly conflicting interests regarding new construction, construction of distributed generation (DG) facilities offers an attractive alternative. Its advantages are a short construction cycle, being less of a burden to the existing transmission network, as it can be located near to the demand, and contributing to prevention of global warming through clean combustion using renewable energy technologies such as natural gas co-generation, natural gas microturbines, or fuel cells. DG using natural resources such as wind and solar generation is also attracting attention.

Under these circumstances, a new set of requirements advents. The first is development of a set of policies for promoting an increased market for power generated from renewable sources. The second is to evaluate the influence of DG from renewable energy sources on the quality of the electricity. The third is establishing the technology requirements for interconnection with the power grid. The fourth is policies for maintenance of grid reliability.

The task of the power grid infrastructure is to transfer the generated electricity from the power plant to the demand site economically, efficiently and reliably. However, if the focus is on low-cost operation, reliability often suffers. If the focus on reliability is too great, resulting in over-investment in equipment, the cost will be a burden to the consumers. In the conflict between these two elements, an appropriate balancing point exists, in theory at least, and should be sought.

We will discuss here the sorts of infrastructure that should be designed and how this should be pursued in the near future in order to maintain high reliability on the energy delivery network under the deregulation of electricity markets.

First, we will discuss the way to use DG effectively in the development of the distribution network and examine the technical issues, challenges and solutions for interconnection of DG to the grid. Then, new concepts of energy delivery systems proposed in Japan will be introduced. One is the "**Demand Area Power System**" proposed by the Central Research Institute of Electric Power Industry (CRIEPI). The second is the FRIENDS (**F**lexible, **R**eliable and **I**ntelligent **EN**ergy **D**elivery **S**ystem), a new electricity delivery system proposed by research groups in Nara, Hasegawa, et al. The third is the "**Super Node Network**" proposed by the National Institute of Advanced Industrial Science and Technology. Also, their characteristics, structures and feasibility will be discussed. Finally, the concept of the "**New Electricity Supply System**" being examined at the Agency of Natural Resources and Energy of the Japanese Government and directions in future research will be introduced.

II. POWER QUALITY AND SUPPLY RELIABILITY UNDER DEREGULATION

Deregulation of power industry is progressing in European countries and in North America. Typically, the conventional integrated supply system is being changed to an unbundled form in which competitors can separately provide generation, transmission and distribution services. Many countries are testing this notion in anticipation of seeing lower power prices through open competition. Nevertheless, the news on the deregulation front has not been entirely good. For instance, transmission reliability is said to have declined dramatically in the United States as a result of decreasing new capital investment in the physical infrastructure of generation and transmission. In the summer of 1996, two large-sized outages occurred in the Western United States. Some reported that a part of the blame could be placed on the drying up of capital investment.

In such a situation, some believe that it is necessary to take the response to improve supply reliability at the load end, in order to prevent outages, for instance, independent power producers (IPP) placing distributed generations close to the load (Fig.1). In the United States, one emerging business is the sale of "unbundled power quality services." The customer pays a premium for the delivery of high-quality power. The idea is to separate customers into two tiers, based on their need for power quality. Customers pay one price for power that is good enough for ordinary use, therefore not necessarily highly consistent in quality of voltage, current, frequency or reliability. Additional fees are added if customers desire better quality power, according to the particular characteristics desired, thus customers are supplied with the type of power that they choose.

In this electricity supply model, it is possible to create a new business offering only a guarantee of uninterrupted electric service. There has been much discussion about how to make this unbundled power quality service business model a reality.

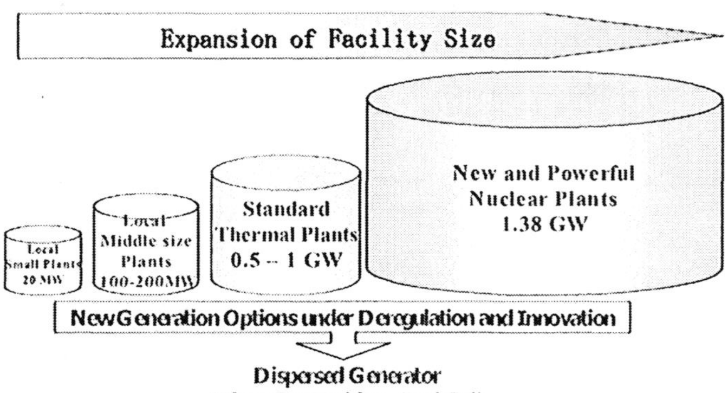

Fig. 1 Distributed Generation in Deregulated Power System

Virtual Utility (ABB), Virtual Power Plant (ENCORP), Custom Power Parks (EPRI, AEP), Demand Area Power System (CRIEPI), FRIENDS and the Super Node Network have been proposed. As is shown in Table.1, these concepts for unbundled power quality systems and their functions a proposed based on application of DG sources, distribute power storage systems, power electronics and informatic technology.

III. FUTURE DISTRIBUTION NETWORK FOR PRACTICAL USE OF DG: DEMAND AREA POWER SYSTEM

Development of the Distribution Network to Effective Use the DC Power Source

The Japanese Government has forecast that the distributic network will be host to about 20,000 MW of distribute generations (DG) by the year 2010 (Table 2). Based c tendencies that we have seen in the past, it is predicted th DG will be unevenly connected to different distributio networks, i.e. a large amount of DG will be connected some parts of the grid and little to the remainder.

It is certainly probable that, in the future, the number of D connections to the grid may be restricted, since it is alread anticipated that they will bring negative influences to th operation of the grid, such as decreases in power quality an difficulties with coordination and protection during outages.

It is also anticipated that the future deregulated pow industry will include interconnected DG generated pow with concern primarily for profit by selling to off-si customers using the grid. This could adversely impact gri operations if a large quantity of power were generated f only a short period and injected into the grid.

CRIEPI proposes a solution for such problems in a ne concept called the "local demand site grid." Elements includ new structures, methods of operation and methods of contr of the grid at the location of power consumption. Fro hereon, I will refer to this concept as simply the "Dema Area Power System."

Table 1. Overview of Custom Power Functions

Function	Processing Overview
Database	Geographic information, facility information, grid structure, substation information, switching status, load recording
Support of facility planning	Facility planning support, load data management, loss computation, outage restoration plan
Support of facility design	Facilities standardization, alternative plan cost comparison, cost management, administration of drawings, contract support
Support of facility installation	Building layout support, building resource plan management, workflow control
Maintenance, security	Maintenance data, maintenance history management, maintenance management planning
Support for operation planning	Outage management, customer complaints processing support, operations planning support
Display, recording, printing	Data input support, document printing, drawing pictures, statistics processing

Table 2 Targeted Installed Capacity of Distributed Generations by the year 2010

Type of New Energy usage	2010
Photo Voltaic	App. 4.82 GW
Wind	App. 3 GW
Waste	App. 4.17GW
Biomass	App. 0.33GW
Use of Sunshine	App. 4.39Gl
Use of Waste and Biomass heats	App. 140Ml & 670Ml
Unutilized Energy including Snow/Ice latent heat	App. 580 Ml
Clean Energy Automobile	App. 3,480,000
Natural Gas Cogenerator	App. 4.64 GW
Fuel Cell	App. 2.2 GW

Characteristics and Architecture of the Demand Area Power System

The effectiveness of the Demand Area Power System will be discussed here, as it is conceived by clarifying its architecture, power system specifications of the necessary component devices as is shown in Fig.2. These are the proposed elements of the Demand Area Power System, the area to which it may be applicable, the architectural design, the basic method of operation and control using information technology and the proposed component devices to be used.

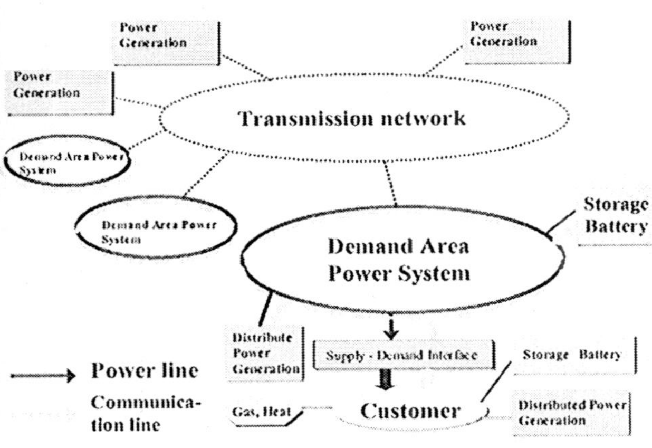

Fig.2 Role of Demand Area Power System in the Future

Effectiveness of the Demand Site Grid

Fig.2 also shows the problems in system operation and control that might occur with the connection of DG sources to the distribution network. It also shows the response measures, the degree to which these correct the problems, and the gain or loss in the utilization rate of the facility for each of the proposed grid architectures. As shown in this figure, the demand site grid structure is the most effective to solve the problems that are anticipated to occur after deregulation. The only remaining challenge is to lower the cost. The specification for the loop controller should include functions and capabilities for it to control grid voltage and current during normal operation.

Conceptual Design of the Demand Site Grid

Fig. 3 defines basic locations for ;

(1) Loop controller to control the voltage and current on the line, and

(2) Supply-demand interface to manage and control DG sources and demand volume.

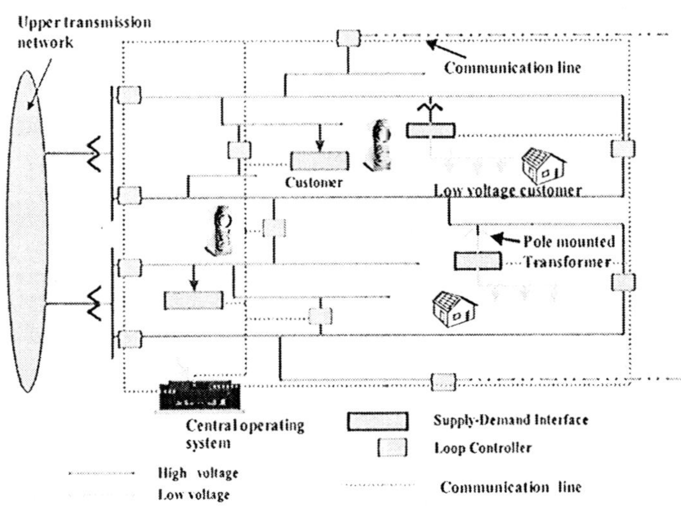

Fig. 3 Structure of Demand Area Power System

IV. ADVANCED ENERGY DELIVERY SYSTEM AFTER DEREGULATION: FRIENDS

New Form of Power Distribution System

Many proposals have been made for the future power distribution system in the post-deregulation power markets. Some would change the grid structure in a flexible way by using various types of equipment. Some consider the supply of high reliability power (unbundled power quality services) and/or power conservation. Some are considering value-added power transmission using highly advanced information technology.

In these future power distribution systems, the following characteristics would be described:

a) Flexibility in grid structure to adjust to normal operations, outages, or scheduled outages during maintenance. This is done with a combination of static-type switches and distributed information processing.

b) Highly reliable power supply, i.e., no outages, utilizing power generated at many different substations,

distributed generation sources, and power storage systems.
c) Option menu service or unbundled power quality service, i.e., a menu of options for power quality from which end users can freely choose the different power quality options from the supplier of their choice.
d) Load leveling and electricity conservation using DG, power storage technology and demand-side management.
e) Improvement of customer services such as through information-providing services.
f) Advanced demand side management.

To realize the above will require DG, distributed power storage equipment, demand side management, power electronics, communications and information processing.

The Flexible Reliable and Intelligent Electrical ENergy Delivery System (FRIENDS) is proposed to realize all of the above (a to f items). The conventionally proposed custom power park offers only functions b, c and d. In other words, the unbundled power quality service is supplied. Items e and f are addressed in the proposal for a Customer Based Bi-directional Telecommunication Network Service.

Features of Proposed New Power Delivery System

Now, the features of the FRIENDS will be introduced briefly. This system may be one example of a future distribution network out of the systems proposed so far.

Let us take an urban area set up with the concept of the FRIENDS. The system of the FRIENDS as shown in Fig.4 has multiple "power quality control centers" located in proximity to the demand sites, as the equivalent of a specific area in the current distribution line. These centers are able to receive generated power from multiple high-voltage transmission lines so that it becomes possible to provide option menu service to consumers in each given locality. In passing, let me mention that the functions of the custom power park are very similar to this part of the system. Power quality control centers can be installed at substations inside the building that receive power from the outside or in the underground portion of the distribution transformer box. In the centers, many assortments of power quality are created. Using a static-type switch also permits flexibility in switching the connections to high- or low-voltage distribution lines.

The power quality control center (QCC) supplies power to low-voltage distribution lines through the transformer using an option menu system. Changes to high-voltage distribution line connections can be carried out with static-type switches inside the power quality control center; therefore, the commodity power can either supply the center or bypass it.

More promising applications of power electronics technologies for power quality enhancement in FRIENDS are:

(a) DC distribution inside the home,
(b) Load leveling using distributed power storage, and
(c) Emergency backup power generators for use during outages.

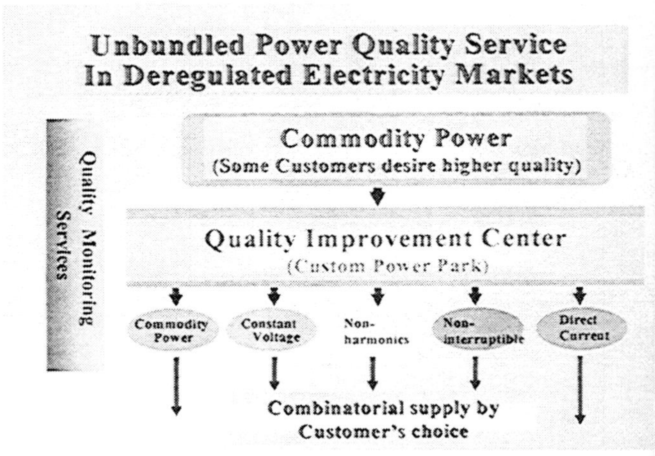

Fig. 4 Functions of Quality Control Center

Customer Information Service

The power quality control center is also networked with optical cable. Therefore, it can control itself or act as a customer information service center doing information processing and information exchange. In the large-scale view operation and demand-site control of the switch and other equipment in the center are handled in tandem by a large control computer at the branch office and a smaller computer at the distribution substation, the center or the consumer site. At the same time, computerized protection and control functions will be provided. The data used in these computers is centrally managed in an integrated database of data from the management, operation, and control of the power transaction system, and consist of data from operations, maintenance, and mapping and the like, and also customer information services as shown in Fig.5.

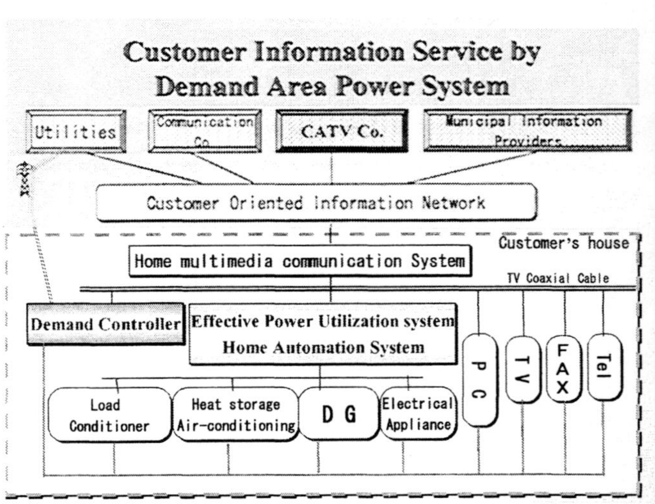

Fig.5 Customer Information Service in the future distribution networks

V. COORDINATED CONTROLLED SYSTEM: SUPER NODE NETWORK

Function of the Super Node Network

It is believed that distributed generation, such as co-generation, and renewable energy generation will be connected to the endpoints of the existing grid in great numbers in the future. It will be difficult to maintain the grid status for long if these continue to be connected to the grid at the current pace; as distributed generation sources are not controlled by the grid operators. Herein lays the necessity of strengthening the functions to control the electric current. However, it is also difficult to respond to this need with the present centralized management scheme of the grid control system.

This is because;

a) DGs come in many different types with many different characteristics,
b) Large numbers of small-capacity facilities are connected to the endpoints of the grid,
c) Renewable energy generation has large fluctuations in output, and
d) It is installed at the end user's site as needed.

It is also the case that even if equipment providing automatic control is connected to each distributed generator, there is a large possibility that there will be interference between the different units when too many of them are operating simultaneously.

This is why the National Institute of Advanced Industrial Science and Technology has suggested the Super Node Network to respond to this need for stabilizing the grid. By grouping nearby installations to exchange control information, the coordinated control actions can be carried out.

Hierarchical Structure of the Super Node Network

The concept of stratification of the exchange of control information and stratification of blocks explained in the previous chapter can be applied to the power grid in the large geographical area. What we call a "super node" is the individual interface devices of power electronics exchanging control information. A super node network is a grid where different types of super-node devices are interconnected. Shown in the earlier figure, a super node network on the grid is stratified according to voltage levels and is divided into blocks following the transmission and distribution line structure.

This network acts in such a way that fluctuations occurring downstream on the grid will have minimal influence upstream. Because of this, each super node device installed at the upper end of the grid must have a large capacity, while nodes closer to the source of the fluctuations require devices with smaller capacity but faster response. Similarly, the storage system must be chosen for the characteristics of capacity upstream and response speed downstream.

VI. POST DEREGULATION POWER SUPPLY SYSTEM FOR PRACTICAL USE OF DG: NEW POWER SUPPLY SYSTEM

Purpose of the New Power Supply System

Meetings of the Electric Power Technology Discussion Group have been held since 1998, under subcontract to the Electric Power Technology Section, Public Benefit Business Division, Agency of Natural Resources and Energy, Ministry of Economics, Trade and Industry (previously called the Ministry of International Trade and Industry or MITI). The name of the committee has changed every year, but the purpose of its meetings has remained the same: to examine the medium-term and long-range direction of technology development in the power industry.

The first discussions were about overall power technology. Subsequently, the discussions have been about the recent partial deregulation of the power market, the future expansion of DG, and rapid advances in IT. Starting in 2001, the theme became issues of technical development needed for widespread use of DG.

In July 2001, *the New Power Supply System* Technology Evaluation Committee was created with experts in DG becoming members. This committee issued a report in March, 2002. The New Power Network System Study Group was chartered to begin in 2002 an examination of the future model of the power supply system proposed in the report (Fig.6).

This committee will:

a) use simulations to develop a concrete model of future systems and analyze its effectiveness,
b) promote the development of individual technologies required for the new systems, while being responsive to societal changes,
c) promote standardization of the technology, and
d) look for ways of cooperating with foreign countries.

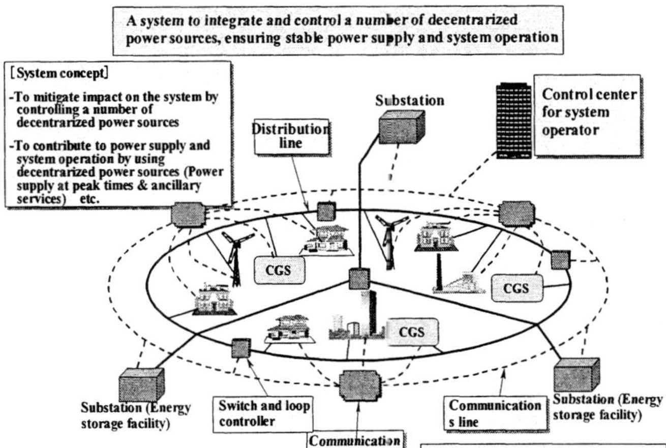

Fig. 6 New Power Supply System (Studied by METI)

Issues and Challenges in Technology Development for the New Power Supply System and a Model for the Future

The search for a new power supply system model of the future where balance and stability is achieved between grid power and the power from distributed generation, with free access to the network for DG. Organizing the issues and technology development challenges by some means such as creating a road map in order to realize the model as is shown in Table 3.

The range of discussion is limited to the technology, whereas examination of the policy issues will be done in tandem by other entities. But the two will influence each other interactively, so that initiative resulting from the technical side will be submitted to the policy-making side and vice versa. Therefore, the discussions in the committee must reflect the policies being enacted

VII. FUTURE POWER SUPPLY NETWORK FOR PRACTICAL USE OF DISTRIBUTED GENERATIONS AND CUSTOMER SERVICES

The global movement for deregulation that is greatly changing the overall form of the electric supply system seems like it is here for the foreseeable future. This is also true for the transformation of the distribution system with highly advanced information technology as well as greater use of power electronics. For the ideas behind the Customer Based Bi-directional Telecommunication Network, Virtual Utility (ABB), Virtual Power Plant (ENCORP), Custom power parks (EPRI, AEP), Demand Area Power System (CRIEPI), FRIENDS and the Super Node Network are significant examples.

The latest technology and deregulation policies ar interacting together to influence the form of the electric powe supply. For instance, power electronics can protect customer at the level of the distribution grid or at the customer site from reductions in power quality that might result from the reduce investment in transmission under deregulation.

Also, the fiber-optic information network in the deregulate world enables unbundled power quality service, online billin systems, and customer information services. Small, high performance, inexpensive computers provide the necessar control more easily than before. Using such equipment customer service can be separated out and offered as commercial product. It is expected that we may see never before dreamed-of business opportunities.

A power supply model for the future emerges from th above elements combined with the expected shortage in th near future of primary energy sources. This model is centere around a customized power distribution grid based on th customer fully utilizing highly advanced power electronics telecommunication and information processing technology.

But under exceptional conditions such at during gri outages, it should be able to suppress the increase of shor circuit current and instantaneously disconnect the sectio where the outage occurred.

The supply-demand interface should be able to gather th data on voltage, current and frequencies at all locations on th grid, including the location where it is installed. Whe outages occur, it must also be able to automatically identif the sections where the outage occurred and where th abnormal voltage occurred. It must also have a logic functio to automatically judge the best way to protect the integrity o the grid in the future energy delivery systems.

Table 3 Issues and challenges in technology development for the future model of the power supply system

	Future model of power supply system	Issues and challenges of technology development
Power distribution system	a. Based on the present infrastructure, the system is to control only grid structure devices, but to allow many DG to be flexibly interconnected. b. Controlling many DG units, this system provides peak power, ancillary services and power supply and contributes to the grid operation.	1. Control technology for grid infrastructure devices – technology to make the grid into a loop and technology to utilize the power storage devices on the grid 2. Control technology for DG sources – independent automatic control technology for DG, control technology by grouping
Demand side system	c. Power supply with differentiated power quality and backup system for customers d. Integrated customer-side power system utilizing IT technologies	3. Unbundled power quality service technology - responding to the different needs using power storage devices and DG; backup technology as protection from the momentary voltage sag 4. Energy management on the customer side with active use of IT

ELSEVIER
IFAC
PUBLICATIONS
www.elsevier.com/locate/ifac

DEVELOPMENT OF AUTONOMOUS DEMAND AREA POWER SYSTEM IN CRIEPI

Hiromu Kobayashi and Satoshi Uemura

The Central Research Institute of Electric Power Industry (CRIEPI)
2-11-1, Iwado Kita, Komae-shi, Tokyo, 201-8511 JAPAN

Abstract: A large introduction of distributed power generation is expected. Under such circumstance, power flow congestion, voltage fluctuation and degradation of safety on distribution systems will occur by the operation of the distributed power generations. As a strong measure of the issues, we have proposed a unique, new power supply system called "Autonomous Demand Area Power System". The system leads free access of the distributed power generation to the grid and makes effective use of the generations using power electronics devices and communication network. The paper described concept, composition and operation measures of the Autonomous Demand Area Power System. *Copyright © 2003 IFAC*

Keywords: Distribution network, Distribution automation, Communication system, Power generation, Voltage control, Power flow, Protection, Safety

1. INTRODUCTION

Centered in urban areas with high power demand densities, the introduction of distributed power generations including gas turbine generation, fuel cell power generation and photovoltaic power generation is expected. Under such circumstances where many distributed power generations are incorporated in the existing power system, power flow congestion, voltage fluctuation and degradation of safety on distribution lines will occur caused by varied output of the distributed power sources. This will make the control of the entire power system operation complicated, and result in a failure to maintain power quality and supply reliability by conventional power system management.

As a strong measure of the issues, we have proposed a unique, new power supply system (referred to as the 'Autonomous Demand Area Power System'). The system leads free access of the distributed power generation to the grid and makes full use of the characteristics of distributed power sources, which feature the effective use of electric power and thermal energy at the demand side, in addition to cost reductions on the power supply side. Thus, both customers and power providers benefit. To facilitate this proposed scheme, we'll try to establish network technology and operation management technology of the autonomous demand area power system, and identify its effectiveness.

2. BASIC CONCEPT OF AUTONOMOUS DEMAND AREA POWER SYSTEM

2.1 Concept of autonomous demand area power system

Fig.1 shows expected future utility power system. The autonomous demand area power system is defined as the segment that includes the distribution system and the secondary system of power supply side. Its facility formation is required to flexibly deal with power flow congestion due to the interconnection of multiple distributed power sources, and increased power demand in urban areas, without any complicated control techniques. This proposed system must maintain stable power supply reliability for a longer period, without frequent strengthening of related facilities even when distributed power sources and/or demand may increase.

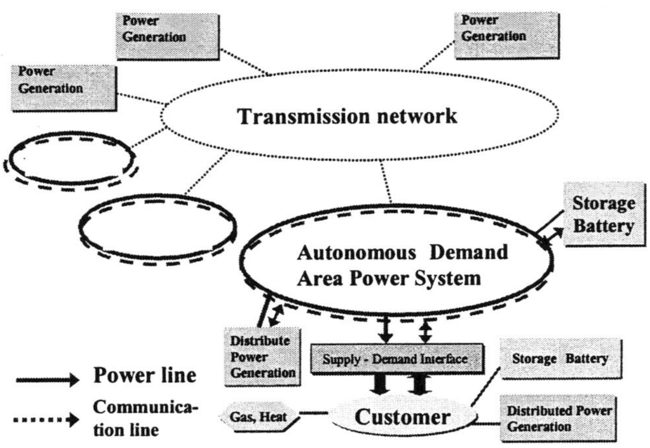

Fig.1 Expected future power system.

To meet these requirements, the loop shaped configuration, aiming at enhanced power supply capability, will be adopted. For areas where particularly large demands are required, the mesh-formed grid configuration will be partly applied. While the autonomous demand area power system can be structured at a low cost by rationalizing its system configuration and maximizing the use of existing facilities /equipment, the use of Loop Power Controllers (LPC) enabling flexible power flow control between loops is considered to be effective for upgrading system reliability.

For operation and control of the autonomous demand area power system, a specific function will be needed for dealing with a balance of energy supply and demand in order to meet the mutual needs of customers and power providers by fully making use of advanced communication networks. This function that we proposed is called Supply and Demand Interface (SDI). Via the SDI, customers learn the current operation status of the power system and rate menus by power providers and retailers, in addition to the power generating condition of their own distributed power sources. Customers can also select the most economical one from among different types of energy including gas, heat and electricity supplied by power providers and retailers in a highly efficient and stable manner. If the power system operation fails, the outage section can be minimized and/or the power flow controlled by using distributed power sources owned by customers, coordinating with adjacent SDI.

Individual SDI is operated in linkage with the central operation control system that manages the entire autonomous demand area power system. The operation control system functions, except for first-aid recovery of failures, to monitor the autonomous demand area power system. By sending data on power transaction rate menus and current power system operation status to individual SDI, it also manages a more efficient operation of the entire autonomous demand area power system and energy conservation for leveling with the whole power system. In short, the operation control system will bring about a more efficient, simplified power system for customers through maximized use of a communication network.

2.2 Configuration of autonomous demand area power system

The autonomous demand area power system is defined as the segment that includes the distribution system (=6.6kV or 22kV) and the secondary system (=66kV). By expanding conventional functions of the distribution line and transformer of distributing substations, several distributing substations are grouped into one unit. This results in flexible power system operation and easier operation control. The proposed system may be in the loop formation, in addition to the conventional tree branch form. LPC for controlling power flow and voltage between loops will be needed. At the supply side, the operation control system and multiple SDIs per several customers will be installed for information exchange between supply and demand sides. Fig. 2 shows the autonomous demand area power system configuration, and communication network structure.

The concepts of the operation control system, the SDIs, and the LPCs, which comprise autonomous demand area power system, are described as follows. The communication infrastructure required for the system is focused on as well.

Operation control system; The operation control system functions to relay data between the SDIs and the upper power system, and to control power flow and restore failures through data exchanges between SDIs. This system aims for control of whole of the autonomous demand area power system.

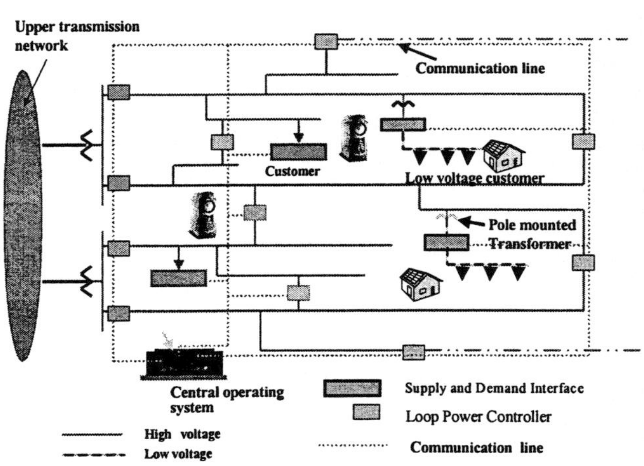

Fig2.Configuration of autonomous demand area power system.

The system collects the data on distribution line

voltage, current and operation of distributed power sources via the SDIs, and judges the current operation status of the concerned autonomous demand area power system. The system changes power routing by sending a power flow control signal to each LPC for power flow leveling within the demand area, and collects customers' metering data. At the time of a failure occurrence, the system detects the faulty section, and executes power interchange for sound sections and restoration by utilizing each SDI and LPC.

Data on real-time rate menus, power system construction, and various customer services are relayed from the operation control system to the SDIs; and, if necessary, commands on direct load control and output control of distributed power sources are sent to the interfaces. When a failure occurs, this system sends the faulty point data to individual SDIs, and transmits a command to return to normal operation to the interfaces when the failure is completely restored.

Loop Power Controller (LPC); The autonomous demand area power system enables free access to distributed power sources, as well as coordinated operation among large-scale power plants, distributed power sources and energy storage facilities. Therefore, the system is required to have flexible facility operation, and its formation will be shifted from tree branch type to loop and/or mesh type.

The LPC, shown in Fig.3, functions to control power flow in the autonomous demand area power system with the loop/mesh shaped configuration. This unit consists of two ac/dc converters linked to direct current (= Back To Back method), and enables the simultaneous control of power flow and terminal voltage at both power line sides. The two power lines can be connected together by the LPC even though voltages and phases between the lines are quite different. As the results, large scale of power network having loop and/or mesh configuration can be composed in demand area. Additionally, because fault current of loop network can be isolated by the LPC, reliability of protection and safety of the network will be able to be maintained. The target capacity of a unit is 500kVA to 1000kVA.

The issues for establishment of the LPC technology are as follows.
- Design detailed specifications for distribution line use.
- Obtain reliability of the function through a long term of demonstration test.
- Miniaturize size and reduce cost. Resolving the issues depends on the development of new semiconductor switching device such as SiC device.

Supply and Demand Interface (SDI); The SDIs, which are installed between the power system and customers, autonomously manage data exchange and energy supply and demand. Fig. 4 shows a location and role of the S&D interface installed between a power system and LV customers. The SDI functions to maximize the benefits of customers and power providers, and to conserve energy and save resources. Customer benefits include minimizing energy costs, obtaining stable energy, a reliable system for protection/security and a power supply to match customer needs. Power provider benefits include an improved facility capacity factor by indirect and direct control of distributed power generations and loads, simplified system operation, and obtained protection /security of the power system.

3. R&D OF AUTONOMOUS DEMAND AREA POWER SYSTEM

We propose on operation technique for the autonomous demand area power system as shown in Fig.5. We also propose a technique to decide the amount of control of each LPC corresponding to the system condition, and develop the program for operation control system. From the calculation result, it is confirmed that a system operator could reflect by the weight to distribution system loss, availability factor and voltage regulation. In the future, we will seek to use optimization techniques to increase the computing and program-solving speed.

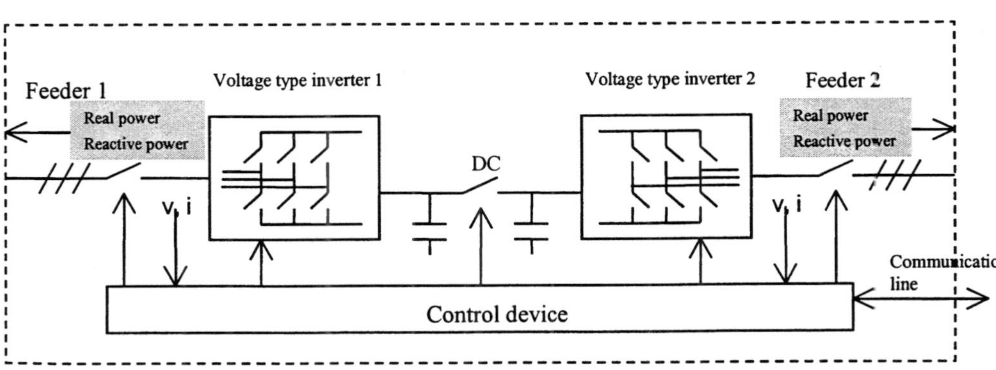

Fig.3. Loop Power Controller adopted back to back method with 2 inverters.

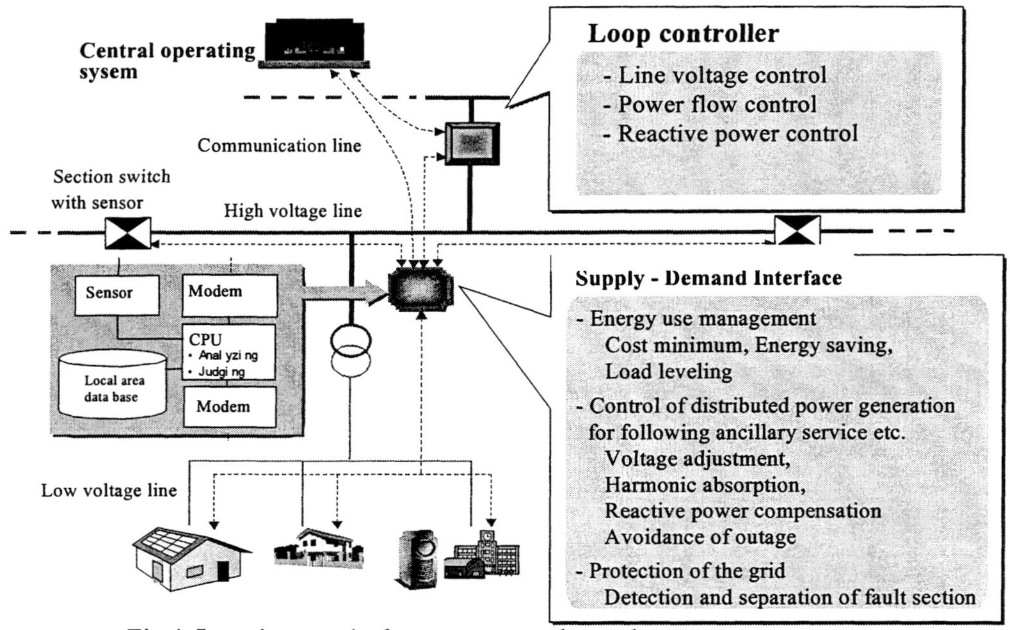

Fig.4 .Location area in the autonomous demand area power system including LPC and SDI.

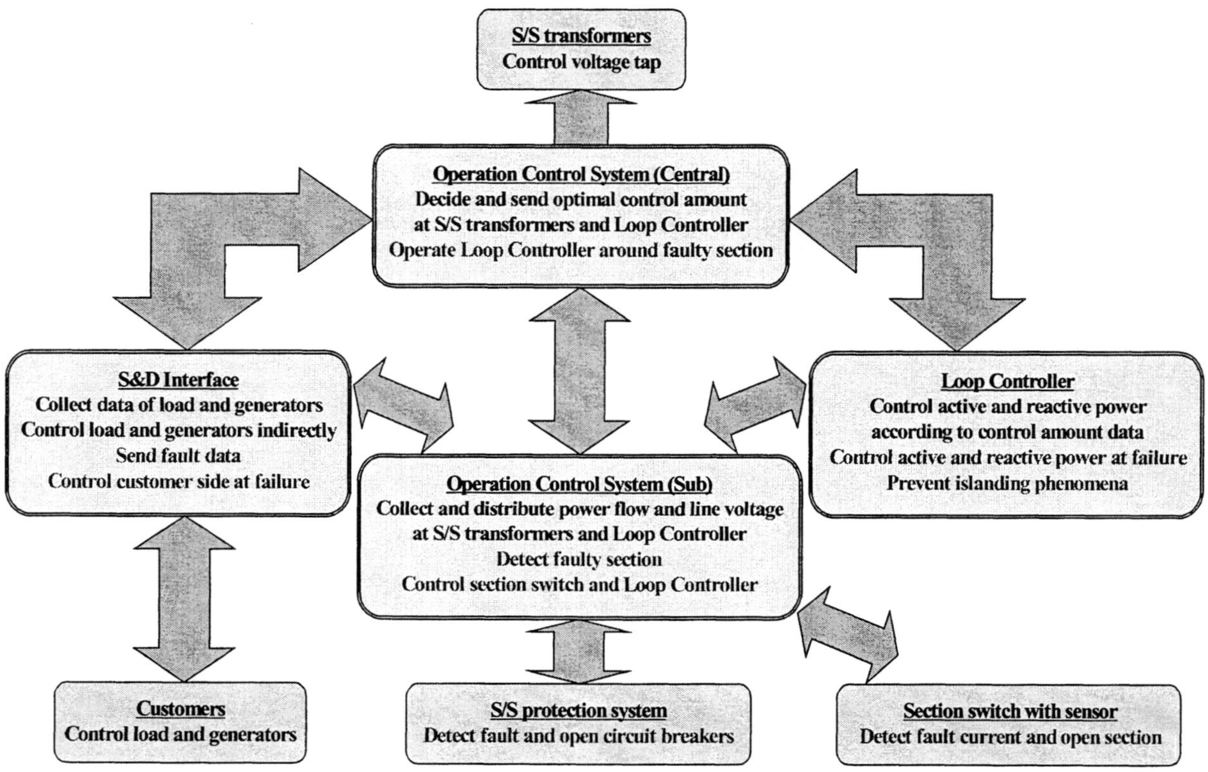

Fig.5. Operation flow for the autonomic demand area power system.

4. CONCLUSIONS

The paper described the Autonomous Demand Area Power System for coping with a large introduction of distributed power generations. Basic composition and operation techniques of the Autonomous Demand Area Power System using the Loop Power Controller (LPC) and the Supply and Demand Interface (SDI) and communication were indicated.

There are many resolved technical and economical issues to be investigated for practical use of the system. We will promote research and development of the LPC, SDI and operation techniques including communication measure, introducing demonstration test at a large scale demonstration test facility in CRIEPI Akagi Testing Center consisting of total 1000kW distributed power generations.

ELSEVIER
IFAC
PUBLICATIONS
www.elsevier.com/locate/ifac

STATE OF THE ARTS OF FRIENDS ON 2003

Koichi Nara, *Ibaraki Univ.,* **Jun Hasegawa,** *Hokkaido Univ.,* **Kiichiro Tsuji,** *Osaka Univ.,* **Toshifumi Ise,** *Osaka Univ.,* **Tsutomu Oyama,** *Yokohama National Univ.,* **Yuji Mishima** *Ibaraki Univ.*

Abstract: FRIENDS(*Flexible, Reliable and Intelligent Energy Delivery System*) is a new power delivery concept which enables to connect distributed power resources to power delivery system and to realize customized power quality services by controlling them. This paper introduces following 4 items of recent technological developments in FRIENDS: (1) an optimal network configuration method of FRIENDS in urban area from a viewpoints of supply reliability, (2) a fault state autonomous restoration method of FRIENDS through multi-agent system, (3) DC loop type QCC(*Quality Control Center*) to realize the customized power quality services for the area where DC power supply is dominating, (4) economical aspect of the customized power quality services. *Copyright © 2003 IFAC*

Keywords: Power system control, Distribution systems, Autonomous control, Optimization problems, Optimal control

1. INTRODUCTION

History of electric power supply started through dispersed and competitive way as supplying power only near the distributed small scale power station. Then, these small power supply networks are interconnected to each other, and wide spread huge monopoly power supply utility have been constructed from a viewpoint of power supply reliability. However, recently, through highly developed distributed generation technology and by introducing competition in power market, the power supply system must be decomposed into dispersed small scale portions and be re-organized again. From the above background, near the customer side, many new concepts to organize the power delivery systems are proposed: such as, Micro-grid, Virtual Power Plant, Power Park (U.S.A.), DISPOWER(Germany), Demand Area Power System and FRIENDS(Japan), etc. FRIENDS(*Flexible, Reliable and Intelligent Energy Delivery System*) was proposed by authors in 1995 (Nara and Hasegawa, 1995), and the detailed conceptual design was presented in reference (Nara, *et al.*, 2000). Therefore, in this paper, recent additional technological developments in FRIENDS are introduced. A brief review of FRIENDS is shown in Chapter2. A network design method with the optimal normal state control and autonomous control operation technique in fault state are explained in Chapter3 and Chapter 4. To realize the customized power quality services, DC type QCC (*Power Quality Control Center*) among many proposed types of QCCs, and economical aspect of the customized power quality services are discussed in Chapter 5 and Chapter 6.

2. BRIEF REVIEW OF FRIENDS

In recent years, many distributed generators (DGs) and energy storage systems (DESSs) are being connected to the power distribution systems, and a lot of problems may emerge: such as power quality, protection, voltage problems, etc. Therefore, as stated in Introduction, many concepts are proposed to overcome the above phenomena or to establish new functions for compensating the vulnerable points of DGs and DESSs.

FRIENDS is a concept which try to attain the following functions by effectively using DGs and DESSs, demand side control, power electronics technologies, modern communication technologies and distributed intelligent facilities, etc.:

(1) Flexibility in reconfiguration of the network,
(2) Reliability in power supply,
(3) Multi-menu services or customized power quality services to allow consumers to select the quality of electric power and the supplier,
(4) Load levelling and energy conservation,
(5) Enhancement of information services to customers,
(6) Efficient demand side management, etc.

A general concept of FRIENDS is schematically shown in Fig.1. As shown in Fig.1, FRIENDS characteristically has so-called *quality control center* (QCC) between distribution substation and customers. In QCC, static switches are equipped to establish flexibility in frequent change of connections of high and low voltage power lines in normal and fault state, and an electric power with various levels of quality is produced for multi-menu services or customized power quality services by using DGs and DESSs allocated in demand sides or in QCCs. A concept of customized power quality services is depicted in Fig.2. Namely, consumers can select qualities of power independently according to their necessity as shown in Fig.2. To attain the customized power quality services or supply reliability, QCC must be supplied power from several substations through power line network as shown in Fig.1. In FRIENDS, QCCs connect to each other as a meshed network. A typical QCC allocation in urban area and an example of network configuration of FRIENDS is shown in Fig.3. In addition, energy exchange among DGs and DESSs through this network together with demand side management, energy conserving measure can be attained.

To realize the above functions, information network and control strategies as well as FRIENDS network configuration and the optimal operation strategies must be developed. The most recent technological developments to realize FRIENDS are introduced in the following chapters.

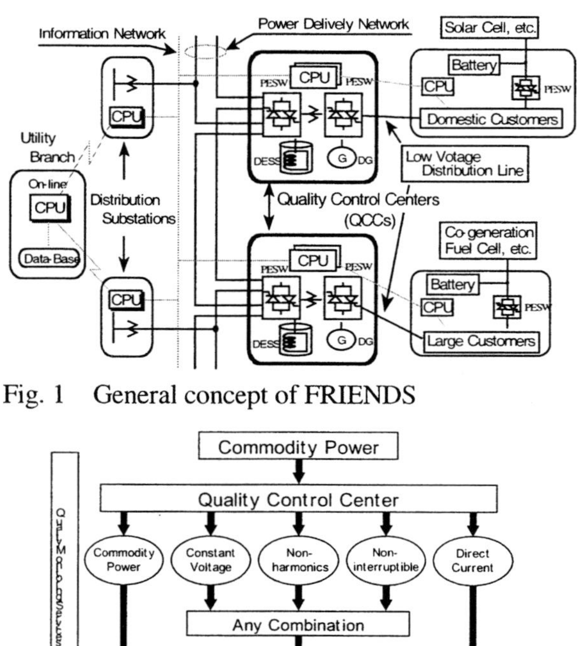

Fig. 1 General concept of FRIENDS

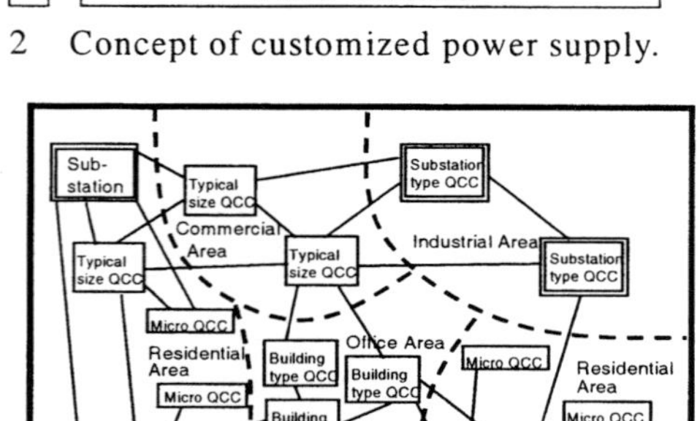

Fig. 2 Concept of customized power supply.

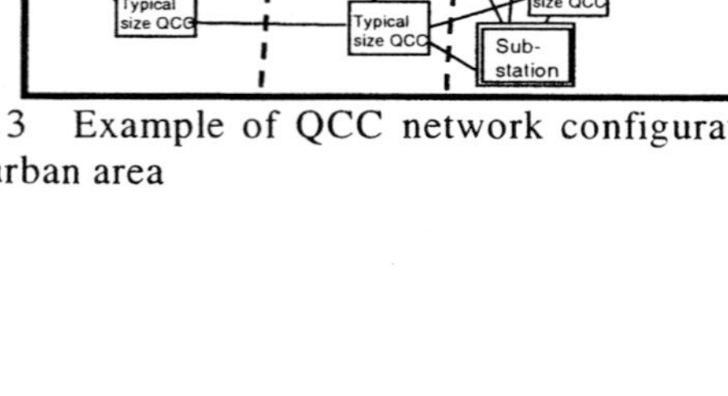

Fig. 3 Example of QCC network configuration for urban area

3. DESIGN OF FRIENDS NETWORK

To simplify the planning problem, an optimal network configuration problem of FRIENDS in urban area from viewpoints of supply reliability, facility installation costs and operational cost is defined to determine where to install power distribution lines and DG units for minimizing total installation cost under the constraints of installable capacity of DG unit and sum of expected power interruption cost (Nara, *et al.*, 2002).

The problem can mathematically be formulated as follows.

[Objective function]

$$\text{Min. } \alpha\left(\sum_{n=1}^{ND}(aX_n + bYN_n) + \sum_{m=1}^{BR} c_m YL_m\right) + \beta\sum_{t=1}^{T} Oloss^t \quad (1)$$

[Constraints]

(DG's maximum capacity)

$$X_n \in \{x^1{}_n, x^2{}_n, .., x^i{}_n, .., x^L{}_n\} \; (n = 1, \cdots, ND) \quad (2)$$

(Expected power interruption cost)

$$\sum_{t=1}^{T}\sum_{r=1}^{FLT}\frac{1}{T} p_r BLCost^{rt} \le \varepsilon \quad (3)$$

where, *ND*: Total number of QCCs, *BR*: Total number of potential lines, X_n: Discrete variable for DG's capacity in n-th QCC, x^i_n: i-th available capacity of DG for n-th QCC (X_n must be selected from x^i_n), YN_n: =1 if X_n is not 0; =0 otherwise, YL_m: 0-1 variable for installation of line m (1: if line m is installed; 0: otherwise), *FLT*: Number of fault cases, *a*: Variable cost for capacity of DG installed at n-th QCC, *b*: Fixed cost of newly installed DG at n-th QCC, C_m: Installation cost of line *m*, $Oloss^t$: Distribution loss at load pattern *t*, *T*: Number of load patterns under consideration, ε : Small constant, α, β : Weighting coefficients.

The objective function (eq.(1)) is to minimize weighted sum of installation cost of power lines and DGs as well as operational cost (distribution losses). $BLCostr^{rt}$ in eq.(3) means the sum of power interruption costs of customers for fault case *r* at load pattern *t*. Also, p_r in eq.(3) means occurrence probability of fault *r*. Namely, eq.(3) means that the expected value of power interruption cost is less than the pre-determined small constant.

For the above problem, a general flowchart of the solution algorithm through tabu search (TS) is shown in Fig.4. In solving the problem, the optimal operation must be found to calculate the distribution losses and the undistributed (interrupted) power cost for every neighborhood solution of TS. To find the optimal operation of FRIENDS for every operating conditions, in a box marked by * in Fig.4, TS and sensitivity analysis are employed to reduce the computational burden.

This solution algorithm is applied to a real-scale urban power delivery system. Fig.5 shows the initial locations of QCCs with the initial network configuration. Parameters for the test calculation are shown in Table 1. The resulting optimal configuration by applying the above algorithm is shown in Fig.6 and Table 2.

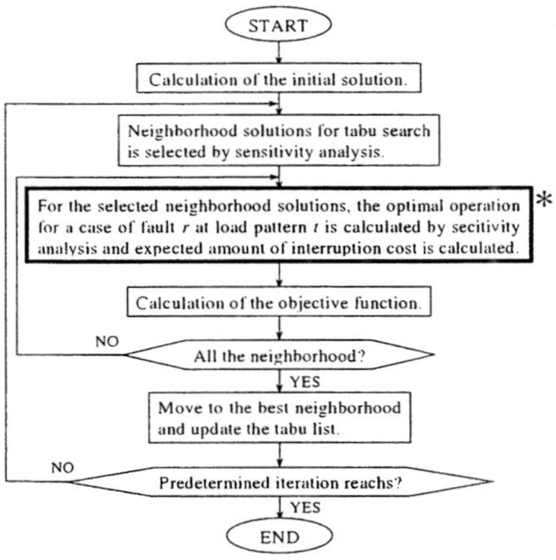

Fig. 4 General flowchart of the solution algorithm.

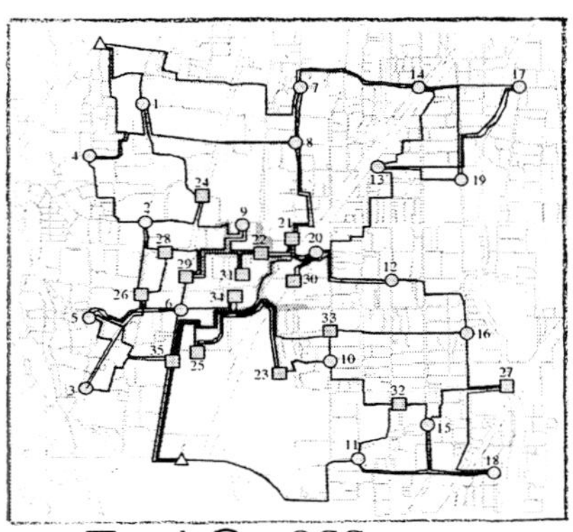

□ and ○ : QCC

△ : distribution substation.

Fig. 5 Initial network configuration

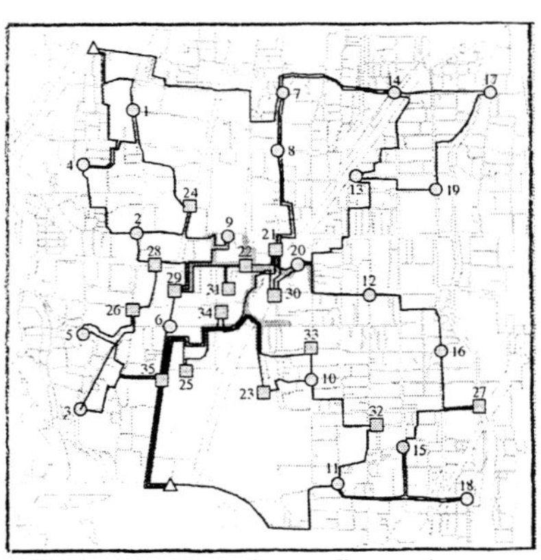

Fig. 6 Optimal network configuration.

Table 1 Parameters for test calculation.

Installation cost of lines [$/km]	70.0
Fault probability of lines [times/km/year]	0.0095
Installation cost of DG(fixed) [$]	2.0
Installation cost of DG(variable) [$]	300.0
Available capacities of DGs [MW]	0.4, 0.8, 1.2, 1.6, 2.0
Coefficients for eq.(1)	α=1.0 , β=1000.0

Symbol "$" means hypothetical money unit.

Table 2 Calculation result.

	Initial	Optimal
Installation Cost [$]	2010.0	1431.3
Distribution Loss [kWh]	34.1	38.43
Expected Interruption Cost [$]	0.00	0.00
Objective Function Value	2044.1	1469.7

4. DISTRIBUTED AND AUTONOMOUS EMERGENCY OPERATION

The fast fault restoration for improving the supply reliability is one of the most important functions of FRIENDS. This chapter presents a method for restoring the distribution systems under the fault situation depressively and autonomously by the multi-agent system (Hara, *et al.*, 2002).

4.1 Interior Structure of QCC considered

The interior structure shown in Fig.7 is considered as a model of QCC. In the QCC, hybrid transfer switches (HTS) for turning on/off the feeders, two PWM-controlled inverters (Inv.1 and 2) and DG as a back-up generator are introduced. By utilizing these devices, it can produce the electricity with three kinds of quality called ordinary quality (OQ), high quality (HQ) and super premium quality (SP) (Hara, *et al.*, 2000). Further, the QCC is assumed to be able to communicate with the neighbor QCCs by means of message passing.

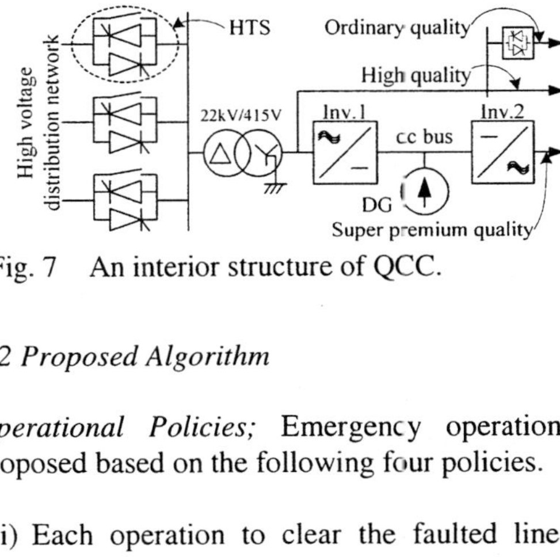

Fig. 7 An interior structure of QCC.

4.2 Proposed Algorithm

Operational Policies; Emergency operations are proposed based on the following four policies.

i) Each operation to clear the faulted line must have priority.
ii) Each agent (QCC) supplies the electricity to its own customer prior to the other agent's one.
iii) Each agent passes the active and reactive powers required by the neighbor agent. If there is not enough reserve of DG to supply the requested power, the agent cooperates as much as possible.
iv) Each agent asks the neighbors to pass the power only when it does not have enough DG capacity.

Example Sequence of the Proposed Algorithm; A model system with four QCCs and two DSSs as shown in Fig.8 was used for numerical simulations. The demand powers of OQ, HQ and SP are 500, 200 and 100kW, respectively. The sequential events after the fault occurrence at the feeder between substation (DSS1) and QCC1 are illustrated in Fig.9, and can be explained as follows: (i) QCC1 detects the fault occurrence by receiving FLT (FauLT) message generated by the fault section detector in QCC1. (ii) QCC1 turns off HTS to clear the fault and then it initiates a stand-alone operation by operating the DG as a backup generator. Since capacity of DG in QCC1 is less than its total demand (800[kW]), QCC1 cuts off its OQ load and sends SRQ (Supply ReQuest) message to two neighbor QCCs. Here, SRQ to the QCC2 has a higher priority to QCC3. QCC2 and 3 propagate SRQ to QCC4 because they also do not have enough capacity of DG to output the requested power. (iii) QCC4 receives two SRQs, however, the SRQ from QCC2 is adopted according to their priorities and SRQ route via QCC3 is overwritten by SRQ via QCC2. QCC4 also propagates SRQ to DSS2. (iv) SP (Supply Possible) message from QCC3 is propagated to QCC1 via QCC4 and 2. QCC4 also sends CSR (Cancel SRq) message to DSS2 in order to nullify the SRQ which was sent before. (v) Finally, QCC1 turns on HTS and a cascade route is constituted. THA (THAnk) message is propagated from QCC1 to QCC3 in order to inform that the SRQ from QCC1 has been terminated successfully.

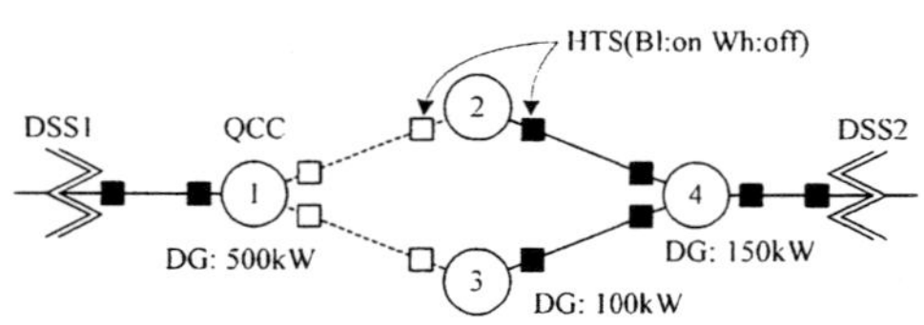

Fig. 8 Four QCCs and two DSSs system.

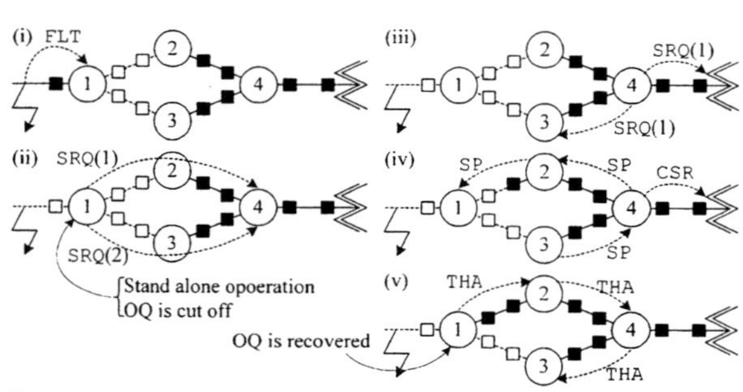

Fig. 9 Sequential events of the proposed operation.

4.3 Quantitative Analysis of Performance

Performance of the proposed operation was investigated through the Monte-Carlo simulation in which the capacity of DG and demand in each QCC, response time of HTS and message delay are modeled as uniform random numbers. The network shown in Fig.10 was employed as a test system and the fault between DSS3 and QCC21 was applied. Fig.11 shows the distribution of the interruption duration obtained through the simulation. This figure shows that very fast restoration (less than 60msec) can be realized by the proposed operation.

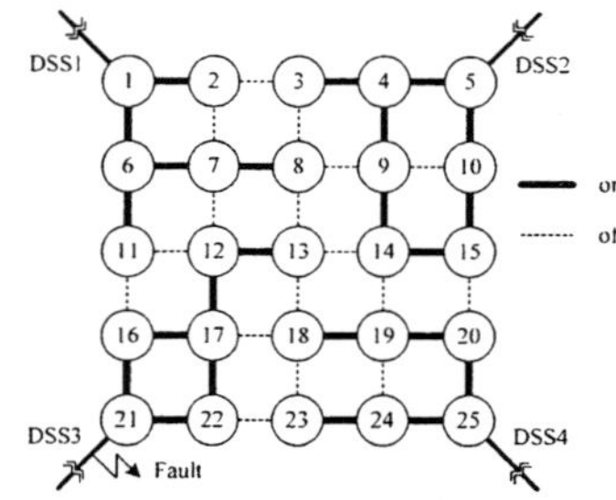

Fig. 10 Twenty-five QCCs and four DSSs system.

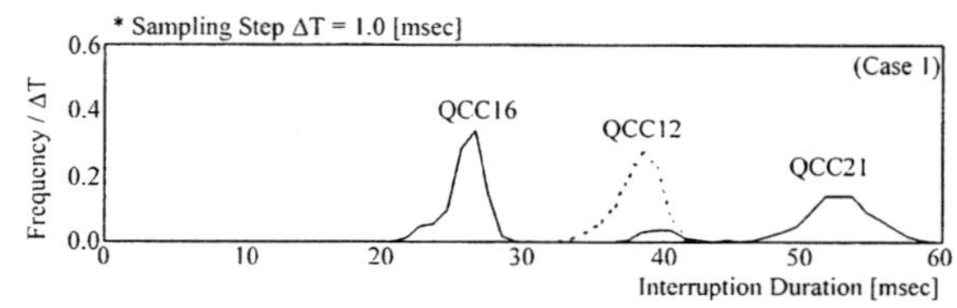

Fig.11 A distribution of the interruption duration.

5. DC LOOP TYPE QCC

The DC loop type QCC studied here has three DC loop lines, that is, the high quality loop, the normal quality loop and the ground loop as shown in Fig.12 (Saisho, *et al.*, 2002). The ground loop is used in common to the high quality and the normal quality line. The high quality line and the normal quality line are connected to AC grid by using bi-directional rectifier, and diode rectifier with PFC (Power Factor Corrector), respectively. AC load are connected to the DC loop by using inverters, and DC loads are connected to that by using DC/DC converters. DGs (Distributed Generators) such as photovoltaic generations, fuel cells, wind turbines and micro turbines, and energy storage system such as secondary batteries and SMESs (Superconducting Magnetic Energy Storage) are connected to the DC loop by using rectifiers or DC/DC converters. Hybrid

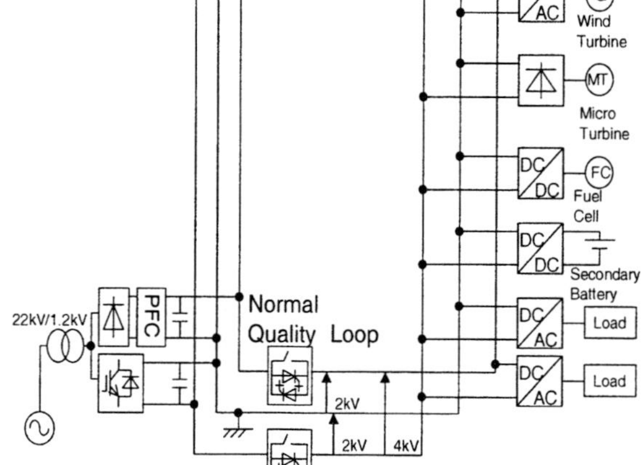

Fig.12 Configuration of DC loop type QCC.

switches composed of a mechanical breaker and a GTO switch are installed in the DC loop in order to separate the faulted part from the system in case of the fault. Superconducting fault current limiter can be used for limiting the fault current before turning off the GTO switch. The voltage of both the high quality and normal quality loop are controlled to be constant by the bi-directional rectifier and the diode rectifier with PFC. Characteristics of this system are summarized as follows.

i) In case of the fault on AC grid, loads connected to the normal quality line of the DC loop can be protected from voltage sags and loads connected to the high quality loop of the DC loop can be protected from voltage sags and interruptions.
ii) Harmonics and reactive power produced by loads do not flow into AC grid.
iii) Rectifiers for the adjustable speed drive system can be omitted, when they are connected to the DC loop.
iv) Losses of the distribution system can be reduced and reliability can be enhanced by the loop configuration.
v) DGs and energy storage devices, which have DC outputs, can be installed easily by using choppers which have simpler circuit than DC/AC inverters, or DC/DC converters which utilize high frequency transformer smaller than line frequency transformer and less switching devices than DC/AC inverters.

Simulation results are shown in Fig.13, which shows the voltage sag case on the AC grid. Due to the voltage sag on AC input, the DC voltage is affected. But the inverter voltage, which shows output voltage of the inverter connected to loads, has no effect of

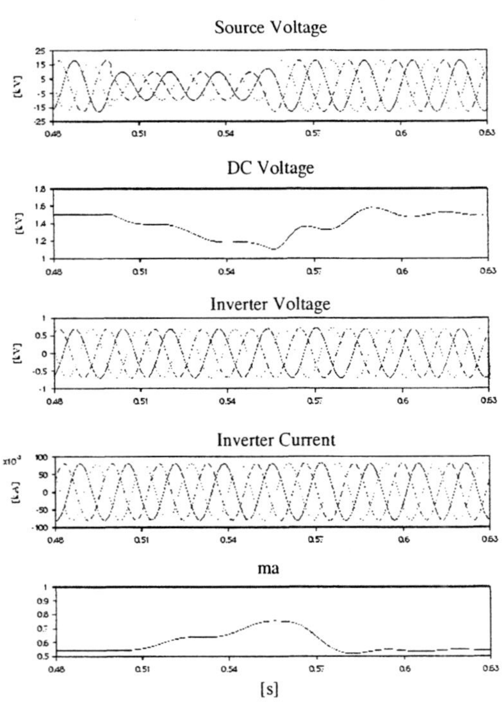

Fig. 13 Simulation result of DC loop type QCC.

voltage sag, and the load can be operated at constant voltage due to the control of amplitude modulation ratio m_a.

This type of QCC is different from other type of QCCs, because the power converters are distributed in a region, on the other hand, power converters are located at a concentrated place in the other types of QCC (Hayashi, *et al.*, 2002). The DC loop type QCC will be suitable where a lot of inverters are used in a region such as industrial park, water pump station and etc.

6. ECONOMIC ASPECTS OF FRIENDS

In order to improve the consumer surplus (economic benefit of consumer), it is important to consider consumer's requirement on reliability and cost. Consumers require high reliability power for limited number of high quality apparatus. In this case, they may accept expensive power. However, consumers do not require high reliability power but cheap power for the rest of apparatus (Oyama, 1999). The required reliability level and quantity of electric energy can be illustrated as Fig. 14.

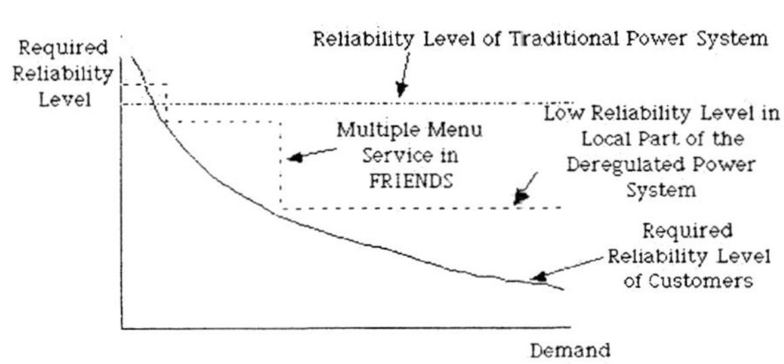

Fig. 14 Demand and supply reliability level.

With the multiple menu service, consumers can choose high reliability power with high price or normal reliability power with normal price. Their choice is based on the consumer utility curve (demand curve).

In this study, the utility curve is obtained based on the questionnaire. It is shown in Fig. 15 (Shimomura and Oyama, 2001).

Consumer surplus can be calculated as follows.

$$(\text{surplus}) = S\,(\text{utility})_i - (\text{payment})_i$$

The subscript i represents the reliability level of power. In this calculation, two levels (h: high reliability and n: normal reliability) are assumed.

The price for high reliability power P_h and normal reliability price P_n are set to recover power purchase cost, QCC cost (converter and storage), and cost for additional distribution lines. It is assumed that a consumer set the ratio of high reliability power and normal reliability power to maximize its surplus.

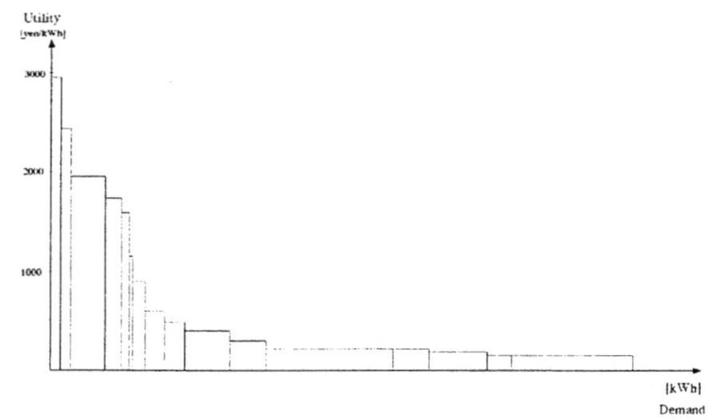

Fig. 15 Approximate form of demand curve.

The relation between consumer surplus and high quality power service availability are shown in Fig. 16. In this case, service availability of utility system is set 0.999. From the figure, consumer surplus with multiple menu power supply is higher than that without it. It is thought that introduction of multiple menu power supply has a merit.

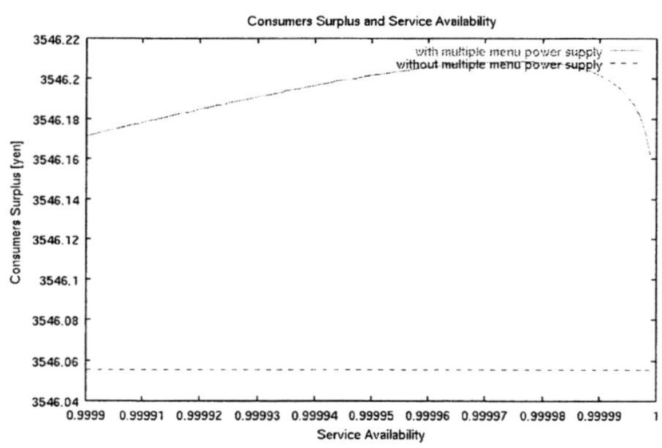

Fig. 16 Consumer surplus and service availability.

7. CONCLUSION

In recent years, many new concepts have been proposed in world wide for the de-regulated power distribution system with distributed power resources. FRIENDS is one of these concepts which tries to attain the customized power quality services by effectively using DGs and DESSs in demand side and QCC. In this paper, some of recent technological developments in FRIENDS research are presented. To realize FRIENDS, more researches are necessary: such as new protection scheme for the power distribution system with DGs, fault state optimal operation to prevent power interruption, inexpensive configuration of QCCs, etc. Although technological break through in power electronics may be a major problem, we also have lots of problems to be solved to realize FRIENDS. The authors expect that many researchers will have interest to the problem to realize FRIENDS.

REFERENCES

Hara R., H. Kita, E. Tanaka, and J. Hasegawa (2002). Emergency Operation Based on Multi Agent Technology in FRIENDS. *Proc. of 14th Power System Computation Conference (PSCC2002)*, Session 03, Paper 4.

Hara R., H. Kita, E. Tanaka, and J. Hasegawa (2000). Interior Structure of Quality Control Center at the Low Voltage Side for the Customized Power Quality Service in FRIENDS. *Trans. of IEE Japan*, **Vol.120-B**, No. 11, pp.1451-1458

Hayashi, Y., T. Ise, and K. Tsuji (2002), AC-type Quality Control Center for FRIENDS -System Configuration and Experimental Results-, *Proc. of the IEEE/PES Transmission and Distribution Conference and Exhibition 2002*, Yokohama, Japan, pp.1191-1195.

Nara, K, and J. Hasegawa (1995), An Advanced Flexible and Reliable Distribution System. *Proc. of EMPD'95*, pp.55-60.

Nara, K, J. Hasegawa, *et al.* (2000), FRIENDS - Forwarding to Future Power Delivery System-. *Proc. of ICHQP*, pp.8-18.

Nara, K., H. Omi and Y. Mishima (2002). Optimal Configuration of New Power Delivery System for Customized Power Supply Services. *Proc. of the Power Systems Computation Conference*, PS19-5

Oyama, T. (1999). Demand Side Prospects of Multiple Menu Service in Power Service with the Quality Control Center. *Proc. of IEEE PES 1999 Winter Meeting*, New York, U.S.A. , pp.422-426.

Saisho, M., T. Ise, and K. Tsuji (2002), DC Loop Type Quality Control Center for FRIENDS - System Configuration and Circuits of Power Factor Correctors-, *Proc. of the IEEE/PES Transmission and Distribution Conference and Exhibition 2002*, Yokohama, Japan, pp.2117-2122.

Shimomura. E., and T. Oyama (2001). A Study on Effect of Multiple Menu Power Supply Systems on Consumers. *Proceedings of International Conference on Electrical Engineering*, **Vol.2**, Xi'an, China, pp.712-715.

ELSEVIER
IFAC
PUBLICATIONS
www.elsevier.com/locate/ifac

THE STATUS OF POWER QUALITY AND/OR DISTRIBUTED GENERATION PROJECTS IN KOREA: ESTABLISHMENT OF GRID INTERCONNECTION STANDARDS

Jae-Eon Kim

School of Electrical & Computer Engineering, Chungbuk National University
48 Gaesingdong Cheongju City, Chungbuk Korea 361-763
jekim@chungbuk.ac.kr

Abstract: National and/or utility grid interconnection standards for interconnecting non-utility distributed generation systems with local utility grid will be essential to their beneficial utilization. In this paper, some activities with respect to establishment of grid interconnection standards of distributed generation in Korea are introduced.

Keywords: Standards, Grid Interconnection, Power Quality, Distributed Generation

1. INTRODUCTION

In Korea, being confronted with the new millennium, major events occurred such as restructuring of the electric power industry, amendment of Electricity Business Act and opening of Korea Power Exchange. Also, investors or independent power producers become to emerge as utility restructuring opens energy market, allowing the customer to choose the energy provider, method of delivery. The energy market forces favour small, modular power technologies that can be installed quickly in response to market signals. Emerging on the scene is a portfolio of micro-grid network or more complex network with distributed generation systems. But national and/or utility grid interconnection standards for interconnecting non-utility distributed generation systems with local utility grid are essential to their beneficial utilization. These interconnection standards are vital to utilities, customers, manufacturers and nation.

In this paper, we introduce restructuring of the electric power industry, penetration levels of renewable energy and cogeneration, and activities for establishing grid interconnection standards of distributed generation.

2. RESTRUCTURING OF THE ELECTRIC POWER INDUSTRY

Act on Promotion of Restructuring of the Electric Power Industry, enacted on December 23, 2000, enabled the division of the generation sector into several companies. The Electricity Business Act, amended on Feb 24, 2001, mandates the establishment of an electricity trade market (Power Exchange) and a regulatory agency (Korea Electricity Commission). Korea Power Exchange (KPX) was established on April 2, 2001, as a non-profit independent organization to facilitate transparent and fair management of the electricity trade market. Its major missions are preparation for launching the future whole sale market, efficient management of power trading, fair and transparent operation of the market and reliable & stable operation of the power system. Two stages of transition were designed for the electricity market in step with each phase of KEPCO's separation and privatization according to the basic plan of the restructuring of the electric power industry(Jan. 21, 1999). CBP(Cost Base Pool) market is the current market structure since April 2, 2001. A TWBP(Two-Way Bidding Pool) market model is being designed and its implementation is scheduled for 2004 after the division of the distribution/retail sector.

In the rule of the KPX, power system operation procedure and criteria are described only for the generator of more than 154 kV class. This means that there are no operation procedure provisions for embedded generators or distributed generators which are mainly connected to the 22.9 kV class distribution systems in Korea.

Table 1 Design for the Electricity Market

Items	1st stage	2nd stage
Market	CBP	TWBP
Competition	Generation	Wholesale
Term	current	After division of distribution/retail sector(2004)

3. RENEWABLE ENERGY AND COGENERATION

As for renewable energy and cogeneration, photovoltaic generation system, wind turbine, gas turbine, gas engine and small hydro-power are considered. These generation systems are connected to electrical distribution lines and operated. Table 2 shows their dis semination status recently.

Table 2 Utilization of New & Renewable Energy And Cogeneration

Items	Dissemination Status
PV system	3,710kW (1999)
Gas Turbine	61 units 148,000kW(2001)
Gas Engine	755 units 277,000kW(2001)
Small Hydro-Power	30 units 42,820kW(2001)
Wind Power	20 units 7,139kW (2002)

On the other hand, several Renewable Energy dissemination projects is being propelled by public organizations in order to attract public attention and to assure the credibility of R&D activities. To expand the renewable energy market, the government supports low-interest bank loans for their producers and facilities.

4. ACTIVITIES FOR ESTABLISHING GRID INTERCONNECTION STANDARDS OF DG

As the introduction of distributed generation systems advances, Korea Electric Power Corporation (KEPCO), Korea Electricity Commission (KOREC), Ministry of Commerce, Industry and Energy are feeling urgently the necessity of Grid Interconnection Standards (GIS) of DG. There were some meetings and two projects associated with this matter, but none is enacted yet. That is, one called as "A Study on Establishing GIS of Gas-cogeneration Systems" conducted by Korea Energy Management Corporation (KEMCO). Another called as A Study on Establishing GIS of Wind Turbine" conducted by Korea Electric Power Research Institute (KEPRI), sponsored by KEPCO. In this section, the GIS draft fromthe two projects is summarized as follows.

4.1 Basic Direction and Purpose

The GIS is to provide a guideline for establishing the formal one that guides government officers, generation utilities, transmission utilities, distribution utilities, electricity service utilities, owners or manufacturers of distributed generation systems and KPX. This GIS contains scope and purpose, definitions of terms, application processes and fees, technical interconnection provisions or requirements, contract provisions or agreement, and tariff provisions.

4.2 Scopes

The GIS is applied to distributed generation systems lower than 20,000 kW which is the maximum capacity of the dedicated 22.9 kV distribution line. For non-dedicated 22.9 kV distribution line, its maximum capacity is limited to 10,000 kW. The types of generators are synchronous machines, induction machines and inverters. For single-phase DG, the maximum allowable size is 5kW or less. Also, for three-phase 380V class DG, it is 5kW to 100 kW or less. For DG more than the above capacity, it is connected to the 22.9kV line through transformer.

4.3 Definitions of Terms

Definitions of terms contain all of technical and non-technical terminologies regarding the GIS provisions.

4.4 Application Processes

DG owner/applicant contacts the dedicated department of the distribution utility (DU) or the electricity service utility (ESU) and submits DG application. Upon receipt of a completed application, the DU has a defined period of time in which to process the application and provide approval to interconnect, approval to interconnect with a list of prescribed changes to the DG design, justification and cost estimate for prescribed changes to DU system, and application rejection with justification. Section 4.5 will help the DG application review. The DU should also provide the various metering options available and discuss the rates available. If the application is approved, the applicant and the DU will enter into a contract or agreement for operating DG. And then, DG will be installed and the commissioning test allowed within a defined period of time by the public authorized inspection agency, which is to inspect whether the installation conforms to national safety codes and local codes.

4.5 Technical Interconnection Requirements

The distribution utility (DU) should review the information provided by the DG owner/applicant to ensure that the proposed installation conforms to the following requirements.

4.5.1 Safety

These requirements regarding safety are intended to ensure that DG is designed and installed in a way that

- is not a hazard to DU personnel or equipment or to other customers.
- does not disturb other customers or degrade the power quality.

That is, disconnection or de-energizing requirements during utility outages, abnormal conditions and system emergency --- islanding detection, visible-break disconnection.

4.5.2 Reliability

System reliability to assure the safety includes power quality and protection requirements.

Power quality requirements are governed by established practices that voltage, flicker, frequency, power factor and harmonics. Deviation from these practices represents out-of-bounds conditions and may require disconnection of the DG from the DU network. The DU service voltage and the DG voltage should be maintained within 220V±13V at the terminal of low-voltage customers. The operation of the DG should not cause excessive voltage flicker on the utility. The frequency of the DG should match the utility frequency varying within 60Hz±0.2Hz so that the two systems can operate in synchronism. At the PCC (point of common coupling), the THD of the voltage shall not exceed five percent and the TDD of the current shall not exceed five percent during normal operating conditions. The power factor at the PCC may be required to keep within lagging 0.85 to leading 0.8 for proper voltage regulations. The maximum voltage unbalance at the PCC shall not exceed two percent during normal operating conditions. The DC component of DG output current should be less than 0.5 percent of the rated rms output current.

Protection requirements to assure the safety may include protection such as overcurrent protection, over/under voltage protection, over/under frequency protection, reverse power protection, transfer trip protection, synchronism checking protection, dedicated transformer, generator disconnect device, or interconnection disconnect device. Besides, some delicate problems should be investigated, that is, islanding, resonant overvoltage, surge, overloading of equipment, feeder reclosing coordination, nuisance break trip, fuse saving, and grounding. The DU and the DG owner/applicant shall be solely responsible for the protection system of the electrical equipment and facilities at their respective sides of the PCC.

4.5.3 Parallel Operating Requirements

Parallel operating requirements are divided into two categories. One is the requirement of the controller of DG, the other is the trip & reconnection requirement for the DU system abnormal conditions. DG may have the capability of parallel operation and black start operation. When the DG is paralleled with the DU system, the voltage regulator shall be capable of providing VARs to the distribution system according to one of the following functions: constant VAR source, constant power factor source, constant voltage source. In the case of real power controller, constant output operation and peak cutting operation can be done.

Abnormal conditions can occur on the DU system, such as voltage disturbances, frequency disturbances, loss of synchronism, feeder reclosing coordination. The trip & reconnection requirement for those should be provided by the DU.

4.5.4 Control & Monitoring requirements

The generation control should take into account coordination with the DU including system protection and operational control concerns. The generation control may be automatically controlled, locally controlled, or remotely controlled. The controls should allow for quick shutdown of the generation in the situation of failure of the level of generation mechanisms or high or low frequency. Also, the power quality state and the behaviours of protection system at the PCC should be monitored through some measurements.

4.5.5 Metering & Communication Requirements

Metering shall be in place to determine the kWh production of the generation source. Additional metering may be required, such as kW demand, kvar, or power factor. Metering shall be of the accuracy according to the electric market rule in KPX. Telemetry may be required in some circumstance to monitor the real time output and other functions of the DG. But it is not required for the DG less than 100 kW that is interconnected to the secondary line of the pole transformer. Communication systems for telemetry and metering should be compatible for interface with the DU.

4.5.6 Data Requirements

Data may be required under certain conditions for purpose of recording transactions and insuring the ongoing safe and reliable operation of DG. Those are divided into real time data, data summaries, and power quality/failure reports regarding operations.

4.5.7 Certification and Testing

The provider of the DG should provide to the DU one-line diagram and test results. Pre-certified equipment may be utilized for interconnection to a DU system without further review or testing of the equipment design by the DU. The use of pre-certified equipment does not automatically qualify the applicant to be interconnected to the distribution system at any point of common coupling. An interconnection review may be performed to determine the compatibility of the DG with the DU system capability at the selected PCC. DG that is not pre-certified shall be equipped with protective hardware and/or software to prevent islanding and maintain power quality. The applicant shall provide the final design and install the protective measures required in sec. 4.5.2 and 4.5.3. The electric provider or DU should review and approve the design, types of protective functions, and the implementation of the installation. They should specify all calibrations for DG owned protection systems that are required herein. Type tested equipment should be certified by a nationally recognized test organization. Pre-certified equipment should first be type tested. The DU reserves the right to witness the testing of installed switchgear, protection systems, and generator and review all such records at all reasonable times. The DG owner/applicant has full responsibility for routine maintenance of the DG, control and protective equipment and the keeping of records for such maintenance.

4.5.8 Approval Process and Data Required

For DG units larger than 100 kW, the following data shall be supplied from the DG owner/applicant to the DU:

- Equipment specifications of generator and protection systems including one-line diagram, protection data, operating characteristics
- Location of the PCC
- Test data
- Synchronizing method
- Maintenance schedules
- Data needed to coordinate the installation with that of the DU
- Expected start up date

For a small DG unit less than 100 kW, there shall be a once-only evaluation of the DG by an independent laboratory. Once the DG is approved, it shall be an acceptable DG that requires no further utility review. The results of the analysis of the independent laboratory shall be made available to the DU. The DU shall approve DG of different sizes within the proper timing. If the DU rejects any implementation, it shall give full reason and explanation.

4.6 Contract Provisions or Agreement

This contract provisions or agreement, which govern the relationship between a DG owner/applicant and the DU, should be embodied in the following items:

- Fees & charge regarding interconnection such as inspection or test fees, fees needed for technical studies and some added system protection equipment, upgrade fees, metering charges for a second meter, standby charges
- Tariff provisions or billing
- Liability and insurance coverage
- Indemnification
- Communication
- Compliance with technical standards
- Allowing the DU on-site inspection
- The handling of costs associated with interconnection
- Obligation of both parties to cure adverse effects
- Access rules
- Dispute resolution

REFERENCES

J.E. Kim (2003), "A Study on Establishing GIS of Gas-cogeneration Systems" sponsored by Korea Energy Management Corporation (KEMCO).

J.E. Kim (2003), "A Study on Establishing GIS of Wind Turbine" sponsored by Korea Electric Power Research Institute (KEPRI).

ELSEVIER
IFAC
PUBLICATIONS
www.elsevier.com/locate/ifac

GENERATION DISAPATCH OF DEREGULATED ENERGY RESOURCES USING STOCHASTIC MODELING

C.S. Chang **Li Wei**

(eleccs, engp1435)@nus.edu.sg
Department of Electrical and Computer Engineering,
National University of Singapore, Singapore 119260

Abstract: Widespread restructuring of electric power industries has led to rapid emergence of distributed energy resources (DERs), which are clustered together to form microgrids, power parks, or virtual utilities. Through advanced control and communication, DERs are becoming more competitive, controllable and flexible. Their impacts should be included in the utility's economic dispatch and unit commitment problems. Utilities solve these problems by minimizing overall cost among all their units, while DERs focus their attention on individual profits. A common framework is proposed here to enable both parties to achieve their respective objectives. DERs with homogeneous cost characteristics are clustered here into several virtual utilities. A stochastic model is proposed to assess the availability of each DER and to decide whether it is profitable for it to participate in the operation. By maintaining a proper co-ordination among all utilities and virtual utilities, the interests of both parties are protected and the overall economy is optimized. The proposed approach is applied to a test system, and numerical results are presented. *Copyright © 2003 IFAC*

Keywords: electric power systems, energy distribution, load dispatching, unit commitment, dynamic programming.

1. INTRODUCTION

Electric power industries are undergoing widespread restructuring from the vertically integrated monopolies to competitive open-market systems. The market opening has allowed the entry of distributed energy resources (DERs), which have significant impacts on the utility's operation. DERs are relatively small and geographically scattered generation units. They offer several desired characteristics such as low capital costs and quick starting. Some units offer bonuses such as combined heat and power, ancillary services and green energy (Willis and Scott, 2000; Borbely and Kreider, 2001; Daly and Morrison, 2001). DERs are increasingly clustered to form microgrids, power parks, or virtual utilities. Through advanced control and communication, these integrated DERs are becoming flexible, controllable and economical (Smallwood, 2002; Hatziargyriou and Meliopoulos, 2002; Kaarsberg and Watson, 2002).

This paper proposes a new approach to accommodate DERs in the utility's operation by minimizing the utility's overall cost of generation, coordinating between load and generation, and allowing DERs to participate profitably. Interests of both the utility and DERs are protected. To achieve these, the paper proposes to modify the traditional approach of unit commitment and economic dispatch by direct accommodation of DERs. The approach is begun by clustering DERs with homogeneous cost and performance characteristics into virtual utilities. The availability of each virtual utility is treated in a stochastic model by considering the failure of DERs as random events. The rating of each virtual utility is discounted accordingly. Participation of each virtual utility is decided according to its profitability.

2. MODELING OF VIRTUAL UTILITY

2.1 Clustering of DERs into Virtual Utilities

Virtual utilities are formed by clustering the DERs, which respond to the buy-back prices from the utility in a homogeneous manner. Each virtual utility k is made up from J_k DERs as follows:

$$P_k = \sum_{j \in J_k} P_j \quad (1)$$

where P_k and P_j are respectively the outputs of virtual utility k and DER j, provided that all DERs are FULLY available. Should the availability be lesser than 100% for any DER, its output becomes a random variable. A method as described in the appendix is proposed to model each DER's availability, the mean value η_k and variance σ^2_k of each virtual utility's output.

2.2 Interpretation of Buyback Price

It is assumed that the utility will benchmark its energy trading with virtual utilities using its marginal cost of power delivery (Schweppe *et al.* 1988). A buy-back price s is therefore established:

$$s = (1 - \frac{\partial P_{loss}}{\partial P_i})\lambda \quad (2)$$

where λ is the marginal generation cost. P_{loss} is the power-delivery loss and P_i is the output of generation unit i.

2.3 Modeling of Virtual Utility Profitability

The profit E_j of each DER j is the difference between two cost components: C_{paid} which is the revenue received from the utility for selling P_j and $C_{cost\ j}$ which is the generation cost of DER j for generating P_j. At each time interval t,

$$E_j = C_{paid \cdot j} - C_{cost \cdot j} \quad (3)$$

where E_j is a function of P_j sensitive to s to be maximized. P_j is subjected to generation limits:

$$P_j^{\min} \le P_j \le P_j^{\max} \quad (4)$$

where $P_j^{\max}$ and $P_j^{\min}$ are unit's maximum and minimum outputs.

As each DER j is paid by the utility at s:

$$C_{paid \cdot j} = s\,P_j \quad (5)$$

$C_{cost \cdot j}$ is expressed as a quadratic function:

$$C_{cost \cdot} = a_{j0} + a_{j1} P_j + a_{j2} P_j^{\,2} \quad (6)$$

To maximize E_j:

$$\partial E_j / \partial P_j = 0 \quad (7)$$

Solving (7) gives the optimum value for P_j

$$P_j = \frac{s - a_{j1}}{2a_{j2}} \quad (8)$$

which relates P_j and s. In cases where P_j is outside its limits, it must be set at the corresponding limit as:

$$P_j = \begin{cases} P_j^{\min} & \text{if } \frac{s-a_{j1}}{2a_{j2}} \ge P_j^{\min}; \\ \frac{s-a_{j1}}{2a_{j2}} & \text{if } P_j^{\min} < \frac{s-a_{j1}}{2a_{j2}} < P_j^{\max}; \\ P_j^{\max} & \text{if } P_j^{\max} \ge \frac{s-a_{j1}}{2a_{j2}}; \end{cases} \quad (9)$$

The optimized profit E_j for DER j is thus

$$P_j^*(s) = \begin{cases} P_j(s) & \text{if } E(P_j) > 0; \\ 0 & \text{otherwise}; \end{cases} \quad (10)$$

The above expressions guarantee the optimal output and profitability of every DER unit. By applying the stochastic model and aggregating their output as in the appendix, the optimal output of each virtual utility is achieved and the overall profit is maximized.

3. MODIFIED ECONOMIC DISPATCH AND UNIT COMMITMENT

The utility has the prime objective of minimizing its generation cost. Traditionally, this involves two problems, namely: unit commitment and economic dispatch (Wood and Woolenberg, 1996). In this paper, each of these problems is modified with the inclusion of DERs.

3.1 Modified Economic Dispatch

Assume a utility with M utility units and connected to K virtual utilities at time interval t. The generation cost F is obtained by economically dispatching these on-line utility units and virtual utilities, as illustrated in Figure 1.

$$F = \sum_{i \in M} C_{cost \cdot i} + \sum_{k \in K} C_{paid \cdot k} \quad (11)$$

which includes the utility generation cost $\sum_{i \in M} C_{cost \cdot i}$, as well as the price paid for purchasing electricity from virtual utilities $\sum_{k \in K} C_{paid \cdot k}$. $\sum_{k \in K} C_{paid \cdot k}$ is calculated by applying (5) to calculate the price paid to each DER j and by applying the stochastic model in the appendix to calculate the price paid to each virtual utility k. The power balance constraint is met in the optimization:

$$\sum_{i \in M} P_i + \sum_{k \in K} P_k = P_D + P_{loss} \quad (12)$$

where P_D is system demand.

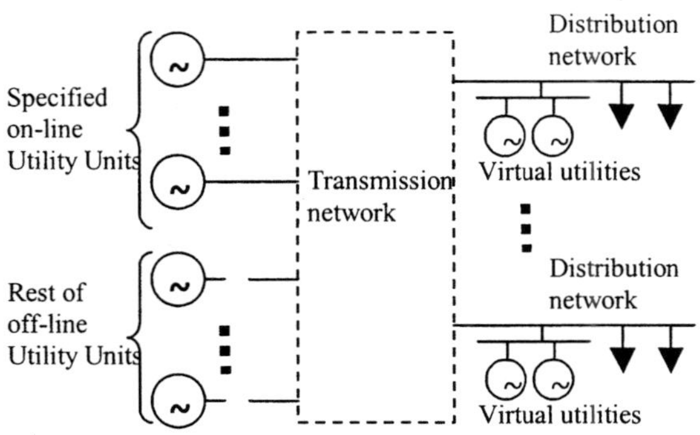

Fig. 1: Economic dispatch of committed utility units and virtual utilities

The utility generation cost $C_{cost\ i}$ is expressed as a quadratic function of P_i:

$$C_{cost \cdot i} = a_{i0} + a_{i1} P_i + a_{i2} P_i^2 \qquad (13)$$

The generation cost F is minimized by successive adjustments of λ as in Figure 2. With the current value of λ, $\sum_{i \in M} P_i$ is calculated by matching each unit's incremental heat rate to λ, and $\sum_{k \in K} P_k$ is calculated by (27) where s is given in (2). P_{loss} is calculated using "B" matrix loss formula (Wood and Woolenberg, 1996). In order to check the convergence of the iteration, (12) is rewritten in terms power mismatch Δ_P as:

$$\Delta_P = (\sum_{i \in M} P_i + \sum_{k \in K} P_k) - (P_D + P_{loss}) \qquad (14)$$

All recent values are substituted in (14) to update Δ_P. Should the mismatch be outside the tolerance, λ is adjusted one-step-up if Δ_P is negative, or one-step-down otherwise. The step size is halved at every iteration. After successive adjustments, the optimum values of λ, and output of each utility and virtual-utility unit are found.

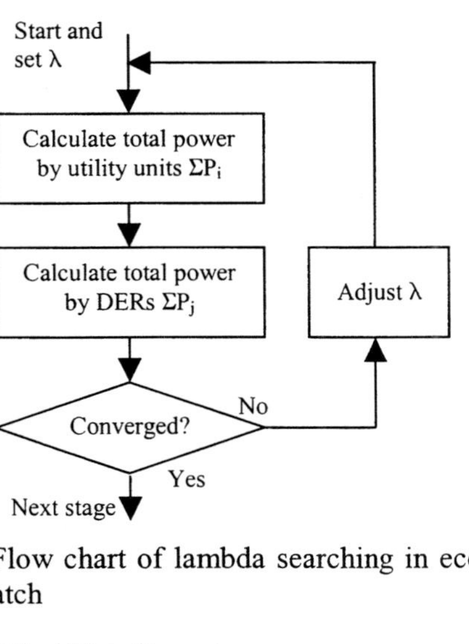

Fig. 2: Flow chart of lambda searching in economic dispatch

3.2 Modified Unit Commitment

The overall generation cost F over T time intervals is found by minimizing the total accumulated generation cost $F_{accu}(T)$.

At each time interval t, a dynamic-programming algorithm (Wood and Woolenberg, 1996) is used to minimize $F_{accu}(t)$ recursively by:

$$F_{accu}(t) = \min[\ F(t) + F_{tran}(t\text{-}1) + F_{accu}(t\text{-}1)\ ] \qquad (15)$$

where $F_{tran}(t\text{-}1)$ is the transition cost from interval t – 1 to interval t. $F(t)$ is the generation cost at interval t, which is minimized in section 3.1.

$F_{tran}(t\text{-}1)$ includes the unit start-up cost. Generator constraints, including unit minimum up time and down time, are considered. For instance, if a unit's previous off time is less than its minimum down time, it will not be committed and the transition cost from its OFF state to its ON state is set to infinite. Virtual utilities are treated as must-run units, having only one state (the ON state). Their outputs would be turned off, should their operation be found unprofitable (Section 2).

4 CASE STUDIES

Tables 1, 2 and 3 describe a utility connected to 8 utility units with a total capacity of 2180MW and 4 virtual utilities (29 DERs) with a total capacity of 142MW. Table 4 gives the 24-hour demand with a peak load at 2046MW and a daily load factor of 0.752.

Table 1: Utility unit characteristics

Utility Unit	P_i^{min} (MW)	P_i^{max} (MW)	Start-up cost($)	Min. up time (Hr)	Min down time (Hr)	Initial state (Hr)	A_{i0}	a_{i1}	a_{i2}
Unit1	200	500	1000	8	6	10	300	7.2	0.0039
Unit2	100	500	950	8	6	10	200	7.5	0.0042
Unit3	150	350	650	6	4	-10	350	7.8	0.00598
Unit4	50	300	550	5	3	10	150	7.9	0.0059
Unit5	50	250	400	4	3	-10	250	8.2	0.0071
Unit6	45	100	250	2	2	-10	200	8.7	0.0067
Unit7	50	100	180	2	2	-10	100	8.5	0.0078
Unit8	20	80	170	2	2	-10	200	9.9	0.0094

Table 2: Virtual utilities and their DER models

Virtual Utility	DER model	Number of units	availability
VU1	Model 1	3	0.98
VU1	Model 4	3	0.95
VU2	Model 2	4	0.97
VU3	Model 1	2	0.98
VU3	Model 3	5	0.95
VU3	Model 4	4	0.95
VU4	Model 3	2	0.95
VU4	Model 5	6	0.94

Table 3: DER unit characteristics

DER model	Number of units	P_j^{min} (MW)	P_j^{max} (MW)	a_{j0}	a_{j1}	a_{j2}
Model 1	5	2	8	16.24	8.46	0.00079
Model 2	4	2	6	12.66	8.48	0.00201
Model 3	7	2	5	10.5	8.75	0.00345
Model 4	7	2	4	8.7	8.74	0.00721
Model 5	6	1	2.5	5.8	8.64	0.02149

Table 4: Hourly load demand (MW)

Hr	Demand	Hr	Demand	Hr	Demand
1	1239	9	1816	17	1615
2	1105	10	1967	18	1667
3	1028	11	2046	19	1717
4	1009	12	1867	20	1905
5	1110	13	1776	21	1697
6	1208	14	1675	22	1526
7	1391	15	1635	23	1429
8	1614	16	1594	24	1296

Two cases are presented here to study the impact of DERs to the system. Delivery losses are not included in both cases. In case A, without considering DER, the utility supplies the consumer load of 36932 MWh at an overall cost of $371775.56 with a spinning reserve of 5% over the 24-hour period. The average cost of generation is 10.07 $/MWh.

With the DER participation in case B, a lower spinning reserve of 2% is assumed. This is because of the short start-up and synchronization times required for DERs. Comparing with Case A, the utility's peak demand is reduced to 1909.6 MW from 2406MW. The utility's plant load factor is now increased to 0.770 as illustrated in Figure 3.

Table 5 shows the simulation results. In Case B, the virtual utilities make a profit of 2.09% (= $361.42/ $17325.36) over the 24-hour period. In addition to its own generating cost of $353155.19, the utility pays $17686.78 to purchase electricity from virtual utilities. The total cost for utility is $370841.97, which is $933.59 or 0.25% less than that of Case A. λ's are lower at most loads for case B, as shown in Figure 4, due to DERs.

Table 5: Generation costs and profits over 24-hour period

	Case A	Case B
Utility Generation (MWh)	36932	35300.87
Utility Generation Cost($)	371775.56	353155.19
Average Utility Cost ($/MWh)	10.07	10.00
DER Generation (MWh)	/	1631.14
DER Generation Cost ($)	/	17325.36
Price Paid by Utility($)	/	17686.78
DER profit ($)	/	361.42
Total Load (MWh)	36932	36932
Total Cost($)	371775.56	370480.55
Average Load Cost ($/MWh)	10.07	10.03

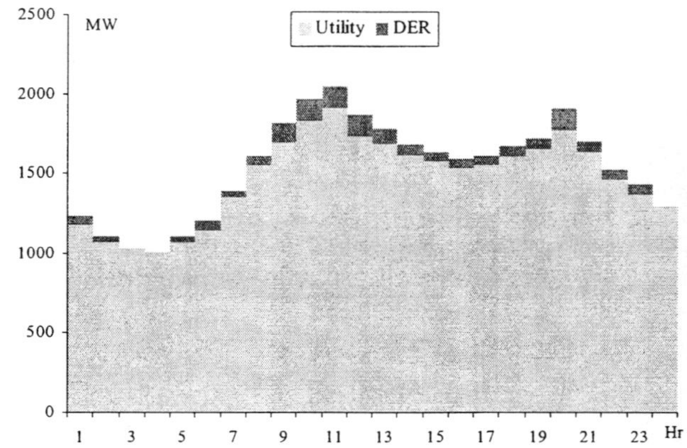

Fig. 3: Generation of utility and DER over 24-hour period

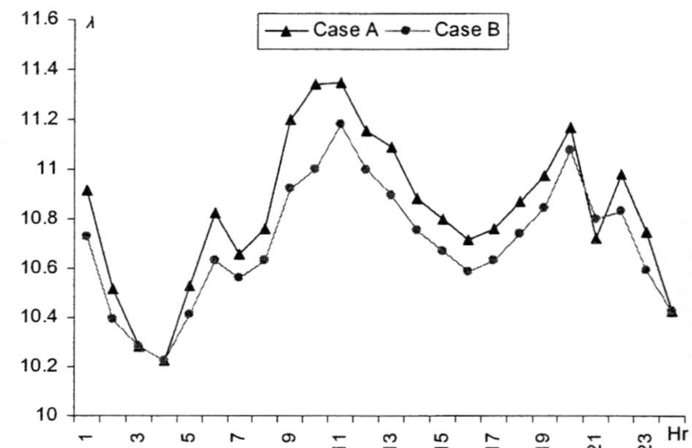

Fig. 4: System lambdas over 24-hour period

5 CONCLUSIONS

This paper presents an approach to accommodate DER generation capacity in the utility's economic dispatch and unit commitment. DERs with homogeneous cost characteristics are clustered into several virtual units, which participate in generation according to their profitability. Availability of the DERs is considered in a stochastic model. Case Studies show the positive contribution of DERs in reducing the overall cost of generation.

6 REFERENCES

A-M. Borbely, J. F. Kreider (2001). *Distributed Generation: The Power Paradigm for the New Millennium*, CRC Press, Florida.

Allen J. Wood, Bruce F. Wollenberg (1996). *Power Generation Operation and Control*, (Second Edition), John Wiley & sons, Inc., New York.

C. L. Smallwood (2002). Distributed Generation in Autonomous and Non-Autonomous Microgrid, *IEEE Rural Electric Power Conference*, pp. D1 - D1_6.

F. C. Schweppe, M. C. Caraminis, R. D. Tabors, R. E. Bohn (1988). *Spot Pricing of Electricity*, Kluwer Academic Publishers, Massachusetts.

H. Lee Willis, Walter G. Scott (2000). *Distributed Power Generation: Planning and Evaluation*, Marcel Dekker Inc, New York.

Nikos D. Hatziargyriou, Sakis Meliopoulos (2002). Distributed Energy Sources: Technical

Challenges, *Power Engineering Society Winter Meeting, IEEE*, **Volume: 2**, pp.1017-1022.

Papoulis A (1965). *Probability, Random Variables, and Stochastic Processes*, McGraw Hill, New York.

P.A. Daly, J. Morrison (2001). Understanding the Potential Benefits of Distributed Generation on Power Delivery Systems, *IEEE Rural Electric Power Conference*, pp. A2/1 -A213.

Tina Kaarsberg, Suzanne Watson (2002). POWER PARKS: A New Solution for Local Energy, *Cogeneration and On-Site Power Production*, January-February 2002, **Volume 3 Issue** 1, James & James (Science Publishers) Ltd, London, UK.

7 APPENDIX

7.1 Modeling of DER Availability

Given the buy-back price s, according to (10) DER j is expected to generate optimal output $P_j^*(s)$ to ensure profitability as permitted by availability. By considering the failures of DER as random events, a stochastic model of the DER is formulated as below. Taking the forced outage rate p_j from the DER's long-term average availability cycle (Wood and Woolenberg, 1996):

$$p_j = \frac{T_{avail \cdot j}}{T_{total \cdot j}} \quad (16)$$

where $T_{avail \cdot j}$ is average time available before failures; $T_{total \cdot j}$ (= $T_{avail \cdot j}$ + $T_{repair \cdot j}$) is mean time between failures; T_{repair} is average repair time.

Therefore the output of DER j is modeled as a random variable denoted as $P_j^R(s)$. In the paper the superscript R is used to denote random variable. $P_j^R(s)$ is subject to Bernoulli distribution, with the probability density function (PDF) $f_j(x)$ being:

$$f_j(x) = (1 - p_j)\delta(x) + p_j \delta(x - P_j^*(s)) \quad (17)$$

where $\delta(x)$ is the impulse function. The mean η_j and variance σ_j^2 of $P_j^R(s)$ is expressed as:

$$\eta_j = p_j P_j^* \quad (18)$$

$$\sigma_j^2 = p_j (1 - p_j) P_j^{*2} \quad (19)$$

7.2 Modeling of Virtual Utility Probability

Since P_j^R (s) is random variable, (1) is rewritten in terms of random variables as:

$$P_k^R = \sum_{j \in J_k} P_j^R \quad (20)$$

where P_k^R the overall output from virtual utility k is a random variable too. The probability distribution of P_k^R is achieved by a numerical convolution using recursive technique, if every unit's output distribution P_j^t is independent of each other. With the recursive technique, the output of DER $P_j^t(s^t)$ is added into P_k^t (s^t) once by one unit. Assume that $f(x)$, the PDF of P_k^R (s) of total u DERs, has been achieved. x is the total output of the u DERs. Now considering to add the output of the $(u+1)$th "N-state" DER, of which in state n the output is $O_{u+1}(n)$ and the probability is p_{u+1} (n), The new PDF $f'(x)$ of P_k^t (s) of the total u+1 DERs can be expressed by:

$$f'(x) = \sum_{n=1}^{N} p_{u+1}(n) f(x - O_{u+1}(n)) \quad (21)$$

where N is taken as 2 representing the ON and OFF states.

After recursively using the numerical convolution (J_k -1) times, all the J_k DERs in virtual utility k are summed up and the $f(x)$, PDF of P_k^R (s) of the overall output of the virtual utility, is achieved. Knowing $f(x)$, the probability distribution of the virtual utility power output is characterized by the mean η_k and variance σ^2_k of P_k^R (s) as:

$$\eta_k = \int P_k^R(s)f(x)dx \quad (22)$$

$$\sigma^2{}_k = \int (x - P_k^R(s))^2 f(x)dx \quad (23)$$

The output distribution each DER within one virtual utility $P_j^R(s)$ is considered to be dependent of each other, as some DERs can be electrically or mechanically coupled. Covariance matrix $\{\mu\}$ is used to describe the dependence between DERs in each virtual utility:

$$\{\mu\} = \begin{bmatrix} \sigma_1^2 & \mu_{12} & \Lambda & \mu_{1J_k} \\ \mu_{12} & \sigma_2^2 & \Lambda & \mu_{2J_k} \\ M & M & O & M \\ \mu_{J_k 1} & \mu_{J_k 2} & \Lambda & \sigma_{J_k}^2 \end{bmatrix} \quad (24)$$

where σ_j^2 is variance of DER j, and μ_{ij} is covariance between DERs i and j. From probability theory (Papoulis, 1965), the mean η_k and variance σ^2_k of the P_k^R (s) are calculated by:

$$\eta_k = \sum_{j \in J_k} \eta_j \quad (25)$$

$$\sigma_k^2 = \sum_{j \in J_k} \sigma_j^2 + 2 \sum_{i \in J_k, j \in J_k, i \neq j} \mu_{ij} \quad (26)$$

where η_j and σ^2_j are the mean and variance of DER j respectively, and their expression are given in (18) & (19). η_k and σ^2_k are functions of the buy-back price s. Therefore, the expected power output of virtual utility k can be expressed as the mean of P_k^R:

$$P_k(s) = \eta_k(s) \quad (27)$$

ELSEVIER
IFAC
PUBLICATIONS
www.elsevier.com/locate/ifac

TUTORIAL ON EVOLUTIONARY COMPUTATION TECHNIQUES FOR POWER SYSTEM OPTIMIZATION

Organizers:
Kwang Y. Lee[a] and Mohamed A. El-Sharkawi[b]

[a]Department of Electrical Engineering, Pennsylvania State University, University Park, PA 16802, USA
[b]Department of Electrical Engineering, University of Washington, Seattle, WA 98195, USA

Abstract: A half-day tutorial session was given on evolutionary computation techniques and their applications to power system optimisation. The purpose of this course is to provide participants with basic knowledge of evolutionary computation techniques, and how they are combined with knowledge elements in computational intelligence systems. Applications to power problems are stressed, and example applications are presented. The tutorial is composed of two parts: The first part gives an overview of evolutionary computation techniques, including fundamentals of genetic algorithms, evolutionary programming and strategies, particle swarm optimization, and ant colony search algorithm, and the hybrid system of evolutionary computation. It also gives an overview of power system applications. *Copyright © 2003 IFAC*

Keywords: evolutionary computation, power system optimization, genetic algorithm, evolutionary strategies, evolutionary programming, particle swarm optimization, ant colony algorithm.

1. INTRODUCTION

Several heuristic tools have evolved in the last decade that facilitate solving optimization problems that were previously difficult or impossible to solve. These tools include evolutionary computation, simulated annealing, tabu search, etc. Reports of applications of each of these tools have been widely published. Recently, these new heuristic tools have been combined among themselves and with knowledge elements, as well as with more traditional approaches such as statistical analysis, to solve extremely challenging problems. Developing solutions with these tools offers two major advantages: 1) development time is much shorter than when using more traditional approaches, and, 2) the systems are very robust, being relatively insensitive to noisy and/or missing data.

The purpose of this course is to provide participants with basic knowledge of evolutionary computation techniques, and how they are combined with knowledge elements in computational intelligence systems. Applications to power problems are stressed, and example applications are presented. The tutorial is composed of two parts: The first part gives an overview of evolutionary computation techniques, including fundamentals of genetic algorithms, evolutionary programming and strategies, particle swarm optimization, and ant colony search algorithm, and the hybrid system of evolutionary computation. It also gives an overview of power system applications.

The second part of the tutorial deals with specific applications of the heuristic approaches to power system problems, such as security assessment, operational planning, generation and reactive power planning, and power plant control. Additional references on modern heuristic optimization techniques are also given.

2. EVOLUTIONARY COMPUTATION

Natural evolution is a hypothetical population-based optimization process. Simulating this process on a computer results in stochastic optimization techniques that can often out-perform classical methods of optimization when applied to difficult real-world problems. This tutorial will provide a background in the inspiration, history, and application of evolutionary computation to system identification, automatic control, gaming, and other combinatorial problems.

The objectives are to provide an overview of how evolutionary computation and other heuristic optimization techniques may be applied to problems within your domain of expertise, to provide a good understanding of the design issues involved in tailoring evolutionary algorithms to real-world problems, to compare and judge the efficacy of evolutionary computation techniques with other more classic methods of optimization, and to program fundamental evolutionary algorithms.

2.1 Genetic Algorithms

Genetic Algorithm (GA) is a search algorithm based on the *conjecture* of natural selection and genetics. The features of genetic algorithm are different from other search techniques in several aspects. First, the algorithm is a multi-path that searches many peaks in parallel, and hence reducing the possibility of local minimum trapping. Secondly, GA works with a coding of parameters instead of the parameters themselves. The coding of parameter will help the genetic operator to evolve the current state into the next state with minimum computations. Thirdly, GA evaluates the fitness of each string to guide its search instead of the optimization function. The genetic algorithm only needs to evaluate objective function (fitness) to guide its search. There is no requirement for derivatives or other auxiliary knowledge. Hence, there is no need for computation of derivatives or other auxiliary functions. Finally, GA explores the search space where the probability of finding improved performance is high.

2.2 Evolution Strategies and Evolutionary Programming

Evolution Strategies (ES) employ real-coded variables and, in its original form, it relied on Mutation as the search operator, and a Population size of one. Since then it has evolved to share many features with GA. The major similarity between these two types of algorithms is that they both maintain populations of potential solutions and use a selection mechanism for choosing the best individuals from the population. The main differences are: ES operate directly on floating point vectors while classical GAs operate on binary strings; GAs rely mainly on recombination to explore the search space, while ES uses mutation as the dominant operator; and ES is an abstraction of evolution at individual behavior level, stressing the behavioral link between an individual and its offspring, while GAs maintain the genetic link.

Evolutionary Programming (EP) is a stochastic optimization strategy similar to GA, which places emphasis on the behavioral linkage between parents and their offspring, rather than seeking to emulate specific genetic operators as observed in nature. EP is similar to Evolutionary Strategies, although the two approaches developed independently. Like both ES and GAs, EP is a useful method of optimization when other techniques such as gradient descent or direct analytical discovery are not possible. Combinatorial and real-valued function optimization in which the optimization surface or fitness landscape is "rugged", possessing many locally optimal solutions, are well suited for Evolutionary Programming.

2.3 Particle Swarm

Particle Swarm Optimization (PSO) is an exciting new methodology in evolutionary computation that is somewhat similar to a genetic algorithm in that the system is initialized with a population of random solutions. Unlike other algorithms, however, each potential solution (called a particle) is also assigned a randomized velocity and then flown through the problem hyperspace. The PSO has been found to be extremely effective in solving a wide range of engineering problems. It is very simple to implement (the algorithm comprises two lines of computer code) and solves problems very quickly.

2.4 Ant Colony Search Algorithm

The Ant Colony Search Algorithms (ACSA) mimic the behavior of real ants. As is well known, real ants are capable of finding shortest path from food sources to the nest without using visual cues. They are also capable of adapting to changes in the environment, for example, finding a new shortest path once the old one is no longer feasible due to a new obstacle. The studies by ethnologists reveal that such capabilities ants have are essentially due to what is called "pheromone trails" which ants use to communicate information among individuals regarding path and to decide where to go. Ants deposit a certain amount of pheromone while walking, and each ant probabilistically prefers to follow a direction rich in pheromone rather than a poorer one. There are several advantages of the ACSA: Distributed computation avoids premature convergence, the nature of positive feedback gives a rapid discovery of good solution, and the use of constructive greedy heuristic helps in finding acceptable solutions in the early stage of the process.

3. SPEAKERS

Mohamed A. El-Sharkawi, University of Washington, USA

Yoshikazu Fukuyama, Fuji Electric Co., Japan

Raul Garduno-Ramirez, Electrical Research Institute, Mexico

Germano Lambert-Torres, Federal Engineering School at Itajuba, Brazil

Kwang Y. Lee, The Pennsylvania State University, USA

Yong Hua Song, Brunel University, UK

4. TABLE OF CONTENTS

Part 1: Theory of Evolutionary Computation

Chapter 6. Ant Colony Search Algorithms in Power System Optimization
Chapter 7. Hybrid Systems - Merging Genetic Algorithms with Other Techniques

Part 2: Selected Applications of Evolutionary Computation

Chapter 8. Power Plant Controls: Feedforward and Feedback Controller Design for Wide Range Operation Using Genetic Algorithm
Chapter 9. Application of Evolutionary Technique to Power System Security Assessment
Chapter 10. Genetic Algorithms for Generation Expansion and Reactive Power Planning
Chapter 11. Evolutionary Algorithm for Operational Planning: Unit Commitment and Economic Dispatch
Chapter 12. Bibliographies on Evolutionary Computation with Power System Applications

5. TUTORIAL BOOKS

Lee, K. Y. and El-Sharkawi, M. A. (Editors) (2003). *A Tutorial Course on Evolutionary Computation Techniques for Power System Optimization*, 2003 IFAC Symposium on Power Plants and Power Systems Control, Seoul, Korea.

Lee, K. Y. and El-Sharkawi, M. A. (Editors) (2002). *Tutorial on Modern Heuristic Optimization Techniques with Applications to Power Systems*, IEEE Power Engineering Society, Publication 02TP160.

6. SPEAKERS' BIOGRAPHIES

Kwang Y. Lee is a Fellow of IEEE. He is a professor of Electrical Engineering at the Pennsylvania State University. He is an Editor of IEEE Transactions on Energy Conversion, Associate Editor of the IEEE Transactions on Neural Networks since 1996, Technical Program Chairman for the 1997 International Conference on Intelligent Systems Applications to Power Systems (ISAP '97), and Program Co-Chairman of the 1989 U.S.-Korea Joint Seminar on Expert Systems for Power Systems. He is active in the Intelligent Systems Subcommittee and Station Operation, Design and Control Subcommittee of the IEEE Power Engineering Society. He is also a Vice Chair of the IFAC Technical Committee on Power Plants and Power Systems. He organized and participated several panels and special sessions in IEEE and other international conferences on intelligent control, distributed simulation, combined research and curriculum development, and undergraduate research. He was also a contributor to "IEEE Tutorial on Artificial Neural Networks with Applications to Power Systems" and "IEEE Tutorial on Fuzzy Logic Applications in Power Systems", and edited an "ISAP Tutorial Book on Current Trend and the State of the Art in Intelligent System Applications to Power Systems". He is the author of over 300 technical publications and proceedings and 15 book chapters.

Mohamed A. El-Sharkawi is a Fellow of IEEE. He is a professor of Electrical Engineering at the University of Washington. He is the Founder of the international conference on the Application of Neural Networks to Power Systems (ANNPS) and the Co-founder of the international conference on Intelligent Systems Applications to Power (ISAP). He is a member of the administrative committee of the IEEE Neural Networks Council representing the Power Engineering Society; Video Tutorial Chair of the IEEE Continuing Education Committee and the neural network council. He is the Founding Chairman of several IEEE task forces and working groups and subcommittees. He is a member of the editorial board and associate editor of several journals. He co-edited an IEEE tutorial book on the applications of neural network to power systems. He organized and taught several international tutorials on intelligent systems applications, power quality and power systems; and organized and chaired numerous panel and special sessions in IEEE and other international conferences. He published over 130 papers and book chapters in these areas, and holds 7 licensed patents. He published recently a textbook entitled "Fundamentals of Electric Drives," Brooks/Cole Publishing, 2000.

Germano Lambert-Torres is a professor of Electrical Engineering and Dean of Research and Graduate Studies at the Federal Engineering School at Itajuba, Brazil. He is the General Editor of Research Technologic Development Magazine,

General Chairman for the first Brazilian Conference on Neural Networks (CBRN'97), General Chairman for the 1999 International Conference on Intelligent Systems Applications to Power Systems (ISAP '99), Program Co-Chairman of the International Congress of Logic Applied to Technology (LAPTEC'00), and Vice-General Chairman for the 2001 International Conference on Intelligent Systems Applications to Power Systems (ISAP '01). He is also a Member of the Internal Program Committee for IASTED International Conference on Power and Energy Systems, International Conference in Evolutionary Computing in Computer, Communication, Control and Power, and WSES International Conference on Simulation. He serves on a number of committees related to intelligent systems in IEEE and CIGRE. He also serves as consultant for many Brazilian and South America Power Industries, such as Itaipu Power Plant. He has more than 300 technical papers published in transactions, proceedings, and book chapters.

Yoshikazu Fukuyama received the B.S., M.S., and Ph.D. degrees in electrical engineering in 1985, 1987, and 1997, respectively, from Waseda University, Tokyo, Japan. He has been working at Fuji Electric Co., Japan from 1987. He was a visiting scientist at Cornell University from 1993 to 1994. His research interests include application of intelligent systems to power systems and power system analysis. He is also interested in general optimization problems. He is a member of IEEE and IEE of Japan.

Yong-Hua Song is a professor of Electrical Energy Systems at Brunel Institute of Power Systems, Brunel University, UK since 1997. He received his B.S., M.S. and Ph.D. degrees from Tchinghua University, Beijing, China. In 2002 he is awarded D.Sc. by Brunel University for his significant contributions to power system knowledge and research. His research interests include power system optimization, control, economics, FACTS and intelligent system applications. He is a fellow of the IEE and a chartered electrical engineer in the UK.

Raul Garduno-Ramirez received his B.S. degree in Electrical Engineering from the National Polytechnic Institute (IPN), Mexico, in 1985, his M.S. degree in Electrical Engineering from Advanced Research Center of the IPN, Mexico, in 1987, and his Ph.D. degree in Electrical Engineering from the Pennsylvania State University. During 1986 he stayed at the National Mechanical Laboratory, Tsukuba, Japan, and has been at the Electrical Research Institute, Mexico, since 1987 where he was been involved in the development of control systems for power plants. Dr. Garduno is a Member of the IFAC Technical Committee on Power Plants and Power Systems. His current interest is in the development of intelligent control systems including advanced controls and process optimization. Dr. Garduno is a member of IEEE, ISA and the National Research System (SNI-Mexico).

ELSEVIER
IFAC
PUBLICATIONS
www.elsevier.com/locate/ifac

NOVEL OPTIMAL NEUROCONTROL FOR A SYNCHRONOUS GENERATOR USING RADIAL BASIS FUNCTION NEURAL NETWORK

Jung-Wook Park[†], Ronald G. Harley[†], and Ganesh K. Venayagamoorthy[††]

[†] *School of Electrical and Computer Engineering*
Georgia Institute of Technology, GA 30332-0250, U.S.A.
Jung-Wook Park (jungwookpark@ieee.org) and Ronald G. Harley (rharley@ece.gatech.edu)

[††] *Department of Electrical and Computer Engineering*
University of Missouri-Rolla, MO 65409-0249, U.S.A.
Ganesh K. Venayagamoorthy (ganeshv@ece.umr.edu)

Abstract: This paper presents the design of an infinite horizon optimal neurocontroller to replace the conventional controllers such as the automatic voltage regulator and governor for the control of a synchronous generator connected to an electric power grid. The neurocontroller design uses the dual heuristic programming (DHP) algorithm, which provides the most robust control capability among the adaptive critic designs (ACDs) family. The radial basis function neural network (RBFNN) is used as the function approximator to implement the DHP. The performances of the proposed optimal neurocontroller are evaluated and its stability issue in real-time operation is analysed. *Copyright © 2003 IFAC*

Keywords: Dual Heuristic Programming, Optimal Control, Radial Basis Function Neural Network, Synchronous Generator.

1. INTRODUCTION

Conventional linear controllers (CONVC) for the synchronous generator consist of the automatic voltage regulator (AVR) and the turbine governor. They are designed to control, in some optimal fashion, the generator around one particular operating point; and at any other point the generator's damping performance is degraded (Venayagamoorthy and Harley, 2001).

Artificial neural networks (ANNs) have been used as an alternative for the CONVC for nonlinear adaptive controllers of various control objectives in the field of electrical power engineering. Especially, the radial basis function neural network (RBFNN) based neurocontroller has recently attracted attention for the control of a synchronous generator (Flynn, *et al.*, 1997; Segal, *et al.*, 2000) and a flexible alternating current transmission system (FACTS) (Dash, *et al.*, 2000) device. This is due to the efficient control capability of the RBFNN neurocontroller to damp low-frequency oscillations as well as the faster convergence speed for identifying of the plant.

Whenever ANNs are used in the control path of a nonlinear power system, the stability issue is an important factor to be considered. Especially, the character of the wide range of operating conditions of a synchronous generator connected to an electric power grid limits the use of the ANN because training of the ANN is based on the gradient descent algorithm. To overcome the issue of instability during the steady state at the various different operating conditions and provide strong robustness for the ANN based controller, the adaptive critic designs (ACDs) technique (Miller, *et al.*, 1990; Prokhorov, *et al.*, 1997; Si and Wang, 2001; Venayagamoorthy, *et al.*, 2002; Werbos, 1992) for the nonlinear optimal control (Bertsekas, 2001; Gregory and Kin, 1992), especially the infinite horizon problem, has been applied lately.

In spite of its functional merit providing a best fit from the input data space, there is no research reported in the literature using the RBFNN as a function approximator to implement the dual heuristic programming (DHP), which has the most robust control capability among the ACDs family.

In this paper, the background of the ACDs with relation to optimal control theory is briefly described. The DHP based optimal neurocontroller (DHPNC) using the RBFNN is designed for the control of the excitation and turbine of a synchronous generator connected to an infinite bus. After training off-line prior to commissioning in real-time operation, the transient and dynamic damping performances for the proposed DHPNC are evaluated with case studies by time-domain simulation.

2. ADAPTIVE CRITIC DESIGNS

Adaptive critic designs (ACDs) technique, which was proposed by P. J. Werbos (1992), is a novel optimization control algorithm based on the mathematical analysis to handle the classical optimal control problem by combined concepts of *reinforcement learning* and *approximate dynamic programming* (ADP).

For a given series of control actions that must be taken sequentially, and knowing the effect of these actions until the end of these sequences, it is impossible to design an optimal controller by traditional supervisory learning methods when ANNs are used. In addition, if the sequences for user-defined cost-to-go function **J** in (1) go infinite in time (so-called *infinite horizon problem*), the Bellman equation using classical dynamic programming (DP) in (2) to minimize/maximize function J has the well-known *curse of dimensionality* problem because the DP prescribes a search which tracks backward from the final step, retaining in memory all sub-optimal paths from any given point to the finish, until the starting point is reached.

$$\mathbf{J}_{\pi}(\mathbf{x}_0) = \sum_{k=0}^{\infty} \gamma^k g(\mathbf{x}(k), \mathbf{u}(k)) \quad (1)$$

$$\mathbf{J}_{k+1}(\mathbf{x}) = \min_{\mathbf{u} \in \mathbf{U}} \left[g(\mathbf{x},\mathbf{u}) + \gamma \mathbf{J}_k(f(\mathbf{x},\mathbf{u})) \right],\ k = 0,1,\ldots \quad (2)$$
$$\mathbf{J}_0(\mathbf{x}) = 0 \text{ for all } \mathbf{x}$$

where k is a discrete time index at each step, $\mathbf{J}_{\pi}(\mathbf{x}_0)$ denotes the cost associated with an initial state $\mathbf{x}_0$, and a control policy $\pi=\{\mathbf{u}_0, \mathbf{u}_1, \ldots\}$, and γ is the discount factor ($0 < \gamma < 1$).

The ACD algorithm requires the use of two "critic" networks, which output an estimate of the total future value of a utility function U arising from the present and past states and the control information, thereby, it learns to approximate the heuristic cost-to-go function **J**(k) in (3).

$$\mathbf{J}(k) = \sum_{p=0}^{\infty} \gamma^p U(k + p) \quad (3)$$

After minimizing the **J** in (3) by the critic network, the control network (the so-called "action" network in the ACD literature) is trained with the estimated output backpropagated from the critic network to obtain the converged weight for the optimal control **u**. The backpropagation calculates the derivatives of future utility with respect to present actions through the critic network. In other words, backpropagation plays the same role as the costate equation in *Pontryagin's Minimum Principle* from the viewpoint of classical optimal control theory.

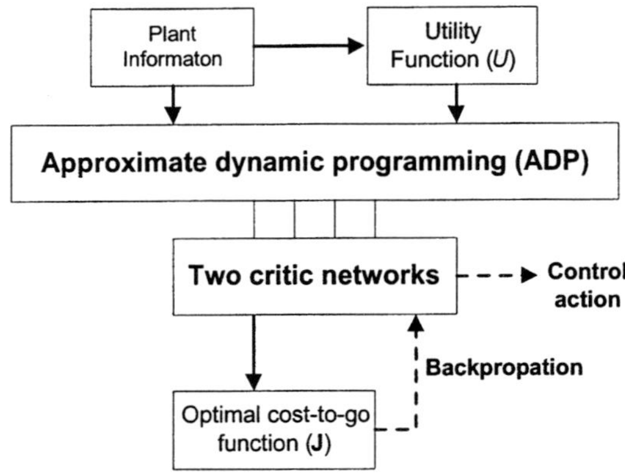

Fig. 1. Approximate dynamic programming (ADP) by two critic networks

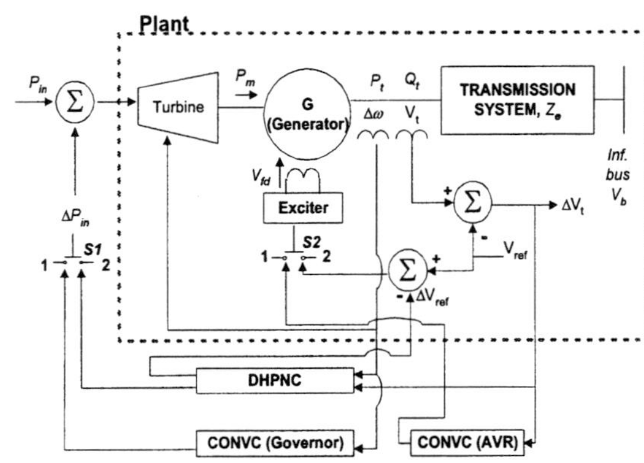

Fig. 2. Plant model used for the control of a synchronous generator connected to an infinite bus

All these explanations are illustrated in Fig.1 and a more detailed description is found in (Park, *et al.*, 2002; Werbos, 1992).

The DHP technique described in this paper uses three different types of RBFNNs, one for the critic, one for the model, and one for the action network implementations. The critic network approximates the derivates of the function **J** with respect to the states of the plant. All the steps taken in the DHPNC design are carried out off-line.

3. DHP BASED NEUROCONTROLLER DESIGN

3.1 *Plant Modeling*

The synchronous generator, turbine, exciter and transmission system connected to an infinite bus in Fig. 2, form the plant (dashed block in Fig. 2.) that has to be controlled. The generator (**G**) with its damper windings is described by a seventh order d-q axis set of equations with the generator current, speed, and rotor angle as the state variables.

In the plant, P_t and Q_t are the real and reactive power at the generator terminal, respectively, Z_e is the transmission line impedance, P_m is the mechanical input power to the generator, V_{fd} is the exciter field voltage, V_b is the infinite bus voltage, $\Delta\omega$ is the speed deviation, ΔV_t is the terminal voltage deviation, V_t is the terminal voltage, ΔV_{ref} is the reference voltage deviation, V_{ref} is the reference voltage, ΔP_{in} is the input power deviation, and P_{in} is the turbine input power.

The position of the switches *S1* and *S2* in Fig. 2 determines whether the DHP based optimal neurocontroller (DHPNC), or the CONVC consisting of exciter/AVR system and turbine/governor system, is controlling the plant. Block diagrams, time constants and gains for the CONVC are shown in (Park, *et al.*, 2002).

3.2 *Radial Basis Function Neural Network*

The radial basis function neural network (RBFNN) shown in Fig. 3 consists of three-layers. The input values are each assigned to a node in the input layer and passed directly to the hidden layer without weights.

The hidden layer nodes are called RBF units, determined by a parameter vector called *center* and a scalar called *width*. The gaussian density function is used as an activation function for the hidden neurons in Fig. 3. Then, linear output weights (vector **V** in Fig. 3) are connected between the hidden and output layers. The overall input-output mapping equation of the RBFNN is as follows.

$$y_i = b_i + \sum_{j=1}^{h} v_{ji} \exp\left(-\frac{\|\mathbf{X} - \mathbf{C}_j\|^2}{\beta_j^2} \right) \tag{4}$$

where **X** is the input vector, $\mathbf{C}_j \in \mathbf{R}^n$ is the jth center of RBF unit in the hidden layer, h is the number of RBF units, b_i and v_{ji} are the bias term and the weight between hidden and output layer respectively, and y_i is the ith output in the m-dimensional space. Once the RBF centers are established, the width of the ith center in the hidden layer is calculated as follows.

$$\beta_i = \left[\frac{1}{h} \sum_{j=1}^{h} \sum_{k=1}^{n} (\|\mathbf{c}_{ki} - \mathbf{c}_{kj}\|) \right]^{1/2} \tag{5}$$

where $\mathbf{c}_{ki}$ and $\mathbf{c}_{kj}$ are the kth value of the center of the ith and jth RBF units. In (4) and (5), $\|\cdot\|$ represents the euclidean norm.

To avoid the extensive computational complexity during training, the batch mode k-means clustering algorithm for centers is initially calculated for the centers of the RBF unit. Thereafter, the pattern mode least-mean-square (LMS) algorithm is calculated to update the output linear weights.

For more detail explanation for the RBFNN, the reader is referred to (Park, *et al.*, 2002; Segal, *et al.*, 2000). By trial and error, twelve neurons for the model network and six neurons for the action and critic networks in the hidden layer are optimally chosen for this study.

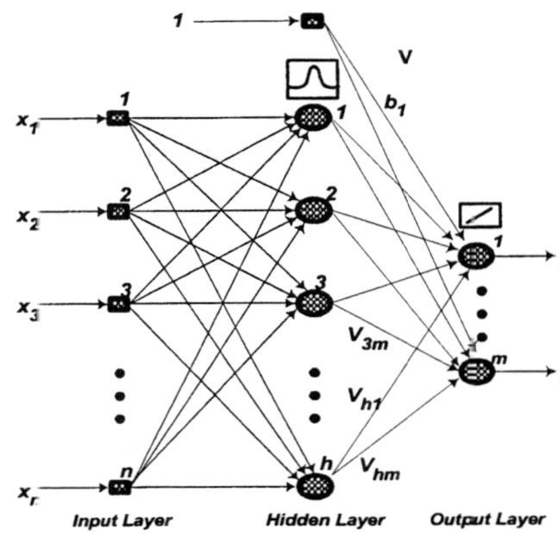

Fig. 3. Plant model used for the control of a synchronous generator connected to an infinite bus

3.3 *Dual Heuristic Programming Neurocontroller*

The configuration for the critic network training in the DHP is shown in Fig. 4. The input reference vector $\mathbf{Y}_{ref}$ into the plant and output vector **Y** from the plant are:

- $\mathbf{Y}_{ref}(t)$, input reference vector to the plant = $[P_{in}(t), V_{ref}(t)]$.
- $\mathbf{Y}(t)$, output vector from the plant = $[\Delta\omega(t), \Delta V_t(t)]$.

The input vector **X** and output vector **Y** for the action and model networks are:

- $\mathbf{X}(t)$ of the action network = $[\Delta\omega(k), \Delta V_t(k) \mid k = \{t\text{-}1, t\text{-}2, t\text{-}3\}]$.
- $\mathbf{Y}(t)$ of the action network = $\mathbf{A}(t) = [\Delta P_{in}(t), \Delta V_{ref}(t)]$.
- $\mathbf{X}(t)$ of the model network = $[\Delta\omega(k), \Delta V_t(k), \Delta P_{in}(k), \Delta V_{ref}(k) \mid k = \{t\text{-}1, t\text{-}2, t\text{-}3\}]$.
- $\mathbf{Y}(t)$ of the model network = $[\Delta\hat{\omega}(t), \Delta\hat{V}_t(t)]$.

The critic network estimates the derivatives of function **J** with respect to the vector of observables $\Delta\hat{\mathbf{Y}}(t) = [\Delta\hat{\omega}(t), \Delta\hat{V}_t(t)]$ (input vector of the critic network) of the plant identified by the model network, and it learns to minimize the following error measure over time:

$$\|E_C\| = \sum_t \mathbf{e}_C^T(t)\mathbf{e}_C(t) \tag{6}$$

$$\mathbf{e}_C(t) = \frac{\partial \mathbf{J}[\Delta\hat{\mathbf{Y}}(t)]}{\partial \Delta\hat{\mathbf{Y}}(t)} - \gamma \frac{\partial \hat{\mathbf{J}}[\Delta\hat{\mathbf{Y}}(t+1)]}{\partial \Delta\mathbf{Y}(t)} - \frac{\partial U[\Delta\mathbf{Y}(t)]}{\partial \Delta\mathbf{Y}(t)} \tag{7}$$

After exploiting all relevant pathways of backpropagation as shown in Fig. 4, where the paths of derivatives and adaptation of the critic network are depicted by dotted and dash-dot lines, the error signal $\mathbf{e}_C(t)$ is used for training to update the output linear weights of the critic network. The mathematical closed form solution of the RBFNN for backpropagation computation, which accounts for the estimated input vector **X** from the future output vector **Y**, based on the chain rule of derivatives, is given in (8)

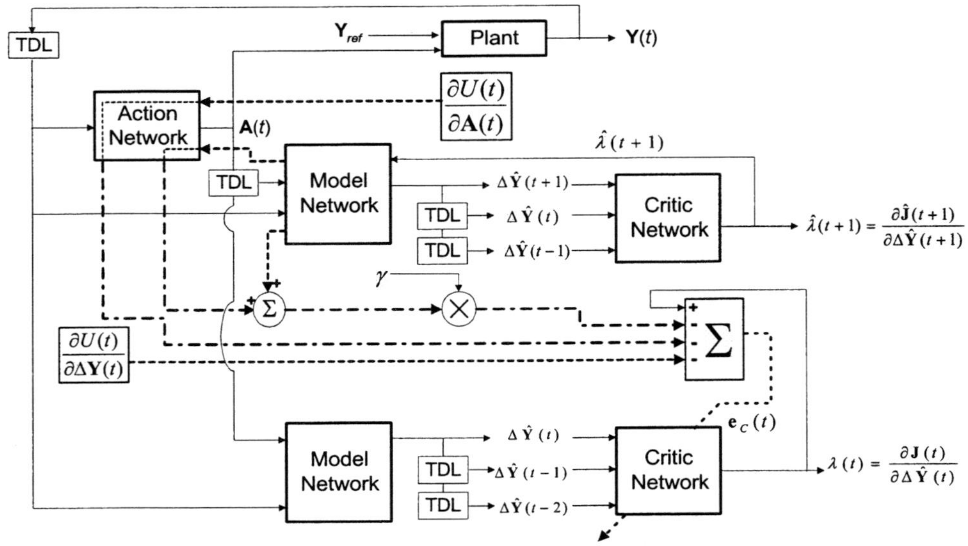

Fig. 4. Critic network adaptation in DHP: This diagram shows the implementation of (10). The same critic network is shown for two consecutive times, t and t+1. The discount factor γ is chosen to be 0.5. Backpropagation paths are shown by dotted and dash-dot lines. The output of the critic network $\hat{\lambda}(t+1)$ is backpropagated through the model network from its outputs to its inputs, yielding the first term of (9) and $\partial \mathbf{J}(t+1)/\partial \mathbf{A}(t)$. The latter is backpropagated through the action network from its outputs to its inputs forming the second term of (9). Backpropagation of the vector $\partial U(t)/\partial \mathbf{A}(t)$ through the action network results in a vector with components computed as the last term of (10). The summation of all these signals produces the error vector $\mathbf{e}_c(t)$ used for training the critic network.

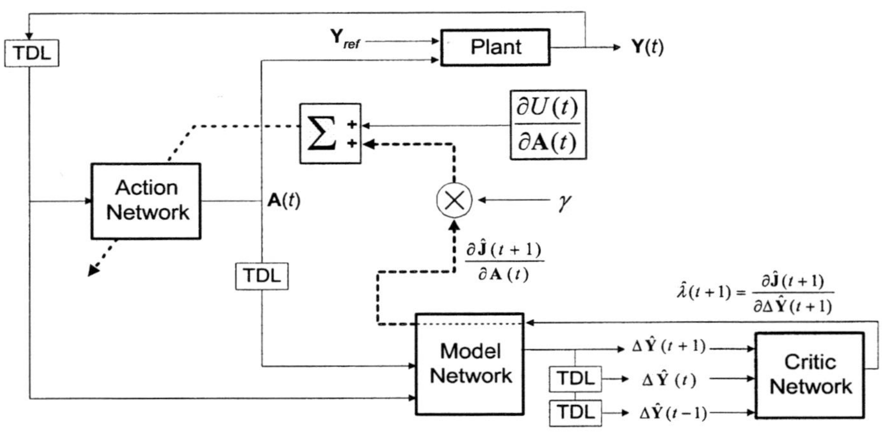

Fig. 5. Action network adaptation in DHP: The discount factor γ is chosen to be 0.5. Backpropagation paths are shown by dotted lines. The output of the critic network $\hat{\lambda}(t+1)$ at time (t+1) is backpropagated through the model network from its outputs to its inputs (output of the action network), and the resulting vector multiplied by the discount factor (γ = 0.5) and added to $\partial U(t)/\partial \mathbf{A}(t)$. Then, an incremental adaptation of the action network is carried out by (12) and (13).

$$\frac{\partial \mathbf{Y}}{\partial \mathbf{X}} = \frac{\partial \mathbf{Y}}{\partial t}\frac{\partial t}{\partial p_L}\frac{\partial p_L}{\partial q_L}\frac{\partial q_L}{\partial p_l}\frac{\partial p_l}{\partial q_l}\frac{\partial q_l}{\partial \mathbf{X}} = \left\{\left(2\sum_{k=1}^{m_l}\frac{\mathbf{C}_k - q_l}{\beta_k^2}\right) f(q_l)\right\}\sum_{j=1}^{m_l} 1\cdot \mathbf{V}_{L,j} \tag{8}$$

where:

- t is target value.
- m_l is the number of neurons in the hidden layer.
- p is the output of the activation function for a neuron.
- q *is* the regression vector as the activity of a neuron.
- L and l denote the output and hidden layer, respectively.
- **C** and β are the center and *width* of the RBFNN, respectively.
- The function f is the gaussian density function defined in the right-hand side in (4) as an exponential form.
- **V** is the vector of output linear weights.

The jth component of the second term in (7) can be expressed using the output of the critic network $\hat{\lambda}_i(t+1) = \partial \hat{\mathbf{J}}[\Delta\hat{\mathbf{Y}}(t+1)] / \partial \Delta\hat{\mathbf{Y}}_i(t+1)$ as follows.

$$\frac{\partial \hat{\mathbf{J}}[\Delta\hat{\mathbf{Y}}(t+1)]}{\partial \Delta \mathbf{Y}_j(t)} = \sum_{i=1}^{n} \hat{\lambda}_i(t+1)\frac{\partial \hat{\mathbf{Y}}_i(t+1)}{\partial \Delta \mathbf{Y}_j(t)} + \sum_{k=1}^{m}\sum_{i=1}^{n} \hat{\lambda}_i(t+1)\frac{\partial \Delta\hat{\mathbf{Y}}_i(t+1)}{\partial \mathbf{A}_k(t)}\frac{\partial \mathbf{A}_k(t)}{\partial \Delta \mathbf{Y}_j(t)} \tag{9}$$

where n and m are the numbers of outputs of the model and the action networks, respectively. By using (9), each component of the vector $e_C(t)$ from (7) is determined by

$$e_{Cj}(t) = \frac{\partial \mathbf{J}[\Delta\hat{\mathbf{Y}}(t)]}{\partial \Delta\hat{\mathbf{Y}}_j(t)} - \gamma \frac{\partial \hat{\mathbf{J}}[\Delta\hat{\mathbf{Y}}(t+1)]}{\partial \Delta \mathbf{Y}_j(t)} - \frac{\partial U[\Delta\mathbf{Y}(t)]}{\partial \Delta\mathbf{Y}_j(t)} - \sum_{k=1}^{m} \frac{\partial U[\Delta\mathbf{Y}(t)]}{\partial \mathbf{A}_k(t)} \frac{\partial \mathbf{A}_k(t)}{\partial \Delta\mathbf{Y}_j(t)} \tag{10}$$

Then, the expression for the weights' update for the critic network using (10) is as follows.

$$\Delta \mathbf{V}_C(t) = -\eta_C \mathbf{e}_C^{\mathrm{T}}(t) \frac{\partial \mathbf{e}_C(t)}{\partial \mathbf{V}_C(t)} \tag{11}$$

where η_C is a positive learning rate and $\mathbf{V}_C$ is the weights of the DHP critic network.

The adaptation of the action network in Fig. 4 is illustrated in Fig. 5, which propagates $\hat{\lambda}(t+1)$ back through the model network to the action network. The goal of this adaptation is expressed in (12), and weights of the action network are updated by (13).

$$\frac{\partial U[\Delta\mathbf{Y}(t)]}{\partial \mathbf{A}(t)} + \gamma \frac{\partial \hat{\mathbf{J}}[\Delta\hat{\mathbf{Y}}(t+1)]}{\partial \mathbf{A}(t)} = 0 \quad \forall t. \tag{12}$$

$$\Delta\mathbf{V}_A(t) = -\eta_A \left[\frac{\partial U[\Delta\mathbf{Y}(t)]}{\partial \mathbf{A}(t)} + \gamma \frac{\partial \hat{\mathbf{J}}[\Delta\hat{\mathbf{Y}}(t+1)]}{\partial \mathbf{A}(t)}\right]^{\mathrm{T}} \frac{\partial \mathbf{A}(t)}{\partial \mathbf{V}_A(t)} \tag{13}$$

where η_A is a positive learning rate and $\mathbf{V}_A$ is the weights of the DHP action network.

Fig. 6 illustrates how the model network (identifier) is trained to identify the dynamics of the plant. The input vector, $\mathbf{A}(t) = [\Delta P_{in}(t), \Delta V_{ref}(t)]$ is fed into the plant with the reference vector, $\mathbf{Y}_{ref}(t) = [P_{in}(t), V_{ref}(t)]$. In off-line training, the input signals of A(t) are generated as small pseudo-random binary signals (PRBSs) with a sampling period of 20 ms. As mentioned before, the output vector of the plant is $\mathbf{Y}(t) = [\Delta\omega(t), \Delta V_t(t)]$, and the model network output is $\Delta\hat{\mathbf{Y}}(t) = \hat{f}(\mathbf{X}(t)) = [\Delta\hat{\omega}(t), \Delta\hat{V}_t(t)]$, where $\mathbf{X}(t)$ is the input vector to the model network consisting of three time lags of system input $\mathbf{A}(t)$ and output $\mathbf{Y}(t)$.

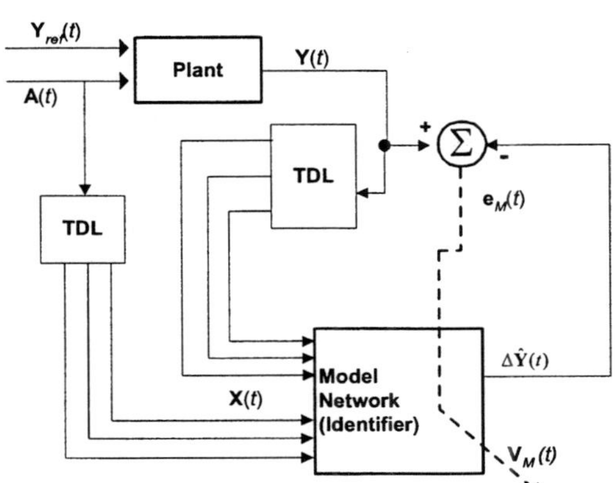

Fig. 6. Training of the model network using the backpropagation algorithm

The residual vector, $\mathbf{e}_M(t)$ given in (14) is used for updating the model network's weights $\mathbf{V}_M$ in (15) during training by the backpropagation algorithm

$$\begin{aligned} \mathbf{e}_M(t) &= \mathbf{Y}(t) - \Delta\hat{\mathbf{Y}}(t) \\ &= [\Delta\omega(t) - \Delta\hat{\omega}(t), \Delta V_t(t) - \Delta\hat{V}_t(t)] \end{aligned} \tag{14}$$

$$\Delta\mathbf{V}_M(t) = -\eta_M \mathbf{e}_M^{\mathrm{T}}(t) \frac{\partial \mathbf{e}_M(t)}{\partial \mathbf{V}_M(t)} \tag{15}$$

where η_M is a positive learning rate and $\mathbf{V}_M$ is the weights of the DHP action network.

4. SIMULATION RESULTS

After training the critic and action network off-line with an acceptable performance, the DHPNC with fixed weights is ready to control the plant during the real-time operation. The performances of DHPNC trained with deviation signals are compared with CONVC for the improvement of system damping and transient stability. Two different types of disturbances are used at an operating point different from the one in which the CONVC is expected to give their best performance, and different from the one where the DHPNC was trained.

First, ±7% step changes in the reference voltage of the exciter (as a small disturbance) is applied to the synchronous generator operating at a steady state condition (P_t = 1 [pu], Q_t = 0.234 [pu], and Z_e = 0.02+j0.4 [pu]). At t=1 s, the 7% step increase is applied. At t=12 s, the 7% step is removed, and the system returns to its initial operating point. The results in Figs. 7 and 8 show that the DHPNC improves the transient system response much more efficiently than the CONVC does with respect to rotor angle (δ) and terminal voltage (V_t).

Next, a severe test is carried out to evaluate the performances of controllers under a large disturbance. At t=0.3 s, a temporary three phase short circuit is applied at the infinite bus for 50 ms from t=0.3 s to 0.35 s for the plant operating at the same steady state condition as in the previous test. The results of this test, comparing the performance of the DHPNC and CONVC, are shown in Figs. 9 and 10. They show that the optimal neurocontroller (DHPNC) damps out the low frequency oscillations for the rotor angle (δ) and terminal voltage (V_t) more effectively than the CONVC.

5. CONCLUSIONS

This research work has made a new contribution by showing the application of the radial basis function neural network (RBFNN) to design the dual heuristic programming (DHP) based optimal neurocontroller for a synchronous generator connected to an infinite bus in a power grid.

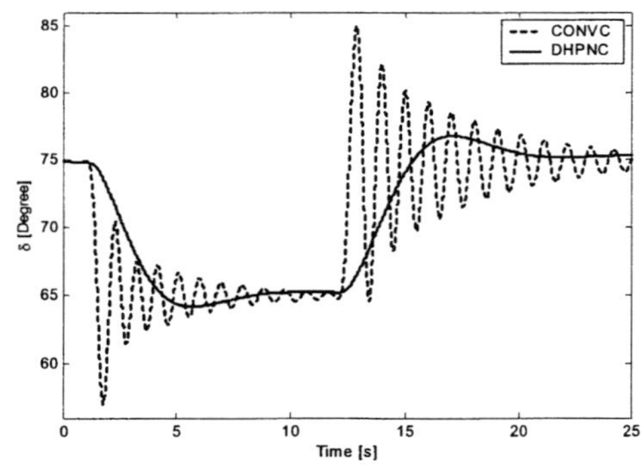

Fig. 7. ±7% step changes in reference voltage of exciter: δ [°]

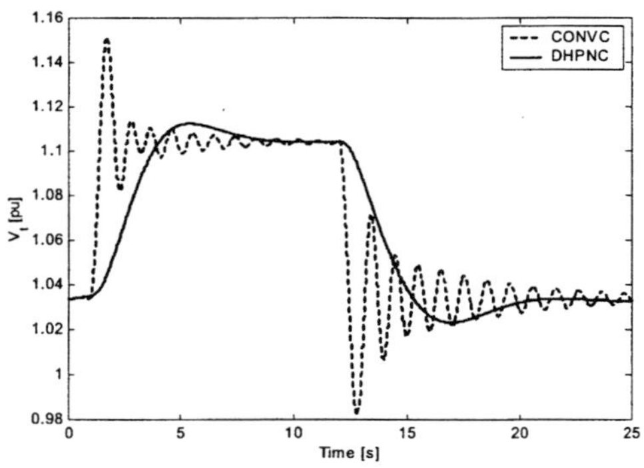

Fig. 8. ±7% step changes in reference voltage of exciter: V_t [pu]

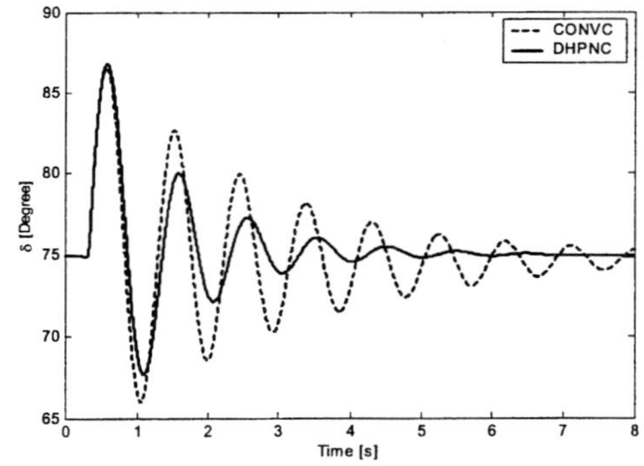

Fig. 9. A 50 ms three phase short circuit at the infinite bus: δ [°]

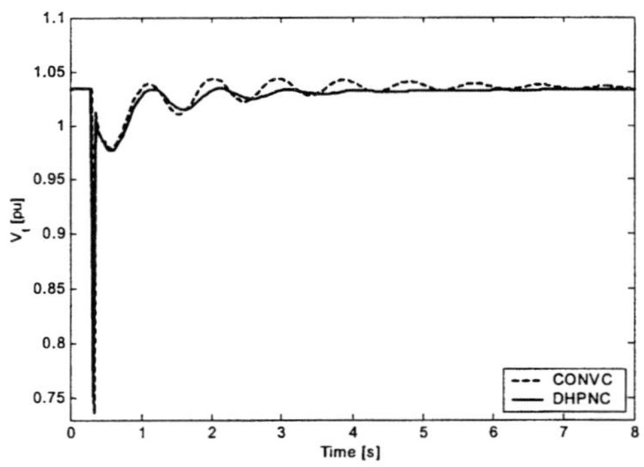

Fig. 10. A 50 ms three phase short circuit at the infinite bus: V_t [pu]

The results show that the DHPNC improves the system damping and dynamic transient stability more effectively than the CONVC. The DHPNC uses the fixed parameters in real-time operation for its model, action, and critic networks, which were trained off-line first. Therefore, the instability issue associated with the neural network based controllers can be overcome, and it provides a robust feedback loop with a powerful dynamic control capability. All these features are desirable and important for optimal neurocontrol technology based on the use of artificial neural networks.

REFERENCES

Bertsekas, Dimitri P. (2001). *Dynamic Programming and Optimal Control.* Athena Scientific, Belmont, Massachusetts.

Dash, P. K., S. Mishra and G. Panda (2000). A radial basis function neural network controller for UPFC. *IEEE Transactions on Power Systems.* 15(4), 1293-1299.

Flynn, D., S. McLoone, G.W. Irwin, M.D. Brown, E. Swidenbank, and B.W. Hogg, (1997). Neural control of turbogenerator systems. *Automatica.* 33(11), 1961-1973.

Gregory, John and Cantian Kin. (1992). Constrained Optimization in the Calculus of Variations and Optimal Control Theory. In: *Handbook of Intelligent Control.* pp. 185-210, D. White and D. Sofge, Eds. New York: Van Nostrand Reinhold.

Miller, W.T., R.S. Sutton and P.J. Werbos (1990). *Neural networks for control.* pp. 67-95, MIT Press, Cambridge, Massachusetts.

Park, Jung-Wook, Ronald G. Harley, and Ganesh K. Venayagamoorthy (2002). Adaptive critic based optimal neurocontrol for synchronous generator in power system using MLP/RBF neural networks. *Proceedings of IEEE Annual Industry Application Society (IAS) Meeting.* 2, 1447-1454.

Prokhorov, D.V. and D.C. Wunsch (1997). Adaptive critic designs. *IEEE Transactions on Neural Networks.* 8(5), 997-1007.

Segal, R., M.L. Kothari and S. Madnani (2000). Radial basis function (RBF) network adaptive power system stabilizer. *IEEE Transactions on Power Systems.* 15(2), 722-727.

Venayagamoorthy, Ganesh K., Ronald G. Harley and D.C. Wunsh (2002). Comparison of Heuristic Dynamic Programming and Dual Heuristic Programming Adaptive Critics for Neurocontrol of a Turbogenerator. *IEEE Transactions on Neural Networks.* 13(3), 764-773.

Venayagamoorthy, Ganesh K. and Ronald G. Harley, (2001). A continually online trained neurocontroller for excitation and turbine control of a turbogenerator. *IEEE Transactions on Energy Conversion.* 16(3), 261-269.

Werbos, P.J. (1992). Approximate dynamic programming for real-time control and neural modeling. In: *Handbook of Intelligent Control.* p. 493-525, D. White and D. Sofge, Eds. New York: Van Nostrand Reinhold.

ELSEVIER
IFAC
PUBLICATIONS
www.elsevier.com/locate/ifac

LOCAL MODEL NETWORKS APPLIED TO NONLINEAR GENERATOR EXCITATION CONTROL

D. Flynn[1] and M.D. Brown[2]

[1]*The Queen's University of Belfast, N. Ireland*
[2]*Atkins Aviation and Defence Systems, Bristol, UK*

Abstract: A multiple model control technique, that incorporates linear control methods within a non-linear structure, is applied to excitation control of a synchronous machine. A non-linear, adaptive generalized minimum variance (GMV) controller is straightforwardly designed to provide automatic voltage regulator (AVR) and power system stabilizer (PSS) roles. The capabilities of the local model network (LMN) controller are then compared with an adaptive GMV controller, illustrating its excellent performance and the transparent nature of the LMN design. *Copyright © 2003 IFAC*

Keywords: local model networks, non-linear control, power system stabilizers, electric generators

1. INTRODUCTION

With electric power systems becoming ever more complex, coupled with increasing economic and operational requirements, advancements in power plant performance and control are continually being sought. Conventional fixed parameter approaches are unable to provide the most effective plant and system control over the full non-linear power plant operating range (Kanniah, *et al.*, 1984). Moreover, tuning of the large number of control loops typically found in a power station can prove to be a demanding exercise. Research studies have suggested a number of strategies, such as adaptive control, that attempt to improve overall control of turbo-generator systems (Wu and Hogg, 1991). The main difficulty, however, with such strategies lies in the robustness of the parameter estimation stage, and robust jacketing software is required if the controller is to work safely in practice.

Non-linear techniques based around neural networks, splines, fuzzy logic, etc., or indeed combinations of the above, have generated considerable interest as convenient and powerful modelling and control tools for excitation control (Kitauchi and Taniguchi, 1997; Shamsollahi and Malik, 1999). However, the non-transparent, black-box nature of these approaches makes incorporation of *a priori* system information difficult. Additionally, theoretical results available from the conventional modelling and control domain can not be easily exploited, so that analyzing behaviour and proving stability are problematic.

Genetic algorithms and other evolutionary techniques have also been examined (Abdel-Magid, *et al.*, 1999). It is recognized that changes in power system loading, network configuration, etc. will have an impact on system behaviour, and in particular the requirements on power system stabilizers (PSS). Consequently, an optimal set of PSS settings are *evolved* to minimize a user-defined cost function, based on the observed response to a range of simulated events. Even assuming that the multi-machine simulation can be considered representative and all plant parameters are known, deciding upon the cost function and the comprehensiveness of the considered scenarios is challenging. Alternatively, adaptation could be incorporated into the non-linear control approaches outlined above, and so create self-organizing fuzzy controllers, self-learning neural networks, etc. (Lown, *et al.*, 1997). However, as with adaptive control, supervision software should clearly be introduced to protect against any anomalous behaviour. This is an issue that has been largely ignored.

An alternative approach that facilitates the use of conventional control techniques within a non-linear context is the local model network (LMN). Within this paper, the principles behind the LM technique are first outlined and explanation is given on how to model the generator excitation system and hence form a transparent, non-linear controller that directly utilizes *a priori* knowledge as part of the design process. The capabilities of the LM controller are then extended to provide an adaptive scheme, which can respond to changes in external power system behaviour. The performance of the LMN controller is compared against a model-based adaptive regulator on a laboratory-based microalternator system for different disturbance rejection scenarios.

2. LOCAL MODEL NETWORKS

The local model network architecture can be considered to be a generalization of a radial basis function neural network, Figure 1 (Murray-Smith and Johansen, 1997). The LMN output, $\hat{y}$, is given by

$$\hat{y} = F(\psi,\varphi) = \sum_{i=1}^{M} f_i(\psi)\,\rho_i(\varphi)$$

The individual M local models $f_i(\psi)$ are functions of the measurement vector ψ, and are multiplied by an interpolation function $\rho_i(\varphi)$, where φ is a function of the current operating region vector. The functions $\rho_i(\varphi)$ are often chosen as normalized Gaussians

$$\rho_i(\varphi) = \frac{\exp(-\|\varphi - \mathbf{c}_i\|^2/2\sigma_i^2)}{\sum_{i=1}^{M}\exp(-\|\varphi - \mathbf{c}_i\|^2/2\sigma_i^2)}$$

where $\mathbf{c}_i$ and σ_i are the centre and width of each Gaussian function. Each $\rho_i(\varphi)$ can be viewed as a model validity function, giving a value approaching 1 when the local $f_i(\ )$ are such that $F(\ ,\ )$ is a good approximation to $\hat{y}$, and a value close to 0 where the local $f_i(\ )$ are such that $F(\ ,\ )$ is not a good approximation to $\hat{y}$. It is customary for the local models $f_i(\psi)$ to be linear in order to take advantage of established techniques. Hence, if $f_i(\psi)$ are of the ARX form, and $y(k)$ and $u(k)$ are the plant input and output at time k, then the LM network represents the non-linear ARX model

$$\hat{y}(k+1) = \alpha_1 y(k) + \alpha_2 y(k-1) + \ldots + \alpha_{n_\alpha+1} y(k-n_\alpha) + \beta_0 u(k) + \beta_1 u(k-1) + \ldots + \beta_{n_\beta} u(k-n_\beta)$$

where

$$\alpha_j = \sum_{i=1}^{M} \rho_i(\varphi(k))a_{ij}, \quad \beta_j = \sum_{i=1}^{M} \rho_i(\varphi(k))b_{ij}$$

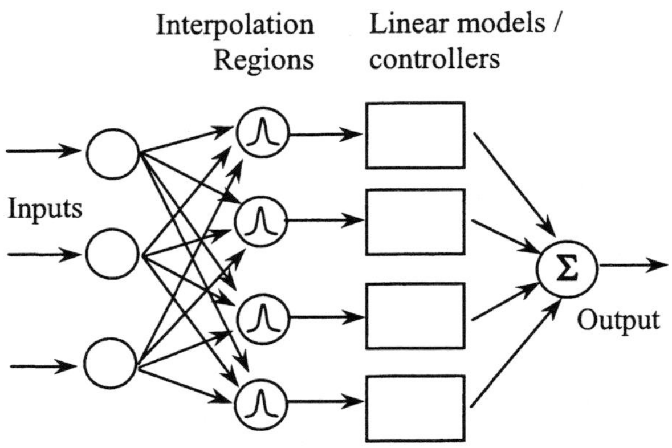

Fig. 1. Local model network structure

such that a_{ij} and b_{ij} are the parameters of the local linear ARX models. Here, α_j and β_j are dependent on the operating point, $\varphi(k)$, to represent nonlinearities in plant behaviour. In contrast with adaptive control schemes, the equivalent α_j and β_j are time varying.

2.1 Generator Excitation Modelling

For the present application, a SISO ARX model of the generator-exciter system can be formed as

$$A(z^{-1})y(k) = B(z^{-1})u(k-k_d) + \xi(k)$$

where $y(k)$ represents the terminal voltage, V_T, of the synchronous machine, and $u(k)$ is the exciter voltage, V_R, at sample time k. k_d is the time delay in an integer number of samples, while $\xi(k)$ is a zero mean white noise sequence disturbing the system. Previous work by the authors and others has suggested that 2nd order ARX models are sufficient to capture the main AVR loop dynamics (Brown, *et al.*, 2002). The operating point of a synchronized alternator is normally defined in terms of its real and reactive power output, P_T and Q_T. Hence, it is convenient and intuitive to select the operating point $\varphi(k)$ as the vector $[P_T(k)\ Q_T(k)]$. Each local model can then be identified for small perturbations about different values of P_T and Q_T.

In order to select M, the required number of local models, a hybrid optimization strategy was implemented, whereby a least-squares cost function was minimized using non-linear optimization for the centres and widths ($\mathbf{c}_i$ and σ_i) of the Gaussian functions, followed by linear optimization of the local models (a_{ij} and b_{ij}) (Brown, *et al.*, 1997). The cost function was formed as the predictive errors of the LMN against an extended training data set. The result of the optimization process was that 5 local models were considered sufficient to cover the operating space. The main objective of this process is to provide an accurate representation when P_T is high, and in the leading power factor region, where the variation in non-linearity occurs most rapidly. The normalized interpolation regions for the models are shown in Figure 2.

3. LOCAL CONTROLLER DESIGN

The local model network has been formed from 2nd order linear models, making it straightforward to design linear controllers for each model. Since these local controllers are interpolated to form a composite control output the response of each controller can be shaped to ensure consistent performance across the entire operating range. In this way, conventional linear control theory can be exploited within a non-linear control framework.

Examination of the individual linear models obtained from the optimization process reveals that some of them are both unstable and non-minimum phase - individual models centred in the leading power factor region. Consequently, the generalised minimum variance (GMV) algorithm (Wellstead and Zarrop, 1991) has been selected as a robust control scheme. So, forming the plant pseudo output $\gamma(k)$ as

$$\gamma(k+k_d)=S(z^{-1})y(k+k_d)+W(z^{-1})u(k)$$

where, $S(z^{-1})$ and $W(z^{-1})$ are weighting polynomials. Modification of the plant output, $y(k)$, can be achieved as follows

$$y(k)=V_T(k)+\lambda\omega(k); \qquad -1<\lambda<0$$

where λ is a weighting factor placed on the rotor shaft speed, $\omega(k)$. Inclusion of the auxiliary signal $\omega(k)$ introduces a power system stabilization (PSS) function to enhance system damping (Kanniah, *et al.*, 1984). $G(z^{-1})$ is then introduced, such that

$$S(z^{-1})=A(z^{-1})+z^{-k_d}G(z^{-1})$$

Since, the local models are of 2nd order, and if it is assumed that $k_d=1$ and $W(z^{-1})=w_0$, it follows that

$$G(z^{-1})=g_0+g_1z^{-1}=(s_1-a_1)+(s_2-a_2)z^{-1}$$

whereupon, the GMV controller is defined as

$$u(k)=\frac{1}{b_0+w_0}\left[-\left(g_0+g_1z^{-1}\right)y(k)-b_1z^{-1}u(k)\right]$$

Controller design is completed by selecting the polynomial $S(z^{-1})=1+s_1z^{-1}+s_2z^{-2}$, and the scalar term w_0. $S(z^{-1})$ can be conveniently chosen as a discrete, stable, 2nd order filter (Brown, *et al.*, 2002). The weighting factor, w_0, permits control signal *detuning*, by trading closeness of reference following against control effort. Since there are 5 local models, suitable $S(z^{-1})$ and $W(z^{-1})$ polynomials can be selected for each controller, tailored to the particular operating region. Each controller output is multiplied by its respective interpolation function, with the weighted signals summed to form the full control signal, which is then subsequently applied to the plant.

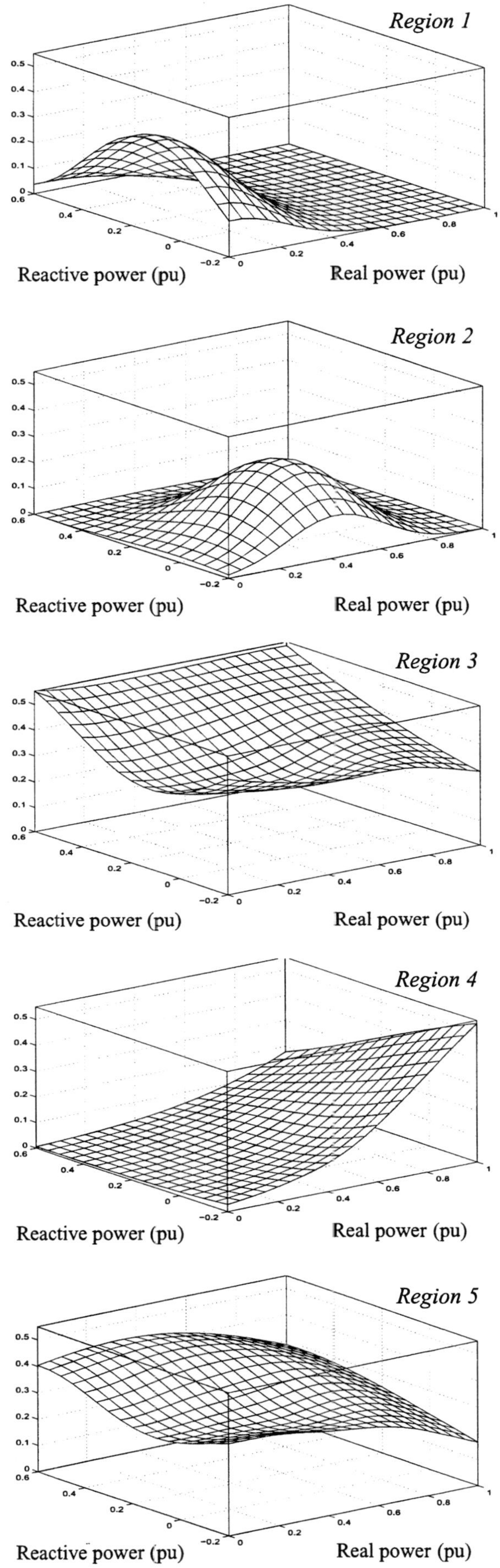

Fig. 2. LMN interpolation regions

4. ADAPTATION OF LOCAL MODELS

The LM structure comprises interpolation functions, designed to characterise the non-linear behaviour of the synchronous machine, and local models or local controllers which are then assigned to each region. It would be possible to adapt both the interpolation function parameters (centres and widths) and the local models themselves (a_{ij} and b_{ij}). However, as the former is largely associated with generator behaviour (variation in P_T and Q_T), and can be assumed to only change extremely slowly with time, the interpolation function parameters may remain fixed.

As previously described the parameters of the local models were determined off-line using linear optimization, based on gathered training data. Subsequently, the respective $A(z^{-1})$ and $B(z^{-1})$ polynomials formed the basis for the GMV controller designs. For on-line adaptation, this process could be repeated, whereby $A(z^{-1})$ and $B(z^{-1})$ are adapted using a recursive least squares algorithm, based on the predictions of the individual models (Wellstead and Zarrop, 1991). Subsequently, controller parameters would be determined. However, since the models themselves are not of direct interest it is more appropriate here to adapt the controller parameters. Hence the polynomials $G(z^{-1})$ and $B(z^{-1})$ are allowed to adapt, dependent on the observed control error.

For an LMN controller, the output is determined as the weighted composite response of the individual local controllers. To enable adaptation, the control error fed back to each parameter estimator is scaled by the interpolation weighting for that particular local controller, i.e. it is assumed that the contribution each controller makes to the final output is reflected in the resulting control error. Also, the local models are only updated if their contribution to the controller output and the control errors exceed an error bound.

Since controller parameters are allowed to adapt on-line, estimator supervision is required, as would be the case for an adaptive controller. It is argued, however, that the requirement here is much less severe as the LMN controller is only required to adapt to changes in external system behaviour. This is in contrast to an adaptive linear controller, which must retune to every change in operating point. Also, despite the LMN controller, as a whole, offering non-linear performance, it is only the local controllers which are adapting and hence supervision can be both transparent and intuitive. Consequently, a variable forgetting factor scheme can be introduced to allow the parameter estimators to track slowly varying process conditions (Isermann and Lachmann, 1985). However, care must be taken to ensure that the estimators do not wind-up due to long periods at constant operating conditions. This can be largely eliminated by minimizing estimator activations, as outlined above, and by monitoring the Kalman gain vectors (Flynn, *et al.*, 1996). As a second measure the

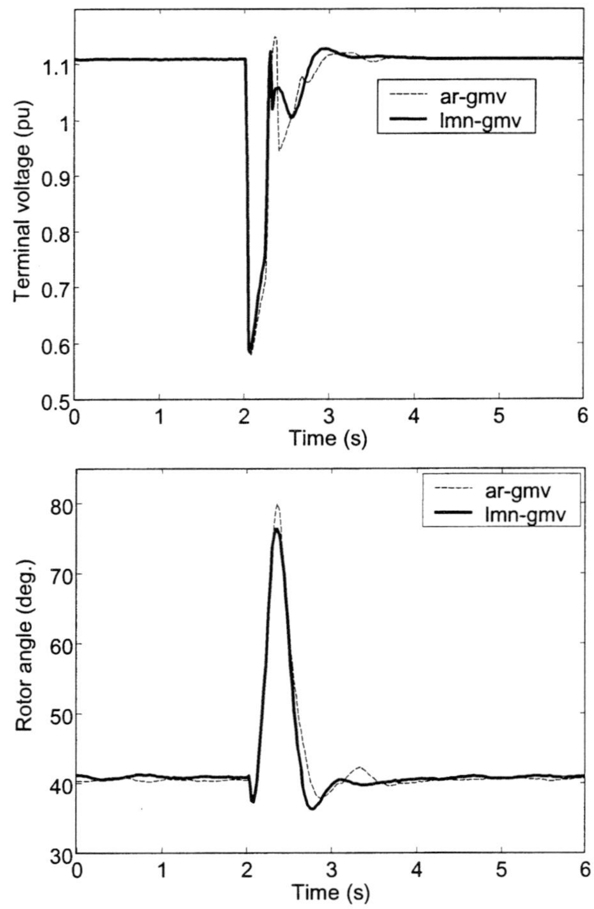

Fig. 3. Terminal voltage and rotor angle (short circuit)

parameter estimators are switched off during severe transients. A further point to note is that during severe oscillations (and hence changes in operating point) an adaptive scheme must rely on the same model, whereas the LMN controller provides full non-linear coverage by constantly interpolating the individual controllers to match the system dynamics.

5. RESULTS

Having outlined the development of a local model network controller, the algorithm was subsequently implemented on a micromachine, synchronized to the N. Ireland power system. Previously, comprehensive tests were conducted in simulation to ensure both the short-term transient and long-term dynamic stability of the system, following both severe and minor disturbances. The micromachine system comprises of a 3 kVA, 220 V, 50 Hz, 4 pole synchronous machine, whose parameters have been selected to match those of a full size machine. The alternator is driven by a separately excited d.c. motor, whose armature current is controlled by an analogue simulation of a three-stage turbine with reheater. The alternator is connected to a delta-star transformer, which has an on-load tap-changing device in seven steps on the secondary terminals. This transformer is connected through a transmission line simulation to the system grid (Wu and Hogg, 1991).

For the purpose of these tests the LMN controller is compared with an adaptive regulator (AR) using the same GMV algorithm as outlined in Section 3, and incorporating an enhanced version of the jacketing software described in Section 4 (Flynn, *et al.*, 1996). It is worth remembering that for the adaptive regulator the model parameters are allowed to continually adapt, and in this way system non-linearities can effectively be captured. Consequently, the adaptive controller is tuned at each operating point, by applying a PRBS input as part of the on-line estimation process, prior to commencing each test. This arrangement contrasts with the LM architecture where the interpolation parameters are fixed, and the relative contribution of the individual models is determined based on the current operating point. The performance of the controllers is illustrated under the following test conditions:

- Three-phase-earth short circuit, duration 250 ms, after 2 s, at the sending end of the transmission line system, at an operating point of $P_T = 0.60$ pu and $Q_T = 0.20$ pu.
- Transformer tap change from position 5 to 4 after 5 s, and from position 4 to 5 after 10 s, at an operating point of $P_T = 0.85$ and $Q_T = 0.15$ pu.

Figure 3 illustrates the terminal voltage and rotor angle responses for both controllers following the 3-phase short-circuit at the sending end of the transmission line. It is clear that both the LMN and adaptive controller provide good damping, although the LMN controller offers an improvement in the initial rotor angle swing and subsequent oscillations. For both controllers the terminal voltage rapidly recovers, following clearing of the short-circuit.

Figure 4 shows the output of the LMN local controllers, which are subsequently multiplied by the weightings of Figure 5. Prior to the fault, the major contribution to the excitation input is from controllers 3 and 5, which can be readily understood by examining the interpolation functions of Figure 2. It can be seen in Figure 4 that the controller outputs vary significantly, confirming the fact that the local controllers have been tuned for different nominal operating points. During the fault, the machine operating point changes rapidly causing highly non-linear behaviour. The relative contributions of the individual controllers changes appreciably, with controllers being phased *in* and *out* at various stages. By contrast, the protection software for the adaptive regulator ensures that the parameter estimator is deactivated during the voltage transients, which effectively *freezes* the identified model.

Tap changes are then applied between positions 4 and 5 of the delta-star transformer, corresponding to a 6% variation in output. Figure 6 illustrates the terminal voltage and rotor angle responses, with the two controllers again providing excellent damping, and fast transient responses. The LMN controller proves

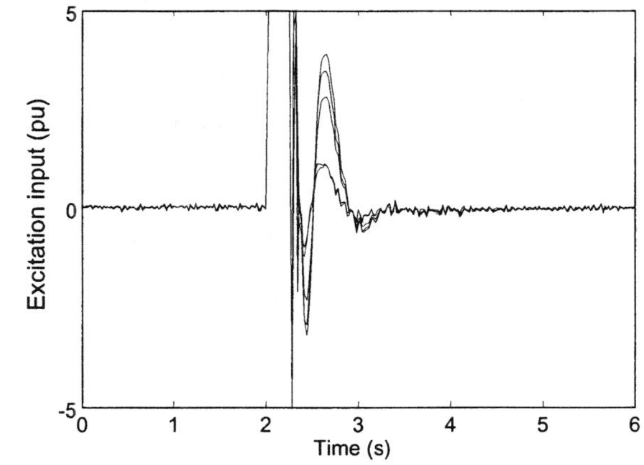

Fig. 4. Local controller outputs (short-circuit)

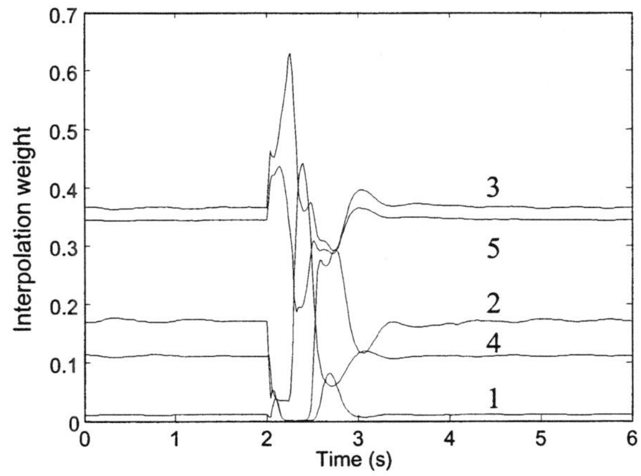

Fig. 5. Interpolation weights (short-circuit)

slightly better at minimizing the rotor swings, particularly following the second tap change. The presence of a PRBS on the AR responses can be seen during the first 3 - 4 seconds. Figure 7 shows the relative local controller weightings. Initially, the majority of the control signal is formed from controllers 3, 4 and 5. During the transient phase, corresponding to an increase in rotor angle, the role of controller 5 is reduced and that of 4 increased, leaving controllers 3 and 4 to provide approximately 35% of the excitation signal.

As a final point, it can be seen that controller 1 does not play a significant role either in steady-state or during transients for either of the presented tests. Examination of Figure 2 shows that the associated model is centred at approximately $P_T = 0.10$ pu and $Q_T = 0.10$ pu. This model thus ensures that the LMN controller is capable of maintaining performance over the entire operating region, however, it is unlikely in practice that the generator will be required to operate at such low output levels.

6. CONCLUSION

A local model network controller has been developed for the excitation loop of a synchronous machine. A major advantage that the LMN structure offers, over other non-linear modelling techniques, is that *a priori* knowledge of the process can be utilized in model structure selection. Hence, using information from previous simulation studies and plant tests, estimates were selected for the non-linear interpolation regions and the structure of the linear models. A hybrid optimization algorithm was then introduced to create a parsimonious structure capturing the system's non-linear dynamics. Finally, individual controllers, based on the generalized minimum variance algorithm, were designed for each of the local models.

The performance of the non-linear LMN controller was then compared on a laboratory micromachine with an adaptive controller adopting the same GMV control strategy. A number of disturbance rejection tests were performed, with the LMN controller providing excellent responses over the generator's entire operating regime. It is worth emphasizing that the *major* aim of the LMN approach is not improved control performance, although this is achieved, but instead to create a framework (representative of the system being controlled) which enables transparent design of a global, *non-linear* control scheme using familiar *linear* techniques. The adaptive controller's performance was comparable with that of the LMN scheme, but requires significant protection software to safeguard operation of the parameter estimator.

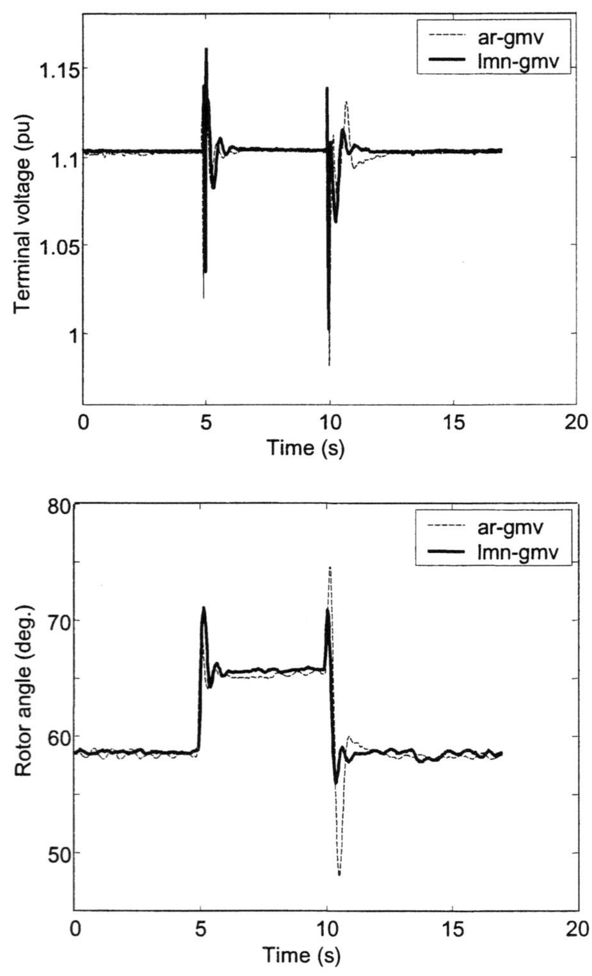

Fig. 6. Terminal voltage and rotor angle (tap changing)

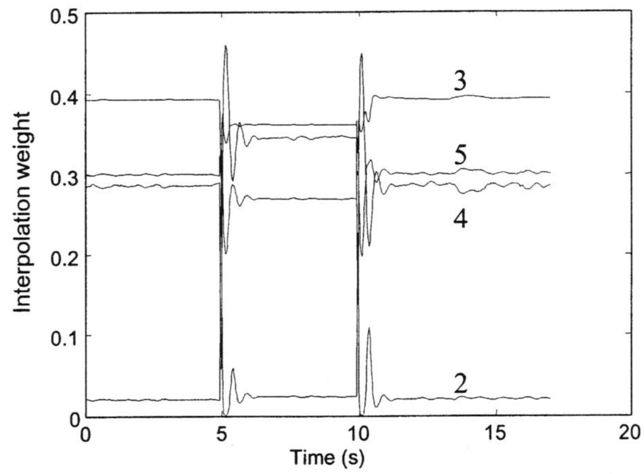

Fig. 7. Interpolation weights (tap changing)

7. REFERENCES

Abdel-Magid, Y.L., M.A. Abido, S. Al-Baiyat and A.H. Mantawy (1999). Simultaneous stabilization of multi-machine power systems via genetic algorithms. *IEEE Trans. PWRS*, **14**, pp. 1428-1439.

Brown, M.D., D. Flynn and G.W. Irwin (2002). Multiple model nonlinear control of synchronous generators. *Trans. Inst. MC*,**24** , pp. 215-230.

Brown, M.D., G. Lightbody and G.W. Irwin (1997). Non-linear internal model control using local model networks. *IEE Proc. Pt. D*, **144**, pp. 505-514.

Flynn, D., B.W. Hogg, E. Swidenbank and K.J. Zachariah (1996). A self-tuning automatic voltage regulator designed for an industrial environment. *IEEE Trans. EC*, **11**, pp. 429-434.

Isermann, R. and K.H. Lachmann (1985). Parameter adaptive control with configuration aids and supervision functions. *Automatica*, **21**, pp 625-638.

Kanniah, J., O.P. Malik, and G.S. Hope (1984). Excitation control of synchronous generators using adaptive regulators. *IEEE Trans. PAS*, **103**, pp. 897-910.

Kitauchi, Y. and H. Taniguchi (1997). Experimental verification of fuzzy excitation control system for multi-machine power system. *IEEE Trans. EC*, **12**, pp. 94-99.

Lown, M., E. Swidenbank and B.W. Hogg (1997). Adaptive fuzzy logic control of a turbine generator system. *IEEE Trans. EC*,**1 2**, pp. 394-397.

Murray-Smith, R. and T.A. Johansen (1997). *Multiple model approaches to modelling and control*. Taylor and Francis.

Shamsollahi, P. and O.P. Malik (1999). Application of neural adaptive power system stabilizer in a multi-machine power system. *IEEE Trans. EC*, **14**, pp. 731-736.

Wellstead, P.E. and M.B. Zarrop (1991). *Self-tuning systems - control and signal processing*. J. Wiley & Sons.

Wu, Q.H. and B.W. Hogg (1991). Laboratory evaluation of adaptive controllers for synchronous generators. *Automatica*, **27**, pp. 845-852.

ELSEVIER
IFAC
PUBLICATIONS
www.elsevier.com/locate/ifac

LOAD CONTROL OF UNIT STEAM-BOILER USING NEUROFUZZY MODELLING TECHNIQUE

Xiang-Jie Liu[*], **Felipe Lara-Rosano**[*],
Marino Sanchez Parra[**], **Raul Garduño Ramirez**[**]

[*]Centro de Ciencias Aplicadas y Desarrollo Tecnologico, UNAM, 04510, Mexico
Mexico City, D.F.
[**]*Electric Research Institute, Cuernavaca, Morelos CP 62490 México*

Abstract: In unit steam-boiler, load control system is to provide continuous active power to satisfy the load demand of power system. The process is characterized by nonlinear and uncertainty. Neurofuzzy modelling based on local mechanism model and dynamic experiments is discussed. A relative fuzzy PI controller is constituted to control the load system in boiler-following mode. The performance of the control system has been verified by simulation and then tested on real-time process. *Copyright © 2003 IFAC*

Keywords: load, fuzzy logic, neural networks, modelling, steam-boiler

1 INTRODUCTION

In unit steam-boiler system, the boiler, turbine and generator combine to adapt to the load demand of power system and maintain its stability. Load control system(Fig.1) plays an important role in the whole unit system and greatly affect the steam-boil running efficiency.

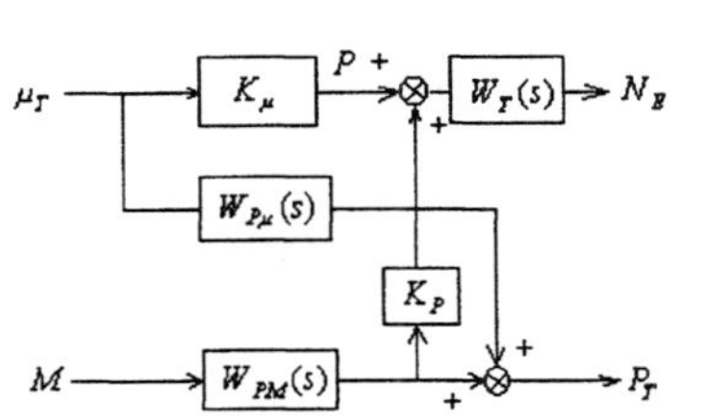

Fig.1 Load system of unit steam-boiler

In figure 1, $W_{PM}(s)$ and $W_{P\mu}(s)$ represent the transfer functions of steam pressure from fuel M and turbine valve opening μ_T respectively. $W_T(s)$ represents the turbine transfer function from steam flow to active power N_E. K_μ and K_P are coefficients.

Steam-boiler turbine system is the complex industrial process with highly nonlinear, non-minimum, uncertainty and load disturbance. There exist two control fashions: boiler-following mode and turbine-following mode. Load-cycling operation between full load and low load leads to the change of operating point right across the whole operating range. Variations in plant variables become quite nonlinear.

An essential tool for the improvement of the control system is to get a valid model. Modelling technique presented obvious significance for steam-boiler system(Astrom and Ecklund, 1972; Irwin, *et al.*,1995). The recent developed neurofuzzy network(Brown and Harris, 1994) could be an effective way to the modelling of such nonlinear system in that fuzzy rules could easily express the expert knowledge in linguistic form while neural networks posses the learning ability which could approximate nonlinear functions with arbitrary accuracy. In the neurofuzzy networks, the whole envelope of process operation is divided into several operating regions, and in each region a local reduced-order model is used to approximate the process(Zhang and Morris, 1995). The neurofuzzy network modelling and control techneque have found application in carbon monoxide concentration system (Tanaka, *et al.*,1995) and fluid bed combustion process(Ikonen & Najim, 1996).

This paper develops a new method of load power

control system using neurofuzzy network modelling method on 300MW unit steam-boiler. AMN(Associate Memory Network) is discussed because it provides a direct link between artificial neural networks and the fuzzy systems. Process input-output data are then used to train the network. With the established neurofuzzy model, a relative fuzzy PI controller is constituted under different operation regions. The performance of the control system has been verified by the simulation process and then tested on real-time load process.

2 MECHANISM DESCRIPTION OF THE PLANT

This paper is concerned with the application of the neurofuzzy network to the modelling and control of the load system of a pulverized coal-firing 300MW unit used for electric power generation. It produces 947 tons of steam per hour at maximum continuous rating. The final superheated steam pressure is 19MPa. The combustion process $W_{PM}(s)$ and $W_{P\mu}(s)$ of load control system described in section 1 is the complex process composes of (figure 2): furnace 1, heated surface 2, superheater 3, steam tube 4 and steam-turbine 5.

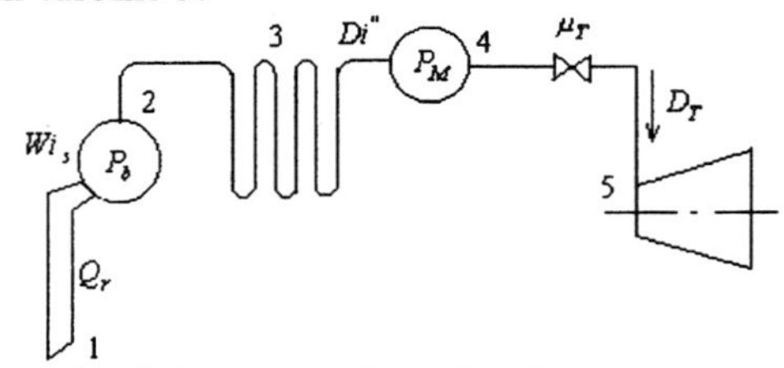

Fig.2 Diagram of combustion process

With mechanism analysis, the combustion process in load system could be expressed in Figure 3.

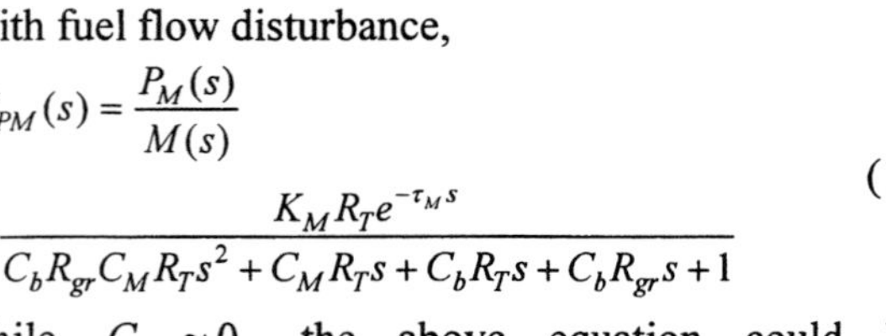

Figure3: The combustion process

With fuel flow disturbance,

$$W_{PM}(s)=\frac{P_M(s)}{M(s)}=\frac{K_M R_T e^{-\tau_M s}}{C_b R_{gr} C_M R_T s^2+C_M R_T s+C_b R_T s+C_b R_{gr} s+1} \qquad (1)$$

while $C_M \approx 0$, the above equation could be approximated by:

$$W_{PM}(s)\approx\frac{K_M(R_T+R_{gr})}{C_b(R_T+R_{gr})s+1}e^{-\tau_m s} \qquad (2)$$

With load disturbance,:

$$W_{p\mu}(s)=\frac{P_M(s)}{\mu_T(s)}=\frac{-K_T(C_b R_T R_{gr} s+R_T)}{C_b R_{gr} C_M R_T s^2+C_M R_T s+C_b R_T s+C_b R_{gr} s+1} \qquad (3)$$

Equation (1)~(3) show that the plant is in essence second-order model and could be approximated by first-order model. However, the representations are extremely local in nature, and do not recognize the non-linear nature of the process. The model parameters R_T, R_{gr} change with different load working condition so that it is in fact nonlinear plant within the whole operation region. Normally variations in a small region around a particular operating point, is reasonable linear. To account for system non-linearities, the entire operating region is, therefore, divided in to a number of local operating zones. In each zone, local linear models of the controller variables are identified using the prediction error identification method. Neurofuzzy networks are designed to represent the different reduced-order model in different load condition.

3 NEUROFUZZY MODELLING OF THE PLANT

3.1 Structure of the neurofuzzy network

The structure of the neurofuzzy network is a B-spline AMN shown in figure 4. The antecedent variables $[x_1, x_2 \cdots x_n]$ are the inputs of the AMN. The output of the network model $\hat{y}(t)$ could be linear combination of local linear models $\hat{y}_i(t)$. In such a way, the network stores and learns information locally and a priori knowledge about the process can be incorporated into the model design.

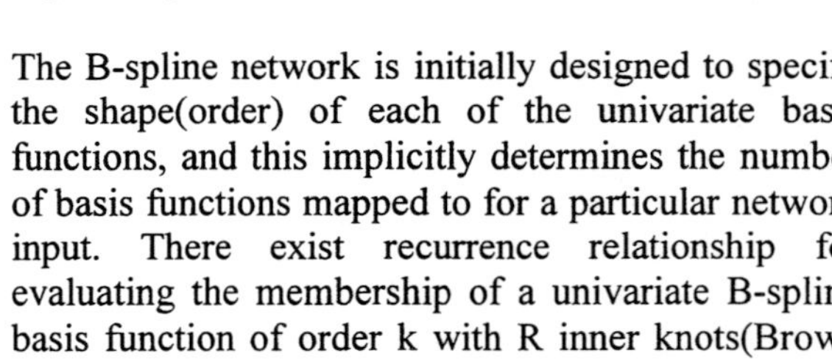

Fig.4: B-spline neurofuzzy network predictive model

The B-spline network is initially designed to specify the shape(order) of each of the univariate basis functions, and this implicitly determines the number of basis functions mapped to for a particular network input. There exist recurrence relationship for evaluating the membership of a univariate B-spline basis function of order k with R inner knots(Brown and Harris 1994) and they have two desirable properties:

1.The basis functions are defined on a bounded support and the output of the basis function is positive on its support , i.e. $\mu_k^j(x)=0,\ x\notin[\lambda_{j-k},\lambda_j]$ and $\mu_k^j(x)>0, x\in(\lambda_{j-k},\lambda_j)$

2.The basis functions form a partition of unity. For any network input, the sum of the outputs of the basis functions is always one, i.e $\sum_j \mu_k^j(x)\equiv 1,\ x\in[x_{\min},x_{\max}]$.

Therefore the neurofuzzy network shown in figure 4 could be expressed as a set of fuzzy production rules:

IF operating condition i (x_1 is positive small, and $\cdots, x_n$ is negative large)

THEN local linear model i :

$$\hat{y}_i(t) = \sum_{j=1}^{n} w_{ij} y(t-j) + \sum_{j=n+1}^{n+m} w_{ij} u(t-k-j+n+1)$$

The multivariate basis function is obtained by the tensor products of the univariate basis functions:

$$a_i(x(t)) = \prod_{k=1}^{n} \mu_{A_k^i}(x_k(t)) \quad (4)$$

where $i = 1,2,\cdots,p$ and n is the dimension of the input vector x(t). p is the total number of memory required in the network given by

$$p = \prod_{i=1}^{n} (R_i + k_i) \quad (5)$$

Thus the desirable properties of the univariate B-spline basis functions are all extended in a natural way to the multivariate basis functions. They are defined on hyperrectangles of size $(k_1 \times k_2 \times \cdots \times k_n)$ and therefore possess a bounded support. The network output is the synthesizing of these p rules:

$$\hat{y}(k) = \frac{\sum_{i=1}^{p} \hat{y}_i(k) a_i(x(t))}{\sum_{i=1}^{p} a_i(x(t))} = \sum_{i=1}^{p} \hat{y}_i(k) a_i(x(t)) \quad (6)$$

3.2 Learning of the neurofuzzy network

Choose the cost function to be:

$$J = \frac{1}{2} e^2 = \frac{1}{2} (\hat{y} - y)^2 \quad (7)$$

where $\hat{y}$ is the network prediction, and y is the real output data. The parameters of the local linear model could be changed in the direction of the negative gradient of J, i.e.

$$\dot{w}_{ij} = -g \frac{\partial J}{\partial w_{ij}} = -ge \frac{\partial \hat{y}}{\partial \hat{y}_i} \frac{\partial \hat{y}_i}{\partial w_{ij}} = -gea_i(x(t)) \frac{\partial \hat{y}_i}{\partial w_{ij}} \quad (8)$$

and

$$\frac{\partial \hat{y}_i}{\partial w_{ij}} = \begin{cases} y(t-j) & j \le n \\ u[t-k-(j-n)] & j > n \end{cases} \quad (9)$$

$g > 0$ is the coefficient determining the learning rate of decrease. $W(0)$ should be chosen such that the parameters could convergent to their real value as soon as possible.

Define the performance index to be

$$J = \frac{1}{N} \sum_{t=1}^{N} \left| \frac{\hat{y}(t) - y(t)}{y(t)} \right| . \quad (10)$$

which will later be used to evaluate the modelling results.

3.3 Plant neurofuzzy network modelling

The I/O data was acquired by the dynamic experiment(Liu, *et al.*, 1999) before the days when the steam-boiler turbine system was to be stopped running for periodic examination.

While coal fuel increase, heat load in furnace increase, so that the heated area increases. This could result in the increment of steam pressure and finally steam flow and steam pressure stay in a relatively higher level (Fig.5). With coal fuel to be constant, the control valve is opened as quickly as possible to yield an increase of output power. The steam flow increases and steam pressure decreases with no time delay. Finally, steam pressure stay in a relatively lower level(Fig.6).

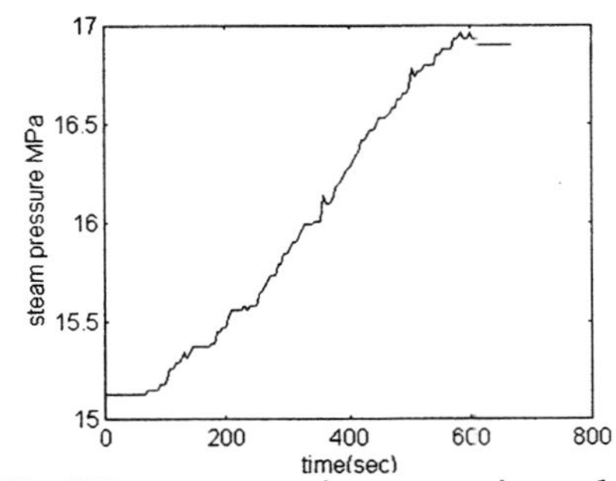

Fig.5 Responses to increment in coal fuel

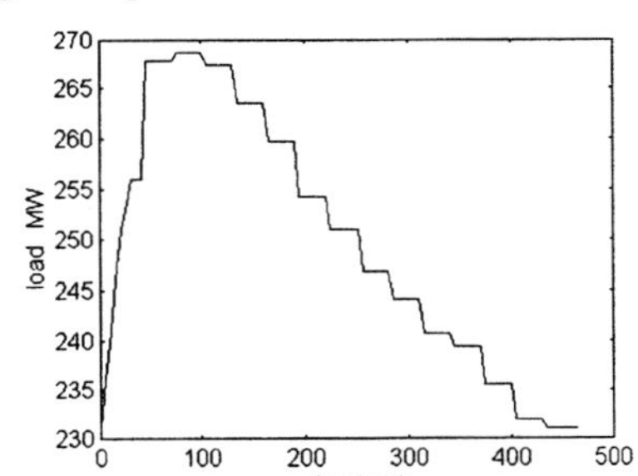

Fig.6 Responses to control valve opening

In the neurofuzzy modelling of the system, load power is the unique antecedent variable because the dynamic parameters in equations (1)~(3) are directly affected by load. In this way, univariate basis functions directly decide the operation region. Load operation was first divided into three based on the operator's former experience, which considers that 300MW is high load, 220MW is medium load and 140MW is low load. Load power lower than 140MW is the process of starting and ceasing of the steam-boiler system, and thus would not be considered in the process of modelling. Local linear model $W_{PM}(s)$ is chosen the form of equation (2) and after training, the identified three-region neurofuzzy model is:

if LOAD is high:

$$y(t) = 0.996 y(t-1) + 0.00415 u(t-120)$$

if LOAD is medium:

$$y(t) = 0.997y(t-1) + 0.00277u(t-120)$$

if LOAD is low:

$$y(t) = 0.998y(t-1) + 0.00217u(t-120)$$

In this model result, the performance index is 0.397. It is not adequate for two reasons:

(1) The local first-order linear model is an approximate model(with $C_M \approx 0$). From the mechanism description equation (1), second-order model represents the model more exactly. Fig.7 shows the fitting curves of the neurofuzzy model using first-order and second-order local linear model under real-time 260MW load condition. It is clear that second-order local linear model represents the complexity of the process more appropriately.

(2) The model representational ability could be increased by dividing the process into more operation regions. Fig.8 gives the five operating regions with language variables "low", "medium low", "medium", "medium high", "high" respectively, which represent more exactly human operator experience.

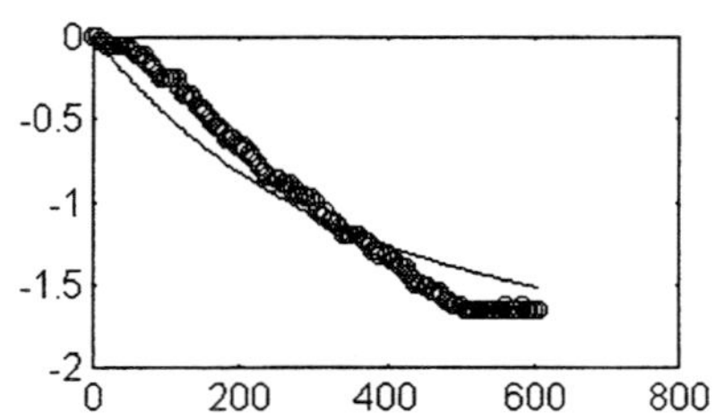

(7-a)

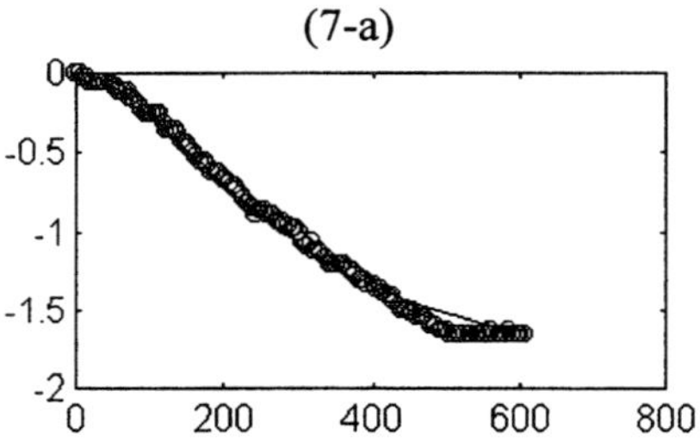

(7-b)

Fig.7 Steam pressure fitting curves using first-order local model (7-a) and second-order model(7-b)

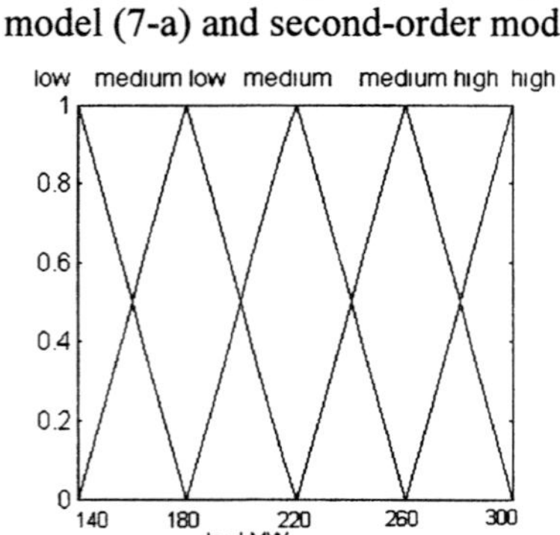

Fig.8 Division of five regions

With second-order local linear model, the model representational ability is increased by dividing the process into five operating regions. After training, the identified five-region model for $W_{PM}(s)$ is:

if LOAD is high:

$$y(t) = 1.9807y(t-1) + 0.9808y(t-2) + 0.0093u(t-120)$$

if LOAD is medium high:

$$y(t) = 1.9842y(t-1) - 0.9843y(t-2) + 0.0063u(t-120)$$

if LOAD is medium:

$$y(t) = 1.9856y(t-1) - 0.9866y(t-2) + 0.00454u(t-120)$$

if LOAD is medium low:

$$y(t) = 1.9886y(t-1) + 0.9887y(t-2) + 0.00323u(t-120)$$

if LOAD is low:

$$y(t) = 1.9901y(t-1) - 0.9901y(t-2) + 0.00245u(t-120)$$

In this case, the performance index is reduced to 0.082. To make a comparison, two other modelling method, namely linear regression model and multi layer perceptron (MLP) model were experimented using the same training data. Linear regression model is presented in the second-order form. MLP uses a fully connected feed-forward architecture of five hidden sigmoid units and a sigmoid output unit. Fig.9 shows comparison of boiler system(solid line in both figure), linear regression model(dotted line), MLP model(dashed line in 9-a) and neurofuzzy model(dashed line in 9-b). It is obviously that linear regression model could not represent the nonlinear plant in the whole operating range. It is only valid around a small region at a particular operation point. MLP could capture the nonlinear system dynamics owing to its ability of approximating nonlinear functions. Nevertheless, it is inferior to that of neurofuzzy model since neurofuzzy model utilizes operation experience and presents the learned information in a more human understandable form.

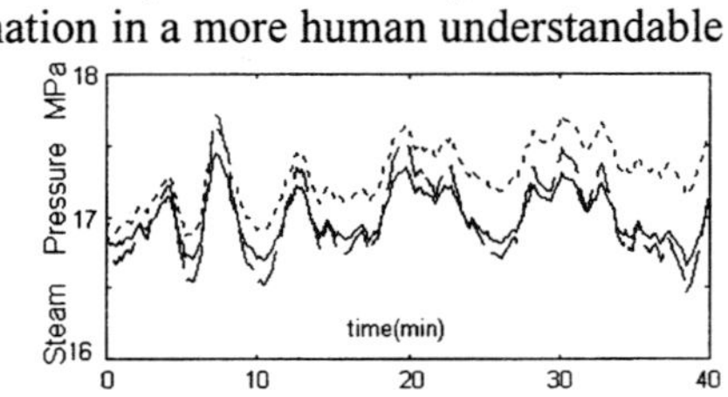

9-a

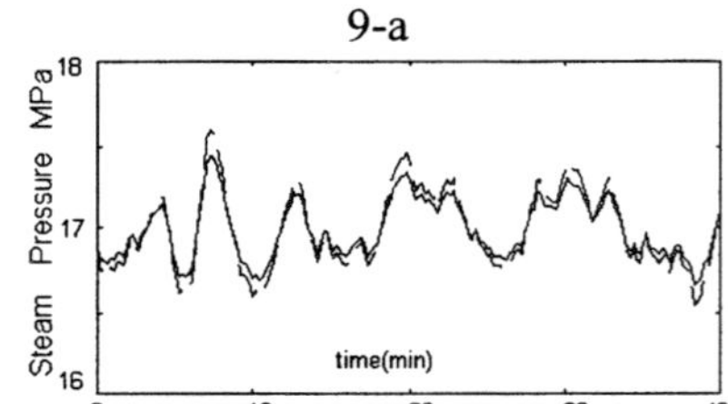

9-b

Fig.9 Model prediction

Neurofuzzy model for load disturbance $W_{P\mu}(s)$ is acquired with the same method.

Still, there exist three kinds of uncertainties causing difficulties to the modelling process:

(1) Even in the stable state with fixed working condition, steam pressure could fluctuate with fuel uncertainty caused by various category and quality of pulverized coal.
(2) Dynamic modelling is based on step response of the process. There is no purely step change in practical situation. The change in fuel delivery is operated manually by operators. And this can be different from one time to another.
(3) Time delay could not be obviously observed at

the beginning due to the fluctuation of the steam pressure. $\tau = 120s$ is the average and approximated time delay.

4 LOAD FUZZY-PI CONTROL ON DCS

The neurofuzzy model of the plant is available for the implementation of a fuzzy controller over the load system in the DCS(Distributed Control System).

In control engineering area, although many advanced methodologies have been developed in the past decades to handle the complex control problems and to achieve better performances, the great majority of industrial processes are still controlled by means of PI controllers. This is because PI controllers assure acceptable performance for a wide range of industrial plants and their usage is well known among industrial operators. It has so far been proved to be particularly effective on first-order or second-order process. Also control methods exist in DCS are mostly PID or PID based advanced strategy. Since the load system has been represented by several second-order neurofuzzy model in local region, PI control strategy could still be used to handle the local model. The adjustment of PI parameters is implemented by fuzzy logic. In this way, the fuzzy logic is a fuzzy supervisor more than a fuzzy controller.

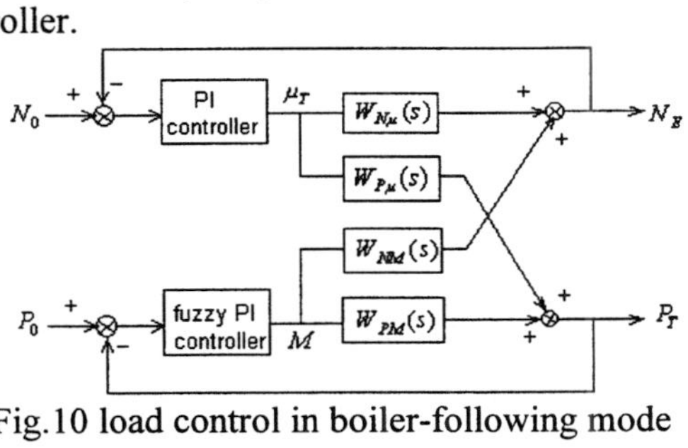

Fig.10 load control in boiler-following mode

The existing scheme for load control is the fixed parameter PI controller, which could obviously be well tuned in one operating point. To substitute it by fuzzy PI controller, the control system structure is depicted as Fig.10. $W_{N\mu}(s)$ is transfer function from control opening change to load power change, $W_{NM}(s)$ is transfer function from fuel change to load power change, i.e:

$$W_{N\mu}(s) = K_\mu W_T(s) \quad W_{NM}(s) = W_{PM}(s)K_P W_T(s) \quad (11)$$

Turbine response is the fast process with time constant $T \approx 9s$ of $W_T(s)$. The turbine controller adjusts the control valve opening by PI controller, so as to satisfy the load power demand quickly. The boiler controller is chosen as fuzzy PI controller, using the similar way to that of neurofuzzy network modelling. The error between the desired steam pressure and the real steam pressure is the input to the boiler controller, the proposed fuzzy PI controller could be depicted as:

IF **LOAD** is "A_i", THEN :

$$u_i(t) = K_{Pi}[e(t) + \frac{1}{T_i}\int_0^t e(\tau)d\tau]$$

A_is are language variables for load division "high", "medium high", "medium", "medium low", "low" respectively. The final output of the controller is calculated as:

$$u(t) = \sum_{i=1}^{5} v_i u_i(t) \quad (12)$$

Where v_1, v_2, v_3, v_4, v_5 are membership functions values for A_is, exactly the same as the fuzzy sets in the modelling process shown in Fig. 8.

In the simulation, proper PI control strategy for each local model could be addressed with model-reference-adaptive-control structure and the reference model is:

$$G_m(s) = \frac{\omega_n^2}{s^2 + 2\xi\omega_n s + \omega_n^2} e^{-\tau s} \quad (13)$$

Let $\xi = 0.7$ $\omega_n = 2$. $\tau = 120s$ is chosen the same as the time delay in the modelling of dynamic experiment. With the cost function defined in equation (7), parameter $\theta = [K_{PI}, T_{iI}]$ could be changed in the direction of the negative gradient of J:

$$\dot{\theta} = -g\frac{\partial J}{\partial \theta} = -ge\frac{\partial y}{\partial \theta} = -ge\frac{\partial y}{\partial u}\frac{\partial u}{\partial \theta} \quad (14)$$

$\frac{\partial y}{\partial u} = w_{(n+1)}$ is the Jacobean of the plant. And

$$\frac{\partial u}{\partial \theta} = \begin{bmatrix} \frac{\partial u}{\partial K_P} \\ \frac{\partial u}{\partial (1/T_i)} \end{bmatrix} = \begin{bmatrix} e(t) \\ \int_0^t e(\tau)d\tau \end{bmatrix} \quad (15)$$

$\theta(0)$ is chosen according to former experience so that the parameters could convergent to their real value as soon as possible.

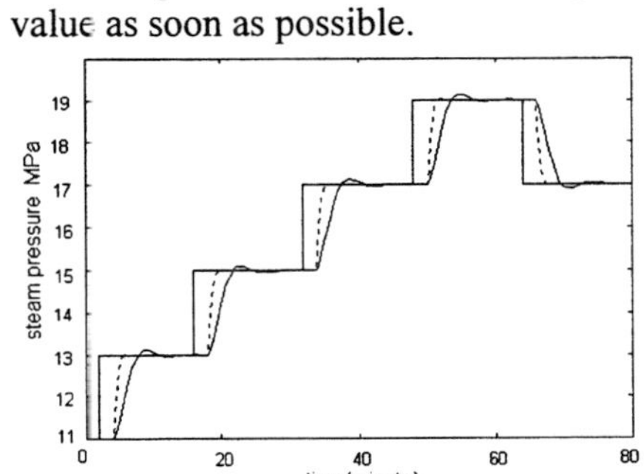

Fig. 11 Simulation of the steam pressure plant

The system is running in a sliding pressure mode, changing in square wave at the five local regions. The closed-loop response curve is shown by the solid line in Fig.11. The dotted line shows the response of the reference model. With continuously updating the parameters, the transient process reaches a good tracking result. In practical situation, it could satisfy the demand of the real steam-boiler system.

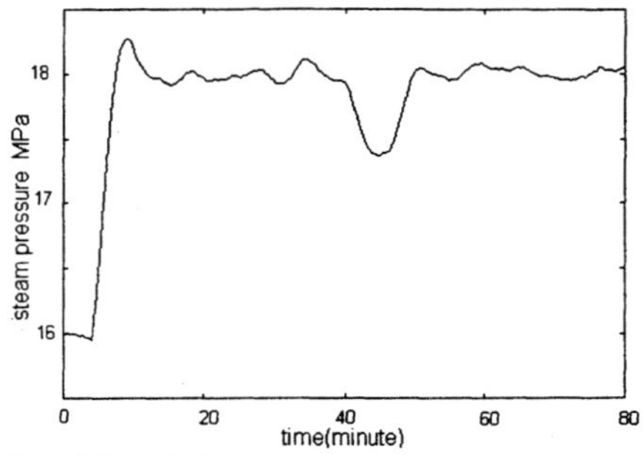

Fig.12 Real-time steam pressure response

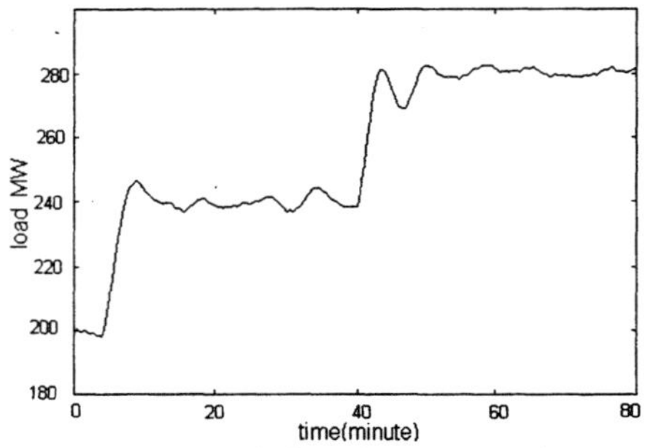

Fig.13 Real-time load response

The proposed method has been tested on real plant in a boiler-following mode. The desired load power N_0 is calculated according to the steam-boiler running condition. Fig.12 and Fig.13 show the response of the load increasing process:

(1) The first step is sliding pressure from 16Mpa to 18Mpa with fixed μ_T and thus load increase from 200MW to 240MW. In transient process, steam pressure overshoot reaches no more than 15%. In steady state, steam pressure fluctuate within $\pm 0.4MPa$ (mainly caused by quality of coal fuel).

(2) The second step is boiler-following load control. While the control valve opening μ_T increases, load increases quickly from 240MW to 280MW by the releasing boiler stored energy. While load begin to decrease, there is compensation of fuel delivery, steam pressure restore to its original level. In this way, steam pressure fluctuates 1.3Mpa to satisfy the load demand of power system.

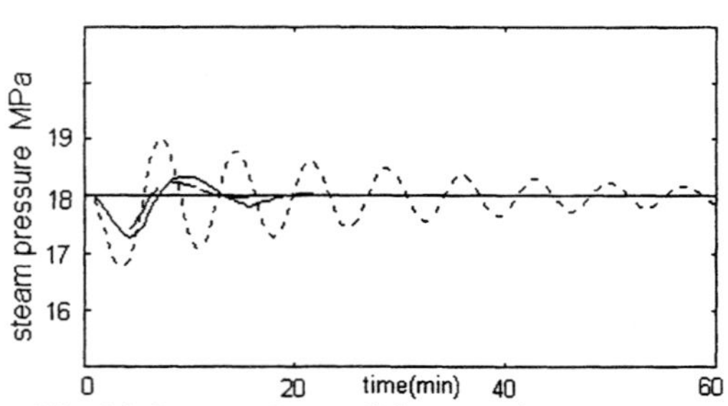

Fig.14 Comparison of the transient process

To make a comparison, the proposed fuzzy-PI controller, gain-scheduling PI controller and fixed PI controller has been simulated on the neuro-fuzzy model under the same load demand condition. Figure14 shows the transient process of steam pressure under 10% load increase condition. The fixed PI controller(dotted line) offers impractical result with large overshot and long setting time, while fuzzy-PI controller(dashed line) and gain-scheduling PI controller(solid line) produce very tight control. Fuzzy-PI controller is a little bit superior to that of gain-scheduling PI controller. This is because the nonlinear model is based on neuro-fuzzy technique and the fuzzy-PI controller is more resource demanding.

5 CONCLUSION

Load control system is the kernel system of unit steam-boiler. Reliable control methods are indispensable in modern power plant. Modelling and control of 300MW steam-boiler load system using neurofuzzy methodology is discussed in this paper. The mechanism model of load system is analyzed. Modelling method using Associate Memory Network is discussed. In AMN, with the dynamic experiments, a complex nonlinear load model is represented by neurofuzzy model containing several local linear models. A relative fuzzy PI controller is simulated on the neurofuzzy model, and then tested on real-time process. The advantage of the method is that it is suitable for improving many industrial plants already controlled by PI-type controllers. It presents significance in imporving the running efficiency of steam-boiler system by overcoming various disturbance and thus shows prospect future.

REFERENCE

Astrom, K.J.,& Ecklund, K. (1972). A simplified nonlinear model of a drum boiler-turbine unit. *Int. J. Contr.*, **16**(1), 139-153.

Brown, M. & Harris, C. J. (1994). *Neurofuzzy adaptive modelling and control.* Prentice Hall.

Ikonen,E. & Najim, K. (1996). Fuzzy neural networks and application to the FBC process. *IEE Proc.-Control Theory Appl.*, **143**(3), 259-269.

Irwin, G., Brown, M., Hogg, B., & Swidenbank, E.(1995). Neural network modelling of a 200MW boiler system. *IEE Proc.-Control Theory Appl.*, **142**(6), 529-536.

Tanaka,K., Sano,M., & Watanabe,H.(1995). Modeling and control of carbon monoxide concentration using a neuro-fuzzy technique. *IEEE Tran. fuzzy system,* **3**(3), 271-279.

Xiang-Jie Liu, Yi-Min Peng & Xiaoxin Zhou(1999). Identification of Boiler Models and Its Fuzzy Logic Strategy. Proc.the 14th IFAC World Congress. Volume O (pp.149-154).Beijing China.

Zhang, J. & Morris, A.J. (1995). Fuzzy neural networks for nonlinear systems modelling. *IEE Proc.-Control Theory Appl.*, 142(6), 551-561.

ELSEVIER
IFAC
PUBLICATIONS
www.elsevier.com/locate/ifac

A HIERARCHICAL MULTILEVEL FLOW MODELING FOR POWER SUBSTATION CONTROL AND RESTORATION

Germano Lambert-Torres **Alexandre Rasi Aoki** **Luiz Eduardo Borges da Silva**

Federal University at Itajubá
Itajubá – MG – 37500-000 - Brazil
Phone: +55-35-3629-1240 – Fax: +55-35-36291187 – germano@iee.efei.br

Abstract: The intelligent system proposed in this paper has been implemented for helping the Piratininga Power and Light's technicians in the system operational occurrences. It allows the operator to go through the ways for problem solving. The use of hybrid systems with artificial intelligence techniques, together with correctional numeric techniques, make possible to solve non-systematic problems and to show to the user the reasoning path used for the problem solving. This last factor is very important in a learning process, one of the aims of this system, once it make possible for the operator to understand the way the specialist in the area uses for solving one problem.

Keywords: knowledge engineering, knowledge-based system, intelligent control, power distribution, and power station control.

1. INTRODUCTION

The increase in the number of blackouts in the Brazilian power electric system (PES) is a consequence of system overload and lack of investments in the electric system. Besides, the long blackout impacts in consumers, economy, and in the PES itself, makes the fast and effective restoring very important (Adibi and Fink, 1994).

One important part of PES is the electric substation, which provides the system the control, transformation and interconnection requirements. The restoring of a substation normal configuration, after a fault occurrence, is made through one human operator intervention, time and again with the help of supervision and control automated system (Valiquette *et al.*, 1991).

Considering the increasing complexity of substations configuration, and that the restoring action is not an automatic procedure, the probability of an human failure, the time required for the execution of the restoring actions is a large one and should be avoided.

This paper comprises the application of Intelligent Planning together with the functional modeling technique Multilevel Flow Modeling (MFM), for the development of a decision support system for helping the substation operator regarding a restoring situation.

The combination of intelligent planning with actual engineering systems have been considered particularly difficult in view of the strain for obtaining sound knowledge in procedures from experts. Besides, this knowledge once obtained remains dependent of dominion and its re-use is scarcely applied in other applications.

The modern control centers are being improved with one more support tool for the decision take that are the Intelligent Systems. These systems have been placed between the old outlets from the several existing programs in the center (state evaluator, contingency analysis, alarms, etc), and the operators, aiming to supply one already screened information and the possible correction procedures for solving the existing problem.

2. FUNCTIONAL MODELING

The aim of this item is to describe briefly the basic concepts of the functional modeling technique MFM (Multilevel Flow Modeling). The MFM functions and objectives definitions will be described, through the construction principles of one MFM model. Also will be shown the basic concepts for the map out of acquired knowledge from one MFM model in one representation of intelligent planning.

In one supervision system it is important to obtain information that explain what the system parts should make (functional qualities), and why they should do it (intentional qualities). This type of information, so called mean-end, is scarcely supplied even non-existent in the classic modeling techniques, which generally explain how the different system parts interact (behavioral qualities).

The functional modeling techniques pick up what functions the system parts can do individually, or together, and how these functions are performed. Multilevel Flow Modeling (MFM) is strongly related to diagnostics, reasoning based upon models and techniques in general for modeling.

MFM uses a very high level of abstraction and is adequate for planning because it allows reasoning regarding the system functions, and an abstraction from the implementation technical details. One of the differences between MFM and other modeling techniques is that MFM extends the survey based upon models with new methods, adequate for the application in complex systems. Another important difference is the functions and objectives definition, while other techniques do not define these concepts explicitly.

3. BASIC CONCEPTS OF MFM

MFM introduces objectives as important and necessary for the representation of objectives of one engineering system. These objectives are achieved if all the system parts work as aimed by the system designer. MFM uses the function concept to catch up the intentions, the functions being the outcome of interactions between the system physics components, and may change according the aimed objectives.

MFM uses two types of functional decomposition, whole-part and means-ends, for the representation of one engineering system, see Figure 1. The decomposition whole-part is used to describe one-function refinements, for presenting more details regarding its realization, or technical implementation. Still one function may represent a complex system using a high level of abstraction; this function may yet be decomposed in sub-functions representing with more details the parts that represent the system. This process generates hierarchic structures with new functions and defined objectives. The decomposition means-ends looks for the function identifications and their connections, so as the function relations with the system objectives.

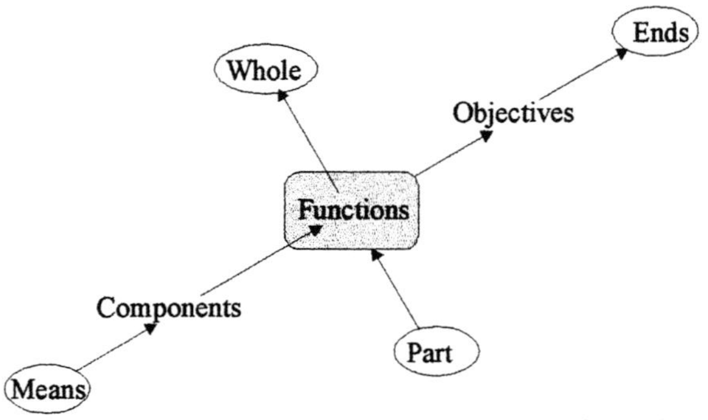

Fig. 1: Functional decomposition axes used by MFM

3.1 Objectives and Functions

The concepts of objectives and functions are important for the representation of engineering system functionality, because all the systems of this type were built with specific purposes. The objectives should catch up the systems purposes, and the functions should express the useful behavior of physic components necessary to reach the objectives

The following are useful concept definitions of objectives and functions adequate to the aims of MFM technique:

Objectives: define the tendency for witch some system activities or its parts are directed.
Functions: define which is the system role, or of its parts, for one or more objectives may be accomplished.

From these definitions it is possible to notice that the objectives and functions concepts are strongly related, once the MFM consider the function as an useful behavior from one system that may be explored, in accordance with its behavior properties, to achieve the desired objectives.

3.2 MFM Structures and Means-Ends Relations

To represent one system's functionality, MFM defines a set of relation's aims-ends that are used for describing the several relationships between functions and objectives:

Source: these functions describe the system capacity for acting as an infinite source of material;

Drain: these functions describe the system capacity to operate as a drain for material;

Transportation: these functions represent the system propriety in supplying a material transfer capacity; the transport function defines one direction for the material flow.

Warehousing: these functions describe the capacity of a system to warehouse material;

Balance Sheet: these functions describe the capacity for a system to provide a balance sheet of material;

Block: these functions represent the propriety of a system to prevent against the transfer of material between two sub-systems.

These functions are adequate for the representation of complex systems functionality that comprises the handling of mass and/or energy. The connection relation link functions through inlet and outlet points, forming a function structure.

One structure MFM is defined as a linked group of functions that process the same type of material. The structures are inter-related by objectives through means-ends relations of the type reach and condition. The reach type relation is used for describing which MFM objectives are accomplished by structure. Usually, one objective is reached by one or more functions. The condition type relation is used for connecting one MFM objective with one function pre-condition.

MFM uses a graphical representation and distinguishes between functions that process energy and functions that process mass as illustrated in Figure 2.

The mass and energy flow structures are built based on a set of functions which are related in complementary pairs, namely source-sink, storage-balance, transport-barrier. The structures are interrelated by goals through two means-end relations: achieve relation and condition relation. The achieve relation is used to describe the MFM goals that a structure achieves. Usually, a goal is achieved by one or more functions. The condition relation is used to connect an MFM goal as a precondition of a function.

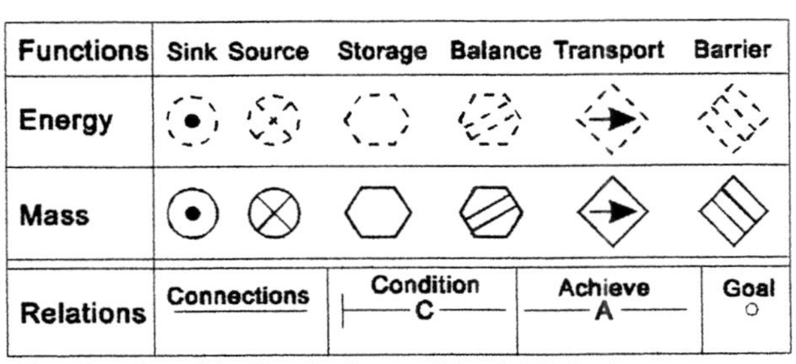

Fig. 2 - Graphic Symbols used by MFM, the means-end decomposition

4. SUBSTATIONS FUNCTIONAL MODELING

The MFM model construction process is highly dependent of the desired abstraction level. A system may be described under several different points of view, and may define several representation levels.

In one substation plant, there are only energy flows, since mass and information flows are not considered. This way, but one representative level will be considered that is the energy flow level.

To construct the MFM model for a substation plant it was necessary to split up the substation in several structures: transmission lines, inlet busbar, transformers, outlet busbar, and loads. For each structure we have a MFM model that fits to one objective, that is the pre-condition of the following structure. In Figure 3, we have a MFM model for the transmission line of one single bar substation.

The MFM model of a transmission line does not include the filter and the lightning arrester, because they are not important for the recovering purposes. The two balances BaN1 and BaN2 represent two diagrams nodes. S11, D1 and S12 are united in transportation TrDS1 because they work as a set and execute the same function.

These MFM functions were defined to reach the objective G1 that is to energize the inlet busbar. For transportation TrDS1 .may work accordingly, the objective GD1 has to be reached by the second structure. It was not used a MFM model for protections, a work with a group of rules giving a Boolean operation condition was used instead. All the forecasted protections for one transmission line and the differential protection frequently used in busbars are included in the second structure shown in Figure 3.

In Figure 4 we have one MFM model for inlet busbar. The function SoB1 is activated by G1, and it is possible to have more than one structure, shown in Figure 3 that reach G1 if we have more transmission lines.

As transformers have different current and voltage levels in primary and secondary windings, transformers were separated in two different functions: the drain function for the primary winding, and the supply function for the secondary winding. Having two transformers T1 and T2, each one reach one objective, G2 and G3, which are energize the transformers secondary windings.

In Figure 5, we have a MFM model for transformer T1. The function SoST1 is conditioned by objective G2. This structure reaches the objective G4 that is energize the outlet busbar. The functions TrDT1 and TrDS1 are similar in their operations.

According to Larsen (1996), the process of a start-up plan development for a MFM modeled plant, may be compared to the problem for establishment of each function in the model.

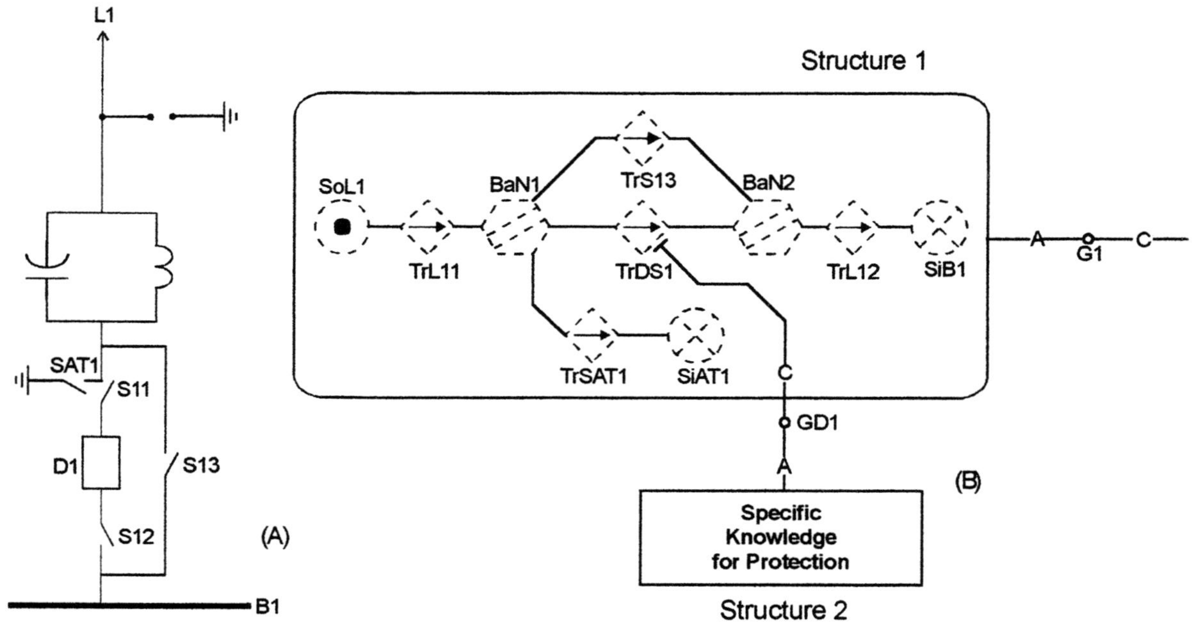

Fig. 3: (A) Diagram and (B) MFM Model for transmission lines

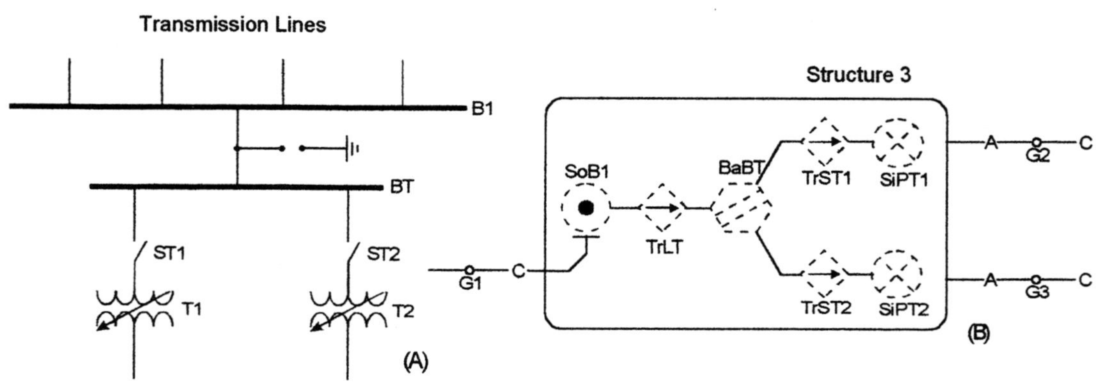

Fig. 4: (A) Diagram and (B) MFM Model for inlet busbar

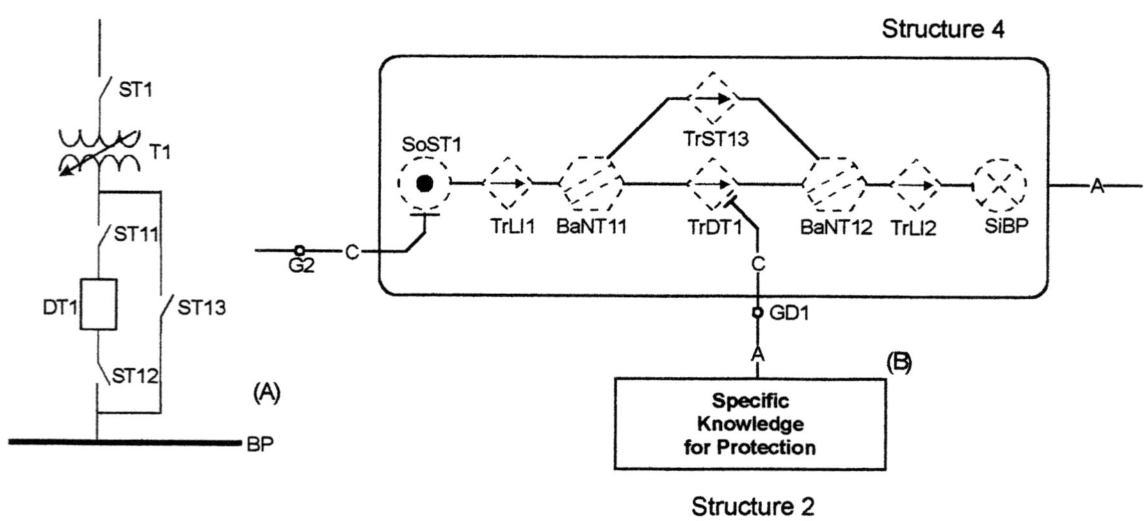

Fig. 5: (A) Diagram and e (B) MFM Model for transformer T1

One function may be set up when both the neighbor functions in the model are in states wanted by the designer and contribute for reaching the operational objectives. Before establishing a function it should be qualified. One function is qualified when it is ready to be integrated with other functions, is that to say, it is able to provide a useful behavior but it is not operating yet.

Based in these concepts, a planning domain was built composed of generalized operators QUALIFY and ESTABLISH, as shown below:

Operator ENABLE <F>

Var:
<F> Function
<NF> Next Function
<PS> Process state
<GD> Goal
Pre:
Enabled <NF>
Goal Achieved <GD>
Structural-pre <PS>
Add:
Enabled <F>

Operator ESTABLISH <F>
Var:
<F> Function
<PF> Previous Function
<A> User Task
<G> Goal
Pre:
Enabled <F>
Goal Achieved <G>
Established <PF>
Activity <A>
Add:
Established <F>

The map out functions is responsible for defining the planning variables in accordance with the acquired knowledge stored in the MFM model.

The topologic structures describe the relation's means-ends MFM. For each function defined in the model, the neighbor functions, previous and posteriors functions are identified and organized in structures.

The process state structures catch up the specific knowledge from a domain. Knowledge must be acquired as a process state when it represents one condition that should be satisfied for supporting the basic behavior of a MFM function.

The solution of a planning problem comprises a sequence of actions that once executed will drive the system for an objective state. These actions come from the domain and generally represent one user action. If one outside action is required for setting up one MFM function, then the knowledge that represents that action should be acquired and stored into the corresponding structure.

5. SUPPORT SYSTEM FOR SYSTEMIC DECISION

The proposed system in four parts: one knowledge system acquisition interface, one modeling system, one map out and one planning system.

The knowledge acquisition interface generates a database from substation protections and equipment. It is possible to work with three different topologies: simple bar, main and transfer bar, and double bar, these three topologies have been chosen for searching solutions to 138-13,8kV substations (Lambert-Torres *et al.*, 1997).

The modeling system provides an intermediate step for acquiring planning knowledge; in this point the database with the MFM models is used for building the planning domain. Beside, this system keeps one structure witch adds and changes easily the knowledge of each MFM function.

The map out between the modeling and the planning system is a set of map out functions responsible for instancing the planning operators and variables. These functions keep independence between both systems. At last, the planning system is responsible for generate a series of complex plans based in the present substation state, according the specified knowledge domain (Lind, 1992).

The aim is to describe shortly the criteria and philosophy of restoring usually used for 138kV substations. The restoring priorities and sequence from equipment will be described, and yet the procedures for the substation restoring after the breakdown and the related occurrence of system disturb.

The prevailing energy model for the present structure of society makes essential the steady supply of electric power, according the established quality standards. The breakdown is considered an abnormal condition, which timing and extension should be reduced in view of the generated consequences and losses. Nevertheless, the electric system defects are inevitable, whether they are generated by outside or inside causes, by electric, environment or human phenomena.

Because the majority of defects are transients, the aim is to minimize the effects and guaranty a steady operation through techniques and equipment, able to provide faster system recovery after partial or total components breakdowns.

The substation recovery is initiated only after the correct and complete elimination of defects is done. During the transient period, the protective relays fill the presence of an abnormal condition in the system operational basic characteristics (voltage, current, frequency, etc) and actuate on the switchgear in a selective way as to isolate the defective component, or the one direct involved in the failure.

The velocity or fastness of the protective relay action aims to reduce the extension of the occurred damage, providing the maintenance of normal operational condition in the healthy parts of the electric system. Nevertheless, depending on substation busbar arrangement, some healthy parts are shut down in view of the lack of adequate switchgear to break up short-circuit currents.

When the fault is transient, once the standard timing for elimination or removal of the cause ends, all the de-energized components may be reactivated in sequence by one recovery system. In the case of a permanent failure, disconnecting switches must be operated in such a way that the minor parts of the system, the one where the failure has occurred, remain de-energized waiting for the maintenance action. While the healthy parts, initially disconnected because of the switchgear limitations from the substation arrangement and other restrictions, should be energized.

It is very important that the restoring system has resources for one effective and selective restoring, soon as the failure be found and identified and the defective branch de-energized, avoiding that the defective circuit branches or the permanent faults actuate through the electric system.

Depending on the extension of fault, only part of the de-energized components or part of the load should be restored. However, there is a restoring hierarchy, as shown in Table 1.

Table 1 - Restoring Hierarchy

START

Sources	TL
Transformation	TRANSFORMES
	AUTOTRANSFORMES
Load	TL, FEEDERS
Control	CAPACITOR BANKS
	REACTORS AND SYNCHRONOUS

END

The operation philosophy and the previous and after fault states of equipment and protective relays, are the starting source of data from the restoring process. Based in these data and after analysis of them, the restoring action begins and will have its sequence defined.

6. CONCLUSIONS

The functional modeling technique FMT has shown to be adequate for acquiring planning knowledge, even though being efficient only regarding the substation topology knowledge. As the knowledge acquisition is based in the FMT model, to have a good model is fundamental.

The mastering of planning based in FMT may be easily understood, though it is necessary the development of specific operators according with the application. The necessary modifications for application in other types of substations is inner to the structures, in the arrangement of the FMT functions, making the maintenance easier and offering flexibility for the decision support system.

As this technique has shown to have a potential for new applications in other substation topologies, the aim is a future implementation of one interface with supervision systems and automated control for electrical substations, making valid the use of this support system for decision together with the automated system.

ACKNOWLEDGMENTS

The authors would like to thank the PRONEX/CNPq/MCT for the financial support to develop this project.

REFERENCES

Adibi, M.M. and L.H. Fink (1994). Power System Restoration Planning. *IEEE Trans. Power Systems*, **Vol. 9**, pp. 22-28.

Lambert Torres, G., G.M. Ribeiro, C.I.A. Costa, A.P. Alves da Silva and V.H. Quintana (1997). Knowledge Engineering Tool for Training Power-Substation Operators. *IEEE Transactions on Power Systems*, **Vol. 12**, No. 2, pp. 694-699.

Larsson, J.E. (1996). Diagnosis based on explicit means-end models. *Artificial Intelligence*, **Vol. 80**, pp. 29-93.

Lind, M. (1992). A categorization of models and its application for the analysis planning knowledge. *Proceedings of the ANP-92 Conference Seminar at Human Cognitive and Cooperative Activities in Advanced Technological Systems.*.

Valiquette, B., G. Lambert-Torres, and D. Mukhedkar (1991). An Tool for Teaching Power System Operation Emergency Control Strategies. *IEEE Trans. on Power Systems*, **Vol.6**, No.3, pp. 1315-1322.

ELSEVIER
IFAC
PUBLICATIONS
www.elsevier.com/locate/ifac

SEQUENTIAL DECENTRALIZED DESIGN OF ROBUST LOAD FREQUENCY CONTROLLERS IN MULTIAREA POWER SYSTEMS

Hassan Bevrani, Yasunori Mitani and Kiichiro Tsuji

Department of Electrical Engineering, Osaka University
2-1 Yamada-Oka, Suita, Osaka 565-0871, Japan

Abstract: A new systematic approach to design of sequential decentralized load frequency controllers for multi-area power systems based on μ-synthesis is described. System uncertainties according to load, practical constraint on control action and desired performance are included in the synthesis procedure. The robust performance in terms of the structured singular value is used as a measure of control performance. A 3-area power system example is presented demonstrating the controllers synthesis procedure and advantages of this method. *Copyright © 2003 IFAC*

Keywords: Power system, LFC, decentralized control, sequential design, μ-synthesis.

1. INTRODUCTION

The sequential control design because of its advantages is the most common design procedure in real applications of decentralised synthesis methods. Sequential design involves closing and tuning one loop at a time. This method is less conservative than independent decentralised design. Because at each design step one utilizes the information about the controller specified in the previous step (Chiu and Arkun, 1992), and, also it is more practical in comparison of common decentralized methods.

The objective of the Load Frequency Control (LFC) is to maintain the frequency and to keep tie-line power flows within pre-specified values. Recently, several decentralized LFC scenarios have been proposed (Hiyama, 1982; Feliachi, 1987; Liaw, and Chao, 1993; Niioka, *et al.*, 1998; Yang, *et al.*, 2002). But all of authors use simultaneous design for a fixed controller structure. Simultaneous design is very difficult numerically for large scale power system, and, it does not provide some of the advantages that are usually the reason for using decentralized control in the first place, such as the ability to bring the system into service by closing one loop at a time, and the guarantee of stability and performance in the case of failures.

In this paper, after introduce the μ-based sequential control framework and pairing inputs and outputs, we will design of a single-input single-output (SISO) controller for each loop. In load frequency controller design for each area, the structured singular value, introduced by Doyel *et al.* (1982) will be used as a synthesis tool and a measure of performance robustness. μ-synthesis is already used to design of sequential decentralised power system stabilisers (Djukanovic, *et al.*, 1999).

This paper shows μ-synthesis can be used for design of multi-area power system load frequency controllers that guarantee robust stability and robust performance for a wide range of operating conditions. Section 2 describes the synthesis methodology. In section 3, the proposed strategy is applied to a three area power system. Section 4 demonstrates the effectiveness of proposed scheme by some simulation results.

2. SYNTHESIS PROCEDURE

In this paper we use the conventional model for each area of multi-area power system, which is widely used by the researchers (Hiyama, 1982; Feliachi, 1987; Liaw, and Chao, 1993; Niioka, *et al.*, 1998; Yang, *et al.*, 2002). Actually, power systems have a highly nonlinear and time-varying nature; however a simplified and linearized model is usually used for load frequency control. The error caused by the simplification and linearization can be considered in robust control strategies (such as μ-synthesis) .

In each area we consider three simple transfer functions according to generator, turbine and power system (rotating mass and load). Fig. 1 shows the block diagram of area 1 from a m-area power system.

The connections of area-1 to other areas are shown in this figure. Referring to Fig. 1, the state space realization of area-i is given by:

$$\begin{aligned} \dot{x}_i &= A_i x_i + B_i u_i + F_i d_i \\ y_i &= C_i x_i \end{aligned} \qquad (1)$$

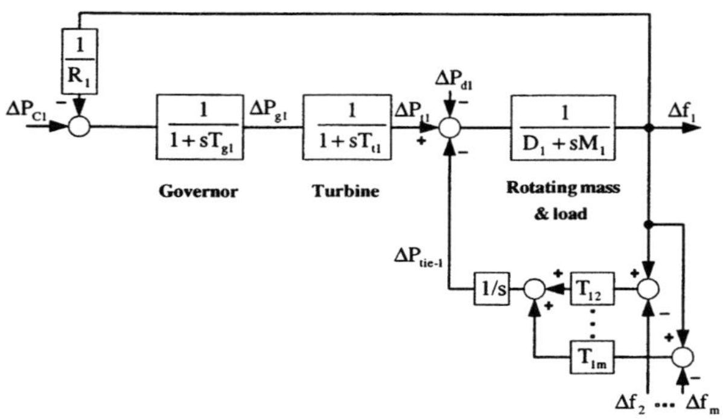

Fig. 1. Block diagram of area-1

The state vector x_i, control input u_i and disturbance input d_i can be defined as follows:

$$x_i = [\Delta f_i \ \Delta P_{ti} \ \Delta P_{gi} \ \Delta P_{tie-i}]^T, u_i = \Delta P_{ci}, d_i = \Delta P_{di}$$

where,

Δf_i : incremental frequency deviation of area- i,

ΔP_{gi} : incremental governor valve position change of generator in area- i,

ΔP_{ci} : reference set point (control input) of area- i,

ΔP_{ti} : incremental output of generator in area- i,

ΔP_{tie-i} : incremental change in tie-line power between area i and other areas,

ΔP_{di} : disturbance in area- i

M_i : equivalent inertia constant for area- i,

D_i : equivalent damping coefficient for area- i,

T_{gi} : governor time constant for area-i,

T_{ti} : turbine time constant for area- i,

T_{ij} : synchronizing coefficient in normal operating conditions between areas i and j,

R_i : drooping characteristic for area-i.

The total real power imported to area-i equals the sum of all inflowing line powers P_{tie-ij} from adjoining other areas i.e.,

$$P_{tie-i} = \sum_j P_{tie-ij} \ . \qquad (2)$$

The real power in per unit transmitted across a lossless line of reactance X_{ij} is

$$P_{tie-ij} = \frac{|V_i||V_j|}{X_{ij} P_{ri}} \sin(\delta_i - \delta_j) \qquad (3)$$

where, P_{ri} is the rated power of area- i, and

$$V_i = |V_i| e^{j\delta_i}, \ V_j = |V_j| e^{j\delta_j}$$

where $V_i = |V_i| e^{j\delta_i}$ and δ_i are the amplitude and the angle of the terminal voltage in area-i.

The main goal in each area is maintaining area frequency and tie-line power interchanges close to specified values in presence of model uncertainties and disturbances. To achieve our objectives and according to μ-synthesis requirements we can modify the each area model as shown in Fig. 2. In comparison of Fig. 1, the inter-area connections are removed and it is considered by ΔP_{tie-i}. This figure shows the synthesis strategy for area-i. It is notable that for each area there are several uncertainties because of parameter variations, model linearization and unmodeled dynamics to due approximation of the rest of power system. However to keep the complexity of the controllers reasonably low, our focus must be concentrated on the most important source of uncertainty. This uncertainty in Fig. 2 is modeled as an unstructured multiplicative uncertainty and W_{ui} represents the related weighting function.

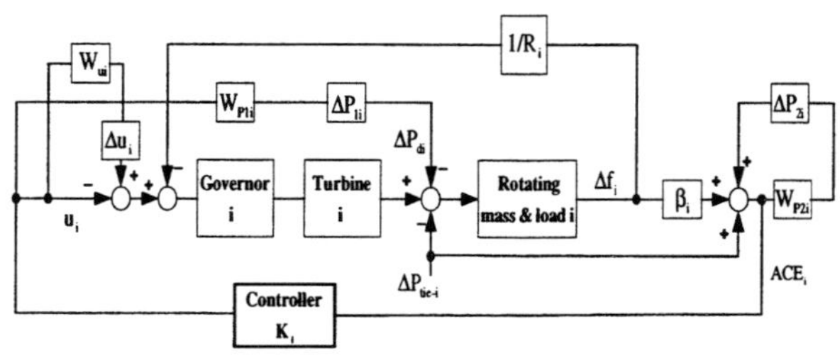

Fig. 2. Proposed strategy for load frequency controller synthesis in area-i.

According to requirement performance and practical constraint on control action, two fictitious uncertainties W_{P1i} and W_{P2i} are added to area model. The W_{P1i} on the control input sets a limit on the allowed control signal to penalize fast change and large overshoot in the control action. The weight W_{P2i} at the output sets the performance goal e.t. tracking/regulation error on the output deviation frequency. Further more it is notable that in order to reject disturbances and to good tracking property, W_{P1i} and W_{P2i} must be such select that singular value of sensitivity transfer function from u_i to y_i in related area be reduced at low frequencies (Bevrani, 1999). Δu_i, Δp_{1i} and Δp_{2i} are uncertainty blocks associated with W_{ui}, W_{P1i} and W_{P2i} respectively.

The synthesis starts with setting the desired level of stability and performance for first area with a set of (u_i, y_i) and chosen uncertainties to achieve robust performance. In order to maintain adequate performance in the face of tie-line power variation and load disturbances, the appropriate weighting functions must be used. The inclusion of uncertainties is adequately allow for maximum flexibility in designing the area closed loop characteristics and the demands placed on the controller will increase.

We can redraw the Fig. 2 as shown in Fig. 3. This figure shows the synthesis framework for i-th controller. G_{i-1} Includes area-i's nominal model, associated weighting functions and scaling factors. As it is mentioned above, the blocks Δp_{1i} and Δp_{2i} are the fictitious uncertainties added to assure robust performance while the block Δu_i models the multiplicative uncertainty associated with local load.

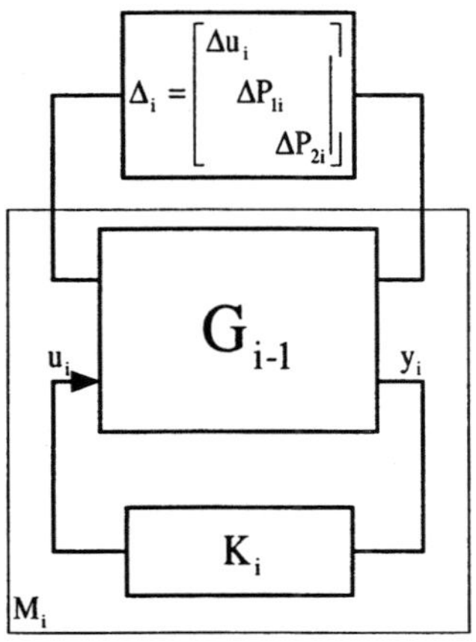

Fig. 3. M-Δ configuration for area-i.

Now in step i, the synthesis problem is designing the robust controller K_i. Based on the μ-synthesis, the robust performance holds for a given M-Δ configuration as it is shown in Fig. 3, if and only if

$$\inf_{K_i} \sup_{\omega \in R} \mu[M_i(j\omega)] < 1. \qquad (4)$$

Using the performance robustness condition and the well-known upper bound for μ, the robust synthesis problem reduces to determine

$$\min_{K_i} \inf_{D} \sup_{\omega} \bar{\sigma}(D M_i(j\omega) D^{-1}),$$

or equivalently

$$\min_{K_i, D} \left\| D M_i(G_{i-1}, K_i)(j\omega) D^{-1} \right\|_{\infty},$$

by iteratively solving for D and K_i (D-K iteration algorithm). Here D is any positive definite symmetric matrix with appropriate dimension and $\bar{\sigma}(.)$ denotes the maximum singular value of a matrix.

When the controller synthesis has been completed, another robust controller is design for a second area with its set of variables, and so on. During design of each controller, the effects of previously designed controllers are been considered. The overall frame work of proposed strategy is given in Fig. 4. It is notable that the block G_0 is assumed to contain the appropriate weighting functions and scaling factors according to Δ_1. The block G_{m-1} includes G_0 and all decentralized controllers K_1, K_2, ..., K_{m-1} designed in previous iteration 1, 2, ..., (m-1) and related uncertainty blocks.

We consider the open loop state-space representation of G_0 as:

$$\dot{x} = Ax + Bu + Fd$$
$$y = Cx + Ed \qquad (5)$$

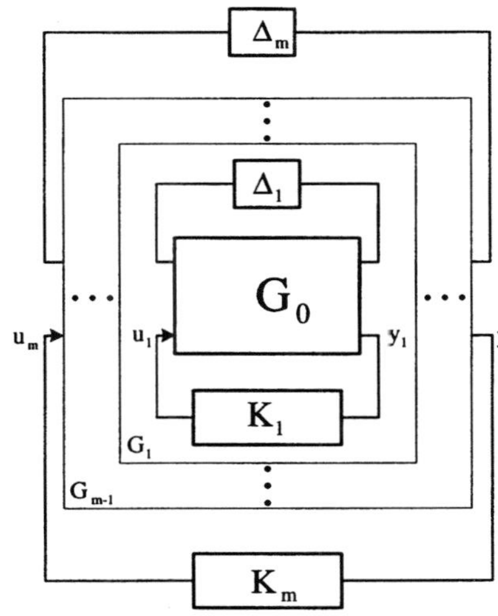

Fig. 4 . Framework for μ-synthesis

In this equations B corresponds to control input, F corresponds to disturbance inputs, C and E correspond to the output measurement which is input to load frequency controller, and,

$$x = [\Delta f_1 \ \Delta P_{t1} \ \Delta P_{g1} \ \Delta P_{tie-1} \ldots \Delta f_m \ \Delta P_{tm} \ \Delta P_{gm} \ \Delta P_{tie-m}]^T,$$
$$d = [\Delta P_{d1} \ \Delta P_{d2} \ \ldots \Delta P_{dm}]^T,$$
$$u = \Delta P_{c1} = u_1 \,; \ y = \beta \Delta f_1 + \Delta P_{tie-1} = y_1 .$$

It should be noted that above equations for open loop system, in each synthesis, must be augmented by including weighting functions and controllers synthesised in previous steps. In each step, μ-controller is designed for one set of input and output variables. When this synthesis has been successfully completed, next μ-controller is designed for an other set of input-output variables, and so on. In every step, the effects of previously designed controllers are taken into account. Therefore by adding one new loop at a time, the closed loop system remains stable at each step. In summary, the proposed method consists of the following steps:

Step 1: Identify the order of loop synthesis.
The important problem with sequential design is the final control performance achieved, may depend on the order in which the controller in the individual loops are synthesized. In order to over come to this problem we must close the fast loops first. Because the loop gain and phase in the bandwidth region of the fast loops is relatively insensitive to the tuning of the lower loops. In other word, for those cases that the bandwidths of the loops are quite different, the outer loops are tuned with the inner (fast) loop in place. It causes the lower number of iterations during re-tuning procedure to obtain the best possible performance. Obtaining the estimation of interactions on each control area behavior to determine the effects of undesigned loops is the other important issue in sequential synthesis procedure. The determination of performance relative gain array (PRGA) and closed-loop disturbance gain (CLDG) methods given in

(Hovd, and Skogestad, 1994) are useful for this purpose.

Step 2: Identify the uncertainty blocks and associated weighting functions according to the first control area input-output set, depending on dynamic model, practical limits and performance requirements.
It is notable that there isn't any obligate to consider the uncertainty in the few parameters, only. Considering the more complete model by including additional uncertainties is possible and causes less conservative in synthesis. However the complexity of computations and the order of result controller will increase.

Step 3: Isolate the uncertainties from nominal area model, generate Δp_{1i} , Δp_{2i} , Δ_{Ui} blocks; and performing M-Δ feedback configuration (formulate the desired stability and performance).

Step 4: Start the *D-K* iteration using μ-synthesis toolbox to obtain the optimal controller, which provides desirable robust performance. This procedure determines the first robust controller.

Step 5: Reduce the order of result controller by utilizing the standard model reduction techniques and apply μ-analysis to closed loop system with reduced controller to check whether or not upper bound of μ remains less than one.
It is notable that the controller found by this procedure is usually of a high order. In order to decrease the complexity of computation in case of high order power system, appropriate model reduction techniques might be applied both to the open-loop system model and H∞ controller model within each *D-K* iteration.

Step 6: Continuing this procedure by apply above steps to other loops (control area input-output set) according to specified loop closing order in step 1.

Step 7: Retuning the obtained controllers to achievement the best performance and check if the overall power system satisfies the robust performance condition using μ analysis.
If the objective is achievement of the best possible performance, the controller which was designed first, must be removed, and then re-designed, but now with controllers which have been synthesized in succeeding steps, because the first synthesis was according to the more conservative state.

The proposed strategy in this paper guarantees the robust performance for multi-area power system, after design of load frequency controllers according to above sequential steps. The advantage of the procedure is it ensures that by closing one loop for a special area at a time, this area gets robust performance; meanwhile the multi-area power system holds its stability at each step. Similarly, during startup the system will at least be stable if the loops are brought into service in the same order as they have been designed (Hovd, and Skogestad, 1994).

3. APPLIED TO THREE AREA POWER SYSTEM

The controllers design approach presented in the previous section, is now applied to a three area power system is shown in Fig. 5, and the nominal parameter values are given in Table 1.

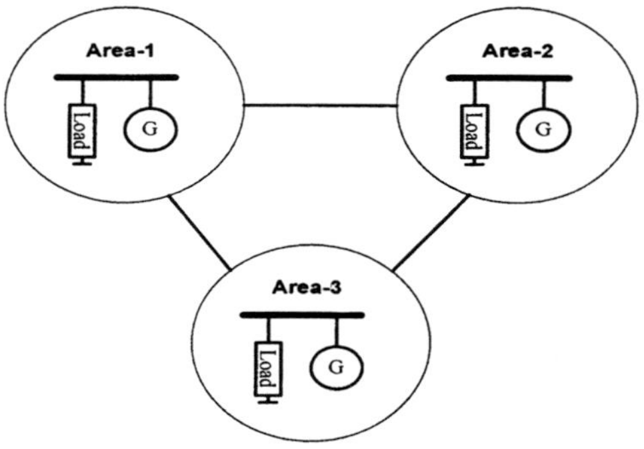

Fig. 5. 3-area power system

Table 1. Three area parameters

Parameter	Area-1	Area-2	Area-3
D_i (p.u.Mw/Hz)	0.006	0.0083	0.008
M_i (p.u.Mw)	0.2	0.167	0.15
T_{ti} (s)	0.3	0.33	0.35
T_{gi} (s)	0.08	0.072	0.07
R_i (Hz/p.u.Mw)	2.4	2.7	2.5
T_{ij} (p.u.Mw/Hz)	0.545	0.545	0.545

A nominal state-space model for this system as a multi-input multi-output (MIMO) system can be constructed as (5), where,

$$x = [\Delta f_1\ \Delta P_{t1}\ \Delta P_{g1}\ \Delta P_{tie-1}\ \Delta f_2\ \Delta P_{t2}\ \Delta P_{g2}\ \Delta P_{tie-2}\ \Delta f_3\ \Delta P_{t3}\ \Delta P_{g3}\ \Delta P_{tie-3}]^T;\ u = [u_1\ u_2\ u_3]^T,$$
$$d = [\Delta P_{d1}\ \Delta P_{d2}\ \Delta P_{d3}]^T;\ y = [y_1\ y_2\ y_3]^T.$$

and $A \in R^{12\times12}$, $B \in R^{12\times3}$, $F \in R^{12\times3}$;

$$A = \begin{bmatrix} A_{11} & A_{12} & A_{13} \\ A_{21} & A_{22} & A_{23} \\ A_{31} & A_{32} & A_{33} \end{bmatrix} \tag{6}$$

$$A_{ii} = \begin{bmatrix} -\frac{D_i}{M_i} & \frac{1}{M_i} & 0 & -\frac{1}{M_i} \\ 0 & -\frac{1}{T_{ti}} & \frac{1}{T_{ti}} & 0 \\ -\frac{1}{R_i T_{gi}} & 0 & -\frac{1}{T_{gi}} & 0 \\ \sum_j T_{ij} & 0 & 0 & 0 \end{bmatrix} \tag{7}$$

$$A_{ij}(i \neq j) = \begin{bmatrix} 0 & 0 & 0 & 0 \\ 0 & 0 & 0 & 0 \\ 0 & 0 & 0 & 0 \\ -T_{ij} & 0 & 0 & 0 \end{bmatrix} \tag{8}$$

$$B=\begin{bmatrix} 0 & 0 & \frac{-1}{T_{g1}} & 0 & 0 & 0 & 0 & 0 & 0 & 0 & 0 & 0 \\ 0 & 0 & 0 & 0 & 0 & 0 & \frac{-1}{T_{g2}} & 0 & 0 & 0 & 0 & 0 \\ 0 & 0 & 0 & 0 & 0 & 0 & 0 & 0 & 0 & 0 & \frac{-1}{T_{g3}} & 0 \end{bmatrix}$$

$$F=\begin{bmatrix} \frac{-1}{M_1} & 0 & 0 & 0 & 0 & 0 & 0 & 0 & 0 & 0 & 0 & 0 \\ 0 & 0 & 0 & 0 & \frac{-1}{M_2} & 0 & 0 & 0 & 0 & 0 & 0 & 0 \\ 0 & 0 & 0 & 0 & 0 & 0 & 0 & 0 & \frac{-1}{M_3} & 0 & 0 & 0 \end{bmatrix}$$

Now we must separate uncertainties and select a set of appropriate weighting functions for each area with a set of (u_i , y_i) as a SISO system in each synthesis. For this example the bandwidths of the three loops are not too different, therefore the order of loop closing is not important. For the problem at hand, we will focus on areas load variation which is narrow but the most important source of uncertainty. A ±30% variation in D_i parameter and $\Delta P_{di} = 0.01$pu (load disturbance) are assumed in each area. In order to modelling the parameter uncertainty as a multiplicative uncertainty, let Goi(s) denote the nominal transfer function from control input to output of area-i and Gi(s) denote the according real model. Then the multiplicative uncertainty can be expressed as

$$\Delta_{ui} = \left|[Gi(s)-Goi(s)]Goi(s)^{-1}\right|. \qquad (9)$$

For example, two sample uncertainties according to $D_{1max} = 1.3D_1$ and $D_{1min} = 0.7D_1$ for area-1 are shown in Fig. 6.

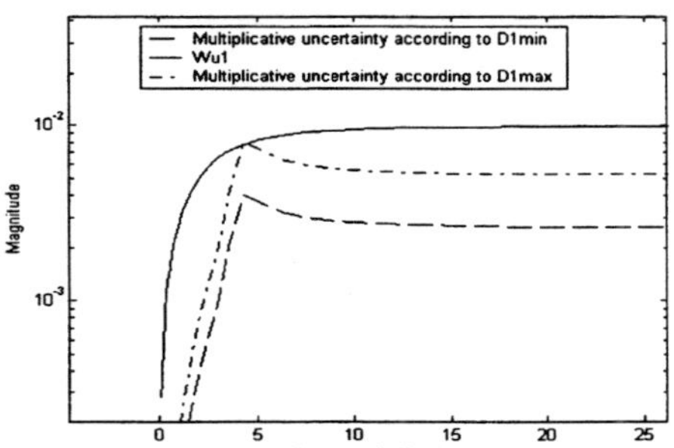

Fig. 6. Multiplicative uncertainty for are-i

In order to reduce the complexity of the controller, we will cover all multiplicative uncertainties due to ±30% variation in D_1 parameter with the following first order weight:

$$w_{u1}(s) = \frac{0.01s}{s+3.5}. \qquad (10)$$

The magnitude Bode plot of this weighting function is shown in Fig. 6. $w_{u2}(s)$ and $w_{u3}(s)$ will be determined, using the same method. W_{P1i} and W_{P2i} in each area are associated with the control effort and tracking/regulation error, respectively. The selection of these weighting functions entails a trade off among different performance requirements, particularly good regulation versus peak control action. As it is referred in previous section an important issue in regard to selection of the weights is the degree to which they can guarantee the satisfaction of design objectives. It will be easy by using μ-analysis, frequently. For the problem at hand a suitable set of weighting functions can be chosen as shown in Table 2.

Table 2 The set of weighting functions for 3-area power system

Area-1	Area-2	Area-3
$w_{u1}(s) = \frac{0.01s}{s+3.5}$	$w_{u2}(s) = \frac{0.04s}{0.9s+2.2}$	$w_{u3}(s) = \frac{0.11s}{1.2s+5}$
$w_{p11}(s) = \frac{0.001s}{10^{-5}s+2.5}$	$w_{p12}(s) = \frac{0.01s}{10^{-4}s+1}$	$w_{p13}(s) = \frac{0.008s}{10^{-6}s+34}$
$w_{p21}(s) = \frac{s+0.1}{13(s+0.001)}$	$w_{p22}(s) = \frac{s+1.05}{60(s+0.05)}$	$w_{p23}(s) = \frac{s+2.2}{33(s+0.002)}$

To reject disturbances W_{P1i} and W_{P2i} are such selected that singular value of sensitivity function be reduced at low frequencies. Further more Table 2 shows that the weight on the control input W_{P1i} chosen close to a differentiator to penalize fast change and large overshoot in the control input. The weight on the error W_{P2i} was chosen close to an integrator at low frequencies in order to get zero steady-state error and good tracking.

Having setup our robust synthesis problem in terms of the standard μ-theory, we used the μ-analysis and synthesis toolbox, to obtain the controllers. A complete set of 3 controllers which meets our objectives are found at the end of fourth D-K iteration. The order of obtained controllers, without model reduction of H∞ controllers during D-K iterations was 16 (K_1), 21 (K_2) and 27 (K_3).The state space realization of the controller has the following form:

$$\begin{aligned} \dot{x}_{Ki} &= A_{Ki}x_{Ki} + B_{Ki}y_i \\ u_{Ki} &= C_{Ki}x_{Ki} + D_{Ki}y_i \end{aligned} \qquad (11)$$

These controllers can be approximated by lower order controllers, using standard model reduction techniques. A balanced realization based on singular values of the observability matrix has been performed (Bevrani, *et al.*, 2002). A vector of the Hankel singular values can be used as an indicator for the acceptable order of result controllers. Using this method the final order of controllers were 2 (K_1), 2 (K_2) and 3(K_3). If we are looking for the best possible performance, the procedure must be continue to retuning obtained controllers; especially the first one that is designed according to more conservative state (Djukanovic, *et al.*, 1999). The re-synthesis of K_1 in this example, gave us a new four order controller.

4. SIMULATION RESULTS

In order to demonstrate the effectiveness of the proposed method, some simulations were carried out. In these simulations, the proposed scenario described in section 2 was applied to the 3-area power system described in section 3. Figures 7 and 8 show the frequency deviation and control action signals in 3 areas, following a step load disturbance of $\Delta P_{d1} = 0.01$pu to area-1.

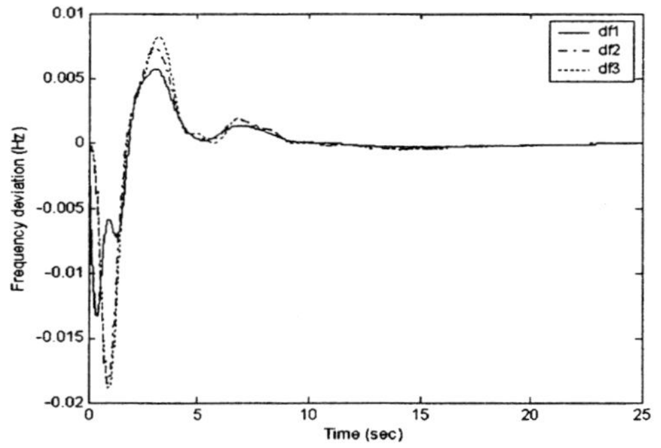

Fig. 7. Frequency deviation following a step load disturbance of $\Delta P_{d1} = 0.01$pu to area-1.

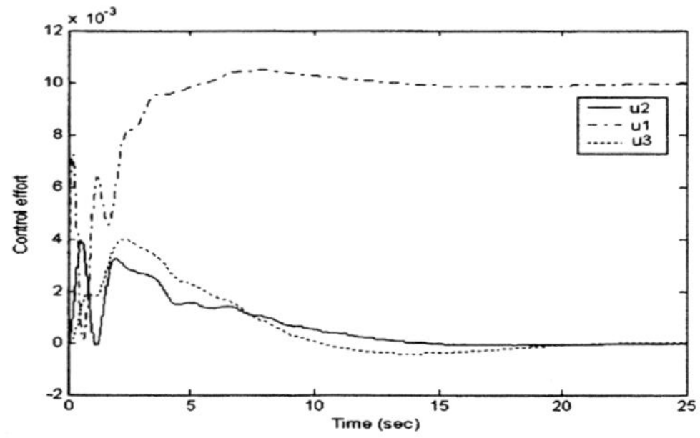

Fig. 8. Control effort signal following a step load disturbance of $\Delta P_{d1} = 0.01$pu to area-1

Fig. 9 shows the frequency deviation following apply a step load disturbance of $\Delta P_{d1} = 0.01$pu to all areas, simultaneously. Finally, Fig. 10 shows the effectiveness of simultaneous 25% increase D_i (i=1,2,3) and a step load disturbance of $\Delta P_{d1} = 0.01$pu to area-1.

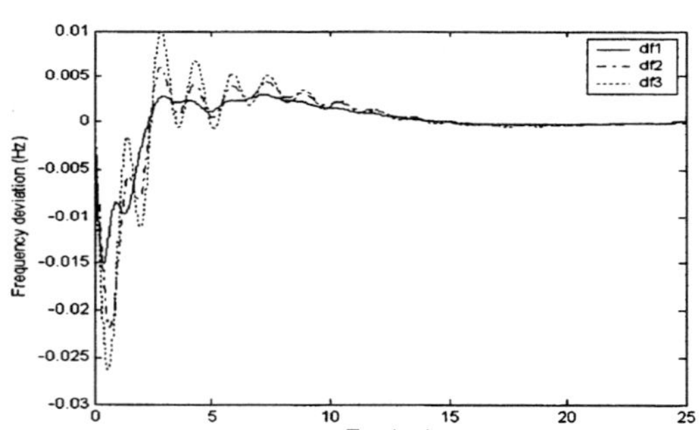

Fig. 9. Frequency deviation following apply a step load disturbance of $\Delta P_{di} = 0.01$pu to all areas.

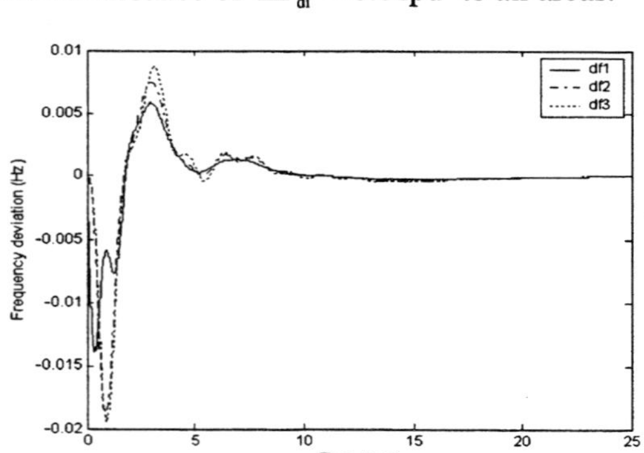

Fig. 10. Frequency deviation following a step load disturbance of $\Delta P_{d1} = 0.01$pu to area-1 and 25% increase in D_i.

5. CONCLUSION

In this paper a new systematic method for robust sequential decentralized load frequency controllers using μ-synthesis in an interconnected multi-area power system has been proposed. At each design step, the information about controllers design in the previous step are taken into account; therefore the method can be less conservative than independent decentralized design and more practical than proposed simultaneous decentralized load frequency controllers. The simulation results demonstrate the effectiveness of the proposed method for solution the LFC problem in presence of uncertainties and disturbances in multi-area power systems.

ACKNOWLEDGMENTS

The authors are grateful to Prof. T. Hiyama from Kumamoto University and Dr. G. Fujita from Shbura institute of technology for their kind discussions.

REFERENCES

Bevrani, H. (1999). Robust load frequency controller in a deregulated environment: A μ-synthesis approach, *Proc. of IEEE Int. Conf. on Control applications*, USA, 616-621.

Bevrani, H., Y. Mitani and K. Tsuji, (2002). Robust low-order load frequency controller in a deregulated environment, *Proc. of 5th Asia-Pacific Conf. on Control and Measurement (APC CM)*, China.

Chiu, M. and Y. Arkun (1992). A methodology for sequential design of robust decentralized control systems, *Automatica*, **28**, 997-1001.

Djukanovic, M., M. Khammash and V. Vittal (1999). Sequential synthesis of structured singular value based decentralised controllers in power systems, *IEEE Tran. On Power Systems*, **14**, 635-641.

Doyel, J.C. (1982). Analysis of feedback systems with structured uncertainties, *IEE Proc., Part D*, **129**, 242-250.

Feliachi, A. (1987). Optimal decentralized load frequency control, *IEEE Tran. On Power System*, **PWRS-2**, 379-384.

Hiyama, T. (1982). Design of decentralised load-frequency regulators for interconnected power systems, *IEE Proc., Part C*, **129**, 17-23.

Hovd, M. and S. Skogestad (1994). Sequential design of decentralised controllers, *Automatica*, **30**, 1610-1607.

Liaw, C.M. and K.H. Chao (1993). On the design of an optimal automatic generation controller for interconnected power systems, *Int. J. Control*, **58**, 113-127.

Niioka, S., R. Yokoyama, T. Ishi, G. Shirai and G. Fujita (1998). Decentralized load frequency control based on H-inf control with observer theory, *Int. Conf. on Electrical Engineering (ICEE'98)*, Korea, 284-287.

Yang, T.C., Z.T. Ding and H. Yu (2002). Decentralised power system load frequency control beyond the limit of diagonal dominance *Electrical power and Energy system*, **24**, 173-184.

ELSEVIER
IFAC
PUBLICATIONS
www.elsevier.com/locate/ifac

MODELLING OF NO_X AND CO EMISSIONS OF A SMALL GAS TURBINE UNIT BASED ON OPERATIONAL DATA AND NEURAL NETWORKS

Francesco Casella, Claudio Maffezzoni

Dipartimento di Elettronica e Informazione - Politecnico di Milano
Piazza Leonardo da Vinci, 32 - 20133 Milano ITALY
E-mail: {casella, maffezzo}@elet.polimi.it

Abstract: Compact models of NO_x and CO emissions of a small gas turbine power plant are discussed, to be used in a plant simulator; the models are based either on semi-empirical correlations or neural networks. Experimental data collected during five months of operation are used to calibrate and validate the models. The results obtained with the different models are shown: semi-empirical correlations fail to accurately describe the observed emissions, while neural-network based models perform well, provided their structure is carefully selected. *Copyright © 2003 IFAC*

Keywords: Power plant modelling, Simulation, Emissions, Neural-network models, NO_x and CO emissions.

1. INTRODUCTION

In recent years, the issue of pollutant emissions such as NO_x and CO from thermal power plants has become more and more important, as environmental regulations impose strict limitations on them. The physical mechanisms which produce these substances are complex, partly unknown and very difficult to model starting from first principles. On the other hand, the availability of compact models, to be used in engineering simulators, is highly desirable.

In this context, research has been carried out on ICARO, a combined heat&power plant installed at ENEA's Casaccia Research Centre (Italy), whose core is a 2 MWe gas turbine unit. Operational data, recorded by the plant SCADA system during five months of operation, have been analysed, in order to obtain reliable models of the plant emission levels.

Initially, semi-empirical models found in the literature have been tested against this data; results, however, were generally quite poor, with only a partial exception. Therefore, the use of black-box models based on neural networks has been investigated, leading to satisfactory results.

The paper is structured as follows: in Section 2 the plant which is the subject of the investigation is briefly described. Section 3 deals with the problem of modeling power plant emissions in general, while in Section 4 the results of applying specific semi-empirical correlation to ICARO test data are illustrated. Models based on neural networks are presented in Section 5, followed by conclusions and further research directions in Section 6.

2. EXPERIMENTAL SETUP

2.1 The plant

The plant considered in this study is a small gas turbine (GT) unit, sketched in Fig. 1; the nominal electrical power produced is 2 MWe. The air from the atmosphere is compressed by a two-stage centrifugal compressor, then fed into the combustion chamber, together with fuel (natural gas) and spray water,

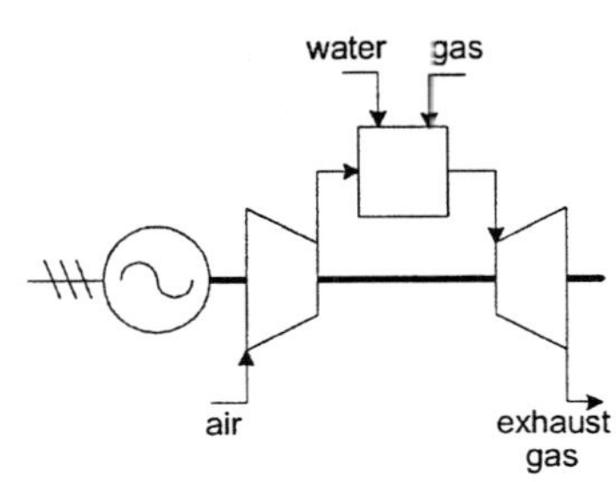

Fig. 1. Plant schematics

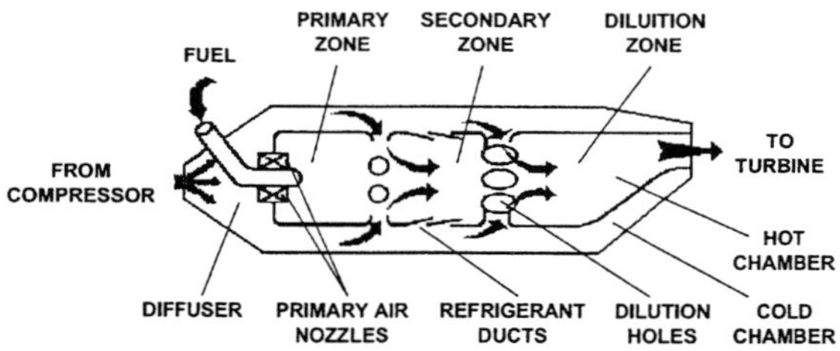

Fig. 2. The combustion chamber

which is used to lower NO_x emissions; the hot gas is then expanded through a two-stage turbine and finally used for water heating in a heat recovery boiler. The control system adjusts the fuel flowrate to achieve the desired electric power production, while the water flowrate is set as a nonlinear function of the measured electric power output. The compressor has a fixed configuration (no IGVs); as a consequence, air pressure and flowrate at the burner inlet show limited variations over the experimental operating range (1177–1351 kPa and 9.95–11 kg/s), with the load changes affecting the outlet temperature (1062–1243 K). A sketch of the combustion chamber is shown in Fig. 2: compressed air enters the cold chamber, from where it enters the hot chamber in three different zones. In the primary zone, sub-stoichiometric combustion takes place; the combustion process is completed in the secondary zone, while further air is introduced in the dilution zone, lowering the temperature to levels compatible with the turbine blades.

The residence time (flowrate/holdup ratio) in the combustion chamber is very short (less than 0.1 s) and much faster than the relevant plant dynamics: for this reason, static models will be investigated.

2.2 Plant simulator

A dynamic plant simulator (Maffezzoni and Aime, 2000) was previously developed. This simulator is based on first-principle thermo-hydraulic models of the plant components (compressor, combustion chamber, turbine), describing pressures, flowrates, temperatures and gas compositions in each control volume. A first attempt was also made at modelling emissions, but the lack of significant experimental data hampered a thorough validation of the model. The model presented in Section 5 has been inserted into this plant simulator. Note that the combustion chamber has been described by a single control volume: further partitioning has proved ineffective in predicting emissions, while being unnecessary for the calculation of the outlet temperature.

2.3 Experimental data

Operating data have been recorded by the plant SCADA system during a time span of 5 months. Atmospheric pressure, temperature, and humidity, compressor discharge pressure and temperature, fuel flowrate and electrical power were recorded every minute, while the average values of $[NO_x]$ and [CO] were recorded every 30 minutes. The air flowrate has been computed from available data using the compressor characteristic; the combustion chamber outlet temperature has been computed by applying basic mass and energy balances and assuming complete fuel combustion.

Around 500 operating points have been selected, according to the following guidelines:

- select intervals corresponding to at least 30 minutes of steady plant operation;
- average all data on 30 minutes intervals, to reduce the effect of noise and random fluctuations;
- select operating points in order to cover the whole operating range (atmospheric conditions and load level) as uniformly as possible.

The emission concentration measurements have been converted to specific production rates (i.e. grams per kilogram of fuel), in order to be consistent with the results in the literature.

The data set has been randomly divided into two parts: the first one (75% of the points) is used to calibrate the models, e.g. by least squares estimation of some parameter, or by backpropagation in the case of the neural networks; the second one (25% of the points) is used as a validation set.

3. MODELING NO_x AND CO EMISSIONS

An overview of the different processes leading to the production of NOx and CO in diffusion flames can be found in Turns (1996). In general, three kinds of models can be used to predict their exhaust concentrations:

1. combustion chamber models based on first-principle equations (3D model of fluid dynamics and reaction kinetics);
2. semi-empirical correlations;
3. black-box models.

Models of type 1. (e.g. Rizk and Mongia, 1993b) are very cumbersome to generate, simulate and calibrate; their elective area of application is the design of combustion chambers. For the purpose of dynamic power plant simulation, models of type 2. and 3. are more appropriate, as they have a much simpler structure and lower computational load. More specifically, the main objective of this research is to reproduce the operation of an existing plant, for which experimental data are available.

4. MODELS BASED ON SEMI-EMPIRICAL CORRELATIONS

Semi-empirical correlations belong to two categories: the first one considers parameters which are inherent to the combustion process, such as the stoichiometric flame temperature or the residence time in the combustion chamber; others instead only employ externally available parameters, such as pressure and flowrates. In general, the predicted concentrations $[NO_x]$ and $[CO]$ (in g/kg of fuel) depend on terms corresponding to the residence time, the reaction rate and on the fuel/air mixing rate.

4.1 Lefebvre model

In 1984, Lefebvre proposed the following correlation (Rizk-Mongia, 1993a), based on experimental work on gas turbines for propulsion:

$$[NO_x] = 0.459 \cdot 10^{-8} \cdot P^{0.25} \cdot F \cdot \tau \cdot \exp(0.01 \cdot T_{st}), \quad (1)$$

where P is the combustion chamber inlet pressure in (kPa), F is the air fraction in the primary zone, τ is the residence time (s) and T_{st} is the stoichiometric flame temperature. In the case under consideration, P is measured variable; τ is easily computed from the air and fuel flowrates and from an estimate of the average temperature in the combustion chamber based on the inlet temperature and on mass and energy balances; T_{st} can be computed from mass and energy balances in case of natural gas fuel, whose chemical composition and heat value are known. Since the actual value of F is not exactly known, and also to better adapt the correlation to the case under examination, a multiplicative constant k has been included in (1), and estimated by least squares on the calibration data ($k = 1.15$).

The results obtained on the validation data-set are shown in Fig. 3: each cross corresponds to a single operating point, its coordinates being the estimated and the measured concentrations in g/kg_{fuel}. It is clear that (1) does not capture the actual formation mechanisms well enough: in fact, most of the estimated values concentrate between 3.7 and 4.0, irrespectively of the actual ones. A possible explanation of this fact is that (1) has been obtained with aero engines, where the variations of pressure and flowrate with the load are much greater than in the case of power generation units.

Lefebvre also proposed a correlation for $[CO]$ concentration:

$$[CO] = 0.333 \cdot 10^{10} \cdot \frac{\exp(-0.00275 T_{pz})}{F \cdot P^{1.5} \cdot F \cdot (\tau - 0.55\tau_{ev}) \cdot (\Delta P/P)^{0.5}} \quad (2)$$

in this case the average temperature in the primary zone T_{pz} has been considered and the residence time must be corrected to take into account the fuel evaporation time τ_{ev}, which is however zero in the case of natural gas. Again a multiplicative constant k was introduced and and estimated via least squares ($k = 22.1$). The results are shown in Fig. 4: also in this case, the correlation cannot reproduce the measured concentrations.

4.2 Rizk-Mongia model I

Rizk and Mongia (1993a) modified Lefebvre's correlations, taking explicitly into account the exponential correlation between the temperature and the reaction rate (as stated by Arrhenius' law), thus proposing the following:

$$[NO_x] = 0.15 \cdot 10^{16} \cdot (\tau - 0.5\tau_{ev})^{0.5} \cdot \frac{\exp(-71000/T_{st})}{P^{0.05} \cdot (\Delta P/P)^{0.5}} \quad (3)$$

$$[CO] = 0.179 \cdot 10^{9} \cdot \frac{\exp(7800/T_{pz})}{P^2 \cdot (\tau - 0.4\tau_{ev}) \cdot (\Delta P/P)^{0.5}} \quad (4)$$

where ΔP is the pressure drop across the combustion chamber. When these correlation are tested on the plant data, the results are comparable to those in Fig. 3 & 4, and thus very far from being satisfactory.

4.3 Rizk-Mongia model II

Another model, proposed by Rizk and Mongia (1993a) starts from the computation of flame temperatures in the primary and secondary zones T_p and T_c. The production of NO_x in the primary zone is given by:

$$[NO_x] = 10^{13} \cdot (P/1.4 \cdot 10^6)^{a(\varphi)} \cdot A(\varphi) \cdot \exp(-71442/T_p) \cdot \tau^{0.64} \quad (5)$$

where $a(\varphi)$ and $A(\varphi)$ are suitably defined functions of the fuel/air equivalence ratio. A similar expression holds for the secondary zone, and the results are then

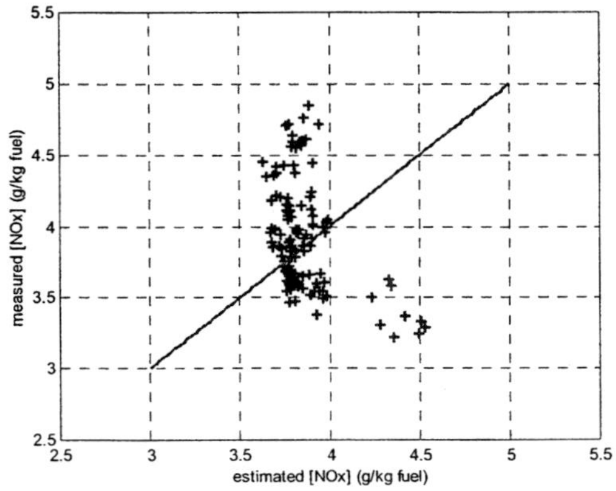

Fig. 3. Results of Lefebvre model (NO_x)

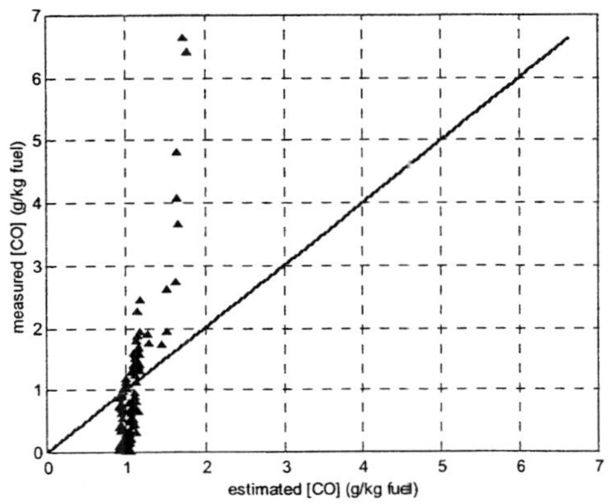

Fig. 4. Results of Lefebvre model (CO)

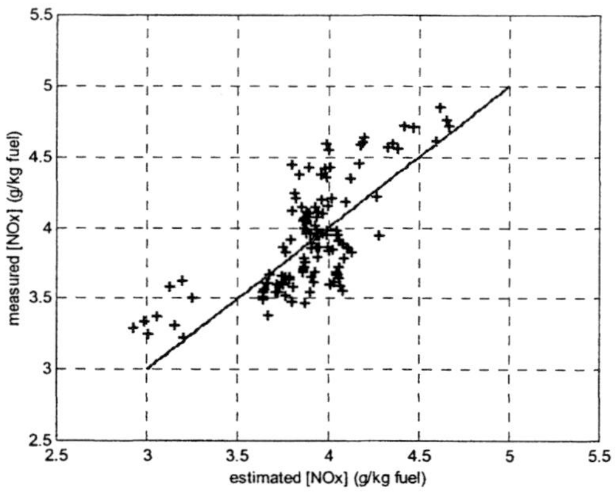

Fig. 5. Results of Rokke model (CO)

added to obtain the total production rate. Also in this case it was not possible to replicate the experimental data consistently.

4.4 Rokke model

This model (Rokke et al, 1993) was obtained with reference to a set of gas turbines for power plants, sized from 1.5 to 35 MW. NO_x concentration was correlated with parameters identifying the operating point, instead of trying to structure the equation according to the physics of the process. The NOx concentration (ppm, in this case) is given by:

$$[NO_x] = 18.1 \cdot P^{1.42} \cdot m_a^{0.3} \cdot \varphi^{0.72} \qquad (6)$$

where P is the combustion chamber pressure (atm), m_a is the air mass flowrate in kg/s, and φ is the fuel/air equivalence ratio. Also in this case, a further multiplicative constant k has been tuned with the calibration data ($k = 1.145$) to better adapt the correlation to the specific plant. The results are shown in Fig. 5: although the accuracy is not very high (the root mean square error, RMSE, is around 7%), this correlation can explain better the full range of measured NO_x concentrations.

5. MODELS BASED ON NEURAL NETWORKS

Since the ability of semi-empirical correlation to explain the experimental data was limited, data-based models were considered. A preliminary analysis showed that the measured emissions could not be accurately explained as linear functions of other macroscopic variables such as pressure, fuel and air flowrate, inlet and outlet temperature. Neural network (NN) models were then adopted. Other applications of NNs to emission modelling in power plants are known (see e.g. Ferretti and Piroddi, 2001, and references therein), even if the specific context is different from the present study.

The NNs considered in this work are standard feedforward networks with a hidden layer (Hertz et al., 1991). The output of each neuron is computed by applying a sigmoidal function to the weighted sum of the inputs plus a bias. The weight matrix W and the bias vector b thus characterise the input-output mapping. Two separate networks are employed to compute $[NO_x]$ and [CO].

The aim of the model is to estimate of the emissions as a function of the smallest possible number of variables local to the combustion chamber. Given the values of the air, fuel and water flowrates w_a, w_f, w_w, the inlet temperature T_i and pressure P, the operating point is completely specified, and thus are the outlet temperature T_o and the values of $[NO_x]$ and [CO]; since the model is algebraic, it can be partially inverted by swapping w_f and T_o. There are thus at least 5 potential inputs for the model; however, it is possible that some subset of them is sufficient to accurately calculate the emissions. Moreover, when the entire GT plant is considered, only three variables can be arbitrarily selected (the atmospheric temperature and pressure, and the power setpoint), all other variables being then fixed by the plant physics or by the control system (relative humidity has a marginal effect during the operating season, characterised by low temperatures). The experimental training data set is then distributed over a 3-dimensional hypersurface. If a NN with more than 3 inputs is trained with this data, it has been found that the performance is severely degraded when the model is used inside the simulator, up to computing completely unreasonable values. This is because the correlation between the inputs in the simulated plant is somewhat different from that of the real plant. For example, $[NO_x]$ is likely to increase with w_f, which is strongly correlated to w_v by the control system; therefore the NN model could ascribe the increase of $[NO_x]$ to the increase of spray water, which is wrong. If the correlation between the two in the simulator is different from that of the training data, this could result in non-physical behaviour of the model. Also, it is clear that even if the burner emission model depends only on local variables, it should be re-trained whenever the entire plant configuration (e.g. the spray water control system, the compressor characteristics, etc.) is changed.

The NN performances with different input combinations have been analysed, and finally w_a, T_i, and T_o were selected as inputs for both $[NO_x]$ and [CO] calculation, since they provide the best performance both in terms of RMSE on the validation data set and in term of robustness when the model is embedded in the global plant simulator.

The input data set is represented on the (w_a, T_o) plane in Fig. 6: the unevenness of the distribution is due to the fact that the plant has never been operated at intermediate load levels. When looking at the (w_a, T_i) plane (Fig. 7), it is apparent that only a limited portion of the subspace is populated by data, since the air flow is correlated to the inlet temperature by the GT characteristics; this is not a problem as long as the NN model in the simulator operates in the same

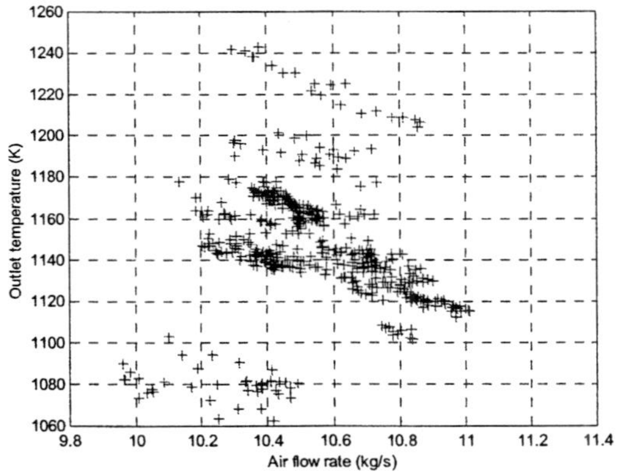

Fig. 6. Experimental data in on the (w_a, T_o) plane

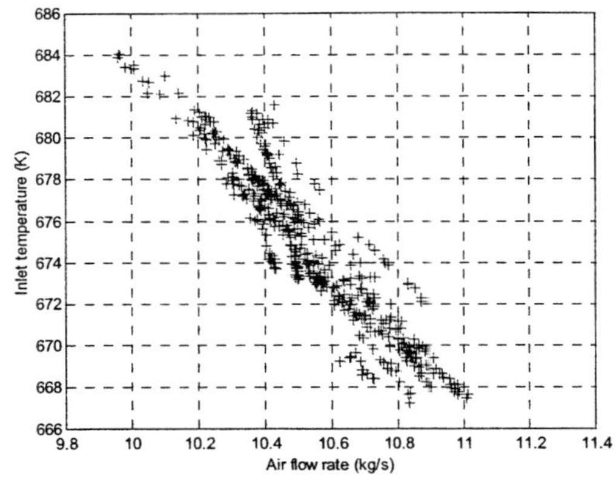

Fig. 7. Experimental data in on the (w_a, T_i) plane

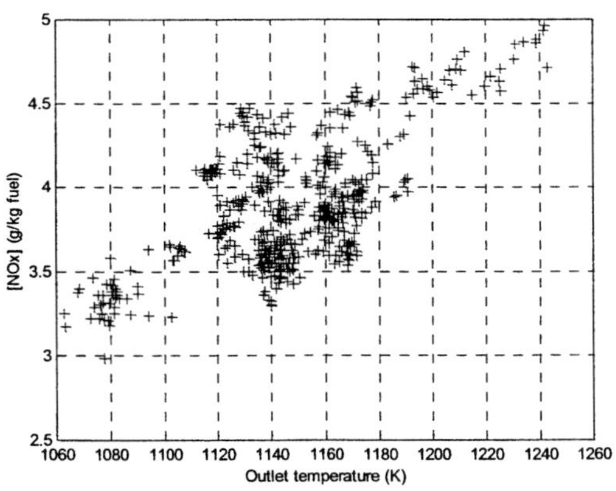

Fig. 8. Experimental data on the (T_o, [NOx]) plane

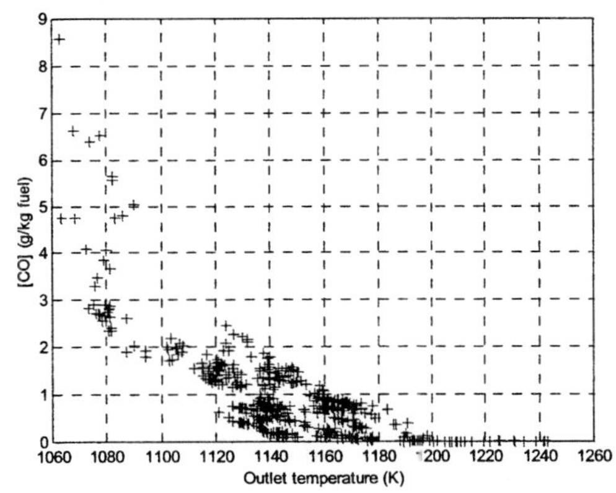

Fig. 9. Experimental data on the (T_o, [CO]) plane

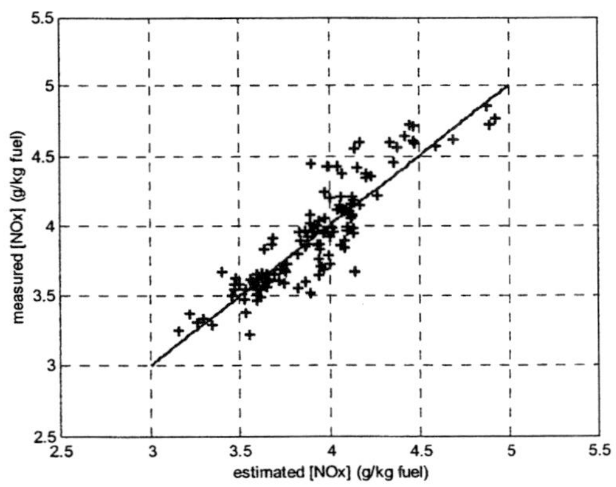

Fig. 10. Results of NN model (NO_x)

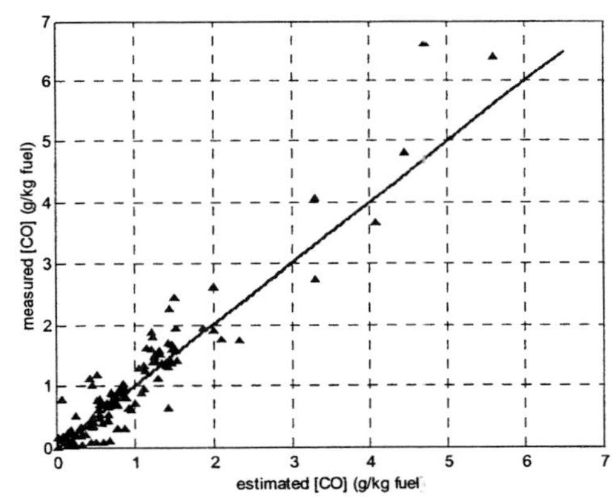

Fig. 11. Results of NN model (CO)

region. If the data is plotted on the (T_o, [NO_x]) plane (Fig. 8) it is clear that NO_x concentration is correlated with the outlet temperature, even if the other two inputs have a substantial effect in discriminating the actual output, particularly in the 1120–1180 K range; the same can be said of [CO] when looking at the (T_o, [CO]) plane (Fig. 9).

The NNs have been trained by a standard backpropagation algorithm on the training data set; the algorithm iterations are stopped at the epoch corresponding to the minimum value of the RMSE on the validation data, to avoid overfitting. It has been observed that the performance varies significantly with the (random) initialisation of weights and biases; therefore, the training phase is repeated many times with different initial conditions, and the weights corresponding to the minimum RMSE are selected.

The number of neurons in the hidden layer has been set to five for both networks; using more neurons slightly improves the RMSE on the validation data set, but degrades the performance when the model is used within the simulator.

The results obtained with the [NO_x] network on the validation data set are plotted in Fig. 10: the RMSE is 0.167 g/kg, corresponding to 4% relative error, which is far better than the results of the best semi-empirical correlation. The results of the [CO] network are shown in Fig. 11: the RMSE is 0.328 g/kg_{fuel} in this case, and the relative error is very large for many

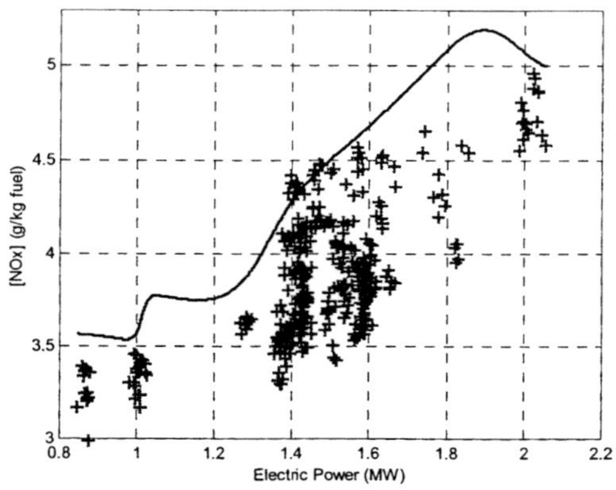

Fig. 12. Simulated NO_x production and training data

operating points. The prediction of CO production is more difficult, probably because it depends on imperfections in the combustion process whose nature and extent are difficult to predict on the sole base of measured variables.

There are many possible causes for the residual uncertainty in both $[NO_x]$ and [CO]: the emission formation process has been modelled as a stationary process over 5 months of operation, but in fact phenomena such as fuel injector fouling or sensor drift could have affected the actual input-output law. Moreover, a constant fuel heat value was assumed in all the computations, but its actual value (which is not measured on-line) may have varied significantly over the set of experimental data points, thus affecting the actual emission output.

It is interesting to understand how the emission model behaves once inserted in the plant simulator. To this purpose, a ramp load change from 40% to 100% of the nominal value has been simulated: the resulting $[NO_x]$ curve as a function of the load has been plotted in Fig. 12, together with the training data set plotted on the same axes. Note that most of the operating points which make up this curve, which correspond to particular atmospheric conditions, do not correspond to any point in the training data set. The typical shape of this curve in commercial GT power plants (Pavri and Moore, 2001) is smooth and monotonically increasing; in our case, this is not true in the intervals 1.0–1.3 and 1.8–2.0 MWe. However, it is apparent that these intervals are void of training data; it should also be noted that the absolute deviation from an ideal smooth curve is less than 0.2 g/kg_{fuel}, i.e. comparable to the network RMSE. In order to avoid such kind of undesirable local behaviour, a more evenly distributed data set would be welcomed: this in turn would require the execution of ad-hoc experimental campaigns on the plant, which unfortunately was not possible in this case.

6. CONCLUSIONS

The problem of obtaining compact models of NO_x and CO emissions of a small GT power plant, for which operational data was available, was considered in this paper. Semi-empirical correlations found in the literature did not lead to satisfactory results. Black box models based on neural networks were then adopted, with good performance. The model has then been used inside a previously developed plant simulator: this required special care in the selection of the model structure and complexity. The proposed approach can be easily applied to any GT power plant, provided that a sufficiently numerous and evenly distributed experimental data set is available. The modelling process is based on the assumption that the emission formation mechanism is repeatable, which could be wrong due to the deterioration of burner and instrumentation performance. This issue should be specifically addressed with ad-hoc experiments to increase the accuracy and reliability of the model.

Another possible application for this kind of models, besides simulation, is their use in virtual sensors, which are particularly interesting as a back-up to temporarily replace off-duty instrumentation, in order to avoid costly plant shut-downs.

ACKNOWLEDGEMENTS

This paper was supported by Centro di Ricerche ENEA, Casaccia (Italy). The authors thank former student Andrea Pontiggia who worked on this problem for his master's degree project.

REFERENCES

Ferretti, G. and L. Piroddi (2001). Estimation of NO_x emissions in thermal power plants using neural networks. *J. of Engineering for Gas Turbines and Power,* **123**, n.2, pp.465-471.

Hertz, J., A. Krogh and R.G. Palmer (1991). *Introduction to the theory of neural computation.* Addison-Wesley.

Maffezzoni, C. and Aime M. L. (2000). Compact dynamic modelling of gas turbine including NO_x emissions. *Proceedings of the IFAC Workshop on Power Plants and Power Systems, Bruxelles.*

Pavri, R. and G.D. Moore (2001). *Gas Turbine Emissions and Control.* GE Energy Services, Atlanta GA.

Rizk, N.K. and H.C. Mongia (1993a). Semianalitical correlations for NO_x, CO and UHC emissions. *Journal of Engineering for Gas Turbines and Power,* **115**, 612-619.

Rizk, N.K. and H.C. Mongia (1993b). Three-dimensional gas turbine combustor emissions modeling. *Journal of Engineering for Gas Turbines and Power,* **115**, 603-611.

Rokke N.A., J.E. Hustad, Berg, S. (1993). Pollutant emissions from gas fired turbine engines in off-shore practice. *ASME Paper 93-GT-170.*

Turns, S. R. (1996). *An Introduction to Combustion: Concepts and Applications.* McGraw-Hill New York.

ELSEVIER
IFAC
PUBLICATIONS
www.elsevier.com/locate/ifac

ANALYSIS ON THE STATISTICAL DIFFUSION OF ELECTRIC MOTORS BASED ON THE CUSTOMERS' END-USE ENERGY SURVEY

Seungchan Chang*, Insoo Kim*, Hag-Sig Son*, Jung-Hoon Kim**

** Dept. of Demand Side Management, The Korea Energy Management Corporation (KEMCO)*
*** Dept. of Electrical Engineering, Hong-Ik University, Seoul, Korea*

Abstract: This paper proposes an analysis technique on the customers' electricity usage of electric motor based on the end-use statistical energy survey. Diffusion of electric motor in the population is estimated using the sample survey results on the customer's installed motor specification and the billing data of electric power consumption. Performing more effective survey and processing end-use field data, internet database software for managing electric motor is developed and used for this survey. This paper describes the statistical sample design for motor survey, development of internet database software for managing electric motor named InterMotorManager V1.0, major results from the motor survey by the categories of electricity usage and the prediction of electric motor diffusion status within the population. *Copyright © 2003 IFAC*

Keywords: Electric Motor, Electric Power Consumption, Demand Side Management, Statistical Energy Survey, Database Management System

1. INTRODUCTION

Electric motors have been widely used for getting a mechanical power, rotational or linear torque in diverse industrial and building areas. Energy consumption due to motor-driven processes accounts for about 54 percent of total annual electrical energy use in Korea, but their specific applications can not be summarized to public on the statistical perspective. In relation to electricity Demand Side Management (DSM), electric motor can be potential candidate for minimizing energy consumption and performing effective load management due to its large share in total electricity usage.

Recently standard induction motor is urged to be replaced by high efficient induction motor which can provide economical benefits to the customers by the reduction of energy use. Energy management is characterized as an ongoing process or continuing effort marked by gradual changes towards an energy efficient outcome. To do this, energy survey or audit is emerged as the first step to achieve an effective performance. The survey results can show where and how energy is being used or wasted. For the in-plant survey, an inventory of energy using equipment should be prepared, showing basic energy use data (usually obtained from equipment nameplates) and indicating typical running time and operating profiles. Without the basic information, it is impossible to tell whether equipment is operating unnecessarily or wastefully, to set standards, or to measure the performance of either an individual piece of equipment, processing line, or plant overall.

Customers' end-use survey should precede all the policy measure related with DSM securing for more cost-effectiveness in the point of public business. The survey procedure, a basic process for getting information on the interested situations, will encompass the management of objective end-use for DSM such as, gathering the field data, monitoring the end-use profiles, constructing the database.

This paper proposes an analysis technique on the customers' electricity usage of electric motor based on the end-use energy survey. Diffusion of electric motor in the population is estimated using the sample survey results on the customer's billing data of power consumption and the installed motor specification. Performing more effective survey and processing end-use field data, internet database software for managing electric motor is developed and used for this survey. This paper describes the statistical sample design for motor survey, development of internet database software for managing electric motor named InterMotorManager V1.0, major results from the sample survey by the categories of electricity usage and the prediction of electric motor diffusion status within the population.

2. THE SAMPLE DESIGN FOR CUSTOMERS' MOTOR ENERGY SURVEY

2.1 Sample Survey Design

The motor energy survey has a finite statistical distribution because the population's total energy consumption can be achieved from the billing data. Diffusion of electric motor in the population can be estimated using the customer's billing data of power consumption and the installed motor specification in the sample. It is mainly because motor energy consumption must be included into the total electricity consumption of customer's billing data.

In the statistical sample design, stratified random sampling is used for selecting the sample customers of electric power. In this paper, two-step stratification is prepared. Utilities' service type and customers' electric power capacity are designated as the main strata. For the sample allocation for the survey on electric power customer, Neyman allocation method is used. In the sample design, the annual energy consumption which is categorized by electricity service type, customer's power capacity, region, electricity usage is used for a basis in comparing the statistical distribution.

Table 1. Strata for the random sampling of customer

Strata a	Strata b
a0 residential (R)	b0 ~ 499kW
a1 general A (G1)	b1 500kW~ 999kW
a2 general B (G2)	b2 1,000kW~4,999kW
a3 educational (E)	b3 5,000kW~9,999kW
a4 industrial A (I1)	b4 10,000kW~
a5 industrial B (I2)	
a6 industrial C (I3)	
a7 agricultural (A)	

Sample allocation for the utilities' service type

$$n_a = n \cdot \frac{N_a S_a}{\sum N_a S_a} \quad (1)$$

where, n_a : No. of the sample customer in strata a

N_a : No. of the population in strata a

S_a : Standard deviation of strata a

Sample allocation for the customers' power capacity

$$n_{ab} = n_a \cdot \frac{N_{ab} S_{ab}}{\sum N_{ab} S_{ab}} \quad (2)$$

2.2. Data Collection Instruments (DCI)

The DCI for motor survey is prepared both paper and internet version. It includes general information of customer, monthly power consumption, installed number and capacity of motor and operational record data for each motor.

General Information of Customer
Sample No., Affiliations, Addresses and contact points, Business type, Main products

Electric Power Consumption Status
Utilities' service type, Contracted power capacity (kW), Monthly billed energy consumption (kWh), coincidence with power system peaks

Electric Motor Installation Status
Number and capacity of installed motor classified by 3-phase induction, single phase induction, high efficient 3-phase induction, DC motor, servo and others

Motor Management Records
Installation profiles: Installed process, name of motor, installed date, motor rewinding, motor type
Nameplate profiles: Capacity (kW), manufacturer, model, voltage rating, pole, enclosure type, insulation rating, efficiency (full, 75%, 50%, 25% load)
Operation profiles: Connected load (fan, pump, compressor, extruder, conveyor, crusher, grinder, blender, planner), axis joint, operation time, regulating load factor, install of adjust speed drive
Measurement profiles: voltage, current, power (kW), power factor, RPM, voltage unbalance, torque (kgfm)

3. ESTIMATION OF MOTOR DIFFUSION

Electric motor diffusion within the population can be estimated by comparing the historical power statistics between the population and the sample. If the sample survey is terminated, motor diffusion status can be obtained using the motor diffusion rate (m_i) with i-typed motor of each stratified domain inside the population.

$$m_i = \frac{\sum\sum(m_{iab} \cdot N_{ab})}{\sum\sum N_{ab}} \quad (3)$$

$$m_{iab} = \gamma \cdot (\sum\sum c_{iab} / \sum\sum n_{ab}) \quad (4)$$

$$\gamma_{ab} = \frac{(n_{ab} / N_{ab})}{(kWh_{n_{ab}} / kWh_{N_{ab}})} \quad (5)$$

where,
a : Utilities' service type index
b : Customer's power capacity index
γ : Scale factor for amending statistical deflection
c_{iab} : No. of i-typed motor within sector a and b
n_{ab} : No. of the sample customers
N_{ab} : No. of the population customers

Computation procedure for estimating electric motor diffusion in the population can be summarized as follows.

1. Compute i-typed motor diffusion rate (or intensity) of each stratified domain from the sample results: $m_{iab} = \sum\sum c_{iab} / \sum\sum n_{ab}$
2. Scale factor computation of each stratified domain by comparing kWh consumed and contracted power between the sample and the population
3. Update motor diffusion intensity with computed scale factor: $\tilde{m}_{iab} = \gamma_{ab} \cdot m_{ab}$
4. Makeup the diffusion intensity of domain a and b escaped in the sample
5. Compute diffused motors within the population

$$C_{iab} = \tilde{m}_{iab} \cdot N_{ab}$$

6. Estimation of total motor diffusion by the region or the usage type (C_{iy}) using the similar step of 1 and 5
7. If $| \sum\sum C_{iab} - \sum C_{iy} | < \varepsilon$, then termination

4. DEVELOPMENT OF INTERNET MOTOR MANAGEMENT S/W: InterMotorManager V1.0

For the effective motor energy survey, internet S/W is developed that consists of users' homepage for local motor management and the master's program for database management. It provides the customer who joined the survey with motor managing tool such as, motor specifications from the manufacturer, electricity rate information, calculation of consumed motor energy, torque, electricity bill when operated with each motor. For the electricity users and motor manufacturers, internet homepage is prepared in the website URL http://www.motorsurvey.or.kr.

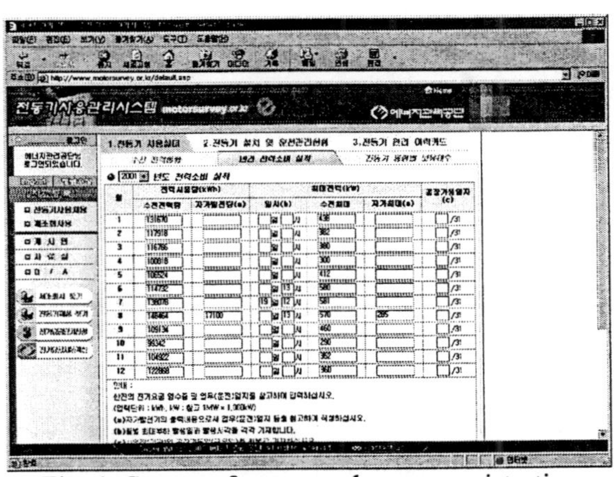

Fig. 1. Screen of consumed power registration

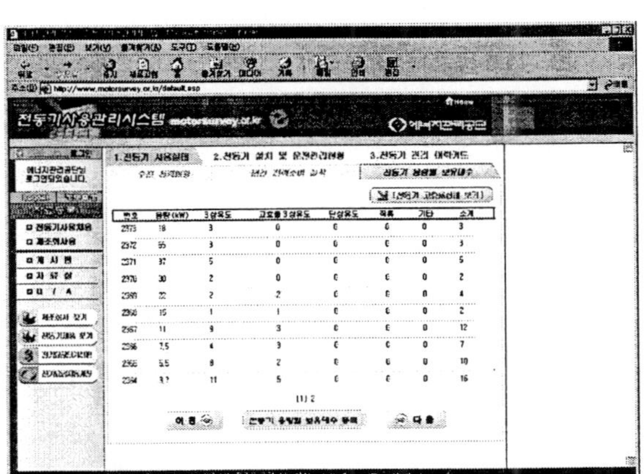

Fig. 2. Screen of customer's motor registration

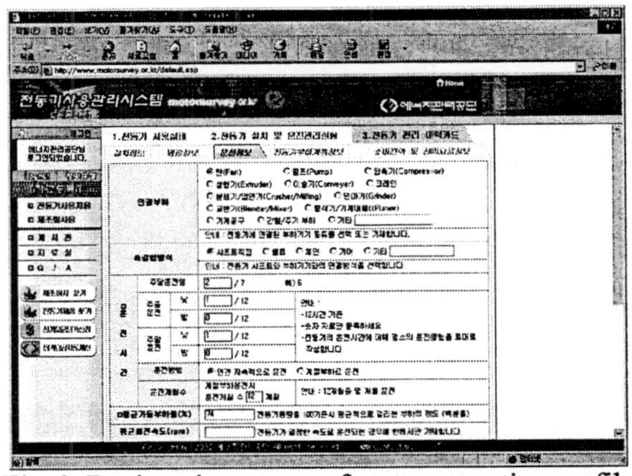

Fig. 3. Registration screen of motor operation profile

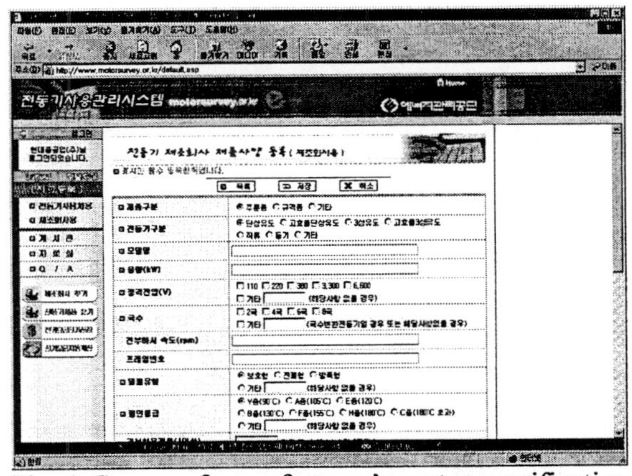

Fig. 4. Screen of manufacturer's motor specification

Electric motor distributions and their energy consumptions can be computed from the sample survey results supplied by the master program based on the electric power statistics in the population.

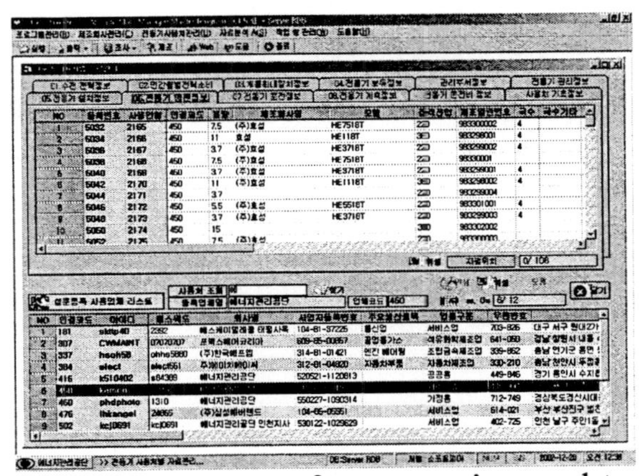

Fig. 5. Master program for customers' survey data

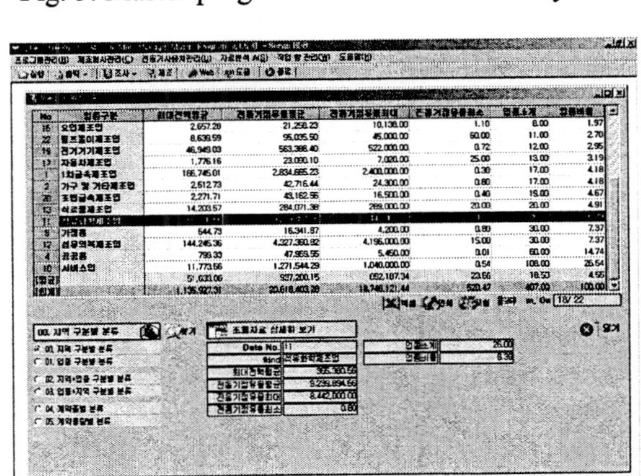

Fig. 6. Analysis screen of motor diffusion

5. CASE STUDY

For the sample survey, three-step investigation of industry and building via post mailing or internet registering, site visiting, and measuring on the installed electric motors was prepared from June to Nov. in 2002. Sample design was performed by the EQ(1) and EQ(2) with consideration of the distribution by the region or the usage type.

Table 2. Results of the sample design for the survey

	a1	a2	a3	a4	a5	a6	a7	Sum
b1	291	71	123	124	88	10	92	799
b2	1,071	222	118	42	105	1,686	19	3,263
b3	1	113	26	3	22	420	1	586
b4		19	2	2	12	317		352
Total	1,363	425	269	171	227	2,433	112	5,000

Replied customers were 780 sites during the survey periods and their annual energy consumptions (on average as of 2001), monthly peak loads (on average) and coincidence of motor demand with the power system peaks summarized 17GWh, 3,875kW and 58.4%, respectively. Electric motor installations within the sample amount to 223,294 sets and their capacity reaches 3GW. Major statistics of motor record from the sample is summarized below.

Installed year on average: 7.7 (motor 5,738 sets)
Enclosure Type: TEFC 3,510 sets occupy 61%, ODP and, Explosion-Proof installed in orders (motor 5,738 sets)
Connected Apparatus: Pump 3,496 sets and Fan 2,366 sets take up 79% and then Compressor, Conveyor, Crane and, Mixer connected in orders (motor 7,462 sets)
Availability (Operation on average): 3,003 hours/ year (motor 8,182 sets)
Load factor for the rating on average: 57.9% (motor 8,179 sets), 73.3%(by the measured data)

Estimating the diffusion of electric motor is applied to the electric power customers in Korea. In Korea, KEPCO supplies electricity and so the population set to KEPCO's customers. As of 2001, electric power statistics described 14,546 thousand customers consumed 239,834GWh annually which has a contracted power as 147.9GW.

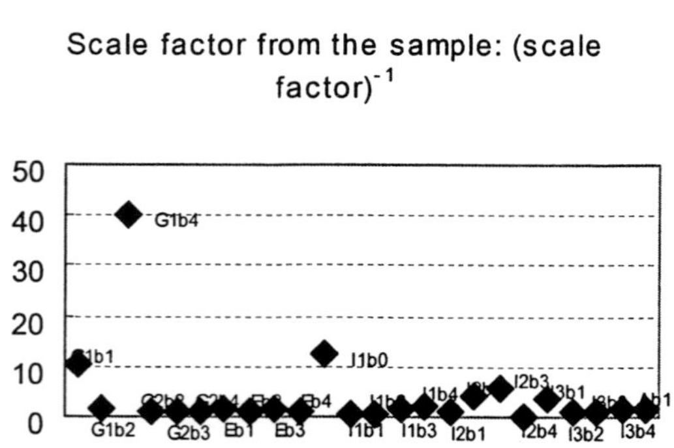

Fig. 7. Scale factor distributions from the sample results

Fig. 7 shows the plots of scale factor inverse. The stratified domain which has large deflections from the sample results when compared to kWh consumption in the population is G1b4, G1b1 and I1b0 in orders. It means that this domain requires more compensation due to lack of surveyed data. In this figure, the domain not plotted is null surveyed data.

In computing the diffusion intensity of domain *a* and *b* that is escaped in the sample, G1b0 sets up from the linear trends of b0~b4 in G1 and G2. Rb0 is applied by 0.01 point of G1b0 and Rb1~Rb4 is applied by 0.8 point of G1b1~G1b4. Eb0 sets up Rb0 and Ab0, Ab2~Ab4 is applied by 0.5 point of related section of G1.

From the estimation results of motor diffusion as of 2001, total installed motors reach 8,446 thousands sets and 76,792MW capacity might be widespread within South Korea. In the point of motor type as shown in Fig. 8, 3-phase induction occupies 95% and then single-phase induction, DC, servo and others are diffused in orders. High-efficient 3-phase induction (for Low Voltage) only possesses 1~1.5% of the total diffusion amounts.

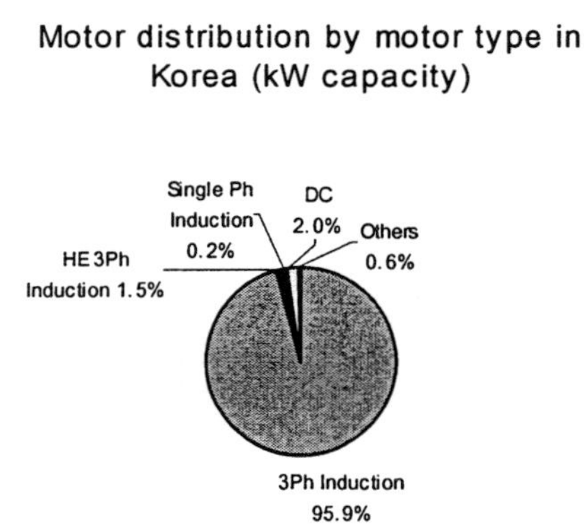

Fig. 8. Motor type distribution inside the population

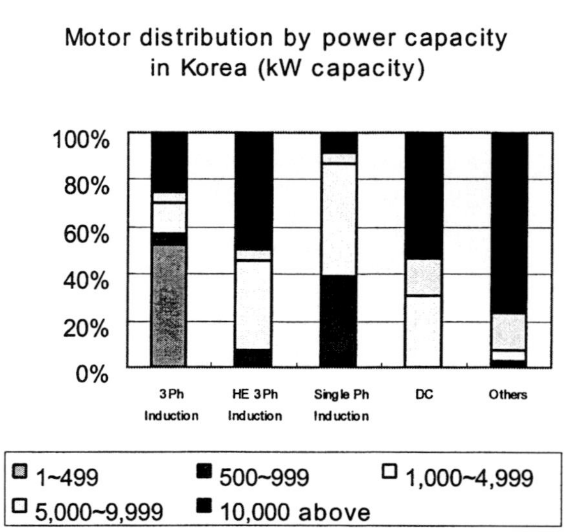

Fig. 9. Motor distribution by the customer's power capacity in the population

In the point of KEPCO's electricity service type as shown in Fig. 9 that represents strata a, industrial has 52%(69% by motor capacity) and, general for commercial and office building has 36%(23% by capacity). Residential-educational-agricultural reaches 12% (8% by capacity). In the point of customer's power capacity shown in Fig. 10, below 500kW shares 62%(50% by motor capacity) and, more than 10,000kW shares 13%(26% by motor capacity). 1,000□4,999kW shares 13% (14% by motor capacity) of the total diffusions.

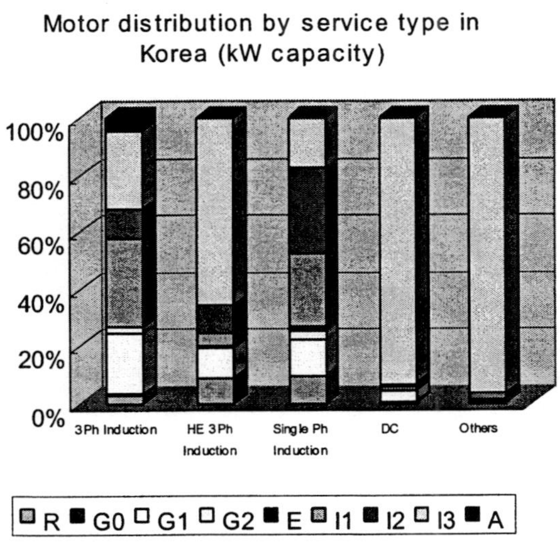

Fig. 10. Motor distribution by the utilities' service type in the population

6. CONCLUSIONS

This paper proposes estimating the diffusion of electric motor from the statistical energy survey. For the effective motor survey and end-use energy management, InterMotorManager V1.0 was developed. Through the development of internet motor management S/W, technical assistance might be possible for internet users on the variety of installed motors and their energy management.

Case studies are performed with the electric power customers inside Korea and we can get the proper information on executing electricity demand side management on electric motor. From the result of this paper, it is great helpful to preparing the frameworks for promoting electric motor high efficiency programs in the perspective of electric power DSM based on the field investigation results of electric motor diffusion.

ACKNOWLEDGEMENTS

The authors specially thank for the cooperation of Korean field workers in the industrial and building sectors who actively participated this energy survey.

REFERENCES

KEPCO (2002), *Electric Power Statistics*, Korea Electric Power Corporation.

KEPCO (2001), *Electric Power Statistics*, Korea Electric Power Corporation.

S. Chang et al.(2002), Database Framework for the Best Practices of Field-oriented DSM Activities in Korea, *ICEE 2002*, **Vol. □**, pp 498~502.

S. Chang(2002), Demand Side Management of Electric Power: The Korean Experiences, International Conference on Power Market, OPCOM Romania.

KEMCO(2002), *Energy Efficiency Statistics*, Korea Energy Management Corporation.

KEMCO(2002), *Guidance on Choosing the Energy Efficiency Products*, Korea Energy Management Corporation.

DOE(2000), *MotorMaster+3.0: User's Guide*, DOE USA.

Seungchan Chang, Sung-Wook Hwang, Hyung-Joon Cho, Balho Kim, Jung-Hoon Kim(1999), Analysis on the Power System Reliability and Production Cost Considering the High-Efficient End-Use Diffusions, *ICEE'99 Proceeding*, **Vol. 3**, pp 394□397.

Chang Seungchan, Kim Jung-Hoon, Hwang Sung-Wook, Kim Balho, Kim Chang-Seob(1998), Development of Demand Side Management (DSM) Monitoring Methodology Considering the Diffusion Process of High Efficient End-Use Appliances, *ICEE'98 Proceeding*, **Vol. 1**, pp 115□118.

Bill Howe et al.(1996), *Drive Power*, E Source Technology.

ELSEVIER
IFAC
PUBLICATIONS
www.elsevier.com/locate/ifac

THE FLUCTUATING REDUCTION METHOD FOR GENERATION POWER OF WIND-PV HYBRID SYSTEM

Jin-Seok Oh*, Sang-Tae Lee*, Seng-Hwan Jun*, Byeong-Duck Ye*, Ji-Yeong Lee**

* *Korea Maritime University*
** *Graduate school of Korea Maritime University*

Abstract: This paper reports the performance of a CB(Circuit Breaker) and converter for Wind-PV (Photovoltaic) system with battery. For this purpose, a fluctuating reduction controller for an electric generation hybrid(wind+PV) system is suggested. The system operates a wind turbine, PV, CB, converter and a battery. Integration of wind and PV sources, which are generally complementary, usually reduce the capacity of the battery. Also, CB controls the overvoltage of the generation system. The objective is to control the operation of the converter and the CB for reducing the power fluctuation. This paper includes discussion on system performance, power quality, fluctuation and effect of the randomness of the wind. *Copyright © 2003 IFAC*

Keywords: Circuit Breaker, Converter, Hybrid, Fluctuation, Potovoltaic, Wind.

1. INTRODUCTION

Wind energy conversion system is based upon the well-known DC link Converter. The wind generator provides current at a variable frequency. This current is rectified onto the DC link using a converter with six diodes.
Generally, the wind turbine generation system has to use a AC-DC converters for an uninterruptible power supply. Fig. 1 shows the schematic diagram of the proposed system for the wind turbine generation which consist of a converter, CB and battery.
The proposed power conversion scheme has a variety pattern of applications in the hybrid generation systems.

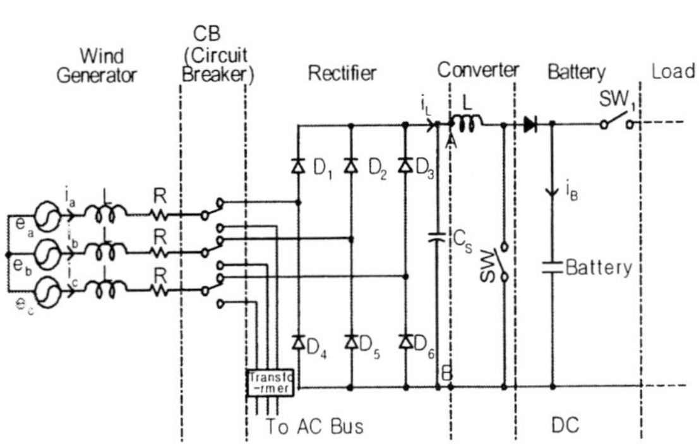

Fig. 1. Schematic diagram of the proposed converter system.

It is also function as an AC to DC converter thus transferring power from the hybrid generator to the battery while the generator supplies power to the converter.
This paper describes the CB control algorithm for overvoltage control and AC-DC converter algorithm for boosting the generation power as booster for generation power applications. The algorithm is developed from an understanding of the energy flow within the system. The purposed scheme can also function as a battery charger and a power supply without any additional power circuit components.
The proposed AC-DC converter with the network is addressed using case studies of voltage fluctuation and harmonics elimination.
In this paper, the impacts on voltage fluctuation can be minimized and furthermore, the network voltage control could also be improved by the proposed operating algorithm.

2. BASIC STRUCTURE AND OPERATING PRINCIPLE

The wind generator provides voltage and current at a frequency directly correlates to the operating speed of the wind turbine.
A converter, typically containing six devices, rectifies voltage and current onto the DC link. A capacitor bank provides filtering, resulting in a voltage stiff DC link characteristic.
The proposed power conversion system converts energy from a variable frequency wind-power to a fixed frequency output as shown in Fig. 1. The system consists of a wind generator, rectifier, converter and CB. These elements discussed below.

2.1 Wind generator

The wind generator is linked to the battery through CB and AC-DC converter. Fig.1 presents wind generating power load to the bus or AC-DC converter according to CB condition. The power captured by the wind turbine may be written as the equation 1.

$$P_{wt} = 0.5\rho A C_P V^3 \quad (1)$$

where, ρ is air density (kg/m³), A is swept area(m²), C_P is coefficient of wind turbine, V is wind velocity(m/s). The torque produced by the wind generator can be written as follows:

$$T_{we} = \frac{3E_{ws}\Phi_{wm}}{2\omega_{wm}L_{ws}}\sqrt{1-\left(\frac{\omega_{we(\lim)}}{\omega_{we}}\right)^2} \quad (2)$$

where, P is number of poles, ω_{we} is (P/2) · ω_{wm}, ω_{wm} is angular shaft speed, Φ_{wm} is flux linked by the stator windings. It is interesting to note that for a given constant voltage E_{ws}, whose value can be obtained through the equation (3) in the terminals of wind generator.

$$E_{ws} = \omega_{we(\lim)} \cdot \Phi_{wm} \quad (3)$$

The rotor speed can be described in the following equation :

$$\frac{d\omega_{wm}}{dt} = \frac{1}{J}\int (T_{wa} - T_{wL})dt \quad (4)$$

Where, J is the inertia, T_{wa} is aerodynamic torque captured from the wind and T_{wL} is the electric torque load. The rotor speed acceleration and deceleration can be described as equation (4). The rate of rotor speed is proportional to the inverse of the inertia and difference between T_{wa} from T_{wL}. The aerodynamic torque is affected by the operating C_P. The voltage fluctuation problem is closer to a steady state problem such as load varying, which is well defined by the real and reactive power distribution.

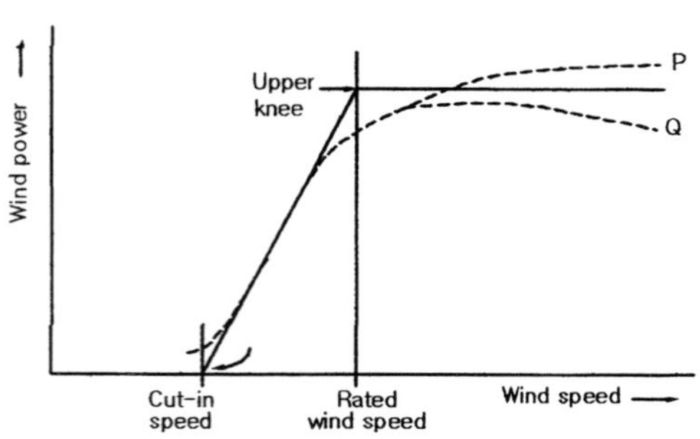

Fig. 2. Relationship between power and wind speed

To understand the effect of fluctuating winds more precisely, consider the theoretical (power/wind speed) curves, for steady winds, given in full line in Fig. 2.

As the wind speed fluctuates about some mean value the power will fluctuate also so that the operating point moves, at varying rates, backwards and forwards along the curve.

When the mean wind speed is near either cut-in speed or the rated wind speed, the true operating point will lie off the curve. Near cut-in it will lie above the curve and near the rated wind speed it will lie below.

The power curve in actual operation will thus be a distortion of the theoretical curve. The amount of the distortion depends upon the amplitude of the wind speed fluctuations. The actual power curve for wind speeds above the rated value may thus lie between the limits indicated by the P-curve and Q-curve in Fig. 2.

The wind speed and power curves are shown in Fig. 3. Fig. 3 shows that the generating electric power is according to the wind speed. It is also noted that the varying wind power can result in bus voltage fluctuation.

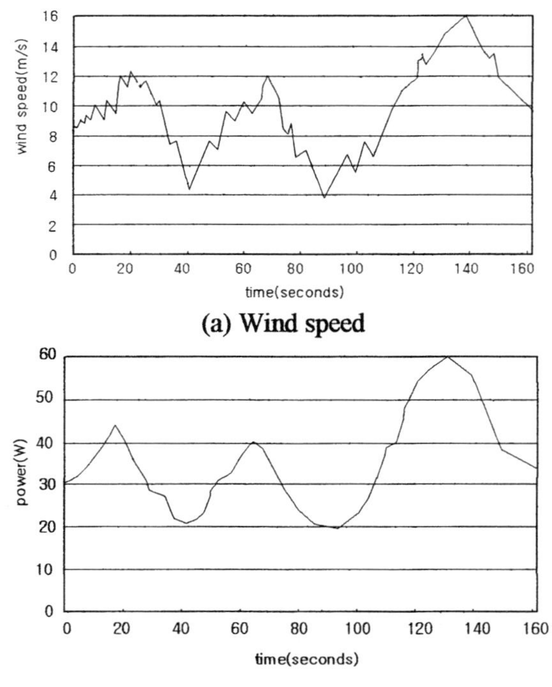

Fig. 3 Wind speed (a) and Output power(b)

2.2 The PV modules

The PV module manufacturer rating is 60W. The PV module power can be calculated by equation (5)

$$P = I_{ph}V_p - I_oV_p \exp\left(\frac{q}{KT}V_p - I_o\right) \quad (5)$$

Where I_{ph} is the cell photo current, V_P is the PV output voltage, I_o is the saturation current, q is the charge of electron, K is Boltzman's constant and T is the cell temperature. The relation between V_P and derivative (dP/dV_p) of output power wit

output voltage can be expressed as

$$\frac{dP}{dV_p} = I_{ph} - I_o \exp(\frac{q}{KT}V_p)[\exp(-I_o) + \frac{q}{KT}V_p] \quad (6)$$

Fig. 4 shows the characteristic curves of equations (5) and (6), and their relationship.

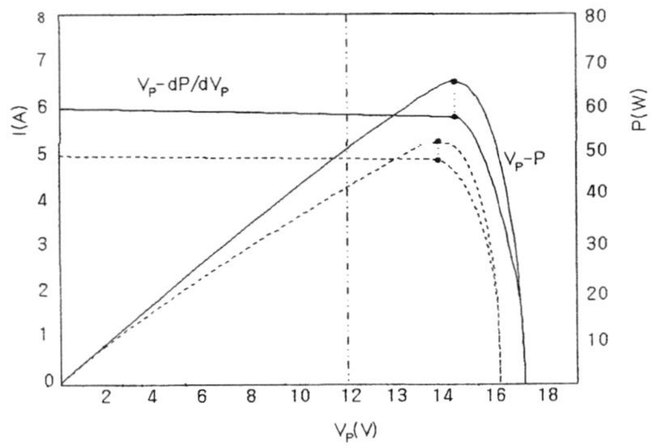

Fig. 4. Characteristic of PV module

It is shown that the MPP(Maximum Power Point) must correspond to such a point where the dP/dV_p is equal to zero. The PV module output voltage can be adjusted by this concept. In this paper, the MPPT is designed by MPC algorithm with $dP/dV_p = 0$.

2.3 Converter

The converter regulates the DC output of wind generator-rectifier unit and PV by switching ratio controller as shown in Fig. 5.

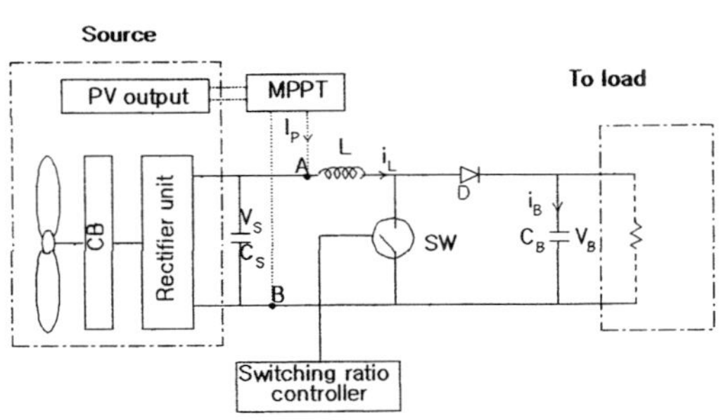

Fig. 5. Block diagram of the generator and converter

We will suppose that the system switches from one state to another. Its equivalent circuits for on-state and off-state are given in Fig. 6.

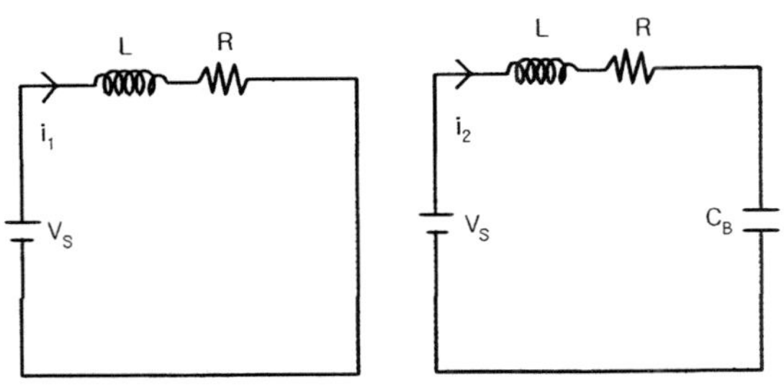

(a) on-state (b) off-state

Fig. 6 Equivalent circuits for the two states

In Fig. 6, the currents i_1 and i_2 are given by

$$i_1 = \frac{V_s}{R}[1 - \exp(-t/\tau)] + I_1 \exp(-t/\tau) \quad (7)$$

$$i_2 = \frac{V_s - V_B}{R}[1 - \exp(-t/\tau)] + I_2 \exp(-t/\tau) \quad (8)$$

Where τ = L / R is the time constant, R is the resistance, V_s is the generating voltage and V_B is the battery voltage.

Equation (7) and (8) are valid for T = $t_{on} + t_d$ for continuous current and $t_{on} + t_d < T$ for discontinuous current. T is the switching period and t_d is the time at which the current becomes zero. From equation (8), t_d is as follow;

$$t_d = \tau \cdot \ln[\frac{1 - A \cdot \exp(-t_{on}/\tau)}{1 - A}] \quad (9)$$

Where A = V_s / V_B. The relation of t_d/τ can be shown Fig. 7.

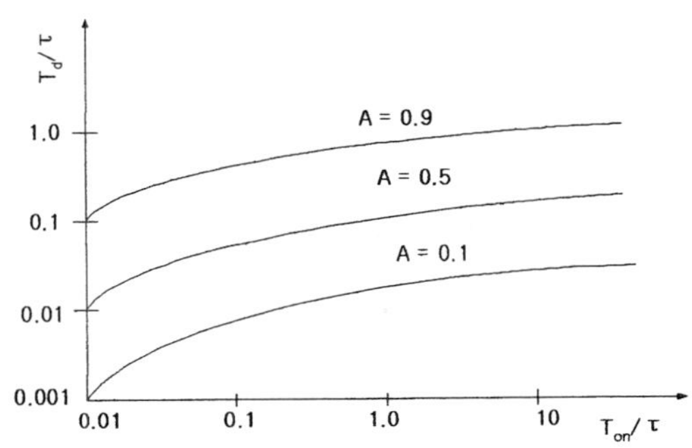

Fig. 7. Relationship between t_{on} and τ

From Fig. 7, these two cases $t_{on}/\tau \langle\langle 1$ and $t_{on}/\tau \rangle\rangle 1$ should be different as equation (10) and equation (11).

$$t_d \simeq \frac{A}{1-A} t_{on} \qquad (t_{on}/\tau \langle\langle 1) \quad (10)$$

$$t_d = \tau \cdot \ln \frac{1}{1-A} \qquad (t_{on}/\tau \rangle\rangle 1) \quad (11)$$

The current i_L for continuous and discontinuous condition is given i_1 and i_2, the system circuit of Fig. 6 can be modeled by the following set of linear differential equations:

$$L\frac{di_L}{dt} = V_s - Ri_L \qquad \text{(on state)} \quad (12)$$

$$L\frac{di_L}{dt} = V_s - V_B - Ri_L \qquad \text{(off state)} \quad (13)$$

This converter has a function of DC-DC boost switching regulator.

The converter switching strategy studied in this paper is a PWM algorithm. The PWM waveform for switching is generated by a modulating wave and a carrier wave. The generation of modulating signal V_M is generated by a difference voltage ($V_S - V_B$)

for the switching condition. The carrier signal V_s is the T-periodic signal. The switching waveforms of converter are shown in Fig. 8.

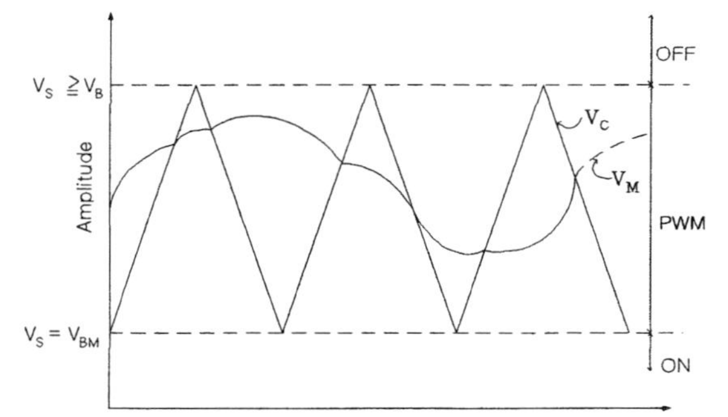

Fig. 8. Switching waveforms of converter

The converter switches from one state to another, whenever the generating voltage V_s is boosted under the control condition $V_{BM} \leq V_s \leq V_B$. Where V_{BM} is a minimum battery voltage for switching condition. The converter has the three operating condition such as OFF, PWM and ON configuration. As can be seen in Fig. 8, the operating condition of ON and OFF is fixed by V_s and V_B.

The switching condition is $V_C - V_M = 0$. The converter switches from on($V_M \rangle V_C$) and off($V_M \langle V_C$) in the PWM area. When the inductor current becomes zero during the off-state, the time is t_d ($\frac{di_L}{dt} = 0$). This converter algorithm includes the PWM strategy and the condition of t_d.

2.4 CB operation

The CB close signal for connecting wind generator to bus is generated by the synchronizing condition such as voltage, frequency and phase. For this purpose, the wind generator voltage is maintained within ±3% of bus Voltage.

The output voltage would be 12Volt with 11m/sec wind speed. The CB control signal can be generated between 10m/sec to 14m/sec.

The synchronizing is performed with the wind generator frequency kept at a high level compared with the bus frequency. In this paper, the CB signal for close is produced with the above condition.

3. Experimental Results

From Fig. 9 (a), (b) and (c), the time evolution of the environmental condition can be observed. Fig. 9 (a) shows the wind speed signal varying between 6m/sec and 14m/sec, with a mean of 10.5m/sec. This waveform is presented the turbulent part of the wind and its variation. The control signal of CB is generated above 11m/sec. Fig. 9 (b) displays the current supply in the converter by the PV system. It can be seen that the PV power is a more static source than wind. Fig. 9 (c) shows the charge efficiency of wind generator and PV that it is estimated with the equation $\{i_L - (i_L - i_B)\} \times 100 / i_L$. Wind-PV hybrid system forms a complementary system and this complementary feature is favorable to system reliability.

The switching action of converter is according to the generation power as shown in Fig. 9 (d). The charge efficiency can be increased by converter and the charging power fluctuation has been greatly reduced. From Fig. 9 (e), the battery voltage keeps a constant voltage with CB and converter. The overvoltage was controlled by using CB for the control output of wind system. Also, the DC link fluctuation can be reduced by switching strategy of converter.

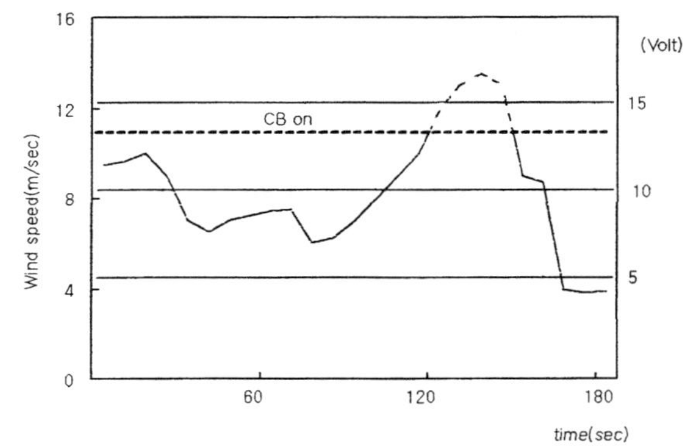

(a)

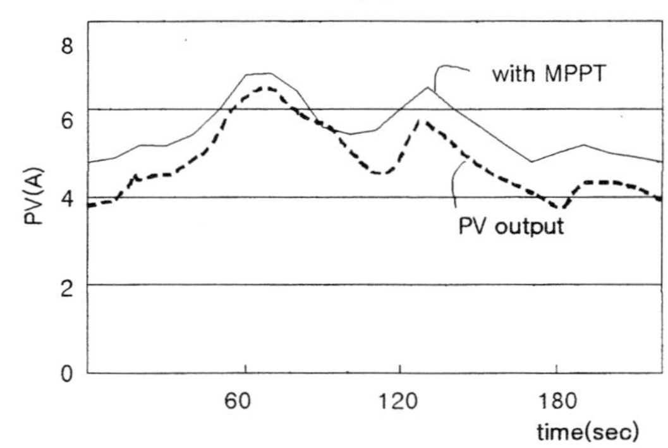

(b)

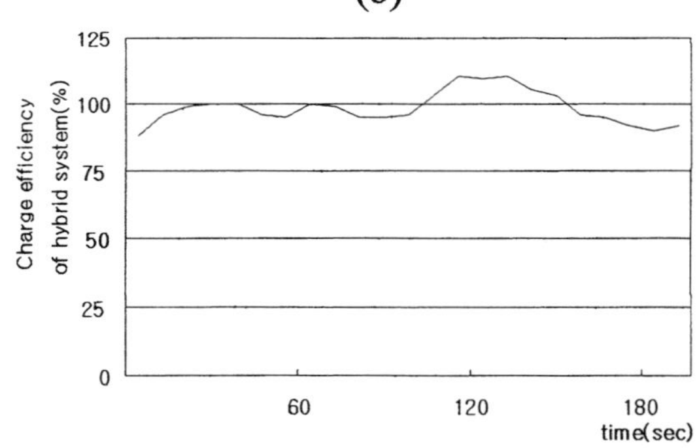

(c)

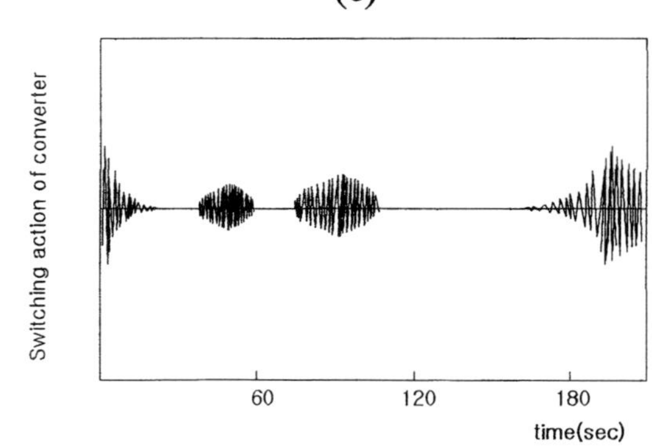

(d)

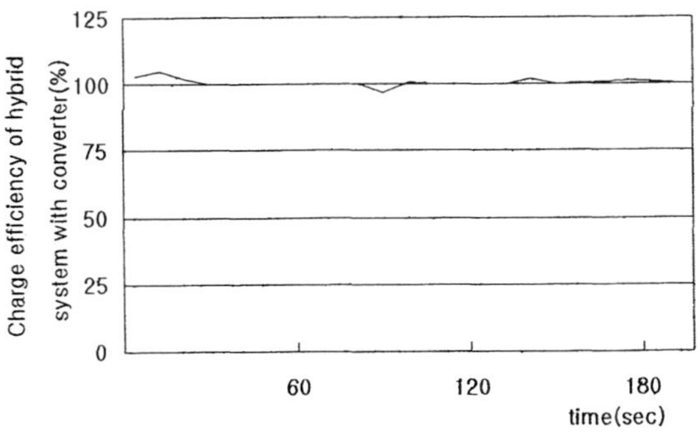

(e)

Fig. 9. Experimental results

4. Conclusion

The wind-PV hybrid system forms a complementary system. Generally, the solar is a more static source than wind, but the wind provides energy during periods of no sunshine. This hybrid system has a high efficiency and reliability.

This paper describes an power control system with CB and converter. The overvoltage of wind system is reduced by CB and DC link fluctuation is controlled by converter. So, the DC bus fluctuation can be reduced with CB and converter. Also, the high efficiency power is achieved by the PWM strategy for converter.

REFERENCES

Abdelali El Aroudi and Ramon Leyva (2001). Quasi-periodic route to Chaos in a PWM voltage -controlled DC-DC boost converter. *Circutts and systems- I*, **Vol. 48,** pp. 967-977.

Edurd Muljadi, Herbert L. Hess and Kim Thomas (2001). Zero sequence method for energy recovery form a variable-speed wind turbine generator. *Energy Conversion,* **Vol. 16,** pp. 99-103.

Francois Giraud and Zyiad M. Salameh (2001). Steady-state performance of a grid-connected rooftop hybrid wind-photovoltaic power system with battery storage. *Energy Conversion,* **Vol. 16,** pp. 1-7.

F. Valenciaga, P. F. puleston, P. E. Battaiotto and R. J. Mantz (2001). Passivity/sliding mode control of a stand-alone hybrid generation system. *Proc.-Control Theory,* **Vol. 147,** pp. 680-685.

Gyu Bum Joung and Jae-Dong Choi, Design of Solar Array Simulator for Spacecrsft, KIEE International Transactions on Electrical Machinery and Energy Conversion Systems, Vol. 12, No. 2, pp. 52-56, June, 2002

Malinowski, Marian P. Kazmierkowski, Steffan Hansen, Frede Blaabjerg, G.D. Marques(2001). Virtual-flux-based direct power control of three-phase PWM rectifiers. *Industry Applications,* **Vol. 37, pp. 1010-1027.**

Peng Wang and Roy Billinton (2001). Reliability beneft analysis of adding WTG to a distribution system. *Energy Conversion,* **Vol. 16,** pp. 134-139.

P. S. Dokopoulos, A. C. Saramourtsis, A. G. Bakirtzis (1996). Prediction and evaluation of the performance of wind-diesel energy systems. *Energy Conversion,* **Vol. 11,** pp. 385-393.

Z. Chen and E. Spooner(2001). Grid power quality with variable speed wind turbines. *Energy Conversion,* **Vol. 16,** pp. 148-153.

ELSEVIER
IFAC
PUBLICATIONS
www.elsevier.com/locate/ifac

FAULT LOCATION AND DIAGNOSIS ALGORITHMS FOR RADIAL DISTRIBUTION SYSTEMS

Seung-Jae Lee, Myeon-Song Choi, Sang-Hee Kang, Bo-Gun Jin
Next-Generation Power Technology Center, Myongji University, Yongin, Korea

Duck-Su Lee, Bok-Shin Ahn
P&C Tech, Anyang, Korea

Abstract: In this paper, a fault location and diagnosis algorithm is proposed. The proposed scheme first identifies fault locations using an iterative estimation of load and fault current at each line section. Then an actual location is identified applying the current pattern matching rules. If necessary, comparison of the interrupted load with the actual load follows and generates the final diagnosis decision. Effect of load uncertainty and fault resistance has been carefully investigated through simulation results, that turns out to be very satisfactory. *Copyright © 2003 IFAC*

Keywords: Distribution system, Power system protection, Fault location, Fault diagnosis.

1 INTRODUCTION

Radial distribution systems are susceptible to faults caused by a variety of situations such as adverse weather conditions, equipment failure, traffic accidents, etc. When a fault occurs on a distribution line, it is very important for the utility to identify the fault location as quickly as possible for improving the service reliability. There have been many researches in the fault location problem for transmission systems. These include a traveling wave-based scheme [1] and harmonics-based scheme [2]. The apparent impedance calculated using fundamental component is the most widely used one [3]. A fault location in the distribution system is not an easy job due to its high complexity and difficulty caused by non-homogeneity of line, fault resistance, load uncertainty and phase unbalance. However the basic approach to calculate the fault location using voltage and current measurement is still same as the transmission system case, that is to calculate the impedance using the fundamental component [4] or harmonics [5].

Dr. Lee is with the department of Electrical Engineering, Myongji University as a Professor since 1988. His main research areas are protective relaying, distribution automation and AI applications to power systems. (sjlee@mju.ac.kr)

Dr. Choi is an associate professor at Myongji University, Korea. His research interest is on power system control and protection. (mschoi@mju.ac.kr)

Dr. Kang is an associate professor at Myongji University, Korea. His research interest is to develop digital protection systems for power systems using digital signal processing techniques.(shkang@mju.ac.kr)

Bo-Gun Jin is now working for his Ph.D. at Myongji University. His main research area is power system control and protection.(sinato@mju.ac.kr)

Duck-Su Lee joined P&C Tech. in 2002, where he is working on the area of protective relaying. (suness@hanmail.net)

Bok-Shin Ahn is President & CEO of P&C Technologies Co.

In this paper, an effective fault location algorithm for the distribution feeder systems is proposed. Higher accuracy has been obtained by updating voltage and current at the load tap or branching point of the feeder and by introducing the load estimation equation that reflects the load current change due to the voltage drop. The load uncertainty has been investigated considering various load models. This paper also proposes current pattern matching rules to determine a single fault location. A single location is identified by comparison of the current pattern with the expected one due to the protective device operation during the fault. Further the load current change due to the circuit interruption is utilized. Effectiveness of the proposed scheme has been proved by a number of EMTP simulations on many realistic distribution systems.

2 ITERATIVE FAULT LOCATION ALGORITHM

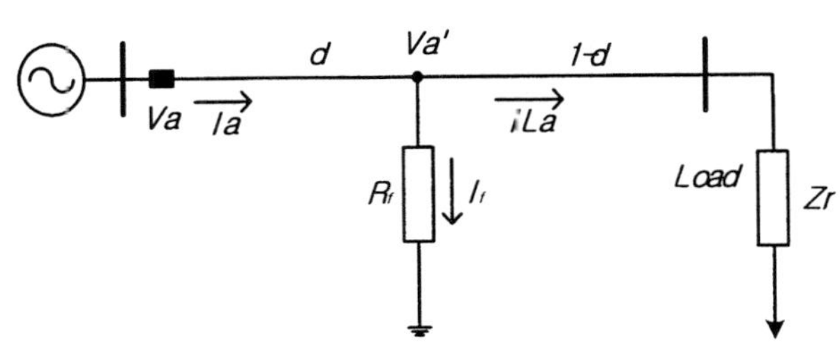

Fig. 1. Simplified distribution feeder model

Consider a phase-to-ground fault on a feeder as shown in Fig.1. The voltage measured at the incoming node is given by Eq.(1).

$$V_a = d \times (Zl_{aa} I_a + Zl_{ab} I_b + Zl_{ac}) + I_f R_f \quad (1)$$

where, V_a: a-phase voltage

I : phase current matrix

Zl : line impedance matrix

I_f: fault current

R_f: fault resistance

d: fault distance

In case of the fault on the single-phase lateral, the voltage equation is given as

$$V_a = d \times Zl_{aa} I_a + I_f R_f \qquad (2)$$

Note that the voltage equation contains three unknown variables – fault distance (d), fault resistance (R_f), fault current (I_f). Taking the real and imaginary parts of the voltage equation and eliminating R_f, the fault distance equation can be obtained.

$$d = \frac{V_{ar} I_{fi} - V_{ai} I_{fr}}{A \cdot I_{fi} - B \cdot I_{fr}} \qquad (3)$$

where, $A = Zl_{aar} I_{ar} - Zl_{aai} I_{ai} + Zl_{abr} I_{br} - Zl_{abi} I_{bi} + Zl_{acr} I_{cr} - Zl_{aci} I_{ci}$

$B = Zl_{aar} I_{ai} + Zl_{aai} I_{ar} + Zl_{abr} I_{bi} + Zl_{abi} I_{br} + Zl_{acr} I_{ci} + Zl_{aci} I_{cr}$

Here the subscripts (r,i) denote the real and imaginary parts. I_f can be obtained from the relation of the load current and fault current.

$$I_f = I_a - I_{La} \qquad (4)$$

The load current during the fault (I_{La}) in Eq.(4) is different from the normal or pre-fault load current due to the voltage drop caused by the fault and it is also an unknown value. An iterative technique to estimate the load current I_{La} has been adopted and the whole fault location process is described as follows:

1) assume I_{La} to be same as pre-fault load current
2) find fault current I_f using Eq.(4)
3) determine the fault distance d using Eq.(3)
4) calculate the voltage at the fault using Eq.(5)

$$\begin{bmatrix} V_a' \\ V_b' \\ V_c' \end{bmatrix} = \begin{bmatrix} V_a \\ V_b \\ V_c \end{bmatrix} - d \begin{bmatrix} Zl_{aa} & Zl_{ab} & Zl_{ac} \\ Zl_{ba} & Zl_{bb} & Zl_{bc} \\ Zl_{ca} & Zl_{cb} & Zl_{cc} \end{bmatrix} \begin{bmatrix} I_a \\ I_b \\ I_c \end{bmatrix} \qquad (5)$$

5) calculate the updated load current I_f using the voltage obtained in step 4)
6) go back to step 2) with new I_f and repeat the process until d converges to a certain value

2.1 Load Current Estimation

Note that step 5) of the fault location algorithm, requires the load current calculation.

A generalized load model [6] in Eq. (6) could be used to estimate the load current.

$$I_{La} = I_{nr} \left| \frac{V_a'}{V_n} \right|^{n_p} + jI_{ni} \left| \frac{V_a'}{V_n} \right|^{n_q} \qquad (6)$$

where, V_n : pre-fault load voltage

V_a' : post-fault load voltage

I_{nr}, I_{ni} : real and imaginary components of pre-fault load current

n_p, n_q : load constant

Here the load constant is determined according to the load characteristic. For constant current load, $n_p = n_q = 0$, for constant power load, $n_p = n_q = -1$, and for constant impedance load, $n_p = n_q = 1$. Usually the load in the distribution system has mixed characteristic and has n_p of 1.5-2.8 and n_q of 2.5-7.0 [7].

2.2 V,I Estimation at Line Section

If the obtained fault distance d is larger than 1, it means the fault is not in that section, but in the following section. So another fault location process should be performed for the next section using voltage and current at the incoming node for the next section. Since the voltage and current measurements are assumed available only at the substation, a way to estimate the voltage and current at the incoming node of each line section is required and described in this section.

Consider a feeder model of Fig.2. The voltage at node k+1 can be obtained using Eq. (7).

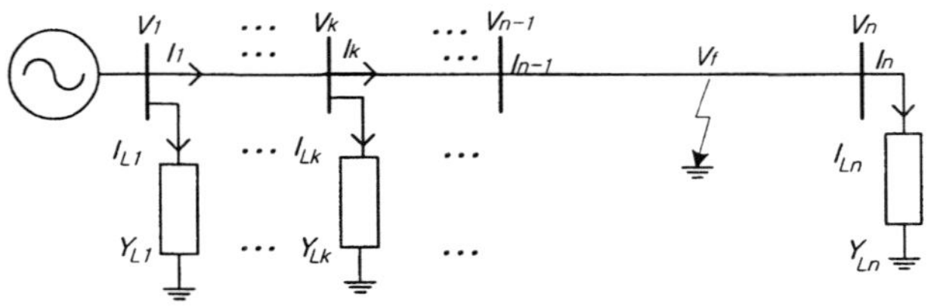

Fig. 2. Single line that simplify supply of electric power system

$$V_{k+1} = V_k - Z_k I_k \tag{7}$$

where, V_k : voltage vector at k[th] node

Z_k : impedance matrix of kth line section

I_k : current vector of kth section

Assuming the constant impedance load, the current flowing out to the k-th load tap can be calculated using Eq.(8).

$$I_{Lk} = V_k \times Y_{Lk} \tag{8}$$

where, I_{Lk} : load current vector of kth load tap

Y_{Lk} : load admittance matrix at k[th] load tap

The current at k-th section, I_k is given as

$$I_k = I_{k-1} - I_{Lk} \tag{9}$$

3 INTELLIGENT FAULT DIAGNOSIS

The fault diagnosis scheme proposed in this study. It first identifies the fault location among multiple candidate locations using current pattern information specified by the protective device operation.

3.1 Diagnosis Based on Current Pattern Matching

In the distribution systems, various protective devices are installed in order to protect the line, transformer and other power equipment. Overcurrent relay, recloser, sectionalizer and fuse are the most commonly used devices and each one has different operating characteristic. So the current pattern during the fault clearing will be subject to the device that is in charge of primary and backup protection. For example, suppose a recloser with sequence setting of 2F2D (2 fast and 2 delay operation) and fuse installed in series as shown in Fig.3. Assume two devices are well set to satisfy the coordination. Then for a fault F1, if all devices have operated as expected, the current measured during the fault clearing would look like Fig.4, which shows a complete recloser operation (2 Fast and 2 Delay).

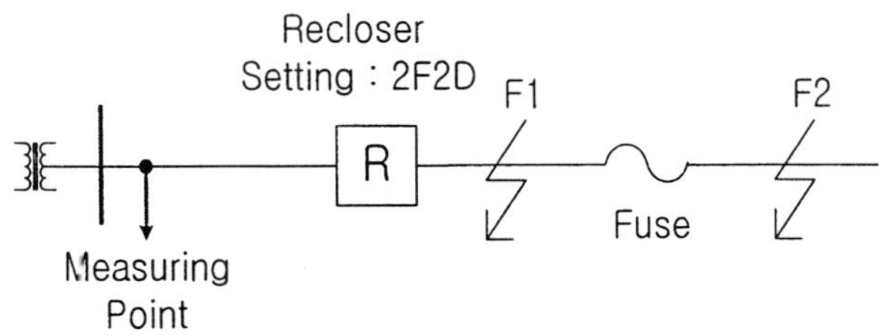

Fig. 3. Recloser-fuse line

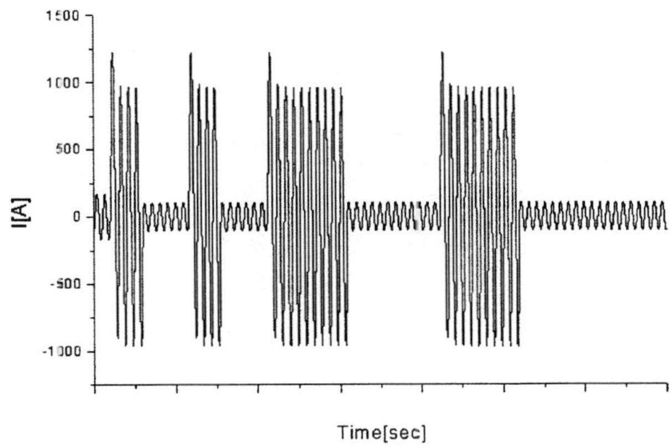

Fig. 4. Current during fault clearing by recloser

However, a fault F2 would generate the current in Fig.5, which shows the first fast operation of the recloser and one reclosing attempt followed by the fuse blow-out during the recloser's delay operation. From this, it can be easily seen that if the current waveform in Fig.4 is given, it means F1 is the fault location. Analysis of such current pattern could identify the protective device type involved in the fault clearing and comparing it with the protective devices that precede each fault candidate, an actual fault location can be determined.

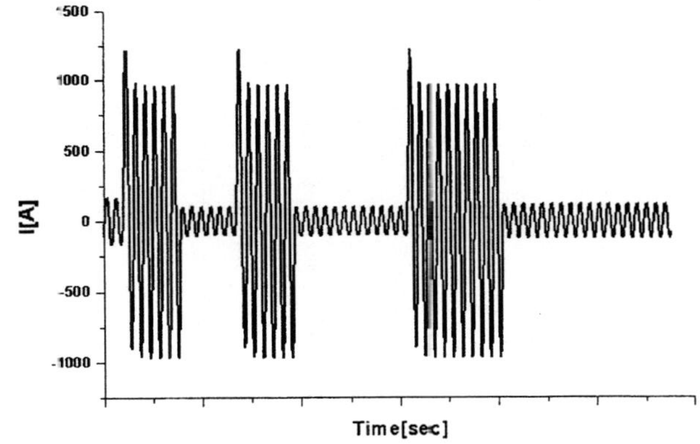

Fig. 5. Current during fault clearing by fuse

Seven different types of protective device combination that could precede the fault and their expected current patterns are listed in Table 1. It has two columns – *device* and *conditions* and each row is the pattern identification rule that reads as 'if *conditions* are met, then the preceding device is *device*'. The waveform pattern is recognized based on the number of reclosing, number of fast and delay operations.

Table 1 Pattern Identification Rules

Protective Devices	Conditions
OCR	① CWF Reclosing number = OCR Reclosing number ② CWF Fast Operation number = OCR Reclosing number + 1
REC	① CWF Reclosing number = REC Reclosing number ② CWF Fast Operation number = REC Fast Operation number ③ CWF Delay Operation number = REC Delay Operation number
OCR – Fuse	① CWF Reclosing number = 0 ② CWF Fast Operation number = 1
REC – Fuse	① CWF Reclosing number = REC Fast Operation number +1 ② CWF Fast Operation number = REC Fast Operation number
Fuse – Fuse	① CWF Reclosing number = 0 ② CWF Fast Operation number = 1
REC – SEC	① CWF Reclosing number = SEC count - 1 ② CWF Fast Operation number = REC Fast Operation number ③ CWF Delay Operation number = SEC Count - REC Fast Operation number
REC - SEC – Fuse	① CWF Reclosing Operation number = REC Fast Operation number ② CWF Fast Operation number = REC Fast Operation number + 1

4 CASE STUDY

In order to show the effectiveness of the proposed algorithm, a 22.9 kV feeder in Fig.6 is taken as a test system that has 21 nodes and single phase and three phase laterals. A node number is indicated along the line with its distance in parenthesis and loads [KVA] are shown in the rectangular box.

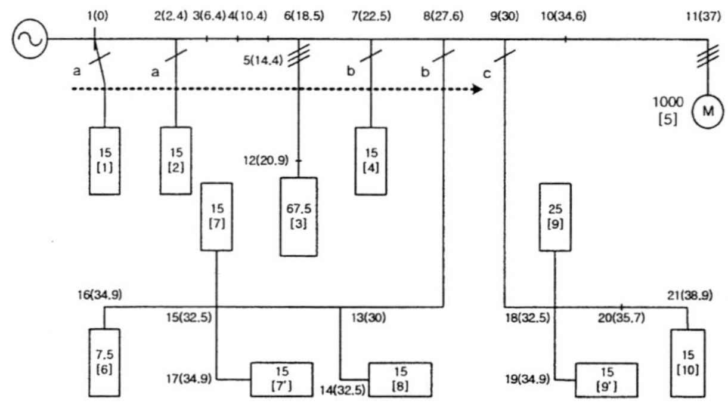

Fig. 6. Model system for fault location

EMTP simulations have been carried out taking 10 fault locations on the thick dotted line in Fig.6 and four fault resistance values, and adopting two different load models. Various parameters include of fault distance, fault resistance, load variation, and load model. The effect of the load uncertainty is also investigated by introducing average 30% variation to the load impedance used in the algorithm.

An estimation error of fault location is calculated using Eq.(10).

$$\%Error = \left| \frac{est.D - real.D}{total.D} \right| \times 100 \tag{10}$$

where est.D, real.D represent the estimated fault location and actual location respectively and total.D denotes the whole line length.

4.1 Fault Location

A. Constant Impedance Load Model with Known Load Impedance

(1) Effect of Fault Resistance

When all loads are constant impedance type and their associated impedance are known, the proposed algorithm has generated the results summarized in Fig.7. The maximum error of 0.6% is observed for a zero fault resistance case while the maximum error of 0.85% is observed for a fault resistance of 50 ohm. It can be seen from the figure that as the fault resistance increases, the estimation error also increases.

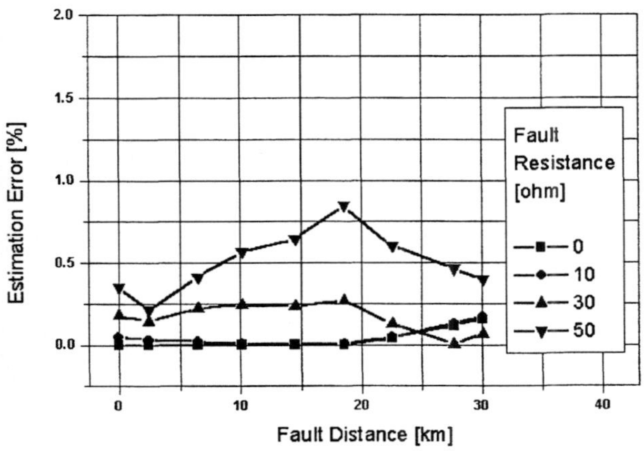

Fig. 7. Fault location error for constant impedance model with known load impedance

(2) Effect of Load Variation

Fig.8 shows the fault location results when there is 30% difference between the impedance used in the algorithm and the real load impedance. It is very interesting to see that load uncertainty does not deteriorate the fault location accuracy at all showing maximum error of about 0.25 % in cases of 30 ohm fault resistance. Even although taking into consideration that 30% load variation is obtained by averaging randomly generated load variation for each load by 30%, it is still very interesting result that shows the robustness and practical power of the proposed algorithm since the exact load impedance can hardly be obtained in a real situation.

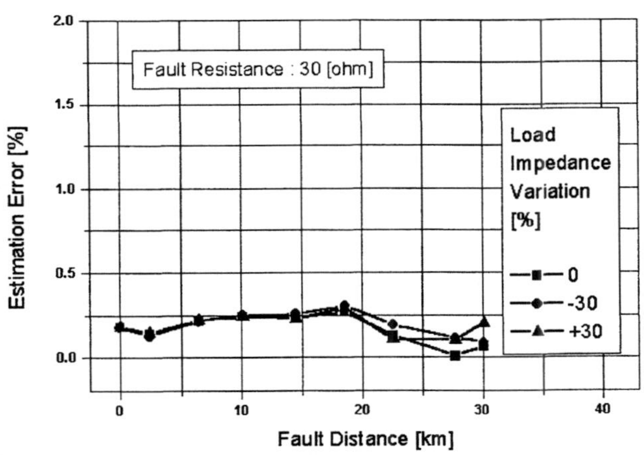

Fig. 8. 30% load variation with R_f=30 ohm

B. Generalized Load Model

(1) Effect of Fault Resistance

A load model represented by Eq.(6) has been adopted with n_p=0.8, n_q=1.5, and its fault location errors for the same system are shown in Fig.9. The same observation can be made for this load model case - the bigger the fault resistance is, the higher the error is. Note that the error is limited within 2.95% for 50 ohm fault resistance. This error might be considered too big to be used for a real application. However in all cases with less than 30 ohm fault resistance, the error is smaller than 1.0% showing enough accuracy for a real application.

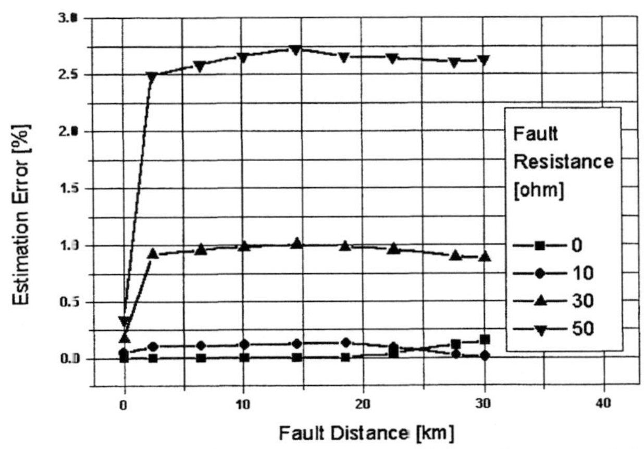

Fig. 9. Fault location error for general impedance model

(2) Effect of Load Variation

For average 30% load variation, again mostly the error is kept within 0.25% showing no accuracy drop for 30 ohm case as can be seen in Fig.10. Some accuracy sacrifice is observed although it is very small in case of 30-ohm fault resistance case, showing maximum error of 1.3%, about 0.3% more compared to no load uncertainty case.

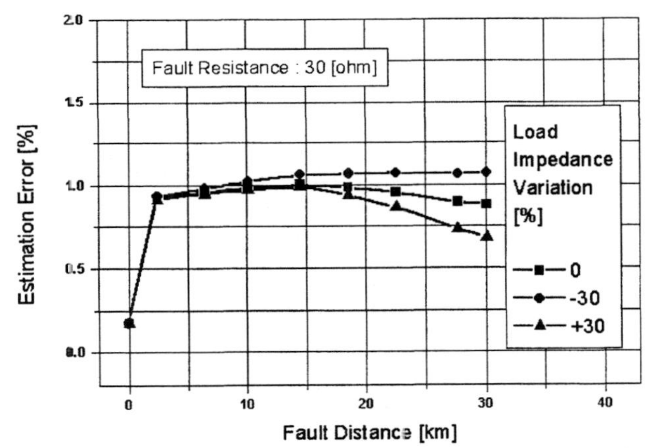

Fig. 10. Error with impedance variation of 30% (R_f=30)

4.2 Fault Diagnosis

In order to show the effectiveness of the proposed fault diagnosis scheme, a feeder in Fig.11 is taken as a test system, which has two reclosers and three fuses. Both reclosers are assumed to have a sequence setting of 2F2D. Simulation for a fault on F2 has been carried out using EMTP. The recorded current is shown in Fig.12. In the current waveform picture, the real time scale is not used because of the limitation on time axis length to be depicted in the paper. The figure indicates the fault clearing by fuse during recloser delay operation following two fast operations, giving apparent three fast operations and two reclosing.

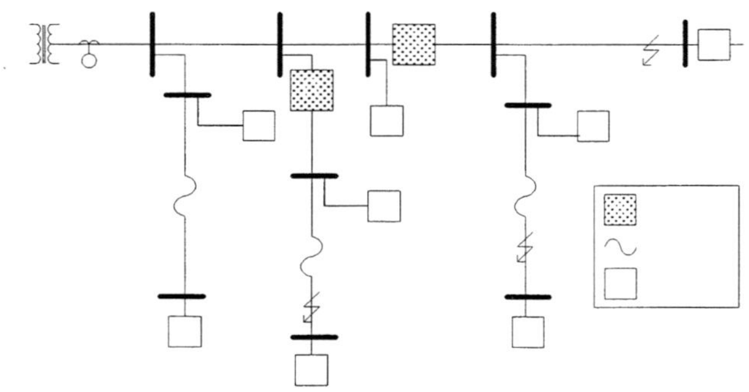

Fig. 11. Example feeder for fault diagnosis

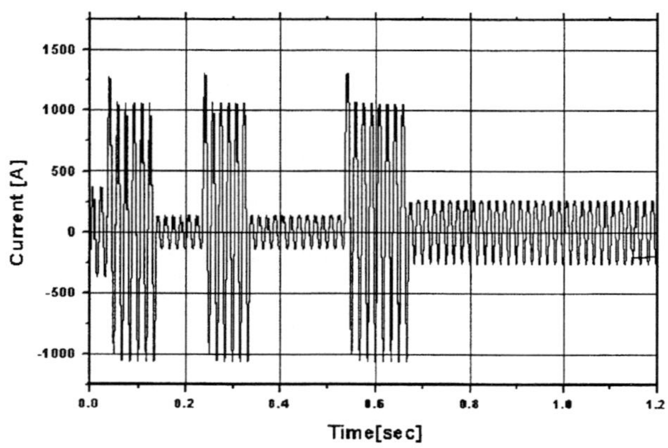

Fig. 12. Fault current waveform

Suppose the fault location algorithm has yielded two more candidate locations F1 and F3 in addition to the actual fault location F2. Note that the preceding protective device pair is OCR-Recloser for F1, and Recloser-Fuse for both F2 and F3. For each fault location candidate, its expected current waveform pattern derived from the pattern identification rules in Table 1 is shown in Table 2. According to this table, the device pattern should be OCR-Recloser and consequently the first diagnosis step based on the current pattern matching has chosen F2 and F3 as fault location. Note that since both have same preceding device pair, they can't be further differentiated using the pattern rules. Now the second step that deals with the load change analysis before and after the fault clearing is carried out. Figure 21 shows 168.17 [A] load current before the fault and 82.70[A] after the first and second fast operations, while 120.02 [A] after the third fast operation, indicating 48.15.[A] load interruption. With the knowledge on the loading (46.77 [A] below fuse2), the one closer to the interrupted loading is selected, which is F2 in this case.

Table 2 Expected operation for each fault candidate

Candidate	F1	F2	F3
Protective device	OCR Recloser	Recloser Fuse	Recloser Fuse
Expected opreation	Reclosing:3 Fast:2 Delay:2	Reclosing:2 Fast:3	Reclosing:2 Fast:3

5 CONCLUSION

This paper proposes an effective fault location algorithm and an intelligent fault diagnosis scheme. The proposed fault location algorithm identifies candidate fault locations using an iterative estimation of load current and fault current at each line section. The diagnosis part determines the actual location by comparing the current waveform pattern with the expected pattern due to operation of the protective devices. If necessary, comparison of the interrupted load with the actual load follows and generates the final decision.

Various simulations changing load model, fault location, fault resistance, and loading have shown practically satisfactory accuracy and high robustness to the load variation of the proposed scheme.

6 ACKNOWLEDGEMENT

Authors would like to thank Ministry of Science & Technology of Korea and Korea Science & Engineering Foundation for their support through ERC program.

REFERENCES

G. B. Ancell, N. C. Pahalawaththa, "Maximum Likelyhood Estimation of Fault Location on Transmission Lines Using Travelling Waves", IEEE Transactions on PWRD, Vol. 9, No. 9, 1994, pp. 680-689.

T. Takagi, Y. Yamakoshi, J. Baba, K. Uemura, T. Sakaguchi, " A New Algorithm of an Accurate Fault Location for EHV/UHV Transmission Lines : Part Ⅰ - Fourier Transform Method", IEEE Trans. on PAS, Vol. PAS-100, No. 3, 1988, pp. 1316-1323.

Yong-Jin Ahn, Myeon-Song Choi, Sang-Hee Kang, Seung-Jae Lee, "An Accurate Fault Location Algorithm For Double-Circuit Transmission Systems", Power Engineering Society Summer Meeting, 2000. IEEE , Volume: 3 , 2000 pp 1344 –1349.

Adly A. Girgis, Christopher M. Fallon, David L. Lubkeman, "A fault location technique for rural distribution feeders", IEEE Transactions on Industry Applications, Vol. 29, No. 6, November/December 1993.

M. E. Hami, L. L. Lai, D. J. Daruvala, A. T. Johns, "A New Travelling-Wave Based Scheme For Fault Detection On Overhead Power Distribution Feeders", IEEE Trans. on Power Delivery, Vol. 7, No. 4, October 1992, pp 1825-1833

Reineri, C.A.; Alvarez, C., "Load research for fault location in distribution feeders", Generation, Transmission and Distribution, IEE Proceedings- , Volume: 146 Issue: 2 , March 1999, Page(s): 115 –120.

P. Kundur, "Power System Stability and Control", McGraw-Hill, 1994, Page(s): 271-275

ELSEVIER
IFAC
PUBLICATIONS
www.elsevier.com/locate/ifac

EVALUATION OF THE RELIABILITY OF DISTRIBUTION SYSTEM TAKING INTO CONSIDERATION CUSTOMER INTERRUPTION COST

Sang-Bong Choi*, Kee-Young Nam*, Seong-Hwan Jeong*, Dae-Kyung Kim*, Kyung-Bin Song**

KERI, Soongsil University***

Abstract: It is raised for methodology to evaluate power system reliability using interruption cost according to power industry is rushed into competition appearance. This paper presents algorithms to evaluate the reliability of power distribution system taking into consideration customer interruption costs. The customer interruption cost is considered as one of the most valuable indices to estimate the reliability of a power distribution system from the customers' standpoint. In the concrete, this paper presents new algorithm that takes into account the load by customer type and failure probability by distribution facilities when calculating the amount of unserved energy by customer type. Finally, evaluation results of unserved energy and system interruption cost based on customer interruption cost are shown in detail. *Copyright © 2003 IFAC*

Keywords: Power System Reliability, Customer Interruption Cost, Unserved Energy, Customer Type, Power Distribution System.

1. INTRODUCTION

In relation to the restructuring of power industry, service reliability of a power system has emerged as a major issue in the power industry recently. In addition, severe competition among energy industries demands energy suppliers to consider the conditions related to service reliability. In other words, as customers can opt for alternative energy source in consideration of price, enhancing service reliability can not be considered as a mandatory strategy necessarily. Therefore, to effectively deal with such issue, it is necessary to find out customers' response to service reliability and interruption costs. In the past, the issue of service reliability in the power industry was focused on ensuring high reliability all the time. However, as cost hike accompanies high reliability, implementing flexible plans in consideration of consumers is a new trend in the industry.

For example, if distribution system facilities are expanded, customers will have more stable power supply due to improved service reliability, which is an advantage, while the facility investment needed will be passed on to the customers as increased electric charges, which is a disadvantage.

As the improvement of service reliability brings the reduction of interruption cost, it is possible to carry out an economic evaluation of a system facility plan from the consumers' standpoint by quantifying the interruption costs following the changes in service reliability(see e.g. Billington et al., 1987; Sullivan, 1996; Nakamura et al., 1999). Therefore, in Japan and other countries, researchers directed their attention to the evaluation of the service reliability of a power system taking into account customer interruption costs. For example, the researchers at Kitami Institute of Technology in Japan suggested a method of evaluating service reliability of a power system taking into account the interruption cost(Nakamura et al., 1997). In general, various facilities is used for a power system including power lines, converters, and switches. And the failure probability of each facility might be different. In addition, the customer interruption costs are different by customer type. But as the method proposed by the researchers at Kitami Institute of Technology does not differentiate these factors and considers them inclusively, accuracy goes down when the interruption cost of a system is calculated. Another method is presented in this paper to overcome this problem. For a distribution system, the interruption cost by customer type was calculated by using the amount of unserved energy. Then a method of adding up the interruption cost by customer type is presented. In addition, a new algorithm that takes into account the load by customer type and failure probability by distribution facilities when calculating the amount of unserved energy by customer type is presented.

2. EVALUATION OF THE CUSTOMER INTERRUPTION COSTS AND RELIABILITY

In recent years, to increase the efficiency of power business through competition and to ensure the customers' choice in power purchase, the opening of power markets at home and abroad is on the rise. As a result, the customers' interest in the soundness between the electric charges and the level of system reliability is increasing. Under this circumstance, when a power company wants to improve its service reliability by reducing interruptions, it is necessary to evaluate how much benefit will be gained by this and present it to the customers. However, as the assessment of customer interruption costs is different from country to country, it is difficult to apply it uniformly. Therefore, gross production index in a macroscopic area was used to evaluate the interruption cost by customer type. Then the amount of unserved energy and the average system interruption time for a sample system were calculated. Finally, service reliability was evaluated in consideration of customer interruption costs. A summarized flowchart is illustrated in Figure 1 below.

2.1 Overview of Methodology

Researches on the evaluation of interruption costs have been carried out in Britain, France, Italy, Canada, the U.S. and Japan since they first started in Sweden. Various methods have been used to evaluate the interruption costs but they can be classified into two. The first type is the macro method, which evaluates the interruption costs in relation to national economy. The second type is the micro method, which calculates the interruption cost by customer type based on the survey of individual customers.

The former, the macro method was conceived on the point that economic loss occurs after power interruption as economic activities are halted, that is justifiable economic value is lost due to power interruption. The simplest method is calculating the interruption cost macroscopically by dividing GNP by total power consumption. A more detailed method is used also. By using tables related to economic activities, this method calculates the interruption cost of each economic sector by dividing the value added obtained in the sector by the power input of the sector. These methods are rough but have merits also. With these methods, it is possible to calculate the interruption cost of a country as a whole or by segment. However, evaluating the interruption cost of individual customers with the values obtained by these methods have some problems.

On the other hand, with the latter, the micro method, it is possible to calculate the interruption costs of customer groups but not individual customers. And the result can be problematic from the objectivity standpoint. As a result, it is necessary to carry out large scale surveys. And actually interruption costs are calculated based on large scale surveys in Sweden, Britain, France, the U.S., Canada and Japan.

2.2 Evaluation of Interruption Cost

As the method of evaluating the customer interruption cost by calculating the gross production index in the region is generally used, in this paper it is applied to calculate the customer interruption costs as follows.

First, according to economic activities, customers are classified into public, service, agriculture, fishery, mining, manufacturing, and residential sectors. Except for the residential sector, the value added was defined as the goods and services produced by economic activities in the country minus the cost of raw materials. And the customer interruption costs were calculated by using the gross domestic production(value added) in each area of public and commercial sectors in the region based on related tables by industry. As for the residential sector, the interruption cost was defined as the value of domestic labor, the [essential activities for subsistence] of the daily activities of a regular household. In other words, the interruption cost for residential customers was calculated as follows. First the amount of housework impeded by power interruption was estimated. Then it was converted into money value based on the estimated value of a full time housewife's monthly labor cost. The data compiled by Korean Women's Development Institute was used as the base of calculation. According to the data, the size of domestic labor force is about 5 million and the monthly value of domestic labor per household is 1.02 million won. And the total yearly cost is 6.1 trillion won. And the customers engaged in service sector were classified into whole sale and retail industry, food and lodging industry, transportation and warehousing industry, communication industry, finance and insurance industry, real estate service industry, and social service and individual service industry. And the customers engaged in public sector constituted the general public administration. And the interruption cost was calculated. Table 1 below shows the estimated values of interruption cost calculated by using power sold and gross production in the area(value added) by customer type in 2001.

Table 1 Customer Interruption Cost by Industry type(As of 2001)

Customer type	Value added (million won)	Power sold (MWh)	Interruption Cost (Won/kWh)	Ratio to the unit electric charge
Public	11,930,465	9,424,826	1,266	11.7
Service	209,719,990	73,303,991	2,861	26.5
Agriculture & fishery	24,526,577	5,985,147	4,098	94.2
Mining	812,699	1,074,284	757	12.3
Manufacturing	182,143,159	128,731,878	1,415	23.0
Residential	68,931,673	39,211,228	1,758	19.2

2.3 Calculation of Average System Interruption Time

As the failure rate by source, the interruption time for repair, and the number of customers experiencing interruption are different, the following formula (1) is used for the calculation of average system interruption time.

$$T_s = \sum_i \frac{t_i \times f_i \times lengths(numbers) \times N_i}{N_T \times 60} \quad ------(1)$$

i = By failure source (powerline, transformer, switch)

N_i = Number of customer experiencing interruption by failure source)

N_T = Total number of customers

t_i = Interruption time by failure source (In minute)

f_i = Failure rate by failure source

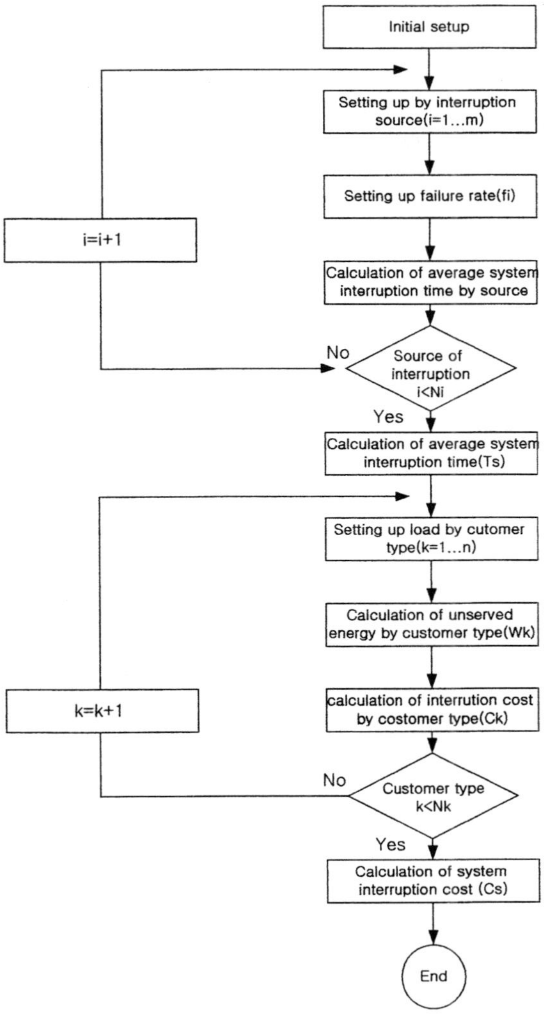

Fig. 1. Flowchart for the evaluation of reliability in consideration of customer interruption cost

At this time, the average interruption time is calculated as in Table 2 in consideration of dispatch time, failure detection time, and average switch changing time.

Table 2 Calculation of Average Interruption Time by Source

Number of Switches	Number of Lines	Number of Switches per Line	Dispatch Time(In Minutes)	Average Failure Detection time (In minutes)	Average Switch Changing time(In minutes)	Average Interruption time (In minutes)
①	②	③ = ①/②	④	⑤	⑥	⑦ = ④+⑤+⑥

2.4 Calculation of the Amount of Unserved Energy

For the calculation of the total amount of unserved energy of a system for the evaluation of service reliability from the customers' standpoint, the amount of unserved energy by customer type needs to be calculated first. By using the load characteristics by customer type in the respective area, the amount of unserved energy by customer type is defined as the following formula(2).

$$W_k = P_k \times T_s \quad \text{-----------------------} \quad (2)$$

Here,

k = Customer of type

W_k = Amount of unserved energy by customer type

P_k = Load by customer type

T_s = Average system Interrupti on time(h)

2.5 Evaluation of Service Reliability consideration of the Interruption Costs

From the amount of unserved energy(Wk) by customer type and the estimate of unit interruption Cost by customer type (Ck,), the interruption cost of a system according to system configuration is calculated as the formula(3).

$$C_s = \sum_k C_k \times W_k$$

Here, ———————————————————(3)

k = Customer type

C_k = Interrupti on cost by customer type

W_k = Amount of unserved energy by customer type

3. CASE STUDY

3.1 Conditions of Case Study

To apply the algorithm that evaluates service reliability taking into account the customer interruption cost presented above, a distribution system of 2 line open loop supply which consists mostly of high voltage customers as shown in Figure 2 was postulated. In this study, for the model system shown in Figure 1, the average system interruption time and the amount of unserved energy by customer type were calculated for the failure by distribution facilities type. Then the level of system reliability was evaluated in consideration of the interruption cost by customer type.

For distribution systems, related regulations require that loads over 100kW should be supplied by high voltage and loads under 100kW should be supplied by low voltage through transformers. Therefore, in this study, we hypothesized that the model system supplies high voltage of over 100 kW to 8 customers and low voltage of under 100 kW to residential customers. In Table 3, load characteristic data including the load amount by switch and customer type are presented.

Table 3 Load Characteristic data of the model power system

Switch number	Load(kW)	Customer type
①	800	Public
②	1,500	Service
③	650	Public
④	3,500	Service
⑤	100	Residential
⑥	2,000	Service
⑦	900	Public
⑧	100	Residential
⑨	750	Public
⑩	3,000	Service

3.2 Results of the Case Study

For the model system, we calculated the average system interruption time in consideration of the failure probability by distribution facilities consisting of power lines, switches, and converters, by using the formula (1) and based on the following assumptions. And the results are shown in the following Table 4.

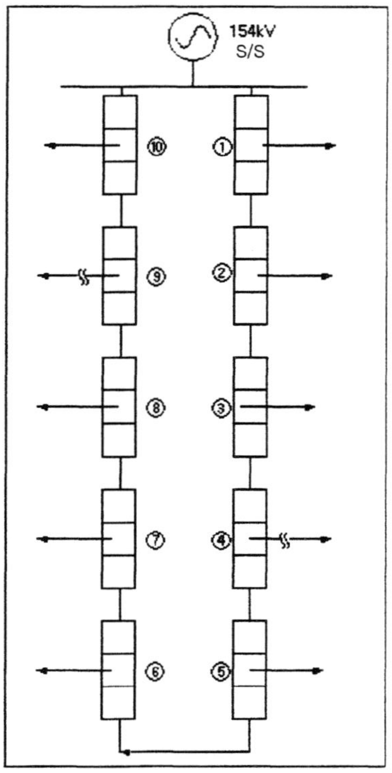

Fig. 2. Configuration of model power system

- ◆ The average switch changing time is 3 minutes per switching station. Also, the time taken to move for repair is considered to be 10 minutes. (Considering the site situation)
- ◆ Calculates the number of lines for the system supply method.
- ◆ Number of average switch stations = Number of switch stations ÷ Number of lines
- ◆ n = (Number of average switch stations ÷ 2^n) ≤ 1
- ◆ Average failure detection time = (n - 1) × 10 minutes /per move
- ◆ Average switch changing time = (2n - 1) × 3 minutes / per switch station
- ◆ Transformer changing time is considered to be 120 minutes.

Table 4 Calculation of system average interruption time

Interruption source	Failure rate	Length (Number)	Total customers	Interruption Time (in min)	Number of Customers Experiencing interruption (annual)
	①	②	③	④	⑤
Lines	0.0129	40	10	24.15	2.5
Switches	0.00145	10	10	24.15	2.5
Transformers	0.00145	10	10	120.0	1.0
Total	-	-	-	-	-

Number of Customers Experiencing Interruption(annual)	Customer Interruption Time (for the year) (In minutes)	Average System Interruption Time
⑥=①*②*⑤	⑦=④*⑥	⑧=⑦/③
1.2900	31.1535	3.11535
0.0363	0.8766	0.08766
0.0145	1.7400	0.1740
1.3408	33.7701	3.3770

By using the average system interruption time calculated with the formula(1), the amount of unserved energy by customer type and the total amount of unserved energy of the system as a whole were calculated with the formulas (2) and (3).

Table 5 Calculation of unserved energy

Customer Type	Load (kW)	Number of Customers	Load per Customer (kW)	Interruption Time per Customer(Hr)	Unserved Energy (kWH)
	①	②	③=①/②	④=Average System Interruption Time*②/60	⑤=③*④
Public	3,100	4	775	0.225	174.48
Service	10,000	4	500	0.225	562.83
Residential	200	4	100	0.113	11.26
Total	13,300	10			748.57

Next, to evaluate service reliability taking into account the interruption cost from the amount of unserved energy calculated in Table 5, the final system interruption cost was calculated by multiplying the customer interruption cost by customer type presented in Table 1. And the results are presented in Table 6.

Table 6 Assessment of system interruption cost

Customer type	Amount of unserved Energy (kWh)	Interruption cost per customer(won/kWh)	Interruption Cost (1,000won/year)
-	①	②	③=①*②
Public	174.48	1,266	220.89
Service	562.83	2,861	1,610.26
Residential	11.26	1,758	19.80
Total	748.57	-	1,850.95

4. CONCLUSION

With the introduction of competition in the power industry, methods of evaluating service reliability have been presented. As a part of such efforts, a method of evaluating service reliability based on the interruption cost by converting the loss customers suffer due to interruptions into money, has been proposed. In this paper, breaking away from the traditional method of power supply, in which the supplier alone decides the level of acceptable service reliability, a method of supply reliability evaluation that reflects customers' side was introduced. In other words, a method of evaluating service reliability based on the interruption cost by converting the loss customers suffer due to interruptions into money, was applied. In addition, to evaluate the interruption cost more accurately, the amount of unserved energy by customer type was calculated in consideration of the failure probability by distribution facilities. Then, the final system interruption cost was calculated by using the interruption cost by customer type.

For the investigation of interruption cost by customer type, various methods are available. However, this study used the calculation method that utilizes the gross production index of a macroscopic area.

REFERENCES

Koichi Nakamura, Susumu Yamashiro(1999). *A Survey Study on Estimation of Customer Interruption Costs.* T. IEE Japan, Vol. 119-B, No. 2.

Koichi Nakamura, Susumu Yamashiro(1997). *A Study on the Estimation of Power System Reliability taking into account Interruption Costs.* PE-97-61.

M.J.Sullivan(1996). *Interruption Costs, Customer Satisfaction and Expectations for Service Reliability.* IEEE Trans. on Power Systems, Vol. 11, No. 2.

R. Billinton, J. Oteng-Adjei, R. Ghaja(1987). *Comparison of Two Alternative Method to Establish on Interrupted Energy Assessment Rate.* IEEE, Trans. On Power Systems, Vol. PWRS-2, No. 3.

ELSEVIER
IFAC
PUBLICATIONS
www.elsevier.com/locate/ifac

OPTIMAL BIDDING STRATEGY IN THE UNCERTAIN ENERGY MARKET

**Pathom Attaviriyanupap, Keita Ogata, Hiroyuki Kita,
Eiichi Tanaka and Jun Hasegawa**

Hokkaido University, Japan

Abstract: This paper proposes an algorithm to determine the optimal bidding parameters for a day-ahead energy market. The algorithm is developed from the viewpoint of the generation company, which has two main objectives. The first objective is to maximize profit while the second objective is to reduce the risk that is induced by the uncertain information. The energy market based clearing price auction will be considered in this paper. The uncertainties of rivals' bidding parameters and forecasted demand are represented using triangular fuzzy number. The evolutionary programming is applied to solve this optimization problem and the results are compared with those obtained from deterministic approach. *Copyright © 2003 IFAC*

Keywords: Bidding Strategy, Deregulation, Energy Market, Evolutionary Programming, Uncertainty.

1. INTRODUCTION

The deregulation and restructuring of the electric power systems have created a competitive open market environment. Deregulation in general often involves unbundling, which in the case of electric utility services represents disaggregation into the basic part of generation, transmission and distribution. These operations became the responsibility of generation companies (GENCOs), transmission companies (TRANSCOs) and distribution companies (DISCOs), with a central coordinator, called an independent system operator (ISO), to balance supply and demand in real time and maintain system reliability and security (Lai, 1998). In the new competitive environment, each market participant searches for their own path to maximize profit. A number of researches have examined the concept of profit maximization (Attaviriyanupap, et al., 2002; Richter, et al., 2000). Bidding strategy is one of those researches that have been studied subsequently so far (David, 1993; Gross, et al., 1996; Hao, 1999; Wen, et al., 2001; Zhang, et al., 2000).

In this paper, the authors propose an algorithm for determining the optimal bidding parameters in the energy market. This method is developed from the viewpoint of GENCO that participates in a 24 hourly energy market based on the clearing price auction system. The proposed algorithm considers the bidding parameters of the other rival suppliers and the predicted demand as a triangular fuzzy number. This means the GENCO has only the possible ranges (not the exact values) of bidding parameters of the rivals and the demand in each hour. This assumption is more reasonable than the deterministic case, which we suppose that GENCO can completely predict the above information. By including the uncertain parameters into the bidding strategy problem using the proposed algorithm, GENCO can determine the optimal bidding parameters, which provide the high enough expected profit within an accepted risk level.

The optimization problem is solved using an evolutionary programming (EP) method taking into account the uncertainties. Test system is used to demonstrate the effectiveness of the proposed method. The simulation results are compared with those obtained from the deterministic approach.

2. ENERGY MARKET

In the energy market, suppliers submit monotonically increasing price curves that describe the energy offered as a function of price. The ISO (or power exchange PX in some market) aggregates the curves into aggregated supply curves for each hour. Then, the ISO/PX determines the point at which the supply curve intersects the forecasted demand. This point sets the market clearing price (MCP) for that hour.

Suppose that a system consist of N_S suppliers. This paper derives the bidding strategy for one GENCO as the first supplier. For the jth supplier, a monotonically increasing linear bid curve $G_{jt} = \alpha_{jt} + \beta_{jt} SP_{jt}$ is submitted to the ISO/PX each hour. Here, SP_{jt} is the power sold by the jth supplier, α_{jt} and β_{jt} are the non-negative energy bidding parameters of the jth supplier. The MCP and the power awarded to each bidder are determined based on the following equations.

$$MCP_t = \frac{\sum_{j=1}^{N_s} \frac{\alpha_{jt}}{\beta_{jt}} + PD_t}{\sum_{j=1}^{N_s} \frac{1}{\beta_{jt}}} \quad , \quad t = 1,...,24 \tag{1}$$

$$TP_{jt} = \frac{MCP_t - \alpha_{jt}}{\beta_{jt}} \quad , \quad t = 1,...,24 \tag{2}$$

subject to

$$TP_{j\min} \le TP_j \le TP_{j\max} \quad , \quad j = 1,...,N_s \tag{3}$$

where MCP_t is the market clearing price for hour t, PD_t is the forecasted demand for hour t, TP_{jt} is the power awarded to the jth supplier for hour t, TP_{jmin} is the lower power limit of the jth supplier, and TP_{jmax} is the upper power limit of the jth supplier.

When the solution to (2) exceeds the maximum limit TP_{jmax}, TP_{jt} is set to TP_{jmax}. When TP_{jt} is less than TP_{jmin}, TP_{jt} is set to zero and the relevant supplier is removed from the problem as a non-competitive participant for that hour.

3. BIDDING STRATEGY PROBLEM

In the deregulated system, suppliers know one part of their load ahead of time from the future market or bilateral contracts. Suppliers can sell the rest energy through the energy market. However, suppliers face a risk in this market because of the uncertain information they have. Bidding strategy is an algorithm to help supplier determine appropriate participation's way. The bidding parameters are selected for each participant so as to maximize the each participant's profit while minimize the risk based on the forecasted demand and expectation concerning rival's bids. The deterministic problem can be formulated mathematically as follows.

The objective function is given by

$$\max \quad PF = RV - TC \tag{4}$$

subject to

$$\sum_{i=1}^{N} P_{it} = Pcon_t + TP_{1t} \quad , \quad t = 1,...,24 \tag{5}$$

$$P_{i\min} \le P_i \le P_{i\max} \tag{6}$$

Here, RV and TC are the revenue and total production cost calculated respectively by

$$RV = RV_{contract} + \sum_{t=1}^{24} \left(MCP_t \cdot TP_{1t} \right) \tag{7}$$

$$TC = \sum_{t=1}^{24} \sum_{i=1}^{N} F(P_{it}) \tag{8}$$

where N is the number of generating units owned by the first supplier, P_{it} is the power generation of generator i in hour t, F_i is the fuel cost function of generator i, P_{imin} is the minimum generation limit of generator i, P_{imax} is the maximum generating limit of generator i,$Pcon_t$ is the amount of power sold on contract at hour t, $RV_{contract}$ is the revenue from contract.

Here, $Pcon_t$ is already determined independent on the market.

The fuel cost is calculated for each unit from

$$F_i(P_i) = a_i P_i^2 + b_i P_i + c_i \tag{9}$$

where a, b and c are unit's constants.

4. BIDDING STRATEGY PROBLEM CONSIDERING UNCERTAINTY

Uncertain parameters to be considered in this paper consist of the demand (PD_t) and bidding parameters of rival suppliers (α_{jt} and β_{jt}, j = 2,...,N_S). These uncertainties are represented with triangular fuzzy number. The fuzzy number is symbolically written as (mean, left spread, right spread). For example, a linguistic declaration such as "power will surely not below 20 MW or above 45 MW, and the best estimate is 30 MW" will be symbolically written as $\widetilde{PD} = (PD, l_{\widetilde{PD}}, r_{\widetilde{PD}})$ = (30,10,15).

The control variables, which have to be determined, in this problem, are the bidding parameters of the first supplier (α_{1t} and β_{1t}). Remember that, these are the crisp values.

When considering uncertainties of demand and rivals' parameters, the MCP and TP_1 became triangular fuzzy number and can be calculated as follows.

1) The mean values of MCP and TP_1 are calculated from (1) and (2) by using the mean values of forecasted demand and rivals' bidding parameters.
2) The possible minimum values of MCP and TP_1 are calculated by replacing the forecasted demand and rivals' bidding parameters with their minimum values.
3) The possible maximum values of MCP and TP_1 are calculated by replacing the forecasted demand and rivals' bidding parameters with their maximum values.

When $TP_{1t} - l_{\widetilde{TP1t}}$ is less than $TP_{1\min}$, $l_{\widetilde{TP1t}}$ should be set to $TP_{1t} - TP_{1\min}$. When $TP_{1t} + r_{\widetilde{TP1t}}$ greater than $TP_{1\max}$, $r_{\widetilde{TP1t}}$ should be set to $TP_{1\max} - TP_{1t}$.

Because $\widetilde{MCP_t}$ and $\widetilde{TP_{1t}}$ are triangular fuzzy numbers, the revenue is also a triangular fuzzy number. It's left and right spreads are calculated from (EL-Hawary, 1998)

$$l_{\widetilde{RV}} = \sum_{t=1}^{24}\left(TP_{1t} \cdot l_{\widetilde{MCP_t}} + MCP_t \cdot l_{\widetilde{TP_{1t}}} - l_{\widetilde{MCP_t}} \cdot l_{\widetilde{TP_{1t}}}\right) \quad (10)$$

$$r_{\widetilde{RV}} = \sum_{t=1}^{24}\left(TP_{1t} \cdot r_{\widetilde{MCP_t}} + MCP_t \cdot r_{\widetilde{TP_{1t}}} + r_{\widetilde{MCP_t}} \cdot r_{\widetilde{TP_{1t}}}\right) \quad (11)$$

After the energy auctions, *MCP* and *TP* are known. The revenue becomes a fixed term in (4). To achieve maximum profit, the supplier must commit the units at the minimum cost. The total production cost can be calculated by minimizing (8). This is done by solving the economic dispatch (ED) problem.

$$\min \quad TC = \sum_{t=1}^{24}\sum_{i=1}^{N} F(P_{it}) \quad (12)$$

subject to (5)

The left and right spreads of total production cost are computed by solving (12) for $TP_{1t} - l_{\widetilde{TP_{1t}}}$ and $TP_{1t} + r_{\widetilde{TP_{1t}}}$ respectively.

The possible minimum and maximum values of profit are calculated from (13) and (14) respectively.

$$PF - l_{\widetilde{PF}} = \min((RV - l_{\widetilde{RV}}) - (TC - l_{\widetilde{TC}}), (RV + r_{\widetilde{RV}}) - (TC + r_{\widetilde{TC}})) \quad (13)$$

$$PF + r_{\widetilde{PF}} = \max((RV - l_{\widetilde{RV}}) - (TC - l_{\widetilde{TC}}), (RV + r_{\widetilde{RV}}) - (TC + r_{\widetilde{TC}})) \quad (14)$$

After the above steps, finally, we design a new objective function as

$$\max \quad \left(w_{PF} \cdot PF - w_l \cdot l_{\widetilde{PF}}\right) \quad (15)$$

where w_{PF} and w_l are weight coefficients.

The objective function expressed in (15) is different from (4) where we assume that we have all correct predicted data in hand. In (15) the risk that induced by the uncertainties of rivals are included. The GENCO solves bidding strategy not only to maximize its profit but also to minimize its risk due to the uncertainties. The high enough expected profit (high *PF*) within an accepted risk level (small value of $l_{\widetilde{PF}}$) is preferred.

5. BIDDING STRATEGY USING EVOLUTIONARY PROGRAMMING

The EP is a computational intelligence method in which an optimization algorithm is the main engine for the process of three steps; namely, natural selection, mutation and competition. It is a stochastic optimization strategy, which places emphasis on the behavioural linkage between parents and their offspring. It operates on populations of real values that represent the parameter set of the problem to be solved over some finite ranges. Each representation is an individual in the EP population. The population is initialized with random individuals at the start of the EP. The EP searches the space of possible real values for better individuals. The search is guided by fitness values returned by the environment. The EP uses two types of rules, named the selection rule and combination rule in its search. The selection rule is used to determine the individuals that will be represented in the next generation. It includes competition in which each individual in the combined population has to compete with some other individuals to get chance to be transcribed to the next generation. On the other hand, the combination rule operates on selected individual to produce new individuals that appear in the next generation. The selection mechanism is based on a fitness measure or objective function values, defined on each individual in the population. The combination rule is used to introduce new individual into the current population or to create a new population based on the current population. This method can provide a high quality solution even the problem is non-convex, and has many local optimums. When EP is applied to the bidding strategy, the method can be described as in Fig. 1.

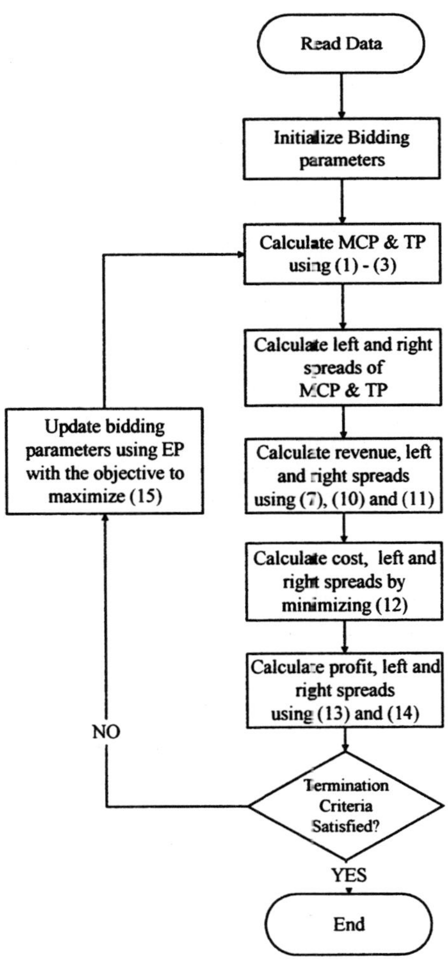

Fig. 1. Bidding strategy using EP

6. SIMULATION RESULTS

In the simulations, the first supplier is assumed to own 10 generating units, and to participate in the energy market with 2 other suppliers. Unit data are given in (Attaviriyanupap, et al., 2002), the rivals' data and demand and their spreads are given in the Appendix. In all cases, we assume that first supplier has a contract to supply 450 MW every hour with the average price 30 $/MWh.

Deterministic case

Firstly, the case in which uncertainties of rivals' bids are ignored is studied. The EP is employed to solve (4). The profit is $226,359. The optimal bidding parameters are provided in Table 1.

Uncertain case

The uncertainties of demand and rivals' bids are included in the problem. The ratio between weight coefficients w_{PF} and w_I has been varied from 1:1 to 1:0.25 with 0.25 step size. The triangular fuzzy numbers that represent the profit for each ratio are shown in Table 2.

From Table 2, it can be seen that the average profit tends to increase when the value of w_I decreases. The possible minimum of the profit, however, becomes smaller. The decision maker has choices when he/she makes the decision. If he/she would like to maximize the expected average profit, the value of w_I should be set to the small value (set w_I to 0 in deterministic case). On the other hand, if the decision maker would like to safe his/her bidding, he/she might select to maximize his/her possible minimum profit (set $w_{PF} = w_I$). However, the possible maximum profit might lower than other cases. The average, possible minimum and maximum profits of each case are shown in Table 3.

The optimal solution in the deterministic case is compared with those obtained from the cases in which the uncertainties are considered. The parameters shown in Table 1 are used and the average profits and their left and right spreads are calculated using (10) – (14). The left and right spreads of profit are $92,882 and $102,054 respectively. With the mean profit $226,359 in the deterministic case, it can be seen that, if the first supplier bids using these parameters, the possible minimum profit is $133,477 (=$226,359-$92,882), while the possible maximum profit is $328,413 (=$226,359+$102,054). The comparison results between deterministic and uncertain case ($w_{PF}:w_I$ = 1:1) are given in Fig. 2. The example of solutions in the uncertain case ($w_{PF}:w_I$= 1:1) is shown in Table 4.

We have then investigated the sensitivity of the bidding parameters of both cases. Firstly, the spreads of demand, α_{jt} and β_{jt} (j=2 and 3) were varied from 10% to 25%. The possible minimum and maximum profits are shown in Fig. 3. Secondly, the mean values of demand, α_{jt} and β_{jt} (j=2 and 3) were changed from original values by 5% and 10%. The results are shown in Fig. 4. According to Fig. 3, the sensitivity of the possible maximum profit from deterministic and uncertain cases is not so different. However, the possible minimum profit from deterministic seems more sensitive than the uncertain case. This implies that considering uncertain parameter provides more stable result. From Fig. 4, the sensitivity from both cases is almost the same.
From both cases, the parameter *PD* seems to affect the left and right spreads of profit more than α and β when the spreads and the mean values of uncertain variables were changed. For both of which the effect of the third supplier was larger than the effect of the second supplier.

Table 1 Solutions in deterministic case

H	$\alpha_{1t}+\beta_{1t}SP_{1t}$	H	$\alpha_{1t}+\beta_{1t}SP_{1t}$
1	$15.06+0.0153SP_{1t}$	13	$21.84+0.0121SP_{1t}$
2	$16.42+0.0103SP_{1t}$	14	$15.66+0.0163SP_{1t}$
3	$17.14+0.0102SP_{1t}$	15	$15.16+0.0151SP_{1t}$
4	$16.67+0.0127SP_{1t}$	16	$15.02+0.0173SP_{1t}$
5	$16.58+0.0127SP_{1t}$	17	$20.26+0.0132SP_{1t}$
6	$15.97+0.0162SP_{1t}$	18	$15.39+0.0129SP_{1t}$
7	$14.68+0.0133SP_{1t}$	19	$17.85+0.0177SP_{1t}$
8	$14.00+0.0149SP_{1t}$	20	$22.04+0.0100SP_{1t}$
9	$15.74+0.0132SP_{1t}$	21	$14.57+0.0173SP_{1t}$
10	$14.75+0.0166SP_{1t}$	22	$20.47+0.0100SP_{1t}$
11	$18.78+0.0136SP_{1t}$	23	$15.45+0.0111SP_{1t}$
12	$14.03+0.0186SP_{1t}$	24	$15.17+0.0169SP_{1t}$

Table 2 Mean profit and left-right spreads

$w_{PF}:w_I$	Mean Profit ($)	Left Spread ($)	Right Spread ($)
1:1	220697	83747	105720
1:0.75	225155	88733	101203
1:0.50	225123	89265	102610
1:0.25	225908	90916	103127

Table 3 Possible minimum and maximum profits

$w_{PF}:w_I$	Mean ($)	Minimum ($)	Maximum ($)
1:1	220697	136950	326417
1:0.75	225155	136422	326357
1:0.50	225123	135858	327733
1:0.25	225908	134992	329035

7. CONCLUSION

The algorithm for determining the optimal bidding strategy for a GENCO in the 24-hour energy market was proposed. In this market, the GENCO submits 24-hourly supply-bidding curves for energy. The market is cleared based on the clearing price system. The profit of GENCO depends on the way he/she bids. The bidding strategy, therefore, is an attractive problem to GENCO. However, because GENCO has no all-perfect information of other rival suppliers and demand for each hour, it is very difficult to bid and to get the profit that he/she expects. These uncertainties in information might cause the loss to GENCO. Certainly, selling the energy through the future market or bilateral contract has very less risk compared with pool-type auction. However, GENCO loses his/her chance to earn more profit from the rest energy capacity he/she has.

The proposed bidding strategy including the uncertainties is more effective than the deterministic one. We demonstrated the algorithm using a test system. And the results showed that the proposed method could provide a high-enough profit within an acceptable risk (acceptable possible minimum profit).

The stochastic optimization method EP was applied to solve the bidding strategy problem. We used this method because it can provide a high quality solution, and gives non-unique solutions, which provide decision maker the choices of decisions. Because we used EP here, it is simple to include constraint such as ramp-rate limit and/or to expand the problem to cover reserve market. These are what we are thinking for the next steps.

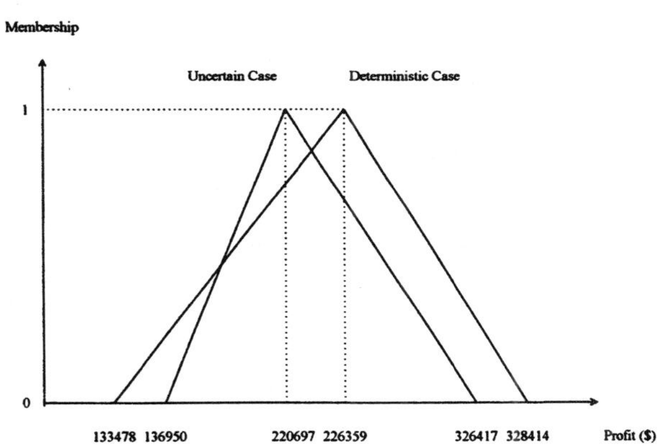

Fig 2. Comparisons of results

Table 4 Solutions in uncertain case

H	$\alpha_{It} + \beta_{It}SP_{It}$	H	$\alpha_{It} + \beta_{It}SP_{It}$
1	$15.69+0.0206SP_{It}$	13	$14.69+0.0156SP_{It}$
2	$15.93+0.0173SP_{It}$	14	$17.07+0.0115SP_{It}$
3	$15.47+0.0198SP_{It}$	15	$17.23+0.0100SP_{It}$
4	$14.00+0.0182SP_{It}$	16	$18.17+0.0111SP_{It}$
5	$14.84+0.0124SP_{It}$	17	$18.74+0.0122SP_{It}$
6	$15.18+0.0147SP_{It}$	18	$17.73+0.0137SP_{It}$
7	$17.64+0.0108SP_{It}$	19	$14.00+0.0145SP_{It}$
8	$17.98+0.0100SP_{It}$	20	$21.76+0.0100SP_{It}$
9	$16.24+0.0117SP_{It}$	21	$19.09+0.0100SP_{It}$
10	$14.00+0.0159SP_{It}$	22	$14.00+0.0139SP_{It}$
11	$14.55+0.0141SP_{It}$	23	$14.46+0.0188SP_{It}$
12	$17.44+0.0115SP_{It}$	24	$14.00+0.0187SP_{It}$

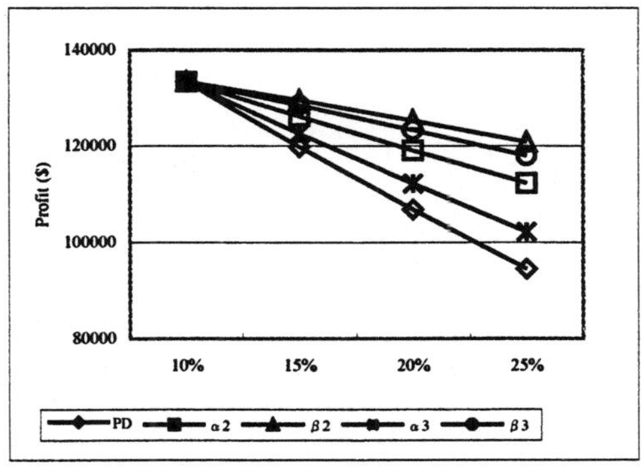

(a) Possible Minimum Profit (Deterministic)

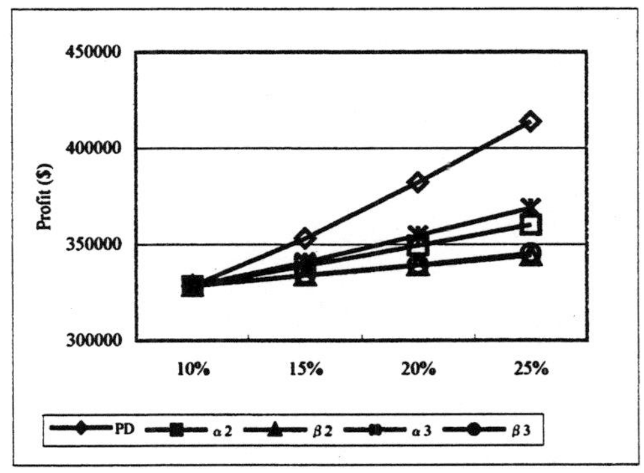

(b) Possible Maximum Profit (Deterministic)

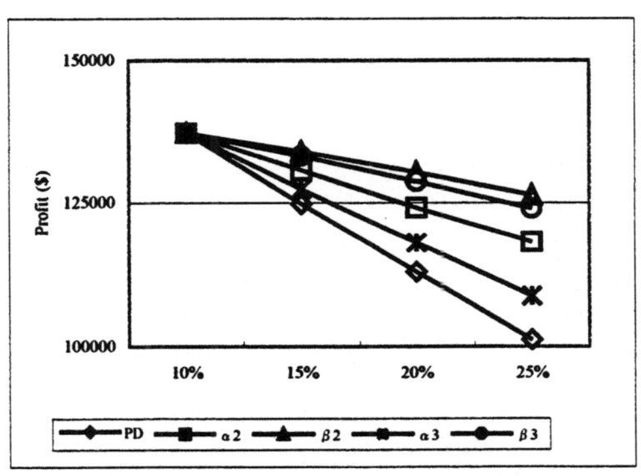

(c) Possible Minimum Profit (Uncertain)

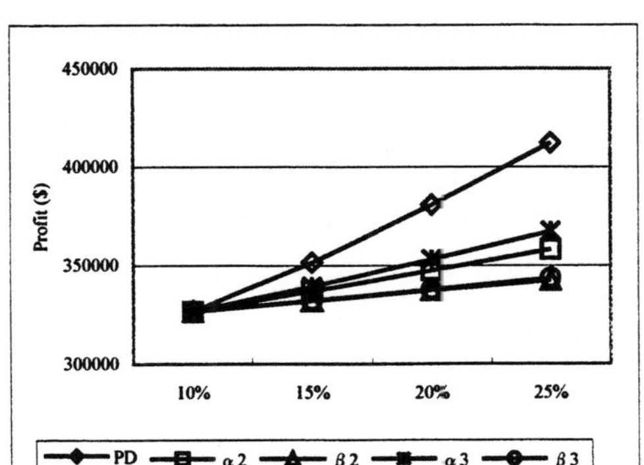

(d) Possible Maximum Profit (Uncertain)

Fig 3. Sensitivity when the left and right spreads changed

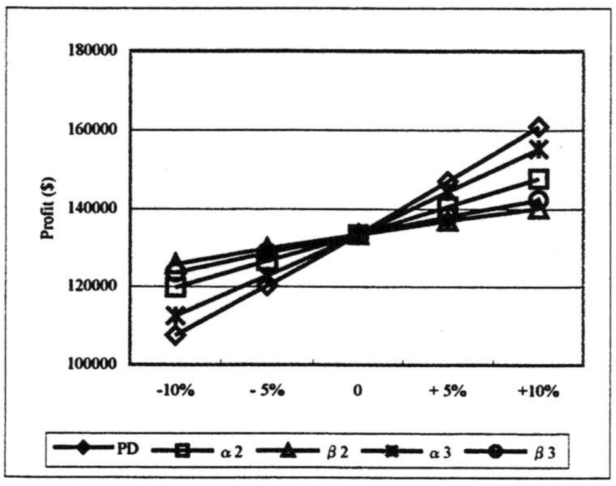

(a) Possible Minimum Profit (Deterministic)

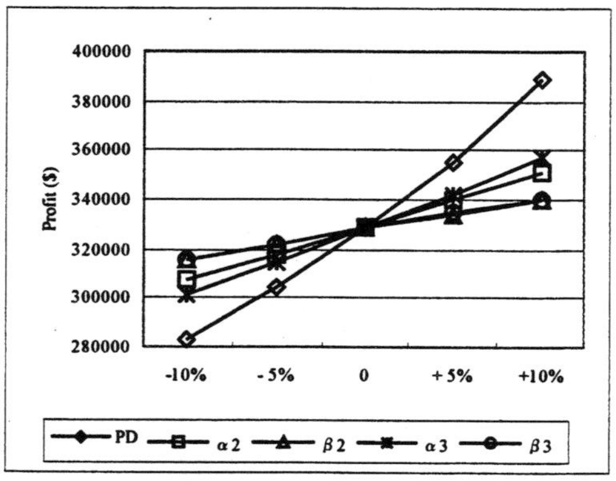

(b) Possible Maximum Profit (Deterministic)

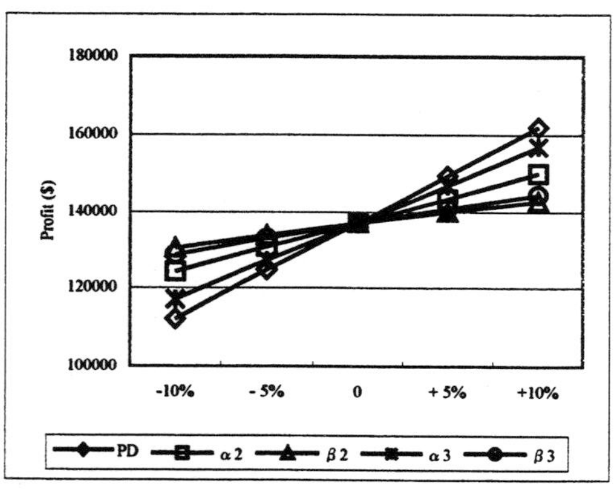

(c) Possible Minimum Profit (Uncertain)

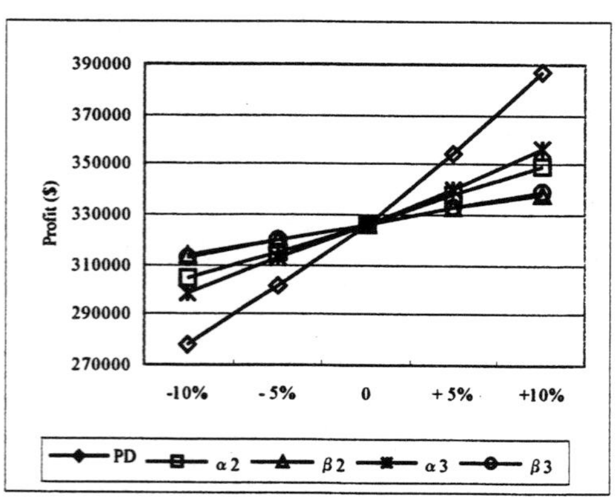

(d) Possible Maximum Profit (Uncertain)

Fig 4. Sensitivity when the mean value changed

8. APPENDIX

<u>Average rivals' bidding parameters and demands</u>

H	Supplier 2		Supplier 3		Demand
	α	β	α	β	
1	16.80	0.0174	16.00	0.0200	810
2	16.76	0.0175	16.30	0.0255	790
3	16.76	0.0175	16.30	0.0260	710
4	17.88	0.0180	16.90	0.0288	1000
5	17.77	0.0210	17.00	0.0285	1120
6	17.85	0.0183	17.18	0.0285	1140
7	18.00	0.0179	17.50	0.0250	1280
8	18.44	0.0179	18.10	0.0200	1670
9	18.42	0.0182	18.55	0.0180	1740
10	19.22	0.0198	19.44	0.0176	2100
11	19.22	0.0210	20.67	0.0176	2210
12	19.00	0.0244	20.70	0.0140	2330
13	19.30	0.0244	22.50	0.0140	2400
14	17.00	0.0244	22.00	0.0132	2290
15	18.46	0.0200	21.79	0.0135	2190
16	20.11	0.0200	23.20	0.0130	2450
17	23.75	0.0190	24.30	0.0127	2770
18	24.37	0.0210	25.60	0.0100	3000
19	24.37	0.0240	26.80	0.0100	3180
20	23.90	0.0240	25.30	0.0100	2890
21	24.00	0.0182	24.30	0.0100	2560
22	20.16	0.0185	22.34	0.0133	2180
23	18.68	0.0196	19.86	0.0136	1320
24	18.56	0.0208	17.20	0.0300	990

* Left spread = right spread = 10% of mean value

REFERENCES

Attaviriyanupap, P., H. Kita, E. Tanaka and J. Hasegawa (2002). A New Profit-Based Unit Commitment Considering Power and Reserve Generating. *Proc. IEEE Power Engineering Society Winter Meeting*, pp. 1311-1316.

David, A. K. (1993). Competitive bidding in electricity supply. *IEE Proceedings: Generation, Transmission and Distribution*, **Vol. 140, no. 5**, pp. 421-426.

EL-HAWARY, M. E. (1998). *Electric Power Applications of Fuzzy Systems*, pp. 38-43. IEEE Press, New York, United States.

Gross, G. and D. J. Finlay (1996). Optimal bidding strategies in competitive electricity markets. *Proc. of the 12th PSCC*, pp. 815-822.

Hao, S. (1999). A Study of Basic Bidding Strategy in Clearing Pricing Auctions. *Proc. of the 21st PICA*, pp. 55-60.

Lai, L. L (2001). *Power System Restructuring and Deregulation – Trading, Performance and Information Technology*, pp. 50-75. John Wiley & Sons LTD, Chichester, England.

Richter, C. W. and G. B. Sheble (2000). A Profit-Based Unit Commitment GA for the Competitive Environment. *IEEE Transactions on Power Systems*, **Vol. 15, no. 2**, pp. 715-721.

Wen, F. and A. K. David (2001). Optimal Bidding Strategies and Modelling of Imperfect Information Among Competitive Generators. *IEEE Transactions on Power Systems*, **Vol. 16, no. 1**, pp. 15-21.

Zhang, D., Y. Wang and P. B. Luh (2000). Optimization Based Bidding Strategies in the Deregulated Market. *IEEE Transactions on Power Systems*, **Vol. 15, no. 3**, pp. 981-986.

ELSEVIER
IFAC
PUBLICATIONS
www.elsevier.com/locate/ifac

A BIDDING STRATEGY DECISION SUPPORT APPLICATION FOR MARKET PARTICIPANTS

JC.Passelergue, B.Marie and C.Galzin

ALSTOM T&D, Energy Automation & Information Business,
9 rue Ampère, 91345 Massy Cedex, France

Abstract: This panel presents the formulation of an decision support application to help the market participants to define their bidding strategy. The tool maximizes the market participant expected profit while keeping his risk exposure under a limit value. The risk exposure is measured using the downside variance principle. *Copyright © 2003 IFAC*

Keywords: Energy Markets, Bidding strategy, Decision support tool, Power system economics.

1. INTRODUCTION

Traders, power producers and industrial energy consumers have today several opportunities to buy and sell energy on power exchanges or balancing markets. This trading diversity will introduce more and more complexity to the market participants in evaluating the potential revenue they can get out of these different markets.

This paper presents an decision support solution to help the user in the determination of a bidding strategy by maximizing his expected profit while keeping his risk exposure below a limit value. The proposed approach takes advantages of the market price forecasts and their standard deviations while taking into account transmission, technical and strategic constraints.

The proposed formulation can be used by all the market participants. Indeed, it can be used by a GenCO trying to sell its remaining energy and/or trying to buy energy in order to reduce its scheduled energy if it is economic to do so. At the same time, it can be used by an industrial consumer (eligible consumer) to buy energy or to sell its previously contracted energy if it is economic to do so. In this paper, by generation bids, we will consider a generation increase for a GenCO and a scheduled consumption decrease for a load. By load bids, we will consider a generation decrease for a GenCO and a consumption increase for a load.

The formulation deals with risk strategy. There is a trade-off to define between risk and reward: to obtain a greater expected profit, one must be willing to take on greater risk. In a more concrete formulation, by putting a higher price on generation bids, we can expect a better payoff, but we increase the risk not to be selected, and consequently to loose a potential revenue. In the proposed solution, the risk exposure of a strategy is given by the downside variation. Using this risk measure, the best bidding strategy can be obtained through a quadratic optimization using market price forecasts while taking in account price uncertainty, maximum risk level and transmission constraints. Market participant then has a view of the global estimated profit on different markets, the probability to be selected for each bid, etc.

2. BIDDING STRATEGY CONSIDERATIONS

In a uniform pricing market, awarded bids are paid at the system marginal price (SMP), which is the price of the last awarded bid. Such a market design is often recognized as providing to the market participants good incentives to not assume a "gaming" behavior. Indeed, with bids paid at the SMP, it could be thought that the best strategy is simply to bid at a price defined directly from the total generation cost (including the fuel costs, the maintenance costs and the crew costs) corrected by a minimum margin factor.

In a pay-as-bid market, the awarded bids are paid at the bid price. Of course, a passive bidding strategy can be applied by the market participant, deriving the bid price from his operation cost. But the benefits of an application helping the bidder to define the "right price" of his bid are easily understandable, because his global profit is directly linked with the bid price. More, such a passive bidding strategy can be risky on the operational point of view, whatever is the market type – pay-as-bid or uniform pricing. When ignoring the market price forecast and the unit ramp constraint, the risk of getting an unfeasible unit schedule after the market clearing is not mitigated.

In a multi-market environment, the problem dimension is even higher. Even if the bidder has a simple strategy to define his bid price, he needs to define on what market place he should submit his bid to maximize his expected profit. As the problem is becoming more and more complex, the need of having a decision support application is becoming more and more obvious.

The proposed decision support solution allows the bidder to define the best bidding strategy, maximizing his expected profit while limiting his risk exposure. Particularly useful in a multi-market or a pay-as-bid single market context, it can also be profitable when applied on a uniform pricing market to mitigate the ramp violation risk. More, applying an inter-temporal optimization, the proposed solution is able to define a strategy consisting on trading at a loss for a given period in order to sell more and get a greater profit during the adjacent periods. This trade-off results from the ramp constraint taking into account.
The following sections presents an decision support solution allowing the bidder to define his bidding strategy through an optimization process.

3. GLOBAL PROCESSING

The bidding study is carried out using quadratic optimization algorithms. Generation and demand bids are considered simultaneously. The results are obtained using a two-step optimization. First, the potential profit loss risk is minimized while trying to reach a minimum expected profit. Then, setting this obtained risk value as a maximum risk level, the expected profit is maximized. This processing is shown in figure 1.

More, this optimization framework allows defining additional linear constraints, such as zonal or area interchange limitations, asset technical constraints. Also, a specific distribution strategy of bids over market places can be defined (defining for instance a minimum level of energy to bid on a given market place).

However, the risk measure has to be relevant to get pertinent results.

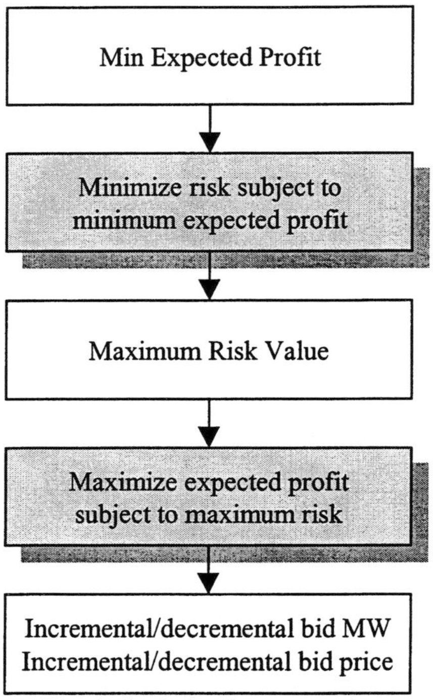

Fig. 1. Optimization processing.

4. RISK MODEL

If the market participant is not awarded in the market, he will not suffer operation losses. However, he will suffer potential profit losses. As a result, the risk measure should represent these potential profit losses (lost opportunity).

In the proposed approach, this risk measure is based on the downside risk determination. The concept of variance fits to the intuitive notion of variation around an expected value. Markowitz introduced this concept in 1956 to deal with the portfolio selection problem (Markowitz, 1959). In his theory, the variance of the total portfolio return is the measure of portfolio risk. Variance allows measuring the risk because it is always nonnegative and no variance is a reflection of no risk. However, the variance penalizes both the high and low returns. In this sense, it fails to capture a reasonable preference for higher returns. The concept of one-sided variance can be derived from the variance concept (Barlow, 1991; Rom, *et al.*, 1994 ; Bjorgan, *et al.*, 1999). It considers deviations either below of some target values (downside deviation) or above of some specific target values (upside deviation).

The downside deviation gives a measure of variation around a specific target value. By minimizing it, the decision support tool will help the user to define the best strategy to minimize the risk of profit losses while considering a minimum expected profit. It should be noted here that this minimum expected profit is defined by the user, and it would vary significantly from a user to an other depending of his risk-loving or risk-adverse behavior.

This measure of the downside risk is an appropriate risk estimate of our problem. In fact, considering figures 2, 3 and 4, we can observe that the maximization of a criterion based on the expected profit can provide a riskier strategy (figure 3). In the same way, the use of variance with its symmetric character introduces a penalization of the positive distances to the minimum profit (figure 4).

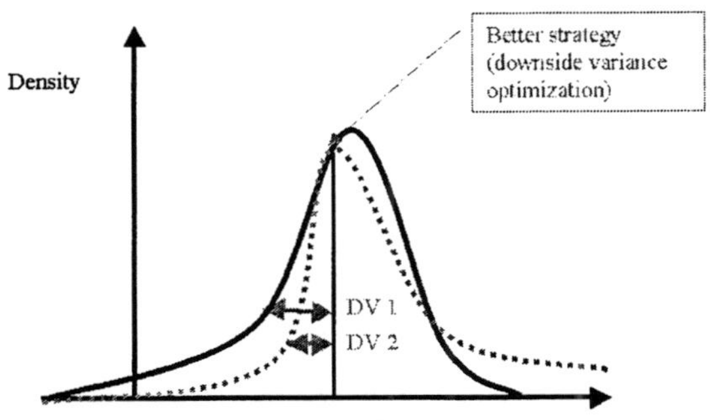

Fig. 2. Downside risk optimization.

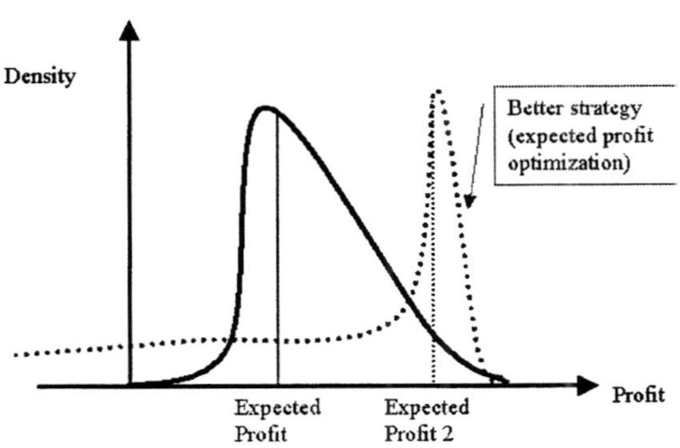

Fig. 3. Expected profit optimization.

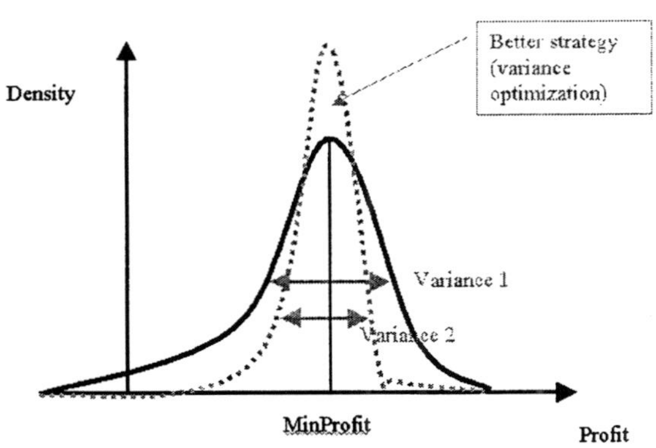

Fig. 4. Variance optimization.

Based on the downside variance formulation, the risk measure, for period h, is given by:

$$\sigma_h^2 = \sum_s Pr_s \cdot \left[min(0, profit_{s,h} - minprofit_h) \right]^2 \quad (1)$$

where σ_h^2 is the risk measure for the period h, and Pr is the probability of having the market marginal price within the price range s. *Minprofit* is the minimum profit defined by the user (in \$) and *profit* the variable (in \$) representing the profit resulting from a bid award in the price range s.

As this formulation is based on price range scenario probabilities, it can be applied both on single-market and multi-market contexts.

In the simplest case, the profit is defined by the difference between the sale revenue and the production cost. Its formulation, for a given market, will depend of the market type: pay-as-bid market or uniform pricing market.

For a pay-as-bid market, the profit from generation will be defined by the sum over all the awarded generation bids (i.e. generation bids with a price lower than the SMP) of the difference between the energy bid price and the energy bid cost, multiplied by the energy bid quantity. The profit from load will be defined by the sum over all the awarded load bids (i.e. load bids with a price higher than the SMP) of the difference between the energy bid cost and the energy bid price, multiplied by the energy bid quantity.

For a uniform pricing market, the profit from generation will be defined by the sum over all the awarded generation bids of the difference between the system marginal price and the energy bid cost, multiplied by the energy bid quantity. The profit from load will be defined by the sum over all the awarded load bids of the difference between the energy bid cost and the system marginal price, multiplied by the energy bid quantity.

However, this profit definition can be altered by additional transmission or market fees. In many cases, these fees can be modeled as a market price bias, or a bid cost bias. For instance, let's assume that the bidder owns two units, the first one in an area A and the second one in an area B. Considering that the market place is located in area A, and a transmission fee from area B to area A exists, when formulating the profit for the unit in area B, the energy bid cost has to be increased by this transmission fee.

5. OPTIMIZATION FORMULATION

As described in the previous section, the solution is obtained through a sequential process.

The first step corresponds to a risk minimization. Given a targeted minimum profit, the risk minimization problem can be formulated as follows:

$$\begin{cases} minimize \ \sum_h \sigma_h^2 \\ subject \ to \\ \quad -Ramp \ constraints \\ \quad -Energy \ bid \ quantity \ constraints \\ \quad -Minimum \ profit \ constraint \\ \quad -Generic \ constraints \end{cases} \quad (2)$$

The ramp constraint modeling introduces a ramp violation variable. This variable will not be null if, for two consecutive scenarios (temporally speaking), the unit generation level varies over the unit ramp capacity. The ramp violation penalty is associated with a penalty price. In case of violation, the global profit is decreased by this ramp violation cost multiplied by the probability of seeing these two consecutive scenarios occur.

The energy bid quantity constraints make sure that the bidder cannot bid more energy than available. However, it can be noted here that a risk-loving bidder could wish to relax such a constraint, considering that the probability to have all his bids awarded is close to zero. This can be reached multiplying by an overgeneration factor the quantity of available energy.

The generic constraints are linear constraints allowing the bidder to set a minimum or maximum limit for the quantity of energy submitted in a specific market. This type of constraint can be used, for instance, to be sure that a minimum quantity of energy will be submitted in each single market (according to the rule of "never put all your eggs in a same basket"). These generic constraints also allow the bidder to deal with inter-area limitations.

The maximization of the expected profit, subject to a maximum non-linear risk measure constraint, would result in a painful, time consuming computation problem. To get round it, the expected profit maximization problem can be formulated as follows:

$$\begin{cases} \textit{minimize} \sum_h \sigma_h^{\,2} - \lambda \cdot \sum_h \textit{ExpProfit}_h \\ \quad \textit{subject to} \\ \qquad -\textit{Ramp constraints} \\ \qquad -\textit{Energy bid quantity constraints} \\ \qquad -\textit{Minimum profit constraint} \\ \qquad -\textit{Generic constraints} \end{cases} \tag{3}$$

with:

$$\textit{ExpProfit}_h = \sum_s Pr_s \cdot \textit{profit}_{s,h} \tag{4}$$

This formulation leads us to simultaneously minimize the risk and maximize the expected profit over the study period. For a risk-avert position, λ would be set as a near-to-zero value, while risk-loving bidders would put a higher λ value. The constraints modeled in this formulation are identical to the previous ones.

6. EXAMPLE

Here is given a simple example to illustrate the variance principle. Let us assume that a quantity of energy of 100 MWh is available for sale at period h on a pay-as-bid market. For simpleness, let's assume that the cost for producing this energy is null. The estimated price distribution on a given market for this period h is as follows:

Price range ($)	Probability for having the clearing price in price range
From 10 to 14	0.2
From 14 to 15	0.8

In such a case, the maximization of the expected profit is formulated as:

$$\textit{maximize}\begin{bmatrix} 0.2\cdot(10\cdot\ q1) \\ +\ 0.8\cdot(10\ \cdot q1+14\cdot\ q2) \end{bmatrix} \tag{5}$$

where $q1$ is the quantity of energy to sell at 10$/MWh and q2 the quantity to sell at 14$/MWh. The maximum expected profit will be given by a sale bid of 100MW at 14$/MWh for an expected profit of 1120$, but with a chance of only 80% to see this bid cleared.

Let's assume now that we want to get a profit not inferior to 1000$. Using the downside risk measure principle, the optimization formulation is:

$$\textit{minimize}\begin{bmatrix} 0.2\cdot[\textit{min}(0,\ 10\cdot q1\text{-}1000)]^2 \\ +\ 0.8\cdot[\textit{min}(0,10\cdot q1+14\cdot q2\text{-}1000)]^2 \end{bmatrix} \tag{6}$$

which results in proposing these 100 MWh of energy at 10$/MWh. The risk measure for this strategy is null. The probability to get less than 1000$ is (theoretically) null.

7. CONCLUSION

This paper presents a decision support solution helping the market participant to define their bidding strategy. Formulated as an optimization problem, the application suggests the quantities and prices to bid in order to maximize the expected profit subject to a risk exposure limit. The risk measure is based on the downside variance principle. This approach can be applied efficiently on single pay-as-bid or uniform pricing markets, and on a multi-market context.

ACKNOWLEDGEMENTS

The authors would like to thank J.Taisne and S.Zahzam, from ALSTOM T&D Energy Automation and Information business, for their contribution to this work.

REFERENCES

Markowitz, Harry (1959). *Portfolio Selection.* Yale University Press, New Haven.

Barlow, W.V. (1991). Asset Allocation in a Downside Risk Framework. *Financial Analysts Journal,* September-October 1991.

Brian M. Rom and Kathleen W. Ferguson (1994). Post-modern portfolio comes of age. *Proceedings of the 4th AFIR International Colloquium,* Orlando, pp. 1285-1291.

R. Bjorgan, C. C. Liu and J. Lawarree (1999). Financial Risk Management in a Competitive Electricity Market. *IEEE Trans. Power Systems, Nov. 1999,* pp. 1285-1291.

ELSEVIER
IFAC
PUBLICATIONS
www.elsevier.com/locate/ifac

MULTI-AGENT BASED STABILIZATION CONTROL USING ENERGY CAPACITOR SYSTEM

Takashi Hiyama **Youhei Matsumoto** **Yuusuke Hara**

Dept of Electrical and Computer Engineering
Kumamoto University, Kumamoto 860-85555, Japan

Abstract: A multi-agent based stabilization control scheme has been proposed for the Energy Capacitor System(ECS) using electrical double-layer capacitors. For the further enhancement of the overall stability of electric power systems, the wide area stabilization control is inevitable based on the information monitored by wide area measurement systems. All the required information is distributed through computer networks for the selection of the site of stabilization control, and also for the decision of control command. To demonstrate the efficiency of the proposed multi-agent based stabilization control, experimental studies have been performed on a 5kVA one-machine infinite-bus laboratory system equipped with a 70Wh (250kJ) ECS. *Copyright © 2003 IFAC*

Keywords: Multi-agent system, electrical double-layer capacitor, energy capacitor system, fuzzy logic control, stability enhancement.

1. INTRODUCTION

Due to the increasing size and complexity of electric power systems, as well as the growth of electric power demand without corresponding infrastructure upgrades, electric power systems are often operated with lower stability margins. Stability improvement is one of the major research subjects for improved power system operation. A new device, the electrical double-layer energy capacitor system (ECS) (Okamoto 1995; Ohshima, *et. al.*, 1998) has been developed for the energy storage.

Switching control of braking resistors is one of the effective means to enhance the overall stability of electric power systems. In the case of the braking resistors, only the absorption of real power is available whenever the braking resistors are switched on; however, the injection of real power is impossible. On the other hand, the absorption and the injection of real power is possible by charging and discharging real power to and from the Energy Capacitor System(ECS) using electrical double-layer capacitors (Hiyama, et. al., 2000: Hiyama, *et. al.*, 2001). In this paper, a multi-agent based stabilization control scheme has been proposed for the ECS. For the further enhancement of the overall stability of electric power systems, the wide area stabilization control is inevitable based on the information monitored by wide area measurement systems. All the required information is distributed through computer networks for selecting the site of stabilization control, and also for determining the control command. A fuzzy logic based control scheme (Hiyama, *et. al.*, 1994: Hiyama, *et. al.*, 1996) is also proposed for the charging or discharging operation on the ECS to enhance the overall stability of electric power systems. The real power flow signal, monitored by the interface agent at the location of the ECS, is utilized to generate stabilization control signals on the supervisor agent connected to the interface agent through the computer network.

To demonstrate the efficiency of the proposed multi-agent based stabilization control, experimental studies have been performed on a 5kVA one-machine infinite-bus laboratory system with a 70Wh (250kJ) ECS. Experimental results clearly indicate the efficiency of the proposed multi-agent based stabilization control scheme even in the existence of the communication delay time through the computer network.

2. FUZZY LOGIC CONTROL SCHEME

The basic configuration of the proposed fuzzy logic control system is shown in Fig. 1 for the charging or the discharging operation of the ECS. The real power flow signal Pe is utilized to generate the stabilization control signal $Uecs$ for the ECS. The proposed controller is implemented by using a personal computer with a DSP board as shown later.

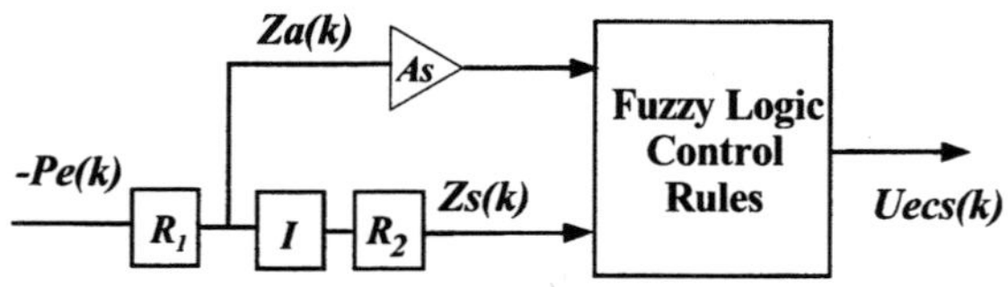

R : Reset Filter I : Integrator
U : Switching Control Signal

Fig. 1. Proposed Fuzzy Logic Control System

The speed and acceleration signals of the generator, derived from the real power flow signal P_e at the location of the ECS through the filtering shown in Fig. 1, are utilized for the stabilization control. Here, Z_a and Z_s are local measures of the acceleration and the speed deviation, respectively, at the location of the ECS. The two dimensional information, i.e., Z_a and Z_s, is utilized to generate the stabilization control signal $Uecs$ to damp the oscillations in electric power systems.

In the fuzzy logic control scheme, the system state is defined by the speed/acceleration state $p(k)$ as shown in Fig. 2. The operating point $p(k)$ is given by

$$p(k) = [\, Z_s(k) \quad A_s Z_a(k) \,] \qquad (1)$$

Then, the radius $D(k)$ is found as

$$D(k)=\sqrt{Z_s(k)^2+(A_s \cdot Z_a(k))^2} \qquad (2)$$

and the angle $\theta(k)$ as

$$\theta(k)=\tan^{-1}(A_s \cdot Z_a(k)/Z_s(k)) \qquad (3)$$

Note that Z_a and Z_s become zero at the final steady state once the study system is stabilized. Further note that A_s is simply a scaling factor for the acceleration signal Z_a.

In Fig. 2, Sector A gives the region where deceleration control is required, that is, charging operation is required for the ECS in order to withdraw the real power from the power system. Sector B is the region where acceleration control is required, that is, discharging operation is required in order to inject real power flow into the power system. The two sectors are defined by using the angle membership functions $N(\theta(k))$ and $P(\theta(k))$ while an additional membership function $G(D(k))$ is needed for the control signal. The function $N(\theta(k))$ gives the grade of charging, and $P(\theta(k))$ gives the grade of discharging. The function $G(D(k))$ determines the level of the control signal. Fig. 3 illustrates the respective membership functions.

A continuous switching control signal is determined by using the following defuzzification function

$$\begin{aligned} Uecs(k) &= \frac{N(\theta(k))-P(\theta(k))}{N(\theta(k))+P(\theta(k))} \cdot G(D(k)) \cdot U_{max} \\ &= [1-2P(\theta(k))] \cdot G(D(k)) \cdot U_{max} \end{aligned} \qquad (4)$$

where U_{max} is the maximum size of the switching control signal. The proposed switching control scheme has three basic adjustable control parameters A_s, D_r, and the overlap angle α. The parameter α is fixed to 90 degrees in this study.

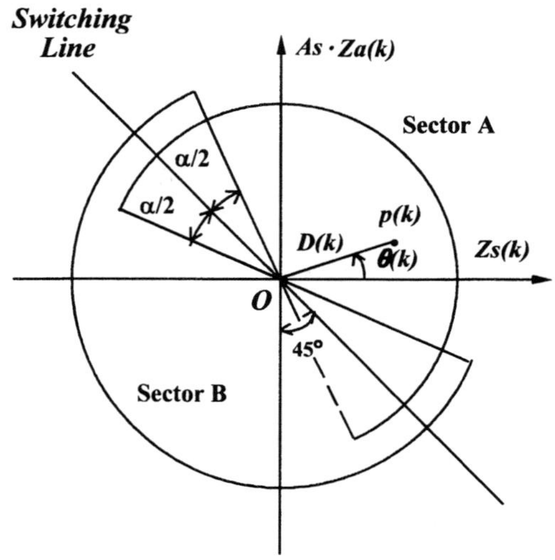

Fig. 2. Phase Plane for Fuzzy Logic Control

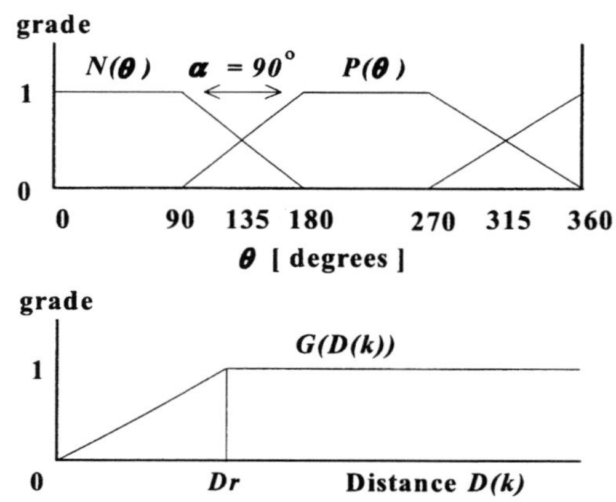

Fig. 3. Angle and Radius Membership Functions

3. LABORATORY SYSTEM

To demonstrate the efficiency of the multi-agent based stabilization control scheme, experimental studies have been performed on a 5kVA one-machine infinite-bus laboratory system shown in Fig. 4. The ECS is connected to the generator terminal through the AC/DC conversion unit as shown in the figure. The generator capacity is 5kVA, and the capacity of the ECS is 70Wh with the maximum charging/discharging power of 3kW. The line switching by the switch S_1 or S_2 is the disturbance for the experimental system.

Two different types of agents are considered in this study: the interface agent which monitors the real power flow at the location of the ECS and also sends the charging or the discharging command to the AC/DC conversion unit. For the interface between the ECS system and the PC based interface agent, the AD/DA conversion is utilized. The charging or the discharging command, which is determined on the supervisor agent using the monitored real power by the interface agent, is sent to the interface agent from the supervisor agent through the computer network. Namely, the computer network is utilized for the interface between the two different types of agents.

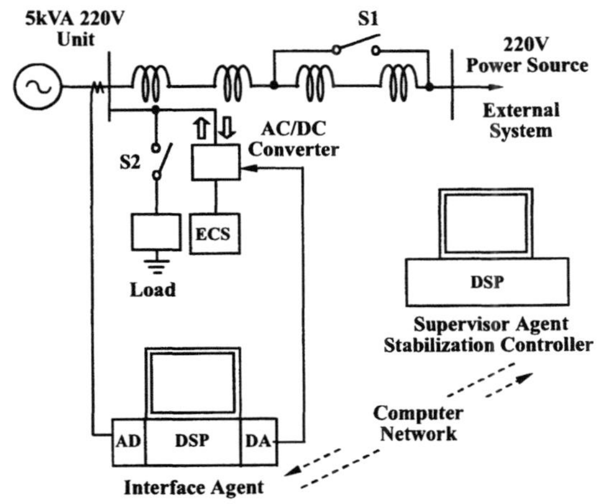

Fig. 4. One-machine Infinite-bus Laboratory System

Fig. 5 shows the overview of the laboratory system. The laboratory system consists of a 5kVA generator driven by a DC motor, a 70Wh Energy Capacitor System with the maximum charging/discharging power of 3kW, a variable load, and the transmission line modules

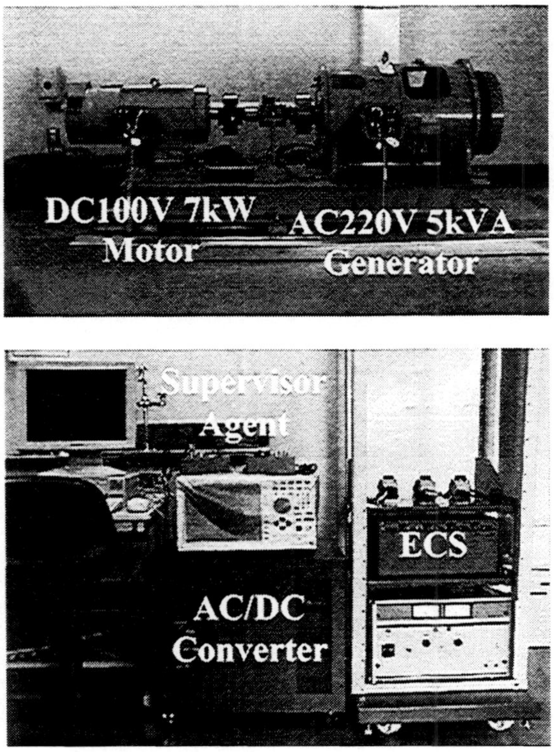

Fig. 5. Overview of Laboratory system

The proposed multi-agent based control system has been developed in the Matlab/Simulink environment together with the Real Time Workshop for its on-line real time operation. Fig. 6 gives the display on the supervisor agent, i.e., the stabilization control system. The DC voltage monitored by the interface agent is utilized to estimate the stored energy on the ECS and also to determine the control action to keep the stored energy level to its pre-specified one. The supervisor agent also has the above voltage control block to maintain the stored energy level.

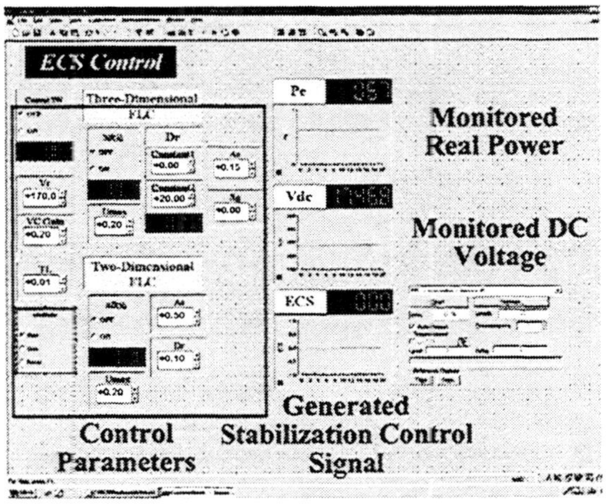

Fig. 6 Computer Display on Supervisor Agent

4. EXPERIMENTAL RESULTS

Typical experimental results are illustrated in Fig. 7 to Fig. 14. In these figures, the real power *Pe*(pu), the reactive power *Qe*(pu), the AC voltage *Vt*(pu), the power from ECS *Pecs*(pu), the ECS terminal DC voltage *Vecs*(pu), and the stabilization control signal *Uecs*(pu) are shown from the top to the bottom.

All the signals are monitored on the interface agents. The stabilization control signal *Uecs* is generated on the supervisor agent based on the real power *Pe* and the ECS terminal DC voltage *Vecs*, both of which are monitored by the interface agent and are transferred to the supervisor agent through the computer network. Here, it must be noted that there always exists the communication delay through the computer network.

Table 1 gives the fuzzy control parameters for the stabilization controllers. The setting of *Umax* up to 0.6(pu) is available on the ECS for the Base KVA of 5kVA. However, *Umax* is specified to 0.2(pu) throughout all the experiments. Namely, the maximum charging or discharging power is specified to 1kW.

Table 1 Fuzzy Control Parameters

Umax = 0.2(pu) Base kVA = 5 kVA	
Direct Stabilization Control:	*Dr* = 0.1, *As* = 0.1
MultiAgent Based Control: (Retuned Parameters)	*Dr* = 01., *As* = 0.35

In Fig. 7 to Fig. 10, the switching of S_1 was utilized to add a disturbance to the laboratory system. For comparisons, the effect of the direct stabilization control is illustrated in Fig. 8. In this case, the stabilization controller consists of just one PC with both the functions of the interface agent and the supervisor agent, therefore, there is no communication delay. As shown in this figure, the stability of the laboratory system is highly improved. In Fig. 9, the same control parameters as those for the direct stabilization control are utilized for the proposed multi-agent based stabilization control system. The stabilization effect is shortly degraded compared with Fig. 9 because of the existing communication delay time. However, the damping of oscillations in the laboratory system is still highly improved. After retuning the control parameters, the stabilization effect achieved by the multi-agent based stabilization control system is almost the same as that by the direct stabilization control as shown in Fig. 8 and Fig. 10.

Additional experiments have also been performed on the laboratory system allowing a different type of disturbance. In Fig. 11 to Fig. 14, the switching of S_2 was utilized to add a disturbance to the laboratory system. The results are almost the same as those for the switching of S_1. Through the retuning of the control parameters, the stabilization performance of the proposed multi-agent based stabilization control system is highly improved with the negligible small difference from the performance of the direct stabilization controller.

Fig. 15 and Fig. 16 show the results of experiments after changing the generator operating point to the rated output 1(pu) from the former setting of 0.6(pu). The multi-agent based stabilization control system highly improves the damping of oscillations

The degradation of the stabilization performance from the communication delay depends on the size of the delay time, therefore, the compensation should be required for the relatively longer delay time. Experimental evaluation of the compensation system is now ongoing on the laboratory system.

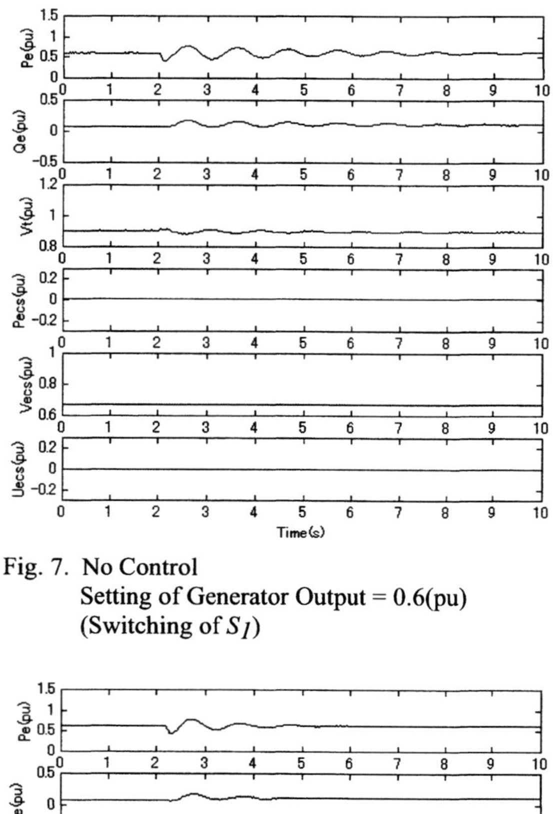

Fig. 7. No Control
Setting of Generator Output = 0.6(pu)
(Switching of S_1)

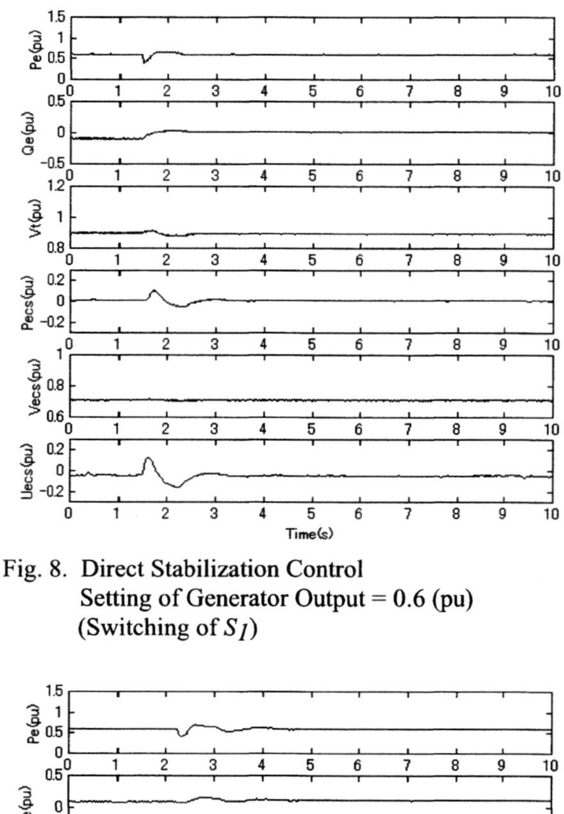

Fig. 8. Direct Stabilization Control
Setting of Generator Output = 0.6 (pu)
(Switching of S_1)

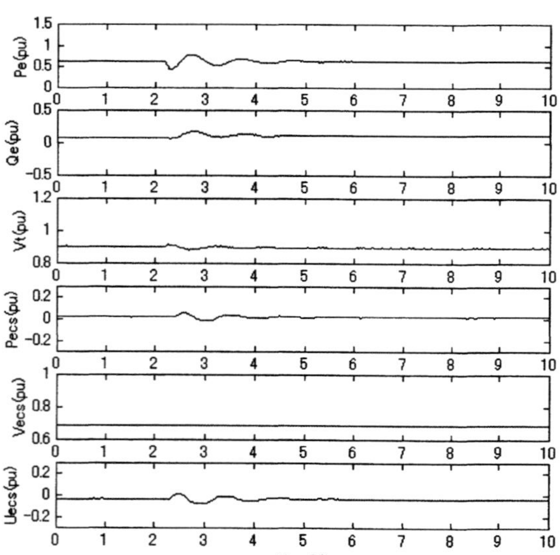

Fig. 9. Multi Agent Based Control
Setting of Generator Output = 0.6(pu)
(Switching of S_1)

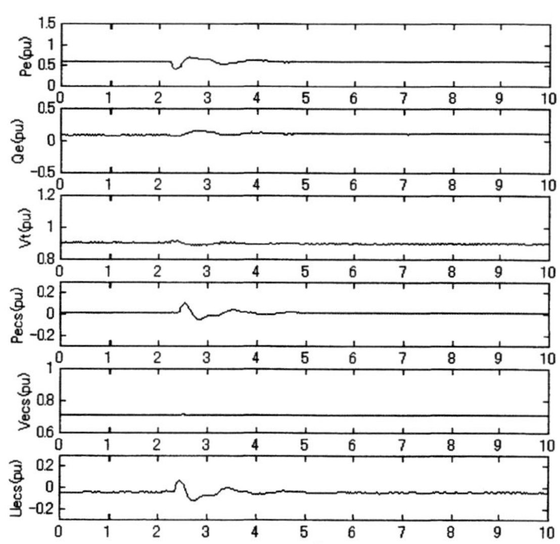

Fig. 10. Retuned Multi-Agent Based Control
Setting of Generator Output = 0.6 (pu)
(Switching of S_1)

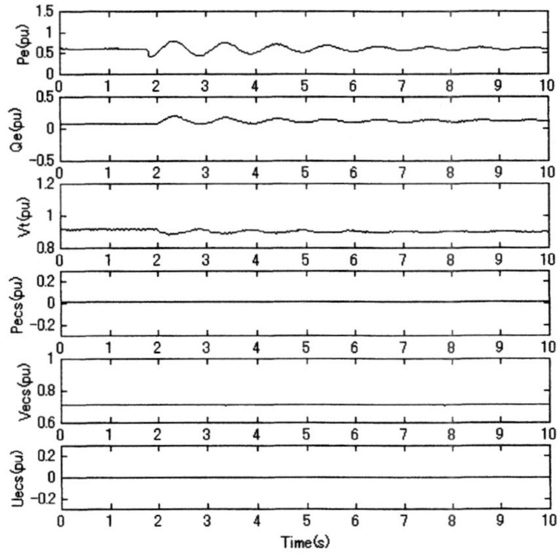

Fig. 11. No Control
Setting of Generator Output = 0.6(pu)
(Switching of S_2)

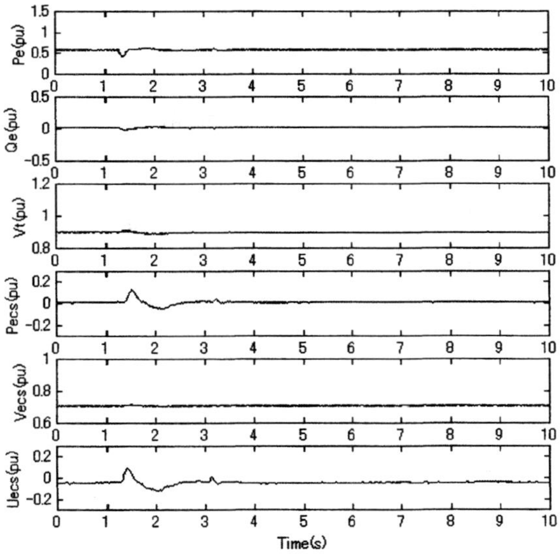

Fig. 12. Direct Stabilization Control
Setting of Generator Output = 0.6 (pu)
(Switching of S_2)

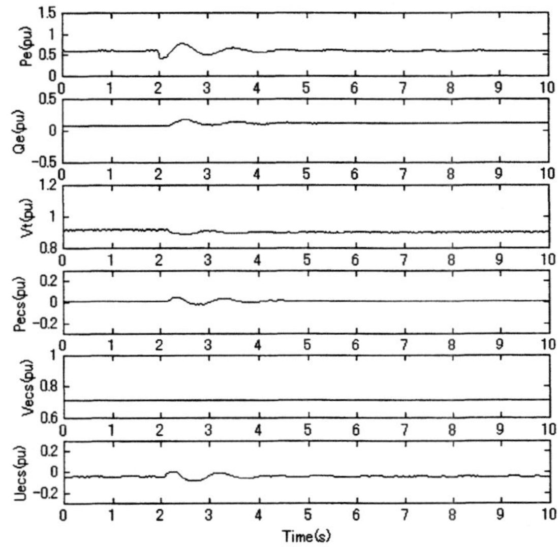

Fig. 13. Multi Agent Based Control
Setting of Generator Output = 0.6(pu)
(Switching of S_2)

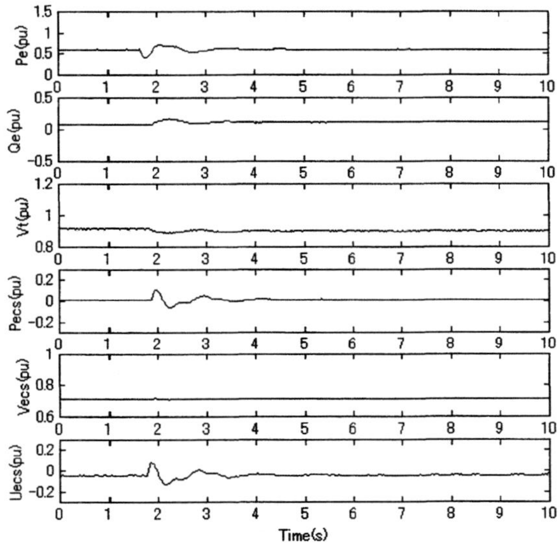

Fig. 14. Retuned Multi-Agent Based Control
Setting of Generator Output = 0.6 (pu)
(Switching of S_2)

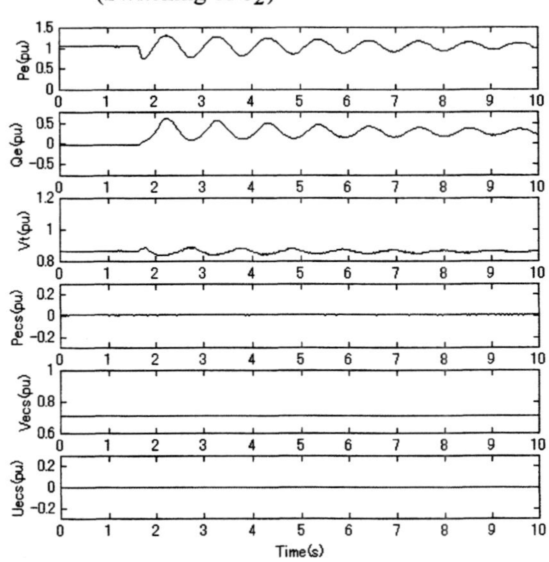

Fig. 15. No Control
Setting of Generator Output = 1.0 (pu)
(Switching of S_1)

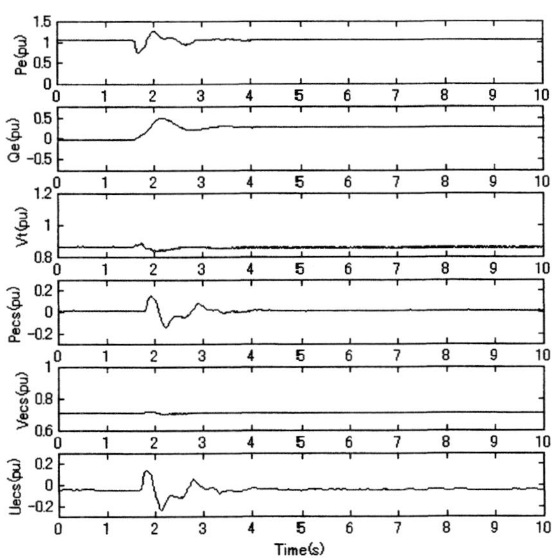

Fig. 16. Retuned Multi-Agent Based Control
Setting of Generator Output = 1.0 (pu)
(Switching of S_1)

5. CONCLUSION

The efficient performance of the proposed multi-agent based stabilization control system has been demonstrated through the experiments on the 5kVA laboratory system with the 70Wh (250kJ) ECS. The damping of the system oscillations is greatly improved by the proposed scheme. The stabilization performance is degraded by the existing delay time of the communication. However, the stabilization performance is easily recovered through the retuning of fuzzy control parameters. For the further studies, experimental evaluation of the compensation system is now ongoing on the laboratory system.

REFERENCES

Okamoto, M. (1995). A basic Study on Power Storage Capacitor Systems. *Transactions of IEE of Japan*, Vol. **115-B**, No.**5**.

Ohshima. M., Shimizu, M., Shimizu, M., Yamagishi, M.and Okamura, M(1998). Novel Utility-Interactive Electrical Energy Storage System by Electrical Double Layer Capacitors and an Error Tracking Mode PWM Converter. *Transactions of IEE of Japan*, Vol. **118-D**, No. **12**.

Hiyama, T., Ueno, D., Yamashiro, S., Yamagishi, M. and Shimizu, M. (2000). Fuzzy Logic Switching Control for Electrical Double-Layer Energy Capacitor System for Stability Enhancement. *Proceedings of the IEEE PES 2000 Summer Meeting*, Vol. **4**, pp.2002-2007.

Hiyama, T., Tomsovic, K., Anami, E., Yamashiro, S., Yamagishi, M. and Shimizu, M. (2001). Experimental Studies on Fuzzy Logic Stabilization Control for Energy Capacitor System. *Proceedings of the ISAP'2001*, pp.394-398

Hiyama, T., Kugimiya, M. and Satoh, H. (1994). Advanced PID type fuzzy logic power system stabilizer. *IEEE Transactions on Energy Conversion*, Vol. **9**, No. **3**, pp.514-520.

Hiyama, T., Oniki, S. and Nagashima, H. (1996). Evaluation of advanced fuzzy logic PSS on analog network simulator and actual installation on hydro generators. *IEEE Transactions on Energy Conversion*, Vol. **11**, No. **1**, pp.125-131.

ELSEVIER
IFAC
PUBLICATIONS
www.elsevier.com/locate/ifac

COORDINATION OF ELECTRIC POWER SYSTEM EMERGENCY CONTROL USING ARTIFICIAL NEURAL NETWORK

Nikolai Voropai*, Dmitry Efimov* and Pavel Etingov**

**Energy Systems Institute,
130, Lermontov Str., 664033 Irkutsk, Russia
Tel.: +7 3952 426 495, Fax: +7 3952 426 796, E-mail: efimov@isem.sei.irk.ru*

***Irkutsk State Technical University,
83, Lermontov Str., 664038 Irkutsk, Russia
Tel.: +7 3952 421 357, E-mail: etingp@mail.ru*

Abstract: The paper describes different principles of emergency control organization in electric power systems. Use of the artificial neural network is considered to form coordinated emergency control actions. The presented way is illustrated on an example of the Irkutsk regional electric power system.

Keywords: Artificial intelligence, Automatic control systems, Backpropagation algorithms, Control actions, Disturbance parameters, Electric power systems, Feedforward networks, Load flows, Transient stability assessment.

1. INTRODUCTION

At present there is no doubt about economical expediency of parallel operation of power plants as component parts of electric power systems (EPSs) and parallel operation of EPSs as component parts of power interconnections. However, the growing size and complexity of such power structures complicate their control. At the same time the risk of occurrence and avalanche-like development of emergencies also increases. That is why emergency control plays more and more important part in prevention of emergencies, termination of their development and restoration of normal operation of EPS. Taking into account a high speed of emergency processes such a control must be largely automatic.

Reliability requirements imply mutual reservation of emergency control device (ECD) actions. For example, at an inadmissible sudden overloading of the transmission line, ECDs of both connected subsystems are to make up and implement control actions (as a rule, disconnection of load and generation) that is to unload the line. Evidently, the uncoordinated actions of ECDs will lead to surplus disconnections. Let us consider another example. At emergency lack of power a part of load can be disconnected automatically, which will decrease the lack of active power, however (if the network capacity is high) can lead to surplus reactive power, inadmissible voltage rise, disconnection of generators and, as a result, to a new increase in power shortage.

The above mentioned and similar examples demonstrate the necessity of coordinated operation of ECDs by control actions and their adjustment.

2. LEVELS OF COORDINATION AND DEFENSE PLANS

Emergency control can be arranged by using two approaches that are based on historically existing priorities of managing and controlling a concrete EPS:

<u>Approach A</u> (most pronounced in Russia and France) is aimed at the priority maintenance of the main grid in operation and the stable parallel operation of EPS elements.

<u>Approach B</u> (typical of the most, although not all EPSs in the USA and Canada) is intended for the priority maintenance of electricity supply to consumers.

Table 1 shows main features, prerequisites and consequences of these approaches.

Within the framework of the mentioned approaches three levels of coordination of ECDs are possible in general: absence of coordination, paired coordination (between two ECDs controlling the same element of EPS, for example, the main transmission line) and centralized coordination (that in turn can be realized either for main lines or the whole EPS). It is worth to note that in the 1990s the EDF experts estimated these levels by the criteria of adequacy, reliability and cost (Faucon, *et al.*, 1995). According to all three criteria the centralized coordination proved to be the best.

Table 1 Two approaches to emergency control

	Approach A	*Approach B*
Main priority	Maintenance of the main grid in operation	Maintenance of power supply to consumers
Technical features of EPS	• Relatively low level of reserving the generation and grid, and therefore substantial tension of load-flow conditions with maximal use of transmission capability of the main grid. • High level of emergency control systems.	• Substantial level of reserving the generation and transmission capability of the main grid, and therefore high level of structural stability of EPS which makes the stability loss improbable even under the heaviest emergency disturbances. • Insufficient development of emergency control systems intended for preventing the initial stability loss.
Legal availability to disconnect consumers	EPS may realize any disconnections practically free.	Forced automatic disconnection of consumers is applied only in rather rare and obvious cases (when an emergency expands with danger for survivability of the whole EPS).
Consequences of using the approach	• Frequent disconnections of EPS elements and short-term limitations of consumer supply. • Effective overcoming of emergency processes leading to heaviest consequences.	• Low probability of limited system emergencies with short-term and small-size limitations of consumers. • Relatively high probability of rare but very heavy emergencies.

The systems of emergency control measures and devices applied to prevent cascading emergencies in the majority of world's countries are called "Defense Plans". The defense plans relate to automatic control, i.e. they are to act against the emergency processes in EPSs, the high rate of which makes impossible human control. According to the EDF data, for example, the defense plan is to act in less than 1.3s after asynchronous run (Faucon, *et al.*, 1995). Several examples of defense plans corresponding to different coordination levels are presented below.

2.1 Defense Plan of French EPS

The defense plan in EPS of France, as was observed by Faucon, *et al.* (1995), performs the following functions:

- Separation of subsystems, where asynchronous mode was detected, from the remaining part of EPS (the measurements of voltage angles are taken in real time at the main substations and transmitted every 50 ms to the central computer). On the base of previous repeated stability calculations, the French EPS is divided into twenty subsystems, between which the asynchronous run is most probable at heavy disturbances. These are the subsystems for which permanent measurements of voltage angles are performed.
- If necessary, load tripping in separated subsystems (there are four levels of automatic frequency load shedding at the medium voltage levels) to prevent disconnection of generators due to frequency fall.

Implementation of this defense plan started after the system emergency in December 1978. Until then (since the 1960s) the use was made of uncoordinated out-of-step protection devices acting at slow voltage swings at the protected line ends. The 1978 emergency demonstrated drawbacks of such an uncoordinated approach.

2.2 Defense Plan of Hydro-Quebec EPS

Due to specific features of this EPS (such as its practical isolation from other EPSs, length of the main grid and very long distance between generation and consumption centers) the EPS disintegration is equivalent to its blackout (there were three severe system emergencies in the 1980s in this EPS). Therefore (in contrast to most other North American EPSs) the main purpose of the Hydro-Quebec defense plan is not to detect an asynchronous mode with fast splitting of the system, but to preserve integrity of the main (735 kV) grid in all conditions (Trudel, *et al.,* 1999). The defense plan includes conventional ECDs installed at substations for:

- Generation and load tripping (for grid unloading),
- Connecting/disconnecting the 735kV shunt reactors,
- Load tripping at fall of the frequency and voltage (the latter has not been realized yet).

The principle of organizing the Hydro-Quebec defense plan is the priority of efficiency over selectivity ("It is not important what has happened, the main thing is to save the main grid"). That is why it includes the volumes of disconnections calculated for the heaviest possible disturbances which were calculated and ranged in advance. One more feature of the Hydro-Quebec defense plan is the absence of mutual coordination in operation of ECDs ("The plan is to be simple, and no external coordination is needed when local devices operate adequately").

2.3 Defense Plan of Brazilian EPS

After the system emergency in March 1999 Brazil started to revise the emergency control system and create a centralized defense plan. At present this plan is at the development stage and includes, as described by Filho, *et al.* (2001):

- Ranging the substations by the probability of heavy disturbances and the degree of their effect on EPS operation. On this basis the priorities for modernization of bus arrangement and equipment and for implementation of advanced ECDs are determined.
- Dividing the EPS into "security zones" (mainly by administrative affiliation).
- System of measurement transmission and centralized remote control (mainly for generation/load tripping) via telephone, microwave, fiber optic and satellite communications.

At present three PLCs (Programmed Logical Controllers) are in trial operation at the Brazillian substations. They are intended for identification of the disturbance reason and location by the measurements and for information transmission to the central PLC. The central PLC in turn is to form a proper control signal for disconnecting the generation and/or load. PLCs are adjusted beforehand on the base of repeated stability calculations.

2.4 Defense Plans of Russian EPSs

Centralized emergency control complexes are installed in several EPSs in Russia (EPSs of Ural, Middle Volga, Taymyr and Norilsk, West Siberia, and Krasnoyarsk-Bratsk-Ust'-Ilimsk power pool) and intended for taking the coordinated control actions aimed at prevention of the system emergency development, maintenance of the steady-state and transient stability and keeping the main grid in operation (Voropai, *et al.,* 1988). The level of an individual EPS is at present the upper level of centralized emergency control. On this level the following functions are performed on-line:

- Tracking the EPS topology (from remote signals),
- Determination of load-flow conditions (from remote measurements),
- Assessment of EPS stability and detection of emergency contingencies, when the system is unstable,
- Determination of necessary control actions for detected cases,
- Formation of signals (in terms of disturbance parameters) for adjustment of local devices.

3. ARTIFICIAL NEURAL NETWORKS AS A TOOL FOR EMERGENCY CONTROL COORDINATION

The main and rather common functions of emergency control complexes are shown in Fig.1. Such tasks of emergency control as on-line assessment of the topology and load-flow conditions, identification of normal operation violations, stability assessment, formation of the control actions and some others, at present can be successfully solved by applying the artificial intelligence.

Among the sections of artificial intelligence the artificial neural networks (ANNs) seem to be most suitable for solving the above tasks due to their following advantages:

- Capability for training and adaptation (this allows the work with inaccurate, incomplete and partly erroneous data),
- Capability for generalization (a well-trained network is able to give adequate responses to input data differing from those used for training),
- Capability for approximation of any continuous functions with a required accuracy,
- Robustness to faults of individual neurons (due to high parallelism of the network),
- Possibility to process a set of input signals and generate a set of qualitatively different responses,
- High speed of response.

Owing to these features ANNs are applied to pattern recognition, classification, simulation of behavior of different objects, etc. In the electrical engineering ANNs are used for studies in many areas: from load forecasts and transient stability assessment to automatic control of various devices in EPS.

In this paper a feed-forward back-propagation ANN is proposed for generation of coordinated control actions. ANNs of that type are to be trained in advance, i.e. are based on the supervised network

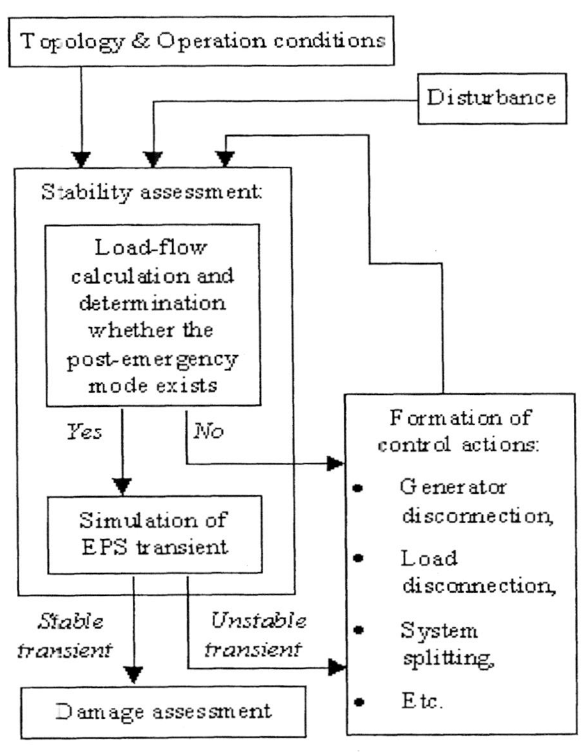

Fig. 1. The structure of emergency control.

paradigm. Fig.2 shows a general principle of the ANN operation.

The type and adjustment of the control action depend on the occured disturbance type and location, and also on the EPS topology and initial operating conditions. Therefore, it is necessary to send that information to the ANN input. The ANN responses (output signals) include information on coordinated control actions, for example, commands to disconnect loads in the deficient part of EPS and/or generators in the surplus part. Then these commands are transmitted directly to starting devices via remote control channels.

The network is trained off-line by the previously chosen representative sample of possible combinations of topologies, operation conditions and disturbances. In the course of training the steady-state and transient stability is analyzed for each case and the necessary coordinated control actions are determined to provide stable operation of EPS with minimal damage for consumers.

4. CASE STUDY

The study was carried out to test the serviceability of ANN to generate control actions on the base of available remote signals and measurements within a centralized emergency control complex.

The regional EPS "Irkutskenergo" was used as a test EPS. The system is characterized by the presence of large hydro-generators and lengthy transmission lines. During the period of winter peak load the operating conditions are tense. The full model consists of more than 200 nodes (including 40 generators) and more than 250 transmission lines and transformers. Fig.3 shows a simplified model of this EPS and main load-flow parameters of the winter peak load.

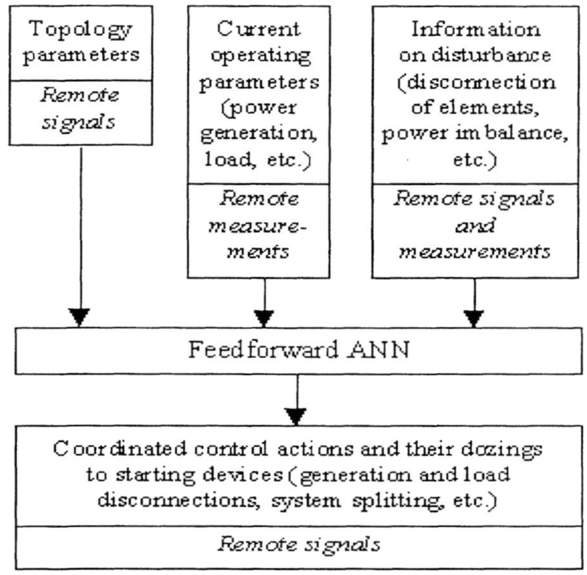

Fig.2. Formation of control actions by ANN.

4.1 Data for the ANN training

The results of load-flow calculations for two initial operating conditions (winter peak and summer minimum of load) and the set of disturbances (including the disconnection of 500kV lines and partial disconnection of generation) were used for the ANN training. Stability was calculated for each condition and each disturbance. If necessary, the control actions (partial disconnection of generation and/or load) were chosen in the expert way to provide stability of transition to post-emergency conditions without surplus disconnections of consumers.

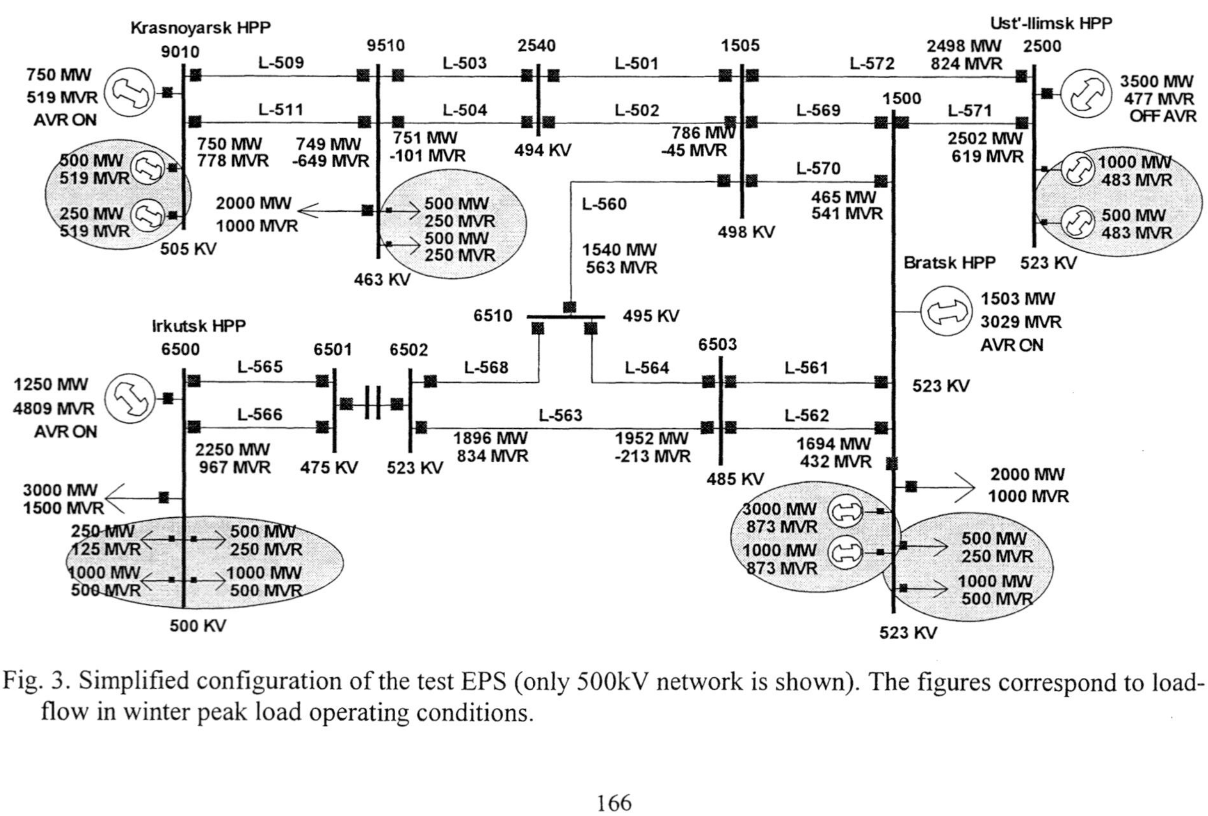

Fig. 3. Simplified configuration of the test EPS (only 500kV network is shown). The figures correspond to load-flow in winter peak load operating conditions.

For instance, at the emergency opening of lines 571 and 572 (see Fig.3), there is no more 500kV connection between the Ust'-Ilimsk HPP (hydropower plant) and EPS "Irkutskenergo". The transmission capability of the 220kV grid is insufficient to cover an imbalance occurred. In this

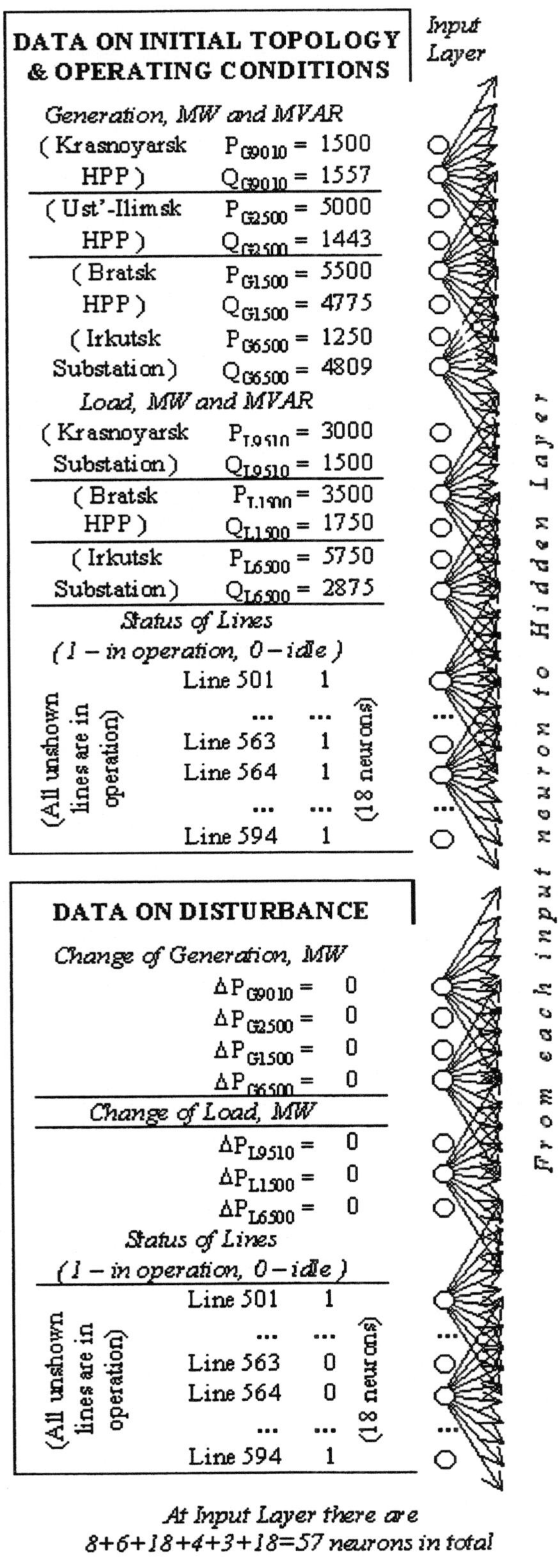

Fig. 4. Make-up of the ANN input data. Figures correspond to the training sample fragment for a certain case (opening of lines 563 and 564 in winter peak load operating conditions).

Table 2 A fragment of the training data set in generalized form (Inputs and Outputs for training of ANN) for winter peak load operating conditions in EPS "Irkutskenergo"

DATA ON DISTURBANCES (Opening of lines or change of generation)	⇒ **CONTROL ACTIONS AND ADJUSTMENTS** (Change of generation ΔP_G & load ΔP_L)
Line 501 or 502 or 509 or 511 or 564	⇒ No actions are needed
Line 571 or 572	⇒ ΔP_{G2500} = -2000 MW
Line 560, 561 or 562	⇒ ΔP_{L6500} = -1000 MW
Line 568	⇒ ΔP_{L6500} = -1500 MW
Line 565 or 566	⇒ ΔP_{L6500} = -1750 MW
Lines 501 and 502	⇒ ΔP_{L9510} = -1500 MW
Lines 509 and 511	⇒ ΔP_{L9510} = -2000 MW and ΔP_{G9010} = -750 MW
Lines 571 and 572	⇒ ΔP_{L1500} = -1500 MW and ΔP_{L6500} = -1500 MW
Lines 561 and 562	⇒ ΔP_{L6500} = -2750 MW and ΔP_{G1500} = -2000 MW
Lines 565 and 566	⇒ ΔP_{L6500} = -4000 MW and ΔP_{G1500} = -3500 MW
Line 503 or 504 or ΔP_{G9010} = -750 MW	⇒ ΔP_{L9510} = -500 MW
ΔP_{G1500} = -2000 MW	⇒ ΔP_{L1500} = -1500 MW
...	⇒ ...

case disconnections of several units of the Ust'-Ilimsk HPP and of a part of the load which is connected to the Bratsk HPP buses are needed.

From the above data a training sample was formed (see its fragment in a generalized form in Table 2).

4.2 The ANN structure and the information make-up

The information delivered to the ANN input characterizes firstly, the current pre-emergency EPS operating conditions (data on active and reactive power of generators and loads) and topology (states of lines) and secondly, the disturbance (changes in active power and changed states of lines). Fig.4 shows the make-up of the input information (the contents of Input Layer neurons) of ANN.

Fig.5 shows the structure of the ANN (hidden and output layers) and composition of output information that is the necessary control action parameters (changes in active power of generation and load). During the training this information is delivered to the ANN output.

On the base of the above data the ANN training was carried out with the use of the conventional backpropagation algorithm (Hertz, *et al.*, 1991).

4.3 The trained ANN testing

Data delivered to the trained ANN input have the

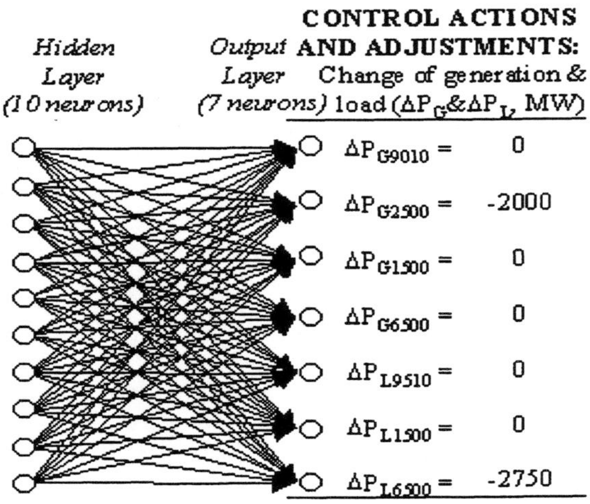

Fig. 5. The ANN hidden and output layers and the output data composition. Control actions correspond to opening of lines 563 and 564 in winter peak load operating conditions.

same structure as during the training. The initial operating conditions were considered to be conditions taking an intermediate place between peak and minimal load conditions of training and a set of operating conditions with some EPS elements under repair. Disturbances that were met and not met during the training were taken as test disturbances. ANN operation results in control actions generated practically instantly as a response to the input data. The ANN output information had the same structure as in the training.

Table 3 shows (in a generalized form) a fragment of input information about disturbances and ANN-generated control actions corresponding to these disturbances.

Transient processes were calculated for each test case both in absence and in presence of the ANN-generated control action. The calculation results showed the following:

- In most of cases in the absence of control actions the loss of stability took place and the post-emergency mode did not exist.
- The application of ANN-generated control actions provided EPS stability support without surplus disconnections of consumers.

The presented results demonstrated good training and generalization abilities of ANN. The trained ANN practically instantly generates adequate responses to the cases including those it has not ever seen before.

5. CONCLUSIONS

An effective way is proposed for generation of the coordinated emergency control actions by the artificial neural network. Effectiveness of this way is confirmed by studies carried out for the real complex electric power system. ANN demonstrated its capability to generalize accumulated knowledge and

Table 3 A fragment of the results of ANN testing (in generalized form)

Test Case	TEST DATA ON DISTURBANCES (Opening of lines or change of generation)	⇒	CONTROL ACTIONS AND AJUSTMENTS Change of generation & load (ΔP_G&ΔP_L, MW)
T-1	Line 560, 561 or 562	⇒	ΔP_{L6500} = -500
	Lines 571 and 572	⇒	ΔP_{L1500} = -1000 and ΔP_{L6500} = -1000
	Lines 563 and 564	⇒	ΔP_{L6500} = -2000 and ΔP_{G1500} = -1500
	Lines 560 and 570	⇒	ΔP_{L6500} = -1500 and ΔP_{G1500} = -1250
	Line 503 or 504 or ΔP_{G9010} = -500 MW	⇒	ΔP_{L9510} = -500
	ΔP_{G1500} = -1500 MW	⇒	ΔP_{L1500} = -1000
	...	⇒	...
		...	
T-M	Lines 569 and 570	⇒	ΔP_{G2500} = -1500
	Lines 571 and 572	⇒	ΔP_{L1500} = -1500 and ΔP_{L6500} = -1000
	Lines 563 and 568	⇒	ΔP_{L6500} = -2500 and ΔP_{G1500} = -1500
	Lines 560 and 570	⇒	ΔP_{L6500} = -1500 and ΔP_{G1500} = -1000
	Line 503 or 504 or ΔP_{G9010} = -750 MW	⇒	ΔP_{L9510} = -1000
	ΔP_{G1500} = -1000 MW	⇒	ΔP_{L1500} = -750
	...	⇒	...

find correct solutions for coordinated actions of emergency control devices in any case. The existing centralized complex of automatic stability control in the regional EPS "Irkutskenergo" can be modernized by applying ANN as an algorithmic base.

REFERENCES

Faucon, O., L. Dousset, J. Boisseau, Y. Harmand and M. Trotignon (1995). Coordinated Defence Plan – An Integrated Protection System. *CIGRE Symp, Helsinki, 1995,* pp. 1-6.

Filho, X. V., L. A. S. Pilottto, N. Martins, A. R. C. Carvalho and A. Bianco (2001). Brazilian Defense Plan Against Extreme Contingencies. *IEEE PES Summer Meeting, Vancouver, 2001,* 6 P.

Hertz, J. A., R. G. Palmer and A. S. Krogh (1991), *Introduction to the theory of neural computation.* Addison-Wisley, Redwood, CA, USA.

Trudel, G., S. Bernard and G. Scott (1999). Hydro-Quebec`s Defence Plan Against Extreme Contingencies. *IEEE Trans. on Power Syst.,* **V.14, No 3,** pp. 958-966.

Voropai, N. I., F. J. Morozov, V. A. Semenov *et al.* (1998). Reliability in the Restructured Russian Electricity Industry. *IEEE Power Eng. Review,* **V.18, No 2,** pp. 8-12.

ELSEVIER
IFAC
PUBLICATIONS
www.elsevier.com/locate/ifac

MULTI-MACHINE POWER SYSTEMS STABILIZING CONTROL USING OUTPUT FEEDBACK EXCITATION SYSTEM

T. Numajiri, G. Shirai
Hosei Univ.
Japan

R. Yokoyama, S. Machi
Tokyo Metropolitan Univ.
Japan

G. Fujita
Shibaura Inst. of Tech.
Japan

Abstract: Electric network is now confronted with market deregulation or overload. Power System Stabilizer (PSS) is used to enhance capacity limits and improve transient stability, and it also make important role in the renewed system. This paper proposes design of excitation control with Fletcher-Powell method, which is non-linear mathematical optimization technique to realize output feedback control. The design procedure and the effectiveness are discussed through 5-machine system models. Scheme of decomposing control matrices for the large network is also discussed for successful application of the method. *Copyright © 2003 IFAC*

Keywords : output feedback method, distributed control, excitation control, power system stabilizer, multi-machine power system.

1. INTRODUCTION

Recently, electric power systems have been complicated against various kinds of uncertainties such as load variation or entry of IPP (Independent Power Producer). Therefore, improvement of electric power system stability has been an important matter. Moreover, long-term disturbance problem associated with long-distance power transmission is also getting attention. PSS (Power System Stabilizer) is a measure device to handle this problem as an economical and effective method. PSS with high capability of long-term disturbance reduction may be achieved by designing it as $P+\omega$ type PSS.

A typical PSS consist of gain and lead/lag compensation function; the design procedure has already been established. However, it can be applied for one-machine infinite bus system, application for multi-machine system is difficult.

Several papers are published to address this problem. For example, GA (Genetic Algorithms)(Zhang and Coonick., 2000) is adopted to design PSS regarding whole system's stability. However, the formulization and algorithm are difficult for a practical application. On the other hand, many paper are discussing multi-variable based generation control, typically LQR (Linear Quadratic Regulator) is used. Yu porposed a standard LQR-based PSS, LQR(Yu and Moussa., 1972) which finds out the control law which was most suitable for evaluation index set up by Yu and specify poles to satisfy the desired characteristic pole assignment method(Yu and Li., 1995) were proposed. But, LQR and pole assignment method are carried out based on state feedback which needs all state variables, since Observer is required, control system becomes complicated, and the form will also be restricted. As a result, in electric power system currently employed as large cooperation system, utilization is difficult for control system by state feedback

Therefore, the method of deciding a control rule by only observable state variables is also proposed (Malik et al., 1985; Feliachi, 1987). This time, the method proposed in this paper is the control law by output feedback using control system consists of only easy to observe state variables are decided by Fletcher-Powell method (William and Athans., 1970, Choi and Sirisena., 1970).

Moreover, dispersed control system using only local information is desirable when generator control system is considered from practical. The Various methods in consideration of interaction between generators have been proposed by dispersed control system. (Folly et al., 1998; Tse et al., 2000; Li et al., 1999; Niioka et al., 2000). In this article, interaction from other-areas is disturbance against the local area, and disturbance reduction control in H^∞control is used, many methods are proposed. Though system decomposition into subsystems seems easy, it involves difficulty because of loop circuits or strong interaction in the network. Moreover, since state feedback has determined the control rule, it is not far from practical use. As this measure, in this paper, multi-machine power systems stabilizing control using output feedback excitation system is proposed. By this system procedure, since only observable information of the local information obtained in each control station is used, it is realizable. And, only control station is divided method in control design realize suitable dispersed control system to power system of at present.

2. OUTPUT FEEDBACK CONTROL

Because of various factors, observing all state variables in an electric power system is difficult. For practical PSS (Power System Stabilizer) design, it is desirable to use only output information. Therefore, output feedback control has been employed.

The output feedback control is represented by state equation (1) and observation equation (2). Where $\boldsymbol{C}$ is the observation matrix, namely only observable state variables are specified among all of state variables.

$$\dot{\boldsymbol{x}} = \boldsymbol{Ax} + \boldsymbol{Bu} \qquad (1)$$

$$\boldsymbol{y} = \boldsymbol{Cx} \qquad (2)$$

Here, matrix $\boldsymbol{x}$ is state vector of n-th system, matrix $\boldsymbol{u}$ is m-th control vector, matrix $\boldsymbol{y}$ is 1-th output vector, matrix $\boldsymbol{A}$ is system matrix ($n \times n$), matrix $\boldsymbol{B}$ is control matrix ($n \times m$), **C** is observation matrix ($l \times n$).

Here, the control input $\boldsymbol{u}$ is given by

$$\boldsymbol{u} = -\boldsymbol{Fy} = -\boldsymbol{FCx} \quad (3)$$

under the evaluation function $\boldsymbol{J}$ is

$$J = \frac{1}{2} tr[\boldsymbol{K}] \quad (4)$$

Here, tr denotes trace for a matrix. Matrix $\boldsymbol{K}$ ($n \times n$) is a solution which fulfills the following formula;

$$\boldsymbol{KA_C} + \boldsymbol{A_C^T K} + \boldsymbol{Q} + \boldsymbol{C^T F^T RFC} = \boldsymbol{0} \qquad (5)$$
$$\boldsymbol{A_C} = \boldsymbol{A} - \boldsymbol{BFC}$$

Here, Q is positive symmetric matrix, R is semi-positive symmetric matrix. In this paper, matrix $\boldsymbol{F}$ ($n \times l$) is used as the optimal feedback gain from (Choi and Sirisena., 1970).

In order to obtain the control input $\boldsymbol{u}$, which makes the evaluation function $\boldsymbol{J}$ minimum, it is necessary to set the gradient of the evaluation function $\boldsymbol{J}$ to zero. The gradient ($m \times l$) is

$$\frac{\partial J}{\partial F} = \boldsymbol{RFCLC^T} - \boldsymbol{B^T KLC^T} \qquad (6)$$

where matrix $\boldsymbol{L}$ ($n \times n$) is a solution which satisfy the following equation;

$$\boldsymbol{LA_C^T} + \boldsymbol{A_C L} + \boldsymbol{I} = \boldsymbol{0} \quad (7)$$

$\boldsymbol{I}$ = unit matrix

Fletcher-Powell method is employed to solve equation (7), in which the calculation is iterated until the solution converges as it satisfies the condition (8).

$$\sum_i \left(\frac{\partial J}{\partial F_i} \right)^2 < 0.1 \quad (8)$$

3. POWER SYSTEM MODEL

One-machine infinite bus model (Yu and Siggers., 1971) can be is linearized as an well-known Heffron-Phillips model (Yu., 1983) by taking pertubatin from nominal operating point. In order to consider stability of a whole system, an extended multi-machine Heffron-Phillips model is also available. Fig.1. shows a multi-machine Heffron-Phillips system model consisting n machines.

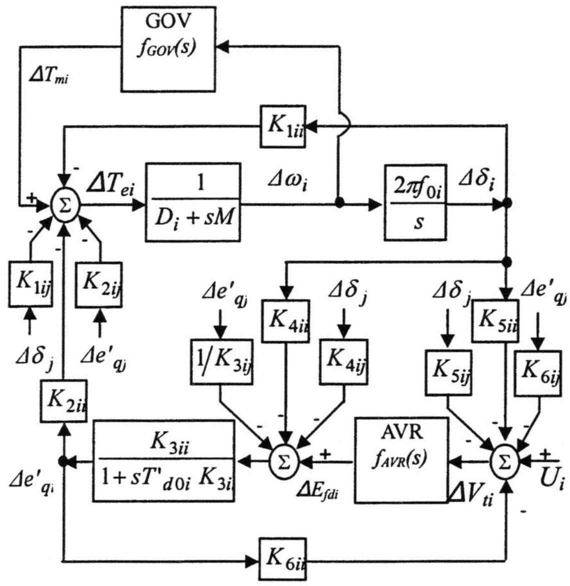

Fig.1. Block diagram of multi-machine power system

In Fig.1, K_{1ii} to K_{6ij} are constants computed from generator characteristic, operating conditions, and system condition. Subscript i is for a constant of the generator, j is for other generator. The following equations are obtained from Fig.1.

$$\Delta\delta_i = \frac{2\pi f_{0i}}{s} \Delta\omega_i \qquad (9)$$

$$\Delta\omega_i = \frac{1}{D_i + sM_i}\left(\Delta T_{mi} - \Delta T_{ei}\right) \qquad (10)$$

$$\Delta e'_{qi} = \frac{K_{3ii}}{1 + sT'_{d0i} K_{3ii}} \left(\Delta E_{fdi} - \frac{1}{K_{3ij}} \Delta e'_{qj} - K_{4ii}\Delta\delta_i - K_{4ij}\Delta\delta_j \right) \qquad (11)$$

$$\Delta T_{ei} = K_{1ii}\Delta\delta_i + K_{1ij}\Delta\delta_j + K_{2ii}\Delta e'_{qi} + K_{2ij}\Delta e'_{qj} \qquad (12)$$

$$\Delta V_{ti} = K_{5ii}\Delta\delta_i + K_{5ij}\Delta\delta_j + K_{6ii}\Delta e'_{qi} + K_{6ij}\Delta e'_{qj} \qquad (13)$$

In this control design stage, $\Delta T'_{ei} \approx \Delta T_{ei} \approx \Delta P_{ei}$ is assumed as the system is operating near the nominal point.

In this paper, PSS (P+ω type) design for a multi-machine system is discussed. Electric output deviation ΔP_{ei} and speed deviation $\Delta\omega_i$ are easy to observe and use as auxiliary input signals for the i-th generator.

4. DISPERSED CONTROL DESIGN

Though present dispersed PSS is allocated locally, from the viewpoint of control design, dispersed control method is not used. For many reasons, such as huge and complex network, introduction of IPP (independent power producer), infrastructure cost of data transmission and its delay, and reliability, a new dispersed control system is desired using only control information.

Here, the following two different methods are used to decompose a power system, and design of the dispersed control system with output feedback is discussed.

4.1 System Decomposition, Type 1

Assuming a power system represented by a state equation (1) is comprised by time-invariant subsystems, a subsystem of the *i*-th area is expressed by

$$\dot{x}_i = A_{ii}x_i + B_{ii}u_i + \sum_{i \neq j} A_{ij}x_j \quad (14)$$

where right hand first term expresses the subsystem, second term is its control input, third term is interaction by other subsystem. The concept is illustrated in Fig.2.

In order to find control F_{ii}, system matrix A_{ii} and control matrix B_{ii} are used. Two methods are adopted to decide these matrices.

Dispersed Control Type A. A_{ii} and B_{ii} are given as diagonal part of matrices A and B respectively. Firstly find full system matrices A and B shown in Fig.2, then decompose to diagonal blocks A_{ii} and B_{ii}, which do not contain interaction from other subsystem.

Dispersed Control Type B. A_{ii} and B_{ii} are given by regarding *i*-th generator as an one machine infinite bus system model. For each subsystem, an one machine infinite bus system model is constructed and these matrices are calculated. Contrary to Type A, A_{ii} and B_{ii} contain interaction from other subsystem

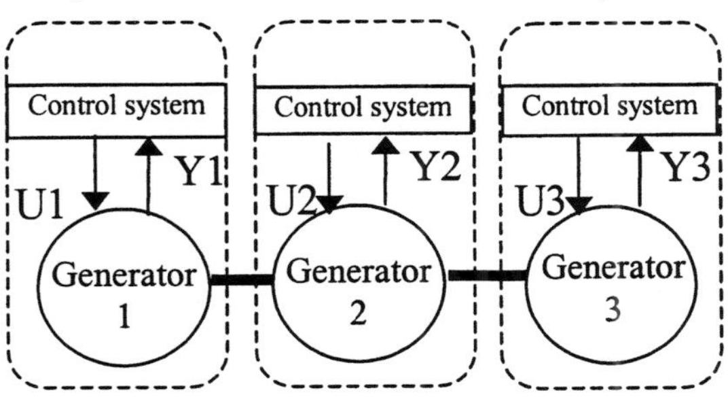

Fig.2. Dispersed Control Type A

4.2 System Decomposition, Type 2
With the methods mentioned above, the decomposition is not so easy if the network forms loop or interaction between subsystems is relatively strong. For this condition, other method to decompose only control station is proposed.

Dispersed Control Type C Assuming a subsystem has an independent control station, decompose the equivalent central control station, then the system state equation of equation (1) may be expressed as

$$\dot{x} = Ax + \sum B_i u_i \quad (15)$$

where right hand first term expresses the whole system, second term is control input from each control stations. The concept is depicted in Fig.3.

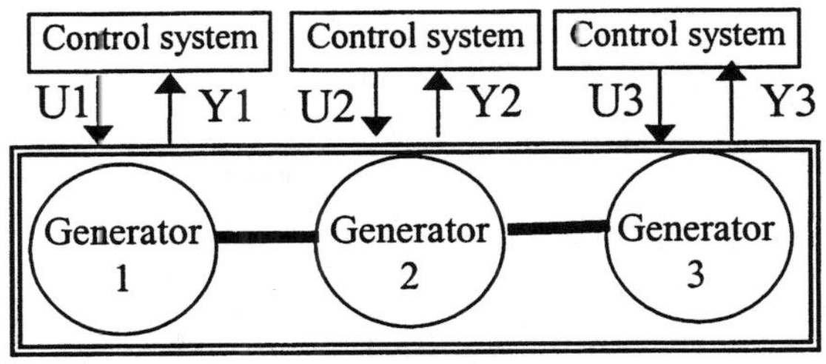

Fig.3. Dispersed Control Type C

The local control F_{ii} is calculated from whole system matrix A and subsystem control matrix B. The whole system including local interaction is considered to the control F_{ii}. This method offers global stabilization of a power system.

5. MODEL STUDY

These three control methods are employed for the design of PSS, and transient simulation is carried out. 5-machine 16-bus model (Bala and Thanikachalam., 1978) is used in the following discussions.

5.1 Study Model
In order to discuss if the proposed methods can be used for complex power systems, 5-machine 16-bus system as presented in Fig.4 is used. 4 control design methods shown in Table.1 are compared. Here, the generator parameter is given in Table.2.

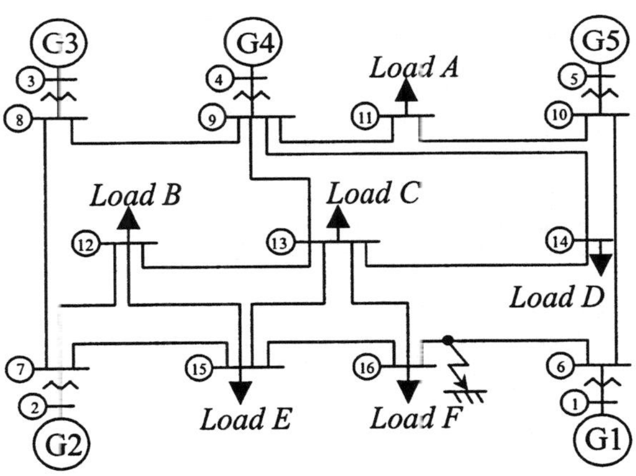

Fig.4. 5-machine 16-bus system model

In this paper, the linearized block diagram is used both for control design and transient stability simulation. The latter requires non-linear characteristic for large disturbance that is important for practical discussion. Therefore, electric output of generator *i*-th is calculated from the following equations.

$$\Delta P_{ei} = P_{ei} - P_{ei0} \quad (16)$$

$$P_{ei} = \sum E'_{qi} E'_{qj} Y_{ij} \cos(\theta_{ij} - \delta_i + \delta_j) \quad (17)$$

where E_q' is absolute of internal voltage of generator *i*-th, Y_{ij} is absolute of equivalent transmission admittance between generator *i*-th and *j*-th, θ_{ij} is phase angle of the admittance.

Furthermore, the non-linearity is considered by calculating ΔV_{ti} from the internal voltage E_q'.

The test disturbance scenario is follows: three-phase ground fault on line 6-16 near bus 16 at 0.1 [s], line open at 0.2 [s], reclosing at 1.0 [s].

Table 1. Dispersed stabilizing controllers

Type	Detail	Observable
A	$\boldsymbol{A_{ii}}, \boldsymbol{B_{ii}}$ Diagonal matrix	ΔT_e, $\Delta\omega_i$ of the area
B	$\boldsymbol{A_{ii}}, \boldsymbol{B_{ii}}$ One-machine infinite bus	ΔT_e, $\Delta\omega_i$ of the area
C	Divided control station	ΔT_e, $\Delta\omega_i$ of the area
D	LQR, ideal state feedback	all state variables
E	Conventional method	$\Delta\omega_i$ of each subsystem

Table 2. Generator parameter

unit		G1	G2	G3	G4	G5
Type		hydro	thermal	thermal	thermal	thermal
Raped capacity	[MVA]	247.5	192.0	128.0	192.0	128.0
M	[s]	23.6	6.40	3.01	6.40	3.01
D	[p.u]	0.01	0.01	0.01	0.01	0.01
T'_{do}	[s]	5.0	5.0	5.0	5.0	5.0
x_d	[p.u]	0.14	0.89	1.31	0.89	1.31
x'_d	[p.u]	0.06	0.12	0.18	0.12	0.18
x_q	[p.u]	0.09	0.86	1.25	0.86	1.25
K_a	[p.u]	10.0	10.0	10.0	10.0	10.0
T_a	[s]	0.2	0.2	0.2	0.2	0.2
K_g	[p.u]	1.0	1.0	1.0	1.0	1.0
T_g	[s]	24.0	7.2	7.2	7.2	7.2
Pc	[%]	3	4	4	4	4

5.2 Simulation Results

In order to discuss the effectiveness of the proposed methods, relative phase angles obtained from transient stability simulation are shown in Figs.5 to 10.

Firstly, crest amplitudes of 1st wave in each case are compared. Without control, it is -20.4[deg], Which is improved as Type A : -19.8[deg], Type B : -18.7[deg], Type C : -16.6[deg], Type D: -14.3[deg], Type E: -17.3[deg]. Moreover, settling times for relative phase angle δ_{21} conversing within ±5 are compared, these are without control : 33.7 [s], Type A : 23.0 [s], Type B : 13.1 [s], Type C : 8.5 [s], Type D : 6.9 [s], Type E: -15.3[deg].

From these results, effect from other area should be taken into account in order to establish dispersed control system as like Type B and C. Especially, remarkable damping is achieved in Type C, because the local controllers are designed regarding the whole system by decomposing only control station, not separating the network. Global stability may be kept even for more complex system. Type A does not contribute any stability improvement. Similar results are also obtained for different fault locations.

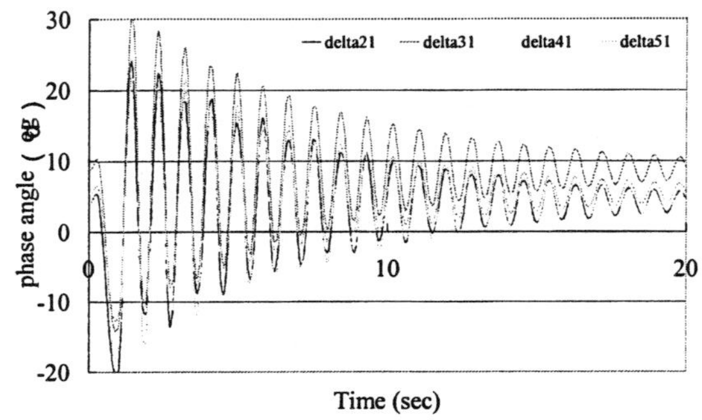

Fig.5. Relative phase angle without control

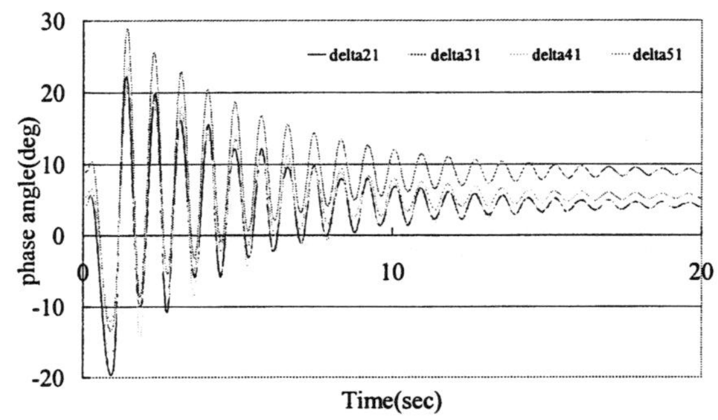

Fig.6. Relative phase angle with Type A

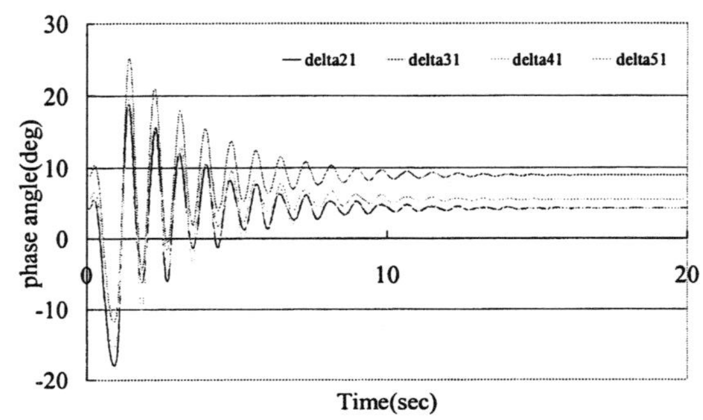

Fig.7. Relative phase angle with Type B

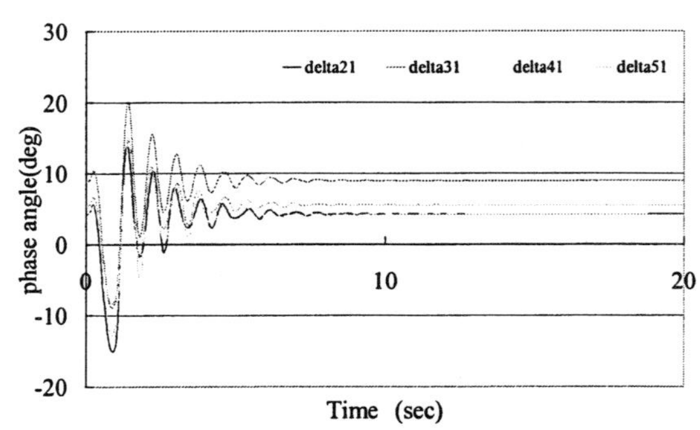

Fig.8. Relative phase angle with Type C

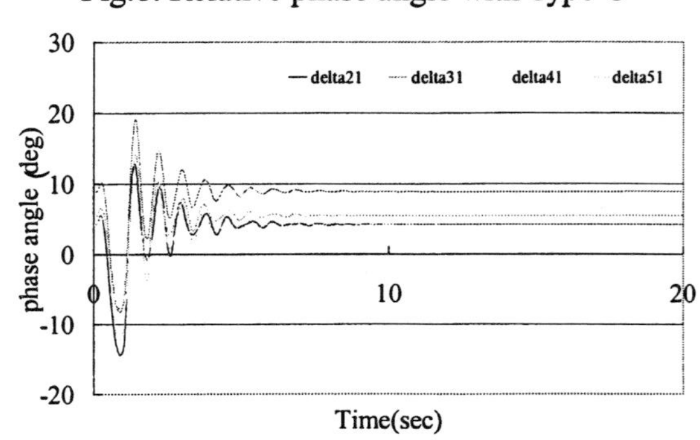

Fig.9. Relative phase angle with Type D

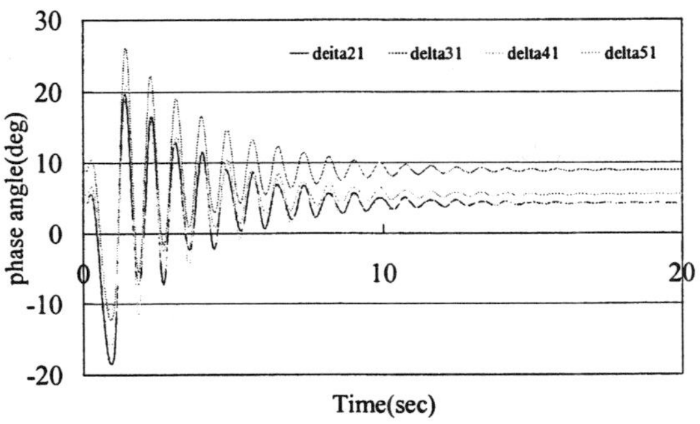

Fig.10. Relative phase angle with Type E

6. CONCLUSION

In this paper, multi-machine power systems stabilizing control using output feedback excitation system is proposed. Practical PSS (P+ω type) is designed in which output feedback control using observable output is used. Transient stability simulation results are compared with LQR as an ideal controller; proposed controller provides preferable damping effect.

Furthermore, in order to apply the proposed method, decomposing only control station, not the network, results in global system stabilization as the control stations are regarding the entire network. Even in each control station only local information is used.

It is concluded the proposed output feedback generation control is an effective method for multi-machine power systems.

REFERENCE

Bala J. L, A. Thanikachalam, 1978
"An Improved Second Order Method for Optimal Load Flow", IEEE Trans. on Power Apparatus and System, Vol.PAS-97, No.4

Choi S. S, H. R. Sirisena, 1974
"Computation of Optimal Output Feedback Gains for Linear Multivariable Systems", IEEE Trans. on Automatic Control

Feliachi A, 1987
"Optimal decentralized load frequency control", IEEE Trans, on Power Systems, Vol.2

Komla A. Folly, Naoto Yorino, Hiroshi Sasaki, 1998
"Improving the Robustness of H_∞-PSSs Using the Polynomial Approach", IEEE Trans. on Power Systems, Vol.13, No.4

Li G. J, T. T Lie, C. B. Soh, G. -H. Yang, 1999
"Decentralized Nonlinear H_∞ Control for Stability Enhancement in Power Systems", IEE Proc.-Gener. Transm. Distrib. , Vol.146, No.1

Malik O. P. et al, 1985
"Decentralized suboptimal load-frequency control of a hydro-thermal power station using the state variable model", Elec Power, Sys, Research. 8, 237-247

Niioka S, R. Yokoyama, G. Shirai, G. Fujita, (2002-1)
'Decentralized Excitor Stabilizing Control for Multimachine Power System', Electrical Engineering in Japan, Scripta Technica Inc, Vol.139, No.1, pp.35-43

Tse C. T, K. W. Wang, C. Y. Chung, K. M. Tsang, 2000
"Parameter Optimisation of Robust Power System Stabilizers by Probabilistic Approach", IEE Proc.-Gener. Transm. Distrib. , Vol.147, No.2

William S. Levine, Michael Athans, 1970
"On the Determination of the Optimal Constant Output Feedback Gains for Linear Multivariable Systems", IEEE Trans. on Automatic Control, Vol.AC-15, No.1

Yu Y. N, C. Siggers, 1971
"Stabilization and Optimal Control Signals for a Power System", IEEE Transactions, Vol.PAS-90

Yu Y. N, A. M. Moussa, 1972
"Optimal Power System Stabilization Through Excitation and Governor Control", IEEE Trans. on Power Apparatus and System, Vol.PAS-91, 1972

Yu Y. N, 1983
"Electric Power System Dynamics", Academic Press Inc, London

Yu Y. N, Q. H. Li, 1995
"Pole-Placement Power System Stabilizer Design of An Unstable Nine-Machine System", IEEE Trans, on Power Systems, Vol.5

Zhang. P, A. H. Coonick, 2000
"Coordinated Synthesis of PSS Parameters in Multi-Machine Power Systems Using the Method of Inequalities Applied to Genetic Algorithms", IEEE Trans. on Power Systems, Vol.15, No.2

ELSEVIER
IFAC
PUBLICATIONS
www.elsevier.com/locate/ifac

SIMULATION OF ISOLATED OPERATION IN FAULT CONDITION BY LOOP POWER FLOW CONTROLLER

Naotaka Okada, Hiromu Kobayashi, Tadao Ishikawa and Kiyoshi Takigawa

Central Research Institute of Electric Power Industry, Tokyo, Japan

Kousuke Kurokawa

Tokyo University of Agriculture and Technology, Tokyo, Japan

Abstract: 6.6kV overhead distribution systems are currently constructed in Japan as radial networks. The load imbalance of each feeder therefore has a negative impact on voltage stability. On the other hand, disconnecting distributed generators with faults causes line overload or under-voltage. Therefore, controlling line voltage and line power flow on the distribution system is critical for the free access of indefinite numbers of distributed generators. We are investigating a new loop or mesh structure distribution system to maximise network capacity. This paper describes a control method for an LPC focusing on 1LG fault states. *Copyright © 2003 IFAC*

Keywords: loop power flow controller, control switching, governor-free operation, current control, frequency control, distribution system

1. INTRODUCTION

In Japan, 6.6kV overhead distribution systems are currently constructed as radial networks. In these systems, each feeder voltage is controlled to within 101+-6V for the standard voltage 100V and 202+-20V for the standard voltage 200V. A line voltage compensator maintains the appropriate voltage for a bank of several feeders. The load imbalance of each feeder therefore has a negative impact on voltage stability.

In the near future, many distributed generators will be connected to the power distribution system. The line voltage will rise over the upper limit of the proper voltage range as a result of the reverse power flow of these distributed generation systems. On the other hand, disconnecting any distributed generators with faults causes line overload or under-voltage. Therefore, it is critical for free access of indefinite distributed generators to control line voltage and line power flow on the distribution system. We are investigating a new loop or mesh structure distribution system to maximise network capacity (H. Taniguchi, 2001; N. Okada et. al, 2002). Though loop or mesh distribution systems generally balance the power flow and regulate voltage, this will increase short circuit current in the distribution system and fault area, the logic being that methods of detecting fault location on loop distribution systems have not yet been established.

We have proposed a loop power flow controller (LPC) using a pulse width modulation (PWM) AC-DC-AC converter to control loop power flow and regulate line voltage in a suitable manner (N. Okada; 2001).

In an ungrounded-neutral system, because the fault current is small at 1LG, isolating the fault section by switch is a possibility. Moreover most faults in distribution systems are 1LG. However, in the case of a fault in the distribution lines, the distributed generation systems have to be disconnected from the distribution line without delay. This is known as the islanding phenomenon. For this reason, distributed generation has a detection relay of islanding operations. Therefore, in cases where the fault section is isolated, it is desirable to continue operation of distributed generations other than the isolated section.

This paper describes a non-interruption control for LPC to separate a 1LG fault section. First we show a mathematical model of LPC based on current control using a voltage type AC/DC/AC PWM converter. Second we propose a method of non-interruption control for LPC that switches from current control to frequency control. We also show a method for the parallel operation required by two frequency control inverters. We confirm by EMTP simulation that the control method can separate the fault section without non-interruption of normal sections. Finally we predict how the islanding protective relay of distributed generations will operate. The results show potential for continuous generation.

2. OVERVIEW OF LOOP DISTRIBUTION SYSTEM WITH LPC

2.1 Conventional Distribution System

6.6kV overhead distribution systems are currently constructed in Japan as radial networks. Fig. 1 shows a radial distribution line route from a distribution substation. A feeder is broken into sections by circuit breakers, closed switches and open switches. The open switches provide flexibility under fault conditions. Under stable conditions, distribution systems are operated as a radial network.

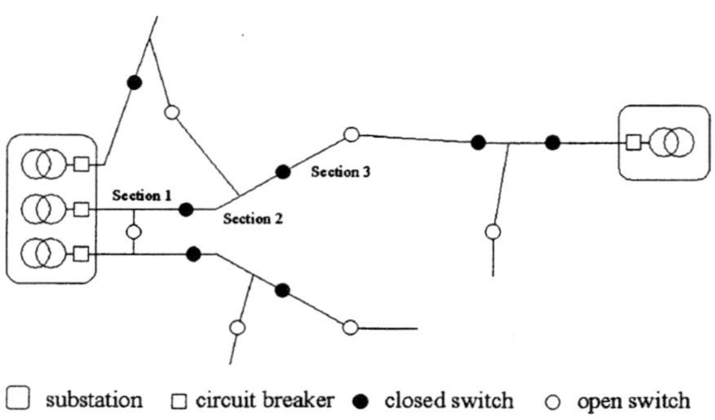

Fig. 1 Radial distribution system

2.2 Loop Distribution System with LPC

A loop system can be easily constructed by closing an open switch. Therefore, we considered the installation points of the LPC to be the open switches shown in Fig. 1.
Fig.2 shows a loop distribution system using LPC.

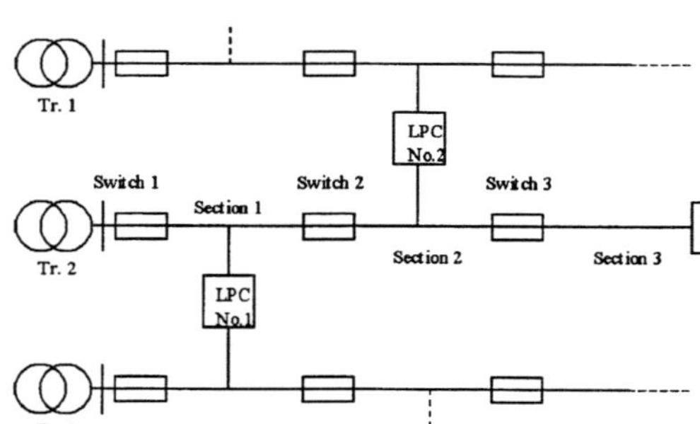

Fig. 2 Loop configuration by LPC

2.3 Loop Power Flow Controller

The basic concept of the loop distribution system using the LPC we propose is as follows (N. Okada; 2001).

A) Aim is free access to a distributed power supply.
B) System responds flexibly to unbalanced load between feeders, and makes effective use of equipment.
C) To enable this, system is constructed in the shape of a loop from a radial.
D) A loop distribution system is installed without altering existing systems such as the protection system, except for loop points.

To achieve all this, we used the BTB (back to back) shown in Fig. 3 as the LPC.

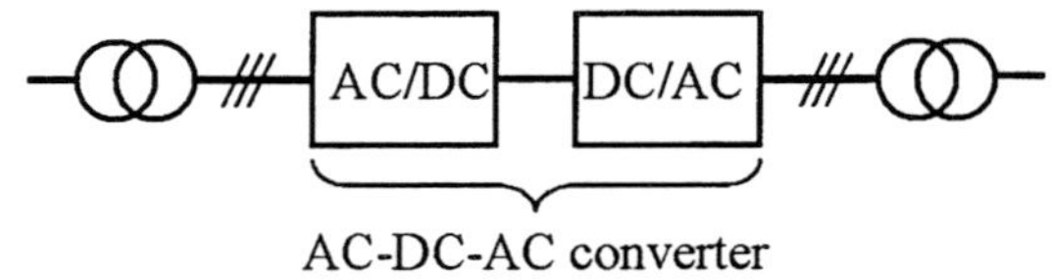

Fig. 3 Configuration of LPC by AC/DC/AC converter

3. MATHEMATICAL MODEL OF LPC

In this section, we show a mathematical model of LPC based on current control using a voltage type AC/DC/AC PWM converter.

3.1 Reactor of Interconnection

The LPC is taken as an AC/DC/AC converter that performs PQ control based on current control with a dq frame (H. S. Kim et al, 1998; Navid R et al., 1995). Fig. 4 shows an ideal AC/DC/AC converter model.

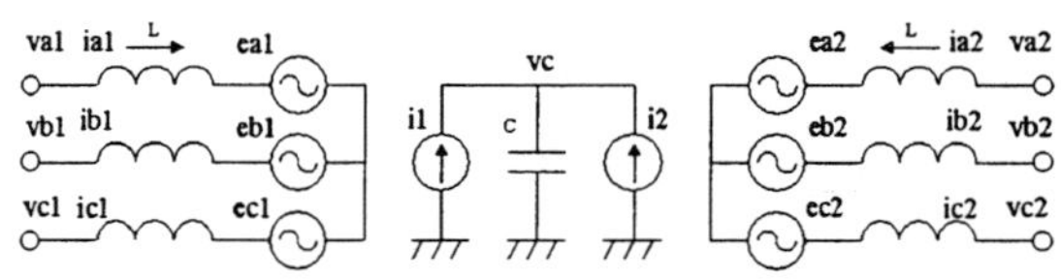

Fig. 4 Ideal AC/DC/AC converter model

Since Fig. 4 is symmetrical, it takes into account the left-hand model.
The relation between the voltage $V_1=\{v_{a1}, v_{b1}, v_{c1}\}$ of the distribution system, the internal voltage $E_1=\{e_{a1}, e_{b1}, e_{c1}\}$, and the reactor current $I_1=\{i_{a1}, i_{b1}, i_{c1}\}$ is shown by the following.

$$L\frac{d}{dt}I_1 = V_1 - E_1 \tag{1}$$

The dq translation is performed for equation (1), and we obtain the following equation.

$$L(p \cdot I - A)I_1^{dq0} = V_1^{dq0} - E_1^{dq0} \tag{2}$$

Where, $I^{dqo}{}_1=\{i^d{}_1, i^q{}_1, i^0{}_1\}$, $V^{dqo}{}_1=\{v^d{}_1, v^q{}_1, v^0{}_1\}$ 、 $E^{dqo}{}_1=\{e^d{}_1, e^q{}_1, e^0{}_1\}$, $p=d/dt$, I is unit matrix and A is

$$A = \begin{bmatrix} 0 & \omega & 0 \\ -\omega & 0 & 0 \\ 0 & 0 & -\omega \end{bmatrix}. \tag{3}$$

3.2 DC Capacitor

The relation between the DC voltage v_c and DC current i_1, i_2 is shown by the following equation.

$$v_C = \frac{1}{C}\int(i_1 + i_2)dt \tag{4}$$

When the effective power that flows in the reactor is taken as P_1, the relation between DC current i_1 and the DC voltage v_c is shown by the following equation.

$$P_1 = v_C \cdot i_1 \quad (5)$$

3.3 Current Control of Interconnected Reactor

Here, current control of the interconnected reactor on the dq frame is performed by the following equation.

$$E_1^{dq} = \begin{bmatrix} K_p + \frac{K_I}{s} & 0 \\ 0 & K_p + \frac{K_I}{s} \end{bmatrix} \left(I_{1ref}^{dq} - I_1^{dq}\right) + \begin{bmatrix} \omega L & 0 \\ 0 & \omega L \end{bmatrix} I_1^{dq} \quad (6)$$

3.4 Power and DC Voltage Control

If the negative phase sequence, zero phase sequence and harmonics component are disregarded with regard to the voltage and current, the effective power of the interconnection reactor is

$$P_1 = v_1^d \cdot i_1^d + v_1^q \cdot i_1^q . \quad (7)$$

Instantaneous reactive power is

$$Q_1 = -v_1^q \cdot i_1^d + v_1^d \cdot i_1^q . \quad (8)$$

Equation (7) and equation (8) may be rewritten for the current to acquire the current reference for the PQ reference.

$$\begin{bmatrix} i_{1ref}^d \\ i_{1ref}^q \end{bmatrix} = \frac{1}{\left(v_1^d\right)^2 + \left(v_1^q\right)^2} \begin{bmatrix} v_1^d & -v_1^q \\ v_1^q & v_1^d \end{bmatrix} \begin{bmatrix} p_{1ref}^d \\ q_{1ref}^q \end{bmatrix} \quad (9)$$

PQ control is carried out by providing the current reference of the equation (9) to the equation (6).
The control of the DC voltage v_c is carried out in accordance with the following equation using the effective power P_1.

$$p_{1ref}^d = \left(K_{PC} + \frac{K_{IC}}{s}\right)\left(v_{Cref} - v_C\right) \quad (10)$$

Similarly, the effective power P_2 of the right-hand converter can perform the v_c control, and do so using both P_1 and P_2.

4. PROPOSED SECTION ISOLATED OPERATION

4.1 Switching from Current Control to Frequency Control

In order to maintain the voltage and frequency of the isolated section isolated from the system of Fig. 2, we apply an LPC that connects the independent section to the other sections. A system is proposed that switches from current control of LPC to frequency control, which maintains voltage and frequency. The left-hand converter 1 of Fig. 4 performs voltage control of the direct current, and we assume that the right-hand converter 2 is switched from current control to frequency control. Switching of control is performed by transposing equation (6) on the following equation.

$$E_2^{dq} = \begin{bmatrix} e_2'^d \\ e_2'^q \end{bmatrix} \quad (11)$$

Where, e'^d_2 and e'^q_2 are considered as the hold value of the value e^d_2 and e^q_2 that is switched, and the equation (6) outputs immediately before.
On the other hand, phase θ is required at the time of dq translation. In the case of a converter that performs an interconnection to a system, the phase can be determined from the voltage of the interconnection point. However, in cases where the section becomes isolated and LPC serves as frequency control, it is necessary to generate phase θ inside LPC. Phase θ at the time of system interconnection is then considered to be the integration of angular velocity ω.

$$\theta = \int \omega \, dt \quad (12)$$

Moreover angular velocity ω is considered as the following equation so that v^d_2 and v_{a2} may be made congruous and it may be set to $v^q_2 = 0$.

$$\omega = \omega_0 + \omega_q$$
$$\omega_q = \int v_2^q dt \quad (13)$$

ω_0 is considered the standard angular velocity. ω_q is the angular velocity for adjustment of $v^q_2 = 0$.
Switching to frequency control from current control is performed by changing to the following equation instead of the equation (13).

$$\omega = \omega_0 + \omega_q' \quad (14)$$

Where, ω_q' is ω_q (equation (13)) immediately before switching.

4.2 Parallel Operation of Two-Frequency Control LPC

If the number of isolated sections is two the LPC needs to be operated in parallel. This requires sharing power flow and control of frequency. Supply-and-demand adjustment by a governor free system is used in the electric power system. The control system of a small-scale isolated power supply system by photovoltaic and the battery is proposed as an isolated power supply using the inverter (N. Okada et al., 1999; N. Okada et al, 2001). Parallel operation by the governor free operation of LPC is then performed. Switching to governor free operation from current control uses the following equation instead of equation (14) simultaneously with switching to frequency control from current control.

$$\omega = \omega_0 + \omega_q' + \frac{1}{M}\int\left(p_2 - p_2' + \frac{\omega_0 - \omega}{r}\right)dt \quad (15)$$

Where, p_2' is the hold value of the effective power flow p_2 immediately before switching; r is a permanent speed regulation, and M is a simulated inertia constant.

5. SIMULATION ANALYSIS

5.1 Simulation Parameters

The simulation was performed using EMTP. The specifications of the 6.6kV distribution system are shown in Table 1. Table 2 and 3 show the specifications of the LPC. Each LPC has the same specifications. Tables 4 and 5 show the specifications of distributed generators.

Table 1 Specifications of distribution system

Item	Specifications
Substation transformer	20MVA
Reactance	7.5%
Line impedance of section 1, 2 and 3	0.2235+j0.4716Ω (1.5km)

Table 2 Specifications of LPC

Item	Specifications
Rated capacity	1000kVA
AC voltage	400V
DC voltage	800V
AC reactor	4.76%
DC capacitor	0.0672 sec
K_P	0.1
K_I	5
K_{PC}	10
K_{IC}	0.1
ω_0	1
M	4 sec
r	4 %

Table 3 Specifications of LPC transformer

Item	Specifications
Rated capacity	1MVA
Voltage	6.6kV/400V
Reactance	2 %

Table 4 Synchronous generator

Item	Specifications
Rated capacity	200kW (250kVA)
Voltage	400V
Inertia constant	3.94 sec

Table 5 Inverter generator

Item	Specifications
Rated capacity	200kW (250kVA)
Voltage	400V
AC reactor	4.76%

5.2 Operation Characteristics of Switching to Frequency Control from Current Control Based on Governor Free Operation of Two-LPC

Fig. 5 shows a simulation model for switching from current control to frequency control, with a load of 500kW (power factor 0.9) in each section. Simulations are performed as follows. (0) Time 0: 1LG fault occurs in section 1. (1) Time 0.1: Detect 1LG fault. (2) Time 0.2~0.3: Current control of LPC is switched to frequency control. (3) Time 0.5: Switch 2 is opened. (4) Time 1.0: Switch 1 is opened. The sequence of isolation is shown in Table 6. In the simulation, 1LG fault is generated at 6 seconds after starting.

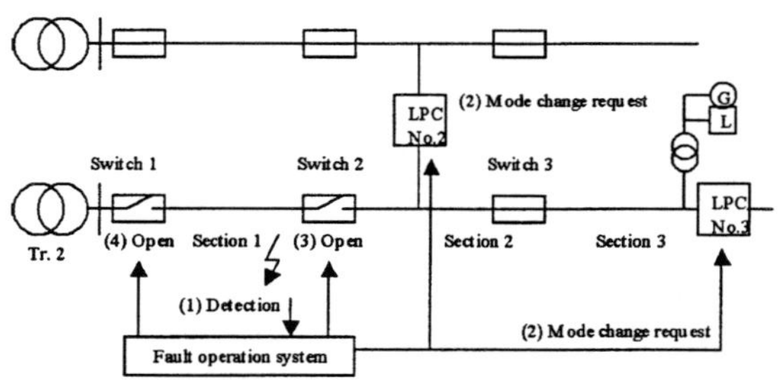

Fig. 5 Simulation model for switching from current control to frequency control

Table 6 Isolation sequence

time (sec)	event or action
0.0	Occurrence of fault
0.1	Detection of fault
0.2 ~ 0.3	Switching to frequency control
0.5	Switch 2 is opened.
1.0	Switch 1 is opened.

Fig. 6 shows the characteristics of switching from current control to frequency control. (a) is the voltage of the zero phase sequence of transformer 2. A 1LG fault is generated at 6 seconds after starting, and switch 1 opens at 7 seconds. (b) is the voltage of section 3. The voltage changes from 0.988pu to 1.018pu at 6.2 seconds. This is generated by controlling P and Q of LPC to make the power flow of switch 2 zero. The power flow change of switch 2 is shown in (d). P and Q of LPC are shown in (e). The detection levels of the over-voltage protection relay of the distributed generation are 1.1~1.2pu. And the detection levels of under-voltage are 0.8~0.9pu. Therefore, the possibility of distributed generation stopping as a result of over-voltage is thought to be low. (c) shows the frequency of the section 3. The levels of over-frequency are 1.01~1.03pu. And the levels of under-frequency are 0.97~0.99pu. The frequency is within these levels. (f) is the frequency change rate at section 3. The frequency change rate is one element of the islanding detection relay with high sensitivity. The detection levels are ±0.05~0.15pu/sec, and detection time is 8~10 cycles. In (f), this value has become 0.18 pu/sec momentarily. However, distributed generation may be assumed to operate continuously, because time is under 2 cycles.

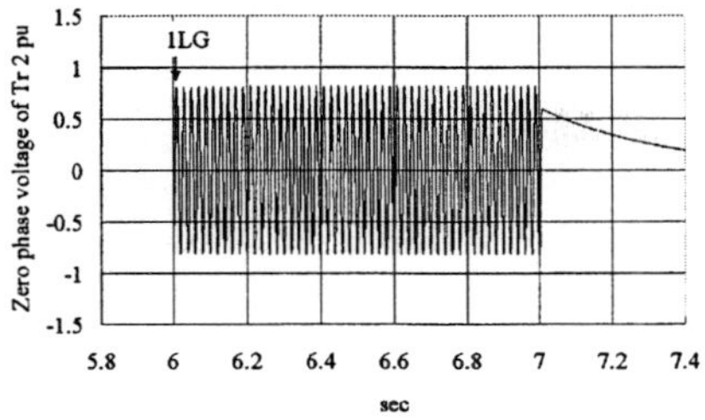

(a) Voltage of zero phase sequence of Tr. 2

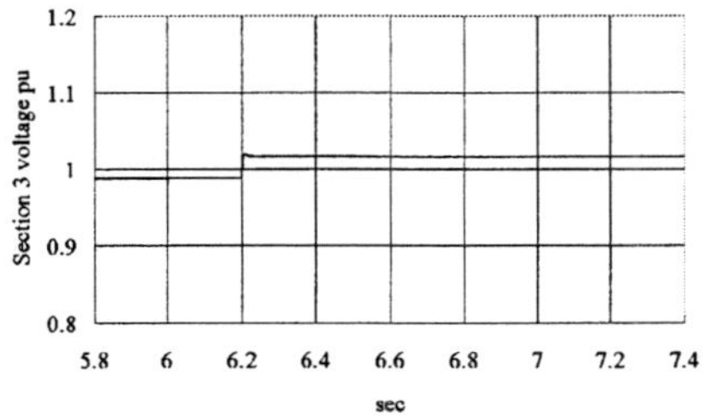

(b) Voltage of section 3

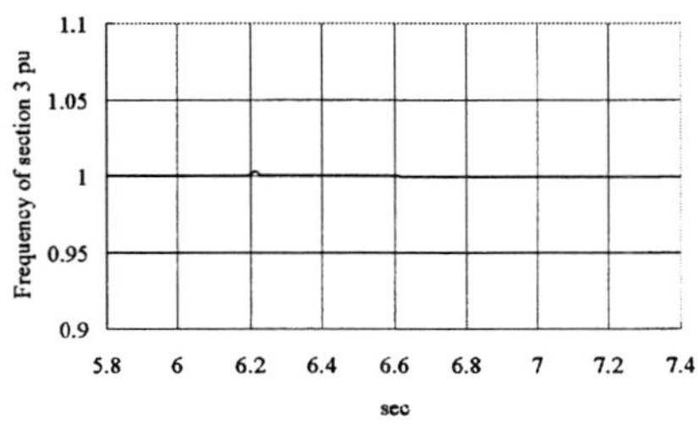

(c) Frequency of section 3

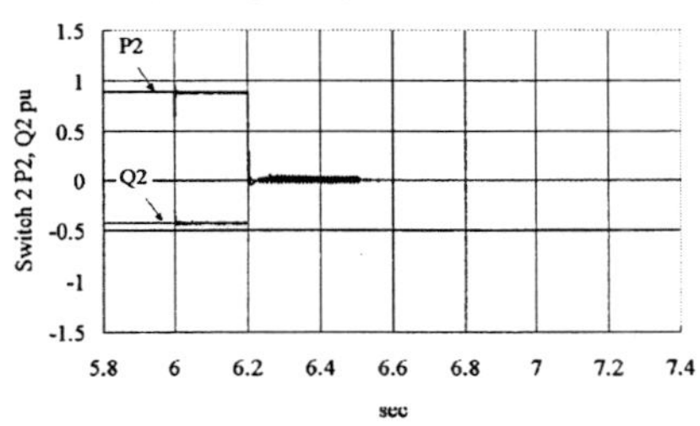

(d) Power flow at switch 2

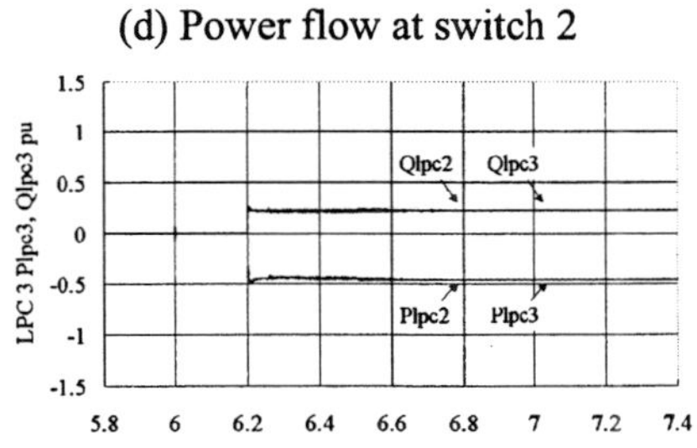

(e) Power flow and reactive power at LPC 2 and 3

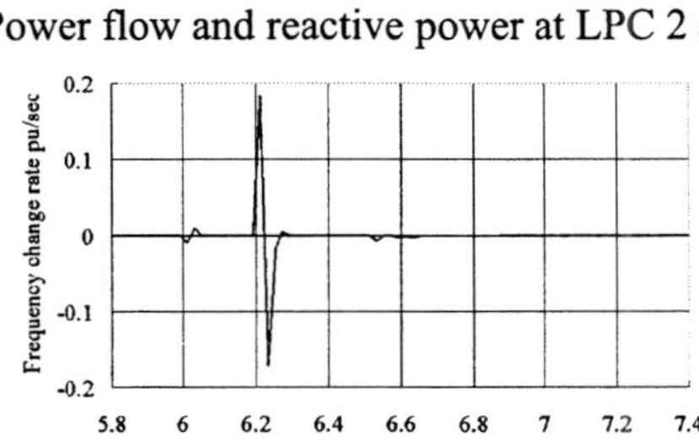

(f) Frequency change rate at section 3

Fig. 6 Characteristics of switching from current control to frequency control

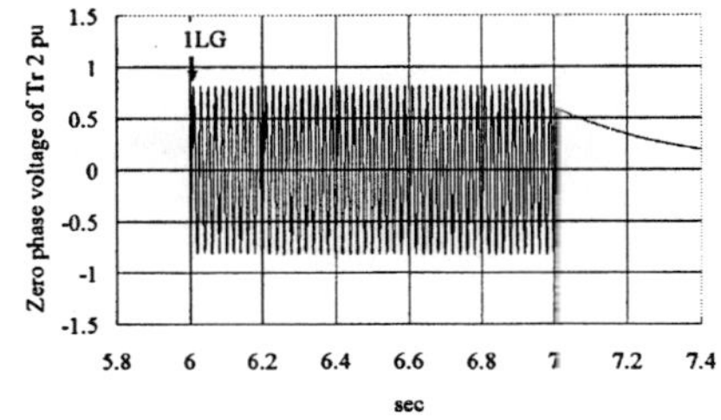

(a) Voltage of zero phase sequence of Tr. 2

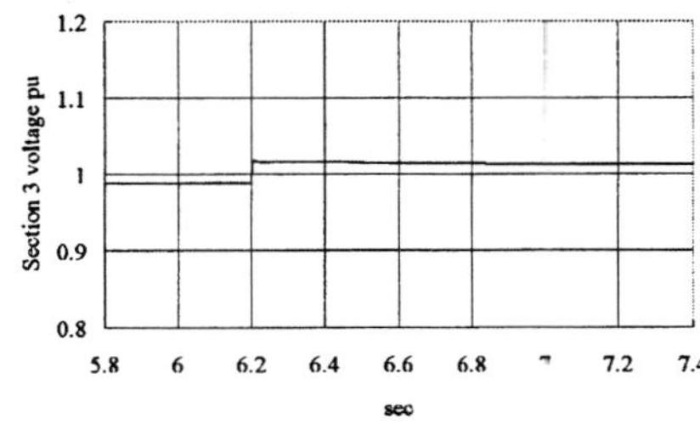

(b) Voltage of section 3

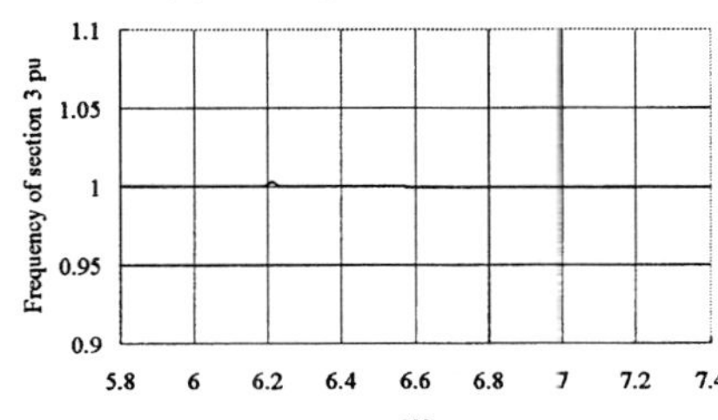

(c) Frequency of section 3

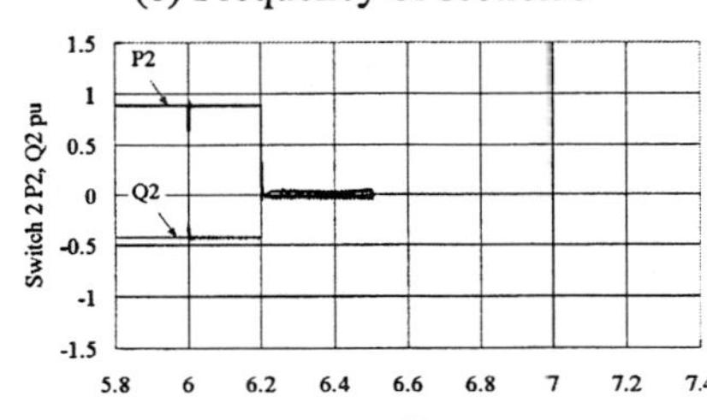

(d) Power flow at switch 2

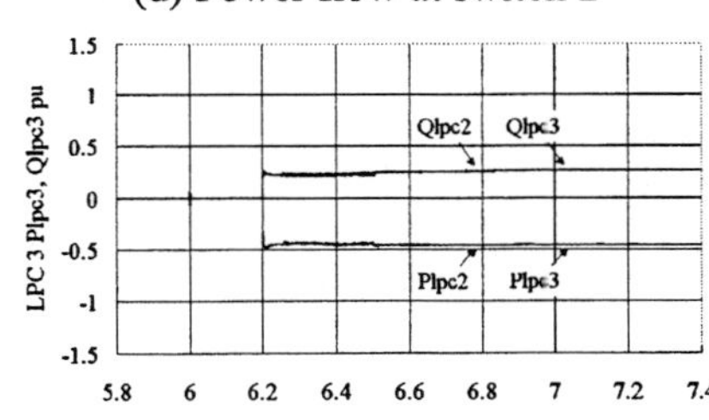

(e) Power flow and reactive power at LPC 2 and 3

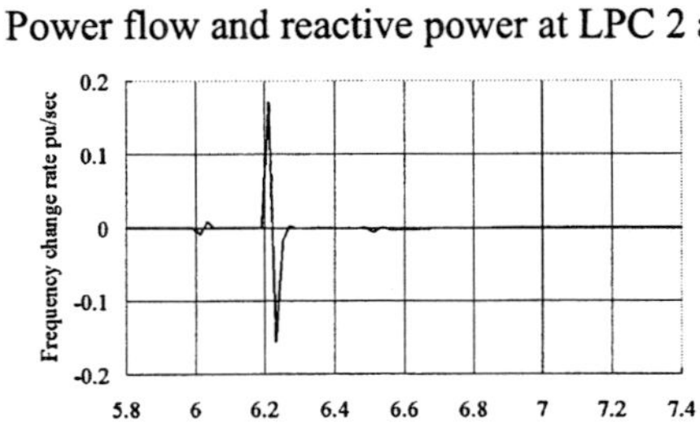

(f) Frequency change rate at section 3

Fig. 7 Characteristics of switching from current control to frequency control with synchronous generator

5.3 Switching Characteristic of Governor Free Operation with Synchronous Generator

Fig. 7 shows the switching characteristics when a synchronous generator connects. The result is thought to be as shown in Fig. 6, and it is believed that distributed generation can continue operation.

5.4 Characteristics of Synchronous Generation Operations

The characteristics of synchronous generator operation are shown in Fig. 8. (a) shows terminal voltage and current. AVR operates by change of terminal voltage, and the maximum current is 1.06pu. This is the tolerance level of the generator. (b) shows the change in PQ of the generator.

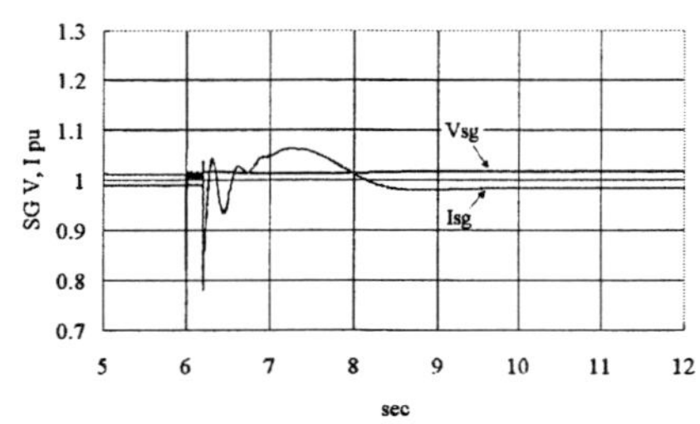

(a) Terminal voltage and current of a synchronous generator

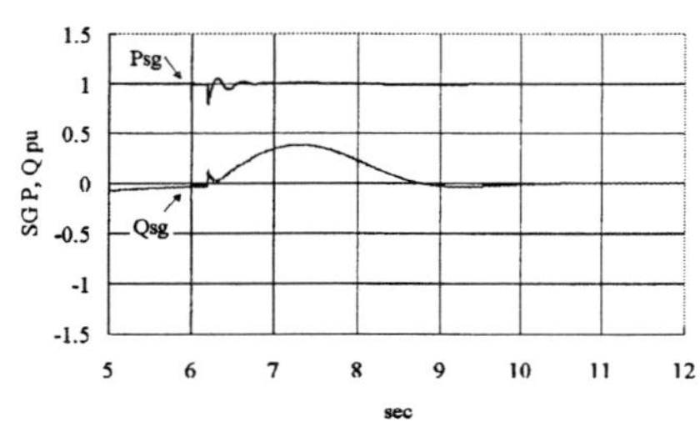

(b) P and Q of a synchronous generator

Fig. 8 Characteristics of synchronous generator operation

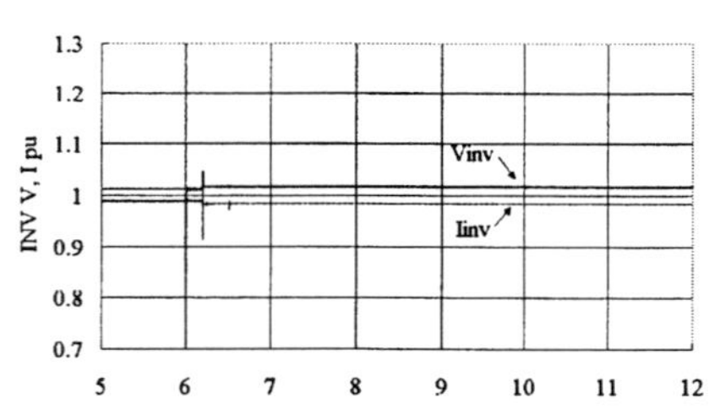

(a) Terminal voltage and current of inverter

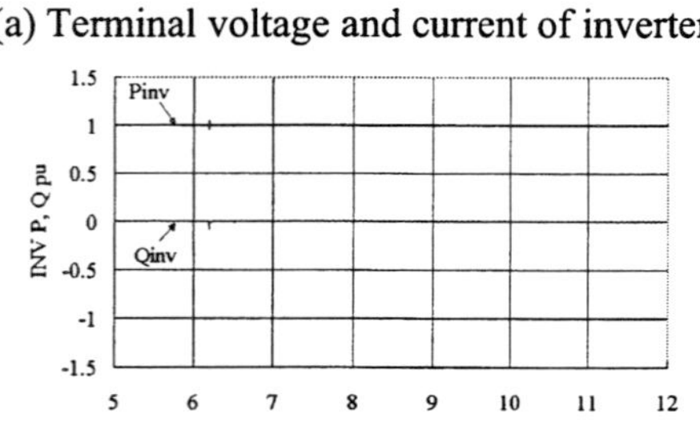

(b) P and Q of inverter

Fig. 9 Characteristics of inverter operation

5.5 Characteristics of Inverter Operation

The characteristics of inverter operation are shown in Fig. 9. Current control of the inverter is high-speed and fluctuation of current (including P and Q) is small compared with the synchronous generator.

6. CONCLUSIONS

We have proposed a method of controlling LPC for 1LG of an ungrounded-neutral system for isolating fault sections by LPC.

We verified switching to frequency control from current control by simulation. Parallel operation of LPC by governor free operation was good.

The simulation analysis appeared to indicate very little possibility that the distributed generation of other sections will stop unnecessarily during isolation of the fault section.

REFERENCES

H. Taniguchi: Dispersed Generation and Power System Control, T. IEEJ, Vol. 121-B, No.9, pp.1065-1068 (2001-9) (in Japanese)

N. Okada et. al, Basic study on installation and control method for loop power flow controller in distribution system, 14 th CEPSI, November, (2002), vol. 3, pp 46-51.

N. Okada, Autonomous Loop Power Flow Control for Distribution System, 7th International Conference on AC and DC Power Transmission, IEE, No. 485, (2001), pp 150-155

H. S. Kim, H. S. Mok, G. H. Choe, D. S. Hyun, S. Y. Choe: Design of Current Controller for 3-Phase PWM Converter with Unbalanced Input Voltage, Proceeding of PESC, fukuoka, (May 1998)

Navid R. Zargari, Geza Joos: Performance Investigation of a Current-Controlled Voltage-Regulated PWM Rectifier in Rotating and Stationary Frames, IEEE Trans. on Industrial Electronics, Vol. 42, No. 4, (August 1995), pp.396-401

N. Okada, K. Takigawa: CONTROL OF LOCAL POWER SUPPLY SYSTEM USING AC MODULES, The Fourth International Conference New Energy System and Conversions,pp.431-434, (1999-June), Osaka, Japan

N. Okada, K. Takigawa: Advanced concept for dispersed power supply system using AC modules, Solar Energy Materials and Solar Cells, volume 67, Numbers 1-4, page 571-581, (March 2001)

ELSEVIER
IFAC
PUBLICATIONS
www.elsevier.com/locate/ifac

DUAL ELECTRIC POWER SUPPLY WITH INCREASING WIND POWER GENERATION, REQUIREMENT FOR AN ADVANCED SECONDARY CONTROL CONCEPT

E. Welfonder* and M. Kurth*
C. Hodurek and H. Radtke****

* *Department for Power Generation and Automatic Control (IVD), University of Stuttgart*
Allmandring 5B, D-70550 Stuttgart, Federal Republic of Germany
Tel. +49 711 685-6214, Fax +49 711685-6590
welfonder@ivd.uni-stuttgart.de

** *Vattenfall, Berlin, Federal Republic of Germany*

Abstract: Based on the example of the East German power system – being coupled to the European interconnected grid - there has been investigated for low load situations
- up to which amount wind power generation can be fed in and transported through the high voltage transmission network as well as
- up to which amount wind power can be overtaken by reducing as far as possibly the power generation of the normal power plant units.

Based on this an advanced secondary control concept (tie line bias control) is introduced for a fair splitting up of the whole renewable power generation, which is fed into the interconnected power system, to the different grid partners. *Copyright © 2003 IFAC*

Keywords: wind power generation, interconnected power system, power fluctuations, secondary control (tie-line bias control).

1. INTRODUCTION

Over the past ten years, the use of renewable power generation units, and in particular of wind power units, has greatly increased. For example, the installed wind power capacity as illustrated in Fig. 1, was already 17.3 GW in 2001 in Western Europe, with alone 8.8 GW of that in Germany. In the course of 2002 the installed capacity is even to increase to 20.9 GW and 11.3 GW respectively (Bundesverband Windenergie e.V., 2002).

The further strong increase to be expected in Germany is based, on the one hand, on the "renewable energy sources act" there (EEG, 2000), according to which up to 9 Ct/kWh is currently paid for wind energy feed-in and, on the other hand, on the planned erection of large off-shore wind parks.

It is true that the electrical energy generated annually with wind energy within the entire interconnected European power system is still relatively low. However, especially in Northern and Eastern Germany the percentage of wind energy generation is at times already considerable, i.e. in off-peak periods with high wind intensities (Luther, 2002). With a further expansion of wind energy generation, operation-related limitations may be required as well on the part of the conventional power plant stations used as within the high-voltage transmission network itself.

This fact and the resulting consequences for power plant and system operation will be explained in the following based on an application-related examination on the example of the VEAG[1] system.

[1] Since 07/01/2002 the German utilities BEWAG, HEW and VEAG have been integrated in Vattenfall Europe.

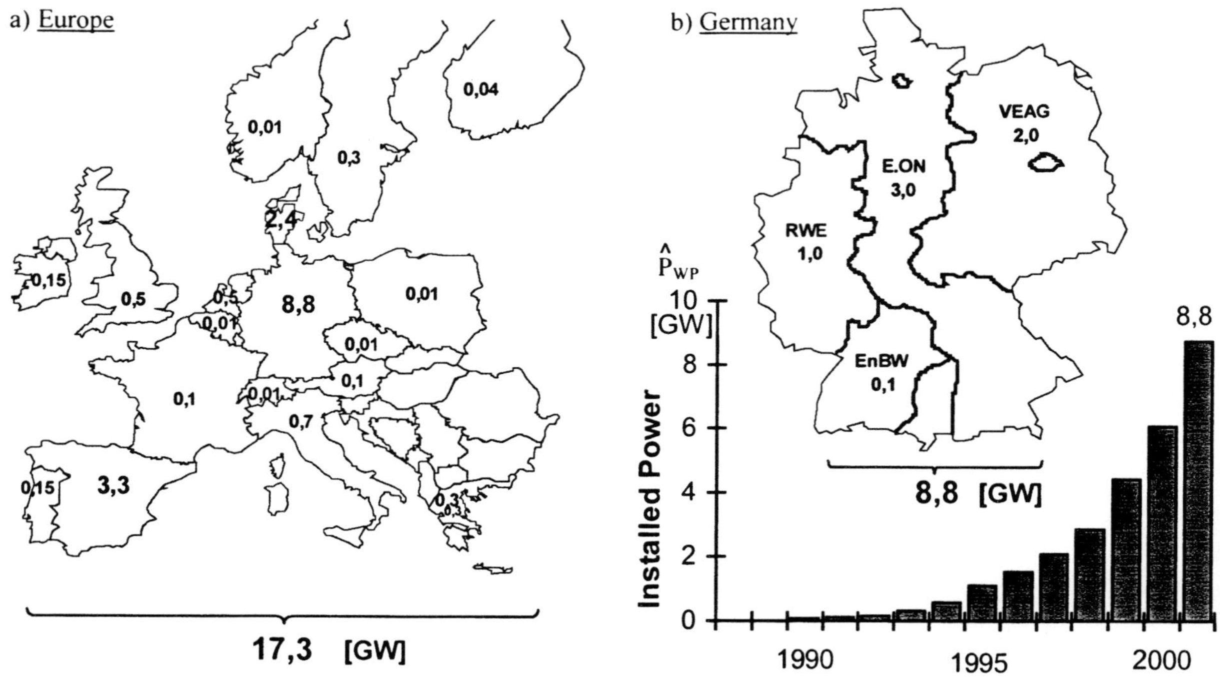

Fig. 1: Installed wind power (state: end **2001**)

2. POWER PLANT AND SYSTEM SIMULATION

The following simulation studies are based on the detailed power system dynamic model of the University of Stuttgart (Welfonder 1987, Kurth 2003), with which the entire UCTE/CENTREL system including the power plant units feeding in and the dynamic loads are simulated in detail.

Fig. 2 gives an overview of the 380/220-kV Transeuropean high-voltage system of VEAG to be simulated. The feeding in of the conventional – non-renewable energy generating – power plants takes primarily place from the Southeast by brown coal fired units being dominant there with nominal powers in the range of 500 to 930 MW.

The large number of wind energy plants that feed into the system from remote locations are taken into account in a simplified manner by 22 regionally located equivalent wind parks with nominal powers in the MW to GW range. Therein 13 equivalent wind parks are feeding in "in the South" of the long VEAG system and five equivalent wind parks "in the North", as well as four future off-shore wind parks, additionally along the Baltic Sea coast.

The amounts of wind power which are fed-in into the VEAG system within the different simulation cases investigated in the following, are illustrated in Fig. 2b.

Within the simulation the grid connection of the equivalent wind parks (Spanner, *et al.*, 1996) is carried out on the 220/380-kV level, s. Fig. 2c. The wind parks are not voltage-controlled and mainly feed in active power. It is assumed that any existing portions of reactive power will be compensated locally. The simulated equivalent wind parks therefore demonstrate the behavior of "negative loads".

In addition, it is assumed that the power fluctuations ΔP_{WP}, which are overlaid on the stationary wind power feed-in in the whole VEAG system, are in the range of ±20%, i.e. the following applies:

$$P_{WP} = \widetilde{P}_{WP} \pm \Delta P_{WP} = (1 \pm 0{,}2)\, \widetilde{P}_{WP}$$

3. REFERENCE CASE

As reference case there serves an off-peak operating state, prevailed on 12/31/2000 at 6:41a.m. within the VEAG system with a load of P_L = 3634 MW and an exchange power of $P_E^{(VEAG)}$ = 1290 MW, see Fig. 3. The high power export is mainly due to the fact that a 930-MW block in Lippendorf is operated on order by VEAG-external companies, i.e. for E.ON and EnBW, and due to this the unit power generated at that time has been exported to these companies.

In the off-peak operating state considered, the largest portion of the power generation was still provided by the conventional power plant units, i.e.

- $P_{sec}^{(VEAG)}$ = 1910 ∓ 95 MW by means of power plant units with secondary control and
- $P_{G,others}^{(VEAG)}$ = 2537 MW by other units with or without primary control.

The remaining power generation of P_{WP} = 477 ± 950 MW was already produced by wind energy plants. This described off-peak operating state serves as the starting point of the following examination scenarios and will therefore be referred to as the basic case (0).

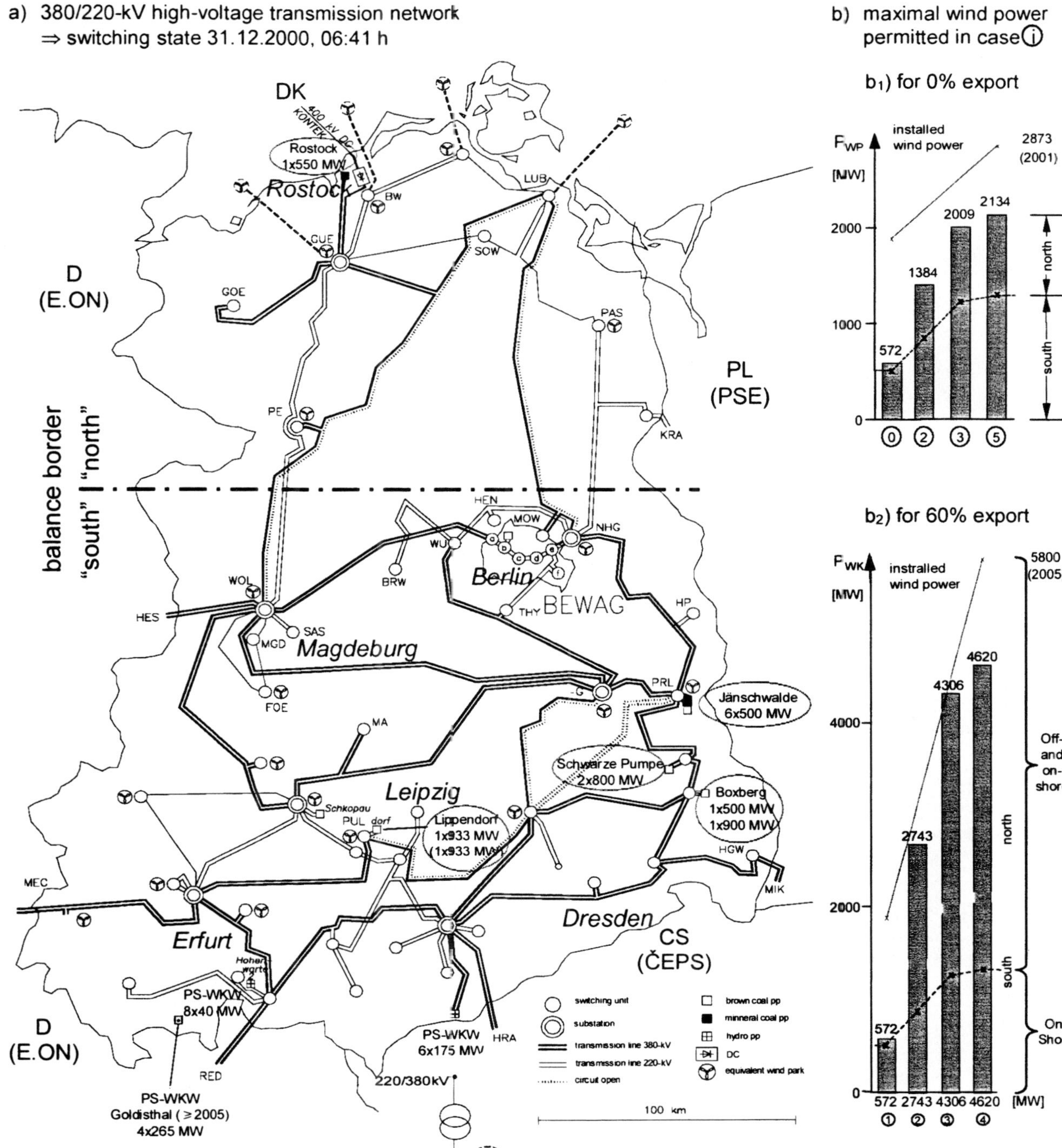

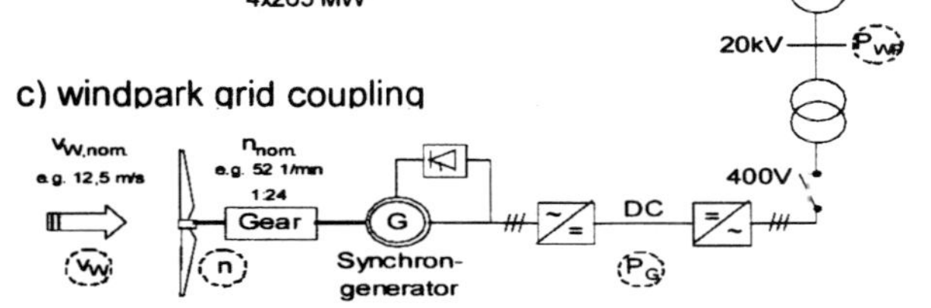

Fig. 2: VEAG-power system with feed-in
- of windparks and as far as necessary
- of conventional power plants

The VEAG system topology which predominates at this point in time is illustrated in Fig. 2a. The geographical assignment can be seen from the entered major cities as well as from the national and system operator borders, drawn in too.

4. EXAMINATION SCENARIO I "WITHOUT WIND POWER EXPORT"

In the first scenario the maximum permissible wind power increase is considered without the possibility of any partial wind power export to the other power system partners with percentually less wind power feed-in.

4.1 Wind-power dependent power plant scheduling

The load and the power export are constant during the entire scenario with $P_L + P_E = 3634 + 1290 = 4924$ MW. As a result of this, the conventional power plant generation must be reduced in accordance with the increasing wind power feed-in. This fact is illustrated in Fig. 4:

Starting from the basic case (0) with a wind power feed-in of $P_{WP}^{(0)} = \tilde{P}_{WP} \pm \Delta P_{WP} = 477 \pm 95 \leq$

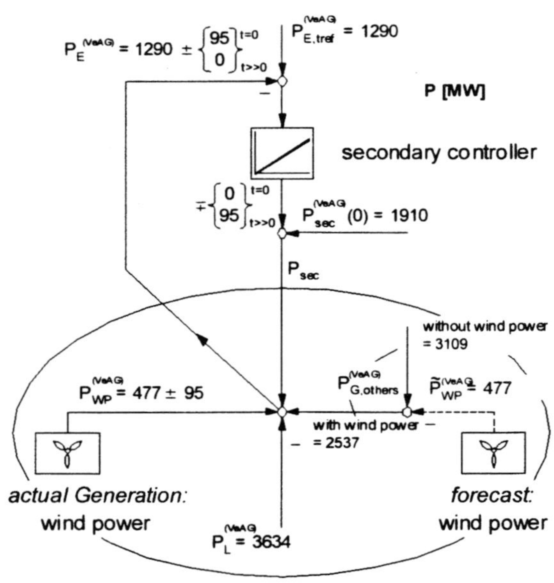

Fig. 3: Summed up load flow situation in the basic case (0)

572 MW, first the power plant units with secondary control will be driven down reaching their lower use limit in duo-operation at $P_{WP}^{(1)} = 694 + 139 = 833$ MW. After then the other power plant units with mainly primary control will be reduced accordingly until these also reach their lower use limit in duo operation at $P_{WP}^{(2)} = 1153 + 231 = 1384$ MW.

A further reduction of the conventional power generation is only possible if the six 500-MW units are switched over from duo to mono operation. With this – VEAG-specific – switchover possibility, now only one of the two steam generators feeds into the shared turbine in duo operation, see Fig. 5.

When switching over the three 500-MW units with only primary control to mono operation, the units with secondary control temporarily regain control leeway which, however, is already consumed again with an increase in the wind power feed-in to $P_{WP}^{(3)} = 1466 + 293 = 1759$ MW. In the case of switching over the 500-MW units with secondary control to mono operation as well, a further increase in the wind power feed-in to a maximum of $P_{WP}^{(5)} = 1778 + 356 = 2134$ MW is possible.

A further increase in the wind power feed-in is not possible within this scenario, as a disconnection of power plant units would:

- impair the availability,
- cause high shut-down/start-up costs and further on
- complicate the problem of voltage regulation in the VEAG system.

4.2 Generator and system behavior

Fig. 5 shows on the example of the wind-power dependent operation of the 500-MW power plant units in the generator graph. Caused by the low inductive load in the case of off-peak system operation, all generators will be operated highly underexcited. To reduce the reactive charging power also all transmission lines, which are not absolutely required in off-peak operation, are switched off, see dotted lines in Fig. 2a.

As the areas for duo and mono operation do not overlap in the generator graph, the generator power of the units with secondary control must be increased before switching over to mono operation. The resulting power excess leads to a temporary increase in the exchange power which, however, is adjusted to its reference value of $P_E = 1290$ MW again within minutes. In Figure 6a this is even carried out within 30

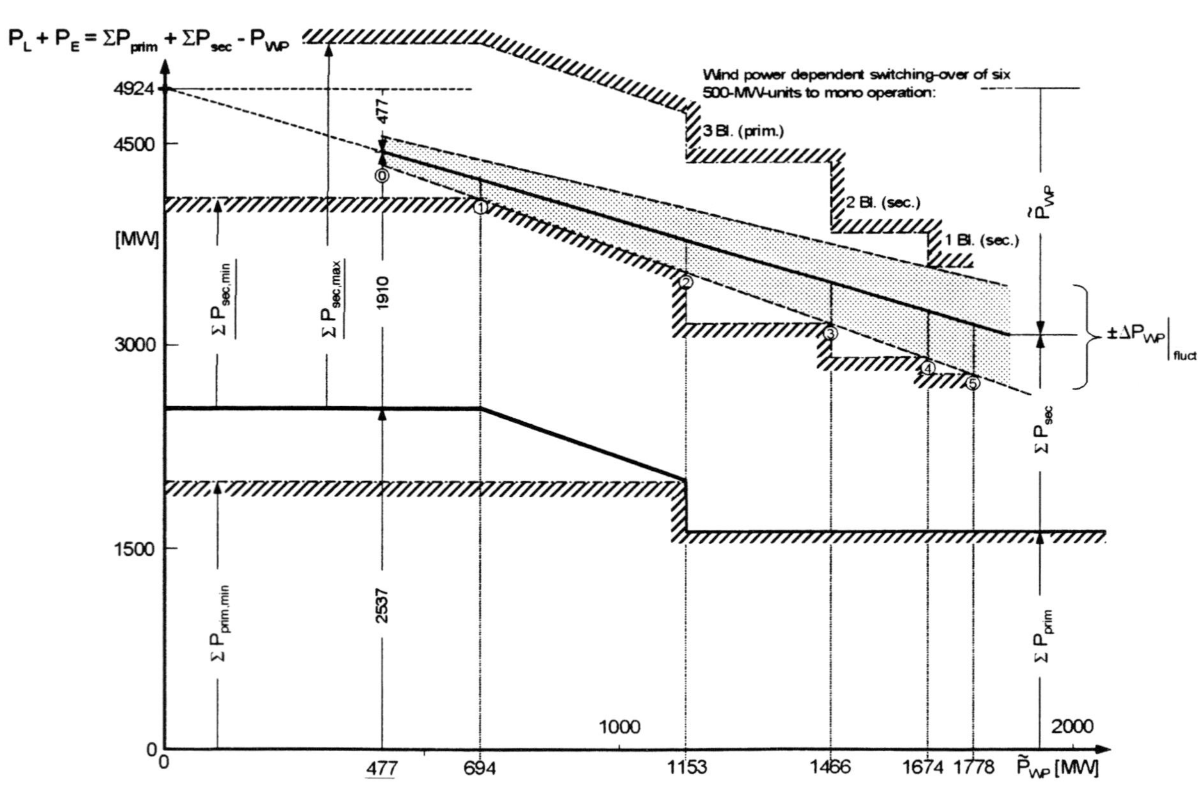

Fig. 4: Summed up real-power balance ⇒ scenario I "without wind power-export

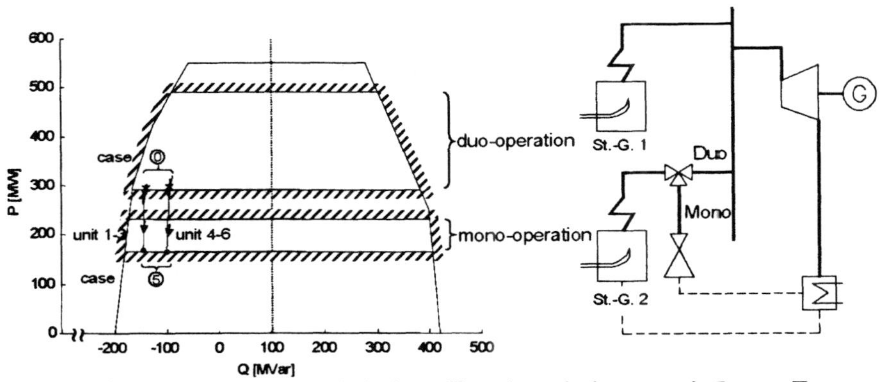

Fig. 5: Generator power graph during off-peak-periods, scenario I resp. II

seconds, as the thermal power plant behavior was simulated - for illustrations reasons - at 10 times the actual speed. As soon as P_E has nearly reached steady state behavior, the simulation of the next case to be examined follows.

As can be seen in Fig. 6b from the transfer power flow between the northern and southern part of the VEAG system, the conventional power plant units feeding-in in the South cover in the basis case the power lacking in the North, too. However, with increasing wind power feed-in the load in the northern part of the VEAG system is covered by the wind energy plants. Therefore the at first negative North-South active power flow increases up to zero. Simultaneously the inductive power losses decrease to zero, which results in a slight increase in the voltage level in the North, see Fig. 6d.

In spite of this, it can generally be stated, that in this examination scenario all voltage curves still run below the critical upper limit of U_{max} = 245 or 420 kV respectively.

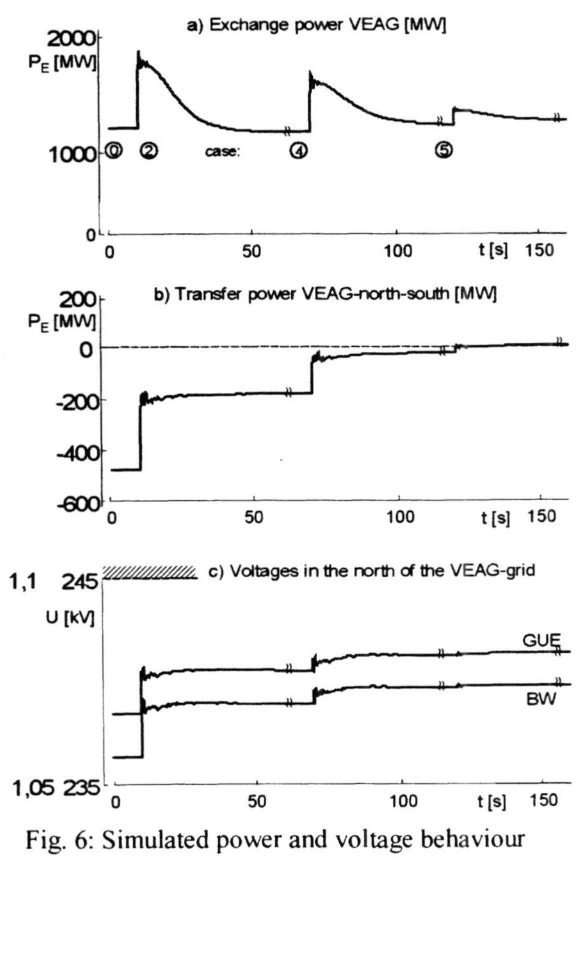

Fig. 6: Simulated power and voltage behaviour

5. EXAMINATION SCENARIO II "WITH PARTIAL WIND POWER EXPORT"

As long as the wind power feed-in is low, it can be compensated by the operator of the power system concerned alone, see Chap. 4. However, with a percental increase in wind power feed-in, it seems to be fair for all power system partners within Germany, or even better within the entire interconnected power system, to participate in the compensation of the regional wind power feed-ins. In this direction tends the following scenario II, in which it is assumed that 60% of the wind power fed into the VEAG system can be exported to other system partners with a lower percentage of wind power feed-in.

In addition the wind power plants to be installed up to 2005 are taken into account, see Fig. $2b_2$. Besides this, scenario II starts from the same off-peak basic case (0). The only slight expansion of power plants and system up to that time is also taken into account, as shown in Fig. 2a. All other system operation specifications, such as transformer staging and voltages at the boundary nodes to the neighboring systems are used unchanged to scenario I.

5.1 Wind-power-dependent power plant scheduling

As can be seen from the summary power balance in Fig. 7, the wind power feed-in can now be increased to a much greater extent, for only 40 % must still be compensated within the VEAG system.

This means that a wind power feed-in up to $P_{WP}^{(1)} = \tilde{P}_{WP}^{(1)} + \Delta P_{WP}^{(1)} = 1136 + 227 = 1363$ MW can be compensated alone by reducing the power generation within the secondary controlled units. From this value the other, mainly primary-controlled units must reduce their generation too. From a wind power feed-in of $P_{WP}^{(2)}$ = 2285 + 457 = 2742 MW this measure is also no longer sufficient, requiring the switchover of the 500-MW units with primary control to mono operation. From a wind power feed-in of $P_{WP}^{(3)}$ = 3067 + 613 = 3680 MW two units with secondary control are switched over to mono operation, and from $P_{WP}^{(4)}$ = 3588 + 718 = 4306 MW the third unit with secondary control is also switched over. This measure 5a enables an increase in the wind power feed-in up to a maximum of $P_{WP}^{(5a)}$ = 3850 + 770 = 4620 MW.

After this completion of the Goldisthal pump-storage power plant station in 2005 it will be possible, instead of switching over the sixth 500-MW unit to mono operation, to run a water turbine temporarily in speed-controlled pump operation and to store excessively generated wind energy in this way. Due to the

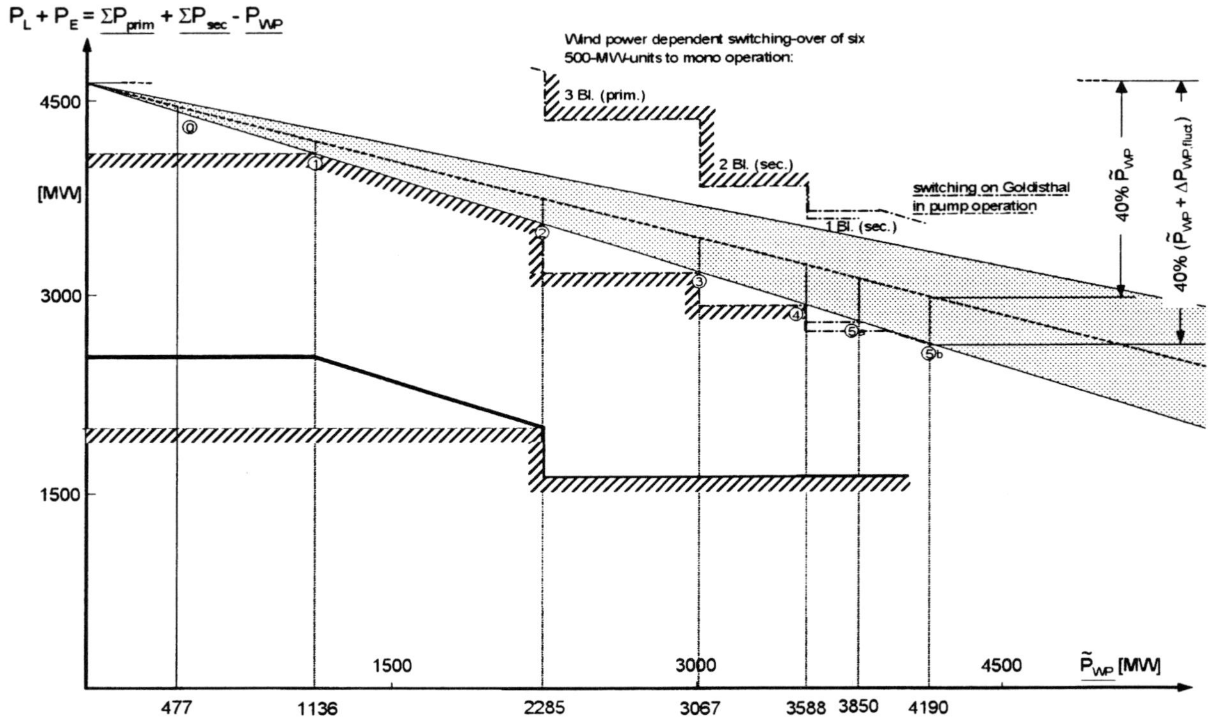

Fig. 7: Summed up real-power balance,
⇒ scenario II with "60% wind power export"

resulting load increase, a further increase in the wind power feed-in up to a maximum of $P_{WP}^{(5b)}$ = 4190 + 838 = 5028 MW is possible. During this process the pump - initially operated at a 190 MW minimum load - is run up with the further wind power increase to a 290 MW full load.

5.2 Generator and system behavior

In both scenarios the behavior of the power plant units illustrated with the generator graphs in Fig. 5 is nearly identical. This is due to the fact that in both cases off-peak system operation is assumed, and that the power generation of the units is reduced each time to the lower limit of their control range.

In contrast to this, the exchange power rises in Scenario II due to the 60 percent wind power export with increasing wind power feed-in P_{WP}, see Fig. 8a. As this power increase is primarily caused from the large off-shore wind park in the North planned in up to 2005, heavy power transport now results in the North-South direction, after the North power draw has already decreased to zero in the course of Scenario I, see Fig. 8b. In conjunction with this the reactive power losses in the transmission system increase again, causing a successive decrease in the voltage level particularly in the northern area of the VEAG system, see Fig. 8c. In the process, the voltage there would – starting with $U^{(0)}$ = 240 kV in the basic case – even drop to 180 kV with a maximum wind power feed-in of $P_{WP|max}^{(5b)}$ = 5028 MW. To prevent a voltage dropping below U = 0.9 U_{Nenn}, compensation coils with a total reactive power of 850 MVAr must be switched off in the North of the VEAG system, see Fig. 8c.

After this switching measure is carried out, the maximum permissible wind energy feed-in on the power-plant side can also just be transmitted on the system side, i.e. by utilizing the entire voltage tolerance range.

A further increase of the wind power feed-in above the maximum permissible power level would have to

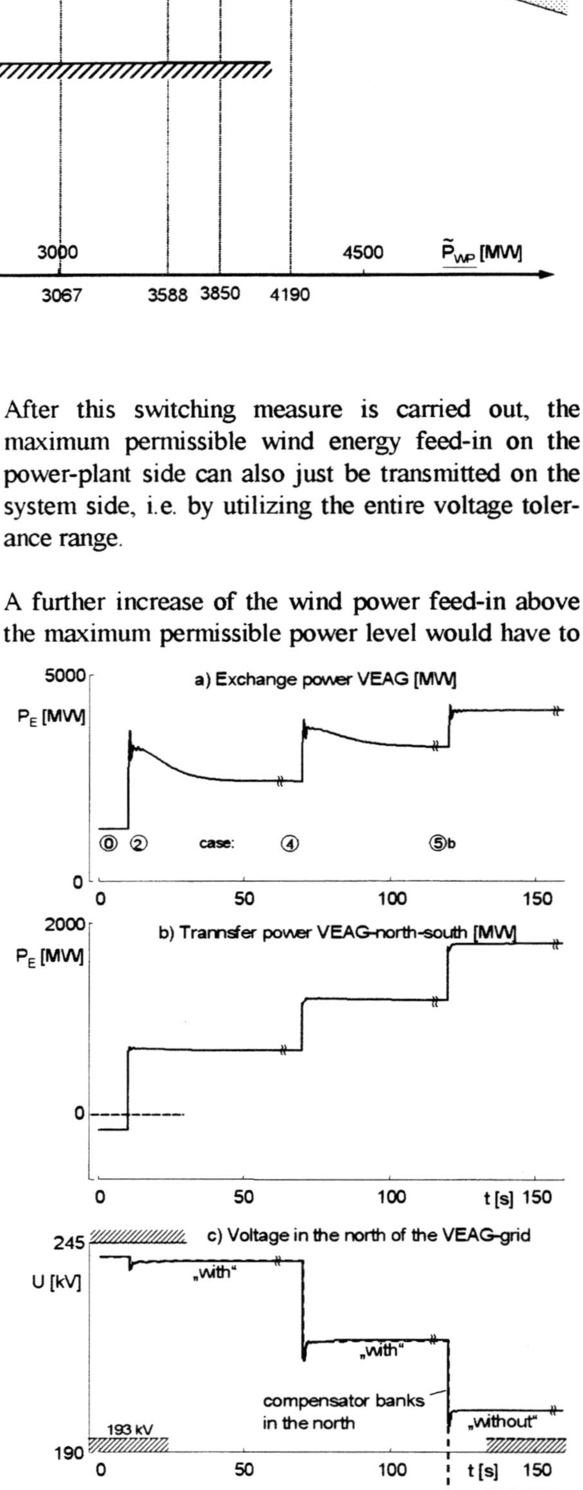

Fig. 8: power and voltage behaviour in scenario II "60% wind power export"

be counteracted with power-limiting adjustment of the rotor blades (e.g. Enercon, 2001) or by temporarily switching off individual wind energy plants. However, this would only prove to be necessary at very high wind intensities, and will therefore probably only be required on a few days of the year.

6. TARGETED NEW SECONDARY CONTROL CONCEPT

However, the measures necessary at power plant stations for enabling a maximum wind power feed-in, do cause disadvantages with regard to energy such as:

- The reduction of the power plant units to off-peak operation results in poorer energy efficiency. This is particularly true for steam power plants, as to be seen in Fig. 9a.
- During mono-operation, i.e. when only one steam generator instead of normally two is connected to the turbine, the second one generally remains in operation for reasons of availability, whereby the generated live steam flow is relaxed via the high-pressure bypass station and must be "disposed of" toward the condenser, provided it cannot be used for steam production or for district heating supply.
- Within hydro power plants pump/turbine storage operation is only possible with a total efficiency of approx. 74 %, see Fig. $9b_2$. Therefore, the temporary storage of wind energy using pump-storage hydro power plants is always combined with power losses of more than 25 %.

Therefore in cases of dual energy supply the measures required on the power plant side should be applied as moderately as possible. This necessitates the participation of all system partners in the compensation of the power generation by renewable energy plants fed in unequally with regard to the time and place. This may also apply to the remote feed-in of fuel cells which, although they are not included in the above plant type, are characterized by a high level of energy efficiency.

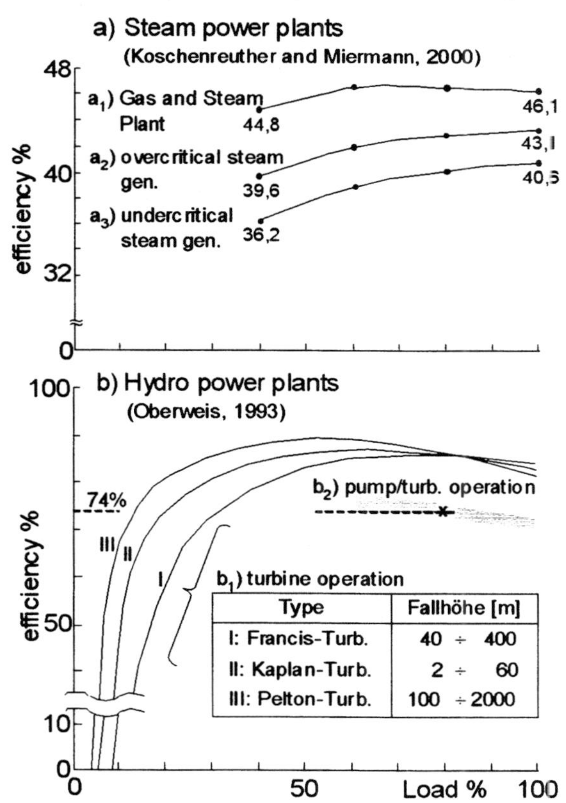

Fig. 9: Part load efficiencies

For this purpose an advanced secondary control concept (tie-line bias control) proves to be required. Therein the regeneratively generated actual power respectively is summarized[2] over the whole control zone or control block and added to the reference value of the respective exchange power P_E.

The regenerative power feed-ins $P_{RE\,i}$ "exported" in this way per each control zone or control block respectively must be gathered to $\Sigma P_{RE\,i}$ in an additional uncontrolled balancing group. The distribution of the whole regenerative power $\Sigma P_{RE\,i}$ among the individual control zones or control blocks respectively should fairly be done in accordance with the respective load share $\alpha^{(i)}$ of the total load ΣP_L, see Fig. 10a and 11a. Only if there should occur bottlenecks, the percentually equal distribution has to be restricted as far as necessary.

The advanced secondary control concept described above is illustrated in Fig. 10 I with the example of the exchange power of the German control block $P_{E\,soll}^{(D)}$ and can be applied correspondingly to the underlaid control zones too, as shown in Fig. 10 II for the Vattenfall control zone. By means of the simple data example the advanced control concept is self explaining.

In both pictures the subfigures a_2 show the classic secondary control concept and the subfigures a_1 and a_3 the share of the connection of regenerative energy generated or predicted respectively for the entire interconnected power system or the corresponding per control block.

Such an advanced secondary control concept is competitively oriented and should therefore also be supported politically, i.e. in the considered case by the European Union. Besides this with a further increase in regenerative energy generation such kind of handling appears inevitable for the entire interconnected power system. For on the one hand, system areas with a high percentage of regenerative energy generation are already approaching the limits of their controllability today in off-peak periods as illustrated in the examination scenario I. On the other hand, the power fluctuations of regenerative generation plants cancel each other out proportionally over the entire interconnected power system. For the moderate wind intensities along the coasts of the Baltic Sea, North Sea, Atlantic and Mediterranean, as well as within Europe generally differ. And the overlaid fluctuations do vary from place to place, and therefore are not correlated with each other (Welfonder, 1990).

[2] When doing so the power feed-in of large wind parks detected by measurement must be extrapolated to the additional share of wind plants being not detected for feeding in remotely.

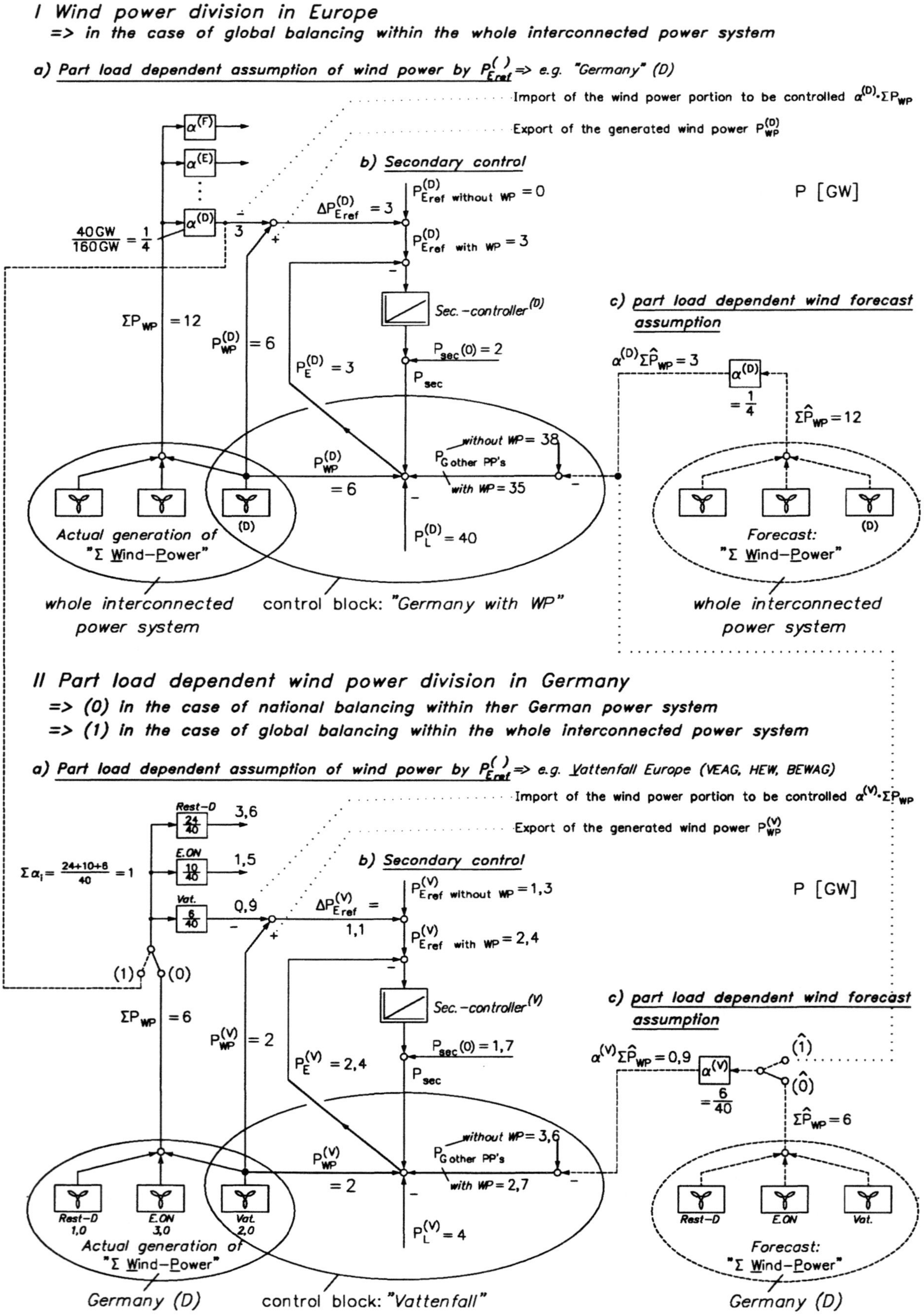

Fig. 10: Advanced secondary control concept, being required

Alone for this reason, the advanced secondary control concept with separate balancing of all regeneratively generated power should be introduced as soon as possible. This is independent of when and how an agreement is reached on the proportional distribution of the whole regeneratively generated and – due to summary detection – smoothed power. Of primary importance is the fact that not each partner in the interconnected power system is forced to compensate the regeneratively generated power fluctuations fed into its system separately (Welfonder, 2002). Applying the advanced control concept also the discussion on the need for energy storage to compensate fluctuating power feed-ins should be defused.

7. SUMMARY AND OUTLOOK

The deregulated electrical energy market recently introduced in Europe with the goal of an energy supply at the lowest possible costs is diametrically opposed to a dual energy supply with the primary use of regenerative energy supply plants.

To illustrate and clarify these facts, a corresponding simulation study has been conducted using the example of the constantly increasing wind energy feed-in into the East German transmission system of the VEAG.

The main results of the study include:
The conventional power plant units feeding into the VEAG system can compensate itself up to 1.75 GW wind power $\pm$ 20% fluctuations, and this by corresponding reduction of the conventional power plant units to minimum generation. However, this is coupled with a poorer energy efficiency of the off-peak operating units.

In the case, that 60% wind power can be exported, up to 4.2 GW $\pm$ 20% of wind power can be fed in and transmitted within the VEAG system. This is, of course, combined with increasing transmission losses.

The power shares fed in regionally by the large number of renewable-energy generation plants at any given time should not be compensated within the individual control zones or control blocks, but should instead be summarily added in a separate balancing group. As a first result of this new concept, regionally occurring uncontrolled wind power fluctuations do already compensate each other proportionally over the entire interconnected power system.

Secondly, the regeneratively generated total power should – in the sense of a competitively-oriented energy supply – be divided up among the individual partners of the interconnected power system in proportions to their load. This kind of distributed wind power compensation should take place primarily before the short and mid-term buying/selling transactions on the energy market will become active, and should only be restricted if bottlenecks occur in the transmission system.

To implement these measures an advanced secondary control concept is required. This is presented in Chap. 6 and explained exemplarily for the German control block and underlaid for the regional control zone of Vattenfall, i.e. the combination of the former German utilities VEAG, HEW and BEWAG.

In a further simulation study, the control behavior of the European power system should be examined taking into account the regenerative power feed-ins for all partners within the interconnected power system.

8. BIBLIOGRAPHY

Bundesverband Windenergie e.V. (2002).
Zahlen zur Windenergie. *http://www.wind-energie.de/informationen/informationen.htm.*

EEG, Renewable Energy Sources Act (2000). *Bundesgesetzblatt I.*

Enercon (2001).
Spec sheet Enercon E-66

Kotschenreuther, H. and L. Miermann (2000).
Leistungs- und Wirkungsgradverbesserung von kohlebefeuerten Anlagen durch kostengünstige Repowering-Konzepte. *VGB-Konferenz "Forschung für die Kraftwerkstechnik 2000", 10-12.Oktober 2000 in Düsseldorf*

Kurth, M. and Welfonder, E. (2003).
Oscillation Behaviour of the Enlarged European Power System under Deregulated Market Conditions.
IFAC Symposium on Power Plants and Systems, Korea.

Luther, M., Radtke, U. (2002).
Operation and Planning of Grids with High Levels of Wind Power.
The World Wind Energy Conference and Exhibitionm July 2002, Berlin.

Oberweis, M. (1993).
Seit 175 Jahren Einsatz von Turbinen in der Wasserkraftnutzung.
http://www.aliai.lu/aliai/rt/rt19993a/rt19993a.htm.

Spanner, M., R. Neifer, and E. Welfonder (1996).
Development and Experimental Identification of Dynamic Models for Wind Turbines and their Fluctuating Power Generation. *VGB PowerTech 76*, **Number 2/96**.

Welfonder, E. (1990).
Control Behavior of Power Systems in the Case of Fluctuating Feed and Required Spinning Reserve. VGB PowerTech 70 (1990), **Number 3 and 4**.

Welfonder, E., Schäfer, Th. and Asal, H.P. (1987).
Control Behaviour of the West European Power System Simulated on the Basis of a Detailed Dynamic Model. *10th IFAC World Congress, July 27-31, Munich.*

ELSEVIER
IFAC
PUBLICATIONS
www.elsevier.com/locate/ifac

VOLTAGE REGULATION FOR DISTRIBUTED WIND GENERATORS

Eric Paalman and Roger Morgan

SP PowerSystems Ltd and Liverpool John Moores University, UK

Abstract: An existing group of wind generators, connected to a weak rural distribution network, has been retro-fitted with an active voltage control system. Detailed measurements have been made of the operation of the voltage controller, and the results used to optimise the target-voltage and bandwidth setting of the tap changer. *Copyright © 2003 IFAC*

Keywords: Wind energy, voltage control.

1. INTRODUCTION

The generation of electricity from wind turbines is of particular interest in the United Kingdom because of its windy climate (British Wind Energy Association 2002). The UK Government has confirmed its enthusiasm for natural energy sources in recent statements (Wilson 2002). Currently only about 550 MW of wind capacity is connected, producing about 0.4% of the national electricity output, but a further 400 MW is expected to become available during 2003, and up to 3 GW may become available by 2005 (Fleming 2002).

The amount of power that can be generated will determine at which voltage level the wind generators are connected. In the UK, large wind-farms are normally connected at 33 kV, while small-scale wind clusters are normally connected into the 11-kV distribution network.

The connection of sources of wind generation requires consideration of many factors (Masters 2002). In particular, connection into 11-kV lines can cause problems of voltage excursions for consumers connected to the same 11-kV line. It is of course possible to reinforce the network, but the cost of doing so may render the project uneconomic. There is considerable interest, therefore, in solving the problems of voltage control where wind generators are connected into weak distribution networks.

The distribution network was originally constructed for power flow in one direction only: i.e. from a central power station, through the transmission lines and the distribution network, to the load centres. The Electricity Supply Regulations stipulate a bandwidth for each distribution voltage level and the networks are designed with on load tap-changers to guarantee that the voltages remain within their set bandwidth. The last point of active voltage control is at the 33 kV to 11-kV transformer, as shown in figure 1. A general discussion on voltage control in power systems can be found in textbooks on electric power such as Weedy (1979).

The voltage is set typically to 11kV +2%. The increment of +2% above the nominal voltage allows for a voltage drop along the distribution line and ensures that low voltage customers receive a guaranteed (single phase) voltage at their terminals of 230 Volt +10% & -6% independent of the load. This technique, together with line drop compensation, is widely used and is very effective in rural areas where power is distributed via long overhead lines. (see figure 1)

Connecting a generator into an existing distribution network can change the load-flow and voltage profile, causing the voltage to rise. If the voltage rise goes outside the statutory limits, (as shown in figure 2), the consumers connected in the vicinity of the generator can receive excess voltage at times when the wind farm is generating. The secondary transformers, 11-kV to 433 Volts, are on a fixed tap and will therefore be determined by the voltage of the 11-kV network.

Several techniques can be used to eliminate or reduce the impact of this voltage rise:

- Restrict the power factor of the generator.
- Upgrade the existing conductor and hence reduce the impedance of the line.
- Reducing the target voltage at the 11-kV bus-bar of the primary transformer.
- Utilise voltage regulators near the point of generator connection.

This paper is concerned with the last of these techniques.

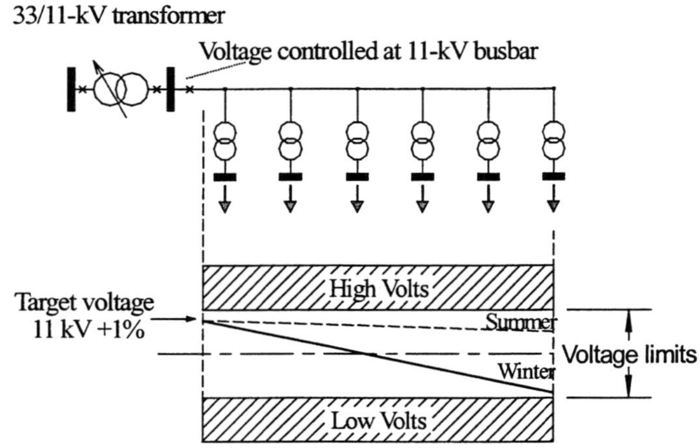

Figure 1 Typical 11-kV system without embedded generation

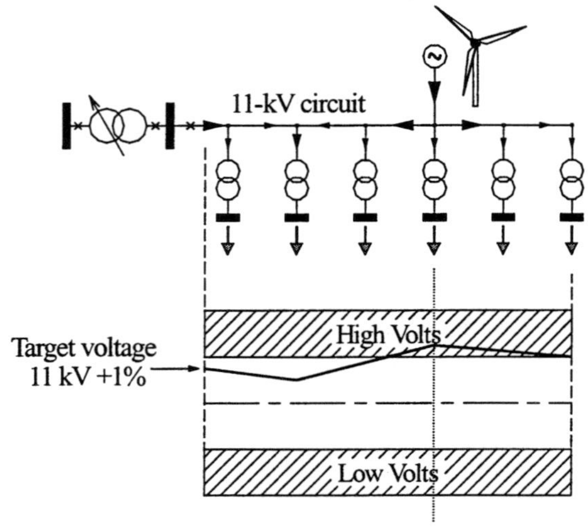

Figure 2 The addition of embedded generation causes unacceptable voltage rise

2. THE INSTALLATION

The installation is situated in a very rural, and very windy, area of North Wales, near the village of Cerrig y Drudion. The electrical load in the area is supplied by an 11-kV overhead line, and consists mainly of domestic and agricultural consumers. Initially a 600 kW wind turbine was installed, and it has recently been upgraded with two further turbines of 850 kW each. To assist with voltage control, a pair of in-line voltage regulators, Cooper VR32, has been commissioned near to the point of connection of the wind farm spur to the 11-kV distribution line. The approach adopted in this study is unique because of the data gathering equipment installed at various strategic locations. This data, which is normally not available, will form a good basis for understanding voltage control and power flow in a practical rural network.

The voltage regulators are auto-transformers with automatic tap changers which can be set to various different regimes of control. A major benefit of these devices is that they can be pole mounted and therefore are easily connected to an existing overhead line. An important factor ensuring simple reliable operation is that the regulators must operate without any dedicated communications equipment but work together with the wind generator and the voltage control at the primary substation. To achieve this each device will have to operate within its own set of pre-defined range of parameters. The voltage regulator is not fast enough to be able to control the voltage variations caused by wind gusting. Instead, the algorithms installed in the control facilities of the wind turbine take care of wind gusting up to 30 seconds. The local voltage controller looks after variations between 30 and 80 seconds. The voltage control at the primary substation is activated after 80 seconds.

3. MONITORING

Monitoring has already taken place at various strategic locations to get a good understanding of the power flow and voltage profile between the wind farm and the existing distribution network. The monitoring will continue in stages to get snapshots of the power flow and voltage control of the system. It is important to take readings at the strategic locations at the same instant in time so the results can be compared in a graph.

There are three strategic places identified at this stage:

1. The 11-kV busbar at the wind farm site.
2. The voltage regulators.
3. The primary substation.

The network operator monitors the information from the primary substation and the information is stored in a computer database. This information can be used to determine the power flow to the main line connected to the wind turbine.

The monitoring equipment for the voltage regulator and the wind farm initially had to be installed temporarily and data was gathered in February 2002, July 2002, and August 2002. The data was gathered in a period of two days continuous monitoring in short intervals, typically readings were taken between 15 and 60 seconds. This gave an accurate picture of the voltage and power flow over the two days. The two days were enough to create a picture showing the various basic conditions, i.e. no wind, windy, no load, heavy load.

Subsequently, monitoring continued over the whole of the winter period of 2002-2003. It allowed consideration of the various operating conditions, varying from zero generation and maximum load, which could be regarded as the normal mode of operation prior to the building of the wind generators, through to maximum generation at times of minimum load, resulting in a major flow of power from the generators back into the sub-station and into the national power network.

4. ANALYSIS

A computer simulation program, Interactive Programme to System Analysis (IPSA), was used to create a model of the network. Real time load data and power flow measurements from the actual network were used for verification. The model was then used to predict the maximum and minimum voltage limits that could be experienced on the network. Optimum settings were calculated and the most effective setting for this scenario is shown in figure 3. The settings ensure tight voltage control when the wind turbine output exceeds the demand in Cerrig y Drudion, i.e. operating in reverse

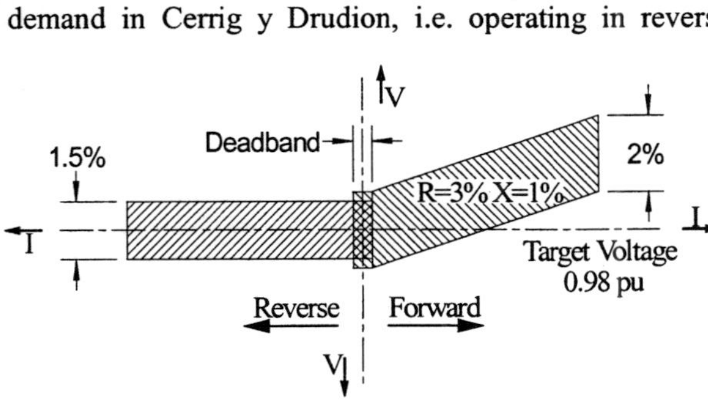

Figure 3 Optimum settings for the voltage regulator

mode. The voltage variations, in reverse mode, are better controlled with a bandwidth of 1.5%. In reverse mode there is no Line Drop Compensation added because the fluctuating power-factor of the circuit. In forward mode, i.e. minimum or no output from the wind turbines, the power-factor is more stable and Line Drop Compensation is added to aid the voltage profile. The power-factor of the system in forward mode is established as 0.95 inductive so an 'R' and 'X' value of 3 and 1 respectively ensures maximum use of Line Drop Compensation. The IPSA studies show that a target voltage of 0.98pu with Line Drop Compensation is the most effective setting at this stage in the project.

5. RESULTS

Results are shown in figures 4-8 (at the end of the paper). Prior to the installation of the voltage regulator, there were several occasions when the voltage exceeded the nominal level by an unacceptable amount. An example is shown in figure 4. After installation of the voltage regulators, with a narrow operating range, results were much improved, as shown in figure 5. Figure 5 shows that whilst the output of the windfarm site has increased from 600 kW to 2300 kW, the voltage rise at the windfarm busbar is acceptable due to tight voltage control at the regulators. The operation of the voltage controller, going both 'down' and 'up', can be seen in figure 6.

A narrow operating range for the voltage controller is clearly desirable from the viewpoint of good voltage control. However, if the bandwidth is set too narrow, there will be excessive numbers of tap changes per day, possibly limiting the lifetime of the tap changer contacts and mechanism. On the other hand, a bandwidth which is set too wide may be insufficiently effective to prevent unacceptable voltage excursions. Measurements were made, therefore, of the number of changes per day as a function of bandwidth. The results are shown in figure 7. There is clearly a major change in the tap-changing behaviour at a bandwidth setting between 1.5% and 2%. Viewed in terms of the lifetime of the units, which are designed for 1 million operations, 25 tap changes per day would represent a lifetime of about 100 years, which is entirely acceptable.

6. CONCLUSIONS

It is evident that the voltage control of the system has been much improved by the addition of the voltage controllers. The main load centre now experiences no voltage excursions outside statutory limits, no matter what the combination of load and wind generator output.

The proposed setting even allows the two newly installed 850 kW generators to run at a less restricted power factor. These new machines are Doubly Fed Induction Generator (DFIG) Machines (Hofmann et al. 1997) which in principle allow a wide range of power factors including capacitive as well as inductive. It appears that the voltage step incurred with the increased power-factor will not be outside the agreed limits. Work is in hand to investigate further the effectiveness of the regulators in such circumstances.

ACKNOWLEDGEMENTS

The authors thank SP PowerSystems for permission to publish. The project was partially funded by the UK Government Department of Trade and Industry. The authors thank Liverpool John Moores University for funding to attend the conference.

REFERENCES

British Wind Energy Association (2002) http://www.britishwindenergy.co.uk/

Fleming, D., (2002) *New Civil Engineer* **21** (November) pp 24-25: Institute of Civil Engineers, London, UK

Hofmann, W., Thieme, A., Dietrich, A., Stoev, A. *EPE'97* Sept 1997 pp 2.723-2.728

Masters, C. L., (2002) *Power Engineering Journal* **16 (1)** (February) pp 5-12: Institute of Electrical Engineers, London, UK

Weedy, B.M., (1979) *Electric Power Systems*, Wiley.

Wilson, B., (2002) *New Civil Engineer* **21** (November) pp 24-25: Institute of Civil Engineers, London, UK

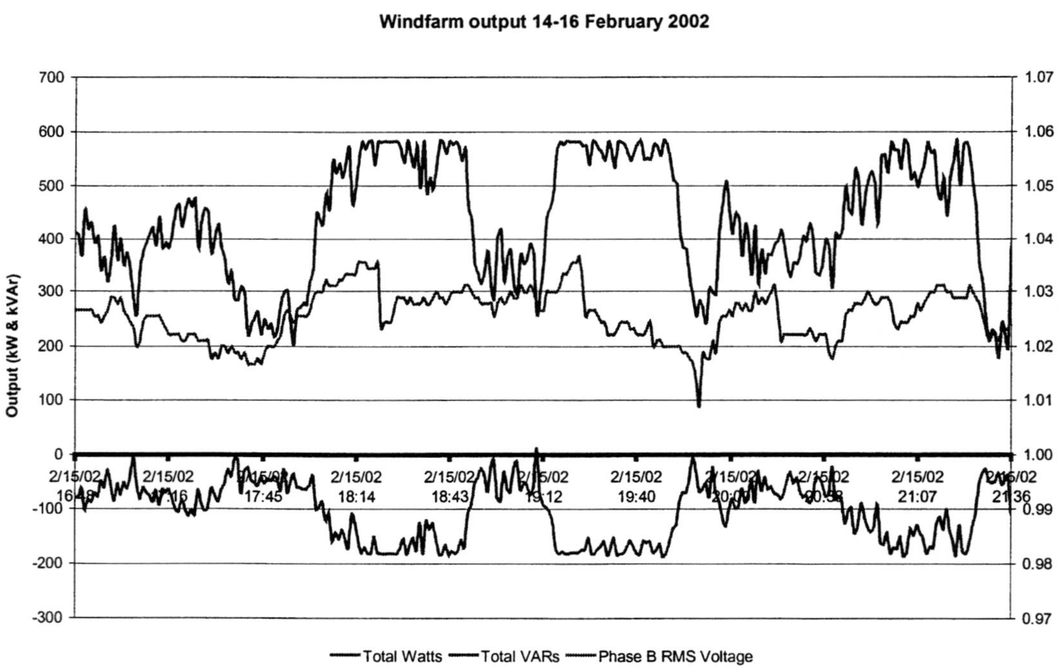

Figure 4 Voltage regulator not in operation.
Upper curve – total watts. Middle curve – RMS voltage. Lower curve – total VARs

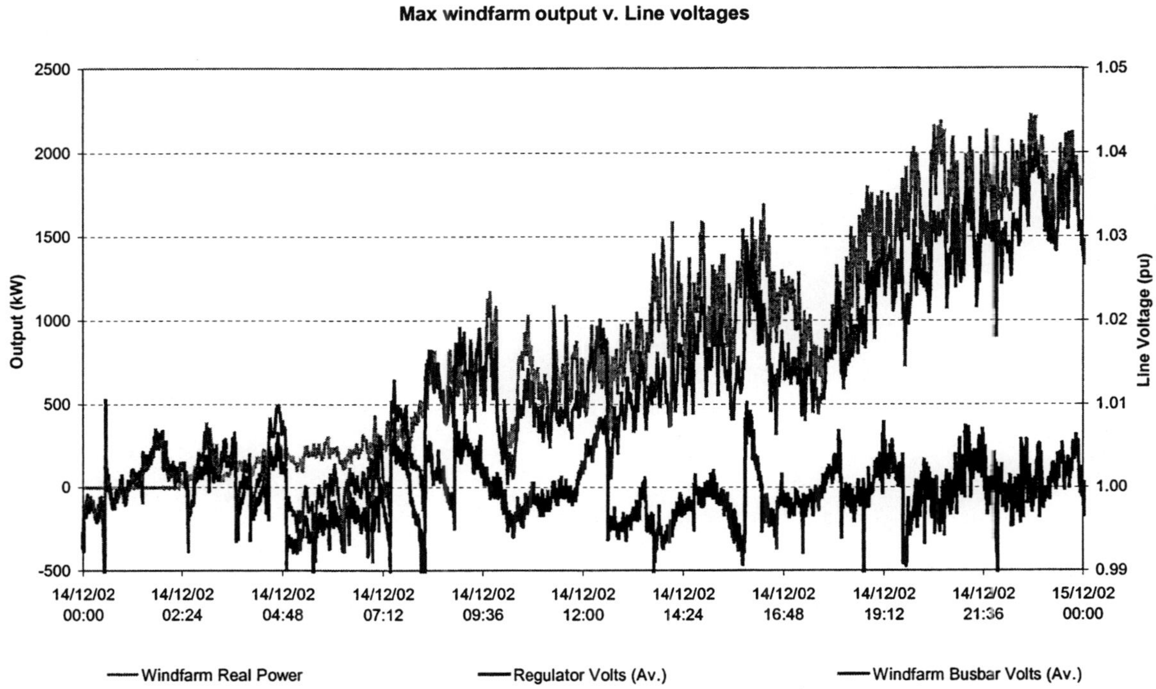

Figure 5 Voltage regulation in operation
Upper curve – windfarm real power. Middle curve – windfarm busbar volts. Lower curve – regulator volts.

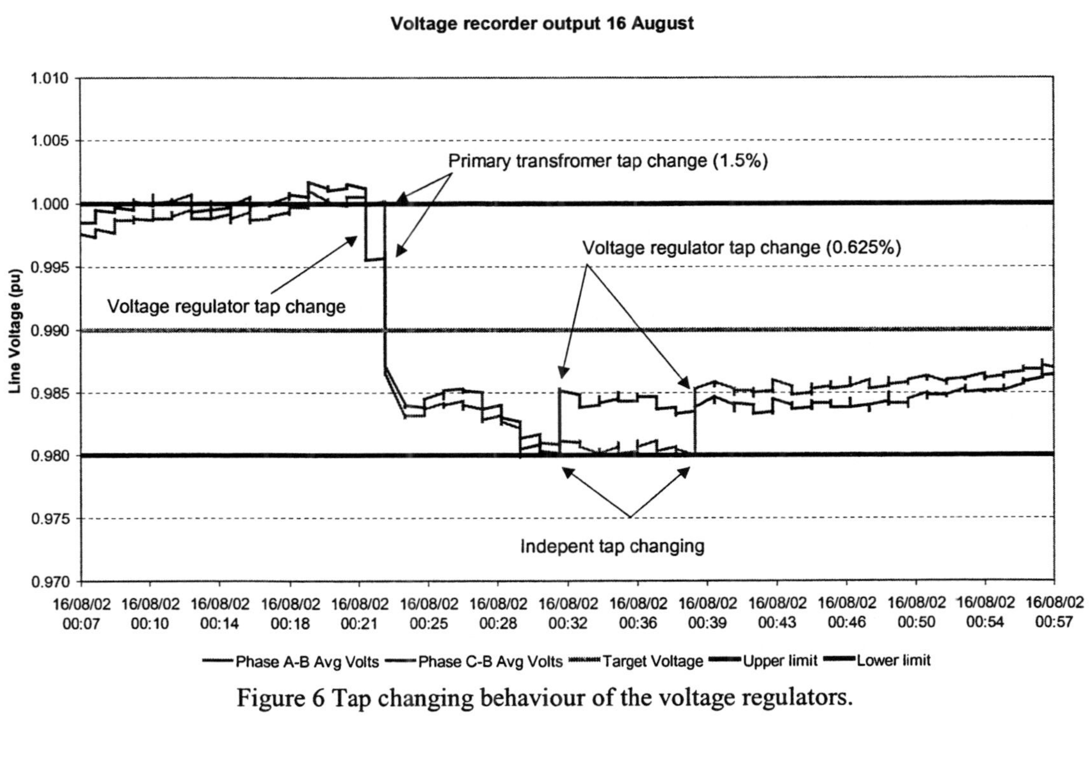

Figure 6 Tap changing behaviour of the voltage regulators.

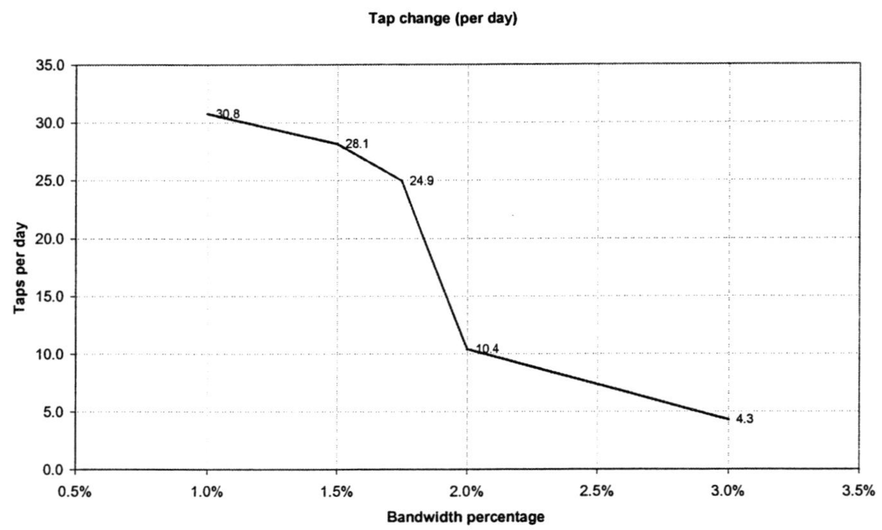

Figure 7. Tap changes as function of bandwidth.

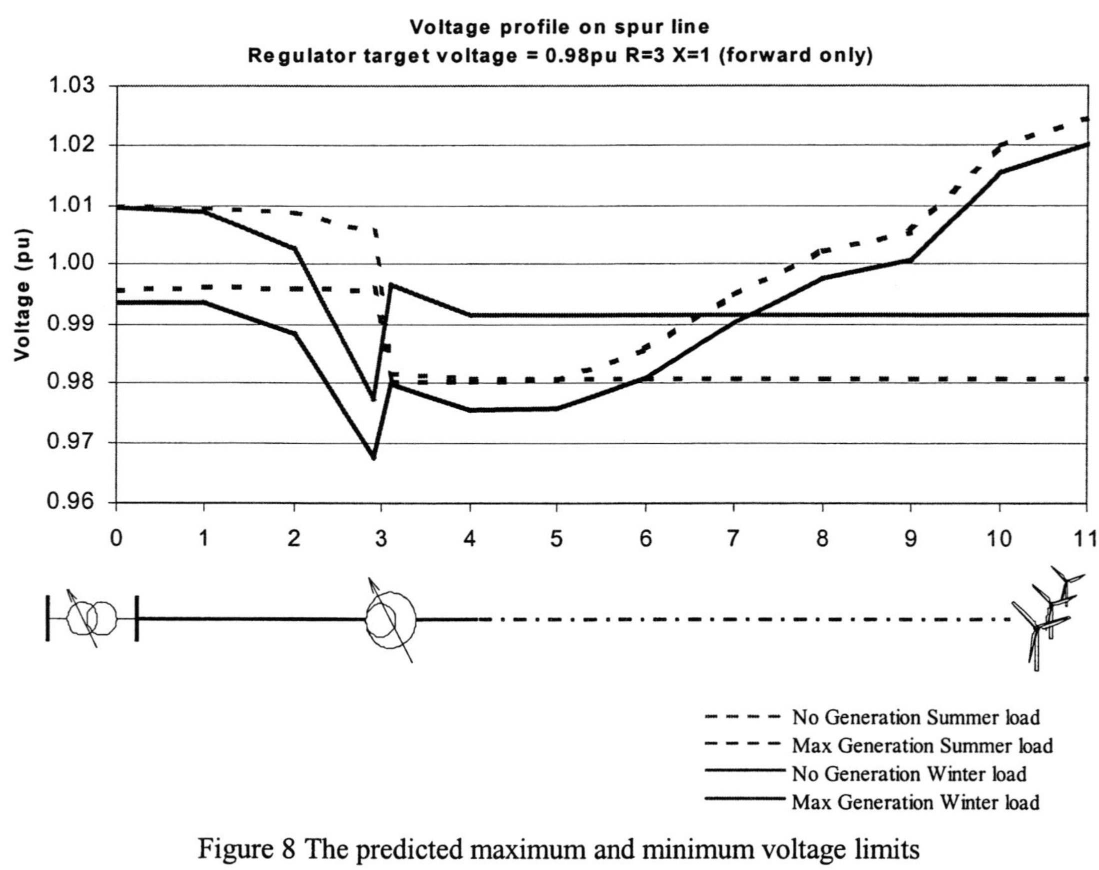

Figure 8 The predicted maximum and minimum voltage limits

ELSEVIER
IFAC
PUBLICATIONS
www.elsevier.com/locate/ifac

VOLTAGE AND OUTPUT POWER STABILIZATION OF WIND POWER GENERATION SYSTEM BY SMES

Tomonobu Senjyu * **Tatsuto Kinjyo** *
Katsumi Uezato * **Hideki Fujita** **

* *University of the Ryukyus, Japan*
** *Chubu Electric Co Ltd., Japan*

Abstract: In recent year, power generation from renewable energy sources is coming up. Particularly, wind power generation is attractive because of it's advantages such as, pollution free, no fuel cost, abundantly available in nature etc. However, the generated power and the bus voltage is always fluctuating and it mainly depends on the wind speed. The superconducting magnetic energy storage(SMES) unit is capable of controlling both the active and reactive power simultaneously and quickly. In this paper, a control configuration for the active and reactive power simultaneous control of SMES which placed at the generator terminals is presented, and H_∞ controller based on Normalized coprime factorization(NCF) approach is applied. Moreover, it is show through simulations, that the controlling sequence of SMES effectively damps out the generator's terminal bus voltage fluctuations and transmission line power oscillations. *Copyright © 2003 IFAC*

Keywords: wind power generation system, superconducting magnetic energy storage, decoupling H_∞ controller, damping control of transmission line power oscillations, compensation of the generator's terminal bus voltage fluctuations, normalized coprime factorization approach

1. INTRODUCTION

From environmental problem or energy resource desiccation problem, etc, introduction of the wind power generation system using natural energy is promoted. From advantageous in respect of cost as a wind generator, many induction generators are adopted. However, in order to carry out through natural energy in terms of wind, the output power and installation terminal bus voltage of a wind generator is always fluctuate with change of wind speed. Moreover, wind power generation system is installed in the end of the distribution line where the distribution line is long in nature because wind power generation system is installed in the intensive wind conditions area. Therefore, an instantaneous voltage sag will become large with increase of the distribution line length by influence of the inrush current at system interconnection[1].

From such a background, the effective interconnection equipment which connects a power system and the wind power generation system is required. Then, this paper is considering introduction of superconducting magnetic energy storage(SMES) as effective interconnection equipment which solves the various problem of the wind power generation system. SMES is capable of controlling both the active and reactive power, simultaneously and quickly by improvement in technical of power electronics equipment. Therefore, various application

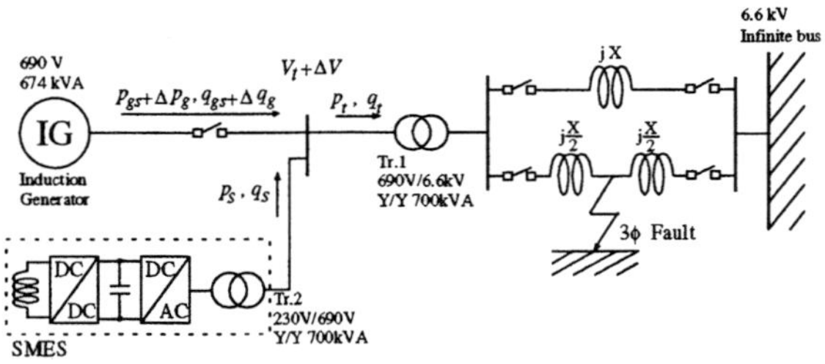

Fig. 1. Wind power generation system.

for SMES to a power system from the merit of the conversion efficiency is expected[2~4]. However, control system design which achieve decoupling control is needed since coherency will arise in the active and reactive power simultaneous control.

In this paper, in the wind power generation system, control system configuration which achieve compensation of the generator's terminal bus voltage fluctuations and the damping control of transmission line power oscillations is proposed. Moreover, a decoupling H_∞ controller is proposed to achieve incoherency for the active and reactive power simultaneous control. In the proposed method, the H_∞ control via Normalized Coprime Factorization (NCF) approach[7, 8] is used to achieve the proposed design idea. A normalized coprime factorization approach has the advantage of selecting weighting function in the H_∞ controller design. Moreover, it achieves controlling a rapid voltage fluctuations and power oscillation by analyzing with an instantaneous value. In order to verify a control performance of the proposed controller, some simulations are performed in the target wind power generation system. The performed simulations are system interconnection, wind speed fluctuation, and three phase to ground fault of transmission lines, respectively. From the simulation results, a transmission line power or the generator's terminal bus voltage at the system interconnection and wind speed fluctuation has been well controlled, and these fluctuations at the ground fault has been similarly suppressed.

2. CONTROL SYSTEM CONFIGURATION

The target wind power generation system which introduced a SMES in this paper is shown in Fig. 1. The induction generator is connected to the infinite-bus-power-system through the tie line of line reactance x. The SMES is connected to the induction generator's terminal through a three-phase transformer. The control objective of SMES is suppressing the fluctuation of generator's terminal bus voltage V_t and is damping the oscillation of transmission line power p_t. Control system configuration of a SMES is shown in Fig. 2. It consist of the shunt device which compose of a three-phase voltage-fed PWM inverter with a switching

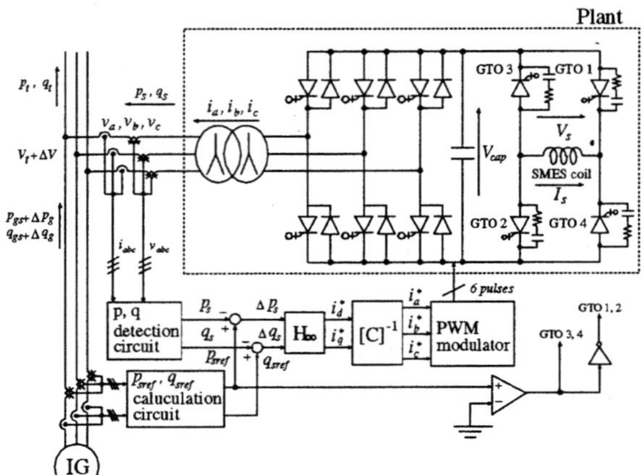

(a) Main circuit.

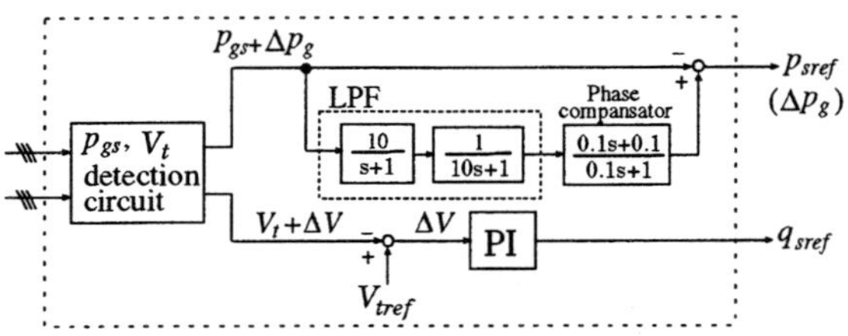

(b) Calculation circuit of the active and reactive power reference p_{sref}, q_{sref}.

Fig. 2. Control system configuration of SMES.

Table 1. DC-chopper operation at transmission line power control.

	GTO1	GTO2	GTO3	GTO4
p_{sref} is negative (charging mode)	ON/OFF	ON	OFF	OFF
p_{sref} is positive (discharging mode)	OFF	OFF	ON/OFF	ON

frequency of 2kHz, and the dc terminals of which are connected to a two quadrant dc-chopper with a superconducting coil and dc-link capacitor. At first, the operation of dc-chopper is explained. As shown in Fig. 2(a), a dc-chopper performs the electric energy charging/discharging controller of the superconducting coil current I_s, and dc-link capacitor is operated as a dc-link voltage source V_{cap}. The command of the electric energy charging/discharging control is determined by positive or negative of the active output power reference p_{sref} of SMES, as shown in Table. 1. The operating function of the shunt device is explained. For the control objective of a SMES, when the fluctuation component of generator's terminal bus voltage ΔV and the oscillation component of output power Δp_{gs} arises by disturbances etc., these fluctuation components is removed by absorbing or releasing the active and reactive power(p_s and q_s) from SMES, as shown in following equation

$$\begin{aligned} \Delta V - \left(K_i + \frac{K_p}{s}\right) q_s = 0 \\ \Delta p_t = \Delta p_{gs} - p_s = 0 \end{aligned} \quad (1)$$

where K_i and K_p is the proportional gain and the integral gain, respectively. The control objec-

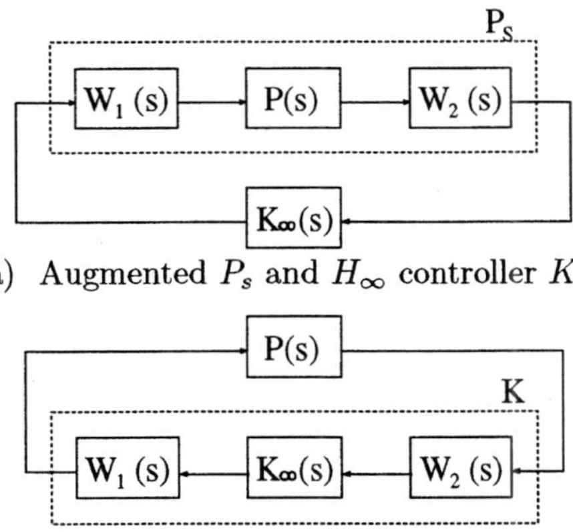

(a) Augmented P_s and H_∞ controller K_∞.

(b) Plant P and robust stabilizing controller K.

Fig. 3. Normalized coprime factorization approach.

tive of the shunt device is the following control of an instantaneous active and reactive output power of SMES(p_s and q_s). As shown in Fig. 2(a), the instantaneous active and reactive output power of SMES is calculated from d-q axis voltage and current obtained by d-q transformation. The difference of the obtained signal and the reference power p_{sref} or q_{sref} is inputted to the decoupling H_∞ controller. The output signal of a decoupling H_∞ controller determines the PWM reference current i^*_{sa}, i^*_{sb}, and i^*_{sc} calculated from d-q inverse transformation. Calculation circuits of the reference active and reactive power (p_{sref} and q_{sref}) is shown in Fig. 2(b), and the determination process of p_{sref} and q_{sref} is described. At first, an instantaneous active and reactive output power of the induction generator $p_{gs} + \Delta p_g$ and the generator's terminal bus voltage $V_t + \Delta V$ are detected. Low-pass-filter(LPF) and phase compensator is the filter which remove the oscillation component. The filter is designed so that an instantaneous active output power p_s of SMES in steady-state could be zero. Inputting a detected signal $p_{gs} + \Delta p_g$ into the filter, the filter outputs the signal removed a oscillation component. The reference signal q_{sref} is determined by the difference of the detected signal $V_t + \Delta V$ and arbitrary reference voltage V_{tref}. When an instantaneous active and reactive output power of SMES(p_s and q_s) tracks $p_{sref}(\Delta p_g)$ and q_{sref}, the oscillation components of V_t and p_{gs} is removed. However, control system design which achieve a decoupling control is needed since a coherency will arise in the active and reactive power simultaneous control. Then, H_∞ control theory is applied to design a 2-input 2-output decoupling H_∞ controller.

3. CONTROL SYSTEM DESIGN

3.1 Plant model

In this section, the decoupling H_∞ controller is designed to achieve incoherency for the active and reactive power simultaneous control. We select the active output power deviation Δp_s and reactive output power deviation Δq_s of SMES as the controller inputs and the select the active component current Δi_d and reactive component current Δi_q as the controller outputs as shown in Fig. 2(a), where Δi_d and Δi_q are the same as that of the PWM reference current i^*_d and i^*_q, respectively. Therefore, the inside of the dotted line in Fig. 2(a) is defined as a plant. The linearized model of the plant is expressed in state-space form as follows;

$$\begin{aligned} \dot{x}(t) &= Ax(t) + Bu(t) \\ y(t) &= Cx(t) \end{aligned} \tag{2}$$

where, u is the system control signal, x is the state variable, and y is the system output, and they are defined by

$$\begin{aligned} u(t) &= [\Delta i_d \;\; \Delta i_q]^T \\ x(t) &= [\Delta p_s \;\; \Delta q_s]^T \\ y(t) &= [\Delta p_s \;\; \Delta q_s]^T \end{aligned}$$

where, Δ shows the deviation from equilibrium point. A decoupling H_∞ controller was designed based on the above state-space representation.

3.2 Normalized coprime factorization approach

Since it is the standard McFarlance-Glover Loop shaping technique, the normalized coprime factorization(NCF) approach applied to the design is described briefly. The NCF method is the technique of performing loop shaping for the gain on a closed loop system using a weighting matrix. Moreover, the NCF approach can solve the problem, even when it is difficult to solve in the mixed sensitivity approach, i.e, when the pole or zero of the augmented plant are close to imaginary axis. Furthermore, there is the advantage in which the selection of the weighting function which satisfies the design specification becomes clearly and easy.

The conceptual of NCF approach is shown in Fig. 3. A plant P is augmented by the proper weighting function W_1 and W_2. The lower limit of γ related to the augmented plant P_S can be obtained by the following equation

$$\gamma_{min} = (1 + \gamma_{max}(XZ))^{\frac{1}{2}} \tag{3}$$

where, $\gamma_{max}(XZ)$ are the maximum eigenvalue of the XZ. X and Z are the solutions of the Riccati matrix equation related to the augmented plant P_S. The robust stability of the system is guaranteed by selecting as $\gamma > \gamma_{min}$.

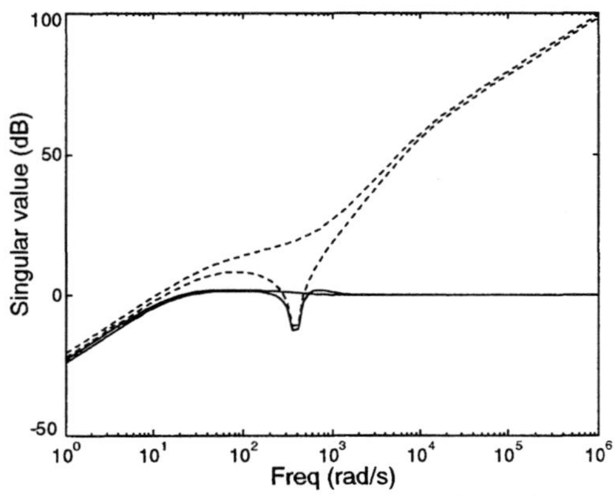

Fig. 4. The design specification (Eq. 4).

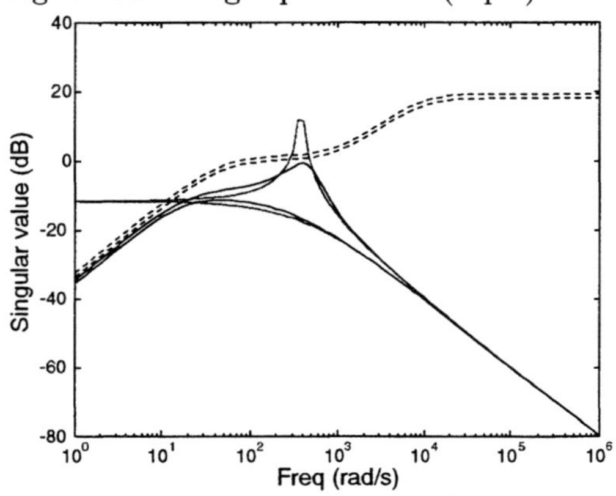

Fig. 5. The design specification (Eq. 5).

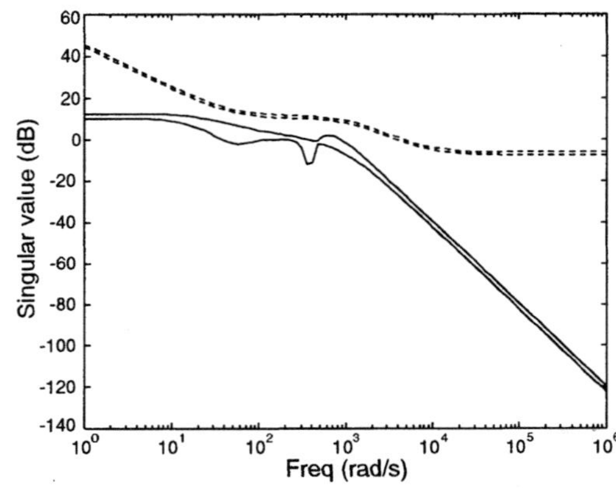

Fig. 6. The design specification (Eq. 6).

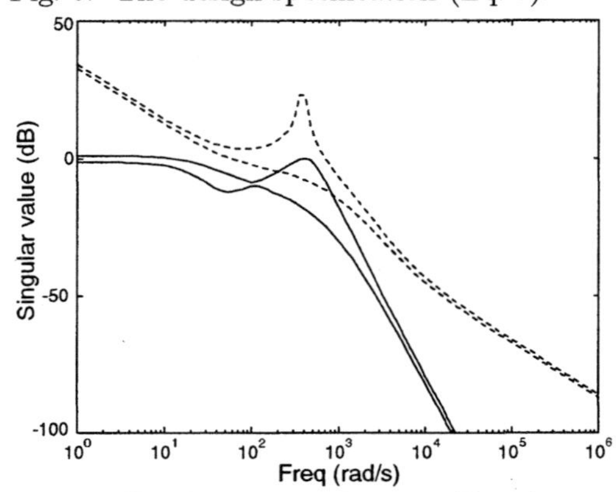

Fig. 7. The design specification (Eq. 7).

Control performance

$$\sigma_{max}((I-PK)^{-1}) \leq \frac{\gamma}{\sigma_{min}(P)\sigma_{min}(W_1)\sigma_{min}(W_2)} \quad (4)$$

$$\sigma_{max}((I-PK)^{-1}P) \leq \frac{\gamma}{\sigma_{min}(W_1)\sigma_{min}(W_2)} \quad (5)$$

Robust stability

$$\sigma_{max}(K(I-PK)^{-1}) \leq \gamma\sigma_{max}(W_1)\sigma_{max}(W_2) \quad (6)$$

$$\sigma_{max}(K(I-PK)^{-1}P) \leq \gamma\sigma_{max}(P)\sigma_{max}(W_1)\sigma_{max}(W_2) \quad (7)$$

where, $\sigma_{max}(\cdot)$ and $\sigma_{min}(\cdot)$ present the maximum singular value and the minimum singular value, respectively.

If the design specifications is not satisfies then the weighting functions W_1 and W_2 are reselected, and design process is repeated until the design specifications satisfies.

3.3 Design of decoupling H_∞ controller

The weighting functions W_1 and W_2 selected in the decoupling H_∞ design is shown in the eqs. (8) and (9), respectively.

$$W_1 = \begin{bmatrix} \frac{0.054s+465}{s+0.001} & 0 \\ 0 & \frac{0.047s+404.34}{s+0.001} \end{bmatrix} \quad (8)$$

$$W_2 = \begin{bmatrix} \frac{4.45s+219.74}{s+1081} & 0 \\ 0 & \frac{4.45s+219.74}{s+1081} \end{bmatrix} \quad (9)$$

In the selection of the weighting functions W_1 and W_2, W_1 is selected to damp the power oscillations, while W_2 is selected to reduce sensitivity in low frequency bandwidth of a plant. By solving the equation (3), $\gamma_{min} = 1.556$ are obtained. In this paper, γ is determined as $\gamma = 2.0$ which guarantees the robustness of the system, since $\gamma > \gamma_{min}$. The design specifications eqs. (4)~(7) is shown in Figs. 4~7. The dotted lines shown in Fig. 5 is singular value plots of the open loop system, and it can confirmed that the resonance point of the open loop system is 10dB in the 400rad/sec neighborhoods. As shown in Figs. 4~7, since the solid line exists in the domain below the dashed line, it can confirmed that the design specifications

Table 2. System parameters.

rated output	600 kW
rated line-to-line voltage	690 V
rated frequency	60 Hz
pole number P	4 pole
inertia coefficient J	687 kg·m^2
stator resistance r_s	0.012 p.u.
stator leakage reactance x_{ls}	0.22 p.u.
rotor resistance r_r	0.015 p.u.
rotor leakage reactance x_{lr}	0.19 p.u.
excitation reactance x_m	7.3 p.u.
line reactance x	0.04 p.u./km
line length	40 km

Table 3. Transformer parameters.

	R_1(pu)	L_1(pu)	R_2(pu)	L_2(pu)	R_m(pu)	L_m(pu)
T_{r1}	0.015	0.03	0.015	0.03	200	200
T_{r2}	0.003	0.04	0.003	0.04	200	200

Table 4. Parameter of PI controller.

Proportional gain	K_p	4200
Integral gain	K_i	2800

eqs. (4)~(7) is satisfied by using above weighting functions. Moreover, it can confirmed that the closed loop system is decreasing the peak value of an open loop system as shown in Fig. 5. Therefore, the designed controller satisfies the robustness of the system.

4. SIMULATION RESULTS

The effectiveness of the proposed control system configurations is demonstrated through computer simulations in terms of the wind power generation system shown in Fig. 1. Each parameters of system, transmission line, transformer, and PI controller in the simulations are shown in Tables 2~4, respectively. The disturbances to the wind power generation system is the transmission line power oscillations and the generator's terminal bus voltage fluctuation at a system interconnection, a wind speed fluctuation and a three phase to ground fault.

4.1 Simulations of system interconnection and wind speed fluctuation

The simulation sequence is shown below:

i) t=1.0sec : system interconnection with a constant wind speed,

ii) 1.5sec≤ t ≤4.5sec : wind speed fluctuation which frequency is 1Hz arises,

iii) 4.5sec≤ t : a constant wind speed.

Fig. 8 shows the simulation results. In Fig. 8(a), the solid line and the dashed line show the simulation results with installing SMES and not, respectively. Fig. 8(a) can be seen that the installing SMES can damp effectively the transmission line power oscillation and can compensate effectively the generator's terminal bus voltage fluctuation as compared to the case of the not installing SMES. From Fig. 8(b), it can confirmed that a SMES is absorbing or releasing the active power of reverse polarity as compared with the transmission line power oscillations, and is outputting the reactive power proportional to the generator's terminal bus voltage fluctuations. Moreover, it can confirmed that the transmission line power is converging to its steady-state value at t = 6sec by the effect of the filter. Furthermore, the output waveform of the active and reactive power of SMES can be seen that the coherency in the active and reactive power simultaneous control has not arisen by applying a decoupling H_∞ controller.

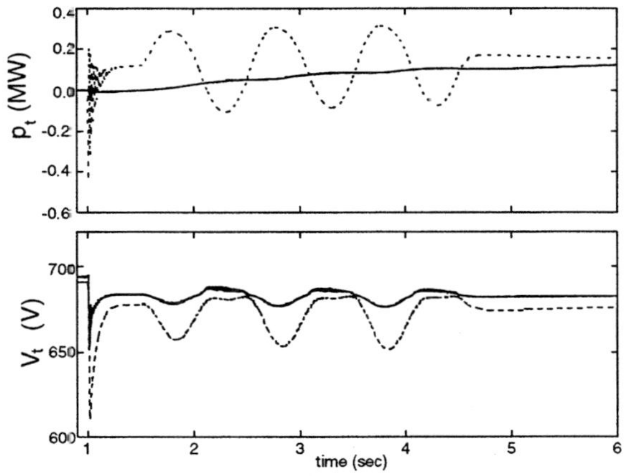

(a) Transmission line power and the generator's terminal bus voltage.

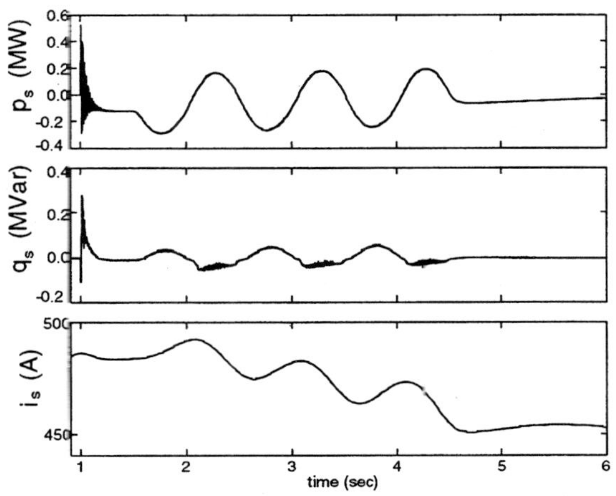

(b) Active and reactive output power of SMES, superconducting coil current.

Fig. 8. Simulation results of system interconnection and wind speed turbulence.

4.2 Simulations of a three phase to ground fault

The simulation sequence is shown below.

i) t=1.0sec : system interconnection with a constant wind speed,

ii) t=7.0sec : a three phase to ground fault is applied to the middle of one transmission line,

iii) t=7.0667sec : cleared by removing the faulted transmission line,

iv) 7.0667sec$\leq t$: the other transmission line is sending power.

Fig. 9 shows the simulation results. The solid line and the dotted line show the simulation results at applying the proposed controller(decoupling H_∞ controller) and PI controller, respectively. As shown in Figs. 9(a) and (b), although considerable degree of interference has arisen after removing fault when a PI controller is applied, it can confirmed that it's decoupling issue for the active and reactive power simultaneous control is attained by applying the proposed controller as compared with a PI controller. Moreover, in the case of the not installing SMES, the transmission line power oscillation is arising, and an instantaneous voltage sag of the generator's terminal bus voltage takes about 0.4sec for converging on a steady-state value, after removing the faulted transmission line. In the case of the installing SMES, it can confirmed that the transmission line power oscillations has been damped by absorbing/releasing an active power from SMES, and an instantaneous voltage sag is converged on a steady-state value for a short time by discharging the reactive power from SMES, as shown in Fig. 9(b).

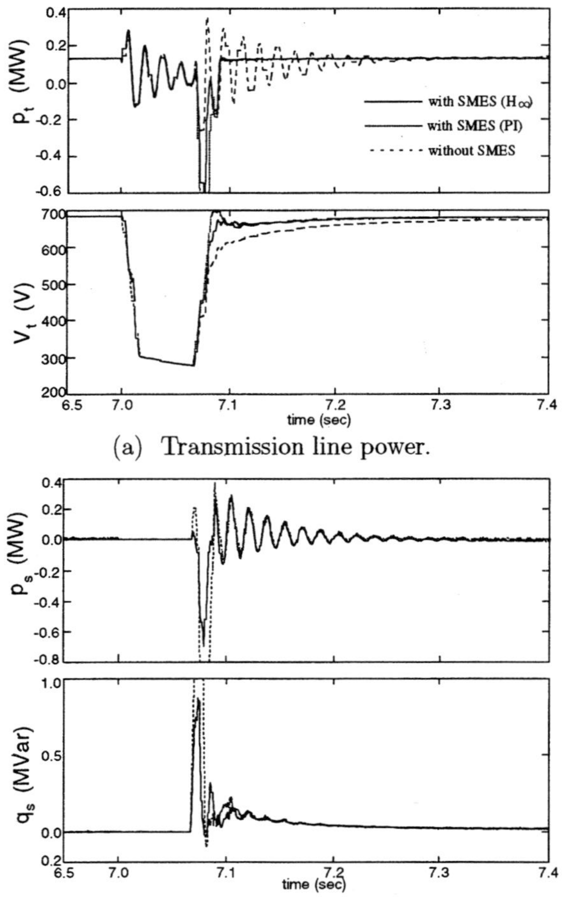

(a) Transmission line power.

(b) Active and reactive output power of SMES.

Fig. 9. Simulation results of three-phase to ground fault.

5. CONCLUSION

In this paper, in the wind power generation system introduced a SMES, control system configuration which achieve compensation of the generator's terminal bus voltage fluctuations and the damping control of transmission line power oscillation was proposed. Moreover, a decoupling H_∞ controller was proposed to achieve incoherency for the active and reactive power simultaneous control. By using the normalized coprime factorization approach, the selection of the weighting function is simplified in the design. It could be confirmed from the computer simulation on that the proposed controller can control effectively the generator's terminal bus voltage fluctuations and the transmission line power oscillation.

REFERENCES

[1] Y. Sasaki, N. Harada, T. Kai, and T. Sato, "A Countermeasure against the Voltage Sag due to a Inrush Current of Wind Power Generation System Interconnecting to a Distribution Line", T. IEE Japan, Vol. 120-B, No. 2, pp. 180-186, 2000.

[2] D. Kamolyabutra, M. Hojo, Y. Mitani, T. Ise, and K. Tsuji, "Experimental Study on Power System Stabilizing Control Scheme for the SMES with Solid-State Phase Shifter(SuperSMES) using local signals", T. IEE Japan, Vol. 119-B, No. 5, pp. 581-588, 1999.

[3] T. Uranaka, Y. Mitani, and K. Tsuji, "Power System Stabilizing Control by a Superconducting Magnetic Energy Storage with a High Speed Phase Shifter", T. IEE Japan, Vol. 115-B, No. 8, pp. 883-889, 1995.

[4] D. Kamolyabutra, Y. Mitani, and K. Tsuji, "Power System Stabilization by SMES with Series Compensator", T. IEE Japan, Vol. 120-B, No. 12, pp. 1615-1612, 2000.

[5] James J. Skiles, Robert L. Kustom, Ka-Pui Ko, Vincent Wong, Ka-Shu Ko, Frederick Vong, and Keith Klontz, "Performance of a Power Conversion System for Superconducting Magnetic Energy Storage(SMES)", IEEE Transactions on Power System, Vol. 11, No. 4, November 1996, pp. 1718-1723.

[6] Byung M. Han, and George G. Karady, "New combined power-conditioning system for superconducting magnetic energy Storage", ELSEVIER Electric Power Systems Research, 37, 1996, pp. 79-85.

[7] I. Ngamroo, Y. Mitani, and K. Tsuji, "Robust Frequency Stabilization Control via Normalized Coprime Factorization Approach", *T.IEEJapan*, Vol.119-B, No.12, pp. 1437-1445 (1999-12)(in Japanese)

[8] D. C. MacFarlance and K. Glover, "Robust Controller Design using Normalized Coprime Factorization plant Descriptions", *Lecture Notes in Control and Information Science*, Vol. 138, Springer-Verlag, 1990.

ELSEVIER
IFAC
PUBLICATIONS
www.elsevier.com/locate/ifac

A CAPACITY ESTIMATION TECHNIQUE OF SOLAR-WIND HYBRID POWER SYSTEM FOR AN URBAN RESIDENTIAL LOAD

Byeong-Gook Kwon, Seung-Chul Lee, Chan-Eom Park

School of Electrical & Electronic Engineering
Chung-Ang University in Seoul Korea

Abstract : In this paper a technique that can estimate the required capacity of a Solar-Wind Hybrid Power(SWHP) system for an unban residential load is presented. The probability distribution of the power output from SWHP system is calculated from the probability density functions(PDFs) of the solar irradiance and the wind speed for each hour. The size of photovoltaic(PV) array and the number of wind turbines are decided so that the total SWHP system output in one day can meet the total daily residential load. The available power for battery system charging can be obtained from the difference between the SWHP system generation and the load. Consequently, the distribution of the battery charging power is calculated by convolving the PDFs of both the SWHP system output and the load. A typical local residential load curve is used for illustration and promssing result is obtained. *Copyright © 2003 IFAC.*

Keywords : solar, wind, hybrid, power system, substitute energy, battery charge

1. Introduction

Solar and wind energies are excellent alternatives to the conventional fossil energies which are both limited and problematic. However widespread utillizations of these clean and unlimited energies their low power density and intermittent availability in contrast to their relatively high installation costs. In particular, becomes of the lack proper capacity signing tools, a rule of thumb estimation of the required capacity normally ends up with an inappropriate installed capacity, which further the return rate for the investment.

In this paper, a technique is proposed that can estimate the appropriate PV array and wind turbine capacities for a typical local urban house load. The technique is based on the probabilities of the solar irradiance and wind speed.

The SWHP system is proposed to utilize the potential compensations between the solar and the wind energies sizing of the battery system is also discussed.

A 10 year data of irradiance and wind speed recorded for every hour of a day are used to compute the PDFs of irradiance and wind speed. The irradiance and the wind speed are considered as random variables studies show that the irradiance has the Beta distribution while the wind speed exhibits the Weibull distribution [1~3]. Normally, the output of a SWHP system is computed by adding the average output of a PV system and the wind system [4~6].

In this paper, we obtained the output PDF of a SWHP system from the convolution the output PDFs of the two systems.

A typical residential load in an urban area is used for illustration. The output of a SWHP system with a battery bank system is used to supply daily load.

Since the battery charging powers are random, the PDF for the battery charging power are derived by convolving the PDFs of the SWHP system output and the load [6,7]. Then the battery charge-discharge curve is plotted using their expected value for each hour.

2. System Configuration

A SWHP system equipped with the battery backup system and operated in parallel with the utility power system network is shown in Fig. 1. The DC power generated from the PV modules and the wind turbine is converted to the AC power through the inverter, and is supplied AC power load.

The surplus power is stored to the battery bank. When solar and wind power generation is insufficient, power is supplied from the battery bank or the utility power line. The battery bank can be discharged up to 40% of its total capacity.

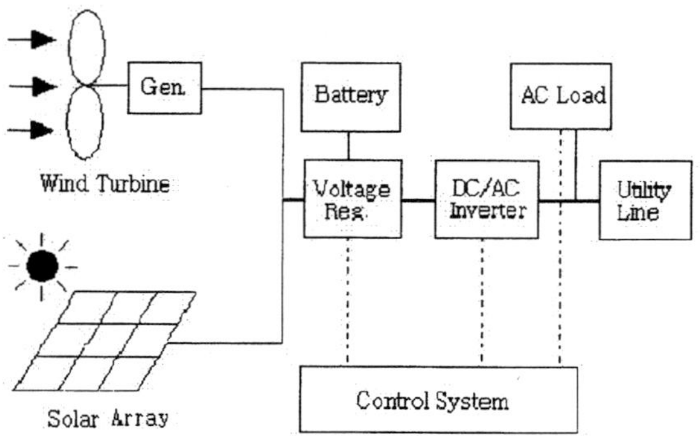

Fig. 1. Solar-Wind Hybrid Power Configuration

3. SWHP System Output Prediction

3.1 PV Module Output Prediction

For the irradiance distribution analysis, the Beta Distribution is used which was verified by Chi-Square Goodness of Fit Test. Then the probability distribution density(PDF) for the solar irradiance can be expressed as

$$f_r(r) = \frac{\Gamma(\alpha+\beta)}{\Gamma(\alpha)\Gamma(\beta)}\left[\frac{r}{r_{max}}\right]^{\alpha-1}\left[1-\frac{r}{r_{max}}\right]^{\beta-1} \quad (1)$$

where, Γ : Gamma function
$r[W/m^2]$: instant irradiance
$r_{max}[W/m^2]$: maximum irradiance
α : shape parameter
β : shape parameter

Then the instant output P_S and the maximum output $P_{S\max}$ of the PV module can be computed from

$$P_S = rA\eta \qquad P_{S_{max}} = r_{max}A\eta$$

$$A = \sum_{m=1}^{M} A_m \qquad \eta = \frac{\sum_{m=1}^{M} A_m\eta_m}{A} \quad (2)$$

where, A : PV array area
A_m : PV module area (m=1,2,...,M)
η : PV array efficiency
η_m : PV module efficiency

From Eq.(2), the PDF of the solar power generation that can be obtained from the PV module of area A[㎡] can be expressed with

$$f_{P_S}(p_S) = \frac{\Gamma(\alpha+\beta)}{\Gamma(\alpha)\Gamma(\beta)}\left[\frac{P_S}{P_{S_{max}}}\right]^{\alpha-1}\left[1-\frac{P}{P_{S_{max}}}\right]^{\beta-1} \quad (3)$$

3.2 Wind Power System Output Prediction

The wind speed distribution is typically modelled with the Weibull Distribution, which can be given as

$$f_V(V) = \frac{k}{c}\left(\frac{V}{c}\right)^{k-1}\exp\left[-\left(\frac{V}{c}\right)^k\right] \quad (4)$$

where, V : wind speed[m/s]
c : scale parameter
k : shape parameter

Fig. 2 shows the power output characteristics of a typical wind turbine.

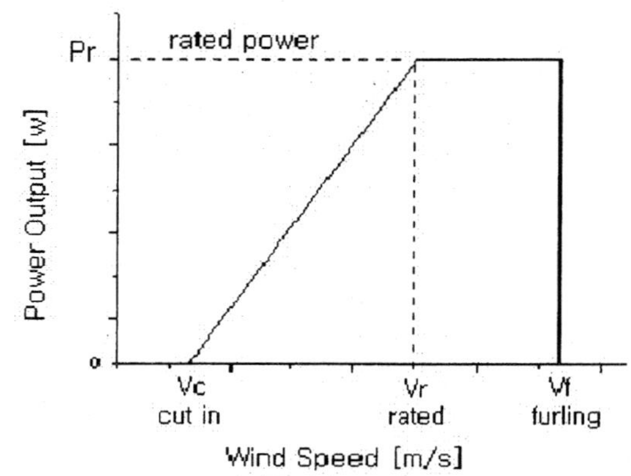

Fig. 2. Wind Turbine Output Characteristics

From the figure, the power output from the wind turbine can be expressed as

$$P_W(V) = \begin{cases} 0 & (V \le V_c) \\ a + bV^n & (V_c < V < V_r) \\ P_r & (V_r \le V < V_f) \\ 0 & (V_f \le V) \end{cases} \quad (5)$$

where, n : order $(V_c \sim V_r)$

$$a = P_r \cdot \frac{V_c^n}{V_c^n - V_r^n}$$

$$b = P_r \cdot \frac{1}{V_r^n - V_c^n}$$

From $f_V(V)$ in Eq.(4), the cumulative distribution of the wind speed can be obtained by integration as

$$F_V(V) = 1 - \exp\left[-\left(\frac{V}{c}\right)^k\right] \quad (6)$$

Therefore the cumulative distribution of the wind turbine output can be expressed with

$$F_{P_W}(P_W)=\begin{cases}1-F_V(V_f)+F_V(V_c), & P_W=0\\ F_V[V(P_W)-F_V(V_c)], & 0<P_W<P_r\\ F_V(V_f)-F_V(V_r), & P_W=P_r\end{cases} \quad (7)$$

where, $V(P_W)=\left[\dfrac{P_W-a}{b}\right]^{\frac{1}{m}}$

By differencing $F_{P_W}(P_W)$ with P_W, the PDF for the wind turbine output can be obtained as follows ;

$$f_{P_W}(P_W)=\begin{cases}[1-F_V(V_f)+F_V(V_c)]\delta(P_W), & P_W=0\\ \left[\dfrac{(V_r-V_c)}{P_r}\right]\left[\dfrac{k}{c_k}\right]\left[V_c+(V_r-V_c)\dfrac{P_W}{P_r}\right]^{k-1}\times \\ \exp\left[-\left(\dfrac{V_c+(V_r-V_c)\dfrac{P_W}{P_r}}{c}\right)^k\right], & 0<P_W<P_r\\ [F_V(V_f)-F_V(V_c)\delta(P_W-P_r), & P_W=P_r\end{cases} \quad (8)$$

3.3 Combined System Output

The PDF for the output of the SWHP system can be obtained by convolving the PDFs shown in Eq.(3) and Eq.(8) as

$$f_{P_h}(P_h)=f_{P_S}(P_S)*f_{P_W}(P_W) \quad (9)$$

4. Optimum Battery Bank Size Calculation

Available battery charging power $C(t)$ is the difference between the SWHP system output and the demand, i.e.,

$$C(t)=P(t)-L(t) \quad (10)$$

Because demand load is also random variable as SWHP system output, need to express by statistical method. In this paper, the each hourly assumed to have uniform between the maximum and the minimum load as

$$f_L(\ell)=\begin{cases}\dfrac{1}{L_{\max}-L_{\min}}, & L_{\min}\le\ell\le L_{\max}\\ 0 & otherwise\end{cases} \quad (11)$$

Therefore, the PDF of the battery charging power for each hour can be formulated as

$$f_C(C)=f_{P_h}(P_h)*f_L(-\ell) \quad (12)$$

Expected value of the battery charging power C becomes positive when battery bank system is charged and negative when discharged.

In this paper, the capacity of the PV array and the wind turbine for satisfying the total daily load is decided simply as

$$P_h=A_S\cdot P_S+n_W\cdot P_W=L \quad (13)$$

where, A_S : PV array area[㎡]

n_W : number of wind turbines

P_S : PV module output [W/㎡]

P_W : wind turbine output [W/ea]

Fig. 3 shows an example of the typical daily residential load pattern of the urban area in Korea[8]. According to energy statistical research result in 2000 year, a monthly electric power energy used per house in domestic urban areas is about 178[KWh] and a daily electric power energy used is about 6300[Wh].

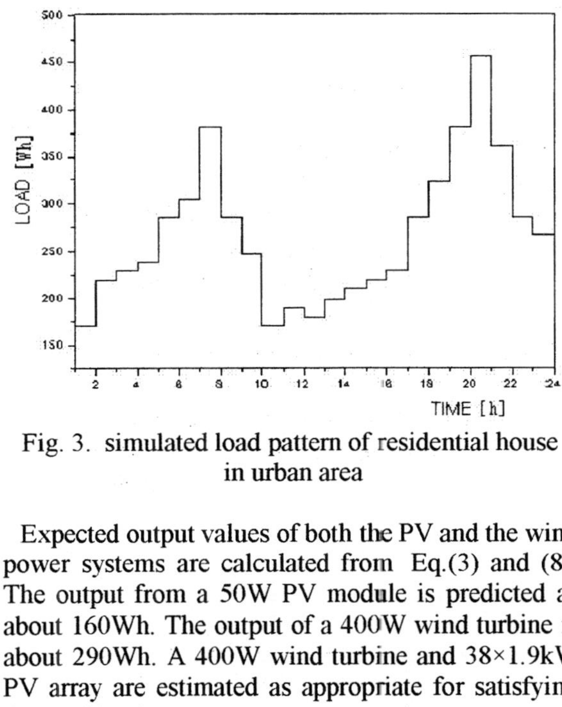

Fig. 3. simulated load pattern of residential house in urban area

Expected output values of both the PV and the wind power systems are calculated from Eq.(3) and (8). The output from a 50W PV module is predicted as about 160Wh. The output of a 400W wind turbine is about 290Wh. A 400W wind turbine and 38×1.9kW PV array are estimated as appropriate for satisfying the total load.

Based on Eq.(12) and for the house load shown in Figure 3, the hourly battery charge-discharge power is computed and is plotted in Figure 4.

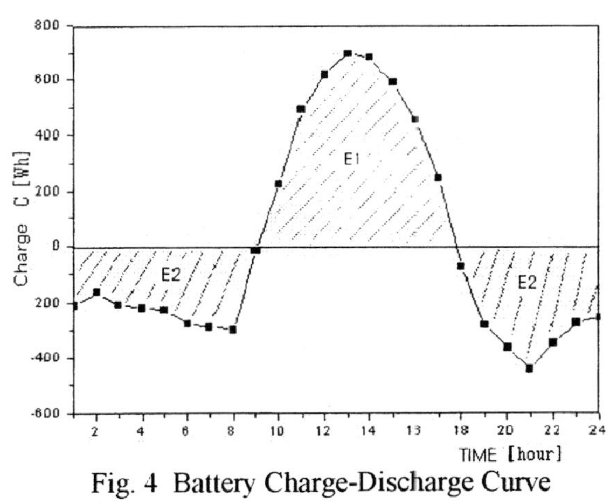

Fig. 4 Battery Charge-Discharge Curve

From Figure 4, Area E1 is the amount of power needed to be discharged from the battery bank system, which is about 4000[Wh]. The Ampere-Hour capacity of the battery bank system is calculated from

$$Battery\ \ Capacity\ [AH] = \frac{C}{V \times U \times R} \qquad (14)$$

where, C : charge energy[Wh]
V : voltage
U : discharge depth rate
R : inverter efficiency

In the Equation (14), with the battery bank system voltage of 12V, the discharge depth rate of 60% and the inverter efficiency of 60%, the minimum capacity of battery bank system is calculated as 620[AH].

Currently, the SWHP system shown in Figure 5 and 6 are being installed on the roof of our engineering college for test.

Fig. 5 System Installation

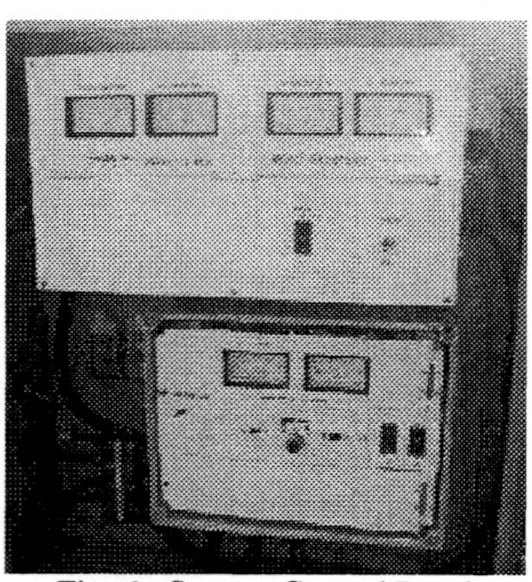

Fig. 6. System Control Panel

5. Conclusion

In this paper a technique that can estimated the appropriate SWHP system capacity for a typical local urban home is proposed. Although the capacity of the SWHP system is computed based on the total daily demand, further study is needed to develop methods that can reflect the price structure of the utility power to maximize the return for the investment for the SWHP system and the effects of the loss of power supply probability(LPSP).

References

[1] Imad Abouzahr, "Loss of Power Supply Probability of Stand-Alone Photovoltaic Systems", IEEE Trans. on Energy Conversion, Vol. 6, No. 1, March 1991

[2] S. H. Karaki, "Probabilistic Performance Assessment of Autonomous Solar-Wind Energy Conversion Systems", IEEE Trans. on Energy Conversion, Vol.14, No.3, Sep. 1999

[3] Bogdan S. Borowy, "Methodology for Optimally Sizing the Combination of a Battery Bank and PV Array in a Wind/PV Hybrid System", IEEE Trans. on Energy Conversion, Vol. 11, No. 2, June 1996

[4] Ziyad M. Salameh, "Photovoltaic Module-Site Matching Based on the Capacity Factors", IEEE Trans. on Energy Conversion, Vol. 10, No. 2, June 1995

[5] Bogdan S. Borowy, "Optimum Photovoltaic Array Size for Hybrid Wind/PV System", IEEE Trans. on Energy Conversion, Vol. 9, No. 3, Sep. 1994

[6] Fadia M.A.GHALL, "Simulation and Analysis of Hybrid Systems using Probabilistic Techniques", Power Conversion Conference-Nagaoka 1997, Proceedings of the, Vol 2, 1997

[7] Imad Abouzahr, "Loss of Power Supply Probability of Stand-Alone Wind Conversion Systems : A Closed Form Solution Approach", IEEE Trans. on Energy Conversion, Vol. 5, No. 3, September 1990.

[8] "Load curve data", Korea Electric Power Corporation, December 2000.

ELSEVIER
IFAC
PUBLICATIONS
www.elsevier.com/locate/ifac

ROBUST MODEL PREDICTIVE CONTROL TO A NONLINEAR BOILER TRUBINE SYSTEM

Li Zhijun Tan Wen Liu Jizhen

(North China Electric Power University, 102206)

Abstract: In this paper, a robust model predictive control algorithm based on infinite norm optimization is applied to a nonlinear boiler-turbine unit. Simulation results show that the designed controller has good tracking, disturbance rejection and robustness against the variations of the operation points due to the plant nonlinearity. Compared with conventional model predictive controller, this control system functions better when the operating point varies. *Copyright © 2003 IFAC*

Keywords: robustness, model-based control, predictive control, nonlinear control system, MIMO, linear programming

1. INTRODUCTION

A boiler-turbine unit is a typical nonlinear MIMO control system, which plays an important role in stable load following, the safety of power plant and fuel saving. Currently, the design of the power plant controllers are mainly based on classical SISO control strategies. Since there are coupling, nonlinearity and model uncertainty in the system, it is difficult to design a controller, which can guarantee good control performance and disturbance rejection by these methods for this system. So in order to ensure power plant work safely and make full use of potential of the boiler-turbine system, multivariable control strategies should be taken.

Some literatures (Cori and Maffezzoni, 1984; Johansson and Koivo, 1984) had reported the application of multivariable control theories to the boiler-turbine system, but these methods usually need an accurate plant model and the designed controllers are very complex. The applications of robust control techniques are reported in another literatures (Kwon W. H. *et al,* 1989; Hwang and Kim, 1995; Tan W. *et al,* 1999). These theories and design methods usually have difficulty in handling input/output constraints. In this paper, a robust MPC (Campo and Morari, 1987) is applied to the same power plant considered in (Kwon W. H. *et al,* 1989).

Model predictive control (MPC) is a receding horizon control strategy. The control action is computed by solving an optimization problem on line in each sampling period. Typically a sequence of predicted control moves will be calculated, but only the first one is implemented. At the next sampling time, the optimization problem is solved again with new measurements, and control input is updated. MPC techniques are widely used in industrial process control practice because the actual control objectives and operating constraints can be represented explicitly in the optimization problem that is solved at each control instant. Many successful MPC applications have been reported in the last two decades (Camacho and Bordons, 1995; Morari and Lee, 1999).

However, conventional commercial MPC only use nominal model, without considering the model uncertainty. For the boiler-turbine system, an accurate model is nearly impossible to build; therefore, the robustness against modeling error is a prerequisite for a practical power plant controller. Robust MPC is a method that combines the merit of robust control techniques that can handle model uncertainty and the advantage of MPC techniques that can deal with constraints.

Moreover, conventional MPC algorithms are only applicable to asymptotically stable system, but drum level is an integrating process. So an improved robust model predictive control algorithm for a multivariable system with integrating subsystem is proposed in this paper.

The rest of the paper is organized as follows. A nonlinear boiler turbine model and model uncertainty analysis are given in Section 2. A robust MPC algorithm applicable to the boiler-turbine system is put forward in Section 3. Controller design and simulation results are given in Section 4. In addition, a comparison with conventional MPC is made too. Finally, conclusions are drawn in Section 5.

2. NONLINEAR BOILER-TURBINE MODEL

A boiler-turbine system can be modeled as a three-input three-output multivariable system (Bell and Astrom, 1987). Three controlled variables are drum stream pressure, electric output and drum level, while three manipulated variables are the fuel actuator position, the control valve position and the feed water actuator position. The main objective of the boiler-turbine control system is to make the electric output follow the load command rapidly while maintaining the drum level and the drum pressure. This nonlinear model is given by the following equation:

$$\begin{aligned}
\dot{x}_1 &= -0.0081u_2x_1^{9/8} + 0.9u_1 - 0.15u_3;\\
\dot{x}_2 &= (0.073u_2 - 0.016)x_1^{9/8} - 0.1x_2;\\
\dot{x}_3 &= (141u_3 - (1.1u_2 - 0.19)x_1)/85;\\
y_1 &= x_1;\\
y_2 &= x_2;\\
y_3 &= 0.05(0.13073x_3 + 100a_{cs} + q_e/9 - 67.975);
\end{aligned} \tag{1}$$

where, the state variables x_1, x_2, x_3 denote the drum pressure (Kg/cm^2), the electric output (MW) and the density of fluid in the system (Kg/cm^3) , the output y_3 is the drum level (m) , the control inputs u_1, u_2, u_3 denote the fuel actuator position, the control valve position and the feed water actuator position respectively. They are normalized and their rates are limited as:

$$\begin{aligned}
&0 < u_1 \le 1, |\dot{u}_1| \le 0.007/\sec\\
&0 < u_2 \le 1, -1/\sec < \dot{u}_2 \le 0.01/\sec\\
&0 < u_3 \le 1, |\dot{u}_3| \le 0.05/\sec
\end{aligned} \tag{2}$$

a_{cs} is the quality factor of steam and q_e is the evaporation mass flow(Kg/s). They are expressed by :

$$a_{cs} = \frac{(1-0.01538x_3)(0.8x_1 - 25.6)}{x_3(1.0394 - 0.0012304x_1)} \tag{3}$$

$$q_e = (0.854u_2 - 0.147)x_1 + 45.59u_1 - 2.514u_3 - 2.096 \tag{4}$$

Table 1 shows different operating points of the system.

Table 1 various operating points

Operating point	1	2	3	4	5
x_1^0	86.40	97.20	108	118.8	129.6
x_2^0	36.65	50.52	66.65	85.06	105.3
x_3^0	469.13	450.81	428	398.23	355.26
u_1^0	0.2090	0.2706	0.34	0.4182	0.5046
u_2^0	0.5520	0.6210	0.69	0.7590	0.8280
u_3^0	0.2555	0.3400	0.436	0.5434	0.6625

In this paper, we will use MPC algorithm based on the step response model of system, which is conventionally called dynamic matrix control (DMC). For the system (1), the step responses are different in different operating points, so we consider the nominal model as operating point 3 in Table 1, and other operating points are considered as modeling uncertainty.

From Table 1, some linearized models are obtained and the step response coefficients in different operating points are depicted as Fig. 1.

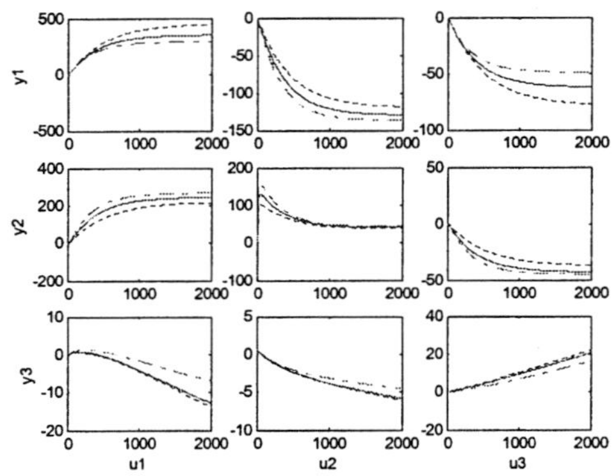

In Fig. 1, Solid line, dotted line, dash dotted line represents step response at operating point 3, 1, 5 respectively. From this figure, it is noted that step response of system varies between dotted line and dash dotted line when the boiler-turbine system runs between operating point 1 and operating point 5.

3. ROBUST MPC ALGORITHM FOR THE BOILER-TURBINE SYSTEM

All MPC algorithms possess three common elements. These elements are: Prediction Model, Objective Function, and control law through optimization.

3.1 Predictive model of the boiler turbine system

For the boiler-turbine system (1), considering the step response coefficients of the *ith* output corresponding to the *jth* input in nominal model :

$$a_{ij} = [a_{ij}(1), \ldots, a_{ij}(N), \ldots, a_{ij}(\infty), \ldots] \tag{5}$$

where $i = 1,2,3$, $j = 1,2,3$,the step response model is employed:

$$y_i(t) = \sum_{j=1}^{3}\sum_{k=1}^{\infty} a_{ij}(k)\Delta u_j(t-k) \tag{6}$$

the predicted value of y at time $t+l$ will be:

$$y_i(t+l\,|\,t) = \sum_{j=1}^{3}\sum_{k=1}^{\infty} a_{ij}(k)\Delta u_j(t+l-k) + d_i(t+l\,|\,t) \tag{7}$$

$$\begin{aligned}
d_i(t+l\,|\,t) &= d_i(t\,|\,t)\\
&= y_{im}(t) - y_i(t\,|\,t) = y_{im}(t) - \sum_{j=1}^{3}\sum_{k=1}^{\infty} a_{ij}(k)\Delta u(t-k)
\end{aligned} \tag{8}$$

where $y_{im}(t)$ is the measurement of y_i at time t, therefore,

$$y_i(t+l\mid t)=\sum_{j=1}^{3}\sum_{k=1}^{l}a_{ij}(k)\Delta u_j(t+l-k)+y_{im}(t) \tag{9}$$
$$+\sum_{j=1}^{3}\sum_{k=1}^{\infty}[a_{ij}(k+l)-a_{ij}(k)]\Delta u_j(t-k)$$

Considering M control action in the future, prediction horizon P, let

$$y_{i0}(t+l\mid t)=y_{im}(t)+\sum_{j=1}^{3}\sum_{k=1}^{\infty}[a_{ij}(k+l)-a_{ij}(k)]\Delta u_j(t-k) \tag{10}$$

$$\Delta U_j(t)=[\Delta u_j(t),\ldots,\Delta u_j(t+M-1)]^T \tag{11}$$

$$Y_{i0}(t)=[y_{i0}(t+1\mid t),\ldots y_{i0}(t+P\mid t)]^T \tag{12}$$

$$Y_{iP}(t)=[y_i(t+1\mid t),\ldots,y_i(t+P\mid t)]^T \tag{13}$$

$$A_{ij}=\begin{bmatrix} a_{ij}(1) & 0 & 0 \\ . & . & 0 \\ a_{ij}(M) & . & a_{ij}(1) \\ . & . & . \\ a_{ij}(P) & . & a_{ij}(P-M+1) \end{bmatrix} \tag{14}$$

so we can derive

$$Y_{iP}(t)=Y_{i0}(t)+\sum_{j=1}^{3}A_{ij}\Delta U_j(t) \tag{15}$$

Considering truncation order N, y_1, y_2 are asymptotically stable processes, so

$$a_{ij}(k+l)-a_{ij}(k)\approx 0, i=1,2, k>N \tag{16}$$

$$y_{i0}(t+l\mid t)=y_{im}(t)+\sum_{j=1}^{3}\sum_{k=1}^{N}[a_{ij}(k+l)-a_{ij}(k)]\Delta u_j(t-k) \tag{17}$$

y_3 is an integrating process, and DMC for integrating process had been introduced in some literatures (Gupta Y. P., 1998; Dai L. K., 1999). For an integrating process unit, the step response is a straight line (ramp) after truncated horizon, so

$$a_{ij}(k+l)-a_{ij}(k)\approx lh_j, i=3, k>N \tag{18}$$

$$y_{30}(t+l\mid t)$$
$$=y_{3m}(t)+\sum_{j=1}^{3}\sum_{k=1}^{N}[a_{3j}(k+l)-a_{3j}(k)]\Delta u_j(t-k)+\sum_{j=1}^{3}\sum_{k=N+1}^{\infty}lh_j\Delta u_j(t-k)$$
$$=y_{3m}(t)+\sum_{j=1}^{3}\sum_{k=1}^{N}[a_{3j}(k+l)-a_{3j}(k)]\Delta u_j(t-k)+\sum_{j=1}^{3}lh_j u_j(t-N-1) \tag{19}$$

From (15)(17)(19), we can derive predictive model of the boiler-turbine system in the nominal model:

$$Y_P(t)=Y_0(t)+A\Delta U(t) \tag{20}$$

where

$$Y_P(t)=[Y_{1P}{}^T(t),Y_{2P}{}^T(t),Y_{3P}{}^T(t)]^T \tag{21}$$

$$Y_0(t)=[Y_{10}{}^T(t),Y_{20}{}^T(t),Y_{30}{}^T(t)]^T \tag{22}$$

$$A=\begin{bmatrix} A_{11} & A_{12} & A_{13} \\ A_{21} & A_{22} & A_{23} \\ A_{31} & A_{32} & A_{33} \end{bmatrix} \tag{23}$$

$$\Delta U(t)=[\Delta U_1{}^T(t),\Delta U_2{}^T(t),\Delta U_3{}^T(t)]^T \tag{24}$$

Considering model uncertainty from Fig.1, we can regard the system as a model family. From

$$(a_{ij}(k))_{\min}\le a_{ij}(k)\le(a_{ij}(k))_{\max} \tag{25}$$

step response coefficients of the system are

$$\hat{a}_{ij}(k)=a_{ij}(k)+\hat{\delta}_{ij}(k) \tag{26}$$

$$a_{ij}(k)-(a_{ij}(k))_{\min}\le\hat{\delta}_{ij}(k)\le(a_{ij}(k))_{\max}-a_{ij}(k) \tag{27}$$

So predictive model can be described as:

$$\hat{Y}_P(t)=Y_P(t)+\hat{\Theta}(t)=Y_0(t)+A\Delta U(t)+\hat{\Theta}(t) \tag{28}$$

where

$$\hat{Y}_P(t)=[\hat{Y}_{1P}{}^T(t),\hat{Y}_{2P}{}^T(t),\hat{Y}_{3P}{}^T(t)]^T \tag{29}$$

$$\hat{Y}_{iP}(t)=[\hat{y}_i(t+1\mid t),\ldots,\hat{y}_i(t+P\mid t)]^T \tag{30}$$

$$\hat{\Theta}(t)=[\hat{\Theta}_1{}^T(t),\hat{\Theta}_2{}^T(t),\hat{\Theta}_3{}^T(t)]^T \tag{31}$$

$$\hat{\Theta}_i(t)=[\hat{\theta}_i(t+1\mid t),\ldots,\hat{\theta}_i(t+P\ t)]^T \tag{32}$$

From equations (20)(26)(27)(28), we can derive maximum and minimum value of $\hat{\theta}_i(t+j\mid t)$ are

$$\underline{\theta}_i(t+l\mid t)=\min_{\Delta u_j\in\Delta\hat{u}_j,\delta_{ij}(k)\in\hat{\delta}_{ij}(k)}\sum_{j=1}^{3}\sum_{k=1}^{l}\delta_{ij}(k)\Delta u_j(t+l-k)$$
$$+\sum_{j=1}^{3}\sum_{k=1}^{N}(\delta_{ij}(k+l)-\delta_{ij}(k))\Delta u_j(t-k) \tag{33}$$

$$\bar{\theta}_i(t+l\mid t)=\max_{\Delta u_j\in\Delta\hat{u}_j,\delta_{ij}(k)\in\hat{\delta}_{ij}(k)}\sum_{j=1}^{3}\sum_{k=1}^{l}\delta_{ij}(k)\Delta u_j(t+l-k)$$
$$+\sum_{j=1}^{3}\sum_{k=1}^{N}(\delta_{ij}(k+l)-\delta_{ij}(k))\Delta u_j(t-k) \tag{34}$$

3.2 Objective function

The objective of robust MPC is that the controller action tries to minimize the objective function for the worst situation. Here, An ∞ type of norm is used as an objective function that is described as:

$$\min_{u\in U}\max_{\theta\in\Theta}J(U,\theta) \tag{35}$$

where

$$J(U,\theta)=\max_{j=1,\ldots P}\left\|\hat{y}(t+j\mid t)-w(t)\right\|_\infty$$
$$=\max_{j=1\ldots P}\max_{i=1,2,3}\left|\hat{y}_i(t+j\mid t)-w_i(t)\right| \tag{36}$$

$w_i(t)$ is reference trajectory of ith output at time t.

3.3 Obtaining the control law using linear programming

The optimization problem of (35) can be reduced to a linear programming problem. Let

$$\mu=\max_{\theta\in\Theta}\max_{i=1,2,3}\max_{j=1,\ldots,P}\left|\hat{y}_i(t+j\mid t)-w_i(t)\right| \tag{37}$$

the problem can be stated as:

$$\min_{\mu,\Delta u}\mu \tag{38}$$

subject to constraints (2) and

$$l\mu \ge Y_0(t) + A\Delta U(t) + \hat{\Theta} - w \qquad (39)$$

$$l\mu \ge -Y_0(t) - A\Delta U(t) - \hat{\Theta} + w \qquad (40)$$

where l=[1,1,..., 1]. We can transform this problem into usual form of linear programming and obtain control action $\Delta u_1(t), \Delta u_2(t), \Delta u_3(t)$ at time t.

4. SIMULATION RESULTS

To test the performance of the controllers, simulations under various conditions are performed using the dynamic model (1) at different operating points. The controller parameters $P = 8$, $M = 3$, the sample time $T = 20$, truncation order $N = 50$. Considering the following situation:

Case 1 A step input disturbance of magnitude 0.1 at every input channel is inserted at $t = 50$ respectively.

Case 2 A step output disturbance of magnitude -10 at *1st* and 10 at *2nd* output channels, 0.1 at *3rd* output channel is inserted at $t = 50$ respectively.

Case 3 A change of operating point from 3 to 1 and from 3 to 5 at $t = 20$ respectively.

In addition, simulations are performed to make a comparison with robust MPC under case 3 using a conventional MPC. Figures 2-7 show the simulation results.

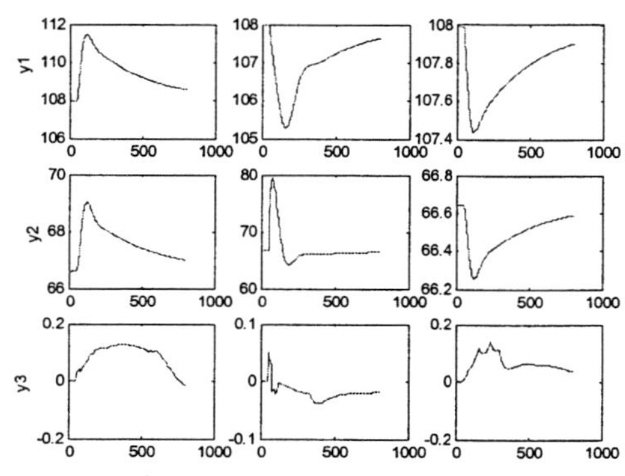

Fig.2 System outputs under case 1

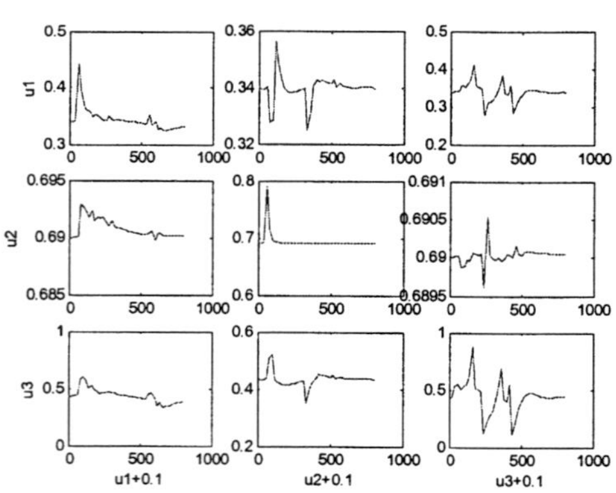

Fig.3 Control actions under case 1

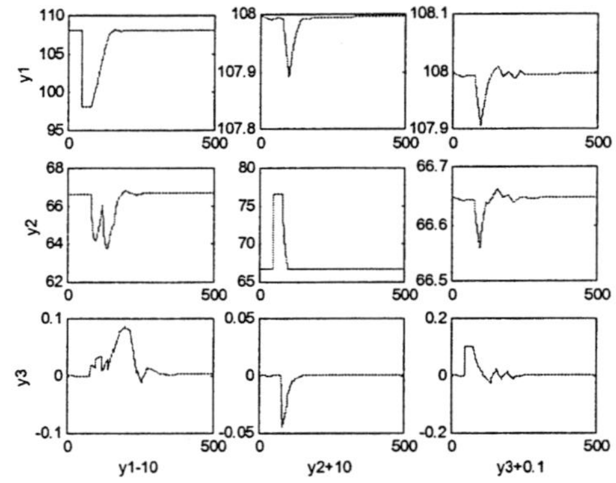

Fig.4 System outputs under case 2

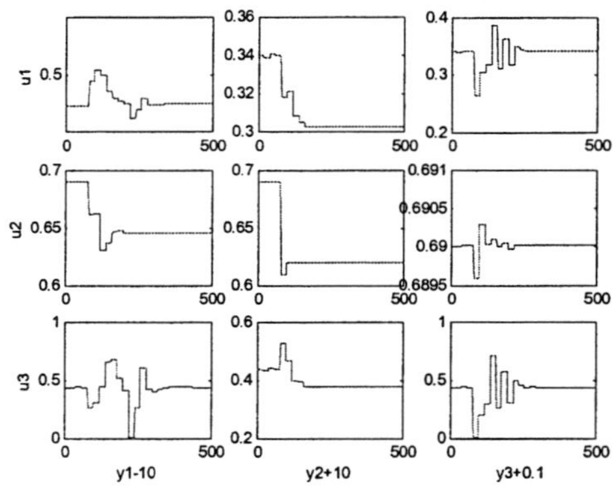

Fig.5 Control actions under case 2

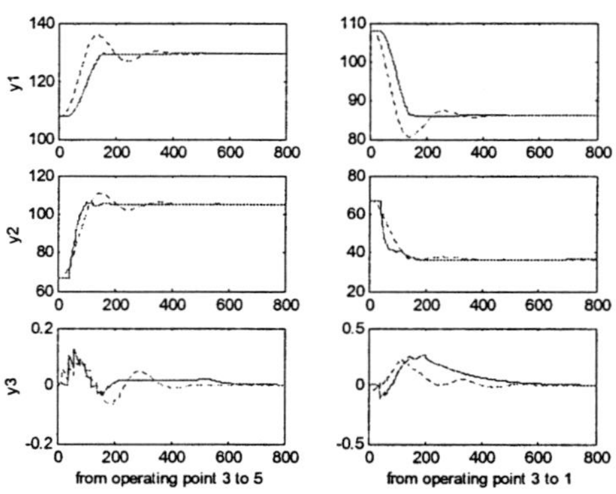

Fig.6 System outputs under case 3
Solid line – Robust MPC
Dotted line – Conventional MPC

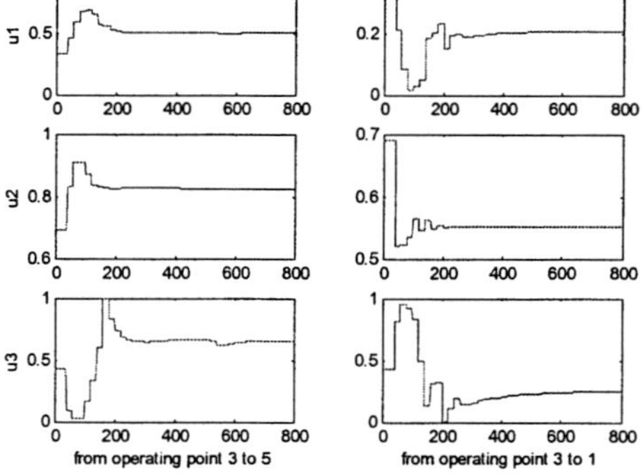

Fig.7 Control actions under case 3

From simulation results, we find that the robust MPC in this paper has the following properties:

1.Robust MPC has the better stability, robustness and tracking ability within input constraints than conventional MPC when the operating point changes.

2.The parameters of robust MPC based on infinite

norm are easier to tune than the ones of conventional MPC because the number of parameter is fewer.

3.The controller has good ability to overcome the output and the input disturbance. But because there is not cost function on control action in ∞ norm objective optimization function, system output responses under input disturbances have a little worse. If using 2 norm objective function and quadratic programming, the ability of the controller overcoming input disturbance will be better.

5. CONCLUSION

In this paper, a robust MPC is designed for a 160 MW fossil fueled nonlinear boiler turbine control system. Some linearized models and their step response coefficients are obtained at different operating points. Half loading point is considered as nominal model, and other operating points as model uncertainty. Due to integrating subsystem of the system, an improved robust MPC algorithm was provided and simulation proved the effective of the controller.

REFERENCE

Bell R. B. and K. J. Astrom (1987), Dynamic models for boiler-turbine-alternator units: data logs and parameter estimation for a 160 MW unit, *Report TFRT-3192*, Lund Institute of Technology, Sweden.

Camacho E. F. and C. Bordons (1995), *Model Predictive Control in the Process Industry.* New York: Springer.

Campo P. J. and M. Morari, Robust model predictive Control (1987), In: *American Control Conference*, Minneapolis, Minnesota, 1021-1026.

Cori R. and C. Maffezzoni (1984), Practical optimal control of a drum boiler power plant. *Automatica*, **20(2)**: 163-173.

Dai L. K. (1999), Dynamic matrix control of integrating system, *Control Theory and Application*, **16(5)**, 744-746.

Gupta Y. P. (1998), Control of integrating processes using dynamic matrix control, *Trans IChemE Part A*, **76(5)**, 465-470.

Hwang C. S. and D. W. Kim (1995), A design of robust two-degree-of-freedom boiler turbine control systems using H_∞ optimization method, *Control Application Proceedings of the 4th IEEE Conference*, Sapporo, 1263-1268.

Johansson L. and H. N. Koivo (1984), Inverse Nyquist array technique in the design of a multivariable controller for a solid-fuel boiler. *Int. J. Control*, **40(5)**: 1077-1088.

Kwon W. H., S. W. Kim and P. G. Park (1989), On the multivariable robust control of a boiler-turbine system, In *IFAC Symposium on Power Systems and Power Plant Control*, Seoul, Korea, 219-223.

Morari M. and J. H. Lee (1999), Model predictive control: Past, present and future, *Computer and Chemistry Engineering*, **23(4)**, 667-682.

Tan W., Niu Y. G. and Liu J. Z. (1999), Robust control for nonlinear boiler-turbine system, *Control Theory and Application*, **16(2)**, 863-867.

ELSEVIER
IFAC
PUBLICATIONS
www.elsevier.com/locate/ifac

NEURO-FUZZY GENERALIZED PREDICTIVE CONTROL OF BOILER STEAM TEMPERATURE

Xiang-Jie Liu*, Felipe Lara-Rosano*,
Marino Sanchez Parra, Raul Garduño Ramirez****

Centro de Ciencias Aplicadas y Desarrollo Tecnologico, UNAM, Mexico City, D.F. 04510, Mexico
***Electric Research Institute, Cuernavaca, Morelos CP 62490 México*

Abstract: A neuro-fuzzy network predictive approach is introduced to design a control system for boiler steam temperature process. While the nonlinear process is modeled by neuro-fuzzy technique containing local CARIMA model, traditional generalized predictive control method can be extended to nonlinear case in a neuro-fuzzy fashion. The proposed method was found to provide a satisfactory performance over the 200MW superheater steam temperature system. *Copyright © 2003 IFAC*

Keywords: neuro-fuzzy networks, nonlinear control, steam temperature

1. INTRODUCTION

Continuous processes in power plant and power station are complex systems characterized by nonlinearity, uncertainty and load disturbance. Applying of traditional control methods encounter great difficulties while the process working condition changes within a large operation range. The effective way is to develop advanced modelling and control strategy on them.

Generalized predictive control (GPC) methods have been proved to be an effective method over linear model and are frequently applied in nonlinear control in an adaptive framework, designing the control law on the basis of a linearised model of the controlled system, which is reestimated periodically to fit the actual operation point with online estimation by recursive identification techniques. The linearised model may be acceptable only in the case where the process is working around the operating point. An adaptive approach may fail if abrupt changes in the operation point are requested by the control specifications or determined by the specific nonlinearity of the system. Self-tuning controller requires continuous on-line identification with injection of additional disturbance into the plant for sufficiently rich excitation of plant dynamics. Failure of parameter estimation algorithm could act to accentuate process disturbances. Problems with unmodelled dynamics and on-line adaption of controllers may not be acceptable in power plants for reasons of reliability(Wu and Hogg, 1991).

Principally, nonlinear process requires nonlinear predictive control for optimum performance. Nevertheless, controller incorporating a nonlinear model may require tremendous computational burden which may disqualify it for on-line application. With specific nonlinear model, the explicit control law may prove a difficult or even impossible task. To incorporate the linear method, Prasad *et al.*(1998) proposed a local model networks based multivariable predictive control, which has been simulated on a 200MW oil-fired drum-boiler thermal power plant.

It is noticed that much of the complex process in power plant could be controlled by manual operation of the experienced operator, which is based on the enough knowledge of the object. This offer effective modelling and control method for automatic control system. The recent developed neurofuzzy network(Brown and Harris, 1994) could be an effective way to the modelling of such nonlinear system in that fuzzy rules could easily express the expert knowledge in linguistic form while neural networks posses the learning ability of approximating nonlinear functions with arbitrary accuracy. In the neuro-fuzzy networks, the whole envelope of process operation is divided into several operating regions. In each region a local reduced-order model is used to approximate the process. Fuzzy logic representing human operator's language expression provides an appropriate means to define operation regions. Thus respectively the local predictive control strategy could be constituted.

This paper develops a nonlinear neuro-fuzzy network modelling and predictive control method and then apply it to the superheater steam temperature control system on 200MW unit steam-boiler. Associate Memory Network(AMN) is used as process predictive model because it provides a direct link between artificial neural networks and the fuzzy systems. Transient process input-output data are then used to train the network. With the established neuro-fuzzy model, a set of relative local GPC controllers are composed under different operation regions to constitute the neuro-fuzzy controller. The performance of the control system has been verified by the simulation process and then tested on superheater steam temperature process.

2 NEURO-FUZZY NETWORK MODELLING

With the following general single-input single-output nonlinear dynamic system:

$$y(t) = f[y(t-1),\cdots,y(t-n'_y),u(t-d),\cdots,\\ u(t-d-n'_u+1),e(t-1),\cdots,e(t-n'_e)]+e(t)/\Delta \qquad (1)$$

where f[.] is a smooth nonlinear function such that a Taylor series expansion exists, a local linear model at an operating point O(t) can be given by the following CARIMA model:

$$\overline{A}(z^{-1})y(t) = z^{-d}B(z^{-1})\Delta u(t) + C(z^{-1})e(t) \qquad (2)$$

where $\overline{A}(z^{-1}) = \Delta A(z^{-1})$, $B(z^{-1})$ and $C(z^{-1})$ are polynomials in z^{-1}, the backward shift operator. They are function of the operating point O(t). Therefore, the nonlinear model could be decomposed into several linear or nearly linear models at different operation points. The total model could be represented by a B-spline AMN shown in figure 1. The antecedent variables $[x_1, x_2 \cdots x_n]$ representing different operation points are the inputs of the AMN. The output of the network model $\hat{y}(t)$ could be combination of local linear models $\hat{y}_i(t)$.

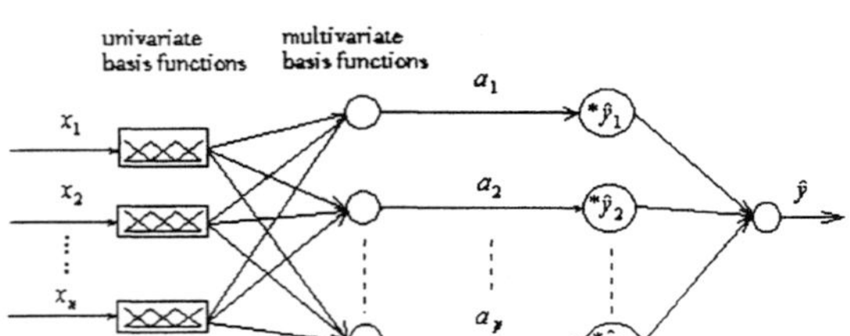

Fig.1 Nonlinear neurofuzzy network model

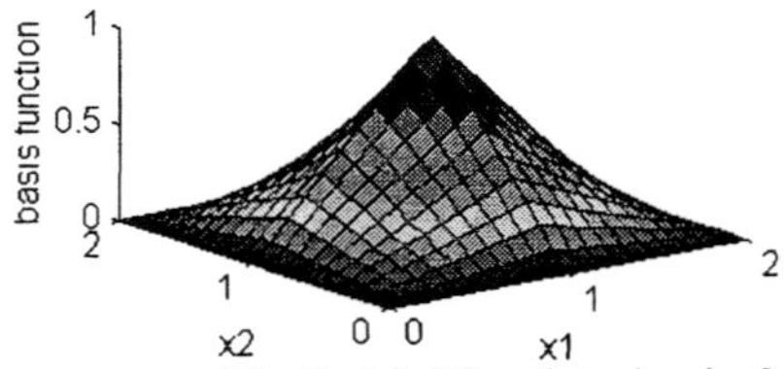

Fig.2 Multivariate basis function

The B-spline network is initially designed to specify the shape(order) of each of the univariate basis functions, and this implicitly determines the number of basis functions mapped to for a particular network input. There exist recurrence relationship for evaluating the membership of a univariate B-spline basis function with R inner knots(Brown and Harris, 1994). Denoting the jth univariate basis function of order k by $\mu_k^j(x)$, jth knot by λ_j and the jth interval by $[\lambda_{j-1}, \lambda_j]$. The basis functions are defined on a bounded support and the output of the basis function is positive on its support, $\mu_k^j(x) = 0,\ x \notin [\lambda_{j-k}, \lambda_j]$ and $\mu_k^j(x) > 0,\ x \in (\lambda_{j-k}, \lambda_j)$. The sum of the outputs of the basis functions is always one, i.e $\sum_j \mu_k^j(x) \equiv 1,\ x \in [x_{\min}, x_{\max}]$.

The neuro-fuzzy network shown in figure 1 could be expressed as a set of fuzzy production rules:

IF operating condition i (x_1 is positive small, and $\cdots, x_n$ is negative large)

THEN local linear CARIMA model i :

$$\hat{y}_i(t) = \overline{a}_{i1}\hat{y}_i(t-1) + \cdots \overline{a}_{in_{\overline{a}}}\hat{y}_i(t-n_{\overline{a}}) + b_{i0}\Delta u_i(t-d) \\ + \cdots + b_{in_b}\Delta u_i(t-d-n_b) + e_i(t) + \cdots + c_{in_c}e_i(t-n_c)$$

or $\overline{A}_i(z^{-1})\hat{y}_i(t) = z^{-d}B_i(z^{-1})\Delta u_i(t) + C_i(z^{-1})e_i(t)$ (3)

The multivariate basis function is obtained by:

$$\alpha_i(x) = \prod_{k=1}^{n} \mu_{A_k^i}(x_k) \qquad (4)$$

where $i = 1,2,\cdots,p$ and n is the dimension of the input vector x(t). p is the total number of memory required in the network given by

$$p = \prod_{i=1}^{n}(R_i + k_i) \qquad (5)$$

Thus the desirable properties of the univariate B-spline basis functions are all extended in a natural way to the multivariate basis functions. They are defined on hyperrectangles of size $(k_1 \times k_2 \times \cdots \times k_n)$ and therefore possess a bounded support. The output is positive inside this domain and zero outside. Fig.2 shows the multivariate basis function formed from two ,order 2, univariate basis function.

The model output is the synthesizing of these p rules:

$$\hat{y}(k)=\frac{\sum_{i=1}^{p}\hat{y}_i(k)\alpha_i(x)}{\sum_{i=1}^{p}\alpha_i(x)}=\sum_{i=1}^{p}\hat{y}_i(k)\alpha_i(x) \quad (6)$$

Choose the cost function to be:

$$J=\frac{1}{2}e^2=\frac{1}{2}(\hat{y}-y)^2 \quad (7)$$

where $\hat{y}$ is the network prediction, and y is the plant output. Training parameter could be changed in the direction of the negative gradient:

$$\dot{\bar{a}}_{ij}=-g\frac{\partial J}{\partial \bar{a}_{ij}}=-ge\frac{\partial \hat{y}}{\partial \hat{y}_i}\frac{\partial \hat{y}_i}{\partial \bar{a}_{ij}}=-ge\alpha_i(x)\hat{y}_i(t-l)$$

$$\dot{b}_{il}=-g\frac{\partial J}{\partial b_{il}}=-ge\frac{\partial \hat{y}}{\partial \hat{y}_i}\frac{\partial \hat{y}_i}{\partial b_{il}}=-ge\alpha_i(x)u_i(t-d-m) \quad (8)$$

where $1\le l\le n_{\bar{a}} \qquad 0\le m\le n_b$ (9)

3 NONLINEAR NEURO-FUZZY NETWORK GENERALIZED PREDICTIVE CONTROL

3.1 The problem formulation with approximation

For the purpose of generalised predictive control, the cost function to be minimized is:

$$J=E\{\sum_{j=d}^{N}q_j[\hat{y}(t+j)-y_r(t+j)]^2\}+\sum_{j=1}^{M}\lambda_j[\Delta u(t+j-1)]^2 \quad (10)$$

q_j and λ_j are respectively the weighting factors for the prediction error and control energy; $y_r(t+j)$ is the ith step reference trajectory, d is time delay taken as minimum costing horizon, N and M are maximum costing horizon for prediction and control variables respectively. The control policy is given by the weighted sum of p sets of local control policies:

$$\Delta u(t)=\sum_{i=1}^{p}\alpha_i\Delta u_i(t) \quad (11)$$

where $\Delta u_i(t)$ is the relative control action in each local region, α_i is the same as that defined in the neuro-fuzzy modelling equation (4) since the contribution of the output estimated by the ith sub-model to the overall process evaluation should be considered the same as that by the ith local control action to the overall system controller. This also meet the idea of T-S model. It is different from T-S method in that the "consequence" does not necessarily contain the "premise" part. With this control structure, the system eliminates the controller switch problem and provides a much more smooth control performance in process operation. With model (6) and control policy (11), equation (10) could be expressed as:

$$J=E\{\sum_{j=d}^{N}q_j[\sum_{i=1}^{p}\alpha_i(\hat{y}_i(t+j)-y_r(t+j))]^2\} + \sum_{j=1}^{M}\lambda_j[\sum_{i=1}^{p}\alpha_i u_i(t+j-1)]^2 \quad (12)$$

With interactions among those subsystems, the minimization of this cost function requires large computing burden. An alternative approximated cost function could be acquired for the simplification with Cauchy inequality,

$$[\sum_{i=1}^{p}\alpha_i(\hat{y}_i(t+j)-y_r(t+j))]^2 \le p\sum_{i=1}^{p}[\alpha_i(\hat{y}_i(t+j)-y_r(t+j))]^2$$

$$[\sum_{i=1}^{p}\alpha_i u_i(t+j-1)]^2\le p\sum_{i=1}^{p}[\alpha_i u_i(t+j-1)]^2 \quad (13)$$

Equation (13) means the sum of the weighted squared errors can be the basis for establishing an upper bound of the original objective function. And

$$E\{\sum_{j=d}^{N}\sum_{i=1}^{p}q_j[\alpha_i(\hat{y}(t+j)-y_r(t+j))]^2\} + \sum_{j=1}^{M}\sum_{i=1}^{p}\lambda_j[\alpha_i u_i(t+j-1)]^2$$

$$=E\{\sum_{i=1}^{p}(\alpha_i)^2\sum_{j=d}^{N}q_j[\hat{y}_i(t+j)-y_r(t+j)]^2\} + \sum_{i=1}^{p}(\alpha_i)^2\sum_{j=1}^{M}\lambda_j[u_i(t+j-1)]^2$$

$$=\sum_{i=1}^{p}(\alpha_i)^2 J_i \quad (14)$$

where

$$J_i=E\{\sum_{j=d}^{N}q_j[\hat{y}_i(t+j)-y_r(t+j)]^2\}+\sum_{j=1}^{M}\lambda_j[u_i(t+j-1)]^2 \quad (15)$$

The derivation procedure shows that the essence of minimization of J is the same as that of J_i. Thus the problem changes to the minimization of J_i in each local model.

3.2 Controller

Equation (15) could be rewritten in a matrix form as:

$$J_i=E\{[Y_i(t+1)-Y_r(t+1)]^T Q_i[Y_i(t+1)-Y_r(t+1)]\} + \Delta U_i^T(t)T_i\Delta U_i(t) \quad (16)$$

The control law by the ith sub-system can be identified as(Clarke *et al.*, 1987):

$$\Delta U_i(t)=(G_i^T QG_i+\lambda)^{-1}G_i^T Q[Y_r(t+1) - F_i\Delta U_i(t-1)-S_i(z^{-1})y_i(t)] \quad (17)$$

where

$$Y_i(t+1)=[\hat{y}_i(t+1),\hat{y}_i(t+2),\cdots,\hat{y}_i(t+P)]^T$$

$$Y_r(t+1)=[\hat{y}_r(t+1),\hat{y}_r(t+2),\cdots,\hat{y}_r(t+P)]^T$$
$$\Delta U_i(t)=[\Delta u_i(t),\Delta u_i(t+1),\cdots,\Delta u_i(t+M-1)]^T$$
$$\Delta U_i(t-1)=[\Delta u_i(t-n_b),\Delta u_i(t-n_b+1),\cdots,\Delta u_i(t-1)]^T$$
$$\varepsilon_i(t+1)=[R_{i1}(z^{-1})e_i(t+1),\cdots R_{iP}(z^{-1})e_i(t+P)]^T$$
$$S_i(z^{-1})=[S_{i1}(z^{-1}),S_{i2}(z^{-1}),\cdots,S_{iP}(z^{-1})]^T$$

$S_i(z^{-1})$ and $R_i(z^{-1})$ are in the Diophantine equation:

$$1=\overline{A}_i(z^{-1})R_{ij}(z^{-1})+z^{-j}S_{ij}(z^{-1}) \tag{18}$$

and:

$$G_i(z^{-1})=B_i(z^{-1})R_i(z^{-1}) \\ =g_{j,0}+g_{j,1}z^{-1}+\cdots+g_{j,n_b+j-1}z^{-(n_b+j-1)} \tag{19-a}$$

$$g_{j+1,i}=g_{j,i}+s_{j,0}b_{i-j} \tag{19-b}$$

$$Q_i=diag(q_{i1},q_{i2},\cdots,q_{iP}) \\ T_i=diag(\lambda_{i1},\lambda_{i2},\cdots,\lambda_{iM}) \tag{19-c}$$

$$G_i=\begin{bmatrix} g^i_{1,0} & & \cdots & 0 \\ g^i_{2,1} & g^i_{1,0} & & \vdots \\ \vdots & \vdots & & \\ g^i_{P,P-1} & g^i_{P-1,P-2} & \cdots & g^i_{P-M+1,P-M} \end{bmatrix}_{P\times M} \tag{19-d}$$

$$F_i=\begin{bmatrix} g^i_{1,n_b} & g^i_{1,n_b-1} & \cdots & g^i_{1,2} & g^i_{1,1} \\ g^i_{2,n_b+1} & g^i_{2,n_b} & \cdots & g^i_{2,3} & g^i_{2,2} \\ \vdots & \vdots & & \vdots & \vdots \\ g^i_{P,n_b+P-1} & g^i_{P,n_b+P-2} & \cdots & g^i_{P,P+1} & g^i_{P,P} \end{bmatrix}_{P\times n_b} \tag{19-e}$$

The optimization control variables are M steps into the future based on the control horizon. Only the first control action is implemented, using a receding horizon principle, i.e,

$$\Delta u_i(t)=d_{i1}{}^T[Y_r(t+1)-F_i\Delta U_i(t-1)-S_i(z^{-1})y_i(t)]$$

where

$d_{i1}{}^T=(1,0,\cdots,0)(G_i{}^TQG_i+\lambda)^{-1}G_i{}^TQ$ is the first row of the matrix $(G_i{}^TQG_i+\lambda)^{-1}G_i{}^TQ$.

4 NEURO-FUZZY MODELLING AND CONTROL OF BOILER STEAM TEMPERATURE

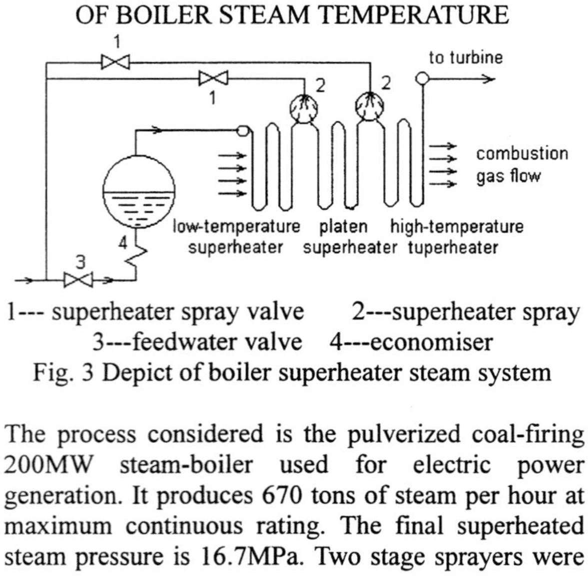

1--- superheater spray valve 2---superheater spray
3---feedwater valve 4---economiser

Fig. 3 Depict of boiler superheater steam system

The process considered is the pulverized coal-firing 200MW steam-boiler used for electric power generation. It produces 670 tons of steam per hour at maximum continuous rating. The final superheated steam pressure is 16.7MPa. Two stage sprayers were applied for the superheater temperature system in a multi-stage control manner to reduce the time constant of each stage and improve the control trait. The rated superheated steam temperature is 540°C.

The task of the superheater in boiler-turbine system is to heat the steam by combustion gas and then send it to turbine. As shown in figure 3, the steam coming from the boiler drum passes through the low-temperature superheater and receives a spray water injection before passing through the radiant-type platen superheater. The steam then receives a second spray water injection before passing through the high-temperature superheater. A properly control system by superheater water spray should maintain strictly the steam temperature within the permitted range(± 10 °C in transient process and ± 5 °C in steady state). Higher temperature could damage the superheater and the high-pressure-turbine by high heat strength. Lower temperature could decrease the running efficiency of the whole steam-boiler. By reducing temperature fluctuations, mechanical stress causing microcracks is diminished, therefore plant life is increased and maintenance cost is decreased. This is obviously the motivation for GPC control, since the variance of the steam temperature is directly related to economic and safety performance.

The superheater outlet steam temperature can be affected by many factors such as steam flow, combustion condition, feedwater temperature, steam enthalpy and combustion gas temperature. Totally speaking, three main factors dominate the superheater outlet steam temperature: load power, gas flow and superheater inlet steam temperature. These are also the reasons why a conventional PID-based control strategy is often unable to perform satisfactorily. Many efforts have been put into developing and testing more advance control schemes on different boiler plants, like adaptive simulation-model-based control of fossil-fuel-fired thermal power plant(Nakamura et al., 1995) and feedforward control based on new measuring techniques on a 256MW Unit(Tommy, 1999). The intention of this section is to develop the nonlinear model of steam temperature with the proposed neuro-fuzzy method based on dynamic experiment of the 200 MW Unit and then constitute the relative nonlinear GPC controllers.

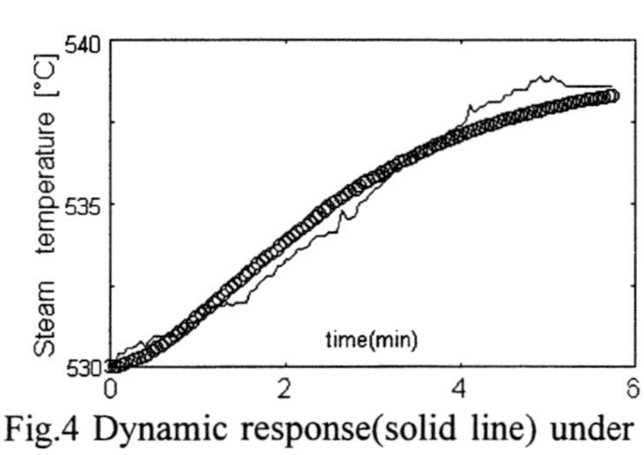

Fig.4 Dynamic response(solid line) under 18% step change in the second stage spray flow and its fitting curve(circled line)

Considering the output "steam temperature θ" with the control input spray μ_θ, dynamic of the steam temperature process could be depicted to be a second order representation (Liu and Zhou, 1999):

$$G(s)=\frac{\theta}{\mu_\theta(s)}=\frac{K_p}{(T_1s+1)(T_2s+1)}e^{-\tau s} \quad (20)$$

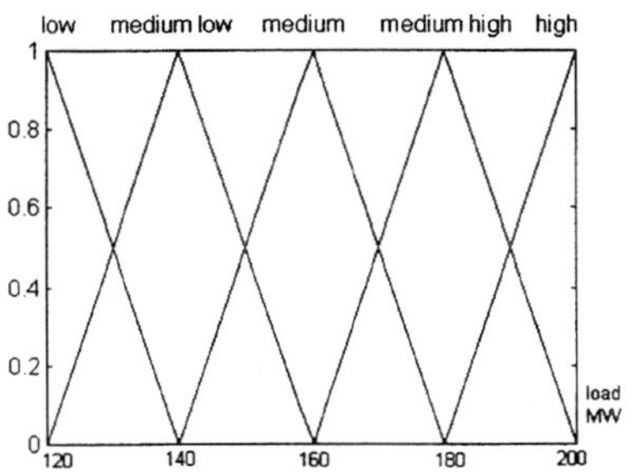

Fig.5 Local division of five membership function

Figure 4 shows real-time dynamic response of steam temperature under 180MW load condition and its fitting curve. The linear model (20) is, however, extremely local in nature, and do not recognize the non-linear nature of the process. Load operation leads to changes in operating point right across the whole operating range. With the neurofuzzy modelling in section 2, load operation is divided into five regions as shown in figure 5. This division is based on the operator's former experience, which considers that 200MW is high load, 180MW is medium high load, 160MW is medium load, 140MW is medium low load and 120MW is low load. Since the length of superheater tube is about tens of meters, there exists time delay from the superheater spray flow to the change of superheater outlet steam temperature. The average time delay $\tau=30s$ is assumed to be uniform and taken as minimum control horizon. Let $C_i=1$ to make $e_i(t)$ an uncorrelated random sequence of zero mean. The identified five-region neuro-fuzzy model is shown in Table 1.

Table 1 Neuro-fuzzy model identification parameters

Load	$\overline{A}_i$	B_i
High	$1-2.8824z^{-1}+2.7682z^{-2}-0.8858z^{-3}$	0.0035
Medium high	$1-2.8940z^{-1}+2.7915z^{-2}-0.8975z^{-3}$	0.0028
Medium	$1-2.9000z^{-1}+2.8100z^{-2}-0.9100z^{-3}$	0.0023
Medium low	$1-2.9130z^{-1}+2.8279z^{-2}-0.9149z^{-3}$	0.0019
Low	$1-2.9200z^{-1}+2.8416z^{-2}-0.9216z^{-3}$	0.0016

Also a global linear model is acquired with the same I/O data in the whole operation region. The real-time open-loop response shows that the neuro-fuzzy model can nearly perfectly describe the nonlinear dynamic behavior but the global linear model can not match the plant. The reason is that linear model could hardly representative nonlinear plant, no matter what kind of identification method has been used.

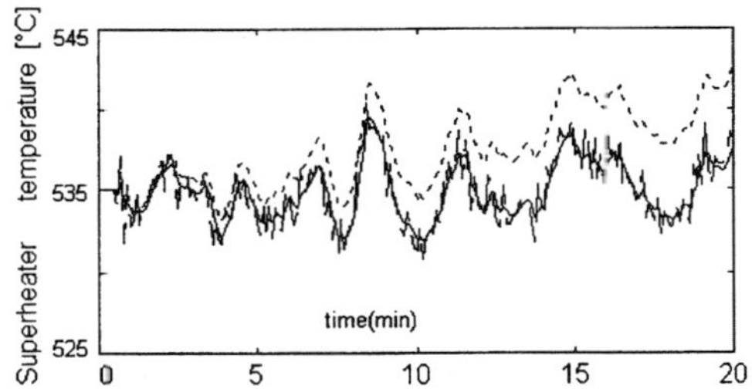

Fig.6 Comparison of boiler system(solid line), global linear model(dotted line) and neuro-fuzzy model(dashed line) responses

Table 2 Neuro-fuzzy GPC control law parameters ($F_i=0$)

Load	$S_i(z^{-1})$	d_{i1}
High	$23.3-45.9z^{-1}+19.6z^{-2}$	0.390
	$32.9-56.1z^{-1}+24.2z^{-2}$	0.452
	$38.7-66.8z^{-1}+29.1z^{-2}$	0.512
	$44.7-78.0z^{-1}+34.3z^{-2}$	0.570
Medium high	$28.0-47.2z^{-1}+20.3z^{-2}$	0.375
	$33.8-57.9z^{-1}+25.1z^{-2}$	0.445
	$39.9-69.2z^{-1}+30.3z^{-2}$	0.520
	$46.2-81.0z^{-1}+35.8z^{-2}$	0.586
Medium	$27.7-47.2z^{-1}+20.4z^{-2}$	0.353
	$33.2-57.5z^{-1}+25.2z^{-2}$	0.423
	$38.9-68.1z^{-1}+30.2z^{-2}$	0.493
	$44.7-79.1z^{-1}+35.4z^{-2}$	0.563
Medium low	$29.4-49.9z^{-1}+21.5z^{-2}$	0.328
	$35.6-61.5z^{-1}+26.9z^{-2}$	0.401
	$42.3-73.9z^{-1}+32.6z^{-2}$	0.478
	$49.8-87.7z^{-1}+38.9z^{-2}$	0.559
Low	$29.9-50.9z^{-1}+22.0z^{-2}$	0.301
	$36.5-63.0z^{-1}+27.6z^{-2}$	0.371
	$43.4-76.0z^{-1}+33.6z^{-2}$	0.445
	$50.8-89.8z^{-1}+40.0z^{-2}$	0.524

The neuro-fuzzy GPC is established based on neuro-fuzzy model. Simulation of steam temperature transient process was conducted under five different load condition. Sampling time and control time are both chosen as 5 seconds. The following parameters: $P=10$, $d=6$, $M=8$, $Q=I$, $\lambda=0.1\times I$, after evaluated in a test and trial mode, are used to achieve an active control. Table 2 lists the control law parameters. Fig.7 shows simulation of steam temperature(dotted line) under set-point tracking using neuro-fuzzy GPC controller.

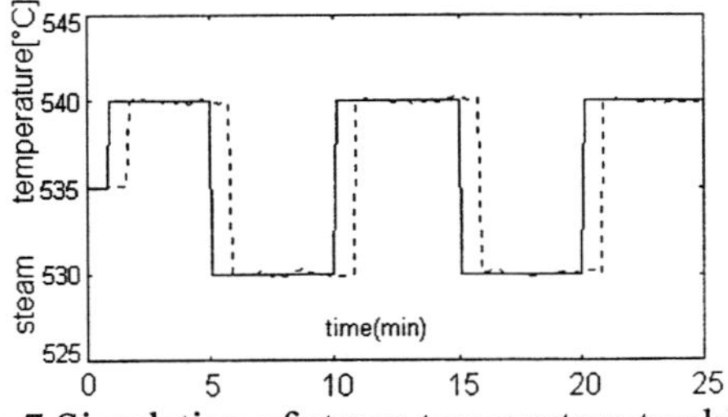

Fig.7 Simulation of steam temperature tracking

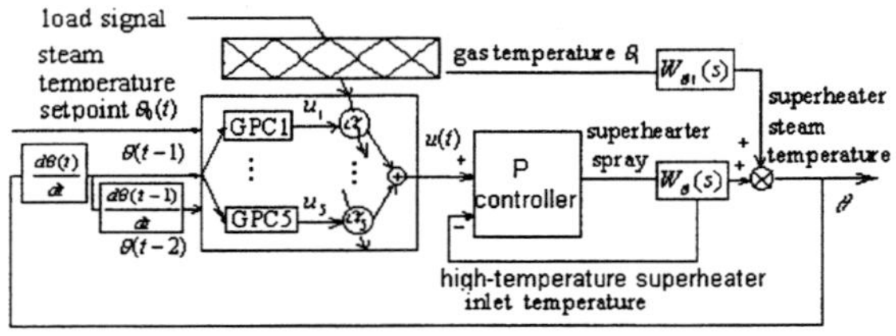

Fig.8 Diagram of boiler control system

The real-time control system is applied in the cascade form (Fig.8). Neurofuzzy GPC controller acts as the principal controller, containing five local GPC controllers. Two controllers are working at the same time, decided by load signal through five triangular membership functions. The subordinate loop adjusts the superheater water spray with ordinary P controller, so as to give a forward response and overcome the inner disturbance quickly. Combustion gas flow acts as feed forward variable(disturbance).

Figure 9 shows the response of the steam temperature under changing load conditions($1\% - 2.5\%$ /min). While the steam temperature fluctuate due to changing load, the system overcome the disturbance in a satisfactory way. Steam temperature overshoot no more than $\pm 7°C$ in transient process. The result is similar to that of (Tommy,1999), which is tested on a 380MW unit, under coal mill stop disturbance.

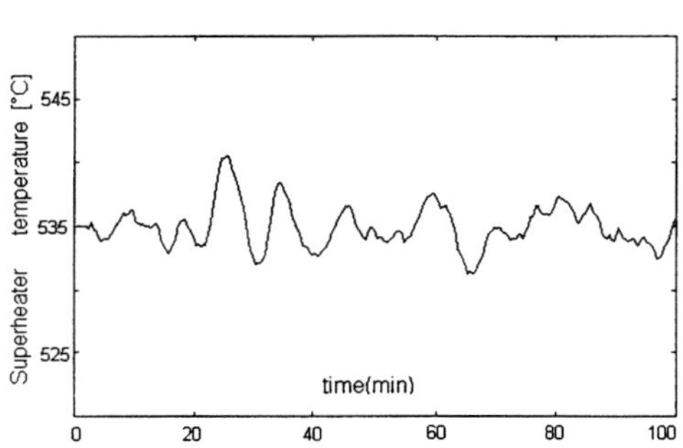

Fig.9-a Response of the steam temperature

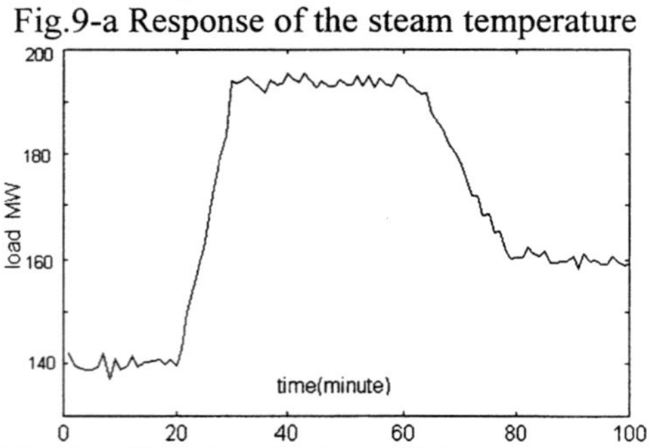

Fig.9-b Changing load condition

To make a comparison, the proposed nonlinear controller has been simulated on both neuro-fuzzy model and on linear model(under 140MW) under the same changing load condition. Figure10 shows that while the load changes at 140MW(t=20min), the responses of steam temperature by the two models are similar. While the load changes at 190MW(t=60min), the response offered by neuro-fuzzy model is superior to that of linear model.

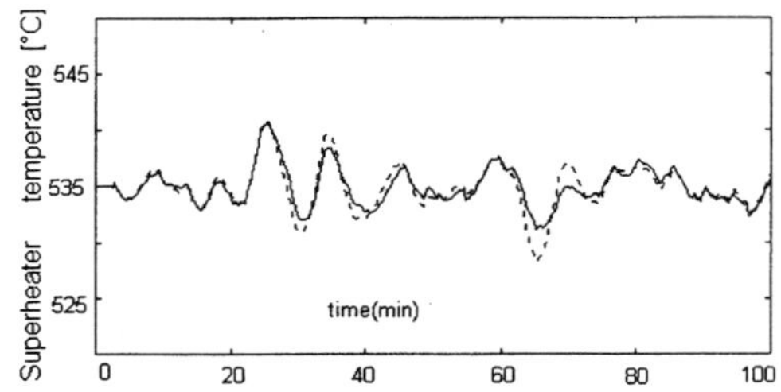

Fig.10 Comparison of the proposed method on neuro-fuzzy model(solid line) and on linear model(dotted line)

5 CONCLUSION

Nonlinear process in power plant can be modeled by neuro-fuzzy modelling technique. The neuro-fuzzy model composed of a series of CARIMA models could represent system nonlinearity perfectly. With the relative GPC strategy, the control system could reach a near optimal structure. The proposed predictive control method is particularly suitable for nonlinear process which can be divided into several local operating regions and thus greatly advance on-line applications. The effort required to develop a neuro-fuzzy predictive model of the specific plant. The advantage is shown in the modelling and control of boiler steam temperature system, which is a typical complex industrial process. In such a way, the proposed method offered an engineering approximation approach for a class of nonlinear industrial system which are formerly difficult to be controlled by traditional method.

REFERENCE

Brown, M. & Harris, C. J. (1994). *Neurofuzzy adaptive modelling and control.* Prentice Hall.

Clarke, D.W., Mohtadi, C. and Tuffs, P.S., Generalized predictive control(1987). *Parts 1 and 2, Automatica*, , **23**, (2), 137-160.

Nakamura, H., Toyoda, Y., and Oda, K.(1995). An adaptive control system for steam temperature of fossil-fuel-fired thermal power plant. IFAC Symp. on Power Plant and Power Systems(pp.201-206). Cancun, Mexico.

Prasad G., E.Swidenbank and B.W.Hogg,(1998). A local model networks based multivariable predictive control strategy for thermal power plants. *Automatica*, , **34**(10), 1195-1204.

Tommy Moelbak.(1999). Advanced control of superheater steam temperatures-an evaluation based on practical applications. *Control Engineering Practice* , **(7)**, 1-10

Wu, Q.H., and Hogg, B.W.(1991). Laboratory evaluation of adaptive controllers for synchronous generators, *Automatica*, **127**(5), 845-852

Xiang-Jie Liu and Xiaoxin Zhou(1999). Identification of Boiler Models and Its Fuzzy Logic Strategy. Proc. the 14th IFAC World Congress. Volume O (pp.149-154).Beijing China.

ELSEVIER
IFAC
PUBLICATIONS
www.elsevier.com/locate/ifac

MODEL PREDICTIVE PRESSURE CONTROL IN STEAM NETWORKS

Yrjö Majanne

Tampere University of Technology, Institute of Automation and Control
P.O. Box 692, FIN-33101 TAMPERE, FINLAND

Abstract: The control task of the industrial power plants typically leads to a complex multivariable control structure with active constraints to be taken care of. **M**odel **P**redictive **C**ontrol, MPC, is applied to control pressure stability in a multilevel steam network. The system is demonstrated in a simulator environment. Besides the ability to control the power plant process successfully, MPC can be used as a convenient tool for analyzing and designing the structure of the steam network. Simulated power plant process controlled by MPC helps to decide the location and the capacity of steam levelling components needed to stabilize the operation of the process. *Copyright © 2003 IFAC*

Keywords: Power plant control, Pressure control, Steam networks, Model predictive control

1. INTRODUCTION

The main task of industrial power plants is to generate and deliver process steam for industrial processes like paper machines, distillation columns, etc. Operation of these plants differs remarkably from that of typical condensing power plants with stable or slowly changing loads. The industrial power plants have to cope with large and fast load changes caused by the operation of the steam consuming processes.

The steam net of the industrial power plant typically consists of several pressure levels. Boilers generate steam to high pressure (HP) header with app. 8 - 10 MPa pressure. Main consumers are connected to intermediate pressure (IP) level with 2 – 3 MPa steam pressure. Low pressure (LP) consumers like preheating and drying processes are connected to 0.2 – 0.6 MPa pressure level. To guarantee a smooth and trouble-free operation of those steam consuming industrial processes, steam pressures and temperatures in different pressure levels should be stabilized as well as possible.

The industrial power plants are typically exposed to large and fast load changes caused e.g. by trips and start ups of the steam consuming processes. The load disturbances are far beyond the control capability of the boilers regulating generation – consumption balance in the steam delivery network. Thus different kind of levelling components and control loops are needed to stabilize steam pressures during the load changes in the steam net.

Requirements of the stable and efficient operation of the industrial power plant lead to a complex control structure consisting of overriding control loops constraining the control actions of basic stabilizing control level. Implementation and tuning of this kind system is difficult and time consuming (Klefenz, 1986),(Majanne, 1999).

Model predictive control algorithms have gained widespread attention during recent years. This is due to many successful applications especially in the chemical industry. (Tinham, 1993), (Warren, 1992), (Oguinnake, 1994), (Oshima, 1994). The principles of the method were developed in the 70's (Richalet, 1976,1978),(Cutler, 1980). An attractive feature of the algorithms is their ability to deal with explicit constraints on system variables Constraints are neglected in the majority of other control algorithms.

MPC can also be applied in fault tolerant control systems (Patton, 1997), (Blanke, 2000). In case of detecting a faulty component in the process, the diagnostic information is used to accommodate or reconfigure the control system to improve the availability of the process. (Majanne, 1993) (Son, 1997).

To implement the MPC system a process model is needed and the control algorithm requires heavy computing. For these reasons the method is not applied widely in the process industry yet. However, nowadays MPC –application packages are available in modern DCS equipment including also tools for process modeling (Metso Automation),(Honeywell). For this reason it is expected that the number of MPC applications will be increased remarkably in the near future.

First in this paper the control problem of steam levelling in the industrial power plant is discussed and the conventional control structure is presented. Next chapter presents the basic principles of MPC. Then an application of a pressure control of a steam network with MPC is demonstrated in a simulator environment. In the next chapter MPC as a tool for process design is discussed briefly. Finally the paper is concluded with comparing the conventional control structure to MPC structure.

2. INDUSTRIAL POWER PLANT

The industrial power plant can be equipped with several different types of boilers with controlled and constant loads. E.g. in Finnish pulp and paper mills the power system typically consists of a constant load chemical recovery boiler and a controlled load multi fuel boiler using e.g. bark, saw dust, peat, and different types of slurries as fuels. Live steam lines of the boilers are connected together with a high-pressure header. Steam pressure in the HP header is typically 8 – 10 MPa.

Steam consumers are operating with lower steam pressures and temperatures. Typical pressure levels are IP–level 2 – 3 MPa and LP– level 0.3 – 0.6 MPa. HP steam is conditioned for the consumers with a back pressure turbine with extraction outlets or with reduction stations. Reduction stations are equipped with pressure reduction valves and temperature control system based on water injection.

The control task in the industrial power plant is to balance the steam generation with the consumption. Steam pressures in different levels (headers) should be stabilized, and the operation efficiency of the plant should be maximized. The importance of the control system is emphasized, because the industrial power plants are typically exposed to large and fast load changes caused by the trips and the start-ups of the steam consuming processes. E.g. a web break on a paper machine may reduce the load of the controlling boiler 20 – 40 % within a few tens of seconds. These

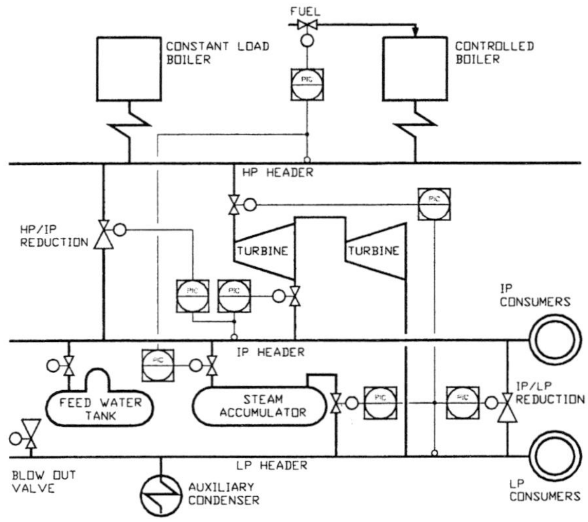

Fig. 1. Typical structure of the industrial power plant with three pressure level steam network.

kinds of disturbances are far beyond the control capability of the boilers and for this reason different kind of load levelling components and control loops are needed.

The function of the load levelling is to smooth the steam consumption by generating artificial load when there is excessive steam in the network and release steam to the network when there is a lack of steam in the network. The process components capable for load levelling are condensing turbine, steam accumulator, feed water tank, auxiliary condenser, and blow out valve.

2.1 Normal operation of the plant

A basic control scheme for the industrial power plant is that the controlled load boiler controls steam pressure in the HP header. Pressures in the IP and LP headers are controlled by the turbine. The extraction flow control splits the turbine inlet flow to IP extraction flow and LP outlet flow. To improve the availability of the power plant the turbine is bypassed by the reduction stations. The reduction stations deliver steam for the consumers when the turbine is not available or its capacity is full. The reduction valves open automatically if the pressures in the IP and/or LP headers drop below the predetermined level.

Compared with the boiler the turbine has much wider control range and much faster control response. In practice this means that the load disturbances caused by the trips and the start ups of the steam consumers could be transferred from the LP and IP headers to the HP header. However, this is not allowed in the full scale, because big disturbances in HP steam pressure can lead to boiler trips caused by the drop of the drum level. Typically the HP header pressure is allowed to fluctuate within ± 5% around the setpoint in order to utilize the steam storage capacities of the boilers. Thus the operation of the turbine controller should be constrained to remain steam pressure in the HP header within the allowed limits.

Also the control actions of reduction stations must be constrained so that pressures in higher level headers won't collapse.

2.2 Load levelling

Steam levelling system can be realized in several ways. The simplest way to stabilize the pressure in IP and LP headers is to blow out the excessive steam when the consumption is rapidly decreased, and keep the blow out valve open until the consumer process is starting up again. From the aspect of energy economics this is the worst concept, because both condensate and energy are lost.

If the turbine is equipped with a condensing section, the steam load can be controlled effectively with production of electric power. Excessive steam can also be stored into the feed water tank by increasing feed water level in the tank. The increased make up water flow increases the preheating steam load. The drawback of this method is that the levelling capacity is quite small and typically it cannot be used as the only levelling component. It is also possible to use the auxiliary condenser to balance the load. It is working in the same manner as the blow out valve, but the condensate is saved. The active use of auxiliary condenser requires that the condenser must be on idle run. This means that cooling water pumps or cooling air fans must be running all the time, which is not energy efficient. The best way to balance the steam load is to use the steam accumulator, perhaps together with the feed water tank.

The steam accumulator is a tank typically connected between IP and LP headers. The accumulator is pressurized between IP and LP levels and half filled with saturated water. If there is too much steam in the network the excessive steam is charged to the accumulator. Charged steam is condensed into the accumulator. Energy released in condensation heats up the water and increases the pressure in the accumulator. The accumulator can store steam until its pressure is equal with the IP header. When there is a lack of steam in the network, the accumulator starts to discharge steam to the LP header. If the disturbance is coming from the LP consumer, the accumulator compensates the disturbance directly. If the disturbance originates from the IP consumer, the turbine control will increase the IP extraction flow and lack of LP steam is compensated by the accumulator. Steam is generated as saturated water is vaporized due to a pressure drop in the accumulator. Vaporization cools the water in the accumulator and drops the pressure. Accumulator

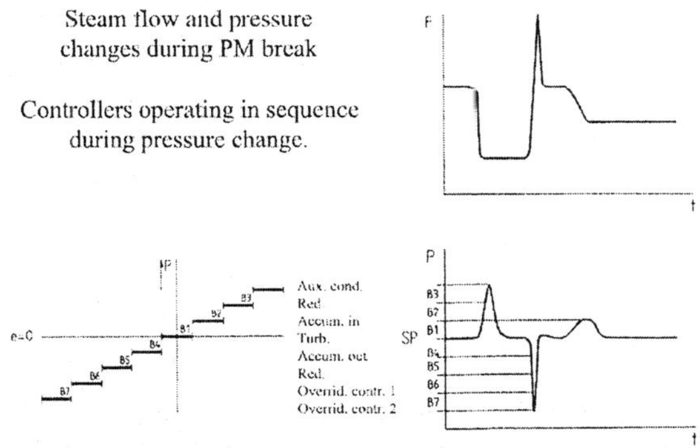

Fig. 2. Sequential operation of the steam levelling components during a web breaks in a paper machine.

can discharge steam to the LP header until the accumulator pressure equals with the LP header pressure.

It is very important to schedule the load balancing actions correctly to achieve good disturbance attenuation with maximal economical efficiency in the network. Fig. 2 shows an example of sequencing the levelling actions as a function of pressure control error. During a web break in a paper machine connected with the IP-net, steam flow decreases and pressure in the header rises. First the turbine tries to control the pressure, but if the control error exceeds the first bias limit, the accumulator starts to charge. If the control error exceeds the next bias level, reduction valve from IP-net to LP-net is opened. If the pressure is still increasing the auxiliary condenser is taken in action.

When starting the paper machine after the break, the flow increases rapidly and the network pressure drops. In the first phase, near the setpoint, the turbine controls the pressure. If the pressure continues dropping the accumulator starts to discharge steam to LP-net. If that is not enough, the reduction valve from HP-net to IP-net opens. If the pressure still drops, the overriding controls constrain the actions of the turbine and the reduction valve preventing HP level to collapse and trip the boilers.

3 MODEL PREDICTIVE CONTROL

The only advanced control methodology which has made a significant impact on industrial control engineering is Model Predictive Control MPC. The main reasons for its success are (Maciejowski, 2002)

- it handles multivariable control problems naturally
- it can take account of actuator limitations
- it allows operation close to constraints, which frequently leads to more profitable operation

Predictive algorithms consist of two tasks: A free response is first predicted over a finite time horizon called the prediction horizon. The free response is predicted assuming that the control variables remain at their current values in the future. The modeling error is taken into account during the prediction. After the free response prediction such control actions are calculated that the actions and the corresponding predicted control errors optimize the selected cost function. The cost function consists of the control efforts, the control errors, and the constraints for control signals and measured process values connected with the control loop. When the control actions are computed, it is usually assumed that only the *M* next control actions are applied. *M* is called the control horizon. Finally only the first control actions are implemented. These steps are repeated at every control cycle (Henttonen, 1996).

4. APPLICATION TO BOILER AND STEAM TURBINE CONTROL

So far applications of MPC in power plants have been demonstrated mostly in simulator environment (Son 1997) (Majanne 1993, 2001) (Hogg, 1990). Applications in real plants are still rare, but there are some for instance about the superheated steam temperature control (Silva, 2000).

In this paper MPC is demonstrated in the simulator environment. MPC is applied to a boiler controlling live steam pressure and a back-pressure turbine controlling extraction pressure and back pressure levels. The constraints interfering in the control actions are integrated in the multivariable MPC-structure. This reduces the number of control loops needed, simplifies the control structure and makes the tuning of the system easier.

Simulator is programmed in Maltlab /Simulink® environment. The simulator library includes nonlinear physical models for typical power plant components like boilers, turbines, etc. The algorithm applied in MPC is a General Predictive Controller, GPC, (Clarke,1987 a,b)). The model of the controlled process is formulated as

$$A(q^{-1})y(k) = B(q^{-1})u(k) + \frac{C(q^{-1})}{\Delta}e(k)$$

$$A(q^{-1}) = I + A_1 q^{-1} + A_2 q^{-2} + ... + A_{na} q^{-na}$$

$$B(q^{-1}) = B_1 q^{-1} + B_2 q^{-2} + ... + B_{nb} q^{-nb}$$

$$C(q^{-}) = I + C_1 q^{-1} + C_2 q^{-2} + ... + C_{nc} q^{-nc}$$

In the demonstration model there is a controlled load boiler generating steam to 10 MPa header. Steam consumers are connected with 2 MPa and 0.5 MPa headers. The consumers are described here as blow out valves. 10 MPa steam is transformed to 2 and 0.5 MPA levels with a back-pressure turbine equipped with extraction outlet. For steam levelling there is a steam accumulator connected between 2 MPa and 0.5 MPa headers. 10 MPa pressure is controlled by the boiler. The Boiler MPC controls fuel feed to the boiler. Controller inputs are 10 MPa header pressure and turbine inlet flow as a feed forward signal. Controller model is identified from the simulated nonlinear process data. Control cycle is 5 seconds and from fuel feed to steam pressure there is a 20 s. delay. The prediction horizon is 300 s. and control horizon is 1 control cycle.

Turbine controller is a 5x2 controller. The controlled variables are 2MPa and 0.5 MPa pressures and control signals are connected to the turbine inlet valve and the extraction pressure control valve between HP and IP sections of the turbine. The three other measurements are the 10 MPa header pressure, the charge and the discharge flows of the accumulator. The 10 MPa pressure is constraining the operation of the controller and the accumulator flow measurements are connected as feed forward signals.

The control actions of the controller are constrained to keep the HP header pressure between the predetermined limits. This means that the turbine controller is not allowed to transfer the load disturbance totally to the HP header. The control cycle is 0.5 s. The prediction horizon is 15 s. and the control horizon 2 s. The controller is configured to constrain the 10 MPa pressure between 9.8 and 10.2 MPa.

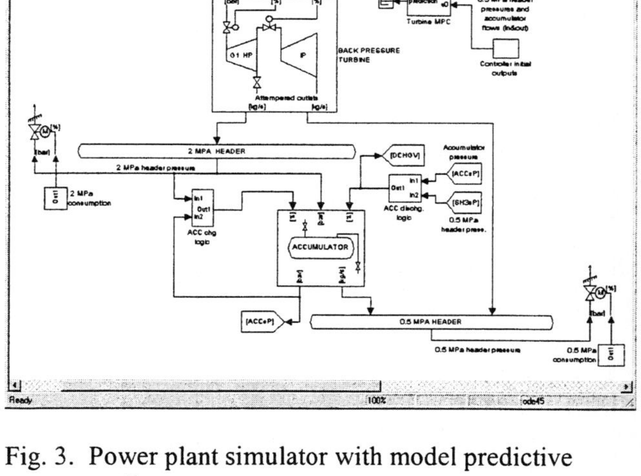

Fig. 3. Power plant simulator with model predictive controllers.

The steam accumulator is controlled by charging and discharging logic. The function of the accumulator is to stabilize steam pressures in 2 and 0.5 MPa headers during the rapid load changes. During the normal operation (no charging or discharging takes place) the accumulator pressure is controlled to 1.2 MPa level. The process simulator with the controllers is shown in fig. 3.

Fig. 4 presents process responses during a load disturbance in the 2 MPa header with the MPC and PI turbine controllers. The boiler controller is equal in both cases (MPC). 2 minutes from the start the load in the IP - level is decreased from 15 to 5 kg/s and 17 minutes after that the load is returned back to 15 kg/s level. The disturbance corresponds with a typical load disturbance caused by a paper machine web break. First the load reduction increases steam pressures in 2 MPa header. The turbine controller transfers part of the disturbance to the 10 MPa header and the accumulator starts to charge steam from 2 MPa header. With MPC the 10 MPa pressure is allowed to fluctuate ± 0.2 MPa. PI controller actuating turbine inlet valve has been limited by an overriding control to keep the 10 MPa pressure between the same limits. With increasing load, pressure in the 2 MPa header drops and the accumulator starts to discharge steam to 0.5 MPa header. The effect of the discharging steam is transferred to 2 MPa header by the back pressure controller. The control results show that MPC can stabilize pressures in IP and LP headers better than the PI controller based system. The accumulator and the boiler are utilized equally in both cases and pressure responses in HP header are almost equal.

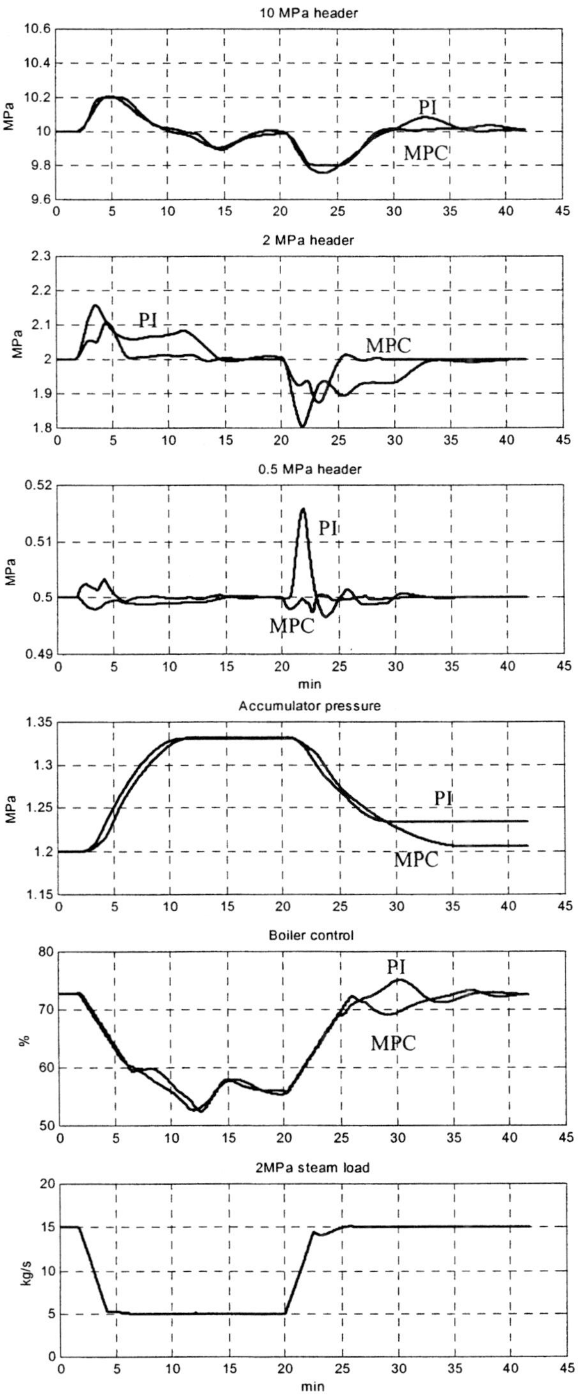

Fig. 4. Process responses for load disturbance with MPC and PI controller based control.

5. MPC AS A TOOL FOR PROCESS DESIGN

A process simulator applied with MPC can be used as a handy tool for process design. When designing the structure of the steam network, the responses for expected load disturbances are simulated with MPC controlled turbine without any steam levelling functions. Simply by changing the pressure controller parameters (weights for control errors in different headers and constraints for pressure limits) the load disturbances can be transferred and shared to different pressure levels. It can be seen easily how much of the disturbance can be transferred to the HP header and how big the response of the rest of the disturbance is in the other headers. This information can be applied to dimensioning the capacity of the steam levelling system.

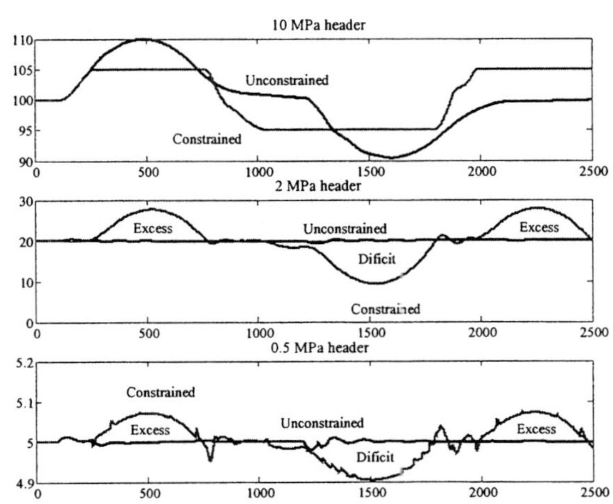

Fig. 5. Simulated effect of disturbance with constrained and unconstrained controller.

Fig. 5 shows an example how MPC can be used in process analysis. First the effect of the disturbance is simulated in a way that turbine controller transfers the disturbance totally to the HP-net. Next the same disturbance is simulated so that HP-net pressure is constrained between acceptable limits of fluctuation. Now it can be seen from the plot how much of that disturbance is left in IP and LP nets and what kind of levelling system is needed.

6. CONCLUSIONS

Model predictive control is a suitable method for controlling processes with strong interactions between process variables (multivariate structure), long delays (predictive) and active constraints (constraints included to the optimization). Industrial power plant is the process having all these features. Conventionally the control of the industrial power plants is realized with individual PI(D) loops with numerous feed forward loops and overriding control actions for realizing the active constraints. This structure leads to a complex solution which is very difficult to test and tune up.

MPC structure simplifies the control structure and makes it easier to commit the system. MPC has "logical" tuning parameters having clear physical meanings like absolute limits and changing speed for process variables. Due to the development of process control equipment, nowadays it is possible to apply this kind of methods even direct in DCS systems. MPC method with process simulators generate also valuable information for process design.

REFERENCES

Clarke, D.W., C.Mothadi and P.S. Tuffs (1987). Generalized Predictive Control - Part I. The Basic Algorithm. *Automatica.* **23** pp.137-148

Clarke, D.W., C.Mothadi and P.S. Tuffs (1987). Generalized Predictive Control - Part II. Extensions and Interpretations. *Automatica.* **23** pp.149-160.

Cutler, C.R. and B.L. Ramaker (1980). Dynamic Matrix Control - A Computer Control Algorithm. *Proc. Automatic Control Conference.* San Fransisco.

Henttonen, J. (1996). Receding Horizon Control Approach to Multivariate Systems. Doctoral thesis. Tampere University of Technology. Publications 192.

Honeywell Inc. Brochures of Honeywell TotalPlant Alcont process automation system.

Hogg B.W. and N.M. El-Rabaie (1990). Generalized Predictive Control of Steam Pressure in a Drum Boiler. *IEEE Transac. on Energy Conversion*, Vol **5**, No 3, September 1990

Klefenz, G. (1986) *Automatic Control of Steam Power Plants.* Bibliographisches Institut, ISBN 3-411-01699-X.

Maciejowski J.M. (2002). *Predictive Control with Constraints.* Prentice Hall, ISBN 0-201-39823-0

Majanne, Y., P. Lautala, J. Henttonen (1993). Autonomous Model Predictive Control System for Power Plant Feed Water Process. *12th IFAC World Congress*, Sydney Australia 1993.

Majanne Y. and K Airola (1999) Handling of Dynamical Disturbances in Industrial Power Plants – Simulation as a part of the Control Design. (in Finnish). *Automaatioväylä*, 7,1999.

Majanne Y., P. Lautala and J. Henttonen (2001). Model Predictive Steam Pressure Control In Industrial Power Plants. *Proc. of the IASTED International Conference on Modelling, Identification and Control.* Innsbruck, Austria, Feb. 19-22, 2001.

Metso Automation Inc. Brochures of Metso DNA process automation system.

Oguinnake, B.A. (1994). On-Line Modelling and Predictive Control of an Industrial Terpolymerization Reactor. *Int. J. Control* , **59** , No 3, pp.711-729.

Oshima, M., I. Hashimoto, H. Ohno, M. Takeda, T. Yoneyama and F. Gotoh (1994). Multirate Multivariable Model Predictive Control And Its Applications to a Polymerization Reactor. *Int. J. Control*, **59**, No 3, pp. 731- 742.

Richalet J., A. Rault, J.L. Testud and J. Papon (1976). Algorithmic Control of Industrial Processes. *Proceedings of the 4th IFAC Symposium on Identification and System Parameter Estimation*, Tbilisi 1976.

Richalet J., A. Rault, J.L. Testud and J. Papon (1978). Model Predictive Heuristic Control: Applications to Industrial Processes. *Automatica* . **14** (1978). pp. 413-428.

Silva R.N., P.O. Shirley, J.M. Lemos and A.C. Goncalves (2000). Adaptive regulation of superheated steam temperature : a case study in an industrial boiler, *Control Engineering Practice*, **8,** 1404-1415

Son W-K, Kwon O-K, Lee M.E. (1997), Fault tolerant model based predictive control with application to boiler systems, *Proc. IFAC SAFEPROCESS'97*, 1240-1245

Tinham, B. (1993). Profitting From Advanced Control. *Control and Instrumentation.* July 1993. pp. 25-29.

Warren, J. (1992). Model Based Control of Catalytic Cracking: *Control and Instrumentation.* July 1992. pp. 33-34.

ELSEVIER
IFAC
PUBLICATIONS
www.elsevier.com/locate/ifac

I&C SYSTEM FOR IMPROVED KSNP (KSNP+) in Korea

Byung-Soo Cho, Hwan-Yong Jung, Sang-Kook Chung

Korea Power Engineering Co., Inc., Seoul, Korea

Abstract: The I&C system design for the KSNP+ utilizes new design approaches and incorporates advanced technologies such as Multiple Loop Configuration, Functional Loop Segmentation, Software-Driven Dynamic Control, etc. The multiple loop controller configuration provides a unique combination of advantages when compared to either a full scale redundant digital control system design or the dedicated single loop controller design. The software-driven control system implemented in the radwaste system design will provide user-friendly and convenient operation environment for the operators. KSNP+ I&C system is expected to enhance the safety and economy through application of advanced technologies. *Copyright © 2003 IFAC*

Keywords: Multiple Loop Configuration, Functional Loop Segmentation, Software-Driven Dynamic Control

A nuclear power plant standardization program is being pursued for the KSNP+, and the first unit is scheduled to start commercial operation by 2008. The I&C system design for the KSNP+ utilizes a new design approaches and incorporates advanced technologies. The KSNP+ I&C design goal is to develop a highly reliable, safe and economical I&C system.

The KSNP+ I&C system is a microprocessor based system to perform both the on-off and analog process control functions. In order to improve the availability, maintainability and reliability of the system, all process controllers are designed with redundancy.

Each loop controller cards are designed to perform multitasking capability or multiple loop controls. In order to take advantage of multitasking capability, plant equipment control loops are grouped and segmented based on plant operational function association, and each group is assigned on a separate and independent loop controller card.

The radwaste control system performs the control and monitoring functions of radwaste process equipment through Operator Interface Stations (OIS). The OIS system is designed and implemented to be a fully interactive software driven real-time on-line dynamic control system.

Figure 1 shows the configuration of KSNP+ I&C system. The main features are as follows:

- Multiple Loop Configuration
- Functional Loop Segmentation
- Software-Driven Dynamic Control (Soft Control) for radwaste control system

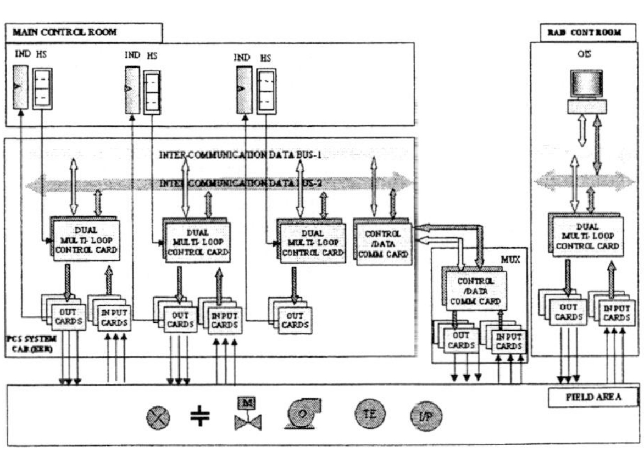

Figure 1. Configuration of KSNP+ I&C system

1. Multiple loop configuration

One of the major advantages of digital control system is the multi-loop control capability which is provided by powerful microprocessors and this multi-loop control capability is widely used in industrial power plants.

However, the safety constraints of nuclear power plants have restricted wide usage of multi-loop control capability for component control. The traditional dedicated control method has been adopted instead, even though a digital control system has become highly reliable.

If the components of nuclear power plant are relevantly grouped based on the functions and configurations of process system, the reliability, functionality and economy can be enhanced by using multi-loop control capability of a digital control system. The number of control loops per a loop controller is decided to maximize reliability, safety, and economy.

KSNP+ has adopted a multi-loop controller approach that would be duplicated with the same functions. Also all the assigned control loops are segmented according to the pre-established guidelines of functional loop segmentation. This approach, the application of multi-loop segmentation and fully redundant controllers, improves the availability and the reliability compared to the conventional dedicated loop controller approach.

2. Functional Loop Segmentation

The analysis of system functional segmentation demonstrates how to improve system availability and reliability by carefully reviewing mechanical flow paths with associated instruments and components and then by establishing a segmentation strategy. In a multiple loop controller design, each controller independently controls a functionally related group of plant components.

Each controller is truly independent of other controllers such that failures, maintenance or testing on one controller does not affect multiple plant functions. Plant components are assigned to controllers in a manner that preserve the redundancy and independence that exist within the mechanical process system being controlled. Components that are mechanically redundant or independent have their associated control functions physically separated into another controller.

This functional segmented grouping is to prevent a failure of one controller from affecting more than one group of equipment. However, in case that splitting of any one group due to a controller capability limitation with respect to the number of associated inputs/outputs, controller response time, etc., more than one controller are provided.

Each loop controllers and associated I/O Module are provided with a dedicated communication interface to the system network.

Where component redundancy exists within a mechanical system, the redundancy is maintained by assigning the redundant component to a separate controller and I/O module. Each I/O module is dedicated for each control loop and independent from other I/O modules, such that failures do not propagate to other I/O modules within a same segmented group.

The functional grouping design, which is the right compromise between economy and reliability, provides a significant increase in MTBF while reducing plant capital requirements simultaneously.

2.1 Power Source Segregation

Plant components and the interfacing instrument loops are divided into the four (4) safety divisions (AE, BE, CE, DE) and two (2) non-safety divisions (AB, BB). Components belonging to each division are grouped and assigned to a unique microprocessor.

Each division AE, BE, CE, DE, AB, and BB will have the required number of groups depending upon its ability to satisfy the design philosophy.

2.2 Functional Group Segmentation

Within the constraints of the division assignments, various functional groupings are created to accommodate the design philosophy. In summary, the functional groupings, which are broken down into group control segments and loop controller component assignments, will support the followings:

- Take advantage of the redundancy of flow paths.
- Take advantage of the redundancy of components.
- Maintain the independence of flow paths.
- Maximize the failure tolerance by factoring the result of the hazard analysis into consideration and the functional groupings are also consistent with:
    - The plant piping and instrumentation diagrams (P&IDs).
    - Fluid system design instrumentation and component control design requirements.
    - Component design requirements.
    - Fluid system design operational requirements.
    - Instrumentation and control requirements for the PCS.
    - Interface requirements for component controls and process instrumentation.
    - System operational requirements.
    - Component technical manuals/system descriptions/specifications.
    - General arrangement drawings.

After initial assignments are made, components are rearranged as necessary to form more compact and cohesive groups as groupings of other systems and components progress. Final group assignments are made based on a review of a system failure mode and effects analysis.

To simplify the discussion in the following functional groups, division SR-A is designated as A1, A2... AN. Similarly the other division groups are designated as follows:

SR-B - B1, B2... BN
SR-C - C1, C2... CN
SR-D - D1, D2... DN
NSR-A - NA-1, NA-2... NA-N
NSR-B - NB-1, NB-2... NB-N

The PCS architecture and the group configuration are shown in Figure 2. Each block represents a group. The divisions are shown to have multiple groups in Figure 2. Each division will have the required number of groups depending upon its ability to satisfy the design philosophy.

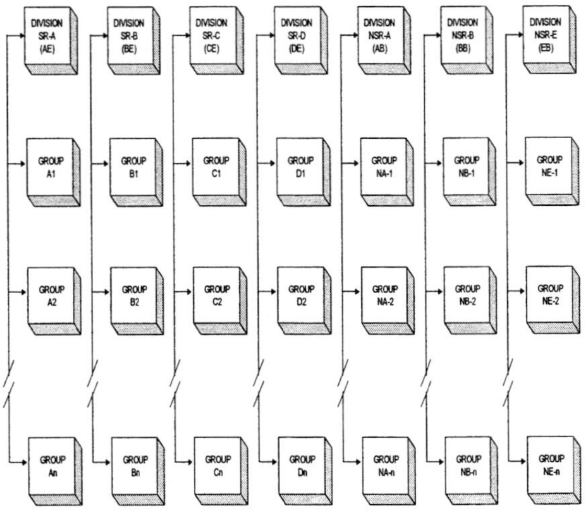

Figure 2. PCS Architecture and Functional Group Configuration

2.3 Group Assignment Methodology

The following is a general procedure for the assignment of the plant systems to PCS subgroup (controller).

- Identify the plant systems controlled by the PCS.

- Identify the electrical classification of 1E or non-1E to define assignment to PCS.

- Define the PCS division assignment for a system.

- Define the system function and identify the system functional independence requirements.

- Assign the system components to PCS subgroup segments to support the requisite system functional independence according to the Functional Group Segmentation.

- Identify the component redundancy and the control power independence requirements.

- Assign the control loops (components and analog loops*) to loop controller as follows:

  - Up to eight (8) control loops per loop controller for safety-related loops.
  - Up to sixteen (16) control loops per loop controller for non-safety-related loops.

2.4 Typical Examples of Functional Group Assignment

The valve A, B, C and D provide seal injection flow to the Reactor Coolant Pumps 1A, 1B, 2A and 2B on Divisions AB, BB, AB and BB respectively as shown in Figure 3. Each valve is assigned to a separate non-safety related group to maintain independence of each RCP. The group assignment is also made consistent with the RCP power train:

- VALVE A RCP-1A Group NA-1 Division (AB)

- VALVE B RCP-1B Group NB-1 Division (BB)

- VALVE C RCP-2A Group NA-2 Division (AB)

- VALVE D RCP-2B Group NB-2 Division (BB)

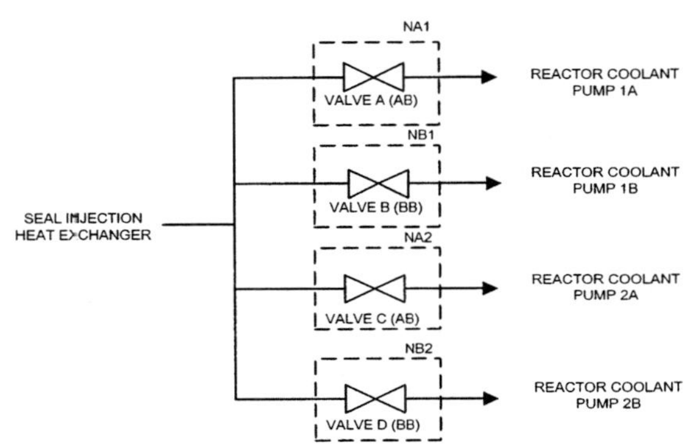

Figure 3. Typical Examples of Functional Group Assignment

The boric acid makeup flow is drawn from the RWT via valve A and the two pumps as shown in Figure 4. Since the valve A, on the main fluid path, provides flow to both pumps which are on Division BB, it is prudent to assign this valve also to Division BB. Since the valve is on the main fluid path it is assigned to group NB-1.

The two pumps are both on Division BB per CLDs and are therefore assigned to separate groups for independence. The pump 1 is assigned to group NB-1 and pump 2 to group NB-2.

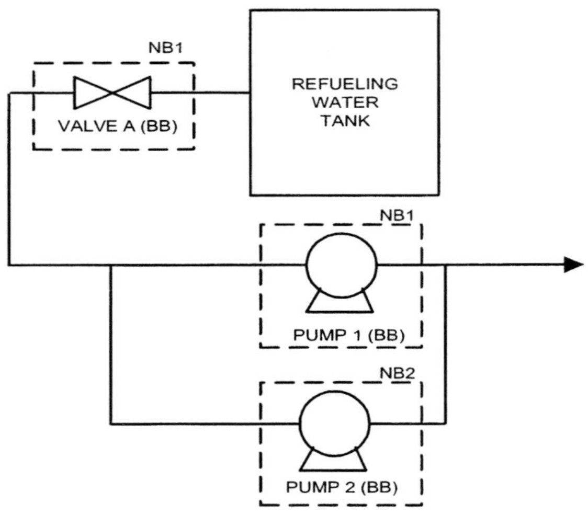

Figure 4. Typical Examples of Functional Group Assignment

3. Soft Control System

The control system for the radwaste systems incorporates a fully interactive software driven real-time on-line dynamic control (herein after called "soft control") system. The soft control system is a completely stand-alone system and it replaces the functions provided by the conventional hardwired dedicated control and monitoring devices.

The soft control OIS is provided with all information and equipment necessary to monitor and control the systems. The soft controls provide process control capability to the operator. The soft controls utilize graphic control faceplates displayed on the Control Function LCD emulating the various conventional physical switches and analog control device functions.

The soft control faceplates are exact graphical replication of the operator console hardwired dedicated control switches, M/A station, other on-off and analog control and indication devices. This is for plantwide uniformity and consistency as well as operator's familiarity. These control faceplates are selectable from system graphics using a trackball.

The system graphic displays consist of graphic representation of the radwaste system. The display of this graphics is in a hierarchical order starting from a high-level system overview down to a detailed low-tier component-level representation based on P&ID, C&ID and CLDs.

The large amount of plant data available to the operator makes it imperative for a means of rapid, logical, convenient and reliable call-up of monitor displays. All displays are selectable for call-up from one or more index or menu displays using cursor control techniques and specific display names.

This scheme does not require keyboard entry of data when selecting a choice from a menu. Many of the displays required for plant operation must allow the operator to enter data values. Data fields that are alterable by the operator are highlighted in some fashion and are preceded by an undisplayed tab stop to permit the operator to tab or use the operator interface device to position the cursor to the beginning of the data field swiftly and accurately.

The displays provided in soft control system are as follows:

- Overview displays
  Overview display provides a summary of the status of plant operation.

- Process graphic displays
  Process graphic displays provide a complete overview and control capability, as applicable, of the process system, with dynamic data clearly indicated such that the operator needs not refer to multiple diagrams to assess system operation.

- Trend displays
  The system is capable of supporting both historical and real time trends. The trends show in a graphic form. All analog points and calculated values addressable by the system are assignable for historical and real-time trend.. All information associated with each trend point in one trend is color-coded and configurable by the operator.

- Barchart displays
  The system have the capability to display horizontal and vertical bar charts. It is possible to assign any analog point to a barchart display.

- Diagnostic displays
  Diagnostic displays provide the operational status of each device in the PCS including the communications network, LCD monitors, keyboards, printers, controllers, I/O cards, and I/O points.

The system includes "select-before-operate" verification procedures for control actions, automatic timed cancellation of operator actions which are not completed, time out conditions for alarms generated for system or operator action as well as procedures for inhibiting control of a device via a tagging operation on the LCD monitor. The operator is able to terminate any sequence of operation at anytime prior to the completion of the final executable action with no adverse affect on the system.

4. Conclusion

The multiple loop controller configuration of the I&C system design for KSNP+ provides a unique combination of advantages when compared to either a full scale redundant digital control system design or the dedicated single loop processor design. Some of the advantages include a significant improvement of the reliability and the availability which can support the improved plant capacity factor while simultaneously reducing the plant construction and O&M (Operational and Maintenance) cost by the minimizing hardware.

Further, appropriate attention to the multiple loop functional design during preparation of plant maintenance procedures can also lead to improvements in operational safety. Clearly, the multiple loop design approach provides the opportunity for prudent and highly efficient use and application of engineering resources. The soft control system implemented in the KSNP+ radwaste system design will provide user-friendly and convenient operation environment for the operators.

KSNP+ I&C system is expected to enhance the safety and economy through application of advanced technologies such as Multiple Loop Configuration, Functional Loop Segmentation, and soft control for radwaste control system.

REFERENCES

SKN 1&2 Project, Segmentation Guideline for Multi-Loop Controller (9-700-J485-001), Korea Power Engineering Company, Korea, 2002.

SKN 1&2 Project, System Functional Description for Plant Control System (9-771-J403-001), Korea Power Engineering Company, Korea, 2002.

SKN 1&2 Project, System Functional Description for Radwaste Control Room (9-753-J403-001), Korea Power Engineering Company, Korea, 2002.

ELSEVIER
IFAC
PUBLICATIONS
www.elsevier.com/locate/ifac

An Improved Human Machine Interface modernization for a Combined Cycle Power Plant

Alfonso Rivas-Suarez **Víctor Alvarez-Cortés**
Carlos E. Uribe-Blanco **Marino Sanchez-Parra**

Electrical Research Institute
Control Systems Division – Process Supervision Department, 29-2. P.O. Box 1-425.
Cuernavaca, Morelos 62001, Mexico.

ABSTRACT

In order to fulfill the Mexican Electric Utility's (CFE) modernization demands, the Electrical Research Institute (IIE) developed at the Process Supervision Department an improved Human Machine Interface (HMI) modernization for a Combined Cycle Power Plant (CCPP) to replace the old one based on VAX workstations with a microcomputer base system using personal computers(PC), maintaining the same functionality and integrating management capabilities as well as processing functions currently performed by a proprietary system. The HMI modernization provided to CFE an integral solution through a single interface allowing the system to interface not only to current proprietary controllers but also to commercial controllers, PLCs or other Intelligent Electronic Device by integrating OPC technology. Data analysis is possible in the improved HMI due to a powerful database back-end engine added to the system. Integrating new technology to legacy systems in order to preserve its life span and be able to integrate modern devices represented a challenge to be resolved. This paper also describes how suitable these new technologies promises were accomplished in the IHM modernization. Improvements to the HMI also included an intuitive color-base navigation in a multi-level model. Improvements in communications involved the replacement of the current PC-based communication concentrators with a pair of high performance servers to improve the overall system flexibility and robustness. Among the more challenging tasks was the integration of a modern commercial redundancy system to work with current proprietary controllers. *Copyright © 2003 IFAC*

Keywords: man/machine interface, microcomputer based system, distributed control, technology transfer, tests.

1. INTRODUCTION

The Combined Cycle Power Plant (CCPP) where the modernization described in this paper was performed, is a generation package formed by three generation units as follows: two gas turbines (GT) generators and its heat recovery steam generators (HRSG) and a steam turbine generator. HRSGs are outdoors while the three turbine generators are indoors.

Generating Units 1 (GU1) and 2a re Gas turbine generators designed to operate using natural gas or gas oil and GU3 is a steam generator (SG) fed with the steam generated in GU1 and GU2.

GU1 and GU2 capacities are 59 MW each and GU3 is 82 MW given a total result of 200 MW installed capacity].

As Garduno and Sanchez (1995,1996) described, the CCPP has a digital distributed control system (DDCS) whose main components are the man-machine interface (MMI) and the all-digital control. The MMI is a set of five workstations where all command and monitoring actions are performed. It is linked with the all-digital control with a gateway based on a PC-based communication concentrators.

A crucial requirement for the HMI system modernization was to maintain the proprietary digital controllers (PDC) achieving monitoring and automatic control functions, actuating directly over the plant elements. Each PDC is backed up by another one with identical characteristics in order to get a highly reliable system.

Currently there has been a trend followed by main vendors of process control system to migrate their product lines to Microsoft Windows platform i on its different versions NT/2000/XP, mainly because is a de facto standard that reduce in a great degree the Total Cost of Ownership, due to the fact that there are different manufactures for PCs, monitors, data acquisition boards, etc, that are totally compatible with this platform, allowing the user to select the best provider without worrying about the devices co-existence.

It was proposed to replace the VAX workstations with Pentium PCs , VMS operating system for Windows 2000 Advanced Server operating system and the Sherryl-Lubinsky graphical editor for a GE commercial software, keeping the same functional scope currently available in the HMI Additionally it was required to integrate alarm management and processing functions as well as historical data collecting and storage, real-time and historical data plotting and the mimic board displayed on screens.

2. NEW HMI SYSTEM ARCHITECTURE

New HMI system developed for the CCPP is based on a client-server architecture with an open hardware platform to provide a powerful automation solution, involving from equipment programming and visualization to batch processing, and supervisory control to the enterprise ERP. Platform selected is also a standard software in industry that permits to concentrate, distribute, control and visualize process information in real-time in all the different enterprise levels, giving to the user an scalable solution -- from a single node to a multi-node where several operators stations share the HMI system information, increasing productivity as well as the Return On Investment.

Main architecture components are:

- 2 servers.
- 11 clients/viewers.
- 15 proprietary digital controllers.
- 2 redundant serial communication networks.
- 1 Fast Ethernet HUB 100Base-TX.

2.1 Servers

The system architecture is based in two high performance servers, operating in a redundant scheme (primary and secondary) in command of concentrating, processing and storing large information quantities to share them through a Ethernet network with a number of viewers.

Main hardware characteristics for each server are:

- Designed for a 24/7 operation (24 hours/7 days a week).
- SCSI disk to meet the rigorous performance demanded by the historical data logging.
- Redundant Hot-Swap power supply.
- Dual Pentium IV processor.
- 512 MB in main memory.
- S-485 Multi-port communication board
- 10/100 mbs Fast Ethernet card.

The software executed on both servers gather data coming from sensors and plant devices, process it and transform it into dynamic objects as text, alarms and graphical displays, allowing the users to have access to information in real-time. Real time information provides the means to better decision-makings avoiding critical situations before they occur, improving quality, productivity and profitability for the plant.

In order to fulfill the established functionality for the system, it was required to install, configure and integrate a commercial software as well as software developed for this application. Commercial software included Windows 2000 operating system, a commercial GUI platform, a Relational Database Management System and development tools for programming in C++ as well as tools for development of Windows based application. Application software developed included the HMI modules and a serial communication manager.

2.2 Servers Functions.

The main functions performed by the servers are:

- Data polling from controllers
- Storage and processing of large amount of information.
- Redundancy capabilities.
- Share information services to clients (i.e. Viewers).

HMI system is executed in a multitasking scheme, where different tasks for data acquisition, historical data logging, process displays to operators, Ethernet s as well as serial communication are all executed in a seamlessly way.

The Serial Communication Manager (SCM) performs data collecting from controllers. SCM also is in charge of controlling the two defined networks called as “Alpha” and “Beta”. Each network has two communication channels for redundancy purposes. SCM application controls configuration, start-up, transfer and stopping of each network as well as the controllers polling and synchronization, managing channel usage and deciding and achieving the channel switching between channels and /or networks.

Both communications networks are connected to proprietary controllers and both networks have the same characteristics and functionality operating simultaneously as “primary” and “secondary”. Information between Servers and controllers is spitted up between both networks, so in case of a failure in a network, the other one is able to keep alive the communication and take control of the whole information transferred.

The SCM application uses a standard protocol for process information exchange (OPC/DDE) to perform the bi-directional information transfer between the controllers and the HMI system.

Redundancy implemented on the servers assures that the system will keep on operating in a case of a hardware failure in the primary server. Redundancy is programmed and configured in such a way that secondary server is ready to automatically handle a seamlessly transfer of all critical functions such as data acquisition, alarms, historical data logging and security operations in the event of a failure without lose of information.

In normal operation, primary server concentrates data coming from controllers and updates this information the secondary continuously, reducing polling load to controllers and improving the response time. In a failure event, secondary server automatically initiate data polling from controllers, assuring that alarms and point values are refreshed without interruption to the operator.

2.3 Viewers

Viewers are connected to server through a star network configuration so the information stored in the server is freely available for operators, engineers, maintenance personnel who interact with the process in real-time as well as the management personnel who analyse on line or off-line historical information regarding the plant behaviour.

2.4 HMI Software characteristics

Some of the features supported in the HMI system are described below:

1) Provides graphical objects (bars, switch, LEDs, tables, charts, trends, etc) to monitor and operate the HMI system as well as the integration of standard graphical objects through OLE/Active technology.

2) Object Library contains industrial elements such as pumps, valves, tanks, etc. as well as the capacity to edit and extend it.

3) Support for creation and embedding of static graphics in GIF, JPG, BMP and TIF format and animation in data displays as well as multimedia content (sound and images in motion).

4) ODBC connectivity to its tags and values.

5) Data acquisition scan rate and refresh on datad isplayed on screens is 1 second per 5000 system variables.

6) Communication drivers to PLCs and DCS and Intelligent Electronic Devices (IEDs) from different manufacturers as well as support for standard protocols such as DDE y OPC.

7) Real-Time automatic data process logging to standard file formats such as TXT and CVS as well as the logging of all interaction between the system and HMI operators.

8) Alarm management that allows the user to classify, acknowledge, organize, filter and display alarms for specific users.

9) Capability to interpret and execute programs using a scripting language.

10) Provides security levels and privileges based on users role.

11) All HMI displays are available to be accessed by an operator via Web using just a web browser as long as permissions and privileges are appropriate.

12) Functions to achieve data capture, evaluation and statistics as well as batch and control sequence to create, configure and execute a conditioned sequence of actions.

3. HMI SYSTEM NAVEGATION

The HMI system allows the operator to locate and access in a fast, efficient and intuitive way all the options and functions available in the system to perform the monitoring and control tasks.

3.1 Multi-Level Navigation

HMI navigation was designed following a multi-level approach where is possible to execute the commands, control actions and data displays pertaining to the level discussed.

Levels managed in the navigation include:

Plant Level

Commands and data displayed on screens have a general scope, i.e. to the whole plant such as the start-up and stop sequence, redundancy control, one-wire diagrams for the plant, plant alarms, etc.

Sub-system Level

Commands and data displayed on screens have a local or subsystem scope. Sub-systems identified in the HMI are:

- Generating Unit #1 (GU1) includes Gas Turbine 1 and HRSG1.
- Generating Unit #2 (GU2) includes Gas Turbine 2 and HRSG2.

- Generating Unit #3 (GU3) is a Steam Turbine

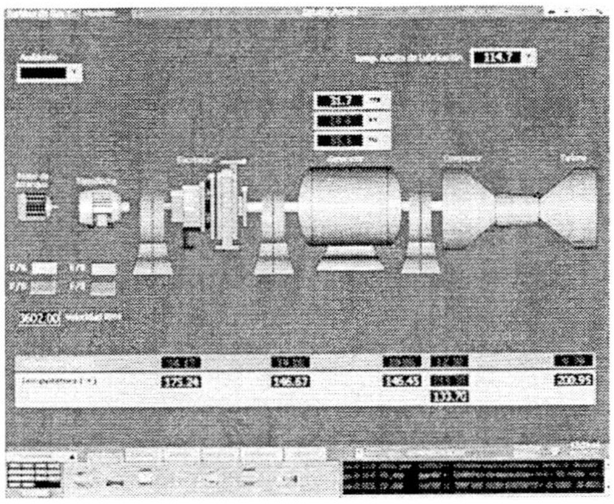

Fig. 1. Gas Turbine in-line monitoring

The HMI Navigation is a logical division the plant into different Generating Units (GUs) which includes sub-systems with defined colours, making easy to the operator the context switching from a general level (plant) to a particular one as well as the context switching between different sub-systems. This scheme provides a safety sense to the operator by providing a clear scope of the commands to be executed as well as the displayed data.

To keep informed the operator regarding its position in the navigation, a "Road-map" display screen is included very similar to those seen in web sites and portals, so that operator can easily locate any display (screen) with a single mouse click.

The HMI navigation layout was designed taking into account that that from any screen display, the operator be able to simultaneously view all the events and alarms generated in the plant in a related way.

HMI screens have defined the following display areas:

<u>Localization Area</u>

Located on the left upper corner shows the operator the displayed data position in the following format: Sub-system | Option | Screen (i.e. Gas Turbine 1 | Models | General Operation).

<u>Main Area</u>

Main area for displaying process diagrams, tabular data, trending, etc. and exhibits about 90% of the displayable area.

<u>Options Area</u>

Available option are: Models (Process Diagrams, Panels (for control), Alarms, Historical data, Trends, Tabular data. According to the level and sub-system a command can have a general or local scope. Each option is identified visually by an icon resembling the screen displayed so its localization is easy for the operator.

<u>Alarms Area</u>

Shows all the alarms generated within the plant and is visible from any display screen regardless of the sub-system or level.

<u>Controllers Status Area</u>

Shows the communication status between the servers and the controllers. It also provides information regarding which controllers are operating and the redundancy status, i.e. which server is operating as primary.

Taking into account that the CCPP operation is achieved by different operator with diverse background and navigation preferences, the HMI navigation was designed to be totally accessible with a mouse device or with a keyboard. However experience has shown that both mechanisms are complementary.

It was also designed an innovative position-oriented mechanism to access all the functions with the keyboard by remembering the icons position and using a few letters (A-to-J) and numbers (0-to-9), eliminating the need for the operator to memorize a large number of keystrokes.

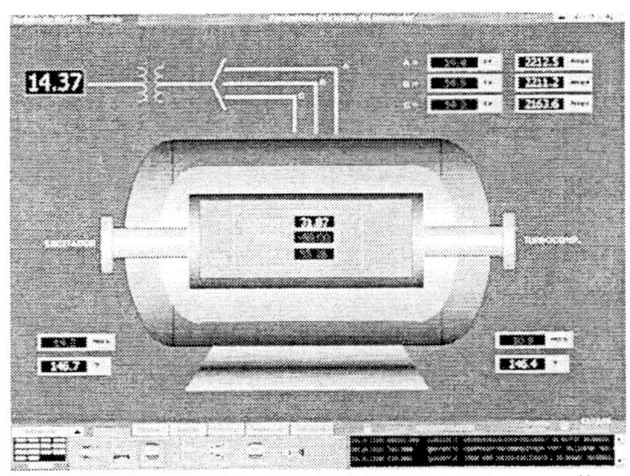

Fig. 2. Gas turbine electric generator on-line monitoring.

3.2 Models (Process Diagrams)

Defined process diagrams to carry out control and monitor functions for the plant. Models options are available for each GU. The default screen for each GU displays the "Current Status" screen display, where the most critical and relevant parameters for the sub-system are shown. Other models screens available are: General Operation, Generator's parameters, Temperatures and Vibrations, etc.

3.3 Alarm management

There are three different alarms displays for each sub-system (i.e. GT1, HRSG1, GT2, HRSG2, ST) classified as analog, digital and both. Alarms screens include information regarding the alarms generated, its date, time, sub-system, alarm ID, state, description and history. Two modes are available to handle alarms: static where a snapshot of the alarms status is taken allowing the operator to deal with them, and the dynamic mode where new alarms are displayed as they are generated in real-time.

3.4 Tabular Data Displays

Displays in a tabular format the field variable values in real-time. Tabular displays are categorized in analog and digital variables using colours to indicate its status. Another developed screens displays tabular information graphically in horizontal bars and alert the maximum value reached by every variable. Maximum values can be reseted in order to refresh information.

4. HISTORICAL DATA LOGGING

Theca pability of a system to monitor and display information and to interact with the system is as important as its capability to store this information to facilitate its analysis to gather knowledge of the process and consequently help in the decision-making to improve the plant quality of service.

The new developed HMI system provides to operators or administrator the required mechanisms to attend the plant behavior in the time through an analysis of the analog and digital signals values acquired from the field devices. To achieve an analysis of the acquired data, the system was equiped with capabilities to store in a massive storage media the acquired data so that later on this information can be retrieved from the HMI in a fast and secure way or from any other application connected to the network. HMI operators can display this stored information in a graphical or grid format. Graphic format shows the historical data of the analog signals values and its behavior in time. This information is useful to verify the behavior of a group of analog signals and its expected trends. Data grids show information in a tabular format similar to those displayed in a MS Excel spreadsheet.

The Relational Database Management System (RDBMS) integrated to the system is a powerful tool for performing the historical data logging in real time. The characteristics of this powerful and versatile RDBMS are:

- Full management of the Relational Model.
- Support for ODBC (Open Database Connectivity).
- Based on open standards since the logged information can be shared and used by other systems.
- High-access speed to log data in order to comply with the required time constrains.
- Capacity to create and store a large number of tables
- Capacity to safety store a great amount of information.
- Capability to schedule the execution of a sequence of task oriented to the database maintenance.

The HMI system carry out the historical data logging for a time frame of several years and logging frequency of every minute and every second. The designed data logging scheme requires that the system administrator define a set of groups and the time frequency to be logged (every minute or second) to the database. Administrator then assigns the variables to be included in each group. The developed system defines two types of groups: first group stores data logged every second and the second group stores every minute. There are 40 groups logging every second and 40 groups logging every minute and every group has an average of 20 variables, giving a total of 1600 variables to be logged.

The system provides mechanisms to perform maintenance activities in a scheduled way, according to the plant requirements Several SQL scripts were coded and configured to keep only two years of historical data, so the database will be able to store information from two years ago to the current date. This capacity will avoid that server's hard disk get filled with obsolete information.

4.1 Real-time trending

The improved HMI system also provides the required mechanisms for operators or administrators to observe the plant behavior in real-time, through the analysis of analog and digital signal values that are being acquired. The analysis is achieved through a graphical tools where is possible to configure a group of variables to be monitored and plotted on a screen to visualize them in real-time. Each group, its variables and its configurations are stored in a file that can be displayed later from the HMI navigation.

A trending tool was integrated to the system and some of the more important features are:

- Capacity to group variables to be monitored.

- Groups can have unlimited number o variables.
- Add and delete new lines to be plotted in any group and modify its color, and style in run-time or off line.
- Capacity display data such as: Title, name of lines.
- Capacity to associate a new line to a variable, current value, engineering units and date.

The same groups of variables are stored for historical data and for real-time trends, so the operator can watch on one screen the historical data and the same variables in the real time trends of the same group of variables.

5. CONCLUSIONS

Benefits.
The developed HMI system replace the whole functionality provided by the two old-proprietary systems and additionally improved several components of the system in the following areas:
1) Capacity to gradually include and replace the proprietary controllers for commercially available devices.
2) Use of inexpensive commercial hardware compared to VAX technology.
3) An improved navigation based on color usage and logical levels, allowing the operator an easy and comfortable operation of the plant via the mouse device, the keyboard or both.
4) Capacity to collect and store a large amount of information for analysis purposes.
5) Information stored is available to other system due to its openness to standards.

Integrating the RDBMS to the system to comply with the time constrains involved a fine-tuning of the RDBMS engine among other problems.

To accomplish a commercial redundancy scheme with the old system required a lot of testing and customization. The redundancy implemented is acceptable but improvements are desirable.

OPC and DCOM proved to be a valid technology to shared and transfer data from the application to the field devices and between the distributed clients (viewers) on the Ethernet network and able to comply with timing demands for a mission critical system.

6. REFERENCES

Garduño R., R. and M. Sanchez P. (1995). Control System Modernization: Turbogas Unit Case Study. *Proceedings IFAC Symposium on Control of Power Plants and Power Systems.* Volume II, pp. 245-250.

Sanchez P., M. and L. Castelo C. (1996). Coordinated Control System Modernization for a Combined Cycle Power Plant., *ISA (International Society for measurement and control). Advances in Instrumentation and Control.* Volume 51, Part 2, pp.1365-1373.

Tieming L., Nyberg T.R., Gang X., 2000. Technologies and Solutions to Integrate Control Networks with Data Networks. Proceedings 4th. Asia –Pacific Conference On Control and Measurement. IEEE.

ELSEVIER
IFAC
PUBLICATIONS
www.elsevier.com/locate/ifac

ADAPTIVE AND INTELLIGENT SYSTEMS FOR GENERATOR MONITORING AND PROTECTION PURPOSES

Waldemar Rebizant, Janusz Szafran [1)]
Seung-Jae Lee, Sang-Hee Kang [2)]

1) Wrocław University of Technology, Poland
Wyb. Wyspianskiego 27, 50-370 Wroclaw, Poland
2) Next Generation Power Technology Center, Myongji University
Yongin, 449-728, Korea

Abstract: New adaptive scheme for signal parameter measurement for generator monitoring and protection is presented. The algorithm is based on the orthogonal components of currents and voltages obtained with adaptive orthogonal filters. The filter data window length and coefficients are adapted according to coarsely estimated signal frequency. As a result, the measurements can be performed with high accuracy in wide frequency band conditions, i.e. also for generator start-up, etc. The measured signals are further used for training of artificial neural networks with the aim to design a robust and effective classifier of generator operation mode. The main attention is paid to the phenomena accompanying pole slipping and out-of-step conditions. *Copyright © 2003 IFAC*

Keywords: digital measurement, adaptive algorithms, artificial intelligence, neural networks, genetic algorithms, generator protection, simulation.

1. INTRODUCTION

Most of the digital measurement algorithms used for power system control and relaying purposes are designed to operate at fixed nominal frequency of input signals since frequency deviations in normal power system operating conditions are very small allowing to get proper accuracy of estimation. There are situations, however, when protections have to operate in serious off-nominal frequency conditions and, if their measuring algorithms are not designed properly, they have sometimes to be interlocked, thus impairing protection system reliability. Examples of such cases are start up (warm up) of a generator or operation mode changing form generating to motoring state (for reversible machines).

In digital protection systems the signals which the decision whether to trip the faulty element is based on are usually determined from orthogonal components of current and voltage phasors obtained by use of non-recursive digital filters. Selective frequency response of the filters and resulting selective features of estimators make them inadequate for application in off-nominal frequency conditions. Contradictory requirements to be selective (suppression of noise) and unselective (to have all-pass unity frequency response) call for adaptive features of the estimators. The idea of adaptive estimation has already been suggested in papers (Winkler and Wiszniewski, 1995; Rebizant, *et al.*, 1999; Moore, *et al.*, 1994).

In this paper a new approach to the problem of signal parameter measurement in off-nominal frequency conditions is presented. The adaptive estimators based on frequency deviation estimation are described. According to the frequency measurement results the orthogonal filters are tuned up to the new frequency band by modification of their data window length and coefficients. The proposed method of adaptive measurement is simple and effective in use.

In the second part of the paper the new intelligent system applying Artificial Neural Networks (ANNs) to protection of synchronous machines against out-of-step (OS) conditions is presented. The out-of-step conditions (loss of synchronism) of a synchronous machine may occur as a result of loss of excitation or during pole slipping. Both effects may on the one hand threaten power system stability and on the other hand cause severe mechanical and thermal stresses to the generator itself (Elmore, 1994). Appropriate protection schemes against OS conditions are applied to avoid the threats mentioned, especially for generators of higher ratings (above 200 MW).

The newly developed and presented here ANN-based OS protection ensures faster and more secure detection of generator loss of synchronism when compared to traditional solutions. To determine the most suitable network topology for the recognition task a genetic algorithm (GA) is proposed (Rebizant *et al.*, 2001). The final ANN structure is found using the rules of evolutionary improvement of the characteristics of individuals by concurrence and heredity. The initial as well as further consecutive populations of ANNs were created, trained and graded in a closed loop until the selection criterion was fulfilled. The results of thorough testing of the designed protection scheme with the signals generated using the Alternative Transients Programme (ATP) are described. The scheme is able to detect but also to predict the coming OS cases with high selectivity. The decision is taken some 500ms-1s earlier than with standard impedance based protection schemes, which gives additional time for appropriate actions to prevent serious generator damage and longer outages for repairs and thus to improve reliability of energy supply to customers.

2. ADAPTIVE MEASUREMENT SCHEME

In this section basic algorithms of signal amplitude measurement as well as their extension for wide frequency band estimation are described. The adaptive approach proposed can be easily applied for measurement of other criterion values used for generator protection (impedance, power components, etc.), as reported in (Szafran, 1999).

2.1 Measurement Algorithms

For measurement of power system signal amplitudes (e.g. voltage $|U|$) the following digital algorithms are usually used:

$$|U|^2 = u_c^2(n) + u_s^2(n) \qquad (1)$$

$$|U|^2 = \frac{1}{\sin(k\omega_1 T_s)}\left(u_c(n-k)u_s(n) - u_s(n-k)u_c(n)\right) \qquad (2)$$

where u_c, u_s are voltage phasors at the orthogonal filter outputs (for instance full-wave Fourier ones), ω_1 is the angular fundamental frequency, T_s is the sampling period and k is a number of delay samples (chosen from the range 1 ... N/4, with N being number of signal samples in one period of the actual fundamental frequency component).

In order to minimize estimation errors in case of application of estimators (1) or (2) for wide fr frequency band measurements (e.g. for generator monitoring and protection), it is necessary to adapt the orthogonal filters to the actual frequency of the signals. It is realized by the following procedure:

- determine the number of samples in one period of input signals N (directly or by signal frequency estimation),
- set the filter data window length to N and modify filter coefficients (i.e. the filter impulse response).

Additionally, in case of estimator (2), the value of k resulting from the currently estimated signal frequency is forwarded to the measurement block, thus on-line updating the previously used delay value. The procedure of adaptive measurement can be represented by the block diagram shown in Fig. 1.

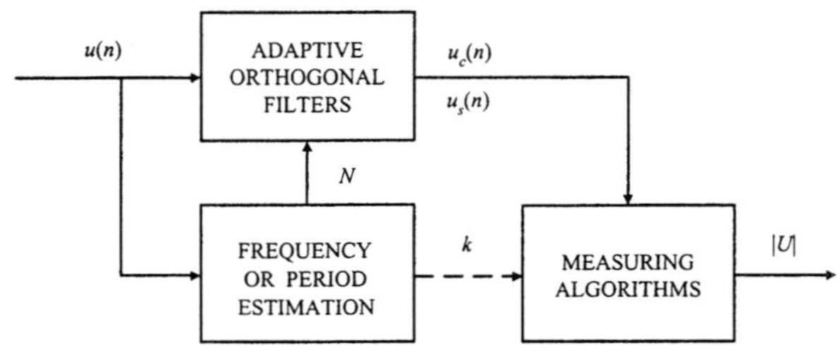

Fig. 1. Block scheme of the amplitude measurement with adaptive orthogonal filters.

2.2 Frequency Deviation - Based Adaptive Scheme

The adaptive estimator presented is based on calculation of the following function (written here for an arbitrary voltage signal):

$$g(\omega) = 0.5\frac{u(n-2k)u(n-d) - u(n)u(n-2k-d)}{u(n-k)u(n-d) - u(n)u(n-k-d)} \qquad (3)$$

where d is certain time delay (d>0).

It can be shown that $g(\omega)$ is proportional to the frequency deviation, i.e.: $g(\omega) \cong -kT_s\Delta\omega$. The value of k is then updated according to the procedure:

$$\text{If}\begin{Bmatrix} g(\omega) < -\delta \\ |g(\omega)| \le \delta \\ g(\omega) > \delta \end{Bmatrix}\text{then}\begin{Bmatrix} \text{increment } k \text{ by } 1 \\ \text{no change of } k \\ \text{decrement } k \text{ by } 1 \end{Bmatrix} \qquad (4)$$

where δ is a discrimination threshold assumed for the expected range of frequency changes. The value of delay k may be any fraction of N, however, once defined, determines clearly the actual value of N. For instance, if k was initially chosen to be equal to N/4, the current value of N is set to be 4 times k. Consequently, the data window length of the filters used is always extended (or compressed) to the actual value of N (for full cycle filters). More details on this approach may be found in (Rebizant, *et al.*, 1999).

2.3 Samples of Simulation Study Results

A number of various test signals have been prepared with MATLAB and ATP programs. Both accuracy and dynamics of the estimators have been tested. In this paper just one example of algorithm operation for ATP generated testing signals is presented.

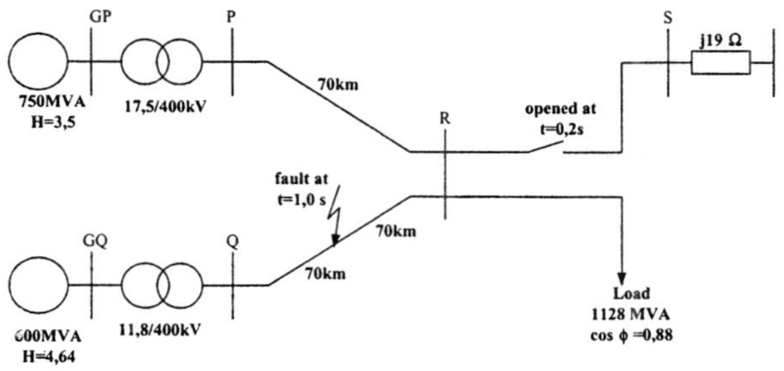

Fig. 2. Test power system modelled in ATP.

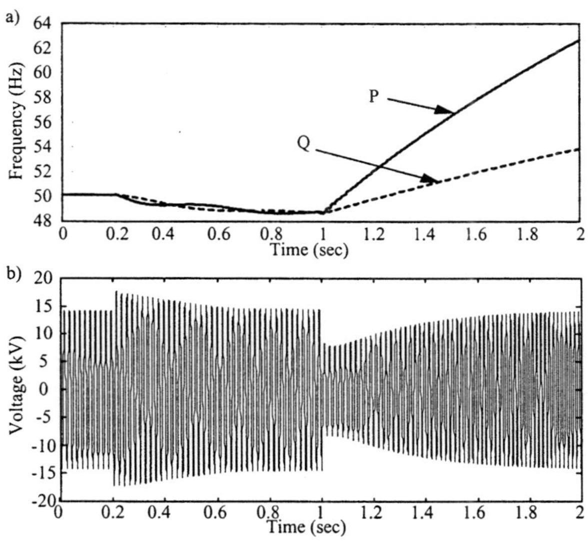

Fig. 3. Transient signals picked up at busbar GP: a) rotor's speed of generators P and Q, b) phase A voltage.

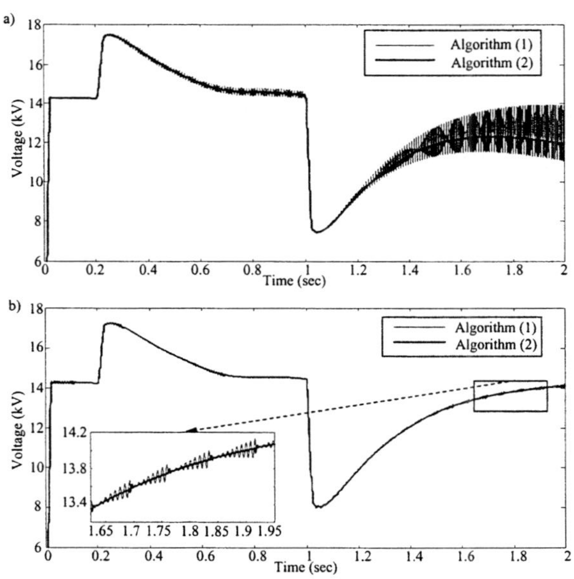

Fig. 4. Amplitude estimation with algorithms: a) non-adaptive, b) with adaptation.

The power system configuration used for tests is shown in Fig. 2. At time t=200ms the switch to the infinite bus is opened (the system becomes isolated). Additionally at t=1.0s a three-phase fault in the middle of the line Q-R was modeled. Transients in generators P and Q rotors' speed as well as the phase voltage at GP busbar are shown in Fig. 3. In consequence of new loading conditions the system frequency decreases (down to 48.5 Hz at the moment of fault inception). Since in this case the fault is not cleared within a reasonable period of time, both generators accelerate reaching speed values far beyond acceptable limits.

The amplitudes of voltage signal measured by non-adaptive and adaptive algorithms are shown in Fig. 4. The biggest differences are seen for larger frequency deviations when the filters used become untuned if the adaptation procedure is off. Then non-adaptive algorithms deliver inaccurate value of signal amplitude (with constant or oscillatory error). Small ripples seen on the measurement curves in Fig. 4b for the case of adaptive estimation are a result of slightly inadequate filter gains between consecutive on-line adjusting of the filter DW lengths.

3. ANN-BASED OUT-OF-STEP DETECTION

In the following sections the ANN design issues are discussed and the genetic method for ANN structure optimisation is proposed. Then the details of developed neural OS protection scheme are described, followed by the results of scheme training and testing with use of ATP signals.

3.1 ANN Design Issues

Artificial Neural Networks represent a modern and sophisticated approach to problem solving for power system protection and control applications. ANNs perform actions similar to human reasoning which rely on experience gathered during so called training. While preparing useful and efficient ANN-based classification/recognition unit, one has to take into consideration at least the following design issues:

- ANN choice (ANN structure type, number of layers and neurones, neurone activation functions, ANN input signals),
- ANN training (training algorithm, initial values of synapse weights and biases), etc.

The ANN input signals have to be chosen in such a way that they provide maximum information suitable for solving of given protection task. Choosing the type of ANN structure and its further parameters is rather a matter of the designer experiences since, unfortunately, there are no general practical rules which could be applied for that purpose. The experimental way with sequential trial-and-error attempts can be followed, however, this may not guarantee the optimal ANN structure to be found.

3.2 Genetic Rules for ANN Design and Optimization

To determine the most suitable network topology a genetic algorithm is proposed. The optimisation procedure commences from a population of artificial organisms called individuals that are defined to describe the topology of the ANN to be optimised. Natural principles according to the theory of Darwin, such as heredity, crossover, mutation, and selection, are used over several generations to develop and improve the characteristics of individuals. The

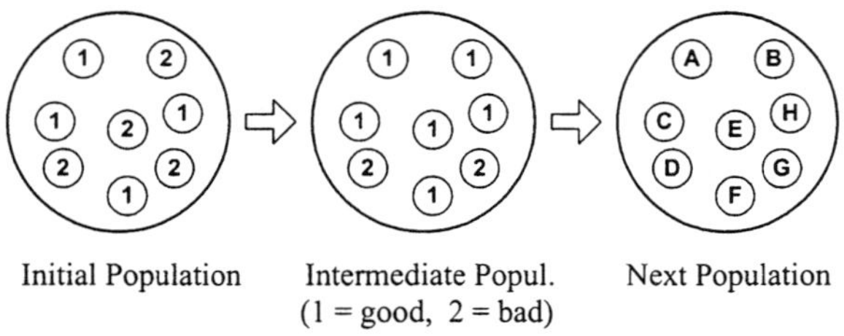

Fig. 5. Principles of the evolutionary process to optimise a network topology.

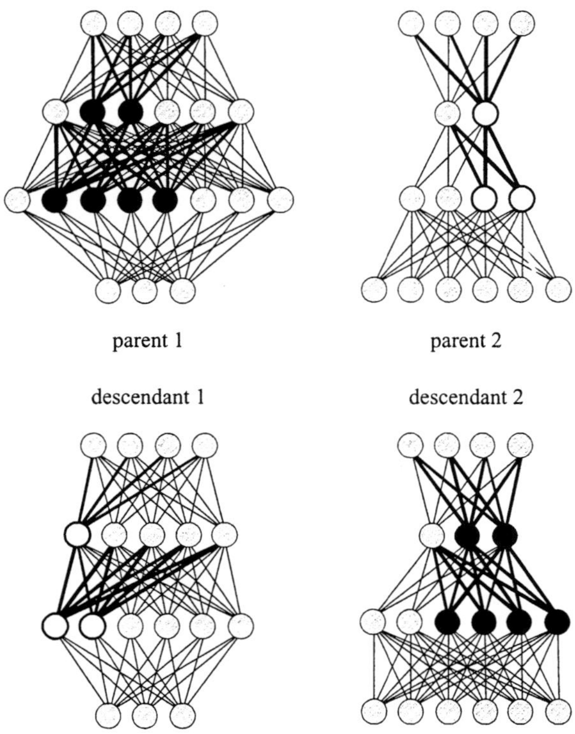

Fig. 6. Crossover of groups of neurones to create new descendent individuals.

selection is made according to the quality of each individual, which can be understood, as the capability to survive. The evolutionary process will end up into an optimum that represents the "stronger" individual, in this case - the most suitable ANN topology.

At the beginning a so-called initial population of neural networks is randomly created. While the number of neurones in the input and output layer are fixed according to the classification problem, the number of hidden layers and the number of neurones in these layers are randomly selected. All the individuals from the considered ANN population are trained with selected typical patterns and validated with all available patterns. This procedure keeps the training time short and emphasises on the generalisation of the network. After the quality of each individual was determined an intermediate population is created where successful individuals are reproduced more likely (Fig. 5).

On this intermediate population several genetic operations can be applied to create a new population. The most important genetic operation is the crossover of two "parent" individuals (Fig. 6) to produce "descendants". A crossover can be an exchange of single neurones, groups of neurones, or whole layers between two ANNs. Crossovers are very important and frequent at the beginning of a genetic algorithm so that a wide variety of different individuals can be produced. Furthermore, mutations can take place, which change randomly chosen genes of the individuals by adding or removing neurones. Another important genetic operation is the recombination, which stands for the creation of a descendant being an identical copy of a parent.

3.3 Implementation of the Genetic Algorithm

The genetic optimisation principles described in previous section have been implemented in MATLAB programme. The initial as well as further consecutive network populations were created, trained and graded in a closed loop until the selection criterion is fulfilled or the prescribed number of generation was reached. The coding of particular individuals was very compact - all the ANNs of given population were represented by the set of the following parameters (genes):

- ANN consecutive number k,
- number of layers l_k,
- vector of layer sizes (number of neurones in the layers) $\mathbf{N}_k$,
- matrix of the connection weights $\mathbf{W}_k$,
- matrix of the neurones' biases $\mathbf{B}_k$,
- quality index Q_k.

The trained population of nets was subjected to quality determination (grading). Various quality criteria may be used to assess the individual nets creating current population. The grading procedure has been organised in such a way that the "best" net (according to the criterion chosen) is always assigned the quality of 1 whilst the "worst" one would have the quality index equal to 0. To avoid accidental cancelling of the best ANN during operations of creating and genetic modifications of intermediate population, the net with quality 1 is always forwarded to the next generation without changes.

At each consecutive optimisation step (for successive population of nets) appropriate changes in the genome of each individual are carried out, depending on the genetic operation applied to the net (-s):

- new number of layers l_k and number of neurones in the layers $\boldsymbol{N}_k$ are assigned,
- elements of the matrices $\boldsymbol{W}_k$ and $\boldsymbol{B}_k$ are cut away, newly established or interchanged between the ANNs.

3.4 ANN Based OS Protection

The general scheme of the neural OS protection scheme developed is shown in Fig. 7. The decision part of the protection is realised with help of an ANN performing typical pattern recognition with appropriately chosen vector of criterion signal samples $\boldsymbol{X(k)}$. The decision (criterion) values have to be previously calculated from available power system signals with use of dedicated digital processing algorithms. The ANN is assumed to produce output equal to 0 for stable patterns and 1 for OS conditions. For the classification purpose a threshold value set to 0.5 is introduced. All the cases for which the ANN output is lower than 0.5 are classified as stable and those for which the threshold is exceeded are recognised as OS cases.

To obtain data for training of ANNs and further testing of the OS protection, the following simple single machine – infinite bus system has been modelled (Fig. 8) with use of the ATP software package. The synchronous machine G1 was connected to the infinite bus system S1 (220 kV) via the block transformer T1 and a 200-km long double-circuit line L1. Within the transmission line L1 a total of 108 symmetrical faults on one of the line circuits were applied, some of them responsible for further developing OS conditions.

Generator output voltages and currents as well as its angular speed were registered in ATP output files. Additional features like voltage/current amplitudes, components of generator power etc. were obtained after digital processing of voltage and current signals. In order to choose the best signals for application as ANN input features, the statistical properties of available signals were determined. The analysis of calculated PDFs allowed sorting the decision signals according to their relative recognition strength. Ultimately, the machine angular frequency deviation $\Delta\omega$ was taken as the most valuable recognition feature in the investigated case. The ANN input vector $X(k)$ was being created on-line from a number of signal samples captured with use of a sliding data window (DW). The DW length was set to 360ms.

The choice of the ANN structure and size for the neural OS protection scheme developed has been done with use of the genetic optimisation procedure. The results given below were obtained for a population of 20 nets trained with a half of available short-circuit patterns and tested with all patterns. Fifty generations of nets were assumed. The investigations have been done for a set of networks trained with the Levenberg-Marquardt algorithm.

The nets constituting the population of individuals were graded according to two different quality indices, i.e. mean square error of the net output Q_{ms} (squared difference between desired and actual output values) and testing efficiency Q_{eff} (percentage of properly recognised OS cases). Consequently, different results of the genetic process have been reached. The "best" nets obtained after 50 generations had 6-1 and 3-1 neurones for the Q_{mse} and Q_{eff} grading indices, respectively. Fig. 9a shows the average remaining error over all ANN topologies of one population as a function of the number of generations. In Fig. 9b the average testing efficiency of the scheme is presented. Smaller values of the mean square error were achieved for Q_{mse} used as grading index, while the highest efficiency resulted

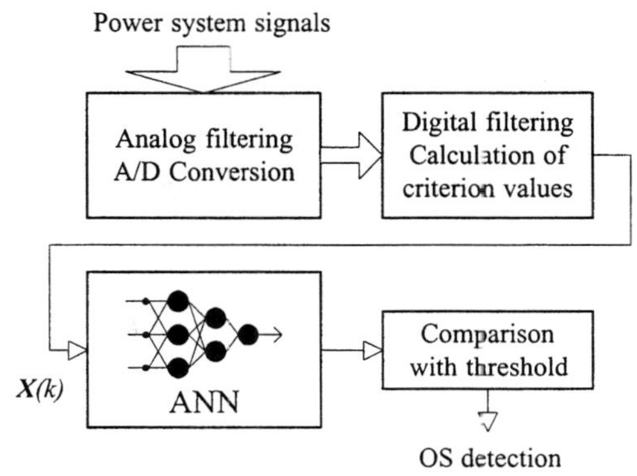

Fig. 7. Neural OS protection arrangement.

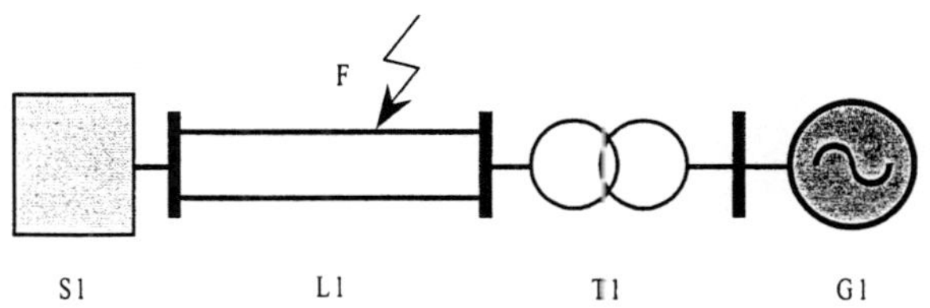

Fig. 8. Test power system modelled.

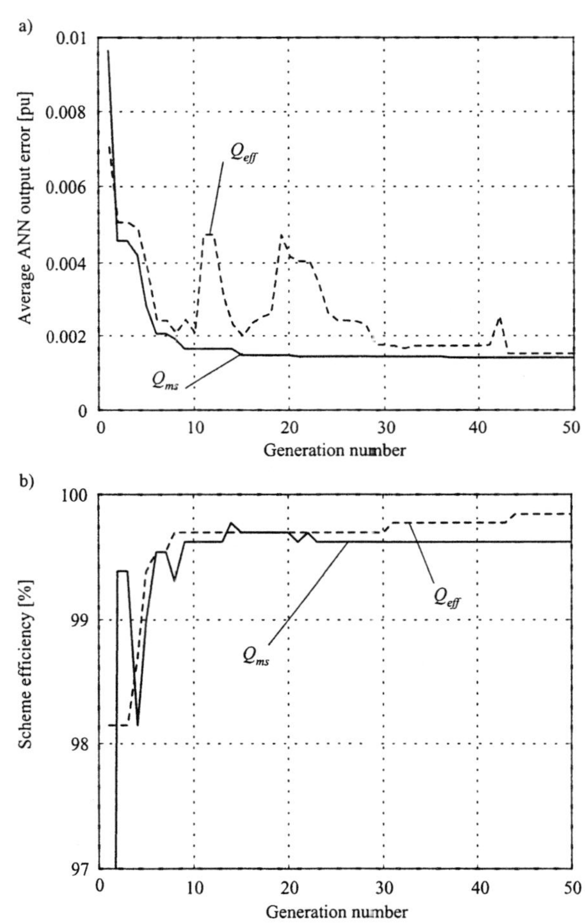

Fig. 9. Optimisation results: a) mean square error, b) recognition efficiency.

from grading the nets with Q_{eff} index. One can observe that monotonic change (decrease/increase) of the mean square error and scheme efficiency was associated with the genetic optimisation process performed with the respective grading indices. The curves for competitive indices, by similar general improvement tendencies, reveal significant oscillatory behaviour. It was found that minimising

Q_{ms} index required greater ANNs and was not always accompanied by good generalisation features. Contrary, smaller nets brought about better OS classification efficiency with non-optimal values of mean square error (well known memorisation vs. generalisation dilemma).

3.5 OS Protection Scheme Testing and Validation

The ANN-based OS protection scheme has been thoroughly tested with ATP-generated power system signals. The scheme displayed high efficiency and very short time of OS detection. It is worth to be mentioned that the values of ANN output error and scheme efficiency shown in Fig. 9 characterise an "average" individual over the whole population of ANNs. The OS protection equipped with the "best" ANNs (graded with quality indices equal to 1.0) displayed 100% selectivity which means that all considered ATP testing cases were correctly classified.

Comparing to other existing impedance-based OS protection devices, a kind of prediction of coming machine instability is performed instead of traditional detection of actually occurring phenomena. The decision was taken within approx. 500ms after fault inception (some 300-900 ms before actual OS appeared), thus leaving enough time for an appropriate action (machine tripping, fast valving) to protect the generator from stresses and preserve the stability of power system. Wide robustness features of the scheme with respect to both various fault types and other synchronous machine ratings have also been confirmed.

4. CONCLUSIONS

In the paper two modern advanced approaches to improvement of digital protection of synchronous generator against OS phenomena are presented. Both signal measurement and decision-making modules of the relay are optimised with introduction of the adaptivity concept and application of artificial intelligence techniques, respectively.

The following features of the adaptive amplitude estimation scheme are worth to be pointed out:

- The proposed scheme is capable of tracking signal amplitude with high accuracy and dynamics even if the system frequency (generator shaft speed) is varying in wide range.
- The adaptive amplitude estimators can be applied in generator digital monitoring, protection and control equipment. This is also valid for other electrical quantities, such as active and reactive power, impedance components or symmetrical components of generator terminal signals.
- Computational complexity of the proposed adaptive measurement scheme is relatively low. Thus practical implementation of the algorithm for simultaneous estimation of various generator signals in real time is utterly possible with currently available signal processors.

The investigations on ANN application to out-of-step protection presented in the paper allowed drawing the following conclusions:

- Promising results in shape of high classification efficiency (OS cases vs normal operation and other cases) have been achieved with application of the designed ANN-based recognition unit.
- Thanks to high sensitivity of the designed neural decision block, the scheme was able to detect or even predict the situations of generator instability basing on tiny symptoms of evolving OS conditions.
- The neural OS detection scheme developed was based on ANNs optimised with of use the genetic procedure. Application of such an approach allowed avoiding non-optimalities usually met with traditional ANN design heuristics and provided most favourable nets for given classification task.
- The obtained neural nets consisted of very few neurones, which allows implementing the scheme on traditional widely available signal processors (no specialised expensive neural chips are needed for implementation of ANNs of such small sizes).

REFERENCES

Winkler, W. and A. Wiszniewski (1995). Adaptive protection - potential and limitation, *CIGRE SC34 Colloqium*, Stockholm, Rep. 34-206.

Rebizant, W., J. Szafran and M. Michalik (1999). Adaptive estimation of wide range frequency changes for power generator protection and control purposes, *IEE-Proceedings on Generation, Transmition and Distribution*, **Vol. 146**, No. 1, January 1999, pp. 31-36.

Moore P.J., R.D. Carranza and A.T. Johns (1994). A new numeric technique for high-speed evaluation of power system frequency, *IEE-Proceedings on Generation, Transmission and Distribution*, **Vol. 141**, No. 5, Sept. 1994, pp. 529-536.

Szafran J., W. Rebizant and M. Michalik (1999). Adaptive measurement of power system currents, voltages and impedances in off-nominal frequency conditions. *Proceedings of the 16th IEEE Instrumentation and Measurement Technology Conference (IMTC'99)*, Venice, Italy, May 1999, Vol. 2, pp. 801-806.

Elmore W.A. (1994). *Protective relaying theory and applications*, Marcel Dekker, New York, US (Elmore (Ed.)).

Rebizant W., J. Szafran, K. Feser and F. Oechsle (2001). Evolutionary improvement of neural classifiers for generator out-of-step protection, *Proceedings of the 2001 Porto IEEE PowerTech Conference,* Porto, Portugal, Sept. 2001, paper PRL1-223.

ELSEVIER
IFAC
PUBLICATIONS
www.elsevier.com/locate/ifac

PARTIAL LEAST SQUARES FOR POWER PLANT PERFORMANCE MONITORING

J. Ritchie and D. Flynn

The Queen's University of Belfast, N. Ireland

Abstract: The problems faced in monitoring power plant performance are outlined and consideration is given to more effective utilization of the data generated on-line by the distributed control system (DCS). Partial least squares (PLS) is identified as a method ideally suited to coping with the large number of correlated and collinear signals available on a power plant, for process monitoring and performance related analysis. Non-linear PLS models are trained using archived data from a local utility to predict quality measures of thermal efficiency, and NO_x and SO_x emissions for generating units dual-firing on both oil and gas. The performance and diagnostic capabilities of the resulting models are examined, illustrating the simplicity of operation of the proposed monitoring tool. *Copyright © 2003 IFAC*

Keywords: partial lest squares, non-linear models, neural networks, performance monitoring, power station boilers

1. INTRODUCTION

The incorporation of distributed control systems (DCS) in power plants for instrumentation and control has become increasingly standard in recent years. The subsequent introduction of sophisticated supervisory layers, to optimise plant performance, has brought both improvements in productivity and significant advances in the manoeuvrability of plant. However, the ease of access to plant-wide signals, arising not only from physical measurements, but also from control loops around the plant and internal features of the DCS itself, potentially complicates the task of monitoring plant operation. The integral alarm systems typically employ univariate methods, with upper and lower thresholds defined for each signal. Particular problems exist for signals which are operating mode or load dependent. Limits are normally set conservatively, so that the operator may be presented with a steady stream of alarms during otherwise normal operating conditions. This problem is magnified when a genuine fault occurs, and the operator is confronted with a flood of alarm messages (Lichnowski, 1999). The root cause of the problem must then be determined and appropriate corrective action(s) implemented.

Monitoring approaches based around a physical model of the plant, which runs in parallel with the real plant, have often been considered (Yoon and MacGregor, 2000). The model should represent both normal operating conditions and specific faults, with individual parameters tuned using available plant data. Clearly, such an approach is best applied to systems with a limited number of inputs / outputs, in order to minimize model complexity. It is also recognized that identification and representation of every possible fault can be extremely time consuming. The designer will require comprehensive knowledge, and access to previous fault records, of the examined system.

An alternative option is to employ structures, such as neural networks, petri nets, etc., to recognize patterns within combinations of alarms and process signals, which can be associated with particular faults (Choi, *et al.*, 1995; Swidenbank, *et al.*, 1998). Detailed knowledge of the relationships between variables is not necessary, although substantial and comprehensive historical data, detailing fault patterns and their diagnosis, is required. However, the *black box* nature of these methods implies that minimal explanation is provided regarding how particular conclusions have been reached. Such approaches also have difficulty in coping with multiple, unrelated faults and it is difficult and time consuming to produce a structure that is complete and up to date.

It is clear, though, that the vast amount of data being collected by distributed control systems could be applied much more effectively. Data mining techniques are ideally suited to dealing with power plant data, which tends to be both highly correlated and collinear. Data mining is often described as the non-trivial process of identifying valid, novel, potentially useful, and ultimately understandable patterns in data. The term comprises a wide range of techniques encompassing easily understood rule bases and decision tree structures to statistical methods and more complex mathematical methods such as fuzzy logic and neural networks. These methods have been applied to a diverse range of applications incorporating chemometrics (MacGregor, *et al.,* 1994), power system security assessment (Huang, *et al.*, 2002) and medical diagnosis (Alonso, *et al.*, 2002).

Typically, models are generated which identify relationships, under normal operating conditions, from historical data. Subsequently, these discovered relationships form the basis for monitoring on-line data, and detecting unusual deviations from the trained behaviour. Techniques are then available to identify and diagnose the possible cause(s) of the problem. Here, the application of partial least squares (PLS) will be examined for power plant monitoring, and a number of extensions to the basic method will be considered. The operational problems faced by operators at a local utility in N. Ireland, Ballylumford power station, will be discussed. The application of PLS techniques will then be demonstrated, and shown to take full advantage of readily available DCS data, and provide an informative monitoring method.

2. NON LINEAR AND RECURSIVE PLS

Principal component analysis (PCA) aims to reduce the dimensionality of a set of interrelated variables, while retaining as much of the variance as possible. This is achieved by identifying latent variables, known as principal components. A reduced set of components may then be selected to represent the majority of the variance in the data. The resulting model characterizes the variation under normal conditions, and so may be used to identify rogue values in future data sets (Dunia, *et al.*, 1996). A useful extension of this technique is offered by partial least squares.

Partial least squares, also known as projection to latent structures, is a robust, multivariate linear regression technique suitable for analyzing noisy and highly correlated data. A model is developed which attempts to explain the variation in the process that is most predictive of the product quality variables.

PLS requires two blocks of data, an **X** block representing samples of independent process variables, and a **Y** block of dependent product quality measurements. Typically, **Y** is measured irregularly, or not as frequently as the **X** block. PLS then aims to provide an estimate of **Y** using the **X** data. If **T** and **U** represent the score matrices for **X** and **Y**, and **P** and **Q** are the associated loadings, then, for, *A*, the required number of components,

$$\mathbf{X} = \mathbf{T}\mathbf{P}^T + \mathbf{E} = \sum_{i=1}^{A} \mathbf{t}_i \mathbf{p}_i^{\,T} + \mathbf{E}$$

$$\mathbf{Y} = \mathbf{U}\mathbf{Q}^T + \mathbf{F} = \sum_{i=1}^{A} \mathbf{u}_i \mathbf{q}_i^{\,T} + \mathbf{F}$$

with the residual matrices **E** and **F** sufficiently small. The above outer relations can be linked by a linear relationship, **B**, between the score matrices of the **X** and **Y** block, namely **T** and **U**, such that,

$$\mathbf{U} = \mathbf{B}\mathbf{T}$$

where **B** is a diagonal matrix. The resulting model however will not be optimal, since the principal components for each block are calculated separately, and thus possess only a weak link to each other. More constructively, the **p** and **q** loading vectors for each component can be determined in sequence using the non linear iterative partial least squares (NIPALS) algorithm. Consequently, at each iteration, the **t** and **u** vector scores for each component can be interchanged, so that slightly rotated principal components are obtained, such that

$$\mathbf{X} = \mathbf{T}\mathbf{P}^T + \mathbf{E}$$

$$\mathbf{Y} = \hat{\mathbf{U}}\mathbf{Q}^T + \mathbf{F}$$

where, $\hat{\mathbf{U}}$ is the estimate of the score $\mathbf{U}$, based on **T**. However, in order to obtain orthogonal **X** block scores it is necessary to introduce a weighting matrix, **W**, as outlined by Geladi and Kowalski (1986). The traditional NIPALS approach requires deflation of both the **X** and **Y** matrices. This may be conveniently replaced by a more efficient kernel algorithm, which only deflates one of the matrices, in this case **Y**.

A number of methods are available for selecting a suitable number of components for the PLS model, balancing parsimony against model accuracy. The most popular methods employ the prediction residual sum of squares (PRESS) statistic, achieved by cross validation of the **X** and **Y** data, which is iteratively separated into multiple training and test data sets (Wold, 1978). Alternatively, examination of the unreconstructed variance of the model has been suggested by Dunia and Qin (1998).

Having developed a model, a number of statistics can be introduced to assess the performance of both the PLS model and the monitored process. So, for example, changes in the squared prediction error (SPE) and Hotelling's T^2 statistic may be indicative of sensor failures, actuator malfunctioning, etc. Threshold limits can be established for each performance indicator, enabling threshold boundaries to be set for typically 95% and 99% confidence levels (Jackson, 1991).

It is also necessary to distinguish fault conditions associated with the process itself. This can be achieved using instantaneous **t** score scatter plots to detect if the process is consistent with normal operating conditions, as defined by the training data. If a process fault is identified, contribution plots may then be constructed to identify signals that have a higher than normal contribution at that time to the observed variation.

2.1 Non-linear Modelling

The above outlined approach creates a linear PLS model, and a simple link between the **T** and **U** score matrices, whereby **U** = **BT**. Alternative structures are, however, required to cope with the dynamic (rather than static) relationships that exist between process variables and the inherent non-linearities which will inevitably be present within an industrial process.

Dynamic PLS (Baffi, *et al.*, 2000) provides a modification to the traditional PLS algorithm, which allows the dynamic aspects of a system to be represented, e.g. relationship between changes in air flow / fuel flow and the variation in NO_x emissions at the stack outlet. This is achieved by creating an augmented data matrix containing lagged values of the process data, which account for time delays in the system. Such an approach becomes essential when the availability of second by second data from the DCS is considered.

In order to cope with system non-linearities, the PLS model may be adapted on-line, in response to changes in say unit load, using recursive PLS methods (Qin, 1998). As new data becomes available it can be included in the model to incorporate slowly occurring system changes. This can be achieved by updating the covariance matrices, as part of the PLS kernel algorithm.

$$\left(\mathbf{X}^T\mathbf{X}\right)_t = \lambda\left(\mathbf{X}^T\mathbf{X}\right)_{t-1} + \mathbf{x}_t{}^T\mathbf{x}_t$$
$$\left(\mathbf{X}^T\mathbf{Y}\right)_t = \lambda\left(\mathbf{X}^T\mathbf{Y}\right)_{t-1} + \mathbf{x}_t{}^T\mathbf{y}_t$$

where $(\mathbf{X}^T\mathbf{X})_t$ and $(\mathbf{X}^T\mathbf{Y})_t$ are the updated covariance matrices at time t, $\mathbf{x}_t$ and $\mathbf{y}_t$ are the new data vectors to be incorporated in the model, and λ is the data forgetting factor.

Before a recursive PLS algorithm can be introduced extensive supervision software will be required. The purpose of the PLS model is to characterize normal plant behaviour, so that deviations can then be noted. It is important to ensure that adaptation only occurs when the process data contains new information (to avoid ill conditioning of the covariance matrices), and, of course, to avoid *learning* fault conditions. The threshold limits for the monitoring indicators will also require adaptation.

Alternatively, a curve fit can be achieved to directly model the non-linear relationship between the **T** and **U** score matrices. This can be conveniently created using a neural network, as the transparency of the resulting model will remain unaffected by this relationship. A radial basis function (RBF) neural network architecture will be discussed here, although a variety of other neural structures would be just as suitable. A RBF network is a single layer feed forward architecture, with neurons in the hidden layer generating a set of gaussian basis functions which are combined by a linear output neuron. Each gaussian function is centred at a point in the input space, determined by methods such as clustering or use of *a priori* knowledge (Broomhead and Lowe, 1988). Individual functions approximate the model in the region around which they are centred. The weights are then linear terms and can be determined using least squares techniques, or singular value decomposition if the data matrix is ill conditioned.

Assuming that the relationship between **T** and **U** is continuous and differentiable with respect to the weighting matrix, **W**, the non-linear mapping can be approximated by a Taylor series expansion. Consequently, following modification of the existing linear algorithms, an error-based updating approach can be applied to iteratively determine the PLS components (Baffi, *et al.*, 1999).

3. DEVELOPMENT OF POWER PLANT MODELS

Ballylumford power station is the largest of three power stations in Northern Ireland, located on the Antrim coast. The power station houses six generating units, 3 x 120 MW and 3 x 200 MW, which were originally designed to operate using heavy residual fuel oil. After a conversion programme between July 1994 and July 1996 they are now capable of firing on either gas or oil. Approximately 2500 sensors record data detailing operation of the plant and its performance. The primary source of information is the DCS which records a range of variables concerning boiler operation for all units. Other variables concerning the feed heating plant, condenser, etc. are available from additional monitoring equipment installed throughout the power station. The majority of these variables are monitored on a second by second basis, although data storage is generally at a lower resolution due to memory limitations.

Given the availability of stored data from the power station for a large number of plant-wide signals, consideration can be given to developing a number of PLS quality models. Thermal efficiency has an obvious dependency on the power output of a unit. However, there are a wide range of contributory factors. For example, effluent dumping by local industry and tidal movements have a significant effect on cooling water (sea) temperature and hence condenser performance. Additionally, power stations are increasingly subject to environmental controls and can incur substantial financial penalties for failure to comply. It is, therefore, of interest to also model NO_x, SO_x and CO emissions generation. Clearly, the nature and level of emissions produced will depend on the type of fuel being used, burner configuration, etc.

While it would be possible to develop a single model for multiple quality variables this approach inevitably implies a degradation in the ability of the model to predict variation in **Y** arising from **X**, due to conflicting objectives. As a result, separate models have been developed for each quality variable considered, with further subdivisions in the models to represent the 120 and 200 MW units, and operation with either oil or gas fuels.

In contrast with model based approaches, which are best suited for systems with a limited number of process inputs and / or outputs, PLS models are able to easily accommodate a large number of variables in the **X** matrix. However, due to the high correlation between signals, the focus on prediction of **Y** and not **X**, as in principal component analysis, and to reduce the computational burden, a carefully considered, reduced set of variables was selected to form the **X** matrix. The signals selected can be distinguished as either being operational or process state variables. Operational variables are under the direct control of the plant operators, and include the number of burners at each level, reheater spray flow, and excess O_2 setpoint. Process state variables that affect the quality variables but can not be controlled are also included, e.g. cooling water temperature, condenser vacuum and gas exit temperature.

3.1 RBF Model Training and Implementation

Initially, linear PLS models were developed with target quality variables of thermal efficiency, NO_x and SO_x - the SO_x model was only required for oil operation. Considering, as an example, the efficiency quality target, a linear model was trained for a 200 MW unit firing on gas, using representative data over the unit's full operating range. The resulting model required 11 components to explain 93% of the variability between **X** and **Y**. However, the objective of a PLS model should be to create a minimal structure to account for the observed variation. This result merely confirms that a power plant is a non-linear system. As such, essential process non-linearities may be hidden in the lower components of the linear model. This possibility can be reduced by creating a number of overlapping linear models that span the entire operating range. However, from a monitoring perspective this will increase the number of SPE, T^2 and score plots which will need to be both created and analysed. The objective of creating an intuitive monitoring system is unlikely to be achieved. This, clearly, leads to the selection of an alternative non-linear structure.

As previously discussed, a number of extensions to the basic PLS algorithm have been developed. Here, a RBF neural network will be used to demonstrate how the non-linearity between the input and output scores can be modelled. Individual PLS models, and RBF networks for each component, were developed using the same training data as before, and for each of the previously listed product quality variables. In this case, the thermal efficiency model for the 200 MW gas-fired unit required only 2 components to explain 93% of the variability in **Y**. The RBF network associated with the first component

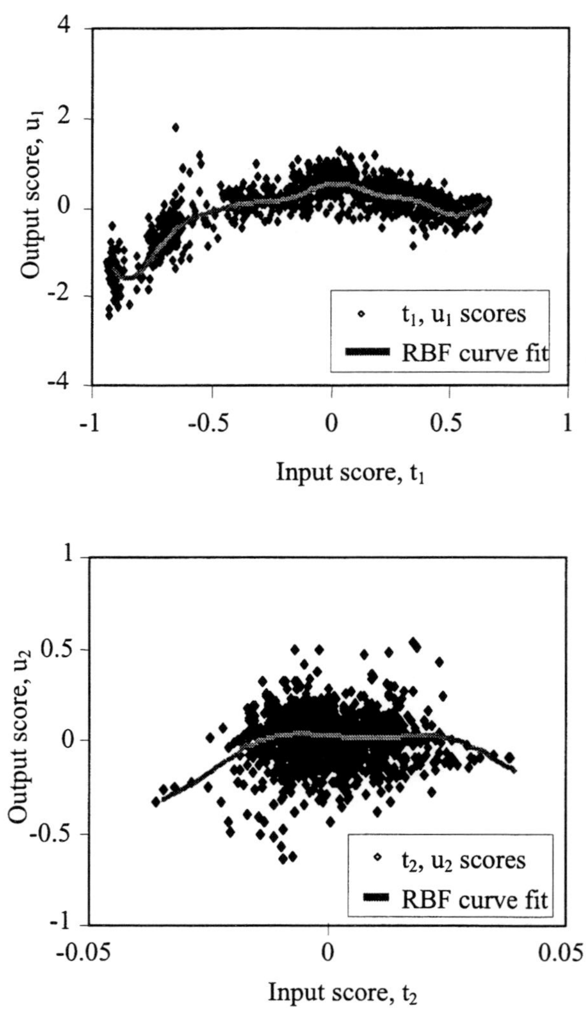

Fig. 1. Thermal efficiency scores scatter plot

contained 23 neurons, while 5 neurons were used in successive component RBF models. The improved modelling of the non-linearity can be demonstrated using scatter plots for the **T** and **U** training data. Figure 1 shows the scatter plots for the first and second components of the efficiency model. The first component is shown to model the majority of the variance leaving little for the subsequent components.

The prediction accuracy obtained from the non-linear modelling technique is shown in Figure 2 for the thermal efficiency model. During the 8 hour test period the unit load varied between 70 and 200 MW. Figure 3 then illustrates the performance of the NOx model for the same generating unit, while the unit load varied between 80 and 200 MW. What is most clear from the figure is the poor quality of the NOx data itself. Measurements are only available every 30 seconds, from alternate side of the boilers, and at a limited resolution. For confidentiality reasons, the vertical scale on all graphs have been normalized.

An RBF PLS model was also developed for a 120 MW unit, while operating on oil, with the goal of monitoring unit efficiency. Here, 2 components were found to explain 95% of the variation in **Y**, with a 16 neuron RBF neural network required for

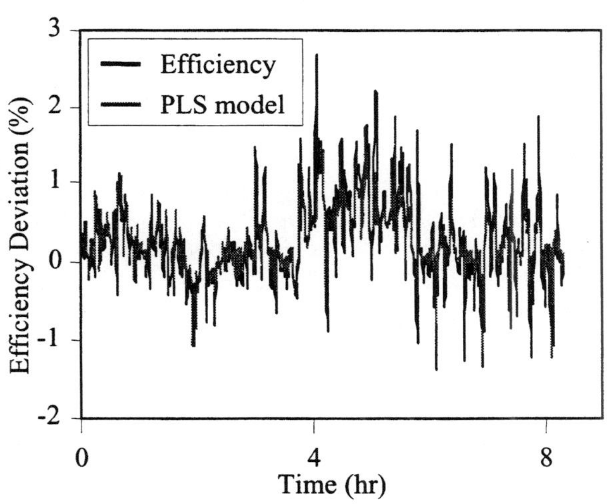

Fig. 2. RBF PLS thermal efficiency model

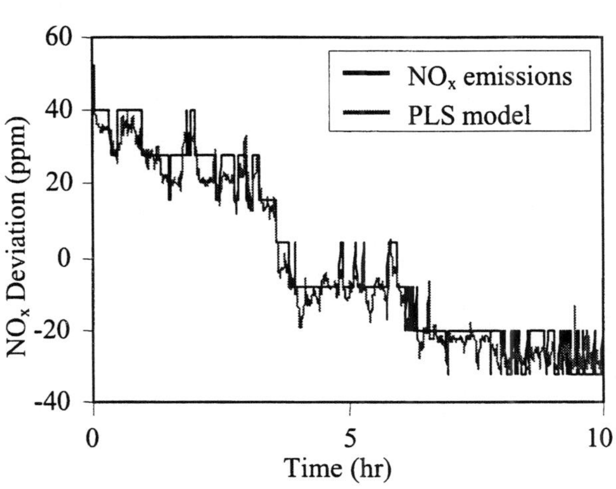

Fig. 3. RBF PLS NO_x model prediction

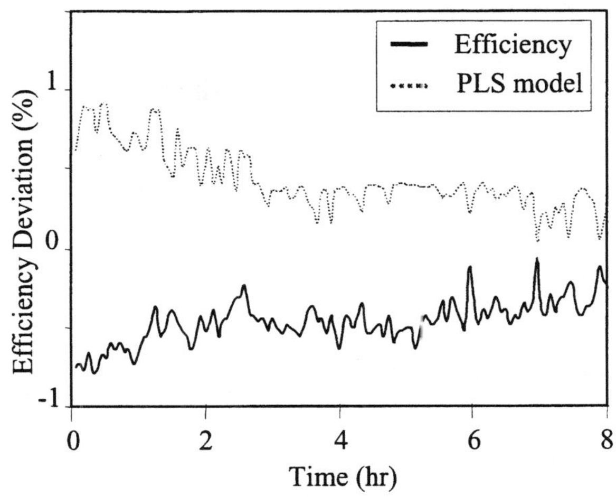

Fig. 4. Monitoring thermal efficiency performance

the first component, and 3 neurons for successive components. Examination of the model under test conditions confirmed the predictive capabilities of the model. Subsequently, the model was used to monitor oil-firing operation in the following weeks. Figure 4 illustrates a significant deviation between the target PLS calculation for thermal efficiency and the value derived from the actual plant. Investigation of the scores plot, Figure 5, shows that the plant has moved away from the normal, trained operating conditions, identified as group *A*, into another region, group *B*. The greater variation can be seen to occur in the t_1 score. So, finally, the contribution of the individual **X** variables to the deviation in the t_1 score can be plotted, Figure 6. The graph shows that signals 22 and 23, cooling water outlet temperatures (A and B sides), have changed significantly, indicating the condenser to be the source of the problem.

CONCLUSIONS

The distributed control system within many power stations provides ease of access to plant-wide signals on a second by second basis, in addition to comprehensive data on historical performance. Data mining, as a generic technique, and partial least squares (PLS) in particular, were proposed as an effective method for utilizing the available data. Non-linear extensions to the basic PLS algorithm were discussed, and subsequently, neural network based models were developed using plant data to monitor the stack emissions and thermal efficiency, for both oil and gas firing, for the 120 and 200 MW units of a local utility.

The prediction capabilities of the various models was then demonstrated. It was also shown that by examination of t score plots and contribution charts that deficiencies in plant performance can be easily highlighted to the operator, and the nature of any problems diagnosed.

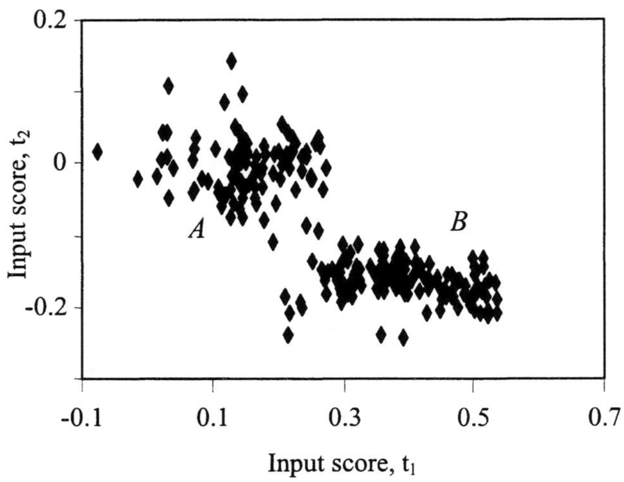

Fig. 5. t_1 v t_2 scores scatter plot

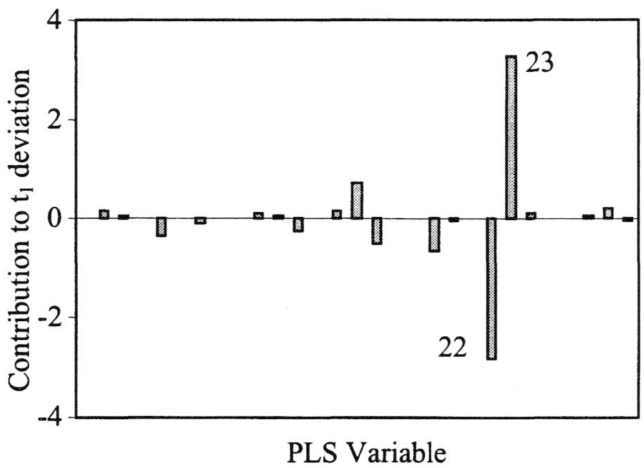

Fig. 6. Contribution of X variables to t_1 score

ACKNOWLEDGEMENTS

The authors wish to acknowledge Premier Power plc for permitting access to their plant, and to the assistance provided by the unit operators and engineering staff at Ballylumford power station. The support of Michael Cregan, The Queen's University of Belfast, is also greatly appreciated.

REFERENCES

Alonso, F., J.P. Caraca-Valente, A.L. Gonzalez and C. Montes (2002). Combining expert knowledge and data mining in a medical diagnosis domain. *Expert Systems with Applications*, **23**, pp. 367-375.

Baffi, G., E.B. Martin and A.J. Morris (1999). Non-linear projection to latent structures revisited (the neural network PLS algorithm). *Computers and Chemical Engineering*,**2 3**, pp. 1293-1307.

Baffi, G., E.B. Martin and A.J. Morris (2000). Dynamic non-linear PLS modelling. *IFAC Advanced Control of Chemical Processes*, Pisa, Italy, pp. 159-164.

Broomhead, D.S. and D. Lowe (1988). Multivariable functional interpolation and adaptive networks. *Complex Systems*, **2**, pp. 321-355.

Choi, S.S., K.S. Kang, H.G. Kim and S.H. Chang (1995). Development of an on-line fuzzy expert system for integrated alarm processing in nuclear power plants. *IEEE Trans. Nuclear Science*, **42**, pp. 1406-1418.

Dunia R., S. Qin, T.F. Edgar, T.J. McAvoy (1996). Identification of faulty sensors using principal component analysis. *AIChE Journal*, **42**, pp. 2797-2812

Dunia, R. and S. Qin (1998). Subspace approach to multidimensional fault identification and reconstruction. *AIChE Journal*, **44**, pp. 1813-1831.

Geladi, P. and B.R. Kowalski (1986). Partial least squares regression: A tutorial. *Analytica Chimica Acta*,**1 85**, pp. 1-17.

Huang, J.A., S. Harrison, G. Vanier, A. Valette and L. Wehenkel (2002). Application of data mining to optimize settings for generator tripping and load shedding system (RPTC) in emergency control at Hydro-Quebec. *Int. Conf. on Probabilistic Methods Applied to Power Systems*, Naples, Italy, pp. 615-621.

Jackson, J.E. (1991). *A user's guide to principal components*, Wiley & Sons Inc., New York.

Lichnowski, A. (1999). Developments in electricity generation plant control management, *People in Control: IEE Int. Conf. on Human Interfaces in Control Rooms, Cockpits and Command Centres*, pp. 90-95.

MacGregor, J.F., C. Jaeckle, C. Kiparissedes and M. Koutoudi (1994). Process monitoring and diagnosis by multiblock PLS methods. *AIChE Journal*, **40**, pp. 826-838.

Qin, S.J. (1998). Recursive PLS algorithms for adaptive data modelling. *Computers Chem. Engng.*, **22**, pp. 503-514.

Swidenbank, E., J. García, D. Flynn, J. Rapún (1998). On-line optimization and monitoring of power plant performance through machine learning techniques. *IEE Control '98*, Swansea, pp. 257-262.

Wold, S. (1978). Cross-validatory estimation of the number of components in factor and principal components models. *Technometrics*, **20**, pp. 397-405.

Yoon, S. and J.F. MacGregor (2000). Relationships between statistical and causal model based approaches to fault detection and isolation. *IFAC Advanced Control of Chemical Processes*, Pisa, Italy, pp. 81-86.

ELSEVIER
IFAC
PUBLICATIONS
www.elsevier.com/locate/ifac

I&C COMMISSIONING RESULTS OF TAEAN #5, 6

Joon-Ho Yoon

Taean TPP, Korea Western Power

Abstract: Taean TPP #5,6 is supercritical one-through coal fired 500MW thermal power plant that adopts two pass, opposite firing boiler, reaction type turbine, which is different from other Korea standard power plant(total : 20 units).
General instrumentation & control commissioning activities were performed to fit controls with the site conditions of the boiler and turbine about 40days.
After we finished tuning over-all control characteristic of boiler and turbine, We could draw optimal main steam temperature and pressure in compliance with load change that are suitable for governor free operation and automatic dispatching system.
This paper covers the principal test activities that were undertaken during commissioning, including Step Response Test, Load Swing Test for 3%, 5%NR/Minute, Load Runback Test, AGC(Automatic Governor Free Operation Test) and ADS(Automatic Dispatch System Test), APS(Automatic Plant Start up Test) *Copyright © 2003 IFAC*
Keywords : DCS, Load Swing Test, Load Runback Test, AGC, APS

Ⅰ. Introduction

○ Control systems are composed of Bailey's INFI-90 DCS to control boiler and BOP, Westing House' WDPF-Ⅱ to control turbine and FGD(Flue Gas Desulfurizer).

INFI-90 DCS system is composed of 1 EWS for designing and configurating control system, 9 OIS for monitoring and controlling system status, INFI-NET for communicating network.

Networks are divided into INFI-NET, Control Way, Expander BUS. Expander BUS transmits data between Multi-Function Processor Module and I/O Slave module which wired field drives.

Control way is a redundant, serial bus running at 1MBaud that allows master modules and INFI 90 plant communication modules to exchange data. INFI-NET is advanced data communications highway.

Operator can get control variable by way of OIS(Operator Interface Station) and can transmit information to field at the same path vice verse.

○ Unit control method adopts BLR-TBN coordination control that controls BLR, TBN and generator all together considering a supercritical one-through BLR.

Unit master is composed of following 3 operation modes according to operation condition.

1.1 Coordination Mode(BLR Master : Auto, TBN Master : Auto)

MW is controlled by TBN Master primarily and BLR Master secondarily. TP is controlled by BLR Master primarily and TBN Master secondarily. In this mode, operator can set MW Demand and in other mode MW Demand follows Actual MW.

1.2 BLR Following Mode(BLR Master : Auto, TBN Master : Manual)

In this mode, BLR Master controls TP error, TBN Master controls MW error. BLR Master in auto follows TBN Master in manual. The Difference between steam flow used in TBN and steam flow generated in BLR bring about the change in TP. BLR Master reacts on that change.

This arrangement is the most responsive but the least stable of the methods of firing rate demand development.

1.3 TBN Following Mode(BLR Master : Manual, TBN Master : Auto)

TBN Master controls TP error, BLR Master controls MW error

TBN Master in auto follows BLR Master in manual. MW is created according to total steam flow into TBN.

This type of control is very stable due to the stability of the turbine valve control of the turbine throttle pressure

※ At the HP Bypass opening, HP Bypass controls TP regardless of operation mode and BLR/TBN Master tracks actual MW.

Unit Master Demand generates MW signal and transmits MW signals to BLR and TBN Master.

BLR Master commands its output to control feedwater, fuel and air master.

TBN Master commands DEHC to open, close governor valve position.

Ⅱ. Commissioning Results of Control System

1. Load Swing Test

To demonstrate that steam temperature going to the TBN can be controlled within allowable limits during rapid load change.

There were several significant Logic Modifications that were executed to improve the tuning characteristics at all loads.

Load Condition	Change Rate	Allowable Steam Temperature Range	Test Result
Between 30 and 50% NR	15MW(3%NR) per Minute	+/- 8℃ (Main Steam)	+/- 8℃
Between 50 and 100% NR	25MW(5%NR) per Minute	+/- 8℃ (Main/RH Steam)	+/- 8℃

- the transient air flow increase/decrease rate
- The fuel compensation logic was modified so that the increase /decrease transient compensation rate would be different at different load rate.

Coal flow signal was time delayed to compensate for actual increase in coal flow into the boiler at both steady state and transient state of the boiler control, and so on.(Refer to Figure 1.2)

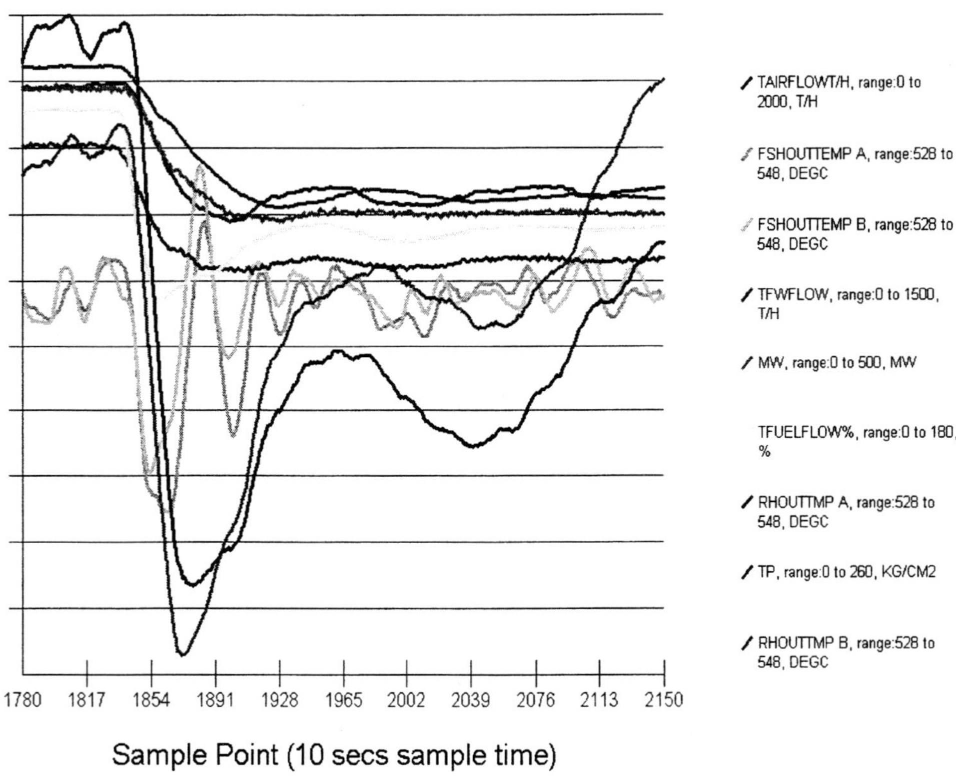

[Figure 1: Load Swing Test of 5%NR/Min]

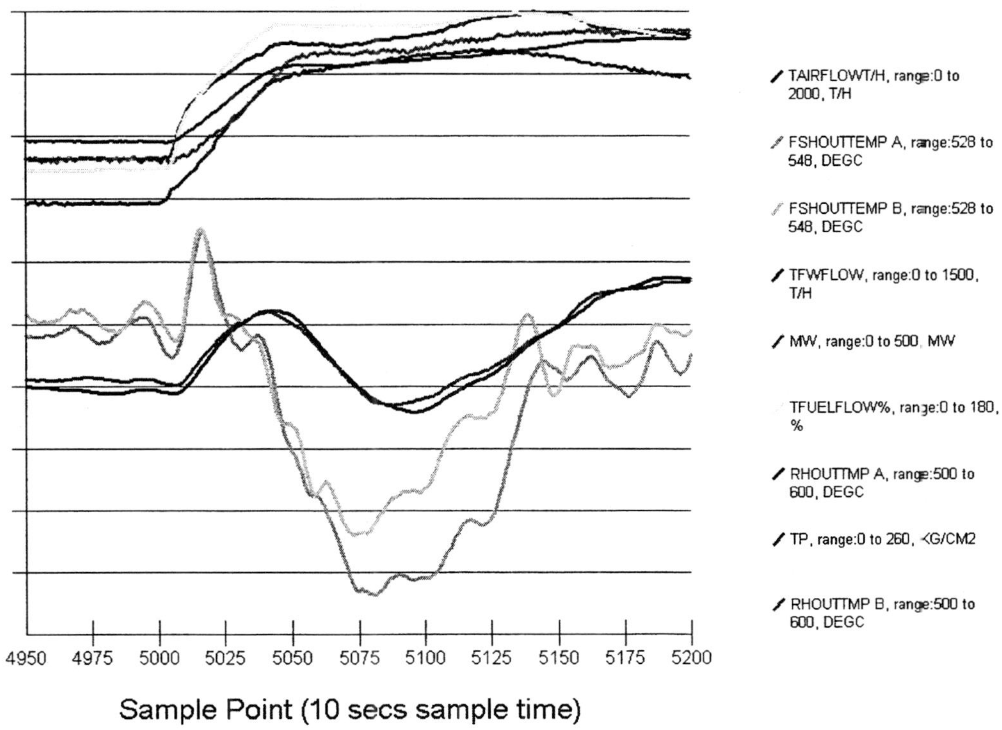

[Figure 2: Load Swing Test of 5%NR/Min]

2. Load Runback Tests

The objective of the load runback test is to demonstrate the capability for rapid, automatic of major boiler auxiliary equipment.

All boiler controls are in automatic mode for the duration of runback test. Therefore, no operator manipulation needs.

As soon as Runback is active, BLR and TBN Master blocks MW and Throttle Pressure error control.

TBN Master will release action of TP error block after 3minutes and BLR & TBN Master will initiate action of Throttle Pressure and MW error control to start actual Coordinate mode control after 15minutes.

During the test, we didn't experience any trip of the boiler that is related to faults in logic configurations.

However, The TBN related control tuning was changed to keep TBN Valve Position 10% less in order to the fast decrease in pressure

Besides, Furnace nose metal Temperature increased due to the heat that is still stored in the boiler metal doesn't get dissipated for more than ten to twenty minutes.

To compensate for the energy that is still reside in the boiler, we have added the 80T/H of feedwater.

Load Runback Test Matrix

Test No	Loss of major Boiler Equipment	Initial Load	Target Load
RB-1	Any FDF	100% NR	55% NR
RB-2	Any CIDF	100% NR	55% NR
RB-3	Any PAF	100% NR	55% NR
RB-4	Any BFPT	100% NR	50% NR

(Refer to Figure 3)

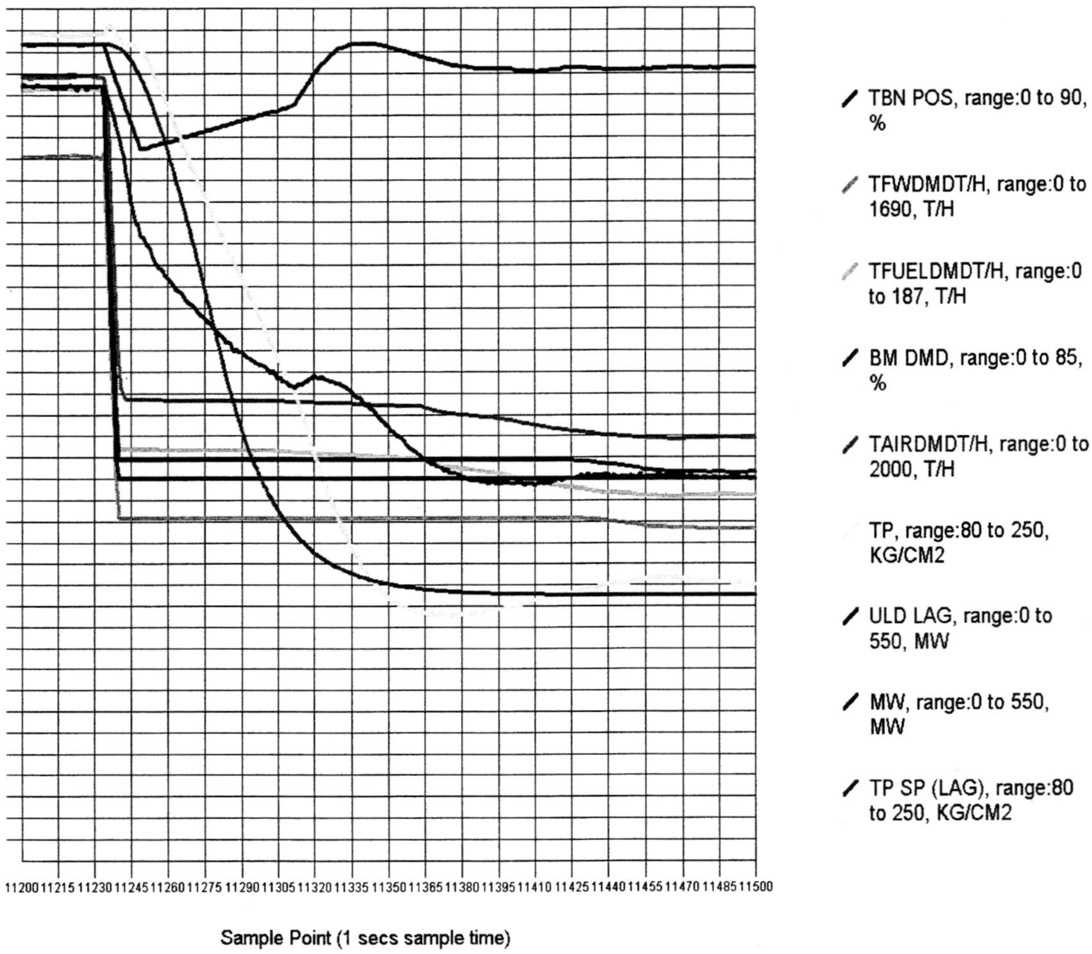

[Fig 3: CIDF Tunback Test]

3. AGC Test

○ AGC(Automatic Governor Free Operation Test)

Operation Spec calls for the 5% Speed Regulation, which is equal to 180RPM speed error to 550MW load change rate.

The test result came out to be better than 5% speed regulation

The reason why we have such a good regulation rate than other 500MW power plant is as followed.

- The turbine is not in sequential mode of operation. So the valve sequential opening nonlinearity is not in this particular boiler at this time

- The MW feedback is seven seconds time delayed. The time delay would increase/decrease the MW demand further up/down and resulted in bigger speed regulation than turbine adjusted regulation rate itself. (Refer to Figure 4.)

Taean #5 Gov Free(Sep 27, 23:00~23:30)

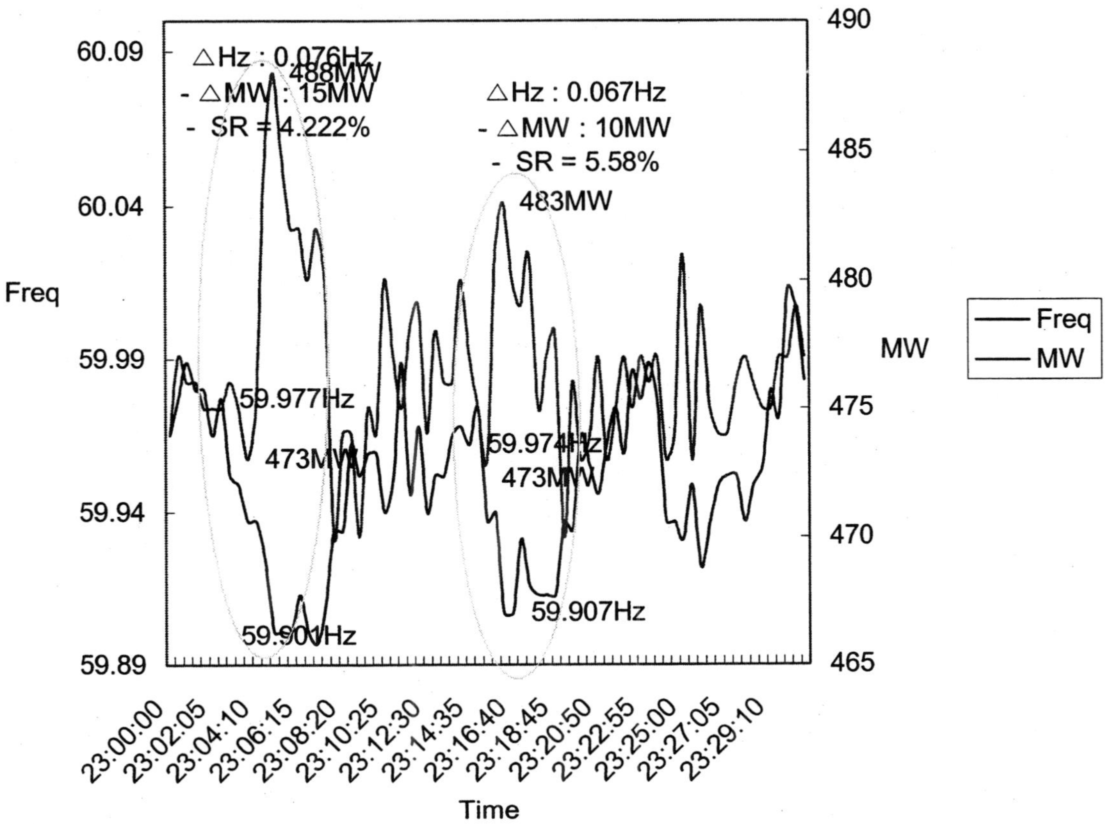

[Figure 4: Governor Free Operation Test]

4. APS(Automatic Plant Start up Test)

The test objective is to demonstrate that the steam generator can be started within the specified times from light-off to full load.

The start-up time gurantees are summarized in below table

Time(minutes) required from light-off to :	Steam to TBN	Full Load	Gurantee
Cold Start - 96 hours after shutdown	120	420	×
Warm Start - 56 hours after shutdown	60	150	×
Hot Start - 6 to 12 hours after shutdown	30	90	○
Restart - 2 hours after full load trip	20	90	×

The automatic progress by plant computer(APS) was performed through the progress of 3 break points after selecting "APS IN" in CRT.
Before the APS test, all MBC, APC, BOP logic was tuned and setup so that the control would wait for the auxiliary equipment to be stabilized before the start of the next step.
We had to modify the originally designed APS logic to adopt to the current Taean Unit #5,6 boiler specific status including mills, warm up burners & fans. Especially for the mills start up related sequence was modified significantly due to the difference between original mill operating philosophy and current site specific mill operating philosophy.
Since the original mill start up logic allows feeder to operate at 5.7T/H, starting up mill at load of 150MW was possible. But the site mill operation is only possible with minimum of more than 18T/H coal rate.
This particular condition makes it very hard for us to operate warm up burner pairs and two mills operating at the same time.
So we have to decide to remove the two mills in service condition from the Load Up 1 completion logic.
(Refer to Figure 5)

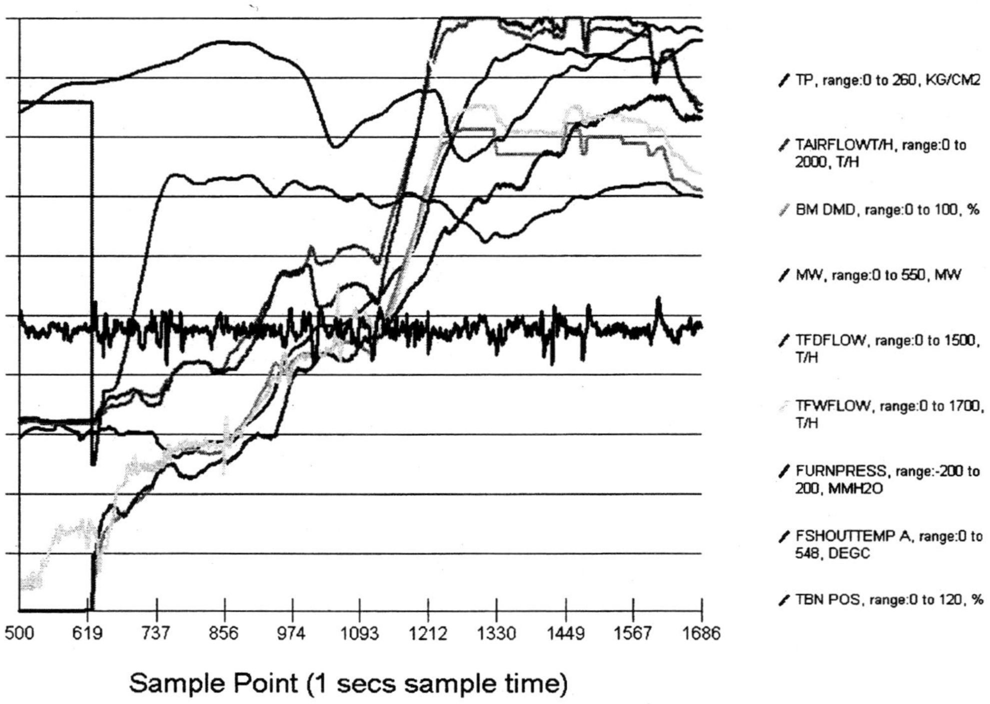

[Figure 5: APS Hot Startup Test]

III. Conclusion

Having vision of 21C Power Industry Leader, we have put our heart and soul into perfect commissioning, technologic independence, growth and development all through commissioning.
As a result, Taean #5 started initial synchronizing 0n July, 20, 2001, commercial operation on Octorber. 16. 2001. Tuning and optimization for control system was performed to fit the site condition of the boiler so that the highest performance efficiency could be drawn relatively in comparision with other Korea 500MW power plant.

Satisfactory Results of commissioning we got are as follows,

- We carried out APC Load Ramp Test tuning for a week and SH/RH steam temperature fluctuation has been improved within operational range at all load changes.
- We didn't experience any trip of boiler during all Runback test.
- We started commercial operation from initial synchronizing at the shortest time.

We have accumulated a lot of know-how and experience regardless of short period.
Now we are going to hand over ours facilities improvement and cases of failure to obtain better construction of following unit.

References

1. Bailey CAD EWS Logic(2000), Bailey Controls Company
2. Field Test Program for Taean #5,6(2001), Doosan Heavy Industries
3. Final Tuning Report(2002), Bailey Controls Company
4. The Control of Boilers(1986), ISA
5. Instrument Engineer's handbook Process Control(1995). Chilton Book Company

ELSEVIER
IFAC
PUBLICATIONS
www.elsevier.com/locate/ifac

STRATEGY OF CONTROL OF EXTENSIVE DISTRICT HEATING SYSTEMS

Jaroslav BALÁTĚ, Bronislav CHRAMCOV, Michal PRINC

Institute of Information Technologies,
Faculty of Technology, Tomas Bata University in Zlín,
Mostní 5139, 760 01 Zlín, Czech Republic,
E-mail: balate@ft.utb.cz, chramcov@ft.utb.cz, princ@ft.utb.cz

Abstract: The District Heating Systems (DHS) are being developed in large cities in accordance with their growth. The DHS are formed by enlarging networks of heat distribution to consumers and at the same time they interconnect the heat sources gradually built. DHS is used in larger cities of some European countries e.g. in Germany, France, Denmark, Finland, Sweden, Netherlands, Czech Republic, Poland and others. Production technology of heat by means of combined production of power and heat (CHP) is an important way to increasing of thermal efficiency of closed thermal loop. The paper shows the system access to the control of extensive DHS - controlled plant. It concerns automatic control of technological string "production, transport + distribution, consumption" of extensive district heating and that is the contribution of this paper. Control by means of advanced control algorithms is a tool (up to now neglect) for decreasing cost of energy and increasing the level of environment protection. *Copyright © 2003 IFAC*

Keywords: District Heating, Control Algorithms, Control System, Hierarchical Control, Combined Heat and Power – CHP, Hydrothermal Power Systems.

1. INTRODUCTION

Cogeneration of Power and Heat is an important way to increasing of thermal efficiency of closed thermal loop. The experiences in design of Control Strategy for Extensive District Heating System in the towns Brno and Prague in Czech Republic are summarised in the paper. It involves the connection of main author's operational experiences gained during many years of his work in the Power and Heating Plant and his further scientific - research activities on the technical university in co-operation with his colleagues, PhD students and with his students. The design of control strategy shows the basic concept of control methods of the district heating system of specific locations in town Brno and in town Prague. Each district heating system has its specific features and therefore it is necessary to create a philosophy of control for each of them. From the point of view of control, this philosophy consists of both general regularities and special features of a specific locality. The idea of a system approach to design of technological string control "production, transport + distribution, consumption" resulted from the specific solution of the way of control in real time and also from short-time preparation of district heating operation in region of Brno, Czech Republic during the last decade of 20th century. The knowledge of operation and experience motivated the author to creating the methods and conception of control of this technological string.

At present it is known from the literature that the problems of optimum control of combined heat and power production (CHP) in sources and also systems of automatic heat consumption control are solved only separately. Very few attentions is paid to analysis of static and dynamic behaviour of heat networks and utilization of these features for operation control of these networks. There are no any works dealing with elimination of transport delay in transport of heat in heat networks. Publications dealing with system approach to control of the technological string as a controlled plant are missing at all. This fact was the motivated cause for solving a new problem.

2. CHARACTERISATION OF CONTROL OF DISTRICT HEATING SYSTEM (DHSC)

The function of DHS is to ensure permanently all economically justified and socially necessary demands for supply of energy to all consumers namely at minimum costs and when respecting further important aspects e. g. of environment. The demands of consumers on the delivered energy can be specified by the following requirements to:

- provide delivery of justified quantity of energy to all consumers according to their needs varying in time,
- provide stable and uninterrupted delivery of energy,
- provide permanent delivery of energy within the limits of specified quality indexes (e.g. temperature of hot water in hot-water piping, steam pressure and temperature in steam piping or possibly frequency and voltage of electric energy at combined production of electric energy and heat etc.)

It is necessary to harmonise the demands of consumers on energy with the requirement of maximum operating economy of the whole DHS i.e. both in the production and delivery of heat with emphasis on complying with ecological indexes in the respective locality.

Economical control of the subsystem of heat energy production consists of two tasks i.e. of *economical distribution of load between separate co-operating sources in circle heat network including economical load distribution between separate production units inside the sources and further also of determination of the suitable composition* of co-operating sources including determination of suitable composition of co-operating units inside the source as well as *in determination of suitable economically based starting and switching off the production units or possibly also the whole sources.*

The important fact is the production technology using *combined production of power and heat.*

Economical operation of the heat distribution network subsystem, heat exchanger stations and consumer's equipment require heat output control in three levels of hierarchy:

- control in the source of heat (on the input into the heat network; or possibly the forced limitation of heat supply);
- control in the transfer station;
- control in the system of consumers.

Economical control in the system of consumers is fully in competence of consumers. Control of heat exchanger stations operation is from the point of view of automatic control quite autonomous; however here it is necessary to think over also further possibilities namely at large consumers.

For heat supply control in DHS the heat supply control in the source of heat is decisive i. e. on the input into the heat network.

The system of district heating - for centralised heat supply is by its character large extensive system. There exists a great number of dependent variables (e.g. relative increases of outputs of separate production units of sources); it is necessary to fix their value always for a certain value of one or more (of vector) independent variables (e.g. change in demand for energy supply to consumers, influence of change in combined production of heat and electric energy, influence of weather conditions, changes of parameters of fuel supplied to the sources etc.).

In this paper there is considered the circular steam heat network supplied by several heat sources. In these sources there are heat exchangers steam-hot water for supplying heat to consumers in autonomous (local) radial networks.

3. BRIEF CHARACTERISTIC OF SPECIFIC LOCALITIES OF DHS

3.1 DHS in Brno

According to the analysis the supposed locality has approx. 400 thousand permanently resident people namely in approx. 164 thousand flats. The total heat input has been 3500 MW_t. This total heat consumption is supplied approx. 30% from Heat Supply System. It is a circle steam network with local radial hot-water networks in each cooperating heat source. Delivery of heat by means of hot-water piping from the nuclear power plant situated in a distance of 42 km was considered (Balátě, 1988).

The following terms are used in further text:

- *District Heating System (DHS)*, system with co-operating Power and Heating Plant and Heating Plants - to be understood as present heat sources and heat networks in the supposed locality,
- *Integrated District Heating System (IDHS)* - to be understood as co-operation of DHS with the heat source in nuclear power plant situated in distance about 40 km; according to the design it had to supply

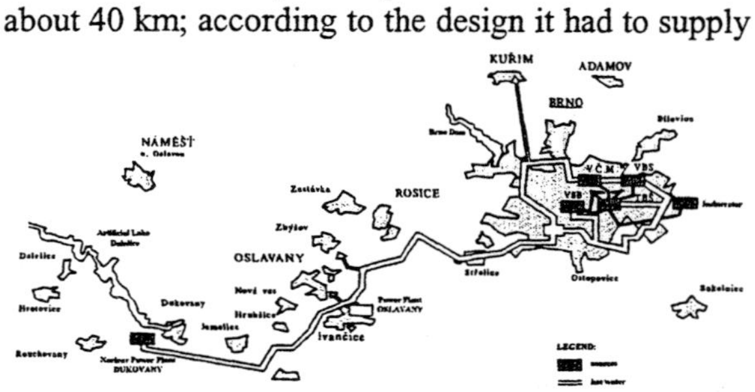

Fig. 1. Integrated District Heating System.

heat using the heat feeder to the blocks of flats in housing estates of the town or possibly to the other connected consumers. Principal scheme of the design to build up an IDHS is on Fig.1. The design including the perspective heat network development in supposed locality and by using the heat source in nuclear power plant is shown on Fig. 2.

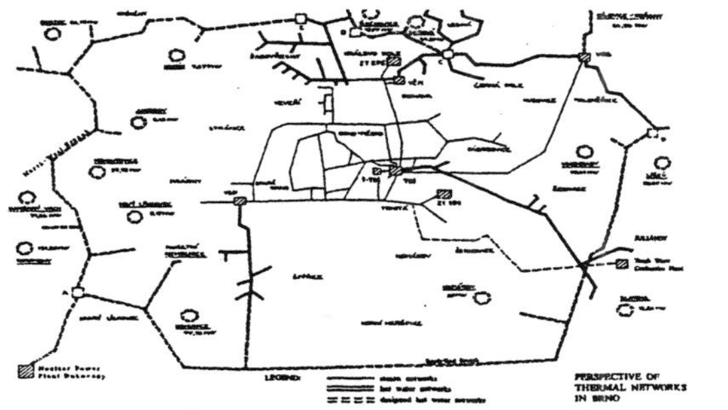

Fig. 2. Perspective of thermal networks

The used abbreviations have the following meaning: TBŠ - Power and Heating Plant, Š - TBŠ - Peak Boilers in TBŠ, further VSB, VČM, VBS - Heating Plants, ZT KPS, ZT 1BS, ZT Zetor – Power and Heating Plants in industrial factories, TKO - Incinerating Plant of Solid Municipal Waste, JE - Nuclear Power Plant.

It is obvious to see from these figures that IDHS forms the original (present) sources which are interconnected by circle steam network and the heat feeder from the nuclear power plant in which had to supply heat in hot water as heat - transfer medium to blocks of flats in new housing estates on the outskirts of town. These housing estates are at present equipped by gas boiler rooms but these should have been reconstructed to exchange stations connected to *north-west and south-east branches of hot water by-pass piping* of the supposed town. The heat sources TBŠ, VSB, VBS and also VČM have their own heat exchangers which deliver heat to their own autonomous hot-water systems. The aim was to interconnect these autonomous hot-water systems with branches of hot-water by-pass piping. Thus the extensive hot-water network had to come into being which would be equipped by sector closures for the possibility to realise changes in network configuration in dependence to the heat consumption in blocks of flats.

3.2 DHS in Prague

It is a hot-water net with installed heat output of 1995 MW_t and el. output of 138 MW_e. The CPH source cooperating from the distance of 36 km is connected to the original DHS. It is an analogical problem as above-mentioned way of control (Balátě, 1996).

4. PHILOSOPHY OF CONTROL METHOD

Philosophy of control is according to author's opinion a basic view of control (i.e. perhaps also by a manual method on a base of operators's experience) and this method (way) of control has to be made objective by means of utilization of objective methods of control supported by control methods using the method of automatic and automated control or using the method of advanced control algorithms.

From the discussion with the operating organisation in locality of DHS in Brno ensued that the control will be realised as follows:

1. Short-time and long-time preparation in dependence to the development trend of weather, co-operation with the superior energy dispatching and with the heat supply dispatching of operating organisation.
2. Operational control in real time in dependence to immediate heat consumption.

The aim of the control has been fixed:

To utilise the source of heat in nuclear power plant as much as possible; to supply the rest of necessary heat from the existing sources of DHS with the requirement of minimum heat consumption in fuel and fulfilling ecological indexes.

For fulfilling the aims of control the control system should ensure activities illustrated by the algorithms of control in the scheme on the Fig. 3. The separate blocks represent activities which concern both the preparation of heat production i.e. covering the predicted course of daily heat supply diagram and control of technological string in real time.

The Incinerating Plant TKO has a specific standing. The task of the incinerating plant is the liquidation of solid municipal waste and therefore the heat production is a secondary product. It will be a duty of IDHS dispatcher to withdraw the heat produced in the incinerating plant and to include it to the bottom part of HSDD in steam i. e. to the area of <u>basic load</u>. The supplied outputs produced in Power and Heating Plants of Factories will be located in the same way.

<u>Long-time and short-time preparation</u> includes the activities specified in paragraphs No. 1, 2, 3 and 4.

<u>Operational control in real time</u> includes the activities according to paragraphs No. 5, 6, 7, 8, 9 and 10. The sequence and reassuming the separate activities is obvious from the Fig. 3.

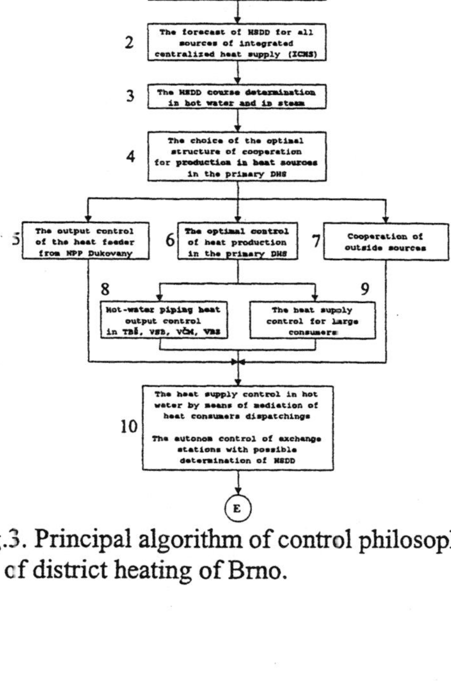

Fig.3. Principal algorithm of control philosophy of district heating of Brno.

5. SUMMARY OF THE METHODS USED FOR SOLUTION OF AUTOMATED CONTROL OF IDHS

The following methods has been or are solved in our laboratory:

1. For heat production:

1.1. Optimisation of heat production in production plant by the method of linear or non-linear programming.

1.2. Making up criterion of regime optimum for co-operation of power and heating plants and heating plants by heat supplying in district heating system namely from the point of view of costs decreasing of produced heat and power and of increasing the environment protection.

1.3. Making up algorithm for selecting optimum operational composition of sources of DHS including elaborating methods of calculating costs of a change of DHS operational composition or operational variant in separate sources.

1.4. Analysis of static and dynamic features of main technologic equipment of DHS sources.

2. For heat distribution:

2.1. Analysis of static features of heat networks and their utilisation for heat distribution control namely:- steam networks, - hot-water networks.

2.2. Making up algorithm for qualitative-quantitative method of control of heat delivery by hot-water piping using continuously calculated prediction of heat supply daily diagram (HSDD). The function of the algorithm is obvious from the Fig.4. (Balátě, 1997; Balátě and Sysala, 1998)

2.3. Automation of operation of consumer's heat exchangers including visualisation of technologic process.

3. For control of the whole DHS:

3.1. Co-ordination of sub-system control of the DHS i. e. of production plants and distribution of heat.

3.2. Utilisation of prediction calculation of course of daily diagram of heat supply for control of DHS.

3.3. Design of function of Heat Supply Dispatching including its co-operation with Heat Supply Dispatchings of operation organisations or possibly with direct large consumers.

3.4. Design of instrumentation of Heat Supply Dispatching by discrete distributed control system covering corresponding hierarchy of control.

The following part of this paper brings near the analysis of the static features of heat networks and their use for heat distribution control and of the prediction calculation of the course of heat supply daily diagram.

6. STATIC ANALYSIS OF HEAT NETWORKS OPERATION

As a tool for the analysis of behaviour of the heat networks, a calculation program for simulation of pressure, through-flow and temperature relations in distributing steam and hot water networks was created. For the calculation of the mass flow in the separate branches of a circular piping system a method called *'circle flow'* was used. It is an analogy with the method of loop current, which is very often used in electrical engineering. The mass flow corresponds here to the electric current, the pressure is proportional to the voltage and the hydraulic resistance corresponds to the electric resistance. If we apply the *1st and 2nd Kirchhoff's Law* to the defined network, we obtain a system of non-linear algebraic equations. The procedure is shown on a simple network having four points of junction and six branches (see Fig. 5)

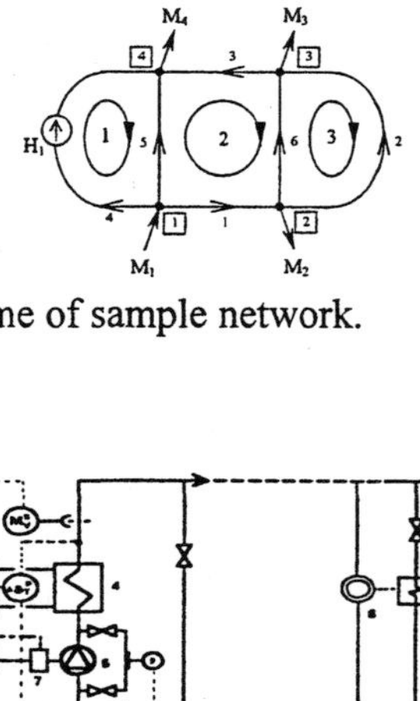

Fig. 5. Scheme of sample network.

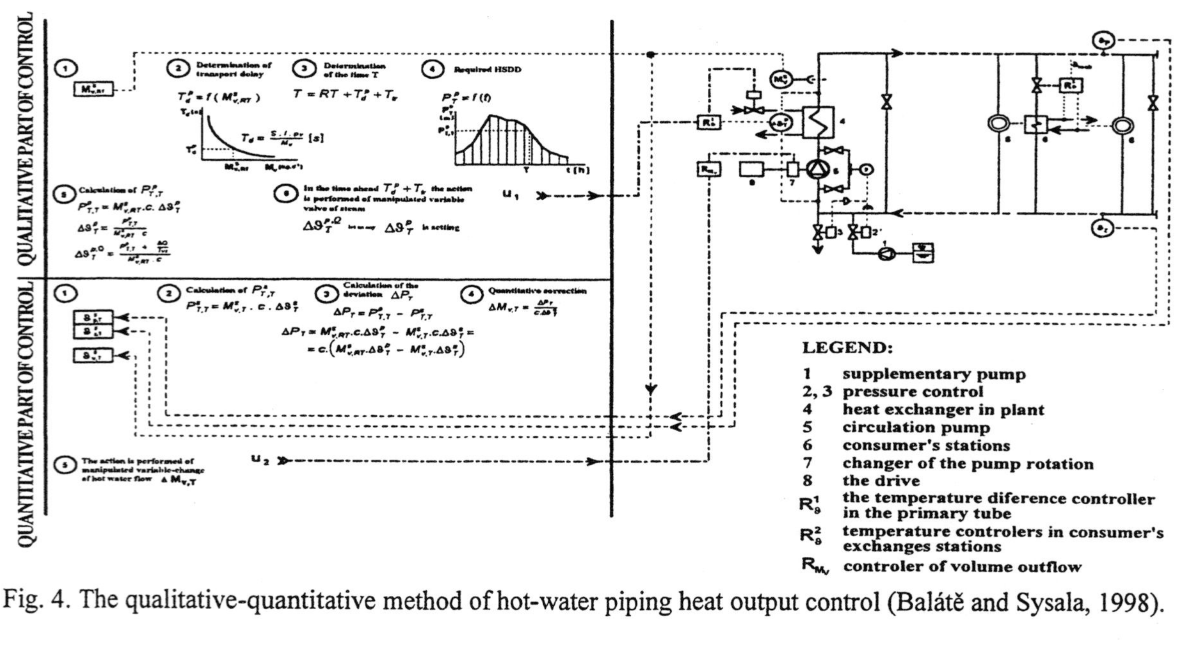

Fig. 4. The qualitative-quantitative method of hot-water piping heat output control (Balátě and Sysala, 1998).

We mark the unknown mass flow of the liquid in the i-th branch by symbol x_i, the taken away/delivered quantity in the i-th point of junction by symbol M_i and the active pressure in the i-th branch by H_i. From 1st Kirchhoff's Law it results: The sum of all flow quantities for a determined point of junction equals zero. Therefore, we can write the following equations of balance:

Point of junction 1 $\quad -x_1 - x_4 - x_5 + M_1 = 0,$ (1a)

Point of junction 2 $\quad x_1 - x_2 - x_6 - M_2 = 0,$ (1b)

Point of junction 3 $\quad x_2 - x_3 + x_6 - M_3 = 0.$ (1c)

The pressure losses in branches Δp arising at flowing of the liquid through the circular piping can be calculated according to the following formula

$$\Delta p = \left(\frac{8L}{\pi^2 d^5 \rho \left(2\log\left(3.72\frac{d}{k}\right)\right)^2} + \frac{8\sum\xi}{\pi^2 d^4 \rho} \right) x^2 = r.x^2 \quad (2)$$

where: L [m] - length of straight sectors of the piping, d [m] - inside diameter of the piping, ρ [$kg.m^{-3}$]- specific mass of the flowing liquid, k [m] - absolute roughness of the inside surface of the piping, $\Sigma\xi$ [-] - sum of the coefficients of the inside resistance of the piping, x [$kg.s^{-1}$] - mass flow through the piping, r [$Pa.s^2.kg^{-2}$] - hydraulic resistance of the piping.

For the chosen loops 1, 2 and 3 (see Fig. 5) we can write according to the formula (2) and according to the 2nd Kirchhoff's Law the following equations for the pressure losses in branches:

Loop 1 $\quad r_4|x_4|x_4 - r_5|x_5|x_5 - H_1 = 0,$ (3a)

Loop 2 $\quad -r_1|x_1|x_1 - r_3|x_3|x_3 + r_5|x_5|x_5 - r_6|x_6|x_6 = 0,$ (3b)

Loop 3 $\quad -r_2|x_2|x_2 + r_6|x_6|x_6 = 0.$ (3c)

To solve the system of non-linear equations represented in our case by equation (1) and (3) we shall use the Newton's iterative method.

The calculation runs on the basis of determined network parameters, source delivery and the consumer's demands. The results of calculation can be shown on the screen in the form of tables or simple column graphs, see Fig.7. Presented data are results of the calculation, which was worked out for the heat network described in the Fig.6.

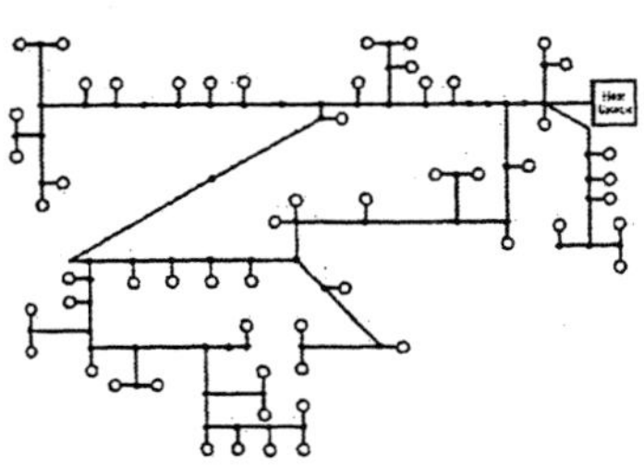

Fig.6. Network topology

The practical importance of the program is not only for design of distributing networks, but above all in operation of district heating systems. The possibility of simulation of relations in the network for various operating conditions, i.e. for instantaneous output of sources or for instantaneous consumptions is very useful for dispatchers in power and heating plants. This can be a contribution not only from technical aspects but also from economic ones.

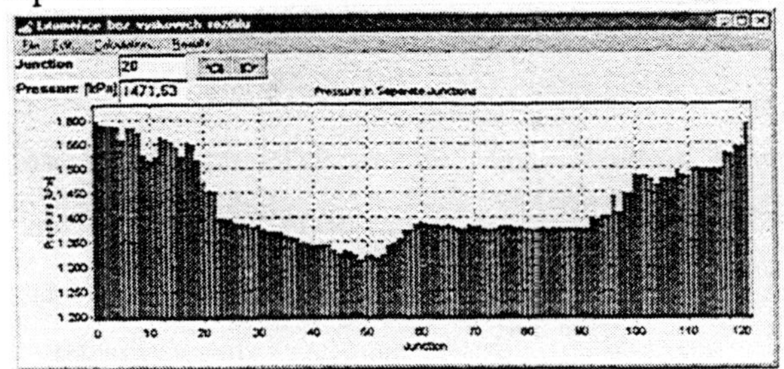

Fig.7. Results presentation in graphs.

7. PREDICTION OF COURSE OF HEAT SUPPLY DAILY DIAGRAM

We can find an application of time series prediction also by the control in the Centralised Heat Supply System (CHSS), especially for control of hot-water piping heat output. The forecast is determined for two ways of using:

1. The predictions of whole HSDD. Forecast of HSDD (see Fig.8) of whole CHSS is utilised for the purpose of heat production control and thus for the purpose of the optimal distribution loading between cooperative production sources and production units inside this sources.

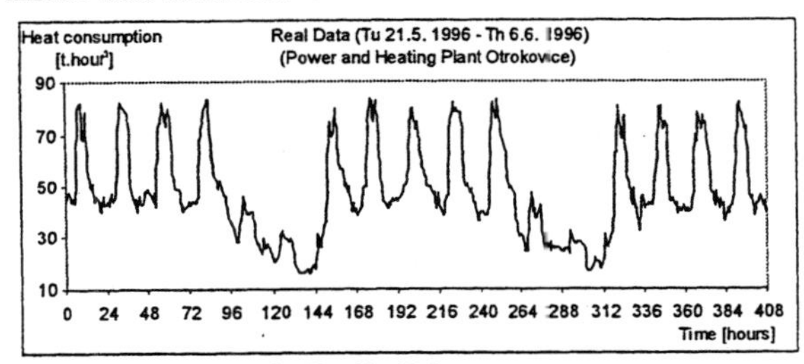

Fig. 8. Heat Supply Daily Diagram.

2. The prediction is determined for continuous acquisition of necessary heat output ahead of the time. It depends on transport delay, namely in the range of 2 up to 16 hours depending on distance of heat sources from consumers, and it is different for each locality.

The methods of forecast. Most of methods which solve the prediction of HSDD are based on mass data processing. But this method has a big disadvantage. It consists of real data out of date. From this point of view it is available to use the forecast methods according to the methodology of Box –Jenkins (Box and Jenkins, 1976). This method works with fixed number of values, which are updated for each sampling period. *The Box-Jenkins (BJ) method.* This methodology is based on the correlation analysis of time series and it works with stochastic models, which enable to give a true picture of trend component and also of periodic components. The course of time series of HSDD contains two periodic components (daily and weekly period). But general model according to BJ enables to describe only one periodic component. We can propose two possible approaches to calculation of forecast to describe both periodic components.

- **The method that uses the model with double filtration**
- **The method – superposition of models**

First we introduce simplified form (5) of general model according to BJ for the next using when there is used substitution in the form (4). We can find more detailed analysis of general model in work.

$$F=\Phi_P^{-1}(B^s)\cdot\varphi_p^{-1}(B)\cdot\Theta_Q(B^s)\cdot\theta_q(B)\cdot\nabla_s^{-D}\cdot\nabla^{-d} \quad (4)$$

$$z_t = F\cdot a_t \quad (5)$$

The method that uses model with double filtration. We can describe model with double filtration through the substitution (4). The model in the form (6) is the result of it.

$$z_t = F\cdot\nabla_{s^*}^{D^*}\cdot a_t \quad (6)$$

where: D – degree of seasonal difference – daily (in equation 4), D* - degree of seasonal difference – weekly, s – daily period (in equation 4), s* - weekly period

It is important to adhere to the general plan for using the method that uses model with double filtration for calculation of HSDD prediction. The model in the form (6) enables to describe the HSDD course (i.e. it describes daily periodic component and also weekly one). It can be used for analysis and prediction of following regular influence of calendar (Saturday, Sunday).

The method – superposition of models. We can use second method i.e. superposition of models for elimination of regular influence of calendar. This method is being used on two models in the form (5). These models are discerned by means of symbols * and **. The time series marked by symbol * is series of values of HSDD outputs in every hour (the sampling period is 1 hour). The time series marked by means of symbol ** is series of values of heat consumption per day (the sampling period is 1 day). The plan of calculating prediction by means of the these method is shown on the Fig. 9.

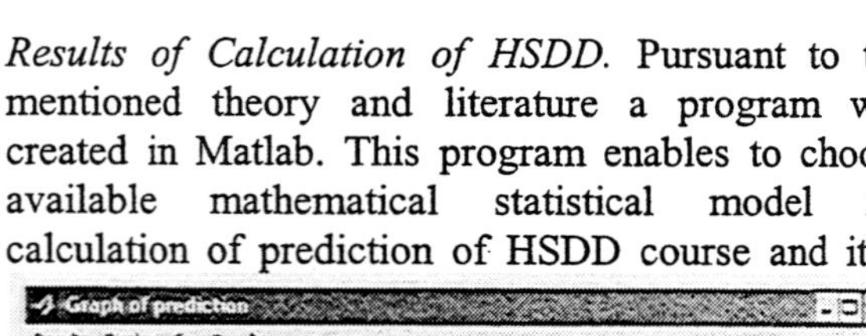

Fig. 9. Superposition of models

Results of Calculation of HSDD. Pursuant to the mentioned theory and literature a program was created in Matlab. This program enables to choose available mathematical statistical model for calculation of prediction of HSDD course and it is used for forecast of HSDD, which is an essential part of control of DHS by means of qualitative-quantitative method. All testing and calculations are based on lot of real data. These data were obtained in specific locality. Examples of results of the forecast calculations are shown on the Fig. 10.

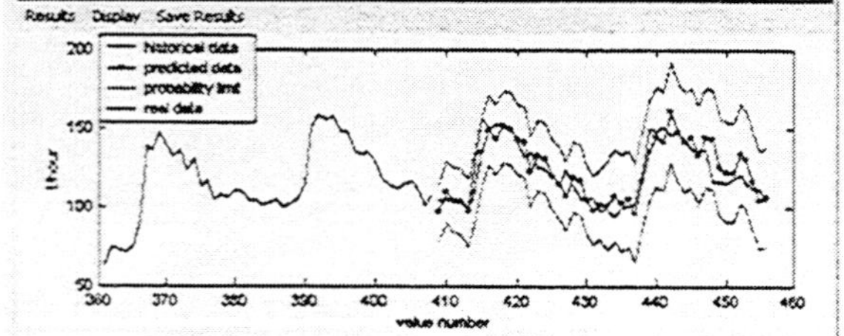

Fig. 10. Results of prediction of HSDD.

8. CONCLUSION

The complicated process of technological string of heat production, transfer and distribution of heat or possibly of heat consumption, as it was mentioned above, requires the system approach when solving the concept of automated control by means of advanced control algorithms. We believe that the intellectual process when determining the philosophy of control can be interesting and useful for the technical public. It is a demonstration of necessity to know technology for designers of control system.

It is necessary to stress that each DHS has its specific features which have to be solved specifically and individually.

We can inform readers that this idea of the presented paper was the base of the proposal in the frame of 5th Framework Programme of EC called: *Cost Effective and Environmental Compatible Operation of District Heating Systems by means of Advanced Control Algorithms.*

Acknowledgements: This work was supported in part by the Grant Agency of Czech Republic under grant No: 101/01/0345 and in part by the Ministry of Education of the Czech Republic under grant No. MSM 281100001.

REFERENCES

Balátě, J. at al. (1988). *The study of district heating control in Brno*, Research work No: VZ-HSZ-070 in Czech, TU Brno, Faculty of Mechanical Engineering, Brno.

Balátě, J. (1996). *District heating in Prague - Design of Philosophy of Control*, Research Work No: KAŘT-PTAS/1996 in Czech, TU Brno, Faculty of Technology in Zlín, Zlín.

Special appreciation with the gold medal of the patent No: 279253 - CZ:

Balátě, J. (1997). *Controle de la production thermique dans circuits de chaleur et méthode utilisée.* 46th World Exhibition of Innovation, Research and New Technology, Brussels EUREKA´97, Brussel, Belgium.

Balátě, J. and T. Sysala (1998). The Way of District Heating Output Control by Means of Hot Water Piping. In: *Preprints: 5th IFAC Workshop on Algorithms and Architectures for Real-Time Control – AARTC ´98Cancun*, Mexico.

Box, G.E.P. and G.M. Jenkins (1976). *Time Series Analysis, Forecasting and Control.* Holden Day, San Francisco.

ELSEVIER
IFAC
PUBLICATIONS
www.elsevier.com/locate/ifac

START-UP SCHEDULE OPTIMIZING SYSTEM OF A COMBINED CYCLE POWER PLANT

Masakazu SHIRAKAWA* and Masashi NAKAMOTO

TOSHIBA Corporation, Industrial and Power Systems & Services Company,
36-5 Tsurumichuo 4-chome, Tsurumi-ku, Yokohama-shi, Kanagawa-ken, 230-0051 Japan
** e-mail: masakazu1.shirakawa@toshiba.co.jp*

Abstract: This paper presents a method to determine the optimal operational parameters of a combined cycle power plant. The proposed method combines a dynamic simulation with an optimization calculation based on the nonlinear programming method. The reason for the optimization of the plant start-up scheduling is being able to reduce the start-up time by keeping the thermal stresses in the thick parts of the heat recovery steam generator and in the steam turbine under their allowable values. *Copyright © 2003 IFAC*

Keywords: Power generation, simulation, dynamic modelling, optimization, nonlinear programming, control engineering.

1. INTRODUCTION

A combined cycle power (CC) plant that combines a gas turbine and a steam turbine can achieve a higher thermal efficiency and lower environmental impact, and have an excellent operation. Therefore, many CC plants have been installed around the world. The CC plants are required to have frequent start-up / shut-downs followed by quick loading in order to regulate power generation rates according to electricity power demands that frequently change. Consequently, the time reduction for start-up without shortening the machine lifetimes has become important for economical fuel consumption. The decisive factors for rapid start-up are thermal stresses caused by temperature gradients in the thick parts of the heat recovery steam generator (HRSG) and in the steam turbine. Important functions of the CC plant control system are to reduce the start-up time and to satisfy the operational constraints.

A start-up schedule for a conventional (simple steam cycle) thermal power plant is usually generated by a mismatch chart using the differential temperature between the steam and steam turbine rotor metal just before the steam turbine starts (Hanzalek and Ipsen, 1966). The conventional control system of the CC plant has basically applied the same method. In this method, turbine design engineers decide the schedule parameters in the mismatch chart during the turbine design stage and install it into the plant control system. Control engineers may adjust these parameters when plant operation problems are encountered during the commissioning operation. This heuristic approach, however, involves design and tuning costs, and it is difficult to obtain the optimal start-up schedule. Moreover, these parameters may not be adjusted during commercial operation after commissioning completion.

Several methods to shorten the start-up time have been proposed. A model predictive control for a conventional thermal power plant has been proposed (Nakai, *et al.*, 1996). This algorithm predicts the future thermal stress in the steam turbine rotor and calculates the optimal future profile of the steam flow rate. The CC plant, however, has many operational restrictions, not only thermal stress in the steam turbine but also loading rates of the gas turbine, temperature gradients of the HRSG, drum water levels of the HRSG and NOx emission from the plant, etc. A fuzzy expert system for a CC plant has been proposed (Akiyama, *et al.*, 1997). To obtain the start-up schedule, this system uses an engineer's experiences in fuzzy rules. It requires additional cost to obtain this engineering knowledge. Recently, nonlinear programming using a plant dynamic model has been studied (Bausa and Tsatsaronis, 2001). This method is applied to a simplified single pressure CC plant model. The model includes many assumptions and simplifications, thus it is not possible to study a practical problem.

This paper aims at proposing a method to determine the practical optimal start-up schedules using a dynamic simulation and nonlinear programming. This method does not require any adjustment of the schedule parameters by engineers. Also, a great deal of labor is not required in order to prepare the knowledge base.

2. COMBINED CYCLE POWER PLANT

2.1 Plant configuration

Many types of CC plants have already been developed. Here we discuss the multi-shaft type CC plant, which consists of two gas turbine units, two HRSG units, and a reheat-type steam turbine unit. The gas turbines and the steam turbine drive the generators. Also, the HRSGs generate steam for the steam turbine using waste heat from the gas turbines. The configuration of this plant is shown in Fig. 1. This plant generates a total output of 625 MW. The gas turbines are of the 1300 deg.C class. The HRSGs are the triple pressure and reheat type with a duct burner. The steam turbine has a 265 MW rated capacity.

2.2 Start-up scheduling method

The start-up process of this plant is shown in Fig. 2. Initially, the first gas turbine starts, purges, fires, and increases its speed toward the rated speed. After

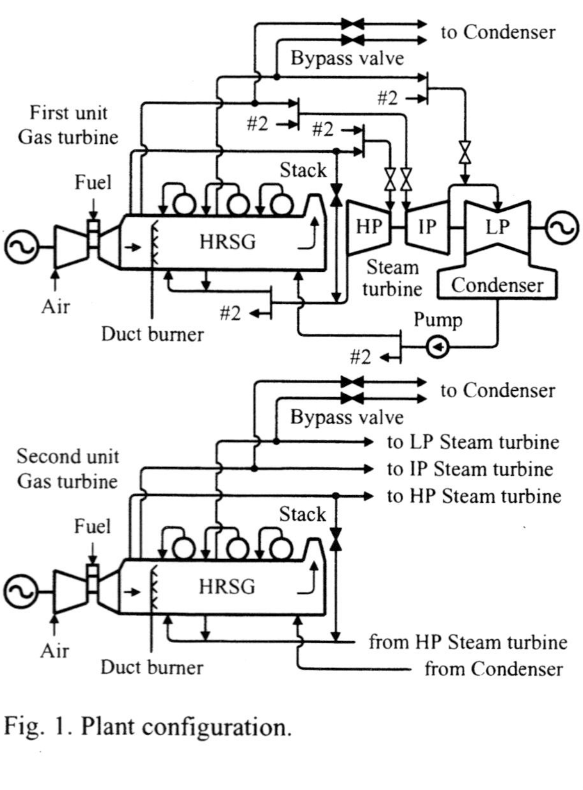

Fig. 1. Plant configuration.

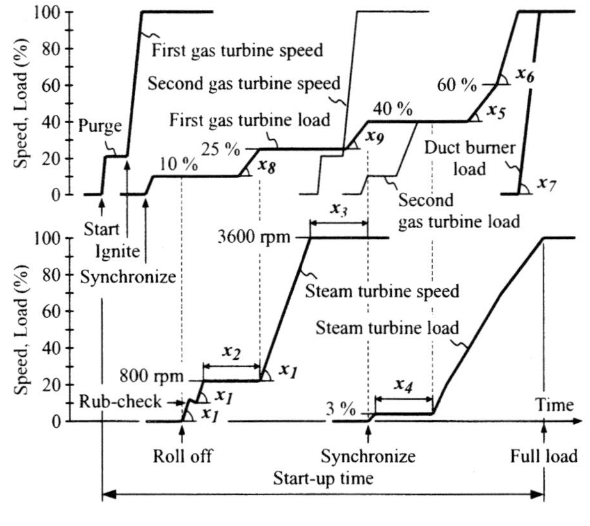

Fig. 2. Plant start-up schedule.

synchronizing, the first gas turbine goes to a minimum load of 10 %. The first HRSG warms at the minimum gas turbine load. When the steam properties have reached the required conditions, the steam turbine starts, and the steam turbine is put into the speed and load control mode. After rub-check, the steam turbine increases its speed toward a predefined speed of 800 rpm at an accelerating rate x_1, and maintains this speed during the low-speed heat soak time x_2. The steam turbine increases its speed again toward the rated speed, and maintains this speed during the high-speed heat soak time x_3. After synchronizing, the steam turbine goes to an initial load of 3 %, and maintains this load during the initial-load heat soak time x_4. During these heat soak operations, the first gas turbine increases its load to predefined loads of 25 % and 40 % at loading rates x_8 and x_9, respectively, so that the steam flow rate should be secured to drive the steam turbine. After completing the initial-load heat soak, the bypass valves of the first unit are closed, then the steam turbine is put into the inlet pressure control mode. The second gas turbine and HRSG are started like the first unit. The second gas turbine goes to a merged load of 40 %, and maintains this load. The bypass valves of the second unit are then closed to merge the second HRSG steam into the first HRSG steam. Afterwards, both gas turbines increase their load again toward the rated load at different loading rates x_5 (40 % to 60 %) and x_6 (60 % to 100 %).

Lastly, after the gas turbines have reached the rated load, both duct burners fire at a loading rate x_7. The steam turbine reaches the rated load with a time delay, due to the delay in the steam generation in the HRSGs.

3. START-UP SCHEDULING PROBLEMS

The problem of the start-up scheduling is to determine the above schedule parameters ($x_1, \cdots, x_9$)

in order to minimize the start-up time while maintaining the operational constraints. The life span of the power plants is severely affected by thermal stresses that develop in the steam turbine and HRSG for reducing the start-up time.

Thermal stresses in the steam turbine occur on the rotor surfaces and rotor bores at the high-pressure (HP) part and the intermediate-pressure (IP) part. A large temperature difference arises between the hot steam and cold rotor metal when steam enters the steam turbine, and also, the temperature difference appears during the gas turbine load-up. As a result, the thermal stresses are developed by the temperature distributions in the rotors. The thermal stress of the thick part of the HP drum is the severest in the HRSG. In general, this thermal stress is estimated by the drum temperature gradient.

If the heat soak times (x_2, x_3, x_4) are extended, the steam turbine accelerating rate (x_1) and the gas turbine loading rates (x_5, x_8, x_9) decrease, then the start-up time becomes longer while the thermal stresses of the steam turbine are reduced, because the differential temperature between the steam and rotor metal decrease. On the other hand, the start-up time is shortened but the thermal stresses of the steam turbine become greater. Also, the temperature gradient of the HP drum depends on the gas turbine loading rates (x_5, x_6) and the duct burner loading rate (x_7).

The plant operational characteristics have been intensively studied using a dynamic simulation method (Maffezzoni, 1992), but it is not easy to determine the optimal operational parameters because it is necessary to iterate the dynamic simulation based on trial and error by the engineer's intuition and experience. Therefore, in a conventional operation method, constant values relating to each start-up mode such as the cold, warm-1, warm-2, hot-1 and hot-2 modes have been used for a plant operation, and these values are predetermined during the plant design stage.

4. CONTROL CONCEPTS FOR START-UP

4.1 Functional structure of the system

Fig. 3 shows the functional structure of the start-up schedule optimizing system proposed in this paper. This system is installed in a plant control system that consists of a computer system, machine control system and operator station (OPS).

The optimum schedule is determined by the optimization with the dynamic simulation. The initial process values for the dynamic simulation are estimated using the cooling curves, and the start-up initiation time comes from the control center on the previous day. Process values are estimated according to equation (1).

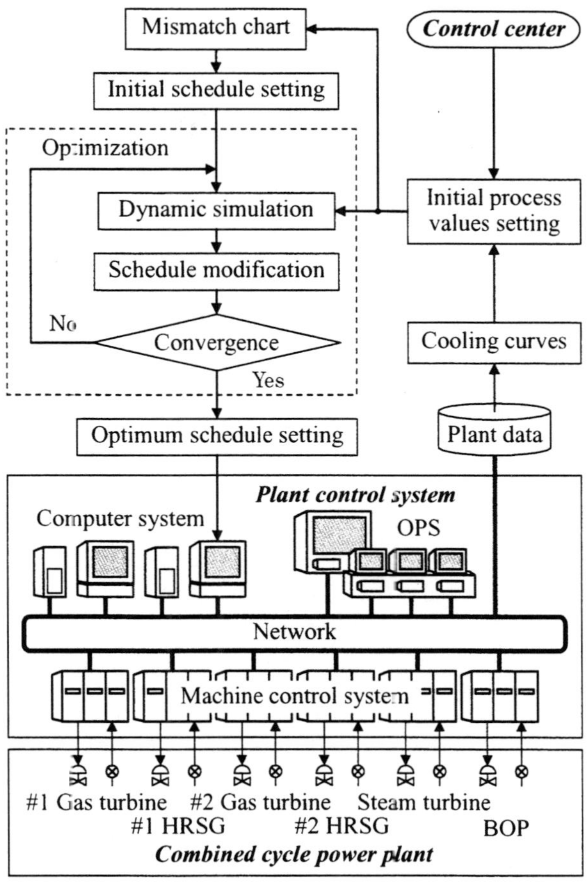

Fig. 3. Start-up schedule optimizing system.

$$P_{su} = (P_{sd} - P_a)\exp\left(-\frac{t_{su} - t_{sd}}{k}\right) + P_a \qquad (1)$$

where P_{su} is the estimated process values at the time of the first gas turbine start, P_{sd} is the process values at the time of the last shut-down, P_a is the ambient conditions, t_{su} is the first gas turbine start time, t_{sd} is the last shut-down time and k is the cooling constants. k can possibly be determined from the actual plant data.

The initial schedule parameters for the dynamic simulation are defined using a conventional mismatch chart. The dynamic simulation in accordance with the initial schedule simulates the plant start-up process. The start-up time and process values are obtained and used for the optimization. A nonlinear optimization method modifies the schedule parameters. The function for convergence judgment decides whether the schedule has reached the optimum point or not. If it does not converge, the modified start-up schedule is sent to the dynamic simulation again. And iterative calculations are automatically occurred until the optimum schedule is determined.

After determining the optimum schedule, the schedule parameters are set in the plant control system. The plant control system operates the CC plant according to the schedule obtained by the optimization.

4.2 Dynamic simulation model

The dynamic simulation model consists of the process equipment model and the control system model. The first principle is adapted to model the process equipment that consists of the gas turbine, HRSG, steam turbine and balance of plant (BOP). The control system is modeled in order to be equivalent to the actual control system. This simulation model is verified by comparison with actual plant data (Funatsu, *et al.*, 1999).

Gas turbine model. The gas turbine responds more quickly than the other equipment. The flow rate and the temperature of the exhaust gas are represented by functions of the fuel flow rate. The gas turbine duct has a time delay because of the large heat capacity.

HRSG model. The HRSG is composed of heat exchangers that have a long time delay when compared to the gas turbine and the steam turbine, and has a significant effect on the starting characteristics of the entire plant. The dynamic characteristics of each heat exchanger are calculated using the mass and energy balance equations as follows.

$$M_m C_m \frac{dT_m}{dt} = Q_{gm} - Q_{mf} \tag{2}$$

$$V_f \frac{d\rho_f}{dt} = G_{fi} - G_{fo} \tag{3}$$

$$V_f \frac{d(\rho_f H_f)}{dt} = G_{fi} H_{fi} - G_{fo} H_{fo} + Q_{mf} \tag{4}$$

where M is the mass, C is the specific heat, T is the temperature, Q is the heat flow, V is the volume, ρ is the fluid density, G is the fluid mass flow rate, H is the enthalpy and t is the time. The subscripts m denote the tube metal, g is the exhaust gas, f is the internal fluid, i is the inlet and o is the outlet.

Steam turbine model. The adiabatic efficiency of the steam turbine is defined as the function of the turbine pressure ratio. The steam flow rate through the steam turbine is calculated using the constant flow coefficient. The turbine power is calculated using the enthalpy difference between the inlet and outlet enthalpies of the steam turbine. No dynamic effect is evaluated in the flow and pressure relation. The thermal stress of the steam turbine rotor is obtained by the temperature distribution, which is calculated by the dynamic model of the heat conduction that divides the rotor into several vertical cylinders.

Pipe model. Since the steam pipe is long and has a large volume, the pressure in the pipe is modeled by the multiple parts and volumes in which the steam is stored.

4.3 Optimization calculation

The start-up problem can be formulated as a function of the optimization problem with operational constraints. Among the schedule parameters $(x_1, \cdots, x_9)$ in Fig. 2, seven parameters $(x_1, \cdots, x_7)$ are selected as optimizing objects and defined as follows.

$$X = [x_1, \cdots, x_i, \cdots, x_7]^T \tag{5}$$

The others (x_8, x_9) are defined by the following relational equations, respectively.

$$x_8 = \max\left(1.0\,[\%/\text{min}], \frac{25-10\,[\%]}{x_2\,[\text{min}]}\right) \tag{6}$$

$$x_9 = \max\left(1.0\,[\%/\text{min}], \frac{40-25\,[\%]}{x_3\,[\text{min}]}\right) \tag{7}$$

The objective function J_T is the start-up time from the first gas turbine start to the plant base load operation. The optimization problem is defined by equations (8), (9) and (10).

$$\min_X J_T(X) \tag{8}$$

$$\text{subject to } c_{mj}(X) \leq c_{lj} \tag{9}$$

$$c_{mj}(X) = \max_{0 \leq t \leq J_T} \left| c_j(X, t) \right| \tag{10}$$

where c_{mj} is the maximum process values during start-up, c_{lj} is its limitations, c_j is the process values, and t is the time. The subscript j stands for specific locations. The thermal stresses of the steam turbine are the HP rotor surface ($j=1$), HP rotor bore ($j=2$), IP rotor surface ($j=3$) and IP rotor bore ($j=4$). The temperature gradients of the HP drum are the first unit ($j=5$) and the second unit ($j=6$).

The operational constraints for equation (9) can be included in the objective function as the following penalty function.

$$J(X) = \alpha_1 \cdot J_T(X) + \sum_{j=1}^{6} \alpha_{2j} \cdot K_j \left(c_{mj}(X) - c_{lj} \right)^2 \tag{11}$$

$$\begin{aligned} K_j &= 0, \quad \text{if } c_{mj} \leq c_{lj}, j = 1, \cdots, 6 \\ K_j &= 1, \quad \text{otherwise} \end{aligned} \tag{12}$$

where α_1 and α_2 are the weighting coefficients.

In order to obtain a feasible solution, each parameter is limited as follows.

$$120\,[\mathrm{r/min^2}] \le x_1 \le 360\,[\mathrm{r/min^2}] \tag{13}$$

$$0.0\,[\mathrm{min}] < x_2, x_3, x_4 \le 90.0\,[\mathrm{min}] \tag{14}$$

$$1.0\,[\%/\mathrm{min}] \le x_5, x_6, x_8, x_9 \le 7.5\,[\%/\mathrm{min}] \tag{15}$$

$$1.0\,[\%/\mathrm{min}] \le x_7 \le 15.0\,[\%/\mathrm{min}] \tag{16}$$

The optimization method used in this paper is the successive quadratic programming (SQP) method (Fukushima, 1986). This method is a kind of nonlinear programming method using the value of the gradient. The gradient of the objective function is approximated as follows.

$$\nabla J(X) = \frac{J(X+\Delta X) - J(X)}{\Delta X} \tag{17}$$

where ΔX is the perturbation of each schedule parameter. $J(X)$ and $J(X+\Delta X)$ can be obtained by the dynamic simulation.

5. APPLICATION RESULTS

Numerical studies have been executed for all the start-up conditions. That is, the initial temperatures of the rotor are from 120 deg.C (standby period is over 100 hours) to 462 deg.C (standby period is 4 hours). A comparison of the conventional method and the proposed method is shown in Fig. 4. This figure shows the relationship between the standby period, the start-up time and thermal stress. Here, the normalized thermal stress shows the maximum value of the four rotors ($j=1$ to 4). For the conventional method (dash dot lines), there are five start-up modes, i.e., the cold, warm-1, warm-2, hot-1 and hot-2 modes that depend on the differential temperature between the steam and steam turbine rotor metal just before the steam turbine starts. The schedule parameters are determined by using the tables of each start-up mode. In this approach, the start-up times become discontinuous. In addition, the start-up times become long because of the thermal stress margins. On the other hand, the proposed method (solid lines) is able to reduce the start-up times by keeping the thermal stresses under the allowable values for all the start-up conditions.

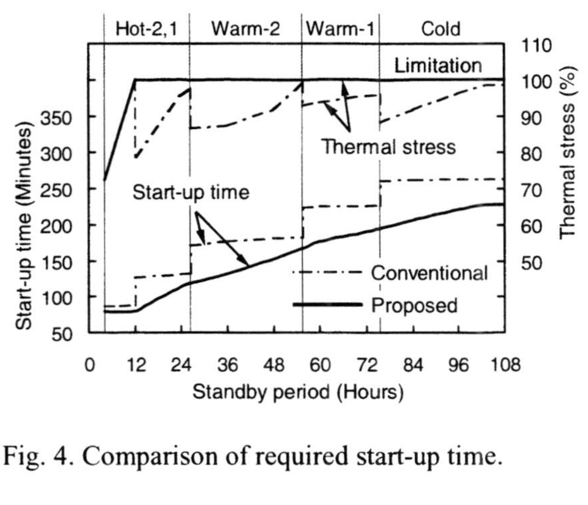

Fig. 4. Comparison of required start-up time.

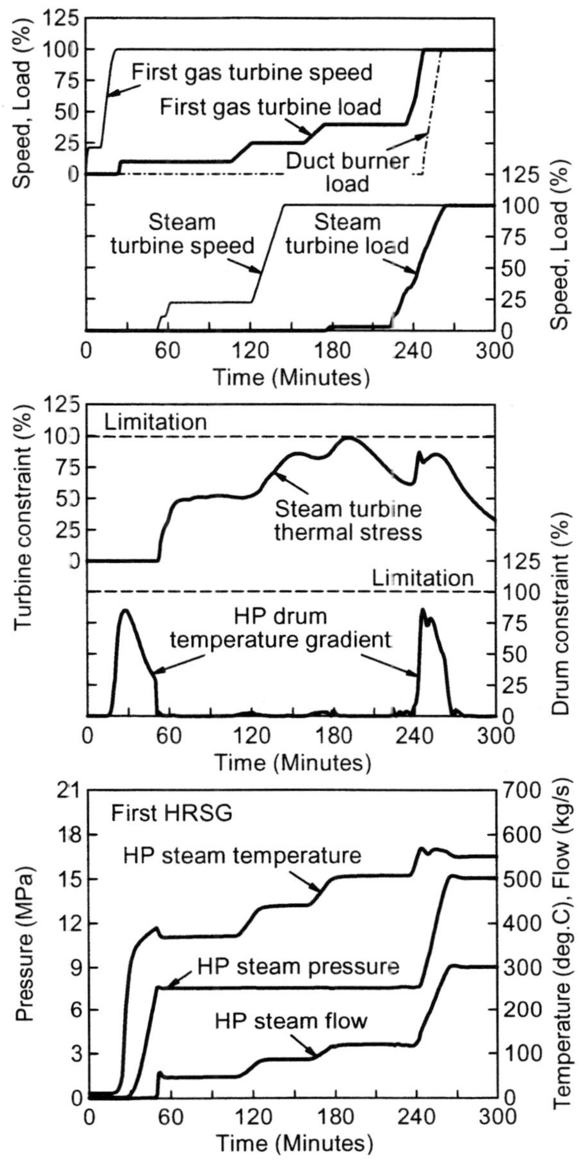

Fig. 5. Conventional method (Initial schedule).

As a typical example, the result of a cold start-up when the initial temperature of the rotor is 120 deg.C (standby period is over 100 hours) is explained. Fig. 5 shows the results of the conventional method. Here, the normalized thermal stress of the steam turbine is the maximum value of the four rotors ($j=1$ to 4), and the normalized temperature gradient of the HP drum is the first unit ($j=5$). The start-up time is 264 minutes. In this case, there are some margins for the operational constraints, and the start-up schedule is not adequate. Fig. 6 shows the results of the proposed method. In this case, the initial schedule parameters were same as the conventional method. The start-up time is 228 minutes. The proposed method reduces the start-up time by 36 minutes compared to the conventional method, and satisfies the operational constraints.

The path to search for the optimal solution is shown in the case of the warm-2 start-up in Fig. 7, where the initial temperature of the rotor is 258 deg.C (standby period is 48 hours). Iterating about ten times minimized the objective function (11), and the optimal solution is obtained. It is necessary to calculate the dynamic simulation eight (number of parameters + 1) times in order to obtain the gradient vector equation (17) for each iteration step. Therefore, the computation time strongly depends on the dynamic simulation. It takes about 2 minutes to calculate the dynamic simulation once using a personal computer (Intel® Pentium® 4 processor 2.20 GHz), so that it needs about 180 minutes to obtain the optimal solution. The computation time depends on the start-up conditions. In a word, the hot start-up condition requires a shorter computation time, and the cold start-up condition requires a longer computation time.

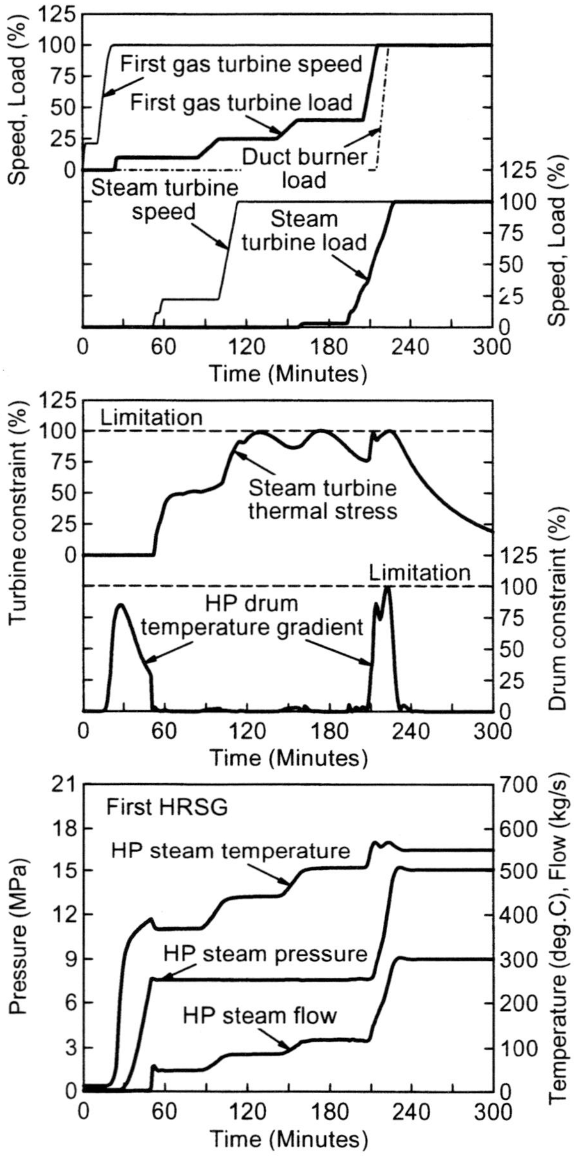

Fig. 6. Proposed method (Optimal solution).

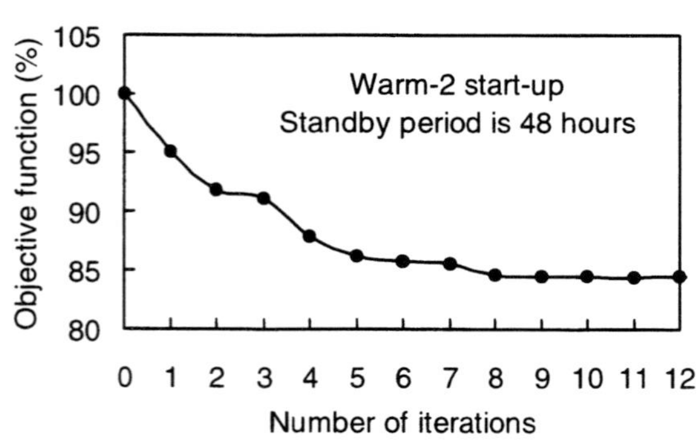

Fig. 7. Convergence history.

6. CONCLUSIONS

A method is proposed to optimize the operational parameters of the combined cycle power plants. The proposed method consists of the optimization calculation and the dynamic simulation.

The proposed method makes it possible to determine the optimal operational parameters by taking into account the dynamic characteristics of the plant.

Future work will add other operational restrictions (e.g., NOx emission from the plant, etc.).

REFERENCES

Hanzalek, F. J. and P. G. Ipsen (1966). Thermal Stress Influence Starting, Loading of Bigger Boilers. *Electrical World*, **Vol. 165**, No. 6, pp. 58 - 62.

Nakai, A., *et al.* (1996). Turbine Start-up Algorithm Based on Prediction of Rotor Thermal Stress. *Proceedings of IFAC 13th Triennial World Congress*, pp. 25 – 30. San Francisco, U.S.A..

Akiyama, T., *et al.* (1997). Dynamic Simulation and its Applications to Optimum Operation Support for Advanced Combined Cycle Plants. *Energy Conversion and Management*, **Vol. 38**, No. 15-17, pp. 1709-1723. Elsevier Science Ltd., U.K..

Bausa, J., and G. Tsatsaronis (2001). Dynamic Optimization of Startup and Load-Increasing Processes in Power Plants - Part I: Method, Part II: Application. *Transactions of the ASME, Journal of Engineering for Gas Turbines and Power*, **Vol. 123**, pp. 246-254.

Maffezzoni, C. (1992). Issues in Modelling and Simulation of Power Plants. *Proceedings of IFAC International Symposium on Control of Power Plants and Power Systems 1992*, **Vol. 1**, pp. 19 – 27. Munich, Germany.

Funatsu, T., *et al.* (1999). An Evaluation of Dynamic Simulation via 1300 deg.C Class Combined Cycle Plant. *Proceedings of ASME PWR-Vol. 34, 1999 International Joint Power Generation Conference (IJPGC-99)*, **Vol. 2**, pp. 735 – 740. San Francisco, U.S.A..

Fukushima, M. (1986). A Successive Quadratic Programming Algorithm with Global and Super-linear Convergence Properties. *Mathematical Programming*, **Vol. 35**, pp. 253 – 264.

ELSEVIER
IFAC
PUBLICATIONS
www.elsevier.com/locate/ifac

Control and Dynamic Simulation for a Combined Cycle Power Plant: The process and control strategies

Marino Sanchez-Parra Yenisse Rudecino-Mendoza

Electrical Research Institute
Control Systems Division – Control and Instrumentation Department, 29-1. P.O. Box 1-425.
Cuernavaca, Morelos 62001, Mexico.

ABSTRACT

The description of the process and the main control strategies for a Combined Cycle Power Plant are presented. The corresponding process models, which were programmed in C language, are the basis for a PC based dynamic simulator of a CCPP that was applied to test the control algorithms belonging to a new Distributed Control System, previously to its commissioning and start up. The control algorithms for the gas turbines, steam turbine, heat recovery steam generators and the condensate system, were carefully tested by means of simulation tests, they were checked for proper operation and tuned while the units were coming back on line after the DCS installation outage. The whole graphical results showing the trend process were collected for analysis. *Copyright © 2003 IFAC*

Keywords: Dynamic tests, simulation, models, control algorithms

1. INTRODUCTION

Combined Cycle Power Plants (CCPP) are becoming increasingly prevalent in the utilities market-place. The reasons for this are numerous, and several are listed in the following section. These units are often required to be very flexible in their operating schedule; it isn't uncommon for CCPP to be required to startup 100 or more times each year. These demands require the implementation of a control system that can facilitate both a rapid and a reliable startup and a profitable power generation. This paper will focus on the control and dynamic simulation tests for a CCPP as a tool for the control strategies validation previous to the plant commissioning and startup.

Overall efficiencies for CCPP are above 50%, with some modern designs approaching 60% or better. Figure 1 shows a simplified flow diagram of this process. The gas turbine (GT) drives an electric generator and a compressor and exhausts into the Heat Recovery Steam Generator (HRSG), which produces superheated steam used to drive a steam turbine (ST) with another electric generator. Many different arrangements are possible, ranging from a single cycle to a more complex designs, such as multiple combustion turbines exhausting into a common HRSG, and more complex HRSGs with multiple drums and/or reheat capabilities. When compared to a conventional steam plant, the popularity of the CCPP can be readily appreciated [Owen,1999]:

- High efficiency, greater than 50%.
- Lower investments costs.
- Lower water requirements. CCPP typically use less than half water required by a steam plant.
- Greater operating flexibility.
- Simplicity of operation, allowing for a reduced staff size.
- Low environmental impact.

Also there are some drawbacks with CCPP; the primary is the requirement for a premium fuel, such as gas or oil.

2. COMBINED CYCLE DESCRIPTION

A combined cycle (CC) process is a kind of marriage between the Brayton and Rankine cycles. The goal is to use the high temperature capabilities of the internal combustion (gas) turbine to increase the working-fluid temperature (air), while using the external-combustion (steam) turbine to reduce the temperature of heat rejection. The GT can be thought of as the topping cycle, while the HRSG and ST is the bottoming cycle [Obleton, 1996].

The combined cycle process transforms the heat energy from gas fuel combustion to electric energy by means of Brayton and Rankine thermodynamic cycles. The integration of such both thermodynamic cycles make possible to get an increased efficiency for the energy transform process. The whole combined cycle process generates energy by means of two GT and a ST. Both GT are identical, and then we will describe the GT process for only one.

The gas fuel energy is transformed to thermal energy thanks to the combustion process developed at the combustion chamber. This is a chemical reaction between the pressured air supplied by the air compressor and the gas fuel. The hot gases produced by the combustion are sending to the turbine, making it rotate at 3600 rpm. Finally, the turbine shaft is coupling to the generator shaft to produce electric energy.

The waste heat gas from the GT stores a high content of heat energy which is unnecessary for the GT. Then to avoid lost, gas is send to a heat recovery system where the heat energy is transferred to water and transformed into steam which drives the ST. Water and steam form a closed cycle which only needs more clean water, also named makeup water, to restore the liquid process leakages. To describe this system we can start at the condenser.

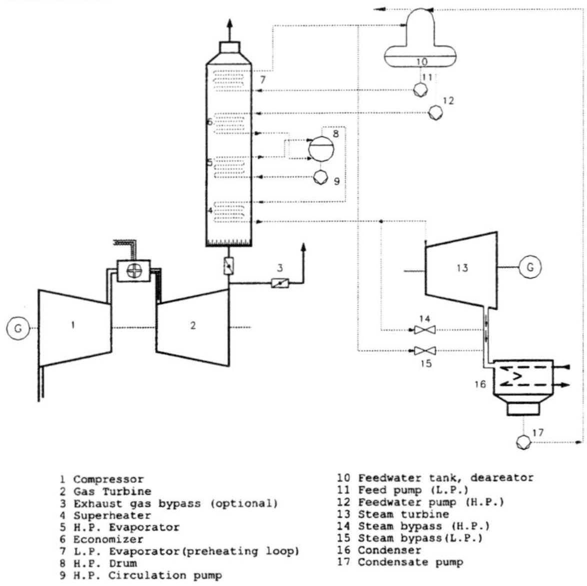

Fig. 1 Simplified Flow Diagram of Combined Cycle Process.

The liquid that is going out the condenser is named "condensate". It is send to a deareator to retire the uncondensed gases with the application of pressurized steam named "pressurization steam". Furthermore it has a bleed valve which helps to reduce the accumulated pressure. Heat transfer from gases to water is made at the HRSG. This equipment has the next four independent heater stages: the low-pressure evaporator (LPE), the economizer, the high-pressure evaporator (HPE) and the superheater (SH). The LPE is topping the tower, where pre-heats the water coming from the deareator and send it to the feed water tank. Up to this point the hot water is known as feed water, which is send to the recovery (economizer), where it is transformed into steam, and to the high pressure drum. At this point liquid is moving away the steam and passes to other stage of evaporation (HPE) and comes back to the drum with more steam.
The saturate steam leaves the top HP drum towards the SH, which is the greatest temperature zone, because at this zone arrive the GT exhaust gases and exists another section called *afterburners* (AB). The AB is for increasing the steam temperature and to get more steam production for the ST and the power generation. The AB raises the CCPP efficiency.
To improve the control of the ST power generation there is a bypassed SH steam line towards the condenser through the main bypass valve. An advantage of this bypass action can be appreciated when the ST is out of service, because the power plant can still continue working with the GT at open cycle with less energy production to protect the heat recovery equipment.
The waste steam from the ST is send to the condenser and finishes its cycle. With the object that the exhausted steam from the ST could be transformed to the liquid face.
For recycling, could water from the cooling tower is passed through the condenser. The most important variables are the condenser vacuum and the condenser level [Sánchez,1994].

3. GAS TURBINE CONTROL & SIMULATION

The GT control for the CCPP encompasses the entire operating spectrum from initial start to final stop. Its main objective is to get the greatest possible electric power generation by means of the compressor air flow regulation and the gas fuel regulation towards the combustion chamber, and supplying the exhaust gases required for the steam production.
The turbo generator is composed by a gas turbine, an axial compressor, the combustion chamber, the starting motor, the inlet-guide-vanes (IGV) actuator, the electric generator and the fuel valves (gas and oil) at the combustion chamber [Garduno and Sanchez, 1995].

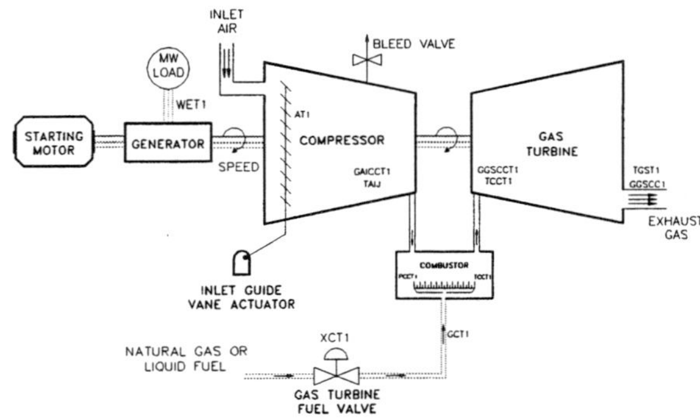

Fig. 2 Gas turbine schematic diagram.

To achieve the desired results, the control actions are the following:

1. Initially the starting motor rolls the unit.
2. As acceleration approaches 25% speeds, the fuel throttle valve pops open and when successful flame is detected, the control system begins to accelerate the GT via the combustion process and the starting motor.
3. The starting motor is off at 65% speed and the unit acceleration progress only by the combustion process.
4. The turbine enters the automatic synchronizing mode as soon as it reaches about 97% speed, and is immediately placed on the line.
5. The instant at the control system detects the main generator breaker closing, raises the fuel valve to immediately pick up minimum load.
6. IGV control maintains 65% opening at startup and 30% when the main breaker closing.
7. At loading the IGV opens from 30% up to 100%.
8. The temperature control and the compressor air flow instabilities control are always active to prevent a failure and they can initiate a fuel valve runback.

For the control system the main variables are:
a) The IGV opening (AT).
b) The GT speed (OMEGA).
c) The generated electric power (WET1).
d) The fuel valve opening (XC).

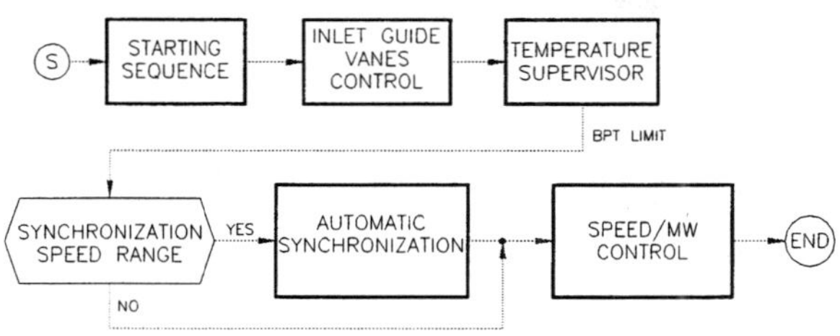

Fig. 3 Gas turbine control functional diagram

4. STEAM TURBINE CONTROL & SIMULATION

The control system of the steam portion of the CCPP extends from cold boilers and zero pressure through maximum firing and overpressure operation. The control system is utilized during initial boiler pressurization and warming of steam lines, with flow piped directly to the condenser through a main bypass valve. As GT firing is increased and exhaust temperatures rise, the control maintains the ST and bypass valve closed while drum pressure gradually builds to a predetermined hot standby level. At precisely the correct throttle pressure the main steam bypass valve is automatically ramped to an initial position to allow steam flow. The bypass valve then modulates to regulate the minimum set point pressure value at the turbine throttle valves while the gas turbine firing continues rising steam temperature to the proper level for matching turbine high pressure rotor parts.

The ST control system includes the ST, two control (throttle) valves to regulate the steam admission to the turbine and the bypass valve for regulating the steam flow towards the condenser (throttle pressure).

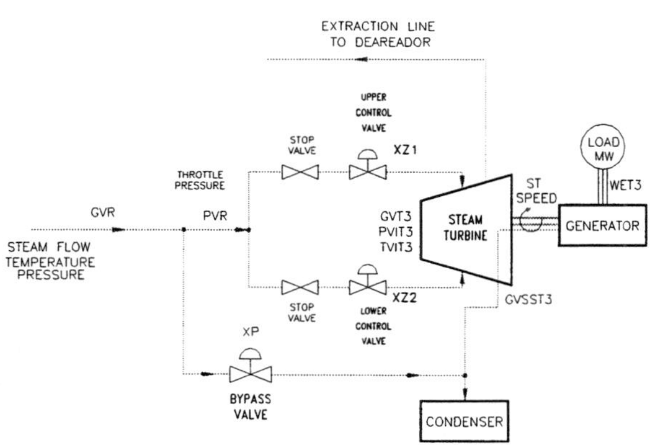

Fig. 4 Steam turbine schematic diagram.

To achieve the desired results, the ST control actions are the following:

1. The speed control mode guides the ST from minimum speed to 100% speed using different speed ramps.
2. The turbine enters the automatic synchronizing mode as soon as it reaches 97% speed, when the field breaker closes. Then is immediately placed on the line.
3. The automatic load control uses MW regulation by a strategy of combined MW feedback and feed forward to regulate dynamic and steady state load.
4. The throttle steam valves have a dynamic response due to a speed/load demand or a throttle pressure runback.

For the ST control the main variables are:

a) throttle pressure
b) ST speed
c) ST electric power

Control modules

The ST main control functions are:

1. Throttle pressure control.
2. Speed control.
3. Load control.
4. Synchronization.
5. Throttle pressure limit (TPL) superviser.

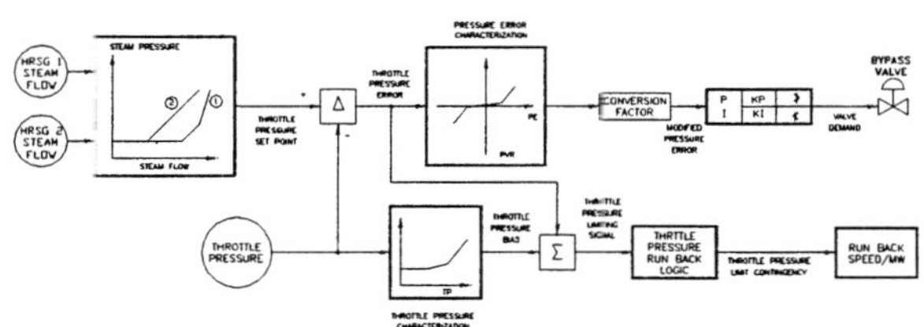

Fig. 5 Throttle pressure control diagram.

Throttle pressure control: the control strategy starts with the characterization of the HRSG steam flow in order to get the throttle pressure set point (PA_PVR). The calculated set point value is compared against the measured pressure variable to get a pressure deviation which will be used as the controller input to get the bypass valve demand (XP).

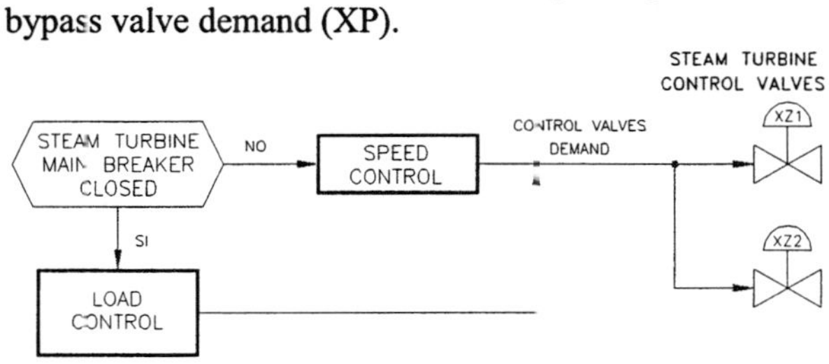

Fig. 6 Steam turbine control functional diagram

5. HRSG CONTROL & SIMULATION

The HRSG has to satisfy the steam demand required by the steam turbine in different operation states which go from turbine acceleration and synchronization to electric power generation.

The HRSG main components are: the heat recovery, the high pressure drum, deareator feedwater-tank and the afterburners. This system links the gas turbine with the steam turbine. At HRSG starts the heat energy transfer to realize the steam generation process. The heat supplied to the heat exchangers belonging to the heat recovery is coming from the GT exhaust gases and from the afterburners combustion, where the superheater is the hottest zone.

For the HRSG control the main variables are:

a) The high pressure drum level.
b) The superheater steam temperature and pressure.
c) The deareator pressure.
d) The fuel flow supplied to afterburners.

The HRSG main control functions are:

1. Feed water control
2. Atemperation control
3. Afterburners control
4. Recirculation water control
5. Deareator pressurization control

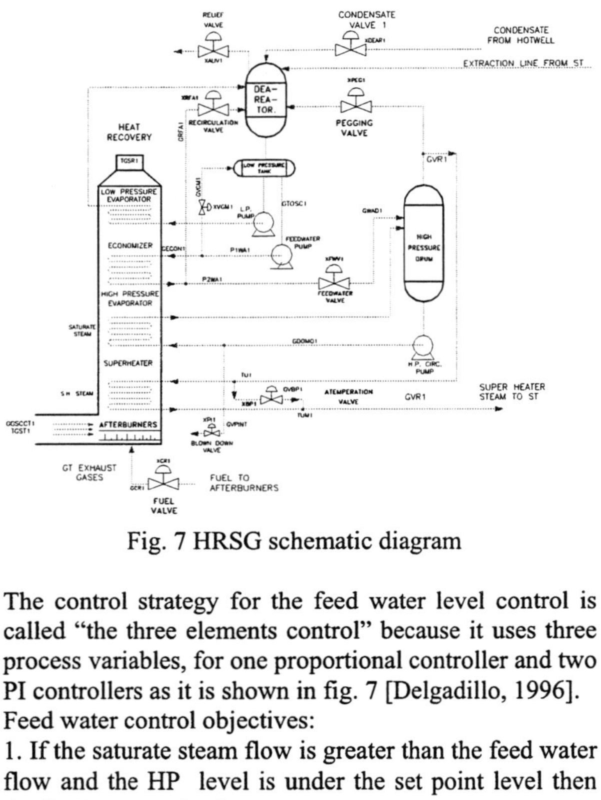

Fig. 7 HRSG schematic diagram

The control strategy for the feed water level control is called "the three elements control" because it uses three process variables, for one proportional controller and two PI controllers as it is shown in fig. 7 [Delgadillo, 1996].

Feed water control objectives:

1. If the saturate steam flow is greater than the feed water flow and the HP level is under the set point level then the feed water valve has to open.

2. If the HP level is up to the set point level then the feed water valve has to close.

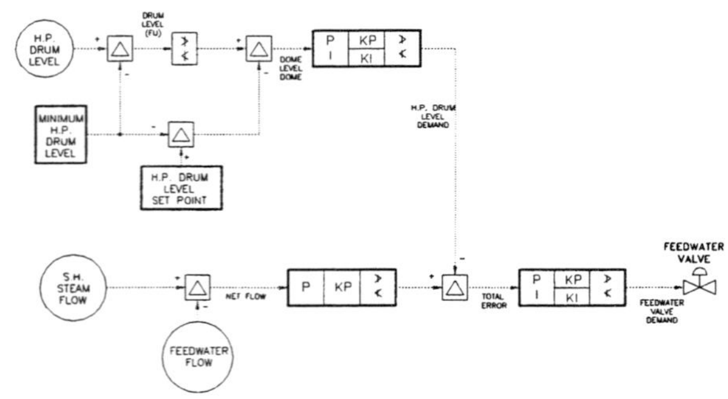

Fig. 8 Feedwater control diagram

Super-heater steam atemperation control.

The super-heater temperature and pressure control is very important because a big increasing for these variables could damage some internal components of the ST. Both the atemperation control and the afterburners control regulate the super-heater temperature.

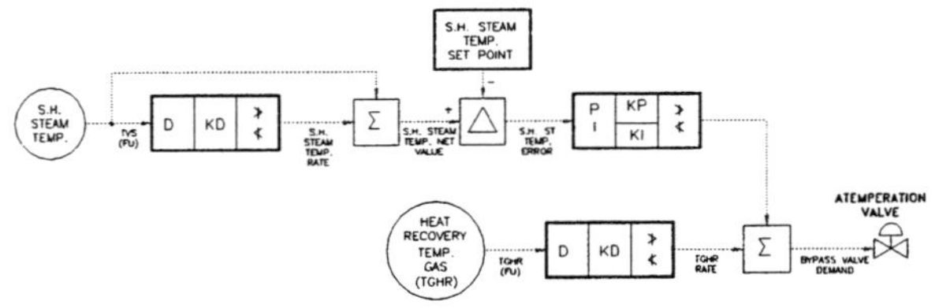

Fig. 9 Atemperation control diagram.

6. CONDENSATE SYSTEM CONTROL & SIMULATION

The condensate system is the process component which has to recover and recycling water from the waste steam to use it again for the steam generation process. This could be done when the ST exhaust steam is condensed and the steam temperature increases, eliminating the air on it. The condensate water is storage to make up the losing water making the process more efficient.

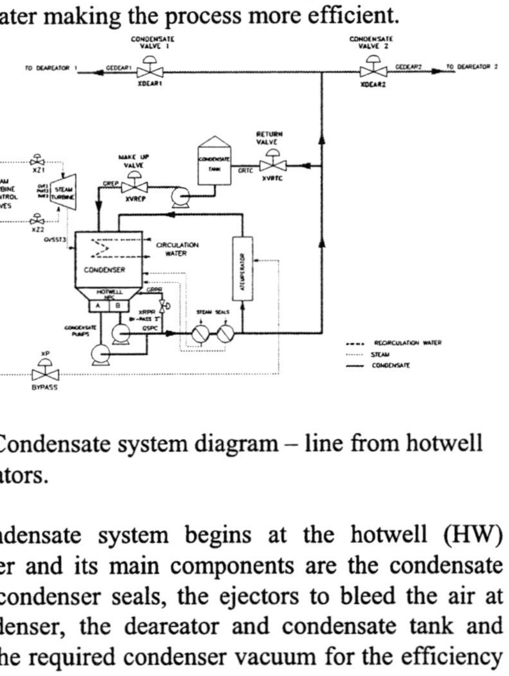

Fig. 10 Condensate system diagram – line from hotwell to deareators.

The condensate system begins at the hotwell (HW) condenser and its main components are the condensate pumps, condenser seals, the ejectors to bleed the air at the condenser, the deareator and condensate tank and getting the required condenser vacuum for the efficiency of the operation.

Condensate control objectives [Sánchez94]:

1. A hotwell high level opens the return valve and the condensate valves, and closes the make-up valve.
2. If the SH steam flow is greater then the condensate flow, then the condensate valves open.
3. A deareator high level opens the return valve and a deareator low level closes the return valve and the make up valve opens.

4. If the condensate flow leaving the hotwell is less than a critical constant value, the control orders the opening of the bypass valve from condenser to hotwell.

7. CONCLUSIONS

The CCPP Dynamic Simulator have been a very useful tool for the evaluation of conventional and modern control algorithms prior to the commissioning and start up in two power plants. Thorough checkout of a DCS could be easily accomplished by testing against an accurate dynamic simulation. Both digital and analog logic could be tested, debugged and tuned. Delivery of control logic that has already been tested also reduce plant down time during system installation. This is an extremely small investment in a DCS retrofit project because controls that have been tested allow the engineer to focus on wiring and connections as the logic has already been scrutinized.

8. REFERENCES

Delgadillo V.,M.A. and J.E. Fuentes G. (1996). Dynamic models for the combined cycle HRSG and Condensate System. *Technical Report* IC-5117 MOD_REC/SC-2.0. Electrical Research Institute (in Spanish).

Garduño R., R. and M. Sanchez P. (1995). Control System Modernization: Turbogas Unit Case Study. *Proceedings IFAC Symposium on Control of Power Plants and Power Systems.* Volume II, pp. 245-250.

Obleton, N. (1996). A strategy for reducing steam turbine startup time in combined cycle applications; *ISA (International Society for measurement and control)Proceedings.* Vol. I, pp. 75-81.

Owen, F.C. and W.H. Harrison (1999); Combined cycle reference plant for today's merchant market. *Proceedings International Joint Power Generation Conference, ASME/ Power Division.* pp. 43-48.

Sanchez P., M., J.E. Fuentes G., , and L. Castelo C. (1994). Digital Control for the Condensate System in a Combined Cycle Power Plant. *Proceedings 1994 International Joint Power Generation Conference, ASME.* "Advances in Steam Turbine Technology for the Power Generation Industry", pp. 227-234.

Sanchez P., M. and L. Castelo C. (1996). Coordinated Control System Modernization for a Combined Cycle Power Plant., *ISA (International Society for measurement and control). Advances in Instrumentation and Control.* Volume 51, Part 2, pp.1365-1373.

GAS TURBINE CONTROL

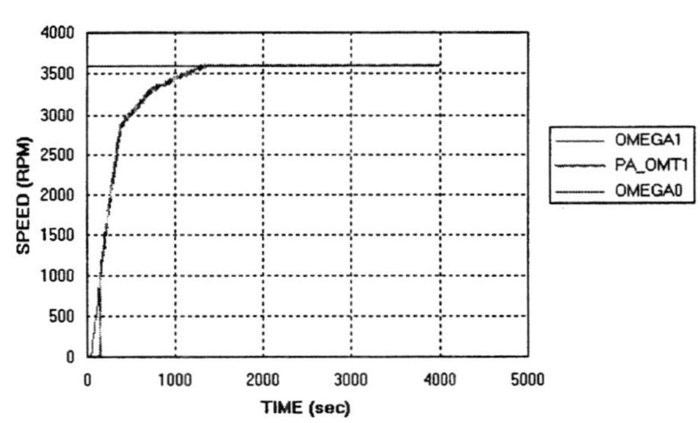

GAS TURBINE CONTROL

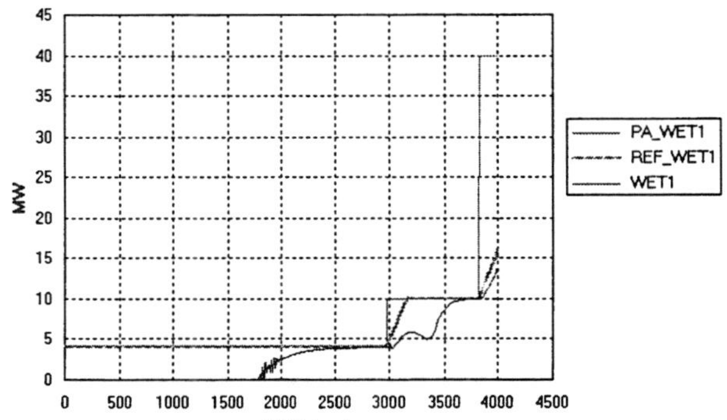

GAS TURBINE CONTROL

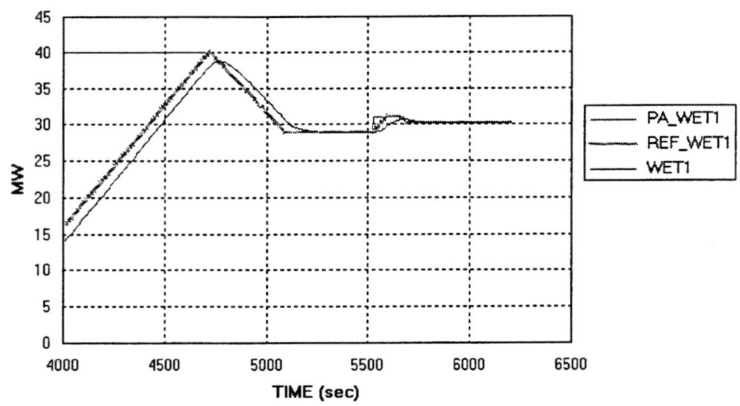

GAS TURBINE CONTROL

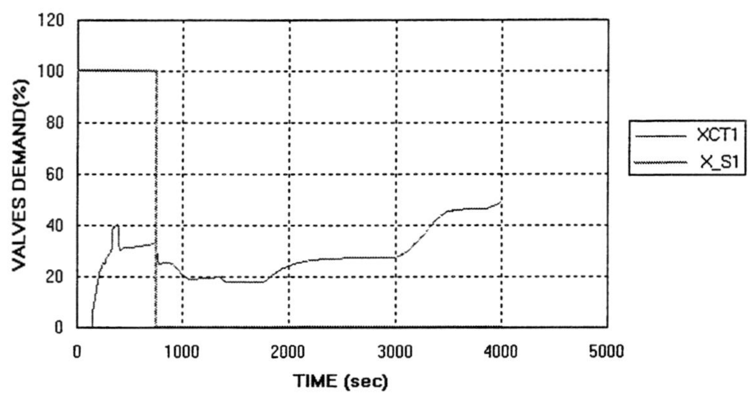

GAS TURBINE CONTROL

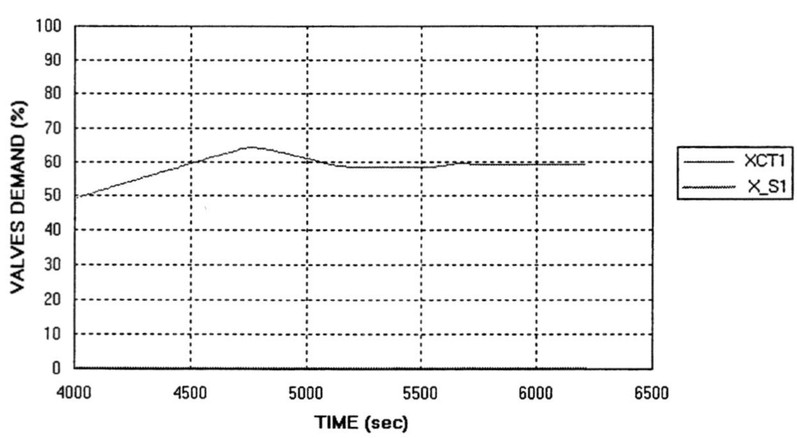

ELSEVIER
IFAC
PUBLICATIONS
www.elsevier.com/locate/ifac

Suggestions to Improve the Reliability of the Fan Actuator

Man-Kwon Han*

Abstract: The fan actuator is the critical part of the Boiler Automatic Control System, adjusting the air flow according to target load. Its malfunction usually leads to the reduction of half load of a power generator or a unit trip, which in turn reduces the Genco profits. As a way to improve its reliability, a power-off device and a repair technique for the mode of its normal operation were suggested. *Copyright © 2003 IFAC*

Keywords : actuators, boilers, controlling elements, logic design, motor control

1. Introduction

In Korea, they adopt the Draft System consisting of the forced draft fan, the induced draft fan, and the primary air fan for the boiler of the 500MW Standard Coal Fired Power Station.

Each unit has six fans installed. The forced draft fan(FD Fan) provides air required for the coal to burn. The induced draft fan(ID Fan) controls the pressure of the furnace and discharges burned gas. And the primary air fan(PA Fan) is in charge of drying pulverized coal and carrying it to the burner. The air volume of the FD Fan and the ID Fan is adjusted by the degrees of the angle of the pitch blade(0~90°) while that of the PA Fan by the changes of the inlet vane location(0~100%). By controlling air volume, the following purposes are fulfilled; one is to have the state of governor free and the other is to meet the demand for the fuel and water used in the boiler, which by the way is in changes all the time. An actuator puts the pitch blade and the inlet vane into motion and is the main subject of this study. The study set out to analyze the cases of its breakdown and their impacts and suggest solutions based on the field experiences and knowledge.

* section chief of the I&C department ,Taean Thermal Power Site Division Korea Western Power Co.,LTD.

2. The Control System of the Fan

The control system presented in the study was for the Unit 1~6 of the Taean Thermal Power Station.

A fan can be divided into two types. One is the axial type that is suitable for supplying large amounts of flux like the FD Fan and the ID Fan, and the other is the centrifugal type that is good for small amounts of flux and high pressure like the PA Fan. Regardless of the types, an actuator alters the position of the pitch blade and the inlet vane according to the demand of the boiler by operating the motor.

2.1 FD/ID Fans Control

The principles involved in controlling the FD/ID Fans are as follows; The boiler DCS sends out a signal for blade position demand according to load(MW) and furnace pressure. The matic controller in the field receives the demand signal and puts the motor into action, which causes the hydraulic device to move and adjust the position of the blade to meet the demand. The potentiometer detects the actual positions of the blade. Once again, the matic controller in the field compares the DCS demand with the actual positions of the blade and then repeats a series of control signals that operate the motor by declination (See the Fig. 1 of the control flow of FD/ID Fans).

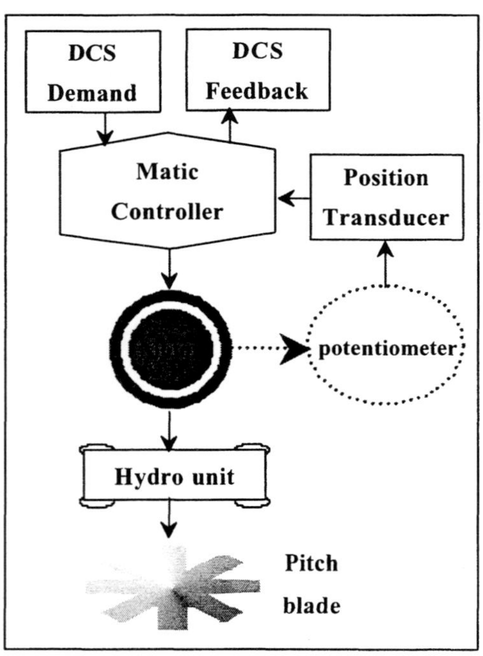

Fig. 1 the control flow of FD/ID Fans

During the control process, there happened many breakdowns in the field matic controller and the potentiometer. The causes of the breakdowns include the installation environment heavy with dust particles and vibration and the repetition of the action in high frequency(30 times/minute).

2.2 PA Fan Control

The PA Fan receives the demand of hot primary air duct pressure that changes according to the speed of the coal feeder and follows the order from the demand. The field controller allows DC 90V to the motor as the demand requires and operates the inlet vanethat is mechanically linked. Being connected to the gear axis of the motor, the two potentiometers detect the actual locations, one of which is sent to the DCS for feedback and the other of which is sent to the field controller. In this way, the actual positions are constantly compared with a demand. The motor is operated by declination with the repeated performance of control signals.

3. Serious Problems of the Actuator

Despite the huge and critical role in the Boiler Automatic Control System, the actuator is low in reliability due to frequent breakdowns.

According to the investigation of breakdowns of the actuator used at the 500MW Coal Fired Power Station, there were total 63 cases in six years from 1996 to 2001 with annual average of 10.5 cases(See Table 1, the breakdowns of each power station).

The major causes of the breakdowns were classified into three. The first one is the field controller, which cannot read the DCS demand and thus maintains the open and close function in a single direction due to the malfunctioning positioner. The second one is the potentiometer, which cannot detect the correct positions of the blade as the contact sides are worn out due to the repeated actions on a certain location. Finally, the controller repeats the open function over and over again as the chip that converts the [kΩ] signal into the [mA] signal at the position transducer is low in quality and sends out lower output than the input.

A breakdown of the actuator results in bigger problems such as load run back(50%) and a unit trip.

Table 1, the breakdowns of each power station

Name of Plant	Cases of the breakdowns
Taean #1~4	14
Hadong #1~2	11
Boryeong #3~6	34
Samcheonpo #5~6	1
Tangjin #1~4	3
Total	63

4. Solutions

Two kinds of approaches in solving a case of actuator breakdown were suggested; one is to cut off the driving power of the motor and prevent more influence of its and the other is to carry out a repair job in the mode of full load operation.

4.1 Development of a power-off device for the actuator

When there are one or more malfunctioning principal parts in the actuator, the open or close function repeats itself. The resulting deviation between the DCS demand and the actual feedback value requires switching the control system from the auto mode to the manual mode and hold on the mode. At the same time, an alarm goes off to make the operator aware of the malfunction and prevent further spread of the impact. In case that the actuator keeps on working even in the manual mode and thus draws the gap between the demand and the actual feedback wider by 15% or more along with the fast changes of the furnace pressure, the source power is shut off to have all the actions on hold. This approach was possible by improving two kinds of the control system, which were the DCS logic and the newly installed magnetic relay inside the MCC(Motor Control Center).

The operating mechanism of the power off device can be described as follows:

It's essential to evaluate an error in the instrument that detects the current locations of the actuator only. If it turns out to be a simple error, there is no needfor a conversion to manual operation or power off. Such a simple error would not impose serious impacts on the boiler's automatic control system.

The power off device is a protective measure to help the stable operation of the power plant. For example, it can play its protective role when the actuator operates irrelevant to the demands of the control system due to a breakdown and thus imposes a serious impact on the boiler's automatic operation. To be more specific, when the gap between the position demand and the position feedback of the FD Fan is more than 10% and the deviation between the total air flow demand and the actual air flow is more than 10% and lasts two seconds or longer, the operation mode of the blade pitch is converted from auto to manual, giving an alarm. The deviation is set 10% since there exists 2-3% of gap between demands and feedbacks under the normal operational situations and the range of bias is ±5%, allowing for the maximum 10%. If the actuator is set in the manual mode and continues to function regardless of the demands, the deviation will rise to 15% or higher from the old 10%. If this situation lasts for more than three seconds, the power to operate the actuator is cut off and thus the actuator stops working, all of which is intended to prevent the shutdown of the generation facilities due to potential spreading breakdowns.

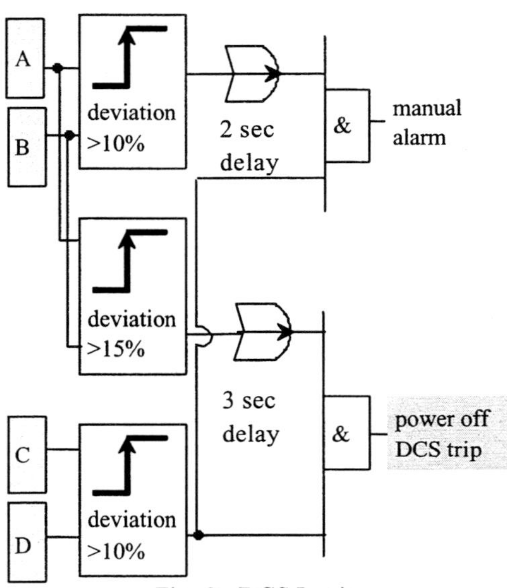

Fig. 2 DCS Logic

Sign	FD Fan	ID Fan	PA Fan
A	position demand		
B	position feedback		
C	Air flow demand	furnace pressure SP	Hot pa duct pressure / furnace pressure Dp SP
D	Total secondary air flow	furnace pressure	Actual Dp

Table 2, description of the sign

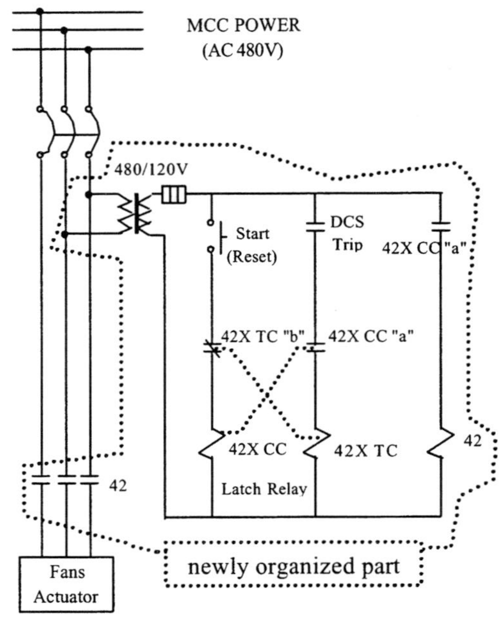

Fig. 3 schematic of the 480V MCC

4.2 Development of a repair technique for the normal operation

The new repair technique was tried for the FD/ID Fans and the PA Fan.

Repairing FD/ID Fans Actuator :

The actuator consists of the field controller, the motor and the gear box. Since the motor and the gear box are combined in one unit, it's necessary to separate them for maintenance purpose during operation. The maintenance efforts are usually for replacing the old potentiometer, which has a high frequency of breakdowns inside the motor. Consisting of the warm and the warm wheel, the gear box cannot operate the pitch blade without the power delivered from outside. It was confirmed that the gear box could be repaired after being separating from the motor. After applying this repair technique to a spare actuator, the results were satisfying and led to further adoption of the technique. In 2002, there were five successful repair cases using the technique.

Repairing PA Fan Actuator:

The motor is combined to the inlet vanes through three links. By concluding both the ends of the middle link with U-bolts, the motor was separated from the inlet vane. Once the malfunctioning part was replaced or calibrated, the operation was turned into the normal state. In 2002, there were two successful repair cases using this approach.

▣ Repairing procedure of the normal operation

- ▶ Governor free off, Load Limit operation
- ▶ Change over of the actuator control mode (Remote ⇒ Local)
- ▶ Links of the Inlet Vane are locked up by the u-bolt
- ▶ Replacement of the trouble potentiometer
- ▶ Calibration of the position

5. Conclusion

This paper provides the description of analyzing a breakdown of the fan actuator, one of the principal control elements in a power plant, and of repairing it without shutting down the power plant. Before the power off device and the technique to do repairing during operation were developed, a breakdown of the fan actuator inevitably led to generating stop or reduced output, which brings enormous losses to the company in this competitive market of power generation. As a matter of fact, it should be every power company's dream to solve any problems while the plant is operating normally. It's such a great pleasure to help their dream come true.

6. References

Katsuhiko Ogata. (1996) *Modern Control Engineering* (Third Edition) Page 207~213

ABB (1995) *Supercritical Steam Generator Instruction Manual* , vol 1.

ELSEVIER
IFAC
PUBLICATIONS
www.elsevier.com/locate/ifac

A STUDY ON THE EFFECTS OF NEW FACTS INSTALLATIONS ON EXISTING DISTANCE PROTECTION RELAYS

**Jongyun Choi*, Youngseong Han*, Inyoung Suh* ,
Byunghoon Chang**, Jongsu Yoon****

**Hyosung Corporation, **Korean Electric Power Research Institute*

Abstract: The UPFC, which is one of the many FACTS devices that are in use in the modern power system, is able to independently control the active and reactive power in the transmission systems by injecting a controlled voltage in series. The series injected voltage by the UPFC can be perceived as a change in transmission line impedance by existing distance protection relays and affect their correct operation. This paper first uses EMTDC models of the UPFC and surrounding power system of the Kangjin substation located in Jeonla-namdo province in Korea to investigate the impedance variation according to different UPFC operating conditions under steady and transient states. Then the operation of the UPFC protection controls, and impedance variations seen by the distance protection relay were reviewed under fault conditions to investigate the possibility of relay misoperation, and the need for relay protection setting modifications.

Keywords: FACTS, UPFC, impedance, distance relaying, protection setting.

1. INTRODUCTION

The UPFC (Unified Power Flow Controller) is a FACTS (Flexible AC Transmission System) controller which can simultaneously support bus voltage and control the flow of power. The UPFC is connected to the transmission grid in parallel and in series. The parallel part has the same structure and function as the STATCOM (Static Compensator), and the series part is similar to the SSSC (Static Synchronous Series Compensator). The first UPFC in Korea has been installed at Kangjin substation located near the southwest coast of the Korean peninsula. The capacity of the installed UPFC is 80MVA (40MVA in series and 40MVA in parallel) and has completed commissioning tests and is expected to begin commercial operation in 2003.

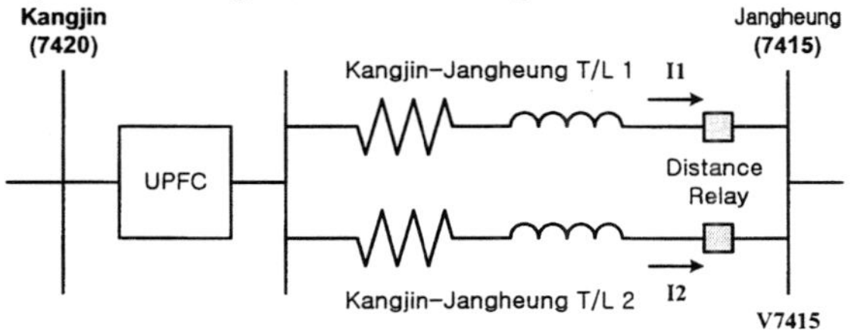

Fig. 1. System diagram between Kangjin and Jangheung .

A conventional distance relay will calculate the distance to the fault location based on the bus voltage and fault current when there is fault at a nearby bus. If the distance is located within a designated zone such as Zone1, Zone2, Zone 3, the fault is removed by operating relays according to the trip time at each zone. When the UPFC is connected in series between the Kangjin and Jangheung transmission lines, the impedance seen by the relays protecting these lines installed at Jangheung substation will include the impedance of the UPFC. Thus the impedance seen by the protection relays may vary according to the UPFC operating conditions (Jung and Kim, 2001; Dash and Pradhan, 2000). EMTDC models that represent the characteristics of the UPFC and surrounding power system were used to investigate the variance in impedance according to UPFC operating conditions at steady and transient states (Yoon, *et al.*, 2001, Han, *et al.*, 2001). Protection operation and impedance changes with bypass switch operation during faults were considered. Impedance changes during faults without bypass switch operation were also confirmed.

2. STEADY STATE IMPEDANCE VARIATION

To investigate the impedance variation seen by the distance relays at steady state conditions on the Kangjin-Jangheung transmission line at Jangheun substation, the impedance values where calculated with the UPFC series inverter injecting a maximum voltage. The corresponding impedance variation (Fig. 3) according to the power flow control range (Fig. 2) is shown.

The impedance range where the distance protection relays trip are shown as triangles concentrated at the center of Fig. 3. The magnitude of the impedance according to injected UPFC voltage is distant from the impedance range that trips the distance protection relays. When the UPFC is operating in steady state, it is unlikely that the UPFC will affect the normal operation of the distance protection relay under any operating conditions.

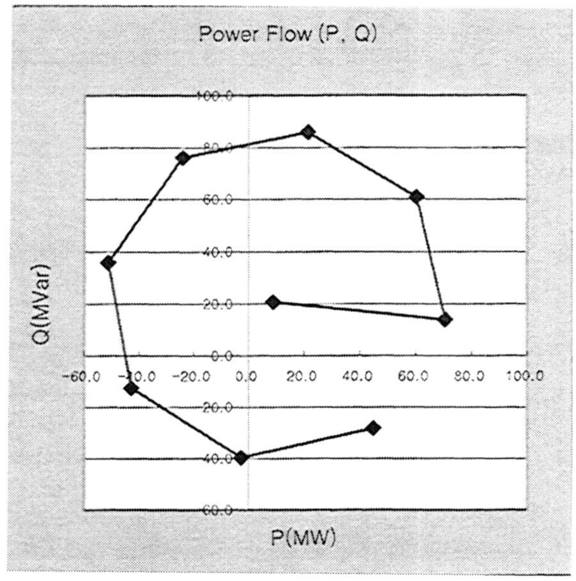

Fig 2. Range of Power Flow Control

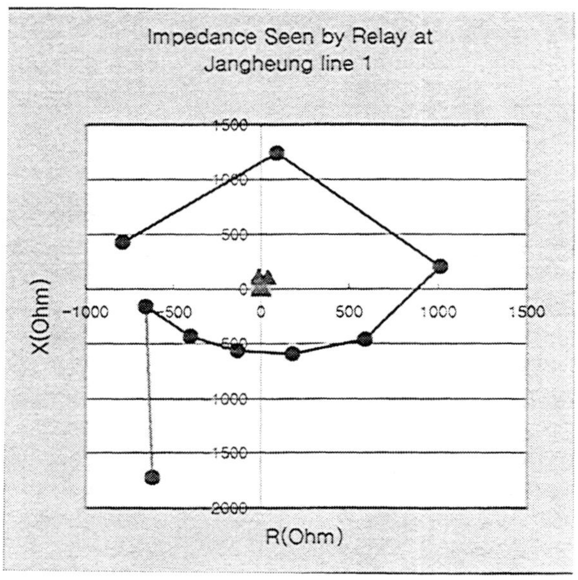

Fig 3. Transmission Line 1 Impedance between Kangjin-Jangheung

3. UPFC PROTECTION

Fig. 4. shows the outline of the UPFC system. When there is a fault in the system, the series inverter and the grid need to be isolated in order to protect the UPFC from the fault current. For this purpose, the UPFC is equipped with series circuit breaker and a thyristor bypass switch. The series circuit breaker always remains closed when the series inverter is not in operation. When the series inverter starts operation the series circuit breaker opens and voltage is injected through the series transformer.

When there is excessive current flowing due to a fault in the system, the series circuit breaker closes. Since the breaker has a mechanical structure it can not close instantaneously and the fault current may cause damage to the series inverter during the short time it takes to completely close the breaker.

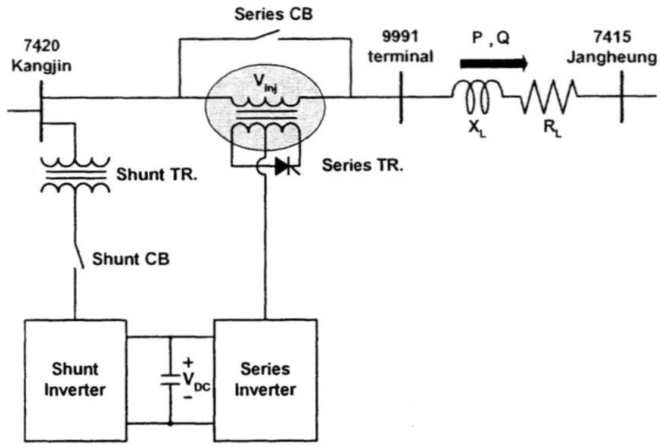

Fig 4. Transmission Line1 Impedance between Kangjin-Jangheung

Thus a static thyristor bypass switch is used to prevent the fault current from flowing towards the inverter while the series circuit breaker is in closing operation. The bypass switch is located at the secondary side of the series transformer, and is turned on/off by a gating signal from the UPFC controller.

3.1 UPFC Protection Sequence During System Faults

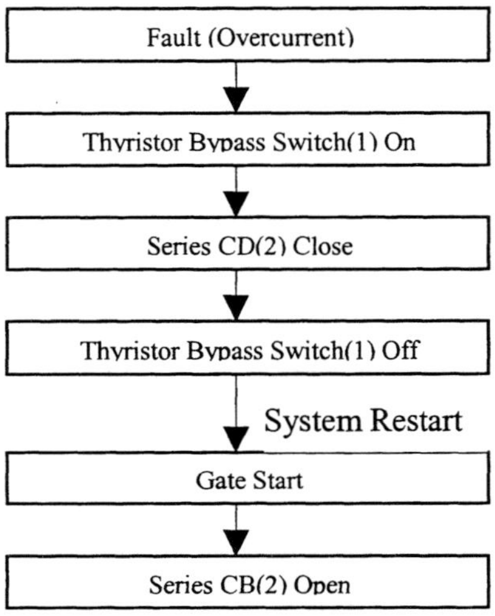

Fig. 5. Protection Sequence During Faults

Fig. 5. shows the UPFC protection sequence during faults. When there is a fault current, the protection control closes the series circuit breaker and Thyristor bypass switch as soon as the fault is detected. The series circuit breaker takes more than 9 cycles to completely close. The Thyristor bypass closes first since it can be turned on within 90 microseconds. The series inverter stops gating as soon as the Thyristor bypass switch is turned on. The Thyristor bypass switch is turned off when all of the transformer excitation current is completely discharged after the series circuit breaker is closed. The UPFC is restarted after the system fault is cleared.

3.2 Impedance Variation

The double line transmission line impedance between Kangjin(7420)-Jangheung(7415) has the following values.

$$R_L = 0.38183(\Omega), \quad X_L = 3.00837(\Omega) \qquad (1)$$

When the protection sequence is in progress, the leakage reactance of the transformer is added in series to the line impedance for approximately 9 cycles when the series circuit breaker is open, and only the Thyristor bypass switch is closed.

$$X_{leakage} = 0.30307(\Omega) \qquad (2)$$

During this period the line impedance is increased by 10%, and only the line impedance will remain after the series breaker is closed.

4. UPFC FAULT SIMULATION

When the Thyristor Bypass Switch of the UPFC operates normally during faults, it is safe to assume that the UPFC will not affect the correct operation of the distance protection relays. Circumstances when the Thyristor Bypass Switch does not operate correctly need to be considered since this could affect the correct operation of the distance protection relays under certain UPFC operating conditions. Thus a EMTDC model of an UPFC was use to verify the operation of the Bypass switch under different conditions (Padiyar, 1998).

4.1 Simulation Process

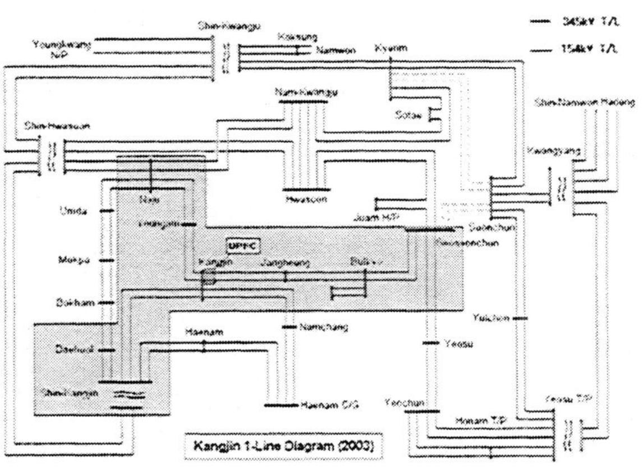

Fig. 6. Power System Diagram of Kangjin and surrounding area.

Two different cases that had the largest impedance during steady state were chosen for the simulations. Fault conditions were generated in different zones, Zone 1 (Kangjin, Jangheung) , Zone 2 (Shin-Kangjin, Youngam, Bulkyo), Zone 3 (Daebool, Naju, Seosoonchun), shown in Fig 6. highlighted in grey. A single phase ground fault and three phase short circuit were simulated for each fault location. All fault conditions were assumed to last for 0.1 seconds.

4.2 Simulation Results

Table 1 shows the operation modes and control commands used in the simulation for both cases

The results of the simulations for Case 1 are shown in Table 2 and Table 3. Case 2 results are given Table 4 and Table 5.

Table 1. Operation Modes and Control Commands used in the Simulation

Case		Operation Mode	Control Commands
Case 1	Shunt Inv.	Iq Control	Iqref = 0
	Series Inv.	PQ Control	Pref = 1, Qref = 1
Case 2	Shunt Inv.	Iq Control	Iqref = 0
	Series Inv.	PQ Control	Pref = -1, Qref = -1

Table 2. Fault Simulation Results (Case 1) – Single Phase Ground Faults

Fault Location	Shunt	Series
Kangjin	Undervoltage, Overcurrent trip	Bypass (Overcurrent)
Jangheung	Overcurrent trip	Bypass (Overcurrent)
Shin-Kangjin	Undervoltage, Overcurrent trip	Bypass (Overcurrent)
Youngam	Overcurrent trip	Bypass (Shunt trip)
Bulkyo	Overcurrent trip	Bypass (Overcurrent)
Daebool	Overcurrent trip	Bypass (Overcurrent)
Naju	Overcurrent trip	Bypass (Shunt trip)
Seosoonchun	Restored to normal after gate stop (Overcurrent)	Bypass (Overcurrent)

Table 3. Fault Simulation Results (Case 1) – Three Phase Short Circuit

Fault Location	Shunt	Series
Kangjin	Undervoltage Overcurrent trip	Bypass (Overcurrent)
Jangheung	Restored to Normal after gate stop (Overcurrent)	Bypass (Overcurrent)
Shin-Kangjin	Restored to Normal after gate stop (Undervoltage, overcurrent)	Bypass (Overcurrent)
Youngam	Restored to Normal after gate stop (Overcurrent)	Bypass (Overcurrent)
Bulkyo	Restored to Normal after gate stop (Overcurrent)	Bypass (Overcurrent)
Daebool	Restored to Normal after gate stop (Overcurrent)	Bypass (Overcurrent)
Naju	Restored to Normal after gate stop (Overcurrent)	Normal operation after gate stop. (Shunt gate-stop)
Seosoonchun	Restored to Normal after gate stop (Overcurrent)	Bypass (Overcurrent)

Table 4. Fault Simulation Results (Case 2) – Single Phase Ground Faults

Fault Location	Shunt	Series
Kangjin	Undervoltage, Overcurrent trip	Bypass (Overcurrent)
Jangheung	Overcurrent trip	Bypass (Overcurrent)
Shin-Kangjin	Undervoltage, Overcurrent trip	Bypass (Overcurrent)
Youngam	Overcurrent trip	Bypass (Overcurrent)
Bulkyo	Overcurrent trip	Bypass (Overcurrent)
Daebool	Overcurrent trip	Bypass (Overcurrent)
Naju	Overcurrent trip	Bypass (Shunt trip)
Seosoonchun	Restored to normal after gate stop (Overcurrent)	Bypass (Overcurrent)

From the simulation results of the two cases, it can be seen that with the exception of Naju, the series Bypass went into operation due to line overcurrent under fault conditions. Thus in other areas where the

Bypass operated instantaneously during faults, the impedance seen by the distance relay was not affected by the presence of the UPFC.

Table 5. Fault Simulation Results (Case 2) – Three Phase Short Circuit

Fault Location	Shunt	Series
Kangjin	Undervoltage Overcurrent trip	Bypass (Overcurrent)
Jangheung	Restored to Normal after gate stop (Overcurrent)	Bypass (Overcurrent)
Shin-Kangjin	Restored to Normal after gate stop (Undervoltage, overcurrent)	Bypass (Overcurrent)
Youngam	Restored to Normal after gate stop (Overcurrent)	Bypass (Overcurrent)
Bulkyo	Restored to Normal after gate stop (Overcurrent)	Bypass (Overcurrent)
Daebool	Restored to Normal after gate stop (Overcurrent)	Bypass (Overcurrent)
Naju	Restored to Normal after gate stop (Overcurrent)	Normal operation after gate stop. (Shunt gate-stop)
Seosoonchun	Normal operation	Bypass (Overcurrent)

The transient fault current flowing in the line between Kangjin-Jangheung during a fault at Naju is not big enough to trigger the operation of the UPFC Bypass switch. If there is a fault at Naju and the distance relays in Zone 1 and Zone 2 fail to operate, the distance relay in Zone 3 between Kangjin-Jangheung needs to go into operation. Since the operation of the relay in Zone 3 is critical, the possibility that this relay might not operate properly during faults needs to be considered. Thus detailed simulations of the effects on distance relays were conducted and reviewed for faults occurring at Naju.

4.3 Impedance Variations During Three Phase Short Circuit Faults at Naju

The impedance variation seen by a distance relay installed at Jangheung Substation during a three phase short circuit fault at Naju was calculated. The impedance calculations were done using the Jangheung bus voltage (V7415) and the currents in each line (I1, I2) as the reference. It was assumed that the UPFC Bypass was not triggered into operation, and the UPFC was in continuous operation. The shunt inverter operation was maintained at Iq control mode, and the series inverter was operated in both the voltage injection and P, Q control mode respectively.
Equations (3) and (4) represent the impedance seen by the relay prior to three phase fault conditions at Naju.

$$R1 = 15.9(\Omega),\ X1 = 104.2(\Omega) \quad (3)$$

$$R2 = 15.9(\Omega),\ X2 = 104.2(\Omega) \quad (4)$$

4.3.1 Voltage Injection Mode

Fig 7 shows the results of the simulation in the voltage injection mode. Figure 7-(a) shows the maximum injected voltage by the series inverter. The prefault (Ppre, Qpre) power flow is shown in Fig 7-(b). The power flow after a three phase short circuit fault at the Naju bus is shown in Fig 7-(c). The impedance seen by the distance relays at Kangjin-Jangheung T/L #1, 2 during the fault is shown in Fig 7-(d).

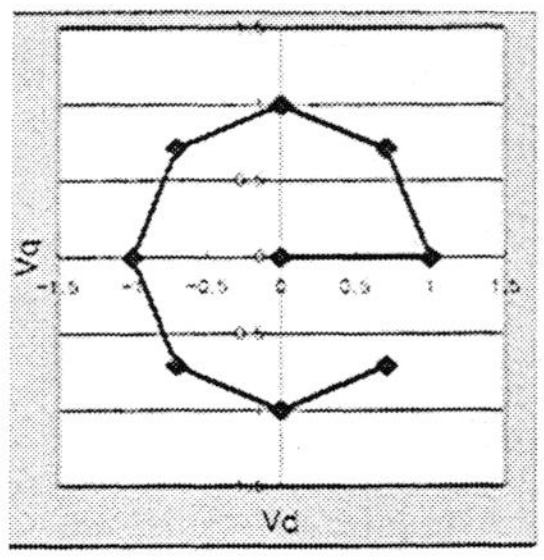

(a) Injection Voltage

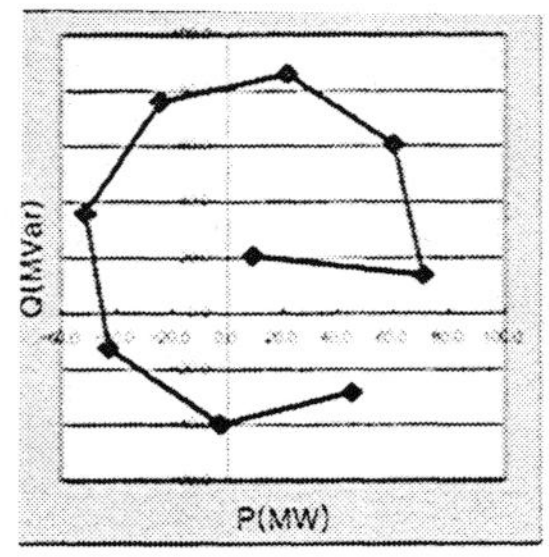

(b) Prefault P, Q

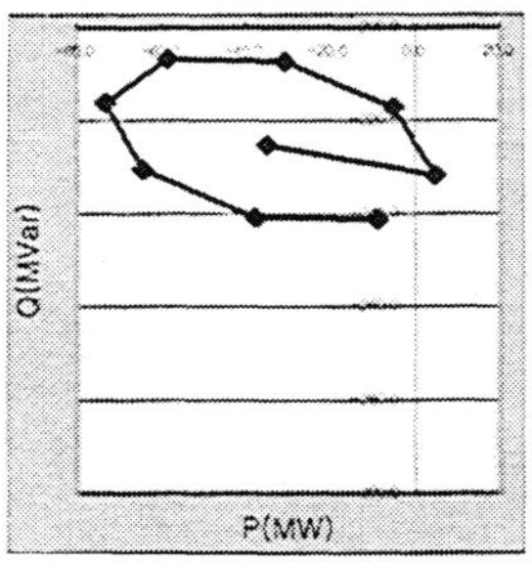

(c) After fault P, Q

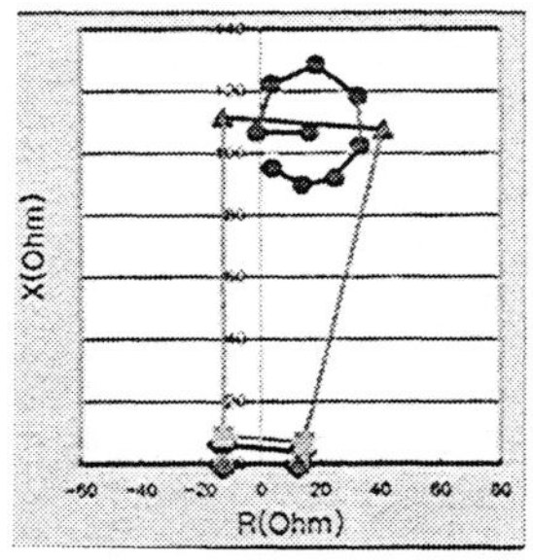

(d) Kangjin-Jangheung T/L-1 Impedance

Fig. 7. Impedance Variation Graph (Voltage Injection Mode).

4.3.2 P, Q Control Mode

Fig 8 shows the results of the simulation during the P, Q control mode. Fig 8-(a) shows the state of the UPFC power flow under maximum controllable power flow command values (Pref, Qref) prior to the fault condition. Fig 8-(b) represents the power flow after a three phase short circuit fault on the Naju bus. The impedance seen by the distance relays at Kangjin-Jangheung T/L #1, 2 during the fault is shown in Fig 8-(c).

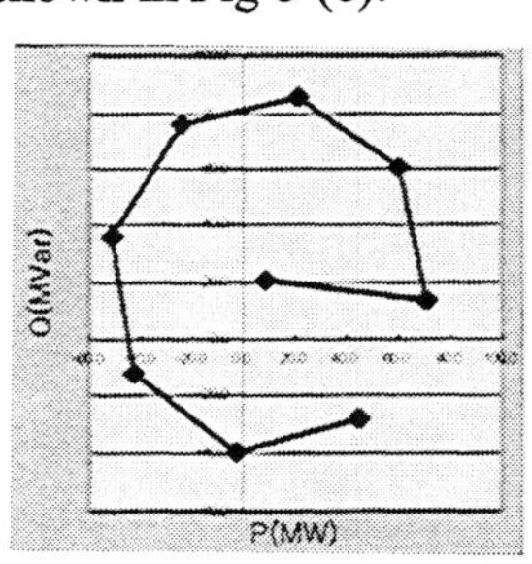

(a) Prefault P, Q

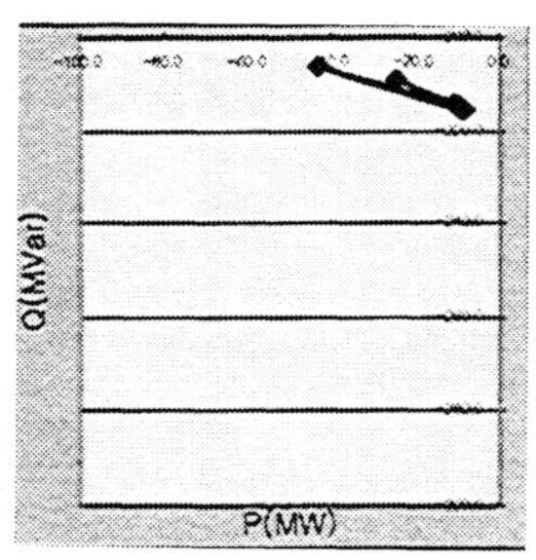

(b) After fault P, Q

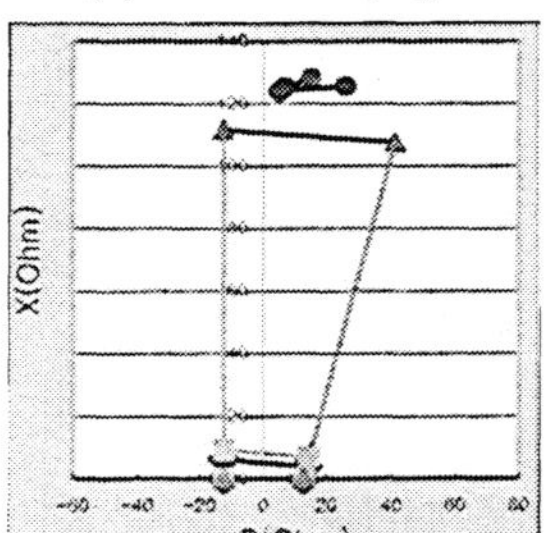

(c) Kangjin-Jangheung T/L-1 Impedance

Fig. 8. Impedance Variation Graph (P, Q control mode)

The results of the simulation in Fig 7 and Fig 8 show that when the UPFC Bypass switch does not go into operation during a three phase short circuit fault at Naju, the impedance seen by a distance relay on Kangjin-Jangheung transmission line at Jangheung substation varies according to the UPFC operating conditions. The impedance variation was largest when the UPFC was in the voltage injection mode. The maximum observed impedance variation was approximately 10% that of the impedance seen with the Bypass switch in operation. When the UPFC is in the P, Q control mode, the injected voltage during faults varies according to the P, Q command values. The magnitude of the injected voltage in most cases was almost the same, and the impedance variation was not as large when compared to the voltage control mode. From the above results, we can conclude that the impedance variation level was not that severe, and will not be the cause of any sudden abnormal relay operation. Some parameter setting changes maybe required for certain cases.

4.4 Impedance Variance during Transient States

The UPFC was assumed to be in steady state operation. A three phase short circuit fault was created at Naju. The impedance variation seen by the distance relay from the Jangheung side was observed. The possibility of abnormal relay operation during the transient state right after the fault due to the impedance seen by the Jangheung side relay having irregularly small values similar to that of Zone1 or Zone2 protection region values was investigated.

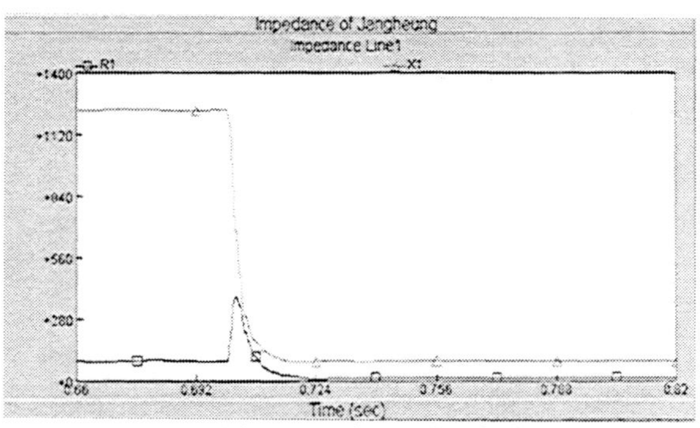

(a) Impedance Variation

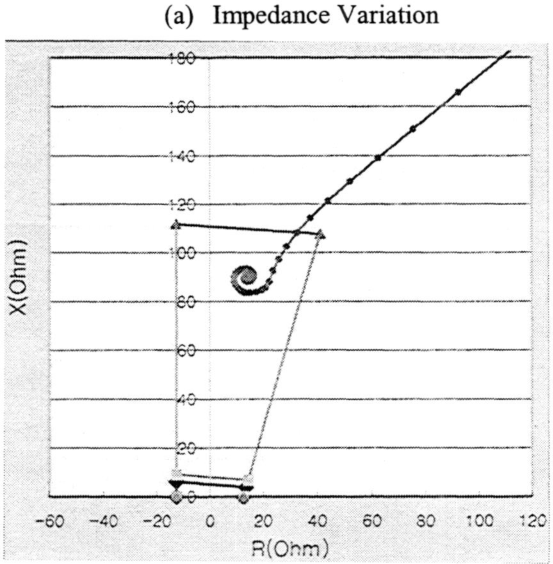

(b) Impedance Trajectory

Fig. 9. Impedance Variation During Transient States

Fig 9 shows the impedance variation of the distance relay on the Janhung side during a three phase short circuit fault at Naju while the UPFC is operating in the voltage injection mode with Vd=0 and Vq=-1. In Fig 3-(b) the impedance trajectory in converges to a point within three cycles (50ms) and the large impedance during the transient state reduces rapidly to within the Zone3 protection region. The impedance variation does not oscillate severely, and the impedance trajectory during the transient state does not fall into the Zone1 or Zone2 protection regions. Thus, the distance relay at Jangheung will not operate abnormally by mistaking the fault at Naju as a fault in Zone1 or Zone2.

5. CONCLUSION

The effects of UPFC voltage injection on distance relays at Jangheung substation were investigated. During steady state operation, the distance relays will not misoperate regardless of the UPFC control state. The UPFC will go into Bypass operation during most of the faults detected by distance relays and will not affect their correct operation. For the Zone1 protection which operates instantaneously, it is recommended that the relay settings be modified to take into account the leakage reactance of the series injection transformer of the UPFC. During faults that do not trigger the UPFC bypass operation such as three phase short circuit faults at Naju, the Zone3 settings need to be modified. The impedance variation seen by the distance relay at Jangheung substation according to different UPFC operation conditions need to be considered.

REFERENCES

Yoon, J., Chang, B., Choo, J., Han, Y., Yoo, I., (2001). "The EMTDC model development of 8-MVA UPFC". *Proceedings of the KIEE PES Spring Annual Conference 2001.* pp 162-165. IEEE PES Korean Chapter.

Han, B., Baek, S., Moon, S., Lee, H., Yoo, I., Chu, J., Jang, B., (2001). "Development of the UPFC Switching Level Simulation Model Using EMTDC". *Proceedings of the KIEE PES Autumn Annual Conference 2001.* pp 15-19. IEEE PES Korean Chapter.

Jung, C., Kim, J., (2001). "A Study for Operation Zone of Distance Relay on Transmission Line Connected UPFC". *Proceedings of the KIEE PES Autumn Annual Conference 2001.* pp 130-132. IEEE PES Korean Chapter.

Dash, P. K., Pradhan, A. K.,(2000). "Adaptive Relay Setting for Flexible AC Transmission Systems (FACTS)," *IEEE Trans. on Power Delivery*, vol. 15, no. 1, pp.38-43, 2000

Padiyar, K. R.,(1998). "Control Design and Simulation of Unified Power Flow Controller," *IEEE Trans. on Power Delivery*, vol. 13, no. 4, 1998

ELSEVIER
IFAC
PUBLICATIONS
www.elsevier.com/locate/ifac

LMI Approach to Decentralized H_∞ Control of FACTS Devices in a Large Power System: Polytopic Representation of Operating Conditions

Malihe M. Farsangi[a], Yong Hua Song[a], and Kwang Y. Lee[b]

[a] Brunel Institute of Power Systems, Brunel University,
Uxbridge, Middlesex UB8 3PH, United Kingdom

[b] Department of Electrical Engineering, The Pennsylvania State University,
University Park, PA 16802, USA

Abstract: A polytopic representation of uncertainties in the operating conditions is introduced for a sequential design of decentralized Flexible AC Transmission Systems (FACTS) controllers for a multi-machine power system. An H_∞-output feedback design procedure using Linear Matrix Inequality (LMI) is applied. Two different types of uncertainties are considered in the design. One uncertainty is due to the interaction from the undesigned loop when evaluating performance, which is bounded in the design of controller using mixed-sensitivity. Another source of uncertainty is considered in the design due to changing in operating point, which is treated by Polytopic Linear Differential Inclusion (PLDI) where various operating conditions are used in the model. The performance of the controllers is verified through non-linear simulations under large disturbances. *Copyright © 2003 IFAC*

Keywords: H_∞ control; polytopic representation; robust control; Linear Matrix Inequality (LMI); FACTS devices; power system oscillation

1. INTRODUCTION

The phenomenon of a poorly damped low-frequency (0.1–2 Hz) inter-area oscillation is very complex as it involves several electromechanical oscillatory subsystems, often comprising several groups of machines distributed over neighbouring utilities. The nature of the interaction makes the damping control design task challenging. The design methodology, therefore, should aim at improving the damping performance of one mode while ensuring the least interaction with other modes. In addition, the controller should guarantee stable operation over the full range of system operating conditions. This type of robust control action has been sought by the H_∞ loop-shaping method (Skogestad and Postlethwaite, 1996). In the study carried out by Farsangi and Song (2001) and in this paper, the H_∞ loop-shaping method is used to arrive at a decentralized controller structure, which provides robust stability and performance over a range of uncertainties of interest. The sequential design of controllers involves closing and tuning one loop at a time, where a loop closed earlier is called a lower-level loop and a loop closed later an upper-level loop. The sequential design guarantees that if robust stability has been achieved after the design of controllers in sequential steps, then the system will, at least, remain stable when loops fail or are taken out of service in the reverse order of how they were designed. For any reason, however, if lower level loops fail, failure tolerance is not guaranteed. This is because by loosing a lower-level loop, the system will face a big change in the state-space model and the controller, which was designed with the lower-loop closed, may not be able to keep the system in a reasonable performance.

While in the previous study by Farsangi and Song (2001) the nominal model is taken linear time-invariant (LTI), in which the H_∞ method is able to consider the uncertainty due to the changes in operating points, failure tolerance is not guaranteed when a lower-level loop fails. In view of this, in this paper, the linear matrix inequality (LMI) technique is used to design an LTI controller valid for all possible operating conditions, including the possible loss of individual loops. A feature of LMI is used to represent uncertainty in which various operating conditions are used in the model. This representation is called a Polytopic Linear Differential Inclusion (PLDI). In this representation, different points of operation constitute the vertices of a region (which is called polytopic), where the system is expected to operate. By ensuring some properties such as quadratic stability for the vertices, it guarantees that all the operating conditions within the polytopic region will possess the same properties.

This methodology is applied to the design of damping controllers for static var compensator (SVC), static synchronous compensator (SSSC) and unified power flow controller (UPFC) placed in a large study system.

In this paper, the problems raised due to the sequential design, are considered in the design. In the next section the problem raised due to interaction between different loops is explained briefly.

2. SEQUENTIAL DECENTRALIZED FEEDBACK CONTROL

Consider the multi-variable plant G with n inputs and n outputs as shown in Fig. 1, where the decentralized controller is assumed to be diagonal, i.e., controllers are designed for each single loop.

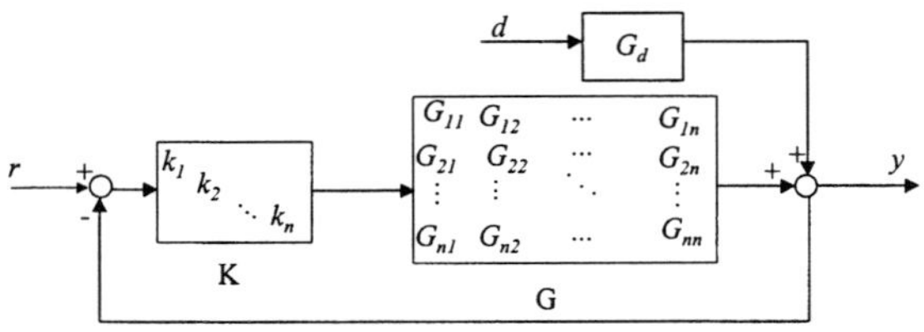

Fig. 1. Decentralised diagonal control of an $n \times n$ plant.

In the figure, r is the input for the overall system, d represents disturbances and y denotes the output response for the overall system when all loops are closed with output error e,

$$e = r - y = Sr - SG_d d \tag{1}$$

where the sensitivity function, $S = (I + GK)^{-1}$, is the transfer function between the reference signal, r and the tracking error, e.

Since in the sequential design, only one loop is considered at a time, closing subsequent loops may change the response of previously designed loops. This problem is raised due to interaction between different loops. Therefore, considering interaction is very important in a successful controller design. The relative interaction in the decentralized control can be presented by the matrix $E = (G - \tilde{G})\tilde{G}^{-1}$ (Grosdidier and Morari, 1986; Hovd and Skogestad, 1994), where matrix $\tilde{G} = diag(G_{ii})$ denotes the matrix with the diagonal elements of G. Then an important relationship for decentralised control is given by:

$$\underbrace{(I + GK)}_{Overall} = \underbrace{(I + E\tilde{T})}_{interactions} \underbrace{(I + \tilde{G}K)}_{individual\ loops} \tag{2}$$

where the complementary sensitivity function, $\tilde{T} = \tilde{G}K(I + \tilde{G}K)^{-1}$, is the closed-loop transfer function of the diagonal system $\tilde{G}$.

Equivalently Equation (2) can be written in terms of sensitivity function as follows:

$$S = \tilde{S}(I + E\tilde{T})^{-1} \tag{3}$$

where $\tilde{S}$ is the sensitivity function for the diagonal system $\tilde{G}$. Note that the sensitivity functions and complementary sensitivity functions for the individual loops are collected in the diagonal matrices, i.e., $\tilde{S} = diag(\tilde{s}_i) = (I + \tilde{G}K)^{-1}$ and $\tilde{T} = diag(\tilde{t}_i) = \tilde{G}K(I + \tilde{G}K)^{-1}$.

Equation (3) shows that $\tilde{S}$ is not equal to the matrix of diagonal elements of S and it also depends on the term $(I + E\tilde{T})^{-1}$, indicating the interaction between different loops. At frequencies where the feedback is effective $\tilde{T} \approx I$ and it can be shown that:

$$(I + E\tilde{T})^{-1} \approx \tilde{G}G^{-1} = \Gamma \tag{4}$$

where, $\Gamma = \{\gamma_{ij}\}$ is the Performance Relative Gain Array (PRGA), known as an indicator for interaction (Hovd and Skogestad, 1994; Skogestad and Postlethwaite, 1996). By substituting Equation (4) into Equations (3) and (1), the following equations can be derived:

$$S \approx \tilde{S}\Gamma \tag{5}$$

$$e = r - y \approx \tilde{S}\Gamma r - \tilde{S}\Gamma G_d d \tag{6}$$

Equations (5) and (6) show that Γ is important when evaluating performance with decentralized control (Hovd and Skogostad, 1994; Skogestad and Postlethwaite, 1996).

To deal with the problem raised due to the interaction, the effect of interaction can be treated as uncertainty in the system. The interaction term $(I + E\tilde{T})^{-1}$ is added to the system as an additive uncertainty (Fig. 2) to express the interaction from the undesigned loop when evaluating performance. This is an estimation of how the undesigned loops will affect the output of the loop to be designed. In Fig. 2, δ denotes the normalized perturbation and $(I + E\tilde{T})^{-1}$ shows the actual perturbation.

3. POLYTOPIC LINEAR DIFFERENTIAL INCLUSION (PLDI)

An uncertain time-variant linear system can be modelled using the Linear Differential Inclusion (LDI) as expressed below (Boyd *et al.*, 1994, Gahinet *et al.*, 1995):

$$\begin{cases} \dot{x} \in \Omega x \\ x(0) = x_0 \end{cases} \tag{7}$$

Any x that satisfies the LDI (7) is called a solution or trajectory of this LDI.

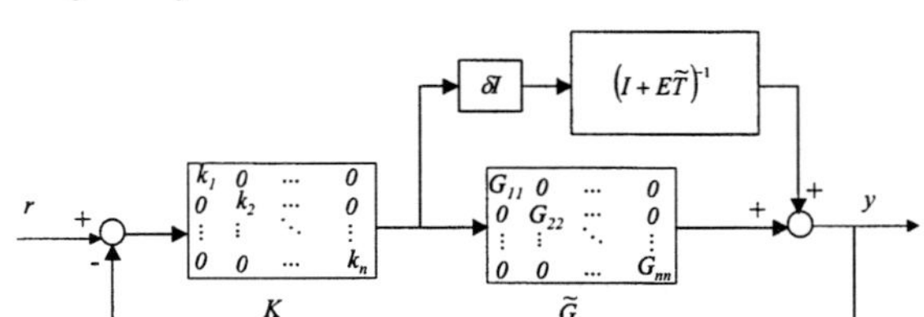

Fig. 2. Block diagram representation of interactions as additive uncertainty.

When Ω in the LDI (7) is a convex subset, described by a convex combination of its k vertices as:

$$\Omega = C_0\left\{\begin{bmatrix} A_1 + j\,E_1 & B_1 \\ C_1 & D_1 \end{bmatrix}, \;\cdots,\; \begin{bmatrix} A_k + j\,E_k & B_k \\ C_k & D_k \end{bmatrix}\right\}$$

where C_o denotes the convex combination, it is named as a polytop. If Ω is a polytop the LDI (7) reduces to a Linear Time-Varying system (LTV) as

$$\begin{cases} E(t)\,\dot{x} = A(t)\,x + B(t)\,u \\ y = C(t)\,x + D(t)\,u \end{cases} \tag{8}$$

For this class of systems, the aim is designing a controller K with the following set of equations:

$$\begin{cases} \dot{x}_K = A_K x_K + B_K y \\ u = C_K x_K + D_K y \end{cases} \tag{9}$$

such that:

- The closed-loop system is stable for all admissible operating points; and
- H_∞ norm of the closed-loop transfer function, does not exceed some level $\gamma > 0$.

For a LTV system, properties such as quadratic stability should be checked to test the feasibility of the design. A sufficient condition for stability of the LDI (7) is the existence of a quadratic function:

$$V(x) = x^T P\, x \tag{10}$$

with a matrix $P > 0$ that decreases along a non-zero trajectory of LDI (7). When such P exists, the LDI (7) is said to be quadratically stable and V is known as a quadratic Lyaponov function. The quadratic stability of polytopic systems can be checked in the Matlab LMI toolbox (Gahinet *et al.*, 1995).

4. STUDY SYSTEM

The system shown in Fig. 3, consisting of 16 machines and 68 buses is considered. This is a reduced-order model of the New England (NE)-New York (NY) interconnected system (Chow, 1997). The first nine machines are representing the New England system generation and machines 10 to 13 represent the New York power system generation.

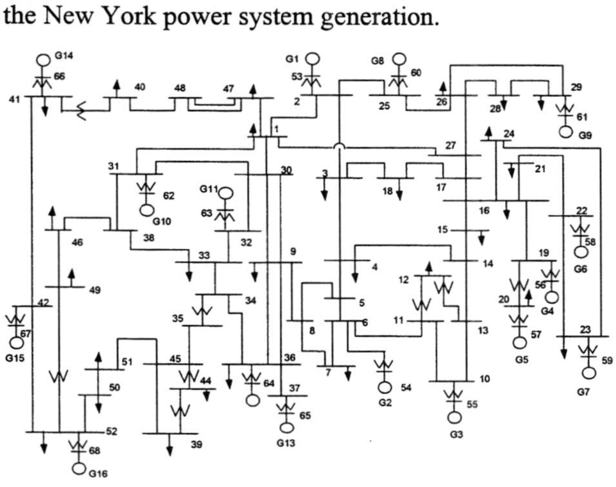

Fig. 3. Single line diagram of the 5-area New England-New York interconnected system.

The last three machines are the dynamic equivalents of three large neighbouring areas connected to the New York power system. The subtransient reactance model for the generators, the first-order simplified model for the excitation systems and the linear models for the loads are used in the study.

The UPFC, SSSC and SVC are placed in the system according to their primary function. For the damping of the inter-area modes, locations for UPFC are selected as bus 8 and line 8-9 and for SSSC and SVC as line 40-48 and bus 7, respectively.

5. DESIGN PROCEDURE

The H_∞ loop-shaping method is used in the sequential design to arrive at a decentralized control structure, which provides robust stability and performance over a range of uncertainties of interest. In a power system, most of the tracking problems are taken care of by the primary controllers such as automatic generation controllers and automatic voltage regulators. Since the objective of the supplementary controller is to damp out system oscillation in a particular frequency band, the design emphasis is different from that of tracking a reference input signal. Zhao and Jiang (1995, 1998) reformulated the H_∞ mixed-sensitivity method for damping purpose as illustrated in Fig. 4, where W_2 is a weighting function that determines the robustness of the closed-loop system and W_3 is a weighting function that determines the damping performance of the closed-loop system. The objective of the controller is to provide additional damping to the system, or equivalently, to reduce the resonance peak of the closed-loop system.

The damping and robustness requirements can be achieved, such that the transfer function of the closed-loop system satisfies the following norm inequality:

$$\left\| \begin{bmatrix} W_2(s)V(s)K(s) \\ W_3(s)V(s) \end{bmatrix} \right\|_\infty < 1 \tag{11}$$

where, $V(s) = \dfrac{y(s)}{r(s)} = \dfrac{G(s)}{1 + K(s)G(s)}$ is the closed-loop transfer function and $\|\;\|_\infty$ denotes the H_∞ norm.

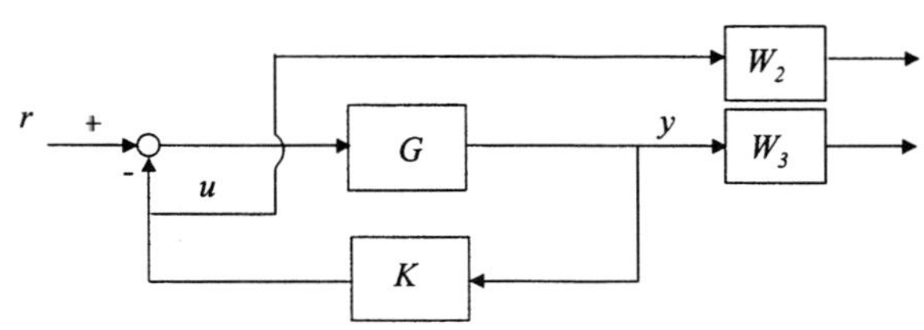

Fig. 4. Block diagram for damping controller design.

The objective of the mixed-sensitivity minimization is to find an internally stabilizing controller that minimizes the H$_\infty$ norm in (11). In the LMI formulation, the same objective is achieved in the sub-optimal sense if and only if there exists a symmetric matrix X_∞ such that (Gahinet *et al*, 1995):

$$\begin{pmatrix} A_{cl}X_\infty + X_\infty A_{cl}^T & B_{cl} & X_\infty C_{cl1}^T \\ B_{cl}^T & -I & D_{cl1}^T \\ C_{cl1}X_\infty & D_{cl1} & -\gamma^2 I \end{pmatrix} < 0,\ X_\infty > 0 \tag{12}$$

with $\left\| C_{cl1}(sI - A_{cl})^{-1}B_{cl} + D_{cl1} \right\|_\infty < \gamma$; and the controller is said to be γ sub-optimal.

The design task can be summarised in the following steps:

- **Step 1**: Reduce the order of the open-loop plant to ease the controller design for each loop;
- **Step 2**: Choose proper weighting functions W_2 and W_3;
- **Step 3**: Consider the extreme operating conditions, where the system is expected to operate;
- **Step 4**: Using the Matlab LMI toolbox to find the matrices in (9) to design a H$_\infty$ LMI based controller sequentially; and
- **Step 5**: Check the degree of stability of the polytopic system by function "quadstab" in the Matlab LMI toolbox.

6. DESIGN OF DAMPING CONTROLLERS FOR THE STUDY SYSTEM

The above procedure is followed to design controllers for SVC (loop 1), SSSC (loop 2) and UPFC (loop 3). In the sequential design the controller is designed by closing one loop at a time. Based on the work carried out by Farsangi and Song (2001), the closing order of loops are: loop 3 (UPFC), loop 2 (SSSC) and loop 1 (SVC).

First, loop 3 with a UPFC is closed and the resulting system with 129 states is reduced to an equivalent model with 15 states using the model reduction technique, which can be found in (Skogestad and Postlethwaite, 1996). The weighting function W_2 is chosen by considering the effect of the interaction from the other loops. The interaction could change the dominant mode in the open-loop frequency response. For each loop, as long as the encountered interaction from the undesigned loops is bounded by W_2, the stabilization of the system is guaranteed.

In order to have a simplified estimate of the effects of closing other loops (the interaction from the undesigned loops, loops 2 and 1), the matrix $(I + E\tilde{T})^{-1}$ is calculated and added as an additive uncertainty term to loop 3. Fig. 5 shows how the interaction from the undesigned loops affects on loop 3. As shown in the figure, the dominant modes are changed due to the interaction from the undesigned loops.

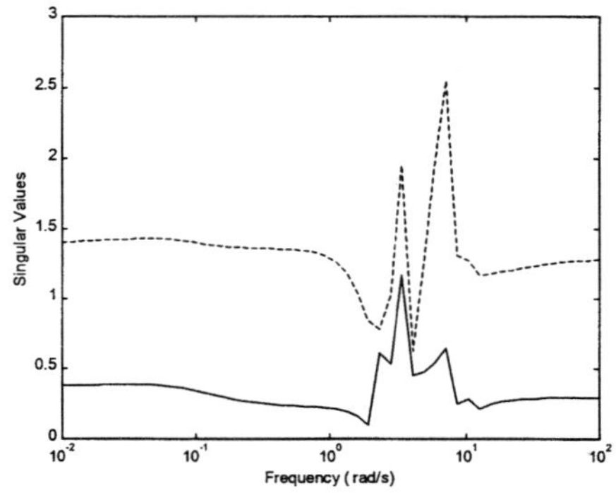

Fig. 5. Loop shape of loop 3: Solid line: without considering interaction; Dashed line: with considering interaction.

The choice of W_2 and W_3 is based on the work carried out by Zhao and Jiang (1998). Based on Fig. 5 the weighting function W_2 is synthesized to have a reasonably wide band to cover the resonant frequency of the system due to the interactions from the undesigned loops. The weighting function W_2 is chosen as follows:

$$W_2 = \frac{s^2 + 38s + 5}{s^2 + 22s + 5}. \tag{13}$$

The weighting function W_3 is selected so that the magnitude of $1/W_3$ at the resonant frequency is smaller than the resonance peak of the open-loop system (shown in Fig. 5, solid line. The resulting weighting function W_3 is as below (Zhao and Jiang, 1998):

$$W_3 = 0.5\frac{s^2 + 120s + 3600}{s^2 + 80s + 1600} \tag{14}$$

Another source of uncertainty to be considered in the design is the changes in operating point, which can be handled by PLDI. In the study system (Fig. 3), tie-line 8-9 between the NE system and the NY system is very week and a large amount of power transfer along this tie-line makes the system unstable. When tie-line 1-2 is disconnected a large amount of power transfers through tie-line 8-9. This condition introduces the largest uncertainty in the system. Also, nominal operation and 10% increasing in the power transfer through the tie-line are considered as another operating points.

By bounding the interaction from the undesigned loop with W_2 and considering different operating point due to the power transfer through the tie-line 8-9 with PLDI, the controller is obtained for loop 3 by using the LMI toolbox is as below:

$$K_{UPFC} = \frac{NUM_{UPFC}}{DENOM_{UPFC}} \quad (15)$$

where

$NUM_{UPFC} = 0.0006\, s^6 - 0.0045\, s^5 + 0.0468\, s^4 - 0.3549\, s^3 + 1.1645\, s^2 + 8.9231\, s + 0.0049$

$DENOM_{UPFC} = 0.005\, s^6 + 0.056\, s^5 + 0.468\, s^4 + 2.71 s^3 + 7.791^2 + 26.145\, s + 11.434$

The quadratic stability is then checked for all different operating points in the polytopic region, which shows the system is quadratically stable.

For designing the controller for loop 2, the order of the plant is reduced to 15. Once again, the term $(I + E\tilde{T})^{-1}$ is added as uncertainty to loop 2 to determine the effect of interaction from other loops. Loop shape of loop 2 is shown in Fig. 6. By considering the uncertainty caused by interaction (shown in Fig. 6), the weighting functions W_2 and W_3 are selected in a similar way as for loop 3.

For loop 2 (SSSC) the same procedure as for loop 3 is followed, except that in addition to uncertainty due to changing in operating point due to power transferred along tie-line, the uncertainty resulting from the failure of loop 3 is also considered. Once again the LMI toolbox is used to design the controller for loop 2 (SSSC) in a form similar to Equation (15), which is given below:

$NUM_{SSSC} = -0.0001 s^6 + 0.0476 s^5 - 0.0892 s^4 + 0.8181 s^3 - 0.4815 s^2 + 2.751 s + 0.006$

$DENOM_{SSSC} = 0.002 s^6 + 0.516 s^5 + 0.1799 s^4 + 0.6887 s^3 + 0.8858 s^2 + 1.769 s + 0.3386$

Once again, the system is quadratically stable for all operating point.

For designing the controller for loop 1, once again the same procedure as for loop 2 is followed except that another extreme condition for the uncertainty due to the failure of loops 2 and 3 is considered. The resulting controller achieved for loop 1 (SVC) in a form similar to Equation (15) is as follows:

$NUM_{SVC} = 0.0012\, s^5 - 0.2857\, s^4 + 0.0075\, s^3 - 3.5897\, s^2 + 3.3284\, s - 0.0004$

$DNOM_{SVC} = 0.006\, s^5 + 0.679\, s^4 + 2.853\, s^3 + 6.698\, s^2 + 18.459\, s + 0.001$

The performance of the controllers designed is evaluated in the frequency and time domains. An eigenvalue analysis is carried out to evaluate the damping ratios for post-fault condition (a three-phase fault occurred at bus 1 in the study system). Table 1 displays the inter-area modes and damping ratios of the open-loop study system. It can be seen from the table that after clearing the fault, one eigenvalue is positive (mode 1) and the system becomes unstable through growing oscillations. By placing FACTS devices, the unstable mode (mode 1) are disappeared. By applying supplementary controllers, the modes have been improved even more.

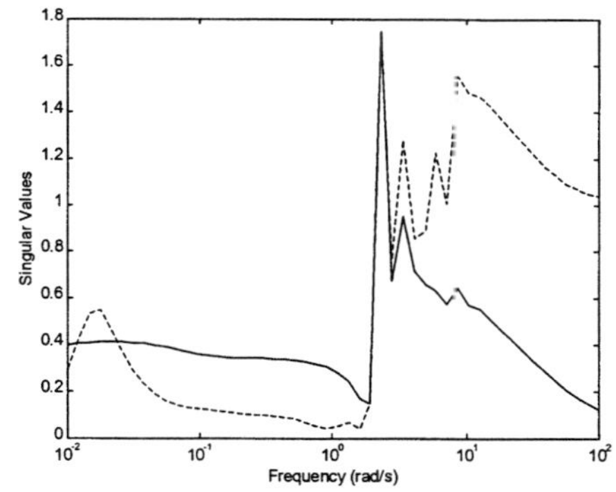

Fig. 6 Loop shape of loop 2: Solid line: without considering interaction; Dashed line: with considering interaction.

A non-linear simulation is necessary to evaluate the performance of the controllers during and after system fault. The variation of machine angles with reference to the machine 15 are computed.

Fig. 7 shows the dynamic behaviour of the study system in terms of machine angles for machines 1, 8 and 14. As shown in this figure, the system oscillations grow in the absence of FACTS devices. The action of FACTS devices stabilize the system oscillations and adding supplementary controllers ascertain damped behaviour.

Also, the responses of the system in different condition of operation, corresponding to failure of loop 2 is demonstrated in Fig. 8. This shows that in the case of losing some of the loops, still the remaining loops can be made robust.

7. CONCLUSIONS

In this paper UPFC, SSSC and SVC are placed in a large power system to damp the inter-area modes. The corresponding damping controllers are designed sequentially. Since in the sequential design, only one loop is considered at a time, closing subsequent loops may change the response of previously designed loops. This problem is raised due to interaction between different loops. To deal with this problem, the effect of interaction is treated as uncertainty in the system, which is bounded by a weighting function in the mixed-sensitivity design method.

Also, this paper presents the advantages of the polytopic representation of uncertainty in the operating conditions for the design of the damping controllers for a large power system. When the polytopic representation is used, various typical operating conditions appear in the model of the system. These different operating points are due to different amounts of power transfers along tie-lines.

The quadratic stability is then checked for all different operating points in the polytopic region. Even in the case of losing some of the loops (loops closed earlier), still the remaining loops can be made robust, guaranteeing the failure tolerance.

Table 1 The effects of FACTS devices on the inter-area modes in the study system.

Mode No.	System without FACTS Devices		System with FACTS Devices		System with FACTS Devices and Supplementary Controllers	
	Inter-area Mode	ζ	Inter-area Mode	ζ	Inter-area Mode	ζ
1	0.0116±1.8127i	-0.006	-0.1043±1.9093i	0.054	-0.3904±2.1812i	0.176
2	-0.0511±2.9195i	0.017	-0.0584±3.0267i	0.019	-0.3139±2.9967i	0.104
3	-0.1427±3.1651i	0.045	-0.1657±3.1733i	0.052	-0.4836±3.6800i	0.130

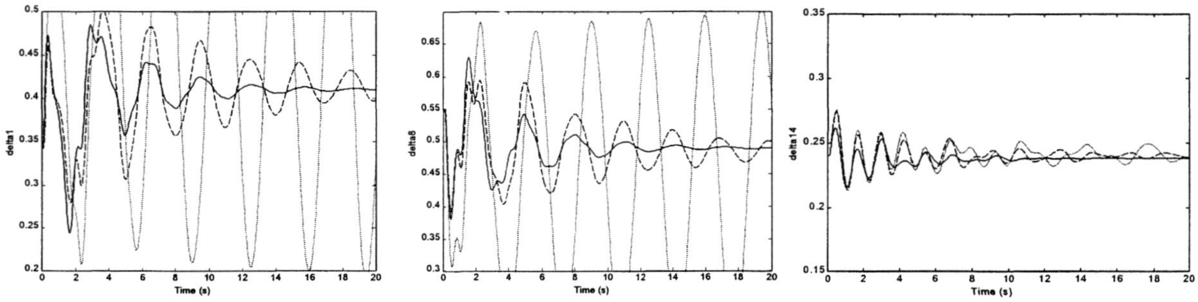

Fig. 7. Dynamic response of the system (machines 1, 8 and 14) following a three-phase fault at bus 1 in the study system: Dotted line: System without FACTS devices; Dashed line: System with FACTS devices; Solid line: System with FACTS devices and with controllers.

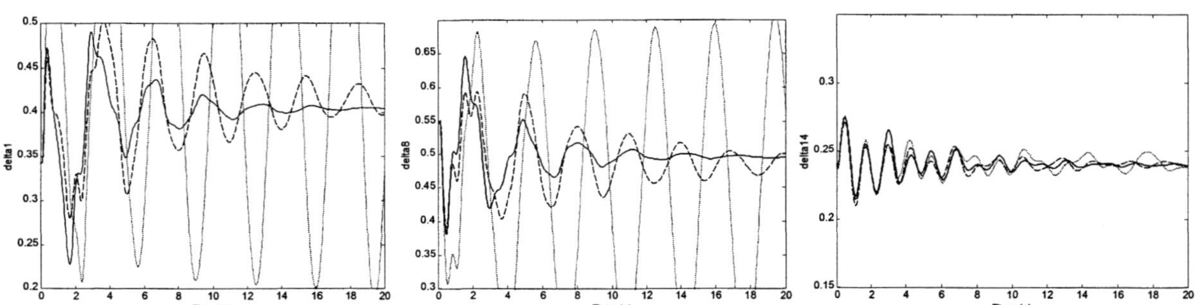

Fig. 8. Dynamic response of the system (machines 1, 8 and 14) after losing loop 2: Dotted line: System without FACTS devices; Dashed line: System with FACTS devices; Solid line: System with FACTS devices and with controllers.

8. REFERENCES

Boyd, S., L. El Ghaoui, E. Feron and V. Balakrishnan (1994). *Liner matrix inequalities in system and control theory*, Society for Industrial and Applied Mathematics, Philadelphia, USA.

Chow J. H. (1997) *Power system toolbox: A set of coordinated m-files for use with MATLAB*, Cherry Tree Scientific Software.

Farsangi M. M.and Y. H. Song (2001) *Sequential decentralised control of FACTS devices in large power systems*" Seventh International Conference on AC-DC Power Transmission, November 2001, pp. 268-273, London, UK.

Gahinet P., A. Nemirovski, A.J. Laub and M. Chilali (1995). *The LMI control toolbox*, The Math works Inc.

Grosdidier P. and M. Morari (1986) *Analysis of interactions using structured singular values*, Proc. the 1986 American Control Conference, **Vol. 2**, pp. 658-663, New York, USA.

Hovd M. and S. Skogestad (1994) *Sequential design of decentralised controller*, *Automatica*, **Vol. 30**, pp. 1601-1607.

Skogestad S. and I. Postlethwaite, (1996) *Multivariable feedback control: analysis and design*, John Wiley.

Zhao Q. and J. Jiang (1995). *Robust SVC controller design for improving power system damping*,

IEEE Transactions on Power Systems, **Vol. 10**, No. 4, 1995, pp. 1227-1232.
Zhao Q. and J. Jiang (1998). *A TCSC damping controller design using robust control theory*, *Electrical Power and Energy System*, **Vol. 20**, No. 1, pp. 25-33

ELSEVIER
IFAC
PUBLICATIONS
www.elsevier.com/locate/ifac

COORDINATION BETWEEN TRANSIENT AND DAMPING CONTROLLER FOR SERIES FACTS DEVICES USING ANFIS TECHNOLOGY

Lijun Cai, István Erlich

Department of Electrical Engineering,
University of Duisburg-Essen, 47057, Germany.

Abstract: This paper deals with the control of the series FACTS (Flexible AC Transmission Systems) devices for the coordination between their transient stability controller and POD (Power Oscillation Damping) controller in multi-machine power systems. The design aspects and their implementation in form of fuzzy-logic coordination controller are presented. Furthermore, ANFIS (Adaptive Neuro-Fuzzy Inference System) is employed for the training of the proposed fuzzy-logic controller. The local signals of the FACTS devices are applied to achieve the coordination objectives. Digital simulations of multi-machine power system subjected to a wide variety of disturbances validate the efficiency of this approach. The proposed control scheme is not only robust, but also simple and easy to be realized in power systems. *Copyright © 2003 IFAC*

Keywords: ANFIS, Coordinated control, Damping control, FACTS, Fuzzy logic, POD controller, Training, Transient controller

1. INTRODUCTION

Nowadays, FACTS devices are used to control power flow and to enhance system stability. Particularly, in recent years, with the deregulation of the electricity market, the traditional concepts and practices of power systems have changed. Better utilization of the existing power system to increase capacities by installing FACTS devices becomes imperative. FACTS devices are playing an increasing and major role in the operation and control of power systems.

Series FACTS devices are the key elements in the FACTS family . They are recognized as an effective and economical means to solve the following problems: steady state power flow control, transient stability control and power system oscillation damping control (Lei, *et al.*, 2000). In this paper, the coordinated control of series FACTS devices is considered.

Usually, the coordination between FACTS transient controller and POD controller is achieved using a pre-set fixed time switch (Lei, *et al.*, 2000). However, the optimal switching time is varied with different disturbances and different operating conditions. In the conventional approach, the switching time cannot be adjusted under different situations. This work focuses on the development of a fuzzy coordination controller between FACTS transient stability controller and POD controller. The switching time is then controlled by means of the fuzzy-logic approach. Furthermore, the parameters of the proposed controller are optimized using the ANFIS algorithm (Shing, and Jang, 1993).

Corresponding Author: Lijun Cai, EAN, University of Duisburg-Essen, 47057, Germany. Email: cailijun@uni-duisburg.de Tel: +49-203-3793994, Fax: +49-203-3792749

This paper is organized as follows: Following the introduction, power system model and FACTS location are considered in section 2. Then in section 3, the conventional FACTS control schemes are introduced. In section 4 and section 5, the structure and the ANFIS training of the fuzzy coordination controller are discussed in detail. The simulation results are given in section 6. Finally, brief conclusions are deduced.

2. POWER SYSTEM MODEL AND FACTS LOCATION

A modified 16-machine 68-bus system (Rogers, 1999) with series FACTS devices, as shown in Fig. 1, is simulated in this paper. Each generator is simulated using 6th order model and the FACTS devices are considered using the power injection model (Cai, and Erlich, 2002).

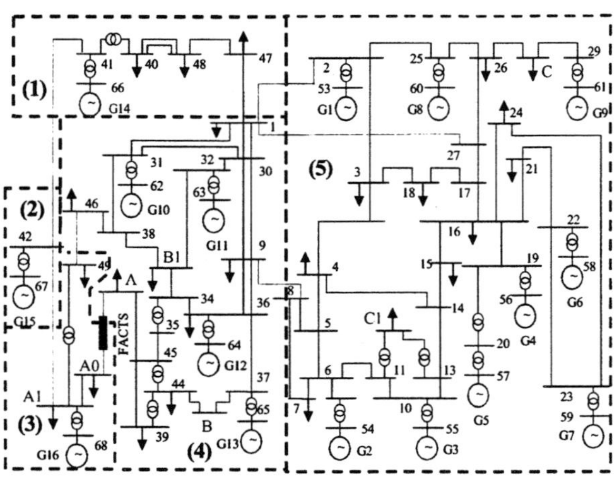

Fig. 1. One line diagram of 16-machine five-area power system

By means of the modal analysis, this system is divided into 5 areas (Rogers, 1999). Because there are three tie lines between area 4 and area 5, the main inter-area oscillation is between area 3 and area 4. The series FACTS devices are located between bus A and bus A0 (on the tie line between area 3 and area 4). The location of FACTS device is determined using the residue algorithm for damping of inter-area oscillations (Wang, 1999).

3. CONVENTIONAL FACTS CONTROLLER

Normally, series FACTS controller focuses on the following three control objectives (Lei, *et al.*, 2000):

1 Steady state power flow control;
2 Transient stability control for improving the first swing stability;
3 Damping control to damp the power system oscillations.

3.1 Steady state power flow controller

In general, the constant line power strategy is used in power flow control (Lei, *et al.*, 2000). The corresponding control loop is shown in Fig. 2, where P_{ref} and C_{ref} are reference values of the active power and the initial compensation value of the FACTS devices respectively. The output C_{FACTS} is the control variable of the series FACTS devices. For instance, for the control of TCSC (Thyristor Controlled Series Capacitor) it is the value of the series capacitance and for UPFC (Unified Power Flow Controller) it is the series injected voltage (Cai, and Erlich, 2002).

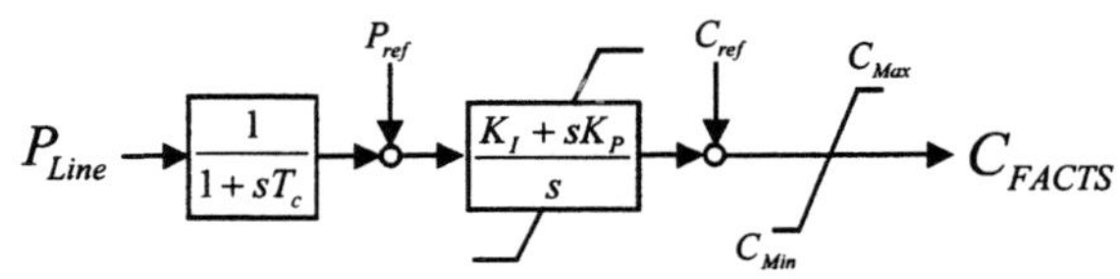

Fig. 2. Steady state power flow controller

3.2 FACTS transient stability controller

Following a large disturbance, the transient controller acts initially to give a maximum compensation level for a pre-set time T_0 (Lei, *et al.*, 2000). The structure of the FACTS transient controller is shown in Fig. 3.

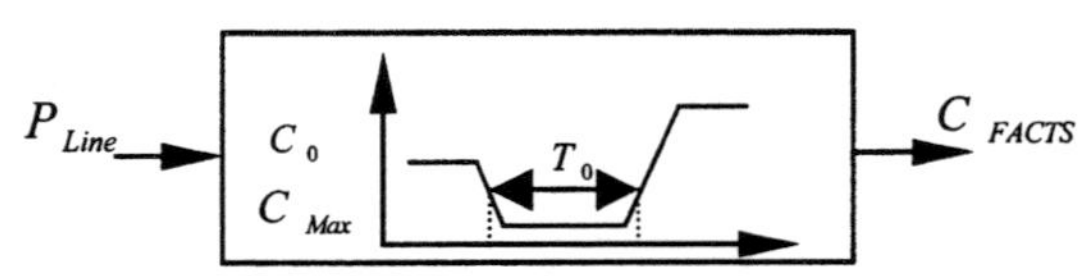

Fig. 3. Transient controller

3.3 FACTS POD controller

In general, the structure of series FACTS POD controller is similar to the PSS controller (Lei, *et al.*, 2000). In Fig. 4, the FACTS POD controller is illustrated. It involves a transfer function consisting of an amplification block, a wash out block and two lead-lag blocks (Lei, *et al.*, 2000; Rogers, 1999).

Commonly, local signals of FACTS devices are always applied for the damping control. In this simulation, the active power through the FACTS devices P_{line} is employed. As stated above, the output is C_{FACTS} which represents the control variable of the series FACTS devices.

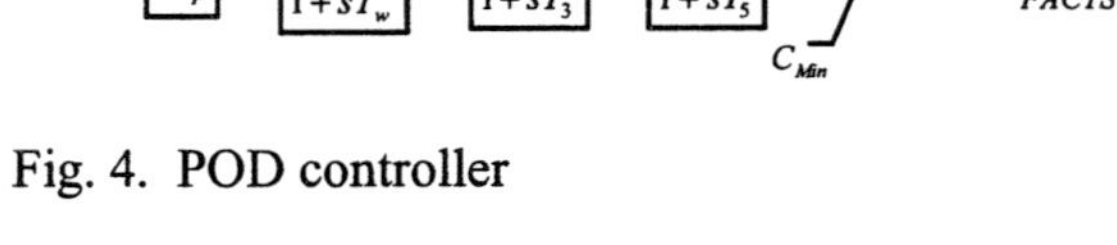

Fig. 4. POD controller

3.4 Conventional coordination method between FACTS transient controller and POD controller

The FACTS transient controller and POD controller achieve different control objectives in different operation states. Conventionally, in order to enhance transient stability and damp the subsequent oscillations, a switching control strategy is always used to coordinate between the two different controllers (Lei, *et al.*, 2000). The switching control scheme is shown as follows:

1 Following a large disturbance, the transient controller acts first to maintain the transient stability of power systems;

2 After the pre-set switching time period T_0, the control is transferred to the POD controller to damp the post-fault oscillations.

The conventional coordination control scheme is shown in Fig. 5. (Lei, *et al.*, 2000)

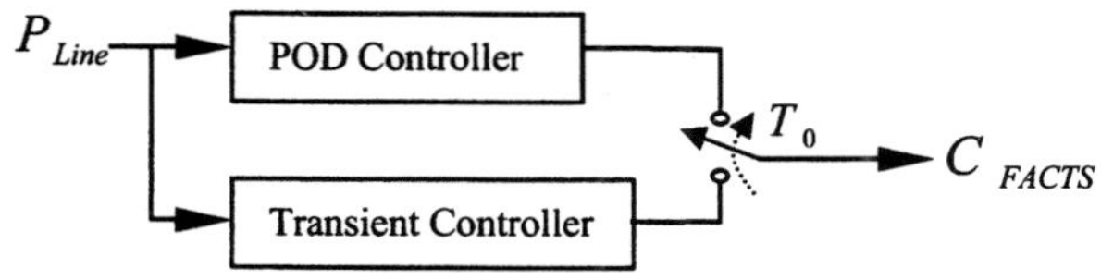

Fig. 5. Conventional coordination between FACTS transient and POD controller

The pre-set switching time T_0 should be reasonably chosen within the transient period. Practically, the exact switching time is determined by trial and error using the transient simulation results. This so determined switching time is physically fixed according to one particular fault sequence. However, power systems may have different disturbances and different operating conditions and therefore the pre-set switching time T_0 may not be suitable for all above situations.

In order to coordinate the transient stability controller and the POD controller under different operating situations and different fault sequences, the fuzzy-logic coordination controller is proposed. The design procedure is as follows:

1 Determination of the structure of fuzzy-logic coordination controller;
2 ANFIS training of the fuzzy-logic controller;
3 Non-linear simulation for verifying the robustness of the proposed controller.

4. FUZZY COORDINATION CONTROLLER

As stated above, the objective is to achieve a good transient behavior and damping performance for the considered system. Therefore the controller must have the following function: By means of the FACTS local signals, design a time-variant switch for the coordination between the FACTS transient controller and POD controller. Moreover, the controller must also react robustly under different situations without knowing fault sequences occurring in the system.

In order to handle the uncertainties of fault sequences and different operating conditions, fuzzy-logic controller is employed. Practically, fuzzy-logic is one of the most successful approaches for utilizing the qualitative knowledge of a system to design a controller. Furthermore, fuzzy-logic controllers do not require the mathematical model of a system. They can cover a wider range of operating conditions and they are robust (Chedid, *et al.*, 2000). In this work, there are two control loops in the fuzzy coordination controller, as shown in Fig. 6, fuzzy-logic loop and protection loop.

The three inputs to the fuzzy coordination controller are: ΔP_{Line}, $\Delta P_e/\Delta t$ and U_A, where U_A is the FACTS terminal voltage magnitude. ΔP_{Line} and $\Delta P_e/\Delta t$ are defined as follows:

$$\Delta P_{Line} = P_{Line}^{(n+1)} - P_{Line}^{(0)}$$
$$\frac{\Delta P_e}{\Delta t} = \frac{P_{Line}^{(n+1)} - P_{Line}^{(n)}}{\Delta t} \tag{1}$$

where

ΔP_{Line}: Active power difference through FACTS device;

$P_{Line}^{(0)}$: The initial active power transferred through the FACTS device;

$P_{Line}^{(n)}, P_{Line}^{(n+1)}$: The active power transferred through the FACTS device at the sample time n and $n+1$;

Δt: The time step between two sample points.

The controller output $\bar{y}$ is the switching signal for the coordination between FACTS transient and POD controllers:

$\bar{y}=1$: Transient controller should be employed;

$\bar{y}=0$: POD controller should be employed.

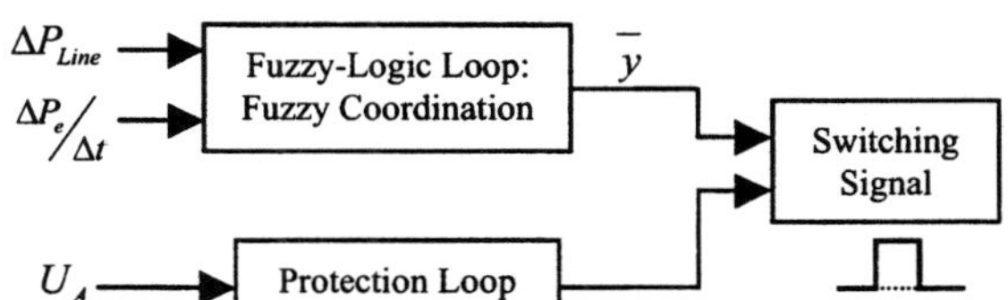

Fig. 6. Fuzzy coordination controller

4.1 Fuzzy-logic loop

The fuzzy-logic loop involves fuzzification, inference and defuzzification.

4.1.1 Fuzzification

Fuzzification is a process whereby the input variables are mapped onto fuzzy variables (linguistic variables). Each fuzzified variable has a certain membership function. In this work, the inputs are fuzzified using three fuzzy sets: B (big), M (medium) and S (small), as shown in Fig. 7.

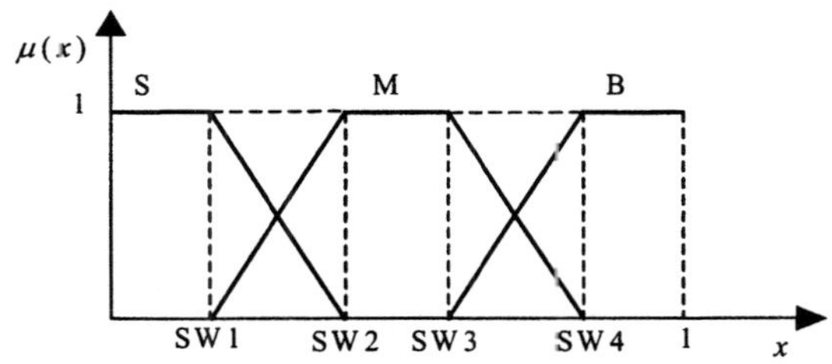

Fig. 7. Membership function

4.1.2 Inference

Fuzzy inference system (FIS) involves fuzzy rules for determining output decisions. The fuzzified input variables are mapped onto the output variables using these fuzzy rules.

In this presentation, the Sugeno FIS (Shing, and Jang, 1993) is employed, because it is able to combine transparency of the rules and the accuracy of the predictions concomitantly. The outputs of the inference system are linear membership functions and the first-order Sugeno fuzzy model has the following form:

if x is A_1 and y is A_2 then $f_i = p_i x + q_i y + r_i$

where in this work, *x* is the ΔP_{Line} and *y* is the $\Delta P_e / \Delta t$ defined as (1). A_1 and A_2 are fuzzy sets in the antecedent, while p_i, q_i and r_i are the consequent parameters (Shing, and Jang, 1993). The fuzzy rules can be obtained from the system operation and the knowledge of the operator. In this paper, the rules are trained using the ANFIS technology.

4.1.3 Defuzzification

The defuzzification process transforms the fuzzy results of the inference into a crisp output. In this work, the weighted average method is employed (Shing, and Jang, 1993).

4.2 Protection loop

The protection loop is to protect the FACTS devices under large disturbances. Its input is the FACTS terminal voltage magnitude at bus A. For instance, under large disturbances, if U_A has a large difference (more than 15%) with the nominal voltage, the FACTS devices will be blocked to protect the power electronic devices.

5. ANFIS TRAINING

In this work, both membership functions and the inference system are optimized using ANFIS technology.

5.1 ANFIS structure

Commonly, the ANFIS has five layers. In layer 1, each node generates membership grades of a linguistic label (Chedid, *et al.*, 2000). In this paper, as shown in Fig. 7, the trapezoid functions are selected. Parameters in this layer are referred to as premise parameters S_1 and they can be trained using the ANFIS learning algorithm (Shing, and Jang, 1993).

Every node in layer 2 is a fixed node and calculates the firing strength of teach rule via multiplication of the incoming signals (Shing, and Jang, 1993):

$$\omega_i = \mu_{Ai}(x) + \mu_{Bi}(y) \tag{2}$$

Nodes in layer 3 compute the normalized firing strength of each rule. The ith node calculates the ratio of the ith rule's firing strength to the sum of all rules' firing strengths (Shing, and Jang, 1993):

$$\bar{\omega}_i = \omega_i \cdot \left(\sum_{i=1}^{n} \omega_i \right)^{-1} \tag{3}$$

In layer 4 there are only adaptive nodes and the ith has the following output (Shing, and Jang, 1993):

$$f_i = \bar{\omega}_i (p_i x + q_i y + r_i) \tag{4}$$

where $\bar{\omega}_i$ is the output of layer 3. $\{p_i, q_i, r_i\}$ are referred to as the consequent parameter set S_2. They can also be trained using ANFIS learning algorithms.

The single node in layer 5 sums up all the incoming signals:

$$\bar{y} = \sum_{i=1}^{9} \bar{\omega}_i (p_i x + q_i y + r_i) \tag{5}$$

The ANFIS structure is shown in Fig. 8.

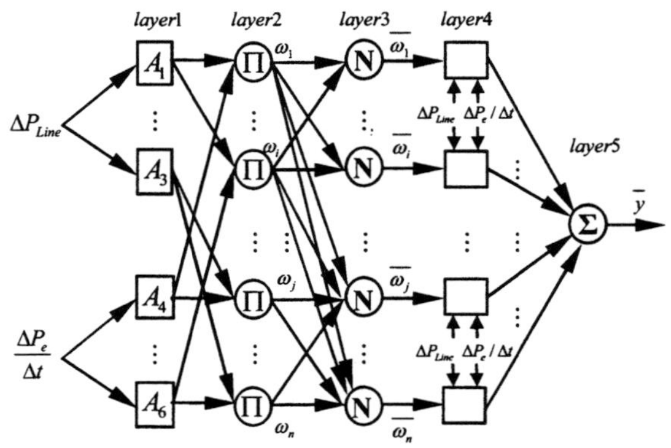

Fig. 8. ANFIS structure

5.2 ANFIS training

The objective of ANFIS training is to learn a good coordination between transient controller and damping controller in presence of uncertainties. The training procedure is achieved based on the batch learning technique, where the tuning of the fuzzy-logic controller is achieved with a back-propagation algorithm using input–output training data set. Considering the computation complexity and the resulting performance, parameters are trained using the gradient descent and the least square estimation (LSE) method (Chedid, *et al.*, 2000).

5.2.1. Fine-tuning of Membership Function

The human-determined membership functions are subject to the differences from person to person and from time to time, and therefore they are rarely optimal in terms of reproducing desired outputs (Shing, and Jang, 1993). In this simulation, since the size of available input-output data set is especially large, the fine-tuning of the membership functions is necessary. Therefore, the learning mechanisms are also employed for the determination of membership functions.

5.2.2. Training of the Fuzzy Inference System

Using (5) and the premise parameters, the overall system output can be expressed as a linear combination of the consequent parameters:

$$\bar{y} = \sum_{i=1}^{9} \bar{\omega}_i (p_i x + q_i y + r_i) = \mathbf{Cz} \tag{6}$$

where **z** is the vector of consequent parameter and C is the matrix of coefficients (Chedid, *et al.*, 2000).

The learning algorithm is composed of a forward pass and a backward pass. In the forward pass, using the available values of S_1, functional signals go forward till layer 4 and the consequent parameter vector **z** are identified by means of the LSE. In the backward pass, the error rates propagate backward and the premise parameters S_1 are updated by the gradient descent (Shing, and Jang, 1993).

5.3 Training data

The training data must cover a wide range of operation and disturbance conditions. Furthermore, they must contain as much information as possible about the examined power systems. Therefore, power system with different fault sequences is simulated to obtain the training data.

The two inputs for the training data: ΔP_{Line} and $\Delta P_e / \Delta t$ can be obtained with the non-linear simulation. The output data for training can be determined by the difference of power angle between the two areas. In this paper, the corresponding center of power angles (COA) (Pavella, *et al.*, 2000) of each area (δ_{Area3}, δ_{Area4}) are employed for the investigation:

$$\delta_{Area3}(t) = \frac{\sum_{k\in Area3} S_k H_k \delta_k(t)}{\sum_{k\in Area3} S_k H_k}$$
$$\delta_{Area4}(t) = \frac{\sum_{k\in Area4} S_k H_k \delta_k(t)}{\sum_{k\in Area4} S_k H_k} \qquad (7)$$

where

H_k — The inertia constant of the kth generator;

S_k — The base power of the kth generator.

The first swing characteristic can be determined using δ_{Area3} and δ_{Area4}.

$$\delta(t) = \delta_{Area3}(t) - \delta_{Area4}(t) \qquad (8)$$

The switching time from transient controller to POD controller is determined at the time for which $d\delta(t)/dt = 0$. This signal is employed only for the training procedure.

The whole training data is composed of four parts: Three-phase short circuits at bus A (Local bus), B (Middle bus) and C (Remote bus) are simulated. Furthermore, the load shedding at bus D (Load) is also simulated to train the fuzzy inference system. Therefore, there are discontinuities between the four parts. The training data is shown in Fig. 9.

Using the training data discussed above, the membership functions and the inference system can be optimized.

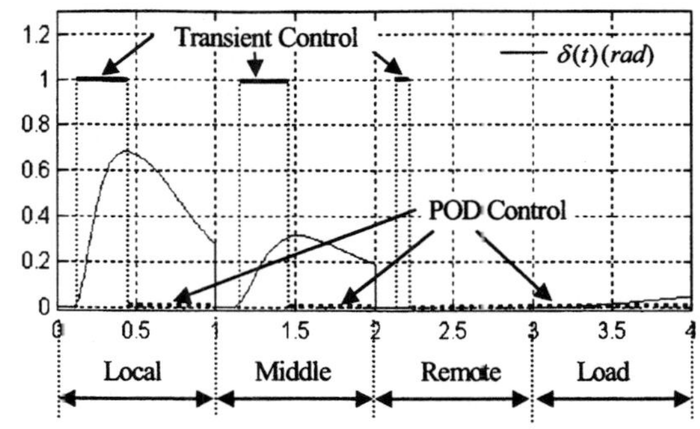

Fig. 9. Training data
—— : Time range for transient controller
······ : Time range for POD controller

5.4 Training results

In this work, ANFIS is trained using 100 epochs and an initial step size of 10^{-4}.

5.4.1. Membership functions

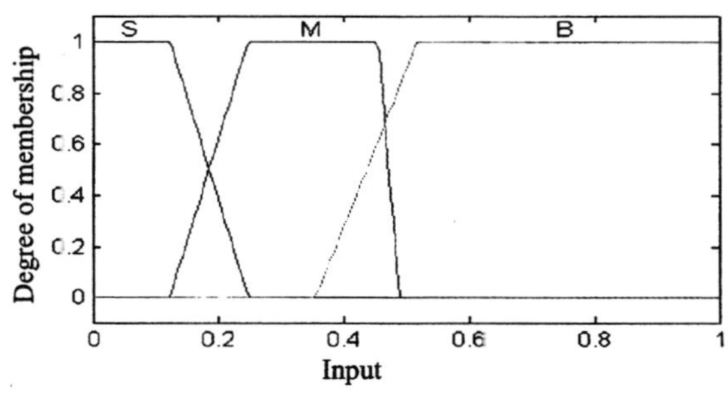

Fig. 10. Membership functions of ΔP_{Line}

The initial membership functions are equally spaced with enough overlap within the input range as shown in Fig. 7. These are typical membership functions to start with for the ANFIS learning.

The final membership functions of ΔP_{Line} are shown in Fig. 10. In order to fit the output of the training data, the initial membership functions are greatly changed. Similarly, the final membership functions of $\Delta Pe / \Delta t$ are also obtained, as shown in Appendix.

5.4.1. Fuzzy inference system

Using the training data, the Sugeno fuzzy inference system is optimized by comparing its output with the objective output data. Fig. 11 shows the most critical training result. A short circuit of 100 ms duration is simulated on the line between bus A0 and a supplementary bus. The near end (supplementary bus) and the remote end (bus A) of the line are cleared at t=0.2 s and t=0.3 s respectively. The difference between the training output and the objective output between t=0.2 s and t=0.3 s is due to the remote end clearance time, where there is an impulse in active power flow through FACTS devices. The other training results (middle bus, remote bus and loss of load) match pretty well the objective output. The detail membership functions and the first-order Sugeno fuzzy model are shown in Appendix.

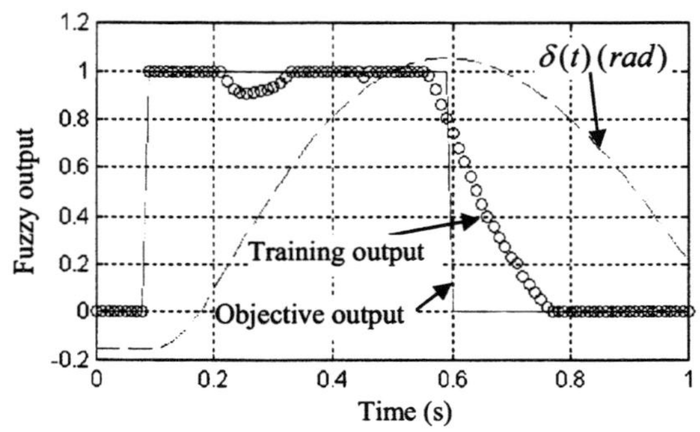

Fig. 11. Training result

6. SIMULATION RESULTS

To verify the performance of the proposed controller, different disturbances are simulated: Three phase short circuits are applied on bus B1 (area 4, middle bus) and bus C1 (area 5, remote bus). The duration of the short circuits are 100 ms and 150 ms respectively. Fig. 12 demonstrates the rotor angle difference between Area 3 and Area 4 in case of using the conventional controller and the fuzzy coordination controller respectively.

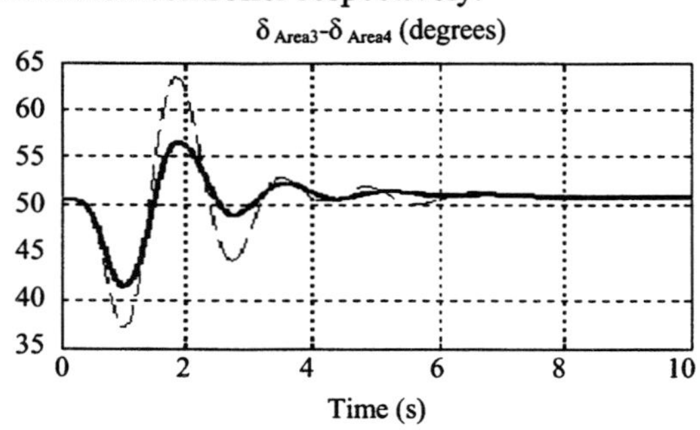

(a) Three phase short circuit on bus B1

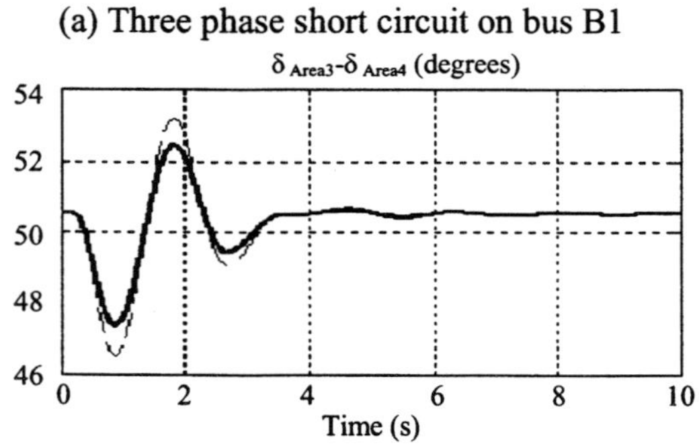

(b) Three phase short circuit on bus C1

Fig. 12. Simulation results
- - Conventional fix-time-switching coordinated control
—— Fuzzy coordinated control

In comparison with the conventional fix-time-switching controller, the examined power system performs better with the proposed fuzzy coordination controller and the dynamic performance is quite improved. Particularly, as shown in Fig. 12 (a), the proposed control scheme leads to significantly better transient behavior under the middle bus disturbances.

7. CONCLUSION

This paper presents a new fuzzy coordination controller for the series FACTS devices to coordinate their transient stability controller and POD controller in multi-machine power systems. Using the training data set, which is obtained by simulating over a wide range of operation and disturbance conditions, the parameters of the proposed controller are optimized using the ANFIS technology. The performance of the proposed controller is verified under different disturbances. Simulation results validate the robustness of the proposed control scheme. Moreover, this approach is also simple and easy to be realized in power systems.

REFERENCES

Cai, L.J. and I. Erlich. (2002). Fuzzy coordination of FACTS controllers for damping power system oscillations. In: *Modern Electric Power Systems Proc. of the International Symposium Wroclaw*. pp. 251-256.

Chedid, R. B., S. H. Karaki and C. El-Chamali. (2000) Adaptive fuzzy control for wind-diesel weak power systems. In *IEEE Trans. on Energy Conversion*, **vol. 15, No. 1**, pp. 71–78.

Lei, X., D. Jiang and D. Retzmann, (2000). Stability inprovement in power systems with non-linear TCSC control strategies. In *ETEP*, **vol. 10, No. 6**, pp. 339-345.

Pavella, M., D. Ernst and D. Ruiz-Vega, (2000). *Transient Stability of Power Systems*, Kluwer Academic Publishers, Boston.

Rogers, G. (1999). *Power System Oscillations*, Kluwer Academic Publishers, Boston.

Shing, J. and R. Jang. (1993). ANFIS: Adaptive-network based fuzzy inference System. In *IEEE Transactions on SMC*, **Vol. 23, No. 3**.

Wang, H.F. (1999). Selection of robust installing locations and feedback signals of FACTS-based stabilizers in multi-machine power systems. In: *IEEE Trans. Power Systems*. **vol. 14**, May 1999, pp. 569-574.

APPENDIX

Table 1 Membership functions

	MFs	a_1	a_2	a_3	a_4
Input1 ΔP_{Line}	S	-0.5548	-0.06164	0.1244	0.2479
	M	0.1226	0.2477	0.4544	0.4875
	B	0.3538	0.5164	1.294	1.788
Input2 $\Delta P_e/\Delta T$	S	-6.75	-0.75	2.00	3.0
	M	2.006	3.00	9.0	10.
	B	8.9997	10	15.75	21.75

Table 2 Linear functions of Sugeno model

i	p_i	q_i	r_i
1	-0.95	0.01	6.716
2	35.211	0.00989	6.010
3	22.1	0.59	0.0558
4	0.367	3.96	2.022
5	24.7	0.0295	3.648
6	-6.815	0.898	0.0815
7	3.12	0.039	6.108
8	0.445	1.97	0.464
9	0.073	0.8322	0.0751

ELSEVIER
IFAC
PUBLICATIONS
www.elsevier.com/locate/ifac

A DYNAMIC SIMULATION ALGORITHM OF UPFC FOR STUDYING POWER SYSTEM OSCILLATIONS

Tae K. Hahn*, **Kwang M. Son****, **Jong-Keun Park***

*School of Elec. Eng., Seoul National Univ.
**Dept. of Elec. Eng. Dongeui Univ.*

Abstract: This paper presents a dynamic simulation algorithm for studying the effect of a Unified Power Flow Controller(UPFC) on the low frequency electromechanical oscillations and transient stabilities. The algorithm is a Newton-type one and gives fast convergence characteristics. The algorithm is applied to an inter-area power oscillation damping regulator design of a sample two-area power system. The simulation results show that the proposed algorithm is efficient and the UPFC is effective for damping inter-area oscillations. *Copyright © 2003 IFAC*

Keywords: FACTS(Flexible AC Transmission Systems), UPFC (Unified Power Flow Controller), Power System Oscillation Damping

1. INTRODUCTION

Recently, Among FACTS devices, the UPFC (Unified Power Flow Controller) is emerging as a promising solution for improving power system characteristics for its high degree of controllability of many power system variables (Gyugyi *et al.*, 1995). Power flow models of UPFC have been developed for the steady-state analysis including power flow control (Fuerte-Esquivel and Acha, 1997; Nabavi-Niaki and Iravani, 1996; Noroozian *et al.*, 1997a)

Current researches on the dynamic modeling and the simulation of UPFC take several approaches. (Nabavi-Niaki and Iravani, 1996) proposed a detailed dynamic modeling of UPFC using switching functions, which is applied to the damping of SSR. A control strategy for SSR mitigation has been proposed, which modulates real power transfer through regulation of the dc link voltage. (Noroozian *et al.*, 1997b) used a power injection model of UPFC, which has been used power flow calculation, to study the effect of UPFC for improving damping of oscillations with energy function based control strategy. (Padiyar and Rao, 1999) developed an iterative scheme that finds UPFC injection currents satisfying network interface conditions. (Arabi and Kundur,1996) proposed a way to consider constraints due to various ratings and operating limits. (Huang *et al.*, 2000) proposed a novel UPFC-network interface technique with dc link capacitor dynamics included. It has been shown that dc capacitor voltage fluctuation during transient is less than 1.5% of the rated voltage for the studied case. (Wang, 2000) studied damping induced by UPFC based on the Heffron-Phillips model of a power system installed with a UPFC. It has also been shown that dc voltage regulator can have a negative damping on the system.

This work was supported in part by the basic research program of the Korea Science & Engineering Foundation under grant No.R01-2000-00269.

This paper proposes a current injection model of UPFC to study the effect of UPFC on the power system low frequency oscillations. The proposed method finds network solutions that satisfy the current balances at the buses including the UPFC bus by forming Newton iteration formula. Since the proposed algorithm is Newton-type, it is conceptually simple and has quadratic convergence characteristics. Since many dynamic simulation programs adopt current balance equation for solving network equations, it can be implemented with a slight modification in the conventional program. The algorithm is applied to the controller design for damping inter-area oscillations in the sample two-area power system model. Case study results show that the proposed simulation algorithm converges rapidly and the designed controller of UPFC is very effective for damping inter-area oscillations.

2. UPFC MODELING

Modeling of the UPFC depends on the range of frequency of concern. The model required for studying low frequency oscillations should faithfully exhibit phenomena of 0.1-2 Hz . The UPFC control systems are developed to satisfy the model proposed in (Gyugyi *et al.*, 1995), which is also shown in Fig.1 (a). Dynamics of the DC link capacitor can be neglected for the above-mentioned frequencies without much error. The series injected voltage can be converted to its equivalent injection current sources as shown in the Fig.1 (b).
The current injection model can be easily adopted for the conventional power system simulation programs, except one difficulty that the injected currents are dependent on the UPFC side voltage, which implies that the current injection of the UPFC should be solved using an iterative technique.

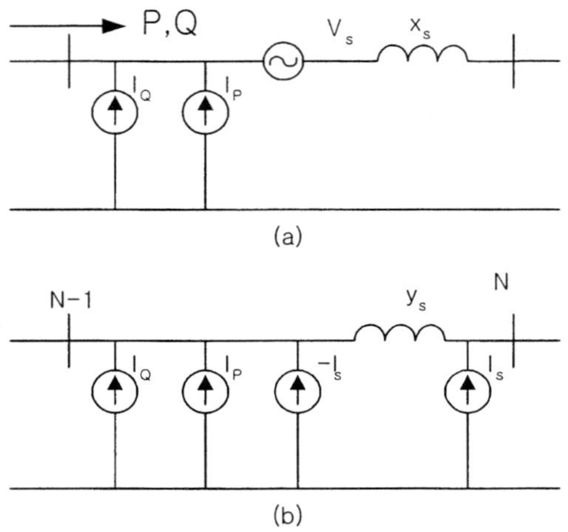

Fig. 1. (a) UPFC model and (b) its equivalent current injection model

3. PROPOSED SIMULATION ALGORITHM

3.1 Dynamic Simulation of Power Systems.

Among various dynamic simulation methods, some dynamic simulation programs adopt the current-balance form for solving network equations of the system with an explicit integration approach. The power system dynamic model can be written as a set of differential equations,(1), and a set of algebraic equations,(2).

$$\dot{\mathbf{x}} = \mathbf{f}(\mathbf{x}, \mathbf{V}_1) \qquad (1)$$

$$\mathbf{I}_1(\mathbf{x}, \mathbf{V}_1) = \mathbf{Y}_1\mathbf{V}_1 \qquad (2)$$

where bold face letters denote the complex quantities, $\mathbf{I}_1$ and $\mathbf{V}_1$ are injection currents and voltage vectors of dimension n ,respectively, and $\mathbf{x}$ is a state variable vector of dimension m. Nonlinear static loads are treated as an injection current dependent on their own bus voltages as shown in (2). So, algebraic equations are, in general, implicit and nonlinear. Equation (2) can be re-written as,

$$\mathbf{F}_1 = \mathbf{I}_1(\mathbf{x}, \mathbf{V}_1) - \mathbf{Y}_1\mathbf{V}_1 = \mathbf{0}\,. \qquad (3)$$

Dividing (3) into real and imaginary parts and then calculating $2n$-by-$2n$ Jacobian gives a Newton iteration formula.

3.2 Proposed Approach

In case a UPFC is installed, the number of nodes becomes N=n+2 and the conventional algebraic equations become

$$\mathbf{I} = \begin{bmatrix} \mathbf{I}_1(\mathbf{V}) \\ \mathbf{I}_U(\mathbf{V}_U) \end{bmatrix} = \begin{bmatrix} \mathbf{Y}_1 & \mathbf{Y}_2 \\ \mathbf{Y}_3 & \mathbf{Y}_4 \end{bmatrix} \begin{bmatrix} \mathbf{V}_1 \\ \mathbf{V}_U \end{bmatrix} = \mathbf{YV} \qquad (4)$$

where $\mathbf{I}_1$ and $\mathbf{V}_1$ are conventional N-2 dimensional complex injection current and voltage vectors in (2), respectively, and $\mathbf{I}_U$ and $\mathbf{V}_U$ are 2-dimensional UPFC injection current and voltage vectors, respectively. The sub-matrix $\mathbf{Y}_1$ is an n-by-n conventional bus admittance matrix including network and generator stators as in (2), and $\mathbf{Y}_2$ and $\mathbf{Y}_3$ are mutual admittance matrices between the UPFC node and the network. $\mathbf{Y}_4$ relates the UPFC inverter side node and the line side node, which are denoted in Fig1. as node N-1 and N, respectively. In order to formulate the Newton algorithm of (4), it can be rewritten as two vector functions in the following form:

$$\mathbf{F}_1 = \mathbf{I}_1(\mathbf{V}_1) - \mathbf{Y}_1\mathbf{V}_1 - \mathbf{Y}_2\mathbf{V}_U = \mathbf{0} \qquad (5)$$

$$\mathbf{F}_U = \mathbf{I}_U(\mathbf{V}_U) - \mathbf{Y}_3\mathbf{V}_1 - \mathbf{Y}_4\mathbf{V}_U = \mathbf{0} \qquad (6)$$

Equation (5) and (6) are complex valued functions with dimensions of N-2 and 2, respectively. In (6), the term $\mathbf{I}_U(\mathbf{V}_U)$ accounts for the dependence of the UPFC injection current on its own voltages, which makes it difficult to incorporate UPFC directly in the conventional dynamic simulation algorithm.

Augmented Jacobian : If we decompose the equation into real and imaginary parts, we can obtain $2N$ equations with $2N$ unknown variables. The dimension of the corresponding Jacobian is $2N$-by-$2N$, having blocks:

$$\mathbf{J} = \begin{bmatrix} J_1 & J_2 \\ J_3 & J_4 \end{bmatrix} \qquad (7)$$

where J_1 is the conventional Jacobian corresponding to (3) and J_2 ,J_3 and J_4 are augmented parts due to UPFC.

The elements of the J_2 and J_3 : The elements of J_2 and J_3 can be obtained in the same way the elements of J_1 are derived. The way to derive J_2, for example, is as follows. Let $\mathbf{Y}_{2,ik} = Y_{2,ik}\exp(j\gamma_{2,ik})$, $\mathbf{V}_{Uk} = V_{Uk}\exp(j\theta_{Uk})$. Then for i=1,2,...,N-2 and k=1,2, the elements of J_2 can be obtained from the following relations. The elements of J_3 can be derived in the same way.

$$\frac{\partial \mathbf{F}_{1,i}}{\partial V_{Uk}} = Y_{2,ik}\exp[j(\gamma_{ik} + \theta_{Uk})] \qquad (8)$$

$$\frac{\partial \mathbf{F}_{1,i}}{\partial \theta_{Uk}} = jY_{2,ik}V_{Uk}\exp[j(\gamma_{ik} + \theta_{Uk})] \qquad (9)$$

The elements of J_4 : In order to obtain the elements of J_4, we need to investigate a more detailed expression of UPFC injection current. From the equivalent current injection model, as in Fig.1, the injection current at the UPFC node can be expressed as follows.

$$\mathbf{I}_U(\mathbf{V}_U) = \begin{bmatrix} \mathbf{I}_P(\mathbf{V}_U) + \mathbf{I}_Q - \mathbf{I}_S(\mathbf{V}_U) \\ \mathbf{I}_S(\mathbf{V}_U) \end{bmatrix} \qquad (10)$$

The shunt inverter injection currents decomposed into two parts. $\mathbf{I}_P$ is in phase with the node voltage so to supply the real power that the series inverter needs. $\mathbf{I}_Q$ is the reactive current component independently controlled by the shunt converter. These two current components should not exceed the limit of the shunt inverter's capability. The series injected voltage is converted to its equivalent current source $\mathbf{I}_S$ in (10). The magnitude and phase of the series injected voltage, $\mathbf{V}_S$, can be independently controlled by the control system. However, since the series injected voltage is superimposed on the shunt inverter side voltage, the phase of the injection current should depend on the phase of the shunt inverter side voltage as shown in (11).

$$\mathbf{I}_S = \frac{\mathbf{V}_S \exp j\theta_{N-1}}{jx_S}, \mathbf{V}_S = V_S \exp j\theta_S \qquad (11)$$

where $\mathbf{V}_{N-1}$ denotes the inverter side voltage, that is, $\mathbf{V}_{U1}=\mathbf{V}_{N-1}$.
In the shunt inverter $\mathbf{I}_P$ is determined from the real power that is injected by the series inverter. From the real power conservation, neglecting losses of the inverter, we can get

$$\mathbf{I}_P = -\frac{\mathrm{Re}\left[\mathbf{V}_S \mathbf{I}^*_{N-1,N}\right]}{\mathbf{V}^*_{N-1}} \qquad (12)$$

The reactive injection current by the shunt inverter is in quadrature-phase with the shunt inverter side voltage in order to inject the reactive power into the power system.

$$\mathbf{I}_Q = I_Q \exp j(\theta_{N-1} - \pi/2) \qquad (13)$$

Using the relations given by (11)-(13) in (10), we can get the elements of 4-by-4 real matrix J_4.
Some comments follow.
1. Although the proposed approach assumes the explicit integration method as a simulation method, it can be easily extended to the case of an implicit integration method.
2. If all loads are constant impedance type, the Jacobian is greatly simplified. So only the 4-by-4 matrix, J_4, need to be calculated for each iteration step as is explained later. The dimension of the Jacobian evaluated at every iteration step depends on the number of nonlinear loads and the number of UPFCs installed.

3.2 A Special Case of Constant Impedance Loads

In case all loads are constant impedance type, then $\mathbf{I}_1$ in (4) is determined from the state variables of generators and excitation systems at every iteration step. So when we solve the algebraic equations at every step, $\mathbf{I}_1$ is given as a pre-determined constant. So equation (4) becomes

$$\begin{bmatrix} \mathbf{I}_1 \\ \mathbf{I}_U(\mathbf{V}_U) \end{bmatrix} = \begin{bmatrix} \mathbf{Y}_1 & \mathbf{Y}_2 \\ \mathbf{Y}_3 & \mathbf{Y}_4 \end{bmatrix} \begin{bmatrix} \mathbf{V}_1 \\ \mathbf{V}_U \end{bmatrix} \qquad (14)$$

Noting that only $\mathbf{I}_U$ is an implicit function of $\mathbf{V}_U$, $\mathbf{V}_1$ can be expressed by $\mathbf{V}_U$ as follows.

$$\mathbf{V}_1 = \mathbf{Y}_1^{-1}(\mathbf{I}_1 - \mathbf{Y}_2\mathbf{V}_U) \qquad (15)$$

Using (15) in (14), $\mathbf{I}_U$ can be expressed as

$$\mathbf{I}_U(\mathbf{V}_U) = \mathbf{Y}_3\mathbf{Y}_1^{-1}\mathbf{I}_1 + (\mathbf{Y}_4 - \mathbf{Y}_3\mathbf{Y}_1^{-1}\mathbf{Y}_2)\mathbf{V}_U \qquad (16)$$

where $\mathbf{I}_1$ is a constant matrix and $\mathbf{V}_U$ can be solved by a Newton-type method if we set

$$\mathbf{Y}_0 = \mathbf{Y}_3\mathbf{Y}_1^{-1}\mathbf{I}_1 \quad , \quad \mathbf{Y}_R = \mathbf{Y}_4 - \mathbf{Y}_3\mathbf{Y}_1^{-1}\mathbf{Y}_2 \qquad (17)$$

Then, (15) becomes the following 2-dimensional complex valued functions. The corresponding Jacobian has the same structure as J_4 except that $\mathbf{Y}_4$ is replaced by $\mathbf{Y}_0$.

$$\mathbf{I}_U(\mathbf{V}_U) - \mathbf{Y}_R\mathbf{V}_U - \mathbf{Y}_0 = \mathbf{0} \qquad (18)$$

So in case all loads are constant impedance type, the algorithm has only to find out the solutions that satisfy the interface conditions at the UPFC node. Once the voltages at the UPFC are found, the voltages at the other nodes are found by (15).

4. CASE STUDY RESULTS

4.1. A Sample Power System Model

A sample system used in the case study is a simple two-area four-machine power system model in (Klein *et al.*, 1991). All generators are 4th order two-axis models(Son and Park, 2000) equipped with 1st order fast exciters. All the exciter parameters are the same as K_A=200(pu), T_A=0.02(sec). The machine parameters are listed in the appendix. All loads are assumed to be constant impedance type.

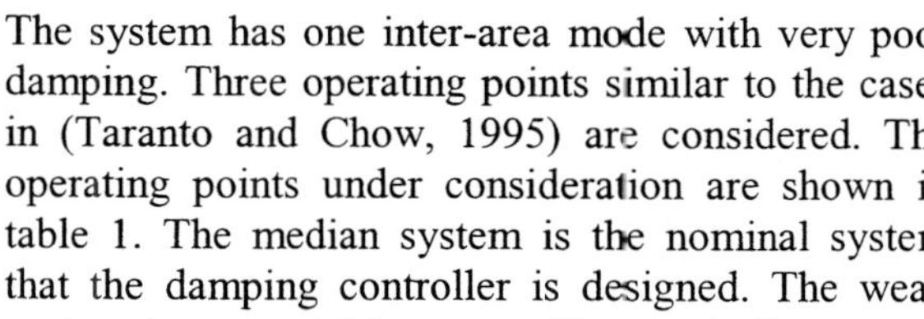

Fig. 2. A Sample Two-Area Power System Model

The system has one inter-area mode with very poor damping. Three operating points similar to the cases in (Taranto and Chow, 1995) are considered. The operating points under consideration are shown in table 1. The median system is the nominal system that the damping controller is designed. The weak system is an unstable case with one tie-line out of service.

Table 1. Operating points under consideration

System	# of ties	Gen/Load (MW)		Damping Ratio(%)
		Area 1	Area 2	
Stiff	3	1120/970	1400/1390	2.82
Median	3	1400/970	1400/1740	0.156
Weak	2	1400/970	1400/1740	-0.175

4.2. UPFC Control System

Fig. 3 shows the structure of the control system of the series part of the UPFC as described in (Rahman *et al.*, 1997). PI-type regulators control the series injected voltage. In order to focus on the series part for its effectiveness for damping, we consider only controlling series injected voltage for damping control.

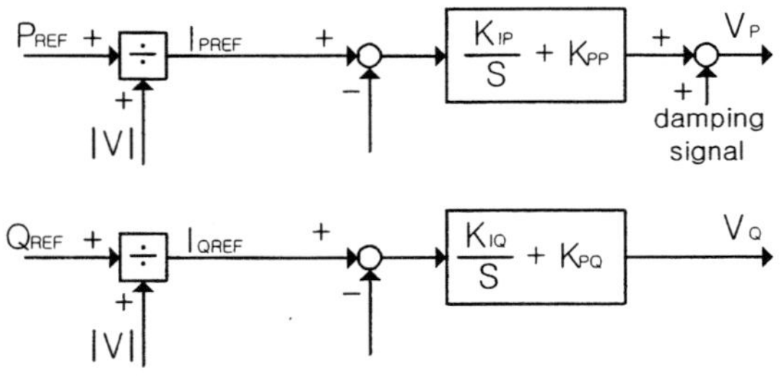

Fig. 3. Control system of UPFC

4.3. Damping Controller

A lead-lag type regulator is considered as a damping controller. As a supplementary input signal, the difference between the real power flow of the line through UPFC and the P_{REF} (reference input of UPFC in Fig.3) is considered.

$$y = K \frac{sT_W}{1+sT_W} \frac{1+sT_1}{1+sT_2} \Delta P_{\text{line}} \tag{19}$$

Fig. 4 shows the loci of inter-area mode with increasing gain, *K*, for various phase lead angles of the damping controller. It can be concluded from the figure that K should be negative or an additional 180° phase lead is needed with a positive *K*. The residue analysis shows that the inter-area mode is nearly in phase with the supplementary input, that is, $\mathbf{V}_P$. So in order to increase damping of the inter-area mode, the negative feedback or phase lead about 180o would be desirable. With a phase lead angle of 15°, *K* is set to be -0.09, which gives damping of the inter-area mode approximately 20%. With the damping control, two local modes of each area are nearly unaffected.

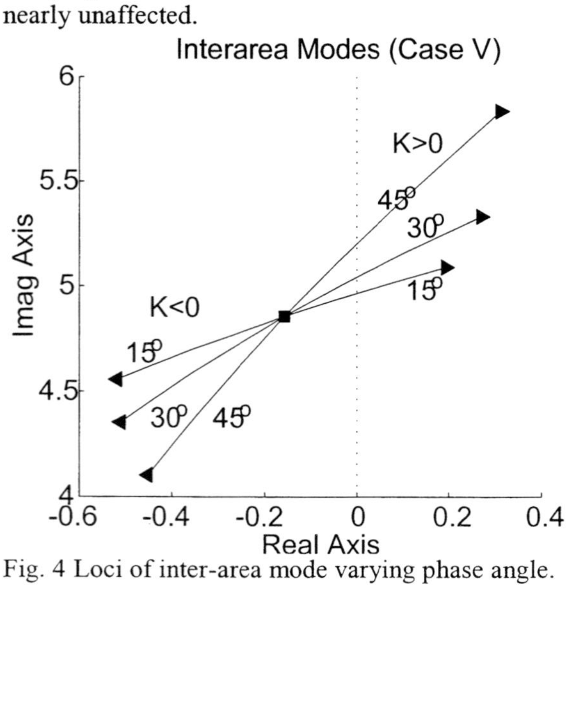

Fig. 4 Loci of inter-area mode varying phase angle.

4.4. Time domain simulations

Time domain simulations are performed for each system with and without damping control for a three-phase six-cycle fault applied to the load bus of area 1. Fig. 5-7 show the rotor angle oscillations between areas, which show that the oscillations are settled down after about 2-4 seconds for all cases. Removing the non-windup limiter decreases the effect of the damping control.

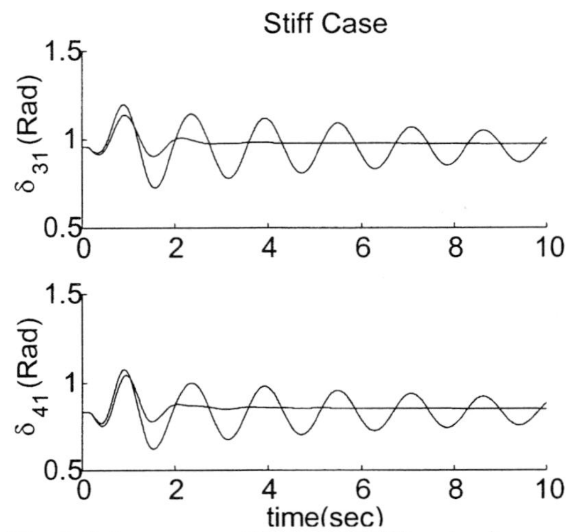

Fig. 5. Inter area oscillations (Stiff system)

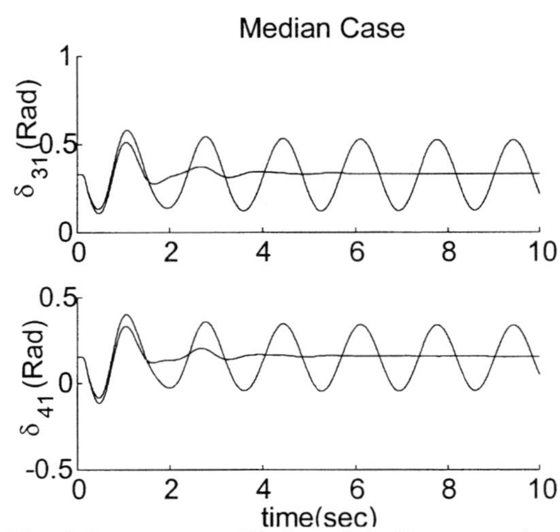

Fig. 6. Inter area oscillations (Median system)

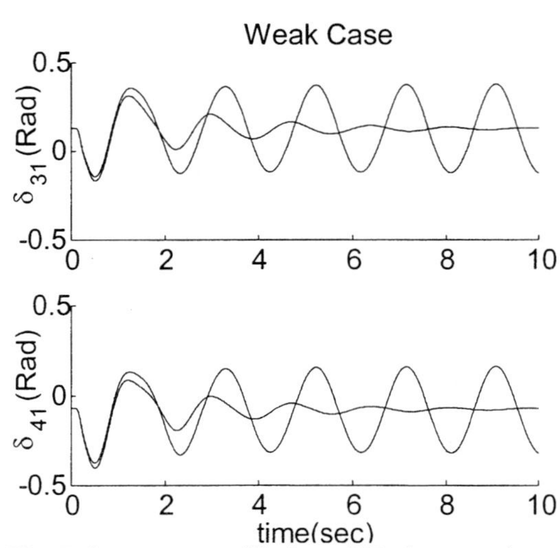

Fig. 7. Inter area oscillations (Weak system)

The limit of series injected voltage is assumed to be 0.05 pu for both $\mathbf{V}_P$ and $\mathbf{V}_Q$. For all cases, K_{IP} and K_{IQ} of the UPFC control system are set to be 0.5. A non-windup limit of ±0.1 is imposed on each integrator.

5. CONCLUSIONS

This paper presents a dynamic simulation algorithm of the UPFC for studying electromechanical low frequency oscillations of a power system installed with a UPFC. Using the proposed algorithm, the characteristics of a damping controller for inter-area oscillations of a sample two-area power system are investigated. The results show that the algorithm is efficient and a UPFC with damping controller is effective for damping inter-area oscillations.

APPENDIX

Table A1 Machine parameters

Generator	1	2	3	4
H	58.5	58.5	55.58	55.58
D	0	0	0	0
T_{do}'	8.0	8.0	8.0	8.0
T_{qo}'	0.4	0.4	0.4	0.4
X_d	0.2	0.2	0.2	0.2
x_d'	0.033	0.033	0.033	0.033
X_q	0.18	0.18	0.18	0.18
x_q'	0.033	0.033	0.033	0.033

REFERENCES

Arabi, S., Kundur, P. and Adapa, R., (2000) "Innovative Techniques in Modeling UPFC for Power System Analysis", *IEEE Trans. On Power Systems*, Vol.15, No.1, pp.336-341.

Fuerte-Esquivel, C.R. and Acha, E., (1997) "A Newton-Type Algorithm for the Control of Power Flow in Electrical Power Networks", *IEEE Trans. On Power Systems*, Vol.12, No.4, pp.1474-1480.

Gyugyi, L., Schauder, G.D., Williams, S.L., Rietman, T.R., Torgerson, D.R., Edris, A., (1995) "The Unified Power Flow Controller: A New Approach to Power Transmission Control", *IEEE Trans. On Power Delivery*, Vol.10, No.2, pp.1085-1097.

Huang, Z., Ni, Y., Shen, C.M., Wu, F.F., Chen, S. and Zhang, B., (2000) "Application of Unified Power Flow Controller in Interconnected Power Systems – Modeling, Interface, Control Strategy and Case Study", *IEEE Trans. On Power Systems*, Vol.15, No.2, pp.817-824.

Klein, M., Rogers, G.J., Kundur, P., (1991) "A Fundamental Study of Inter-area Oscillations in Power Systems", *IEEE Trans. On Power Systems*, Vol.6, No.1, pp.914-920.

Larsen, E.V., Sanchez-Gasca, J.J., Chow, J.H., (1995) "Concepts for Design of FACTS Controllers to Damp Power Swings", *IEEE Trans. On Power Systems*, Vol.10, No.2, pp.948-956.

Nabavi-Niaki, A. and Iravani, M.R., (1996) "Steady-State and Dynamic Models of Unified Power Flow Controller (UPFC) for Power System Studies", *IEEE Trans. On Power Systems*, Vol.11, No.4, pp.1937-1943.

Noroozian, M., Angquist, L., Ghandhari, M. and Andersson, G.(1997a), "Use of UPFC for Optimal Power Flow Control", *IEEE Trans. On Power Delivery*, Vol.12, No.4, Oct. pp.1629-1634.

Noroozian, M., Angquist, L., Ghandhari, M. and Andersson, G., (1997b) "Improving Power System Dynamics by Series-Connected FACTS Devices", *IEEE Trans. On Power Delivery*, Vol.12, No.4, pp.1635-1641.

Padiyar, K.R. and Rao, K.U., (1999) "Modeling and Control of Unified Power Flow Controller for Transient Stability", *International Journal of Electrical Power & Energy Systems*, Vol.21, No.1, pp.1-11.

Rahman, M., Ahmed, R. Gutman, R.J.O'Keefe, R.J. Nelson, J. Bian, (1997) "UPFC Application on the AEP System: Planning Considerations", *IEEE Trans. On Power Systems*, Vol.12, No.4, pp.1695-1701.

Son, K.M. and Park, J.K., (2000) "On the Robust LQG Control of TCSC for Damping Power System Oscillations", *IEEE Trans. On Power Systems*, Vol.15, No.4, pp.1306-1312.

Taranto, G.N. and Chow, J.H., (1995) "A Robust Frequency Domain Optimization Technique for Tuning Series Compensation Damping Controllers" *IEEE Trasn. On Power Systems*, Vol.10, No.3, Pp.1219-1225.

Wang, H. F., (2000) "A Unified Model for the Analysis of FACTS Devices in Damping Power System Oscillations – Part III: Unified Power Flow Controller", *IEEE Trans. On Power Delivery*, Vol.15, No.3, pp.978-983

ELSEVIER
IFAC
PUBLICATIONS
www.elsevier.com/locate/ifac

INTERFACING AN EXISTING ENERGY MANAGEMENT SYSTEM TO A NEWLY ESTABLISHED MARKET OPERATIONS SYSTEM

Hyo-Sang Lee, David L. Rosa, Pradip Kambale, Rick Schaal

Korea Power Exchange Kangnam-Ku, Seoul135-791, Korea
KEMA Consulting, 4377 County Line Road, Chalfont, PA 18914
KEMA Consulting, 7575 Golden Valley Road, Suite 300, Golden Valley, MN 55427
Alstom-ESCA, 11120 NE 33rd Place Bellevue, Washington 98004

Abstract: To support the deregulated energy markets, electric utilities and marketing entities are implementing new market operation systems to provide the necessary bid/offering applications, scheduling algorithms, and after-the-fact settlement tools. Utilities are integrating existing Energy Management Systems (EMS) that already provide some of the required tools and data with these new marketing operation systems. This integration work requires careful coordination of the interfaces between the Market Operating System and EMS. This paper identifies issues associated with these interfaces and presents guidelines leading to successful implementation. *Copyright © 2003 IFAC*

Keywords: Energy Management Systems, Computer Interfaces, Databases, Implementation, and Software Tools.

1. INTRODUCTION

The deregulation of the electricity industry is occurring all over the world, forcing previously vertically integrated electric utilities to split their organizations into several business units, as required by government initiated electricity market laws. Deregulation now requires management, employees and consumers to view the electric power generation and supply business from different perspectives. With all of the dramatic changes taking place in the electric industry, the task of managing the electric power system has universally become much more complex. Electric utilities throughout the world today have System Operators managing their power system using an Energy Management System (EMS) for the optimum control of both the generation capacity and the transmission network. The creation of an electricity market environment now requires System Operators to incorporate newly established market rules and procedures into their normal and emergency operating procedures. In addition, the new electricity market dictates the need for new data and tools to support the electricity market being managed by a newly formed group called Market Operators.

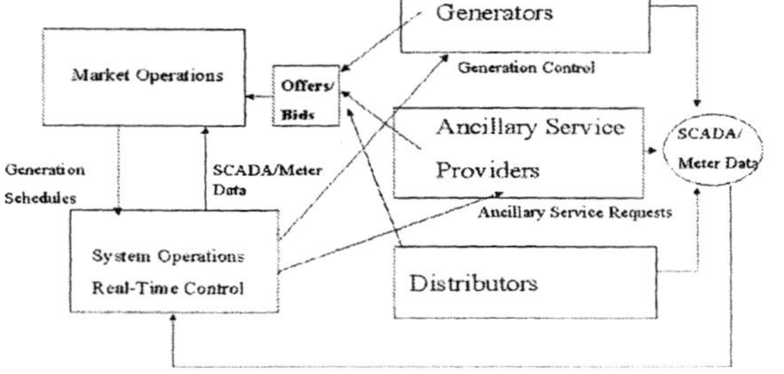

Figure 1: Overview of Market Operation and System Operation

The goal of the Market Operator is to ensure that consumers obtain the best price of electricity while the goal of System Operator is to deliver electricity reliably and securely to the same consumers. Figure 1 shows a high-level overview of the market operation and system operation.

The Market Operator, similar to the System Operator, needs a complete set of computerized tools to evaluate, schedule and operate the electricity market. These tools are typically provided by the Market Operation System (MOS) which is comparable to the EMS for System Operators. Some tools and data needed for market operation (for example, system load forecasts, SCADA data, State Estimator, etc.) are already present in the EMS. Other tools (such as bid/offer processing, bid price generation scheduling, after-the-fact settlement, etc) are unique to the MOS. The overlap of common functions and data exchange requirements between MOS and EMS, coupled with the unique application sets required in each system creates various issues that range from operational planning to the implementation of the MOS in relationship to the Energy Management System.

This paper identifies issues involved in interfacing the MOS to the EMS and presents guidelines leading to a successful implementation.

2. IMPLEMENTATION ISSUES AND RECOMMENDATIONS

Energy Management Systems, now in use for decades at most electric utilities, have become a critically important tool for the System Operator. Although core applications of an EMS include Supervisory Control and Data Acquisition (SCADA), and AGC, other advanced applications and functions including real-time and study Network Analysis and optimization applications, are fairly common. Yet EMS systems do vary in terms of functionality that is required to meet the unique and specific needs of the individual utility company. The EMS traditionally supported both the economic scheduling of generation and the operation of the transmission network. This approach worked well for vertically integrated utilities and provided an integrated set of applications to the System Operator in support of the mission to maintain secure and reliable operation while emphasizing the maximization the economic performance of the collective set of generation. As electricity markets are implemented, the responsibility to provide the secure and reliable operation of the transmission network and energy supply is typically retained by system operations staff while the competitive price scheduling of generation becomes the responsibility of market operations staff. The Market Operator then is tasked with the scheduling and rescheduling of generation to ensure that the constraints of the electrical network are not violated and its system network security is maintained. An after-the-fact settlement application in the MOS then reconciles actual generation output for final monetary disbursements.

It is well known that the basic core functions typically implemented in the EMS do not provide adequate tools for operations staff in the market system environment. Utilities migrating to a new deregulated electricity market are faced with a series of alternatives for providing the computerized systems needed for both the market operation and the transmission network operation:

- Purchase a new system capable of supporting both the market and power systems operations.
- Separately purchase a new market system and a new EMS that will operate in an integrated fashion.
- Purchase a new market system and integrate it with the existing EMS.

Few utilities have been able to use the first alternative, as one of the byproducts of deregulation typically results in a separation of the generation scheduling activity and the transmission network operation. As it relates to the second alternative, most utilities are unable to purchase two new systems because recent and/or historical expenditures do not economically justify replacement of the existing EMS. Accordingly, most utilities are faced with the third alternative of integrating a newly procured market operation system with an existing energy management system.

The process of specifying and implementing a MOS requires the challenging preparation of defined market rules and procedures for all market participants to follow. These market rules and procedures dictate all aspects of the market: from qualifying market participants and establishing a bidding process, detailing the method in which the power system must operate during various normal and emergency operating conditions, to the process in which the after-the-fact settlement takes place. The task of market and system administration then becomes to identify the tools, namely hardware and software that will meet this challenge.

One of the primary tasks in this process is to analyze the case where one system is going to meet all requirements for both market and network operations. In this case an interface application to outside market participants is still needed, which presumably incorporates the use of a separate system. This approach requires the EMS to deviate from its traditional role. Therefore, it is critically important to precisely identify what functionality is required by the EMS and what functionality is required by the MOS, in order to fully support the electricity market. Additionally, it must be determined as to what functionality and data sets are common between the MOS and the EMS systems so overlap issues can be avoided or exploited. The question then becomes: can the requirement be met by duplicating common functions and data sets within two separate systems or by specifying that it be performed in the system where it is most appropriate and then exchanging the needed data between the MOS and EMS?

From a systems perspective, the obvious choice is to support functions only once and to do so within the most appropriate system. From a support aspect, the educational background and skills of the staff required to support a MOS and an EMS are uniquely different for each system, so it makes more sense to place functionality in the system where experts are most available. This same concept should follow with an evaluation of the system maintenance effort, which will be required when deciding on the overall implementation approach of the MOS and its interface with an existing EMS.

Considering the effort and expertise that is required to operate a MOS and an EMS within a market environment, the reasonable approach is having two independent systems: one for system operations and one for market operations. The common data shared between the two systems is then exchanged as dictated by market rules and procedures according to industry standards. As a point of emphasis, data exchanged between the systems must utilize common industry-standard protocols and data formats. Today's implementations strive to minimize or even eliminate data formatting issues by adapting standard data formats such as the Electric Power Research Institute's (EPRI) Common Information Model (CIM). Requiring EMS and MOS vendors to implement data exchange in an industry standard format not only benefits the market system but also aids the utility in exchanging data with any other system or applications in the future. This new requirement, however, will dictate software modifications or upgrades be performed if the EMS system is already in place and does not support standards such as the CIM. The focus here is to explore implementation of data protocol standardization as much as possible to minimize interfacing issues for the current implementation and as well as future system changes or upgrades. This is the objective of an open systems design as it enhances the ability for hardware and software functions to be added to existing platforms without restriction or expensive customizations.

3. BASIC EMS-MOS DATA EXCHANGE

The data required in the MOS from EMS consists of SCADA data, generation dispatch data and solution results of Network Applications. In addition, compliance monitoring data is sent to EMS if compliance monitoring is performed in the EMS.

In the MOS, the real-time SCADA data is used to determine network topology, normal/abnormal conditions, equipment outages. The real-time information such as branch flows, bus voltages is used to determine congestion and abnormal conditions in the power system network. The settlement processes used generation dispatch signals to calculate payment to generators.

EMS needs the following data from MOS: generator and transmission facility ratings, outages, maintenance and planning schedules, ancillary service information, and generation schedules.

Where file format requirements are necessary in the data exchange of the above items, the format specification should be very clear and consistently and globally applied to all interfaces involved. For example, if an interface requires an FTP type export/import file format specifying a unique id for each record, such as an outage list exported from the EMS, make sure the unique id structure for other similar FTP data exchanges are non-conflicting and specified in the same way.

4. APPLICATION OVERLAP BETWEEN MOS AND EMS

The EMS system typically has functions, which support real-time operations, while the MOS system requires functions, which support scheduling functions. In a regulated environment the objectives of both functions are done and supported by the EMS.

To minimize the cost of the MOS, existing functions within EMS should be used whenever possible and feasible. The trade-off is always between the cost of purchasing and integrating new functions within the MOS versus the cost of developing and interface and/or modifying an existing EMS. Also one needs to evaluate expertise required on the MOS side to maintain and execute these functions. If the expertise already resides with the System Operations staff, certain cost advantages might be present.

The EMS functions that are required in the MOS are Load Forecasting, Generation Scheduling, and transmission security.

Load forecast is needed in the MOS to ensure that the bids and offers are consistent with predicated load. What may be different in the MOS environment from the EMS environment is a requirement in the forecasting periodicity and time intervals. Since the load forecast application already functions with an established database in the EMS, it can appropriately be maintained in the EMS. However, it will need some level of modification in order to meet market requirements involving periodicity and/or time intervals.

The generation schedules, calculated by Generation scheduling software, have to satisfy the market rules. In this case, the effort required to modify the EMS scheduling function may be significant. To comply with market rules and to keep the market operation decisions independent from real-time system operations, it is advisable to implement a separate Generation Scheduling function within the MOS system.

The network application analysis functions assist the System Operator in evaluating transmission security of the power system network. On the other hand the study version of these network applications are required to evaluate contingencies for new generation schedules and determine possible remedial solutions. The network applications are required both in EMS and MOS but different modes, namely, real-time and study mode. The MOS function can use the results of State Estimator (SE) provided by the EMS instead of maintaining a separate SE function. Making EMS network analysis functions, such as Contingency Analysis and Optimal Power Flow, available to the Market Operator will depend on the Market Operator's capability to move necessary data into the EMS environment and then modify and run studies in a study mode without affecting the real-time system operations. Again the use of these EMS functions for market operation use should be evaluated for each unique market environment.

5. EMS SOFTWARE MODIFICATION TO SUPPORT MARKET OPERATION

System and market operations are dependent on each other for data, application results and operator actions. The Market Operator after processing bids from market participants provides optimum generation schedules, which should be followed. These generation schedules are typically for different time horizons and are updated frequently from day ahead to 5 minutes intervals depending on market rules. These schedules, however, need to be interfaced with the EMS AGC application so that AGC can use these schedules to calculate generation signals that need to be sent to power plants. Likewise, MOS function should be able to read and use EMS data. This dependency requires modifications to existing EMS software as well as the MOS software being implemented. The successful integration and operation of the EMS and the MOS within the deregulated market environment is not risk free. Risks associated with this integration are discussed in the next section.

6. IMPLEMENTATION RISKS

In the deregulation environment, organizations that are purchasing market systems expect to acquire the latest technology, which includes the most recently developed hardware platforms and software application functions. The vendor(s) selected to implement the market system may or may not be the same vendor(s) that implemented the EMS system. If the two systems were purchased from the same vendor one can expect minimum interface issues involved during implementation, assuming that the existing system supports the present industry interface standards. In this situation the same vendor for both systems has the advantage of knowing the software, data structures, hardware linkages with the availability of in-house experts to resolve problems which can substantially reduce the implementation risks and costs.

When the MOS and EMS are acquired from different vendors, resolving software, data and hardware issues should be given high priority since significant challenges are certain in the implementation of the market system. To minimize the potential risk, interface requirements must be detailed as much as possible and specified as a critical item within the specification. This requirement will give a clear idea to the selected vendor regarding the scope of the work. The vendor should be required to implement an approach to manage and develop interface software using non-proprietary software to minimize potential conflicts and delays in the project.

For the utility, if the EMS system is managed by a different department than the department responsible for the delivery of the MOS, it is equally important that both departments coordinate and work cooperatively together in understanding the capabilities, requirements and the needed interface between both systems. This situation should be amplified if a different vendor is providing the MOS than the vendor who provided the EMS. Working in this fashion will minimize potential conflicts and project risks. It will also help to alleviate concerns with respect to the different operating philosophy and constraints of the two departments.

In spite of a utility's best attempts, the specification may not be perfect and will require some level of change, correction or clarification. Because of this, there will be occasions where direct dialog between the two vendors will be mandatory. Accordingly, the competitive nature of each vendor must be understood so the utility can make suitable provisions to accommodate the vendors' needs and manage the potential project risks. Proprietary and intellectual property concerns will exist among the two vendors so special emphasis in safeguarding their concerns will be an important consideration. One way to mitigate issues is to provide suitable working and testing environments created in a manner that will allow a vendor to do their work without impacting the work or schedule of the other vendor.

For the testing phase, it is also a good idea to define a DMZ (demarcation zone) between the two vendors and the systems they have provided and to clearly establish their roles and technical responsibilities on their perspective side of the DMZ. Defining a clear line of responsibility at the system level will yield an added benefit of being able to isolate and more quickly identify system problems should they occur both before, during, and after the implementation system is completed.

Finally, it is critically important that the implementation phase include a transition plan that allows the utility and the vendors to move from the Site Acceptance Testing and Market Trial testing phases into final production as seamlessly as possible. This is especially important when you consider that the EMS may need to operate for more than a year before the Market system and its interface to the EMS is brought on-line in a production mode.

7. SYSTEM MAINTENANCE

Maintenance tasks for the MOS and EMS systems alike include the continuous updating of the power system model and its related displays and databases. This is driven by changes that normally occur within the life cycle of the electric power system along with typical hardware and software maintenance, and software upgrades, as new releases become available. The power system model used within the MOS and EMS alike should be identical or at least, generated from the same source database. It follows that since the two systems must represent the same power system, they must be able to exchange data based on an identical power system model that electrically matches at every interfacing point. As the power system changes through the addition of new equipment, lines, substations, etc., system support personnel update the power system model to provide the latest system configuration to System Operator. The MOS also needs to use the latest power system model to reflect recent changes in the actual power system for an accurate presentation of information to the Market Operator and market participants. The question then is how can the power system models within two separate systems be efficiently updated to guarantee that System and Market Operators are doing their work based on the same, identical model? Should this effort be duplicated? The same question is applicable to software updates and system upgrades. Furthermore, how will the power system data be translated in order to serve as an input to EMS and MOS systems, which have been provided by different vendors? If two different power system data sources are maintained, how will data be synchronized in order to support the concept of a single power system model? These questions and others should be asked and scrutinized during specification stage so that the selected vendor is fully aware of the scope and can propose software, which will satisfy the requirements. In addition, these same questions should be revisited during the implementation phase to insure that the issues, their associated risks, and potential alternatives are well understood

From the utility's perspective, it is important to have a single source database for the power system model. Additional work, of course, will be required for the vendor(s) or system support staff to translate the data from that single source database appropriately so it can be imported/exported and utilized properly within the MOS and EMS systems. The same principle would apply to any software changes and system upgrades required during the systems' natural life cycle and/or in the modification or addition of market rules.

A set of detailed procedures need to be captured in the form of a change management process to keep both systems continuously updated and synchronized where and when appropriate. The coordination of these changes will require special effort, which can be done by using properly coordinated system administration tools and methods.

8. Common Information Model

Many major difficulties in establishing a data exchange between an EMS and other systems can be overcome by implementing the industry standard for power system models and databases. EPRI has proposed the Common Information Model (CIM) to define power system model data that is used to facilitate data exchanges. The international standards organization, IEC, is also working on developing standards based on the CIM.

Deregulation is quickly expanding world wide and since EMS and MOS systems will most likely be from different vendors, it is becoming more important than ever before to standardize the exchange of real-time data and market data.

By implementation of industry standards for data modeling and data exchange, the significant effort and cost required for data maintenance and integration of applications from various vendors can be minimized.

9. CONCLUSION

In the Market Operating System design and implementation process, the task of interfacing a MOS to an existing EMS system is difficult and challenging. This task can be considered low risk given the use of existing standards and experience in interfacing such systems. However, the actual experience within the electric utility industry proves that data exchange and interface issues can pose a significant risk to the successful implementation of the MOS if they are downplayed or managed improperly. In this paper, several solutions were suggested to minimize the risk associated with a MOS and EMS data exchange interface implementation. By carefully analyzing the interface requirements between the MOS and EMS systems, it is possible to determine and specify the interfaces and the necessary requirements during the specification stage with a high degree of confidence. Doing so provides a well-defined scope for the various vendors, the developers of these interfaces and responsible utility staff thus eliminating the greater risks to the MOS implementation project.

ELSEVIER
IFAC
PUBLICATIONS
www.elsevier.com/locate/ifac

PROTECTION, CONTROL AND MONITORING SYSTEM WITH WEB INTERFACE FOR SMALL DISTRIBUTION SUBSTATIONS

Alexander Apostolov, Ph.D. **Rick Schaal**

ALSTOM T&D EAI
Los Angeles, CA, USA *Bellevue, WA, USA*

Abstract: The paper describes in detail the functional and communications architecture of an integrated system for small distribution substations. It analyses the requirements for protection, control and power quality monitoring and disturbance recording at the substation and distribution feeder level. It describes the distribution of different functions between the IEDs used by the system. *Copyright © 2003 IFAC*

Keywords: protection, control, monitoring, automation, systems

1. INTRODUCTION

Because of requirements for reduced engineering, commissioning and maintenance costs, as well as the availability of Internet access to substation information, utilities are looking for new solutions based on today's Intelligent Electronic Devices (IED) and substation integration systems. At the same time they have to consider the need for power quality monitoring and recording of different abnormal system conditions, that might affect sensitive customer loads.

A Web-enabled protection, monitoring and control system is designed to provide a pre-engineered solution for a complete, cost-effective solution, in order to meet the needs of municipal or rural distribution substations. The system is based on modern off-the shelf technology and industry standard communications. It includes three main components:

- Multifunctional protection IEDs
- Multifunctional power quality monitoring or disturbance recording IEDs
- Web-enabled substation HMI (Human Machine Interface)

The paper describes in detail the functional and communications architecture of the integrated system. It analyses the requirements for protection, control and power quality monitoring and disturbance recording at the substation and distribution feeder level. It describes the distribution of different functions between the IEDs used by the system.

The role of different substation HMI modules related to power quality monitoring and recording are presented later in the paper. Web-enabled functions in the HMI and the IEDs are discussed as well.
Built-in analysis functions in the IEDs are combined with off-line analysis modules in the HMI in order to provide complete coverage of all required primary and backup functions.

2. WEB-ENABLED SYSTEM FOR PROTECTION, MONITORING AND RECORDING IN DISTRIBUTION SUBSTATIONS

Substation integration and automation is the direction that the utility industry is going. However that process is not as fast as expected. This is due to a great extent to the fact that these systems are expensive and require significant engineering effort at different stages of the integration. The development of pre-engineered solutions and the optimal use of multifunctional IEDs of different

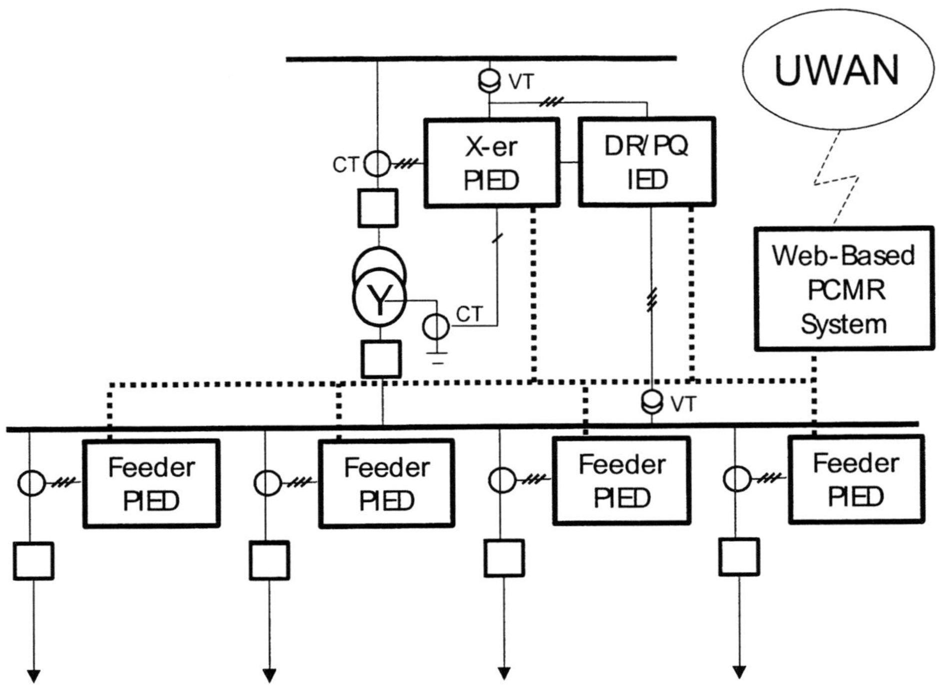

Fig. 1 Simplified diagram of a Web PCMR system

types will result in significant reduction in the cost of such integrated systems.

A first step in this direction is the development of Web Based Protection, Control, Monitoring and Recording Systems for small distribution substations with a single transformer and 2 to 4 distribution feeders. Figure 1 shows the one-line diagram of such a substation with the location of CTs, VTs and the multifunctional IEDs.

For new substations the base configuration would use microprocessor relays for feeder protection that provide a complete set of the basic protection functions implemented in existing substations with electromechanical or solid state relays, including four-shot autoreclosing.

If the utility would like to only partially upgrade the protection, monitoring and control functions and keep the existing reclosing devices, a microprocessor feeder protection relay without reclosing can be used.

The two configurations are basically identical and are offered to support both existing and new distribution substation installations. The two relay types can also be mixed in the configuration allowing migration to newer protection equipment and services.

The base protection schemes shown above can be expanded or enhanced through the selection of the Protective Intelligent Electronic Devices (PIED) used on the high side of the transformer.

The Web PCMR can be pre-assembled for use as a turnkey solution. Enhanced protection options can easily be added into the base system. The basic system is designed to support standard industry communication protocols such as Modbus RTU or DNP 3.0. The RS-485 Modbus data communications network allows the Substation PC to poll the relays in a Client/Server manner. Data is retrieved from the IEDs in real-time and stored in the PCM system memory resident SCADA database.

The PCM system can be remotely accessed by either modem connection or Ethernet LAN. The PCM system can interface externally to other systems or operators through either the Modbus or DNP 3.0 protocols, ICCP (TASE.2) network, or just an Internet Explorer 4.0 or above browser. If desired, an optional Local User Display is also available. The local user display includes a 15" monitor, keyboard, and trackball.

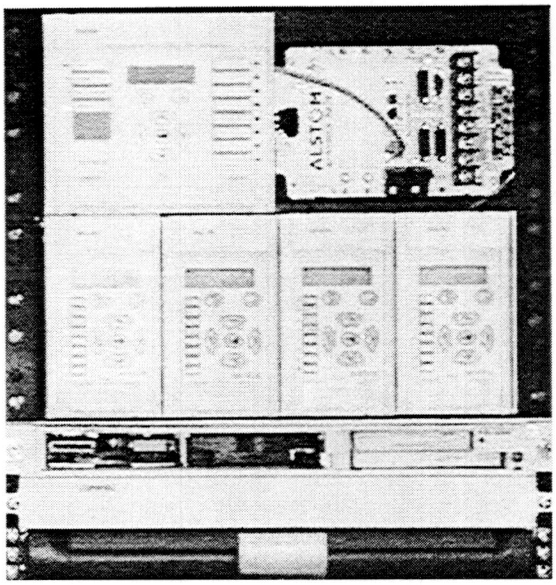

Figure 2. 19" rack mounted WebPCM system for a small distribution substation (1 transformer and 4 feeders)

The WebPCM system is designed to be installed into a standard 19" rack (see Figure 2). The rack can either be provided with the system or customers may already have space available in existing racks or panels. A typical WebPCM configuration is shown below in Figure 2.

The IEDs installed on the high side of the transformer can be of different types, depending on the requirements of the user regarding power supply, integrated control or advanced functionality. The configurations can be tailored to meet the specific needs of the utility in a very cost-effective manner.

The distribution of protection, control, monitoring and recording functions between the different IEDs used in the integrated system is very important in order to ensure the most cost effective and at the same time complete solution.

Since protection and monitoring/recording devices are designed with different primary functionality, they should be selected in such a way, that the requirements for primary and backup devices for each function are available.

Obviously, the primary protection functions will be performed by the protection IEDs on the transformer and the distribution feeders.

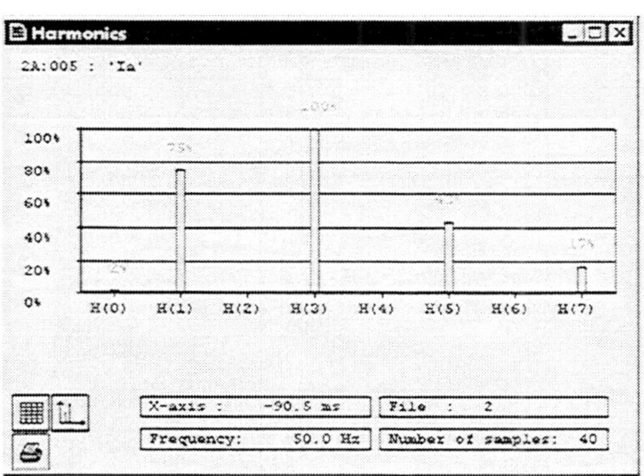

Figure 3. Results from the off-line harmonic analysis of feeder protection IED

From the definitions of power quality events in the beginning of the paper it is clear that most of them are related to voltage measurements. Understanding of the distribution of power quality events (especially voltage sags caused by short circuit faults on transmission lines) throughout the system is very important in order to improve the performance of the protection systems and reduce the effect on users connected to the distribution system. That is why the power quality monitoring and recording IED should have two sets of three phase voltage inputs – one set monitoring the voltages at the transmission level and the other the voltages at the distribution level (see Figure 1).

Power interruptions at the transmission level, voltage sags, swells, unbalances, variations, frequency variations, harmonics, the K-factor for the transformer are all implemented in the power quality monitoring and recording device.

It also provides waveform, disturbance and trend recording of user selected set of system parameters.

The main recording and monitoring function for feeder faults or other abnormal power system conditions should be performed by the feeder protection IEDs. Since the faulted feeder and the transformer are in series, backup recording for feeder faults can be provide by the monitoring recording device on the transformer and the transformer protection IED.

Since the feeder IEDs do not have individual feeder voltage transformers, the detection of power interruptions of the individual feeders is based on monitoring of the status of the breaker auxiliary contacts. Operations of the breaker are recorded as events and time-stamped with 1 msec resolution. Based on this data, off-line analysis can be performed to classify momentary, temporary or sustained interruptions.

Off-line analysis of the waveform records from the feeder IEDs can also be used to determine the harmonic content of the measured feeder currents. Figure 3 shows in graphical form the results from the analysis of a waveform recorded by a distribution feeder relay.

3. HUMAN MACHINE INTERFACE

One of the main problems with any integration system is the development of the Human Machine Interface. This is achievable for the small distribution substations under consideration, because they have a pretty much standard configuration. The HMI is configured to represent the distribution substation as shown on Fig. 4.

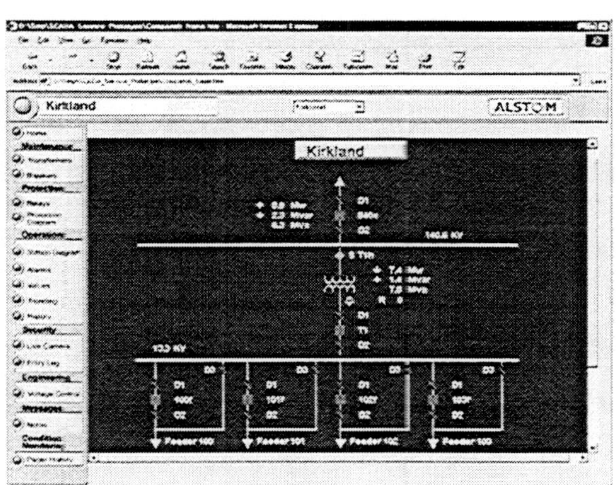

Figure 4. Substation one-line diagram in a Web browser window

Fig. 4 shows the substation one-line diagram from the Web browser interface.

Still, the user may have requirements for customization in order to develop a "utility standard" HMI. This requires the availability of engineering tools such as a display builder. That allows cutomization of displays and graphical representation of data and alarms. The displays may be interconnected through hot links for navigation between views. Any custom solution will require a high premium, thus giving the user an incentive to use the pre-engineered product.

These Display Builder tools are user friendly and support a drag and drop display to database linkage tool which allows easy connection of icons to database values such as status, analogs, and counters.

3.1 Web Browser Interface

The Web PCMR system is also equipped with a Display Builder that is quite unique. When a display is built and linked to the database two viewable images are automatically generated. One of the images can be locally displayed on an optional user display unit and does not require a browser to be viewed. The second image is created using HTTP, which means that the display may be viewed with Microsoft's Internet Explorer from a remotely connected PC. These displays may be viewed locally on one of the two Ethernet connections or remotely, using the optional 56 Kbps Modem for external access via phone line.

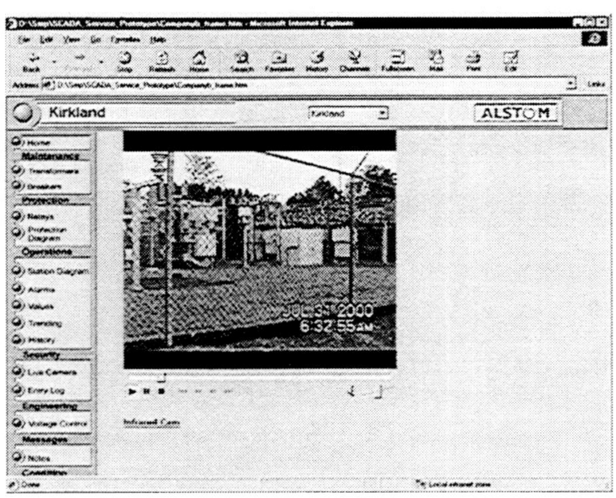

Figure 5. Substation camera view from the Web browser interface

Fig. 5 shows a page with a view of the yard from a substation camera (in the center of the page). On the left side of the page are the controls that allow the user to select other screens such as the substation one-line diagram, load profiles, the communications architecture, alarm and event logs, etc.

The extraction of the different types of records for power quality or any other system event analysis can also be performed over the Internet using the built-in FTP server in the power quality monitoring and recording IED.

4. CONCLUSIONS

The Web-enabled system for protection, control and monitoring in small distribution substations is based on the integration of multifunctional protection and monitoring/recording devices and a Web-enabled HMI.

The different IEDs are integrated into a hierarchical system. The IEDs are connected to the substation computer, with the IEDs performing preliminary local event detection, alarm, logging and recording at the lower level and the substation computer performing the centralized monitoring and analysis at the substation level.

Web access is provided to all HMI functions, as well as for extraction of waveform or disturbance records from the built-in FTP server in the power quality monitoring and recording IED.

- Does not require any additional equipment to be installed, commissioned and maintained in order to provide primary and backup functionality.
- Is part of a pre-engineered, cost effective, complete solution for Web enabled protection, control, monitoring and data acquisition in distribution substations.
- Allows faster analysis of the cause of voltage sags, swells, interruptions, etc. at the distribution feeder level based on the available relay operation information.

ELSEVIER
IFAC
PUBLICATIONS
www.elsevier.com/locate/ifac

NETWORK COMPUTING TERMINAL AS AN INTELLIGENT INFORMATION TERMINAL FOR POWER SYSTEMS

**Yasuhiro Kurosawa, Katsuhiko Sekiguchi, Atsushi Takeuchi
Shigeki Katayama, Hiroyuki Maehara, Masanori Shimada**

TM T&D Corporation, Tokyo, Japan

Abstract: The IT has been progressing rapidly, and all devices can be connected to networks, in order to realize anytime and anywhere computer-supported environment. In such situation, we propose the architecture of NCT (Network Computing Terminal) as an information terminal for power systems, and show some application examples. The NCT exactly acquires the phenomenon generated in power systems as digital raw data, edits it autonomously, and cooperates with other NCTs or PC servers if necessary. It can be applied for broad uses, such as fault analysis, measurement of power quality, and remote maintenance servicing. *Copyright©2003 IFAC*

Keywords: Power system, Information terminals, Networks, Measurement, Monitoring, Internet

1. INTRODUCTION

Recently it is necessary for the electric power utilities to save cost and utilize power system facilities efficiently, because of the progress of deregulation and the introduction of market competition. On the other hand, the requirement to the power quality and reliability has become higher as electricity has been essential for our life surrounded by precise electronic equipment, information equipment and a lot of appliances. Social systems are becoming more sensitive, not only to power system outages but also to voltage dips and sags that can have a serious effect on many industrial and commercial operations.
To survive in this new environment, electrical power companies have to realize cost reduction, high efficiency operation and high quality of power supply. The implementation of information-oriented systems to collect widely distributed data and to process valuable information is becoming increasingly important for accomplishing above requirements.

2. APPLICATION OF THE INTERNET TECHNOLOGIES TO POWER FACILITIES

In response to these trends, the authors have proposed and developed a new information terminal for power systems using Internet technologies. In general, the application of Internet technologies can help to achieve the following desirable features:

- Easy access from any location whether local or remote
- Single point management of information (data is gathered in one location but can be accessed from many locations)
- Information sharing supported by open protocol
- Flexible solutions, from a basic low-cost system to a sophisticated high-level system can be provided using off-the-shelf technologies
- Scalable system configuration (future extensibility)
- Plug and play
- Various support services are available using standard Internet tools

By applying Internet technologies and implementing Web server functions in Network Computing Terminals (NCTs), the NCTs can become data terminals in substations, thereby delivering various types of raw data outside the substation through the Internet or intranet systems. By accessing each NCT and server through a browser, the raw or processed data from several substations can be collected at any location. The proposed system offers the following advantages, enabling the effective supervision and operation of substations, as well as providing a means to save labor in operation and maintenance.

- Substation information can be readily observed at different substations or remote locations using a general-purpose browser.
- Information on measurement, monitoring, control, equipment status and protection, which has been handled separately in the past, can be easily integrated.
- Remote maintenance and equipment monitoring can be readily performed using the information acquired from the data.
- Fault analysis and power system observation based on wide area data becomes possible.
- Documents can be converted into electronic format and accessed by utility personnel.
- Can be easily applied in all substations, even those without numerical IEDs such as numerical relays or bay control unit, by applying NCTs with data acquisition and Internet communication functions.

3. BASE TECHNOLOGIES ADOPTED IN PROPOSED SYSTEMS

The most prominent feature of the proposed system is the application of Internet technology. This is applied at individual equipment level so that each NCT can handle power system data in real time and be connected to a network as a server that is autonomous in all respects. In other words, each NCT functions as an information terminal that delivers power system information and solutions.

Fig. 1 shows the architecture of the new proposed system. The architecture separates the NCT into a Core area and a Web area. The Core area performs real-time measuring for the power system. The Web area is connected with other IEDs, servers and browsers through a communication network. The software of the Web area is described in *Java*, which enhances conformity with the network environment and enables interoperability between different equipment types. This is facilitated by its features of hardware independence, object-oriented language and safety. The Web area is realized by hardware installed with a *Java* virtual machine (JavaVM).

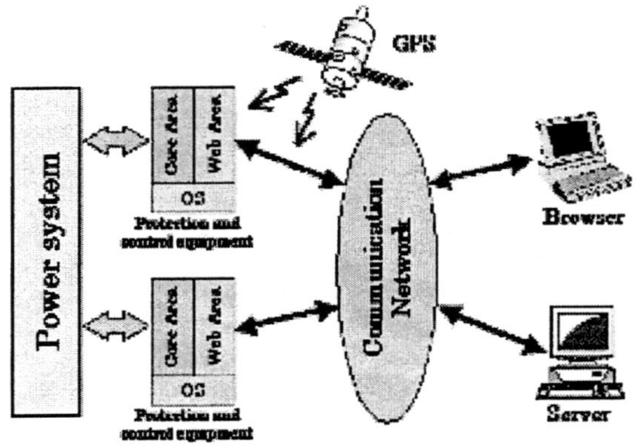

Fig. 1. Basic architecture of the new proposed system

The NCT is an information terminal for power systems, compactly embedding real-time and Web functions, and may also be referred to as a "web-top IED". Fig. 2 provides an outline of the NCT and Fig. 3 shows a functional block diagram. The NCT has a dual CPU (CPU-R and CPU-N), digital input and output circuits, analog input circuits and serial interfaces. CPU-R and CPU-N perform processing for the Core and Web areas respectively, providing status information and analog quantities in real-time. This data can be delivered to the Web after time-tagging as necessary, synchronized to the GPS clock. The NCT can thus be used in a wide range of fields as a data acquisition terminal. It can also be used as an interface unit to connect existing non-digitized equipment to the Internet or intranet systems.

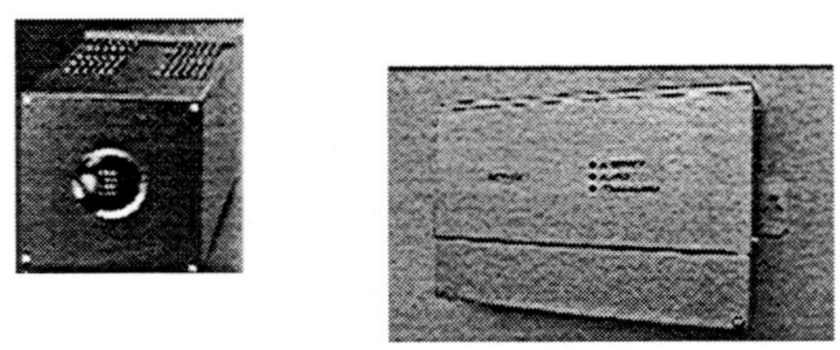

135W, 133H, 245D (mm) 151W, 230H, 45D (mm)

Fig. 2. Outline of the NCT

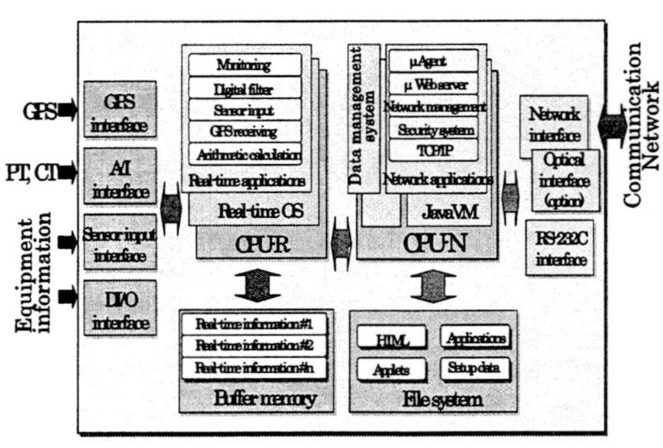

Fig. 3. Functional block diagram of the NCT

4. EXAMPLES OF PRACTICAL SYSTEMS USING THE NCT

Application of Internet technologies to various systems in a substation enables greater efficiency in the operation and maintenance of power facilities, and makes it efficiency to support the operation and analysis of power systems, and monitor power quality. Intranet systems using NCT devices are already in commercial operation in Japan. Some practical implementations are described below.

(1) Power quality and security monitoring (Type-A)
- PQVF measurement
- Voltage sag detecting
- Higher harmonics measurement
- Phasor measurement

(2) Fault location and fault analysis (Type-A)
- Web-based fault locator
- Web-based disturbance recorder

(3) More efficient operation and maintenance of power facilities (Type-GX applied)
- Power equipment monitoring

(4) Remote operation and monitoring system
- Mobile agent technology

4.1 Power Quality Monitoring Functions

NCTs can calculate electrical values for power quality and transmit these data to a PC via a network.

PQVF measuring

It measures and records the active power (P) and the reactive power (Q) on individual power transmission lines, voltage (V) of the busbar and frequency of the voltage (F). Measurement of PQVF is activated manually, by a timer or automatically upon detection of deviation or fluctuation in the voltage, frequency or active power.

Voltage sag detecting

When voltage sag is detected in the voltage of the busbar, this function records the onset time, duration and the lowest voltage.

Higher harmonics measurement

Analyzes and records higher harmonics in the voltage of the busbar. The function is launched manually or using a timer.

Acquisition and recording of PQVF and higher harmonics of more than one substation at the same time are made possible using a timer, synchronized with GPS. Measured PQVF data etc. downloaded on a personal computer is displayed on the screen using display software (viewer). Fig. 4 shows the screen with PQVF data on display.

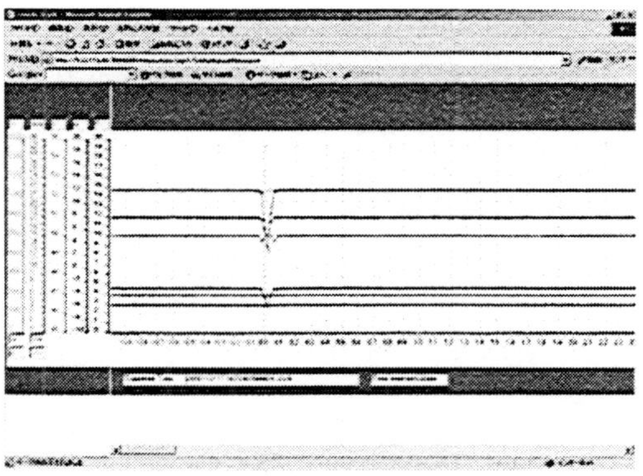

Fig. 4. Display of the viewer

Phasor Measurement System

To monitor dynamic characteristics, in particular the transient stability of a power system on a real-time basis, a phasor measurement system is useful, which collects data including the amplitude and phase angle of the voltage, and active power flow from multiple locations on the power system, and extracts the dynamic characteristics of the system from the data early and with a high degree of accuracy (Phadke 1993).

The phasor measurement system described below measures the phase difference between different locations in a power system using the Phasor Measurement Unit (PMU), constructed using the NCT (Toyoda, et al. 2000).

Fig. 5 shows the architecture of the Phasor Measurement System. The NCT performs A/D-conversion of the input voltage with 16bit, sampling frequency of 4,800Hz or 5,760Hz, and calculates the amplitude and phase of the input voltage by the phasor operation with the one-sec pulsed signal received from the GPS satellite for reference, and saves the results.

Fig. 6 shows a waveform of the actual difference of phase angle. In this case, PMUs (PMU-1 and -2) were installed at two locations separated by several hundred kilometers, and the voltage phase collected through the telephone line and simultaneously sampled. A difference in phase angle of 38-40 degrees between PMU-1 and PMU-2 was found, including a frequency component of around 0.57Hz as swing.

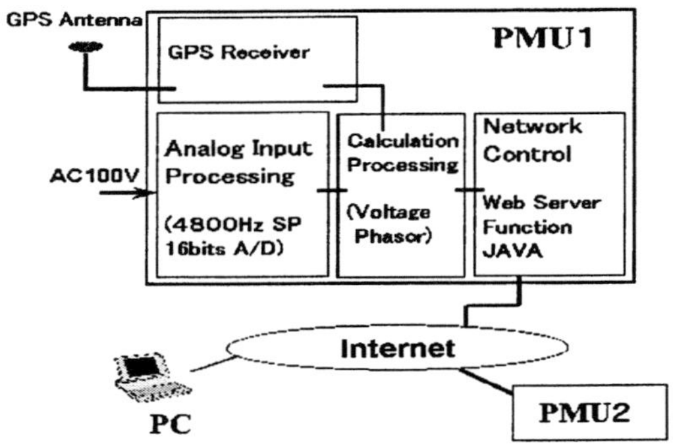

Fig. 5. Architecture of the Phasor Measurement System

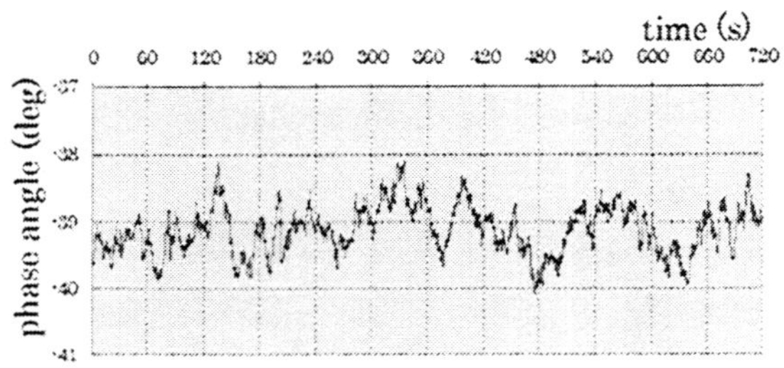

Fig. 6. Transition of the difference of phase angles

PMUs have been installed at several universities in Japan and have been utilized for the analysis of the power fluctuation among the different locations. Fig.7 shows these locations of PMUs.

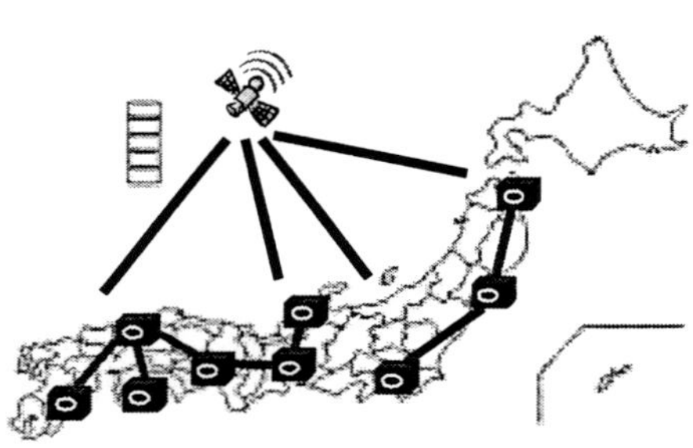

Fig. 7. The locations of PMUs in Japan

4.2 Fault Location and Fault Analysis

Web-based Fault Locating System

We have developed a Web-based fault locating (FL) system for transmission lines, and by the proposed system, good results have been obtained for actual faults in the power system.

The possibility of building a low-cost, high-performance FL system by combining Intranet technology with real time processing technology and excellent results have been obtained in actual power system faults. Table1 shows the examples of the results in actual power system faults.

Table1. The examples of the fault location results in actual power system

	Date	Fault	Result	Faulted Point	Error
1	Jul21, 2001	2L-c	6.9km	6.8km	0.1km
2	Jul22, 2001	1L-c, 2L-ac	6.6km	6.0-6.3km	0.3km
3	Aug.10, 2001	2L-b	2.2km	2.6km	-0.4km
4	Aug.10, 2001	1L-b	4.4km	4.4km	0km
5	Aug.10, 2001	1L-b, 2L-bc	5.6km	4.4km	1.2km

Web-based Disturbance Recording and Analysis System

The Web-based disturbance recording and analysis system is a system for collecting information on the electric quantities and operations of protection and control equipment across a wide area of the power system, and displaying and analyzing status by synchronizing the time (Kudo, et al. 2001).

In this system, the instantaneous values of power system quantities are time-tagged at the NCTs where the data is acquired. The time-tagging clock signal is provided from the GPS system. Data is then transferred to a server. Fig. 8 shows an example of the browser screen.

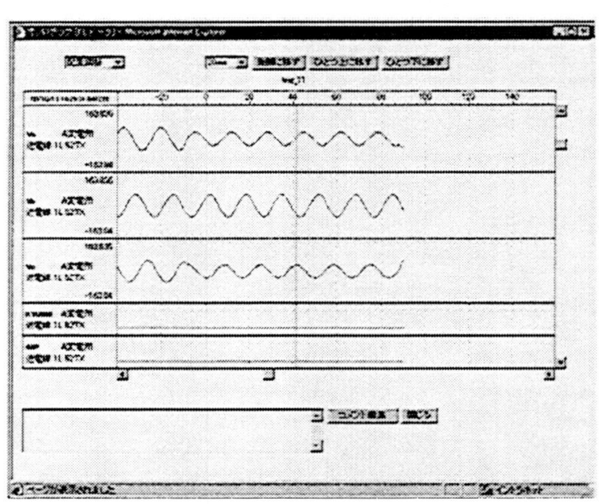

Fig. 8. An example of waveform for fault analysis

4.3 Power Equipment Monitoring System

Maintenance servicing of switchgear, transformers and other power system equipment has generally been performed using the periodic check system, involving regular site visits and checks on equipment. Recently however, a condition-based maintenance (CBM) approach has been attracting attention. This method regularly monitors the condition of plant and detects symptoms of failure in advance to optimize maintenance and service intervals.

Various monitoring techniques are used for condition-based maintenance, such as SF6 gas density monitoring, partial discharge monitoring and monitoring of circuit breaker contact wear. By adding the NCT to the power system equipment, data on condition-based maintenance can be obtained through an intranet system, thereby allowing equipment monitoring and data analysis on a browser in a remote maintenance center. This enables savings to be made in the manpower required to visit and check power system equipment and allows the mapping of planned maintenance servicing. Fig. 9 shows an example of a browser screen being used in power equipment monitoring.

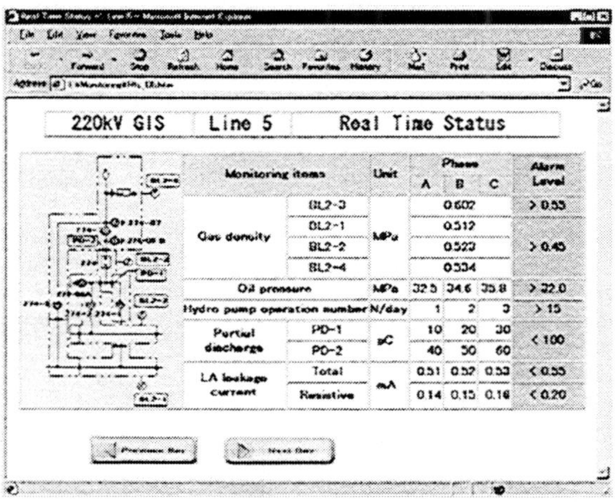

Fig. 9. Power equipment monitoring screen

4.4 Remote Operation and Monitoring System Using Mobile Agent Technology

By having individual protection and control equipment function as Web servers, these protection and control devices can be operated easily and maintained from a remote location using a general-purpose browser. To upgrade operation and maintenance functions, a mobile agent moves flexibly between a browser personal computer and the protection and control equipment and performs transactions inside the equipment. Some types of mobile agent have also been developed. Fig.10 shows a remote operating and monitoring system built with agent technology

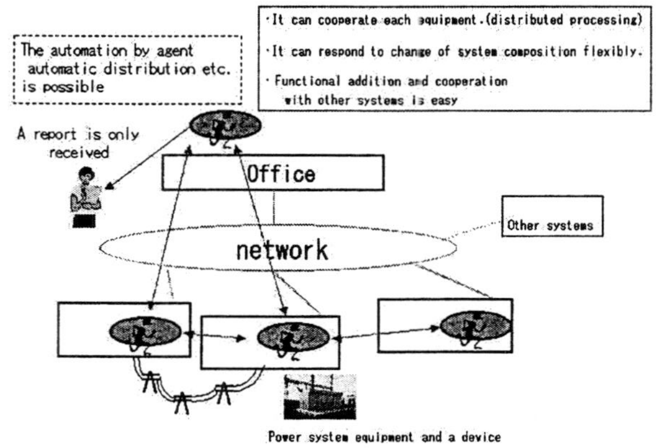

Fig.10. The remote operating and monitoring system built with the agent

5. CONCLUTION

The efficiency of various tasks is generally improved by application of Internet/internet technologies. We have developed the NCT - hardware to provide improving the efficiency of maintenance services and operation of the power facilities, and reducing the cost of acquiring system information. We are now concentrating on practical use of these applications employing the above results. The real-time information at individual equipment level of power systems is directly accessible using the NCT. Because the NCT has autonomous configuration with the Web server function, it is possible to construct systems capable of responding flexibly to the installation of additional application software and increase of facilities and components to be integrated.

We have a conviction that increases in efficiency of operation and maintenance of the protection and control equipment and power facilities, and reduce information-gathering cost by using proposed systems that adopted the NCT in.

REFERENCES

A. G. Phadke(1993), "Synchronized Phasor Measurement Power Systems," *IEEE CAP*, 10-15

David G. Hart, William Peterson, David Uy, Jochen Scheneider, Damir Novosel, Ray Wright (2000), "Tapping Protective Relaying for Power Quality Information," *IEEE CAP*, 45-49

J. Toyoda, et al. (2000), "Development of a High Accuracy Phasor Measurement Unit using the GPS and Internet technologies," *in Proc. 2000 T.IEE Japan Power and Energy Society Conf.*, No. 162

K. Hamamatsu, H. Watanabe, K. Sekiguchi, R. Tsukui, K. Igarashi and P. Beaumont (2001), "A New Approach to the Implimentaion of Intranet-based Measurement and Monitoring," *in Proc. 2001 IEE Developments in Power System Protection Conf.*, No. 479

K. Kudo, Y. Morita, A. Takeda, T. Funaba (2001), "Development of the internet/intranet applied power system measurement and fault location system," *in Proc. 2001 T.IEE Japan Power and Energy Society Conf.*, No. 266

K. Sekiguchi, et al. (2000), "Power System Protection and Control System Based on Intranet Technology - Concept and Basic Architecture," *in Proc. ICEE2000*, E-2-04

K. Sekiguchi, et al. (2000), "Power System Protection and Control System Applying Internet Technology - System Architecture and Applications," *in Proc. ICEE2000*, P-5-12

N. Yamamoto, Y. Tsuyuguchi, Y. Ishihara, M. Kato, Y. Shirota (2001), "Application of the Intranet-technology to a Protective Relay," *in Proc. 2001 T.IEE Japan Power and Energy Society Conf.*, No. 265

Z. B. Shukri et al. (2000), "Experience in Integrating Protective Relay Communication and Protocol Needs in Substation Automation Environment for the Grid Protection Integration and Monitoring Center," *in Proc. CIGRE Session - 2000*, 34-105

BIOGRAPHIES

Yasuhiro Kurosawa received the BS and MS degrees in electrical engineering from the university of Tohoku, Sendai, Japan, in 1974. He joined Toshiba Corporation in the same year and has been engaged in the development of protective relays and relay apparatus. Currently he is Senior Chief Engineer of Protection & Control Systems Division in TM T&D Corporation. He is a member of IEEJ.

Katsuhiko Sekiguchi received the BS and MS degrees in communication engineering from the university of Tohoku, Sendai, Japan, in 1979 and 1981 respectively. He joined Toshiba Corporation in the same year and has been engaged in the development of protective relays and network computing applications for power system protection & monitoring. He is Chief Engineer of Engineering & Manufacturing Department in TM T&D Corporation, Protection & Control Systems Division. He is a member of IEEJ and IEEE.

Atsushi Takeuchi received the BS degree in electrical engineering from the university of Tokyo, Japan in 1980. He joined Toshiba Corporation in the same year and has been engaged in the engineering of protection & control systems. He is now Manager of Planning & Coordination Department in TM T&D Corporation, Protection & Control Systems Division. He is a member of IEEJ and IEEE.

Shigeki Katayama received the BS and MS degrees in applied mathematics and physics engineering from the university of Kyoto, Japan in 1997 and 1999, respectively. He joined Toshiba Corporation in the same year and has been engaged in the network computing development on power systems. He works for TM T&D Corporation now. He is a member of the Operations Research Society of Japan.

Hiroyuki Maehara received the BS and MS degrees in electrical engineering from the university of Nagoya, Japan in 1990 and 1992 respectively. He joined Toshiba Corporation in the same year and has been engaged in the development of monitoring and diagnostic systems for heavy apparatus. He is Deputy Manager of Research & Development Department in TM T&D Corporation, Protection & Control Division. He is a member of IEEJ.

Masanori Shimada received the BS degree in electrical engineering from the university of Fukuoka, Japan. He joined Toshiba Corporation in the same year and has been engaged in the design and engineering of protection & control systems. He is Deputy Manager of Engineering Department in TM T&D Corporation, Protection & Control Systems Division. He is a member of IEEJ.

* TM T&D is a new joint company of Toshiba Corporation and Mitsubishi Electric Corporation in transmission and distribution business.

ELSEVIER
IFAC
PUBLICATIONS
www.elsevier.com/locate/ifac

ENHANCEMENT OF DATA HANDLING IN INTEGRATED POWER SYSTEM ANALYSIS PACKAGE, *NAMED IMPACT*, USING INFORMATION TECHNOLOGIES

Yasuyuki Tada, [*,1] **Takeshi Yamada,** [*]
Tsuyoshi Takazawa, [*] **Yumiko Takahara** [**]
and Takashi Shishido [**]

[*] *Tokyo Electric Power Co.*
[**] *TEPCO Systems Inc.*

Abstract: This paper describes a new method to enhance data handling and reduce effort in power system analysis. The power system is very large and incorporates many pieces of equipment. To reduce effort in power system analysis, it is necessary to increase data handling efficiency. However, conventional power system analysis applications incur a complicated data format requiring experienced operators and/or planners. On the other hand, many useful data-handling methods are provided by information technology (IT). In this paper, a graphical user interface and database function are applied to data handling and are installed in an integrated power system analysis package, named IMPACT.

Keywords: Power Systems Analysis, Graphical User Interface, Database, Stability Analysis, Power Flow Calculation

1. INTRODUCTION

Many power system analytical packages are presently available throughout the world. Historically, the conventional power system analysis applications are coded in FORTRAN language so that analytical data must be expressed by complicated data formatting. Operators and planners, therefore, must be highly experienced.

For example, bus and branch ID codes are necessary in defining power system data. The performance of the power system network depends on network configuration. Power system analysts determine a preferable network configuration by changing all the bus switchgear statuses in the network. When changing the bus status, the analytical data must be modified to all the ID codes by hand coding, requiring immense human effort, in a conventional power system analysis program environment.

This paper discusses a new data-handling method we developed, in which recent information technology (IT) is applied to power system analysis. The integrated power system analysis package is named IMPACT, an abbreviation for Integrated Multi-purpose Package of Advanced Computational Tools for power system engineering. A graphical user interface (GUI) and database function are applied to IMPACT to evaluate the advantages of using IT for power system analysis.

The effectiveness of the GUI was reported in a recent paper (Tada, *et al.*, 2000). However, the reported GUI is deployed to all data in the memory,

[1] 4-1 Egasaki-cho, Tsurumi-ku, Yokohama 230-8510, Japan, tada@rd.tepco.co.jp

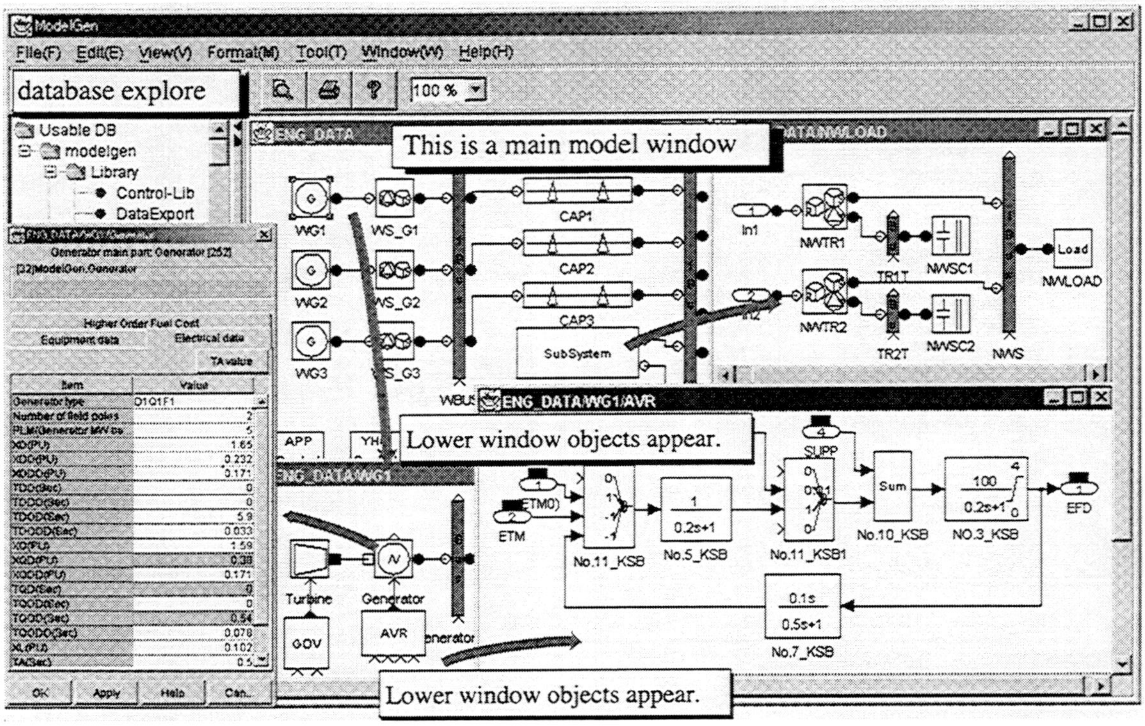

Fig. 1. Example of ModelGen's Window

thereby preventing treatment of a large power system. Also, this is not a preferable method for data sharing for all analysts. This paper discusses the effectiveness of combining a GUI and a database which are compatible with the Internet.

2. IMPACT

We developed IMPACT to reduce human effort in power system analysis, a paricularly important goal considering that in a deregulated environment, the number of cases to be analyzed will increase. It is well known that a graphical user interface (GUI) is preferable for the management of system data. IMPACT has a powerful graphical user interface, named ModelGen. Fig. 1 shows an example of the ModelGen's window. The components of IMPACT include ModelGen, many advanced applications, a post-processing interface and numerous useful utilities. Fig. 2 shows the outline of IMPACT.

2.1 ModelGen

In the IMPACT environment, all power system analytical data is input by the functional ModelGen interface.

In general, power systems are so large that the graphics window provided by a PC is too small to input the power system data in its entirety without some contrivance. ModelGen employs a sub-block concept, which is a child window lying under the main window. The data contained in a sub-block is part of the total system data. The power system data can be grouped into small sections based on the same geometrical region. Power networks also have a hierarchical structure. For example, power stations consist of a generator, turbine and controller. The sub-block concept is, therefore, a useful solution to inputting power system data using a GUI. There are no constraints in creating sub-blocks for any categorized data subset. As sub-blocks appear in the main window as a rectangular box, the drawing area is reduced. When a sub-block is double-clicked, detailed sub-block data appears and any of the data may be modified. Sub-blocks can be defined recursively.

The ModelGen window can be used not only for data input, but also for displaying analytical results such as the initial power flow state.

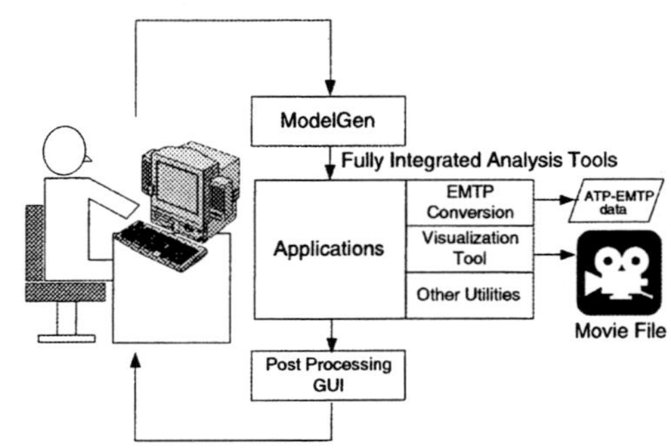

Fig. 2. Outline of IMPACT

Bus Configuration The performance of a power system network depends strongly on the network configuration, which is decided by the switchgear status of the buses. In conventional power system analysis programs, the user must define a logical bus ID to create node and branch data. Such data maintenance is extremely labor intensive, very complicated and vulnerable to mistakes. Furthermore mistakes made during this procedure are, in most cases, also undetectable. Thus, it is preferable that the task of bus ID management is not carried out by a person. The ModelGen can express any bus configuration as a single line diagram, as shown in Fig. 3. Therefore, logical bus ID management by the user is unnecessary even when the switchgear status of the bus is changed.

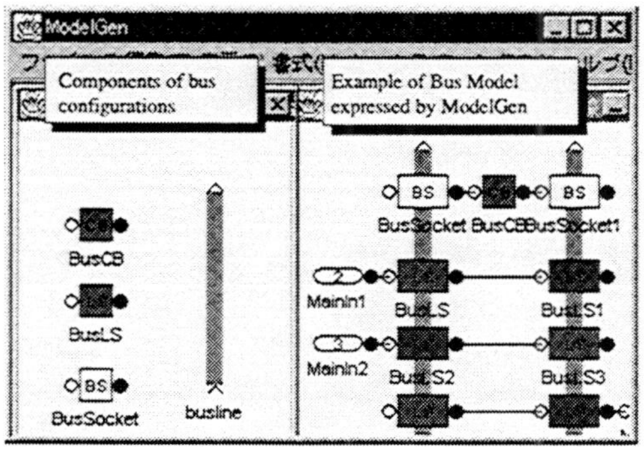

Fig. 3. Bus Configuration Window

Controller data Any controller data such as AVR, Gov. and PSS can also be input by the GUI as shown in the lower-right of Fig. 1. IMPACT provides a controller parts library to arbitrarily express any controller. Fig. 4 shows the parts of the controller library window. Controller data, which is derived from the arbitrary block parts and which is used frequently, can also be input to a new user-defined library.

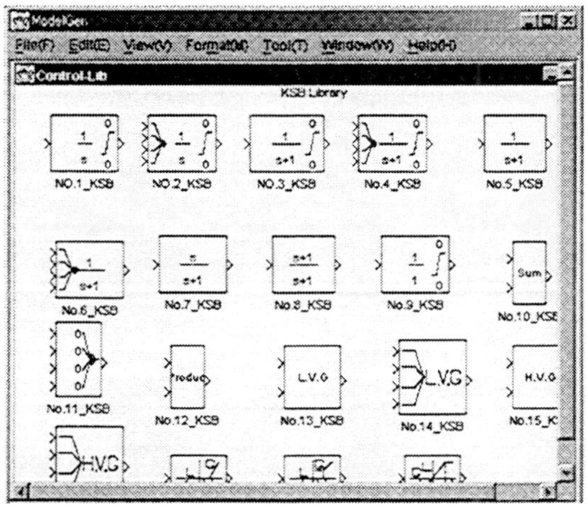

Fig. 4. Example of ModelGen's Window

To design a controller, CAD software such as Matlab/SIMULINK is very important. In the IMPACT environment, controller data input from ModelGen can be converted to Matlab/SIMULINK (Release 12) file format. Using this function, AVR frequency response would be able to easily apply Matlab/SIMULINK commands.

Data is converted from actual measurement to analytical data Certain equipment test data can not be directly used for analysis. For instance, users are required to convert the test data of a three-winding transformer into three-branch analysis model data, even though the data conversion logic has already been defined. Such conversion is extremely cumbersome and the validation of analytical data is difficult. ModelGen supports automatic data conversion logic enabling the test data to be directly used in the IMPACT environment.

2.2 Applications of IMPACT

IMPACT includes many useful applications, as described above and shown in Fig. 5. Since all the applications provide advanced, important functions required to evaluate the performance of a power system, an essential matter in the planning or operation of the system, more value is gained by combining the applications. This section introduces each application.

An important point to note is that ModelGen data should be prepared to run all of the applications because it automatically converts the data for each application. In this way, multidirectional analysis is easily achieved. (All IMPACT applications allow 300 generators, 3,000 nodes and 5,000 branches.)

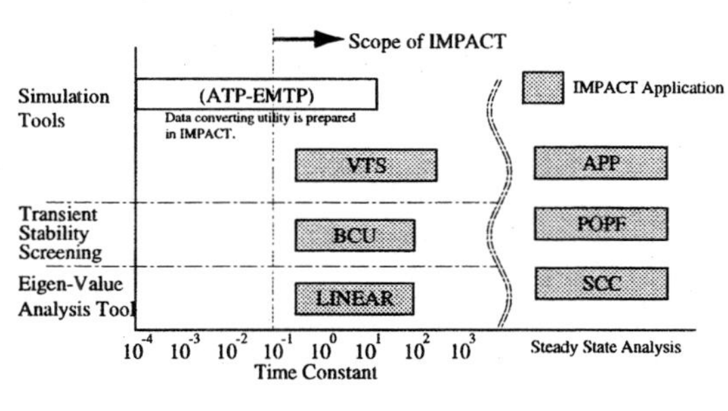

Fig. 5. Target Region of IMPACT Applications

APP Advanced Power flow Program - APP is a tool used to calculate the power flow of a system. APP provides advanced facilities such as voltage sensibility analysis of the power system in the steady state, or P-V curve drawer. These are available for verification of voltage stability.

VTS Variable Time-step Simulation program - VTS is a tool used for analyzing power system phenomena at a time scale of several seconds or minutes and is used to validate the stability or performance of controllers such as speed governors or the PSS. It should be noted that VTS is capable of automatically changing the simulation time step during the simulation. Calculation time is, therefore, efficiently reduced. VTS applies an r.m.s. value.

BCU Boundary of stability-region-based Controlling Unstable equilibrium point - BCU is a screening tool used for transient stability analysis. It screens out unstable cases from assumed contingencies without time-domain simulation. With the use of this tool, the simulation time spent on contingency studies for the planning or operating stage could be reduced to approximately 10% (Chiang *et al.* 1995; Tada *et al.* 2002).

LINEAR Linearization and Eigen-Analysis Recipes - LINEAR is a tool for calculating eigen values, which linearize the dynamic parameters in a power system. It aids in control system design such as that for the PSS, to accomplish desirable small-signal stability.

POPF Powerful Optimal Power Flow program - POPF provides optimal operating conditions for power system equipment with respect to fuel cost, transmission loss, injected reactive power or a combinations of these. POPF aids the power system operator and can also be used for predicting the power system condition for cases in which there are certain unknown operating parameters, for instance, in the planning stage. This application applies MINOS (Murtagh *et al.* 1998), which is an optimization tool developed by the Systems Optimization Laboratory, Department of Operations Research, Stanford University.

SCC Short Circuit Current calculation program - SCC calculates short circuit currents for single-phase and/or three-phase faults, which could be a critical factor in network planning or operations. In addition, feasible reconfigurations can be checked to reduce short circuit currents.

ATP-EMTP data conversion In the same way, the ModelGen data can be converted for ATP-EMTP, which is presently one of the most reliable simulation tools by which to analyze electromagnetic phenomena. Using this function, it is easy to compare instantaneous value analysis and r.m.s. value analysis, although it generally requires considerable manpower. The key technique to realizing this utility is that ModelGen converts controller data to the TACS code of ATP-EMTP. This conversion function is valuable for ATP-EMTP users or those who intend to use it.

2.3 Numerical Example

To address effectiveness of IMPACT, this paper provides a simple model system, shown in Fig. 6. The most important characteristic of model data is the existence of a series capacitor. The turbine model of GEN1 has a three-mass system. The condition of sub-synchronous resonance (SSR) is satisfied.

Single-mass Model Case The detailed mass system is not preferable for analyzing transient stability. The ModelGen converts detailed mass data to equivalent single-mass data. Fig. 7 (a) shows the simulation result of VTS, then this result is the r.m.s. value analysis result. Fig. 7 (b) shows the ATP-EMTP result. The ATP-EMTP data conversion utility is used to create ATP-EMTP data from GUI data to reduce the time required for data handling. The results are almost the same even though the simulation methods are different. The centralization of data handling is very important in maintaining the adequacy of analytical results without increasing human labor.

Three-mass model case Using a detailed mass model, SSR phenomena can be analyzed in the IMPACT environment with ATP-EMTP, as shown in Fig. 8. In the most common method for analyzing SSR, many complicated data conversions are necessary. However, in the IMPACT environment, additional tasking for this calculation from the transient stability study is not necessary because the data specification is well defined and irregular data is not introduced by the ModelGen interface. Integrated data handling has the potential to reduce the labor required for power system analysis.

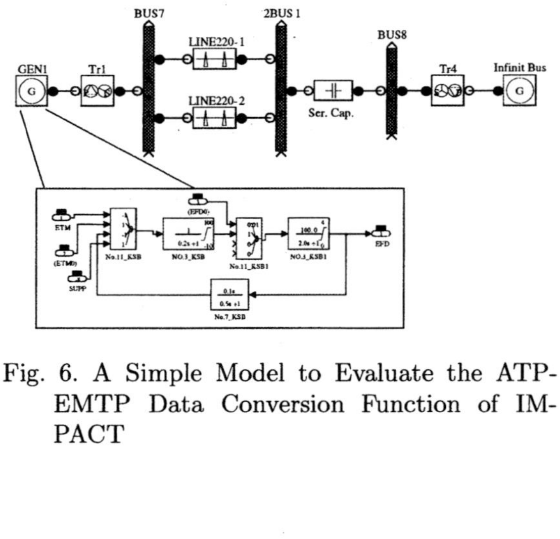

Fig. 6. A Simple Model to Evaluate the ATP-EMTP Data Conversion Function of IMPACT

3. DATABASE FUNCTION OF IMPACT

Although the GUI is very useful and powerful, the management of large volumes of power system data is impossible because the GUI requires enormous memory. In IMPACT, the database function solves this problem. The employed database system is PostgreSQL (http://www.postgresql.org/), which is freeware.

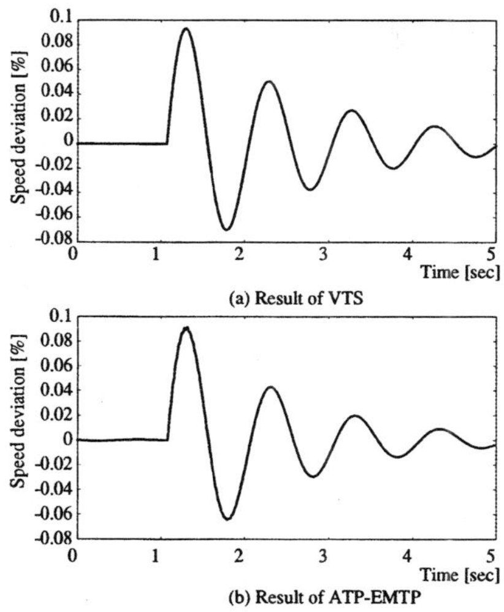

Fig. 7. Calculation Results of VTS and ATP-EMTP with Single-mass Model

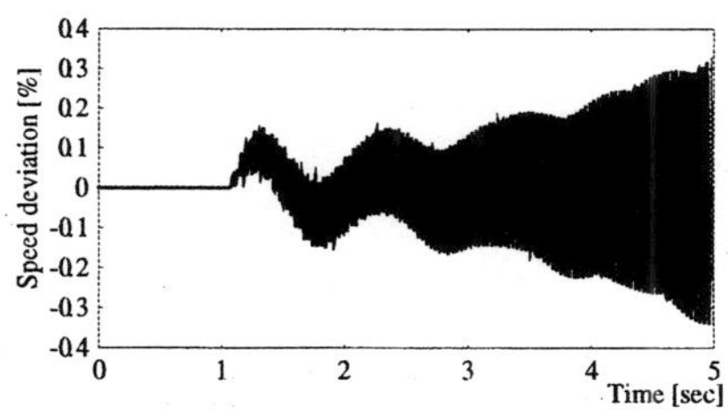

Fig. 8. Calculation Result of ATP-EMTP with Three-mass Model

3.1 *Basic Function of the Database*

Fig. 9 shows a diagram of the ModelGen database function. The basic functions are as follows,

(1) Allows the handling of very large volumes of power system data. There are no practical limitations with respect to size.
(2) Allows multi-user access of the same data via TCP/IP protocol. Therefore, data sharing is realized via the Internet and/or Intranet environment.
(3) XML is used for data transactions between servers and clients to enable highly flexible data processing.
(4) The ModelGen database can be treated as an equipment database.

In general, the relationship between the security of the database and operationality is a trade-off. ModelGen provides original user authentication methods. ModelGen has only three access rights: an administrator, the power-user and the user. When a user logs on to the ModelGen server, any user name is available and right is selected with the correct password. In this method, user administration is unnecessary, thereby simplifying database operations.

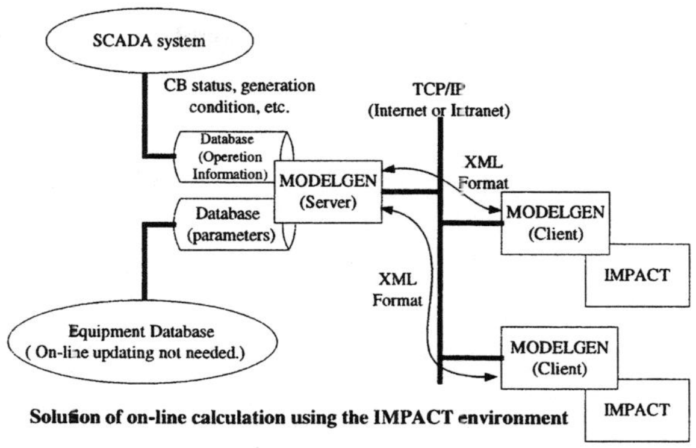

Fig. 9. Example of ModelGen's Window

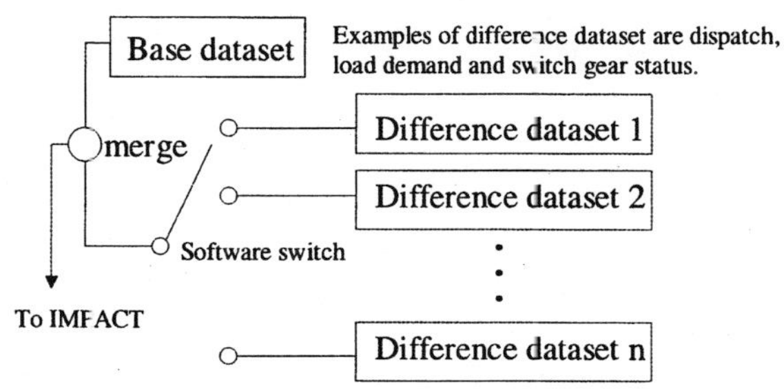

Fig. 10. Example of ModelGen's Window

3.2 *Optional Data Handling Functions*

While GUI is very convenient, mouse operation is onerous for power users. Also, while a synoptic table of the same equipment is useful, the data inputting dialogue of ModelGen cannot express it. To solve these problems, ModelGen provides a function in which CSV format data is outputs for the synoptic table and is also directly input to the database table.

3.3 *Management of Difference Data*

In power system engineering, comparison of the performance, which in some cases differs by only a few parameters, is important. In conventional analytical systems, even though the difference data is minimal, complete analytical data is prepared. The IMPACT database can efficiently manage difference datasets, as shown in Fig. 10. The importance of this function may increase in a deregulated environment because cases of parameter studies would increase.

4. CONCLUSIONS

This paper discussed the importance of the data handling of large power systems for power system analysis. Several solutions to enhance data handling we are proposed. The effectiveness and disadvantages of the GUI for inputting all power system data were discussed first. An improved data handling method based on GUI was then developed, and the necessity for a database for handling the data of a large power system was discussed.

Future tasks for improving the IMPACT system are as follows;

(1) Studies to realize connecting the IMPACT environment to the SCADA system
(2) Using the IMPACT environment for training simulation
(3) Creating a function for Matlab/SIMULINK data conversion to ModelGen data
(4) Creating a control language to manage all IMPACT functions to reduce labor for repetitive calculations such as N-1 criteria confirmation

In summary, we plan to establish an IMPACT environment for all power system analytical calculation interfaces.

REFERENCES

Y. Tada, A. Kurita, M. Masuko, Y. Takahara, K. Koyanagi (2000), "Development of an Integrated Power System Analysis Package", *IEEE PES. PowerCON 2000*, Dec. 2000, Perth Australia.

H.D. Chiang, C.C. Chu and G. Cauley, "Direct Stability Analysis of Electric Power Systems Using Energy Functions: Theory, Applications and Perspective", *Proceedings of the IEEE*, Vol. 83, No. 11, November, 1995, pp. 1497-1529.

Y. Tada, A. Kurita, Y. Zhou, K. Koyanagi, H.D. Chiang, Y. Zheng (2002), "BCU-guided Time-Domain Method for Energy Margin Calculation to Improve BCU-DSA System", IEEE/PES T&D Asia Pacific 2002, Oct. 2002, Yokohama Japan.

B.A. Murtagh and M.A. Saunders, MINOS 5.5 User's Guide, Technical Report SOL 83-20R, System Operation Laboratory, Stanford University, Stanford, CA, Jul. 1998.

ELSEVIER
IFAC
PUBLICATIONS
www.elsevier.com/locate/ifac

NEXT GENERATION NATIONAL- WIDE HYDROPOWER INTEGRATED SCADA SYSTEM.

Young-Gue Kim ; Woo- Bog Song; Nam- Bin Cho *

Hydropower Facilities Dept.
Korea Water Resources Corporation
Daejeon, KOREA

Abstract: Korea Water Resources Corporation (KOWACO), the only organization specialized in water resources management in Korea, is planning to establish a centralized control system for a related operation, in the central control room, of decision making as to water management and electric power generation.
This plan is intended to convert the existing on-site operation system in each multi-purpose dam across the nation into a central operation system in the Corporation's head office, for which Generation Integrated Operating System (GIOS) will be established control center in the Corporation's head office and modernized nine (9) power stations, substations (RDAC). This article discusses realization of GIOS, especially real-time operation of hydropower plants and a specific architectural features for system communication. *Copyright © 2003 IFAC*

Keywords: Algorithms, Automatic control, Back-up. Communication networks
Computer application,Control,Database,Hydro generations
Power station control

1. INTRODUCTION

The Corporation's major facilities in the rivers with the multi-purpose dams are located in Han River, Nakdong River, Geum River and Sumjin River zones: △ Han River zone having Cheongju Dam and Soyang Dam, △ Nakdong River zone with Andong Dam, Imha Dam and Hapcheon Dam, △ Gum River zone with Daecheong Dam and Yongdam Dam and △ Sumjin River zone with Juam Dam. With the establishment of GIOS, generation facilities of multi-purpose dams in each water zone will be related across the nation as water management and control facilities, which will be important national resources. Power generation and water gate facilities for the multi-purpose dams are as follow:

M.P dam	MW	No. Of gates	Remarks
Soyang	100×2	5	
Andong	45×2	4	Pump PH
Daecheong	45×2	6 10	
Chugju	100×4 6×2	6 20	
Hapcheon	50×2 0.6×2	5 5	
Juam	11.3×2	5	
Imha	22.5×2	4	
Namgang	6.3×2	5	
Yongdam	1.1×2 0.12×2	7	

2. TELECOMM. NETWORK

Telecommunication network of the Corporation is made up of the super highway information network and ATM-type backbone network between the head office and major cities across the nation. In this project that will be introduced by phase, the Corporation's ATM back-bone network will be used, and separate exclusive lines will be established from the major cities to the multi-purpose dams, whereby a reliability of the data transmission to the multimedia that are required by the system will be achieved. As an exemplary case, HOCC of the head office and RDAC of Daechung Dam will be linked on the exclusively leased lines, and the existing telecommunication network of the Corporation is illustrated in the Attachment "GIOS Telecommunications Networks".

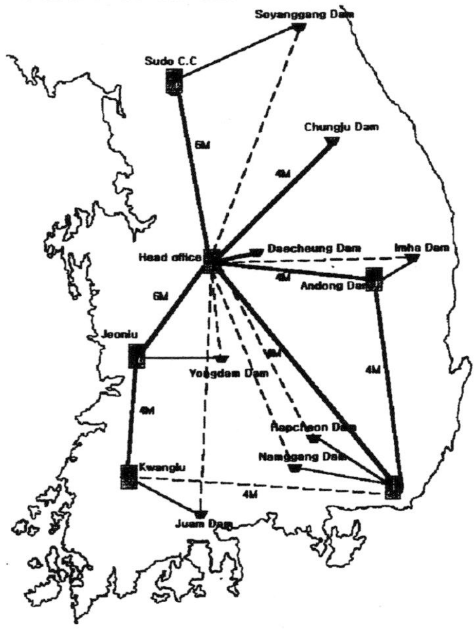

Fig. 1. GIOS Network configulation

3. CONFIGULATION OF GIOS

The Corporation's GIOS, a real-time control system to be established for a related management of water resources, is composed of two layers, namely a central control system (HOCC) for the nation as a whole and on-site system to be installed in each multi-purpose dam (RDAC), where control is made, in principle, at HOCC over the server of each RDAC.

- HOCC
- RDAC
- RTU
- Relation system

(power generation business information system, real-time water management system, etc.)

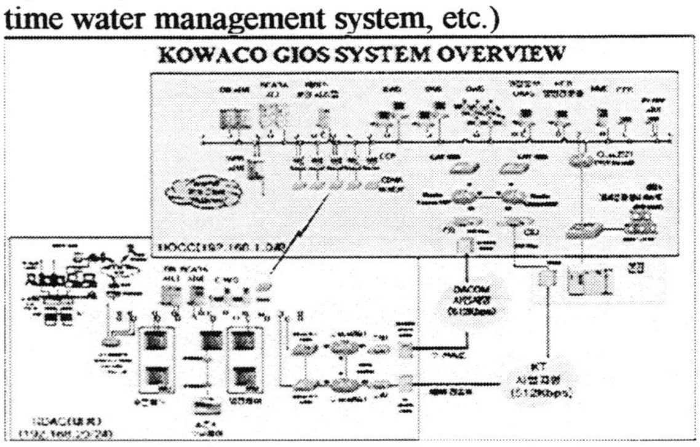

Fig. 2. GIOS System configulation

3.1 GIOS Functions

3.1.1 HOCC functions

HOCC should consist of a related system such as HARDWARE and SOFTWARE, and has functions, performances and other additional functions as hereunder.

- Document acquisition and control (DAC)
- Man-machine relation (MMI)
- Water gate control
- Operation report
- HDR
- Work permit processing
- On-line data base maintenance (ODBM)
- Real-time data saving and backup function
- On-line graphic display maintenance (OGDM)
- Data communication with related system
- Various technical calculations
- Simulator for training operation personnel
- Development and test of Software
- Management of off-line database
- Maintenance of Hardware

3.1.2 RDAC Functions

HOCC should consist of a related system such as HARDWARE and SOFTWARE, and has functions, performances and other additional functions as hereunder.

- Document acquisition and control (DAC)
- Man-machine relation (MMI)
- On-line graphic display maintenance (OGDM)
- Various technical calculations
- Development and test of Software
- Management of off-line database
- Maintenance of Hardware
- Real-time data saving function

3.2. COMM. INTERFACE

3.2.1 OVERVIEW

The main communication path between our HOCC and the 1st stage project site (Daechung, should use the leased line and underwater construction of ATM backbone network. The secondary path should be configured with consideration of CDMA network or satellite communication. For Andong Dam and Imha Dam that is the new RDAC installations, separate routers and CSU equipment for communication with HOCC should be provided. Also, the bidder should submit suggestions for optimal configuration, switching methods, and performance improvement for secondary path.

3.2.2 HOCC-RDAC COMMUNICATION

3.2.2.1 Configuration of main comm. Path

The bidder should suggest equipment with the equivalent or better specification as below and implementation method for performance improvement by using the leased line and underwater construction of ATM backbone network.

(1) ROUTER (HOCC)

- Processor Speed: 225MHZ
- Flash Memory: 32MB or higher
- DRAM: 128MB or higher
- Interface: 10/100Mbps Ethernet 2ports

(2) CSU (5 Sets for HOCC)

- Features: Voice, data and video transmission with E1 class (2,048 Mbps) 4 wire full-duplex mode
- Specifications: Compliant to ITU-T G.703, G.704, and G.732
- Control: Various loop control using CCS and CAS format
- Interface: Provides V.35 or RS-449 interface

3.2.2.2 Secondary path configuration (CDMA)

The Secondary path configuration via CDMA or satellite communication by automatic or manual switching in case of main path of GIOS

(HOCC ↔ RDAC) failure.

3.2.2.3 Communication method

(1) The communication should be compliant to IEEE802.3 Standard Ethernet specifications between HOCC and RDAC.

(2) The CDMA wireless communication should be compliant to the latest IS-95C specifications with cell phone technology.

4. MAIN APPLICATION of GIOS SYSTEM

4.1 HGC(Hydropower Generation Control

(1) The hydroelectric power generation schedule by KOWACO (Korea Water Resources Corporation) is currently being implemented by KPX (Korea Power Exchange) through the use of Hydrothermal Coordination (HTC) Program. The HTC formulates daily schedule on an hourly basis for all medium- to large-scale generators that transmit electricity to the electric power hierarchy of Korea, including generators that belong to Korea Water Resources Corporation (KOWACO). The power generation schedule for KOWACO is currently being delivered to operators of MPD via phone. The Automatic Power Generation Control (AGC) software of KPX is taking advantage of the daily schedule as an economic base-load for generators of KOWACO participating in AGC and these generators have a capacity of at least 50MW.

(2) At the moment, the HTC Program of KPX is utilizing annual hydroelectric power generation schedules planned on a monthly basis for the following year by KOWACO as input data.

(3) Once GIOS is implemented, KOWACO will transmit daily energy schedules to KPX at least once a week for the following 4 weeks through KPX and HOCC-KPX data links and according to the Medium-Term Hydroelectric Power Generation Schedule (MTHS). Using MTHS according to HTC will be a more appropriate and realistic economic base-load for generators of KOWACO than using the current annual schedule. When the Short-Term Hydroelectric Power Generation Schedule (STHS) is completed as planned after the plan for comprehensive water management room is completed, the most optimal and comprehensive water resource management system will have been created as it will be possible to make modifications on a real-time basis and in place for base-load created for HTC.

4.2 WORK PERMISSION PROCESSING (WPP)

WPP system is designed for supervisor of GIOS to approve, after discussing appropriateness with the relevant departments, each permission request from the plant for the matters such as mal-function, inspection, maintenance that interrupt the normal operation of a power station. Interactive software(s) will be provided to search for, make and save the permission. It is recommended for a bidder to suggest the package of programs that meets the following specifications if available. The suggested program should be described in the offer and have the structure of sever/client.

(1) The function of work permission tasks in a maintenance system is composed of indicator to make maintenance work description for a task style of each specific device. Maintenance work explanation may be saved as a file for maintenance work explanation to be used for the function of on-line work permission.

(2) Maintenance work should give explanation for a progressively and logically itemized order as well as the safety inspection to operate the facilities, to perform their maintenance and to tag them.

The face-to-face consulting with MPD operator should be described whenever the principal procedure ends. Procedure in emergency to re-operate the facilities should be included but it is not necessary to be converse order of the interrupting procedure.

(3) A HOCC operator can update maintenance explanation files based on the results recorded on work history file.

(4) Memory area of maintenance work explanation should be sufficient enough to perform repetition of it at least 10,000 times and the specific requirements of memory will be agreed in case of SOW.

(5) An operator can make a monthly report of maintenance work as well as a special report to his managers by using a report maker. Work permission files of maintenance system and work recording file can be sorted and linked with each other for the usage of reports At least, It is possible to sort maintenance by the type of facilities, facility ID, or the required works.

(6) The specific month's summary of spare stocks and materials used in a work-recording file should be available.

4.3. POWER PLANT APPLICATION

(1) This application processes information necessary in operating directly for a generator, its facility, a power transmission facility and etc, such as the current status of facilities, expected information, and calculation.

(2) It should be made up of sever/client structure and interactively provide information to operator W/S. Additional function should be proposed if it is necessary.

4.3.1 Hardware specifications

- CPU: Pentium IV of 1.7GHz or higher
- RAM: 1GB(800MHz ECC RDRAM) or higher
- HDD: 40GB or higher
- CD-ROM: 32X or higher
- FDD: 3.5"
- NIC: 64 PCI dual port 10/100 in a slot
- MONITOR: 18" TFT, 1024 by 768 pixels

4.3.2 Analyzing operation status

(1) Resultant generation analysis by output/efficiency

Generating time by output/efficiency is calculated cumulatively to make data for a fixed period of time.

(2) Output/efficiency trend
Maximum output/efficiency and the current output/efficiency should be demonstrated in one trend to compare each other and also the past trend should be demonstrated.
(3) Calculation of available generation volume by water level
Calculation of available generation volume by water level at specific power output
(4) Expectation of generation volume by used flow quantity
Expectation of generation volume is available as used flow quantity and generating time by power output and the current water level are inputted.
(5) Calculation of the expected water level
Calculation of the expected water level is available as inflow quantity, used water quantity; duration and the current water level are provided.
(6) Expectation of generating time durability by power output according to the used water quantity
(7) Fault recording
Printing sequence of event report in case of heavy fault and printing the trend of the fixed period of time before/after accident.

4.3.3 Efficiency analysis of a generator
It can calculate a generation and a used flow of each generator through the statistical processing function and analyze the efficiency of a generator to guide it according to the optimized efficiency curve It also notify a warning to a operator if it is out of the appropriate bounds.
- Analyzing generating output of each generator and a correlation curve of used water quantity by water head
- A functional relation curve of each generator between time and efficiency
- A functional relation curve of each generator between generation and efficiency

4.3.4 Regulating pond age control
The application software provides outflow rate estimation and opening degree of the gate in response to water level based on the data including inflow rate, water capacity and discharge rate of regulating pond age, and water level.

4.3.5 Auto control and auto recovery
It can give the guidance for current condition and recovery for operator to handle them promptly and safely in case of auxiliary power switching and system failure. It also contains sequence to reduce its procedure as possible.

4.3.6. Maintenance Program
Maintenance system is designed to use a workstation interactively. In addition, backup of magnetic tape and facilities for file search should be available.
- Maintenance schedule
- Tasks for work permission
- History of equipment and failure
- Inventory of spare stocks
(1) Maintenance schedule routine of the control system is to make the annual maintenance plan for water turbine generator, switch gear, control equipment, a preliminary equipment, relay, and others necessary in operating each multipurpose dam. It should consider recommendations for maintenance from the operating department.
(2) A maintenance operator should be able to maintenance re-plan based on countermeasures against unexpected power failure and recommendations for maintenance work from a MPD supervisor. The re-plan should be made promptly by acquiring some data from an operator and other data from the existing plan.
(3) Functions of maintenance plan may include availability of printing in a line printer and a new report should be made in case of re-planning.
(4) Each maintenance plan report should be classified into a separate plan for each operating department.

5. CONCLUSION

The newest technology of the remote concentrated control method of the Hydropower facilities widely spread has described above.
The main functions of the GIOS system at each node has built at KOWACO have followed bellow.

Fig. 3. GIOS Control Centre

- DB Server collect and store the information from SCADA server on real time base, and exchange the data with application server, and back-up DB have updated automatically.
RDB between Headquarter and Site always synchronized.
- SCADA server is configured dual for redundancy to avoid any obstacles while it is processing jobs in the case of automatic switched to Bach-up system as it failed.
Systems have the flexibility of performance and spare room for future expansion after RDAC system implemented, and they have functions of booting for all network node of GIOS system, interruption, status of network, managing and monitoring of node configuration.
- Training simulator support to trainer. It can be scanning real working, imaginary trip for optimise plant operation and take back system in on-line.
- Direct control panel can be monitoring and control remote RTU independently and directly control for server in control centre.

6. REFERENCES

Korea Water Resources Corporation (KOWACO) (2002).
Technical Specifications of GIOS Project **Volume1,2,3,4.**

ELSEVIER
IFAC
PUBLICATIONS
www.elsevier.com/locate/ifac

FEASIBILITY ANALYSIS OF CONTROL-LOOP INTERACTION COMPENSATION FOR A FOSSIL-FUEL POWER PLANT

Raul Garduno-Ramirez

Electrical Research Institute
Division of Control Systems, GCI
Cuernavaca, Mor. 62490 Mexico

Kwang Y. Lee

The Pennsylvania State University
Department of Electrical Engineering
University Park, PA 16802 USA

Abstract: Fossil-fuel power plants are complex systems currently being required to take on load-following duties with full-range variations in power generation. Most of their control systems consist of decentralized PI control loop configurations supplemented with ad-hoc compensation schemes, intended to minimize the effects of interaction among the control loops caused by the non-linear coupled process dynamics. This paper analyses the viability of an inverse decoupling compensation scheme to lessen control loop interaction to ease wide-range power generation control throughout the power plant operating space. The proposed compensator introduces compensation factors among the control signals, which can be systematically determined from an equivalent process gain matrix that conveys all required information about control loop interaction. *Copyright © 2003 IFAC*

Keywords: Power plants, Wide-range operation, Decentralized control, Control-loop interaction, Relative gain array, Interaction compensation, Steady-state decoupling.

1. INTRODUCTION

Late years have witnessed increased participation of fossil-fuel power plants in load-following duties to satisfy the large load demand changes imposed by market conditions. This practice demands wide-range capability in both, power generation and power rate of change, from any power unit. Variations from the minimum to the maximum generation level in short time periods are not rare anymore. This conceptually simple requirement imposes high physical demands on power unit equipments, and leads to conflicting operational and control situations.

Most control systems at fossil-fuel power plants consist of decentralized control configurations based on conventional PI controllers. These schemes were designed under the assumption that the plant might be separated into several single-input-single-output processes. This assumption is only valid during normal operation at base load, where the process characteristics are nearly constant, linear, and non-interacting. Conversely, these conditions do not hold for wide-range load-following operation regimes, where traditional control schemes, designed and tuned for regulation and disturbance rejection, but set-point tracking, may decrease process performance due to the nonlinear and coupled plant dynamics, which change with the point of operation.

Decentralized control schemes have been enhanced with many different compensation schemes to reduce, or to eliminate, the effects of control loop interaction to enhance power plant operation (Taft, 1987; Dimeo and Lee, 1995). Current compensation schemes have been developed through various decades of research and practical experience; most designs are industrial property of power plant manufacturers. With a few exceptions (Ray and Majumder, 1985), the design of most compensation schemes is an ad-hoc process that strongly relies on the designer's experience, nor the resulting schemes are good for wide-range operation.

This paper analyses the feasibility of a compensation scheme to diminish control loop interaction to ease wide-range power generation control throughout the power plant operating space. The structure of the proposed interaction compensator is derived from the multivariable case for disturbance rejection through feedforward control, which is equivalent to a simple inverted decoupling scheme. Compensation factors are introduced among the control signals, preserving the direct control paths of the original uncompensated decentralized control scheme. These factors can be systematically determined from an equivalent process gain matrix that conveys all required information about control loop interaction, as established by the relative gain array (RGA)-based interaction analysis technique (McAvoy, 1983). Section 2 presents the control loop interaction compensator as an inverted decoupler obtained from a feedforward disturbance compensator. Section 3 introduces the control loop interaction compensator for a power plant. Section 4 examines the realization feasibility of the interaction compensator throughout the power plant operating space. Finally, Section 5 concludes this paper.

2. INVERTED DECOUPLING

2.1 Single Loop Disturbance Compensation

A basic feedback control loop is typically represented as in Fig. 1, with transfer function:

$$Y = [1+GG_c]^{-1}\,[GG_cR + G_iW] \tag{1}$$

where R, W, and Y are the Laplace transform of the reference, disturbance, and output signals, in that order, and G, G_c, and G_i are the plant, controller, and disturbance transfer functions, respectively.

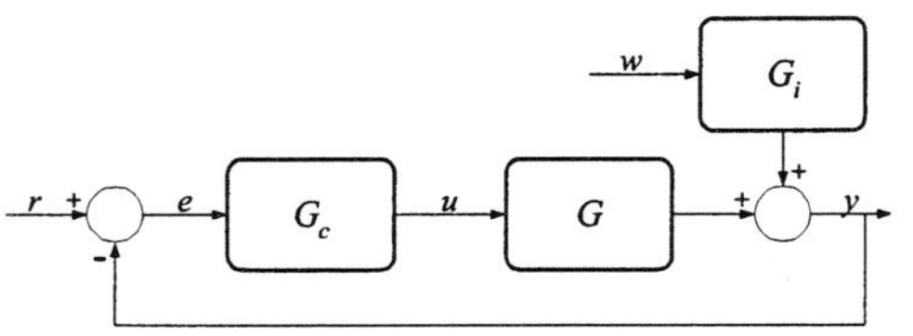

Fig. 1. Feedback control loop.

It can be shown that this control structure does not provide the means for load disturbance rejection. It only provides compensation of model uncertainties through the design of G_c. The fundamental control loop can be extended through feedforward control to achieve load disturbance rejection as shown in Fig. 2, with transfer function:

$$Y = [1+GG_c]^{-1}\,\left[(GG_c)R + (G_i + GG_d)W\right] \tag{2}$$

where G_d is the disturbance feedforward control transfer function. Feedforward control action makes faster corrections than feedback control, acting before any deviation of the measured variable takes place. Theoretically, G_d can be designed to suppress the effect of load disturbances as:

$$G_d = -G^{-1}G_i \tag{3}$$

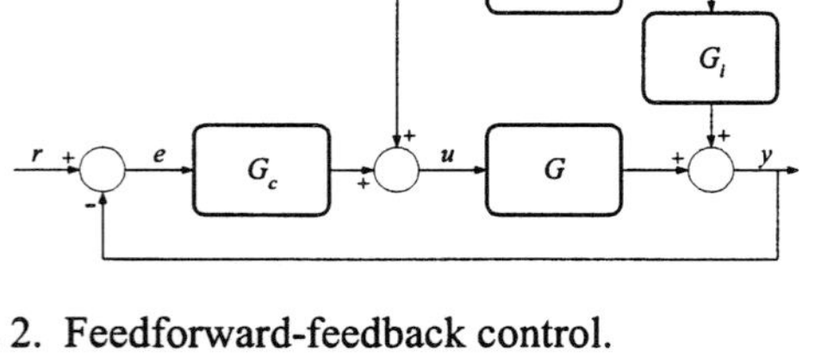

Fig. 2. Feedforward-feedback control.

2.2 Multiple Loop Disturbance Compensation

In industrial control systems, most load disturbances consist of loop interaction due to the process coupled dynamics. The single feedforward-feedback control loop scheme can be extended to compensate for any number of loop interaction effects in a multivariable process. Thus, transfer function (2) becomes:

$$Y_k = [1+G_kG_{ck}]^{-1}\left[\sum_j (G_kG_{ck})R_j + \sum_{j\neq k}(G_{ijk} + G_kG_{djk})U_j\right] \tag{4}$$

where j = 1, ..., k, ..., m is the control loop number. The transforms Y, R, and U correspond to the output, reference, and control signals, respectively. G_k is the process transfer function in loop k, G_{ck} is the controller transfer function in loop k. G_{ijk} and G_{djk} are the interaction, and disturbance feedforward transfer functions from loop j to loop k, respectively.

All control loops with the form in (4) can be arranged in a scheme as shown in Fig. 3 for a two-input-two-output (TITO) system, where the transfer functions for feedforward disturbance compensation form the control loop interaction compensator.

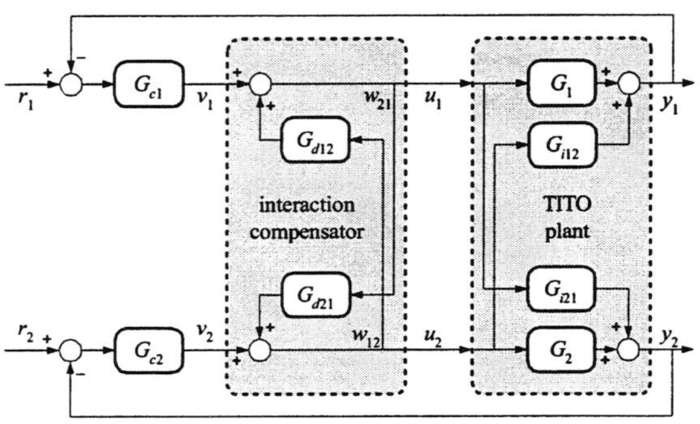

Fig. 3. TITO control loop interaction compensation.

2.3 Inverted Decoupling

Decoupling of multivariable processes includes three schemes: ideal, simplified and inverted (Luyben, 1970), which augment the plant dynamics to appear as a set of independent plants that can be individually controlled. The feedforward interaction compensator is structurally equivalent to an inverted decoupler. However, it should be kept in mind that the objective of the interaction compensator is not to accomplish perfect decoupling, but to decrease the interaction among control loops to lessen the control workload on the feedback controllers, so that the originally decentralized control scheme will effectively perform throughout the plant operating space.

As a basis to design the interaction compensator for a fossil fuel power plant, the inverted decoupler is first explained for the TITO linear plant defined by:

$$\begin{aligned} y_1 &= P_{11}u_1 + P_{12}u_2 \\ y_2 &= P_{21}u_1 + P_{22}u_2 \end{aligned} \tag{5}$$

where y_1 and y_2 are the plant outputs, u_1 and u_2 are the inputs, and P_{11}=G_1, P_{12}=G_{i12}, P_{21}=G_{i21}, and P_{22}=G_2 are the plant input-output transfer functions. In the inverted decoupler structure (Fig. 3), the plant inputs,

u_1 and u_2, are related to the controller outputs, v_1 and v_2, through:

$$u_1 = v_1 + D_{12}u_2 \quad , \quad u_2 = v_2 + D_{21}u_1 \tag{6}$$

where D_{12}=G_{d12} and D_{21}=G_{d21} are the decoupler transfer functions, which are to be designed in such a way that the apparent plant (decoupler and plant) appears as two independent systems of the form:

$$y_1 = P_{11}v_1 \quad , \quad y_2 = P_{22}v_2 \tag{7}$$

Substituting (6) in (5), taking (7) into account, and solving for the compensator transfer functions yields:

$$D_{12} = -\frac{P_{12}}{P_{11}} \quad , \quad D_{21} = -\frac{P_{21}}{P_{22}} \tag{8}$$

In the inverted decoupling case, both the decoupler transfer functions (8) and the apparent plant transfer functions (7) have the simplest expressions.

3. INTERACTION COMPENSATOR DESIGN

According to the inverted decoupler design in Section 2, the design of the interaction compensator is based on the mathematical model of a fossil-fuel power plant. The essential dynamics of a 160 MW oil fired drum-type power plant, for wide-range simulations, have been remarkably captured through a third order nonlinear model (Bell and Astrom, 1987):

$$\begin{aligned}
\frac{dE}{dt} &= \left((0.73u_2 - 0.16)P^{9/8} - E\right)/10 \\
\frac{dP}{dt} &= 0.9u_1 - 0.0018u_2P^{9/8} - 0.15u_3 \\
\frac{d\rho_f}{dt} &= (141u_3 - (1.1u_2 - 0.19)P)/V_t \\
q_e &= (0.85u_2 - 0.14)P + 45.59u_1 - 2.51u_3 - 2.09 \\
\alpha_s &= (1/\rho_f - 0.0015)/(1/(0.8P - 25.6) - 0.0015) \\
L &= 50(0.13\rho_f + 60\alpha_s + 0.11q_e - 65.5)
\end{aligned} \tag{9}$$

where the inputs are the positions of valve actuators that control the mass flow rates of fuel (u_1 in per unit), steam to the turbine (u_2 in per unit), and feedwater to the drum (u_3 in per unit). The three outputs are electric power (E in MW), drum steam pressure (P in kg/cm^2), and drum water level deviation (L in m). The three state variables are electric power, drum steam pressure, and steam-water density (ρ_f), α_s is the steam quality, and q_e is the evaporation rate (kg/sec). A linear state-space model can be obtained at any suitable operating point defined by $x_o = [E_o\ P_o\ \rho_o]^T$, $u_o = [u_{1o}\ u_{2o}\ u_{3o}]^T$ and $y_o = [E_o\ P_o\ L_o]^T$ (Garduno and Lee, 2002):

$$\dot{\hat{x}} = A\hat{x} + B\hat{u} \quad , \quad \hat{y} = C\hat{x} + D\hat{u} \tag{10}$$

where $\hat{x} = x - x_o$, $\hat{u} = u - u_o$, and $\hat{y} = y - y_o$ are the state, input, and output vector deviations, in that order. The required transfer matrix model can be obtained from the linear state-space model (10) as:

$$\hat{Y} = T\ \hat{U} = \left[C(sI - A)^{-1}B + D\right]\ \hat{U} \tag{11}$$

where $\hat{Y}$ and $\hat{U}$ are the Laplace transform of the output deviation, $\hat{y}$, and the input deviation, $\hat{u}$, s is the Laplace complex variable, and T stands for the system transfer matrix. The elements of the transfer function matrix are found to be:

$$\begin{aligned}
T_{11} &= [A_{12}B_{21}s] / d(s) \\
T_{12} &= [B_{12}s^2 + (A_{12}B_{22} - A_{22}B_{12})s] / d(s) \\
T_{13} &= [A_{12}B_{23}s] / d(s) \\
T_{21} &= [B_{21}s^2 - A_{11}B_{21}s] / d(s) \\
T_{22} &= [B_{22}s^2 - A_{11}B_{22}s] / d(s) \\
T_{23} &= [B_{23}s^2 - A_{11}B_{23}s] / d(s) \\
T_{31} &= [D_{31}s^3 + (B_{21}C_{32} - D_{31}(A_{11}+A_{22}))s^2 + \\
&\quad ((A_{32}C_{33} - A_{11}C_{32})B_{21} + D_{31}A_{11}A_{22})s - \\
&\quad A_{11}A_{32}B_{21}C_{33}] / d(s) \\
T_{32} &= [D_{32}s^3 + (B_{22}C_{32} + B_{32}C_{33} - D_{32}(A_{11}+A_{22}))s^2 \\
&\quad + (A_{11}A_{22}D_{32} - A_{11}B_{22}C_{32} + A_{32}B_{22}C_{33} - (A_{11} \\
&\quad + A_{22})B_{32}C_{33})s + (A_{11}A_{22}B_{32}C_{33} - \\
&\quad A_{11}A_{32}B_{22}C_{33})] / d(s) \\
T_{33} &= [D_{33}s^3 + (B_{23}C_{32} + B_{33}C_{33} - D_{33}(A_{11}+A_{22}))s^2 \\
&\quad + (A_{11}A_{22}D_{33} - A_{11}B_{23}C_{32} + A_{32}B_{23}C_{33} - (A_{11} \\
&\quad + A_{22})B_{33}C_{33})s + (A_{11}A_{22}B_{33}C_{33} - \\
&\quad A_{11}A_{32}B_{23}C_{33})] / d(s)
\end{aligned} \tag{12}$$

where

$$d(s) = s(s - A_{11})(s - A_{22}) \tag{13}$$

3.1 Equivalent Gain Matrix

The interaction compensator is designed as a static inverted decoupler that uses the steady-state gains of the plant transfer functions instead of the dynamic plant transfer functions in (12):

$$K = \lim_{s \to 0} T(s) \tag{14}$$

From (14) and (12), the elements of matrix K are:

$$\begin{aligned}
K_{11} &= \lim_{s\to 0} T_{11}(s) = \frac{A_{12}B_{21}}{A_{11}A_{22}} \\
K_{12} &= \lim_{s\to 0} T_{12}(s) = \frac{A_{12}B_{22} - A_{22}B_{12}}{A_{11}A_{22}} \\
K_{13} &= \lim_{s\to 0} T_{13}(s) = \frac{A_{12}B_{23}}{A_{11}A_{22}} \\
K_{21} &= \lim_{s\to 0} T_{21}(s) = -\frac{B_{21}}{A_{22}} \\
K_{22} &= \lim_{s\to 0} T_{22}(s) = -\frac{B_{22}}{A_{22}} \\
K_{23} &= \lim_{s\to 0} T_{23}(s) = -\frac{B_{23}}{A_{22}} \\
K_{31} &= \lim_{s\to 0}\left(-\frac{A_{32}B_{21}C_{33}}{sA_{22}}\right) = \lim_{s\to 0}\frac{\gamma_{31}}{s}
\end{aligned} \tag{15}$$

$$K_{32} = \lim_{s \to 0}\left(\frac{A_{22}B_{32}C_{33} - A_{32}B_{22}C_{33}}{sA_{22}}\right) = \lim_{s \to 0}\frac{\gamma_{32}}{s}$$

$$K_{33} = \lim_{s \to 0}\left(\frac{A_{22}B_{33}C_{33} - A_{32}B_{23}C_{33}}{sA_{22}}\right) = \lim_{s \to 0}\frac{\gamma_{33}}{s}$$

where γ_{31}, γ_{32}, and γ_{33}, are appropriately defined for the equalities to hold. The steady-state matrix K is undetermined because of K_{31}, K_{32}, and K_{33}, which reflect the integrative process dynamics of the drum water level control loop, necessary to achieve wide-range operation. This problem makes it difficult to directly quantify the control loop interaction based on the relative gain array (RGA) method (McAvoy, 1983), as would be desired, using the Hadamard product (element by element) of matrix K:

$$\Lambda = \left(K^{-1}\right)^T * K \tag{16}$$

Fortunately, interaction analysis based on the RGA shows that, in these cases, the interaction information provided by the process steady-state gain matrix can also be retrieved from an equivalent gain matrix, which can also be used to design the compensator. Matrix K may be written as:

$$K = \lim_{s \to 0}\begin{bmatrix} K_{11} & K_{12} & K_{13} \\ K_{21} & K_{22} & K_{23} \\ \frac{\gamma_{31}}{s} & \frac{\gamma_{32}}{s} & \frac{\gamma_{33}}{s} \end{bmatrix} \tag{17}$$

then,

$$K^{-1} = \lim_{s \to 0}\begin{bmatrix} L_{11} & L_{12} & L_{13}s \\ L_{21} & L_{22} & L_{23}s \\ L_{31} & L_{32} & L_{33}s \end{bmatrix} \tag{18}$$

with matrix L appropriately defined. Then, the RGA is obtained from (17) and (18) taking the Hadamard product, canceling s, and taking the limit:

$$\Lambda = \begin{bmatrix} L_{11}K_{11} & L_{21}K_{12} & L_{31}K_{13} \\ L_{12}K_{21} & L_{22}K_{22} & L_{32}K_{23} \\ L_{13}\gamma_{31} & L_{23}\gamma_{32} & L_{33}\gamma_{33} \end{bmatrix} \tag{19}$$

that would be calculated the same with the equivalent steady-state gain matrix, given as:

$$K = \begin{bmatrix} K_{11} & K_{12} & K_{13} \\ K_{21} & K_{22} & K_{23} \\ \gamma_{31} & \gamma_{32} & \gamma_{33} \end{bmatrix} \tag{20}$$

3.2 Compensator Design

The control loop interaction compensator design procedure for the fossil-fuel power plant requires that the product of the equivalent steady-state gain matrix K and the inverted decoupling matrix D be equal to a decoupled steady-state gain matrix M:

$$\begin{bmatrix} M_1 & 0 & 0 \\ 0 & M_2 & 0 \\ 0 & 0 & M_3 \end{bmatrix} = \begin{bmatrix} K_{11} & K_{12} & K_{13} \\ K_{21} & K_{22} & K_{23} \\ \gamma_{31} & \gamma_{32} & \gamma_{33} \end{bmatrix}\begin{bmatrix} 1 & D_{12} & D_{13} \\ D_{21} & 1 & D_{23} \\ D_{31} & D_{32} & 1 \end{bmatrix} \tag{21}$$

where D_{ij}, i=1,2,3, j=1,2,3, are the decoupling factors, or interaction compensation factors, to be designed, and M_1, M_2, and M_3 are the decoupled steady-state gains for the power, pressure, and level control loops, respectively. The interaction compensator is shown in Fig. 4 (Garduno and Lee, 2000).

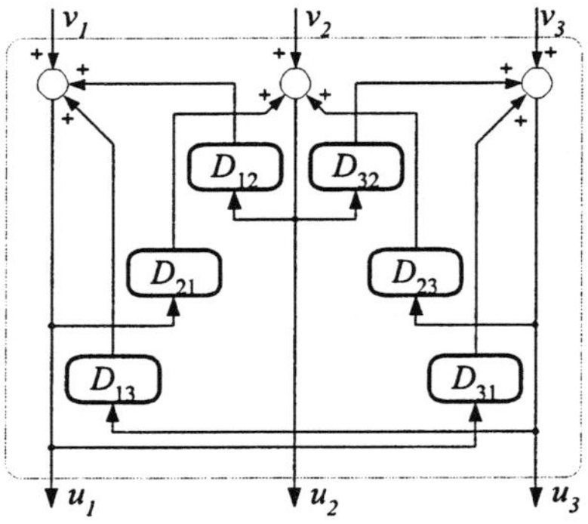

Fig. 4. Control-loop interaction compensator.

Carrying out the product on the right hand side of (21), equating to zero the off-diagonal elements, and solving the resulting system of equations yields the desired interaction compensation factors:

$$\begin{aligned} D_{21} &= \frac{K_{23}\gamma_{31} - K_{21}\gamma_{33}}{K_{22}\gamma_{33} - K_{23}\gamma_{32}} \\ D_{31} &= \frac{K_{21}\gamma_{32} - K_{22}\gamma_{31}}{K_{22}\gamma_{33} - K_{23}\gamma_{32}} \\ D_{12} &= \frac{K_{13}\gamma_{32} - K_{12}\gamma_{33}}{K_{11}\gamma_{33} - K_{13}\gamma_{31}} \\ D_{32} &= \frac{K_{12}\gamma_{31} - K_{11}\gamma_{32}}{K_{11}\gamma_{33} - K_{13}\gamma_{31}} \\ D_{13} &= \frac{K_{12}K_{23} - K_{13}K_{22}}{K_{11}K_{22} - K_{12}K_{21}} \\ D_{23} &= \frac{K_{13}K_{21} - K_{11}K_{23}}{K_{11}K_{22} - K_{12}K_{21}} \end{aligned} \tag{22}$$

Therefore, in steady-state the apparent power plant will appear as three decoupled processes with gains M_1, M_2, and M_3:

$$\begin{aligned} E &= M_1 v_1 = K_{11} v_1 \\ P &= M_2 v_2 = K_{22} v_2 \\ L &= M_3 v_3 = \gamma_{33} v_3 \end{aligned} \tag{23}$$

4. COMPENSATOR FEASIBILITY

The procedure outlined in Section 3 calculates the compensation factors at the point of operation where the state-space and the transfer matrix models in (10) and (11) were found. Wide-range operation requires knowledge of the compensation factors along any predefined power-pressure operating policy in the power plant operating space, defined by the set of all feasible points of operation. These points lie between the upper and lower pressure limits shown in Fig. 5, which also shows a constant-pressure and a variable-pressure operating policy.

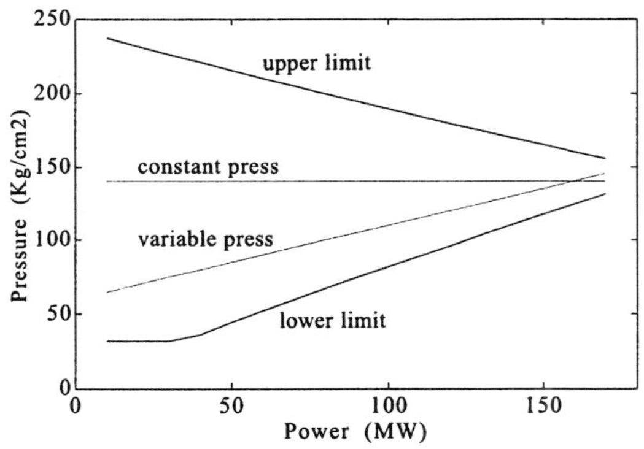

Fig. 5. Power plant operating space.

Operation above the constant pressure policy is not used because of low process efficiency. Therefore, the zone of interest for investigation is that between the constant pressure and the lower limit policies. Tables 1 and 2 present the decoupling factors along the (E, P) constant pressure and lower limit policies, found with (23). This shows that the compensation factors can be calculated without any trouble in the operating zone of interest. In the next two sections, the degeneracy and sensitivity of the equivalent plant gain matrix is demonstrated (Garduno and Lee, 2002).

Table 1. Constant pressure compensator factors

E	P	D_{12}	D_{13}	D_{21}	D_{23}	D_{31}	D_{32}
20	140	-1.13	0.17	1.42	0.0	1.56	-0.31
40	140	-0.55	0.17	1.42	0.0	1.56	-0.09
60	140	-0.36	0.17	1.42	0.0	1.56	-0.02
80	140	-0.27	0.17	1.42	0.0	1.56	0.01
100	140	-0.21	0.17	1.42	0.0	1.56	0.03
120	140	-0.17	0.17	1.42	0.0	1.56	0.05
140	140	-0.14	0.17	1.42	0.0	1.56	0.06
160	140	-0.12	0.17	1.42	0.0	1.56	0.07
180	140	-0.11	0.17	1.42	0.0	1.56	0.07

Table 2. Lower limit compensator factors

E	P	D_{12}	D_{13}	D_{21}	D_{23}	D_{31}	D_{32}
20	32	-0.04	0.17	7.13	0.0	1.78	0.01
40	36	-0.02	0.17	6.20	0.0	1.76	0.02
60	52	-0.03	0.17	4.19	0.0	1.70	0.03
80	67	-0.04	0.17	3.17	0.0	1.66	0.03
100	82	-0.06	0.17	2.55	0.0	1.63	0.04
120	96	-0.07	0.17	2.14	0.0	1.61	0.05
140	110	-0.08	0.17	1.84	0.0	1.59	0.06
160	124	-0.09	0.17	1.62	0.0	1.57	0.07
180	138	-0.10	0.17	1.44	0.0	1.56	0.07

4.1 Numerical Feasibility

Designing the interaction compensator is equivalent to solve the system of linear equations:

$$y = Ku = Mv \tag{24}$$

for the process input vector, u, in terms of the control signal vector, v:

$$u = K^{-1}Mv = Dv \tag{25}$$

Clearly, if K is degenerate, decoupling could be very difficult to achieve, and if K is not full-rank decoupling can not be achieved. Degeneracy of the equivalent plant gain matrix K can be assessed through its condition number, which is defined as the ratio of the largest to the smallest singular value of the equivalent plant gain matrix:

$$cn = \frac{\overline{\sigma}\ (K)}{\underline{\sigma}\ (K)} \tag{26}$$

The larger the condition number (>100), the poorer the numerical conditioning of matrix K, that is, the larger its degeneracy. Fig. 6 shows the condition number along the constant pressure and lower limit policies. Clearly, it can be seen that the gain matrix has no degeneracy problems.

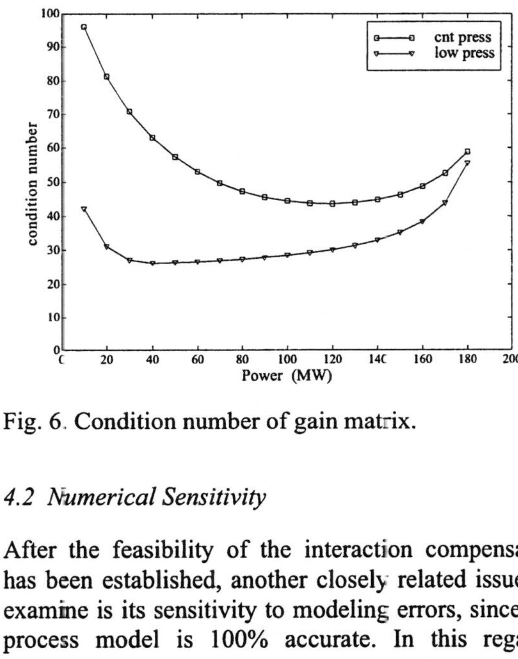

Fig. 6. Condition number of gain matrix.

4.2 Numerical Sensitivity

After the feasibility of the interaction compensator has been established, another closely related issue to examine is its sensitivity to modeling errors, since no process model is 100% accurate. In this regard, consider an error ΔK in the estimate of the steady-state gain matrix:

$$y = (K + \Delta K)\ u \tag{27}$$

The compensator D, designed in terms of K, yields:

$$y = (K + \Delta K)\ Dv = (K + \Delta K)K^{-1}Mv \tag{28}$$

where the amount of error introduced into y due to the model mismatch is given by:

$$\Delta y = \Delta K K^{-1} Mv \tag{29}$$

that can be rewritten using the definition of the inverse of a matrix as:

$$\Delta y = \frac{\Delta K Adj(K) Mv}{|K|} \tag{30}$$

The output error Δy is inversely proportional to the determinant of the gain matrix, $|K|$. Fig. 7 shows $|K|$ along the constant pressure and lower limit policies. Clearly, it can be seen that $|K|$ is large enough to produce very small output deviations due to modeling errors.

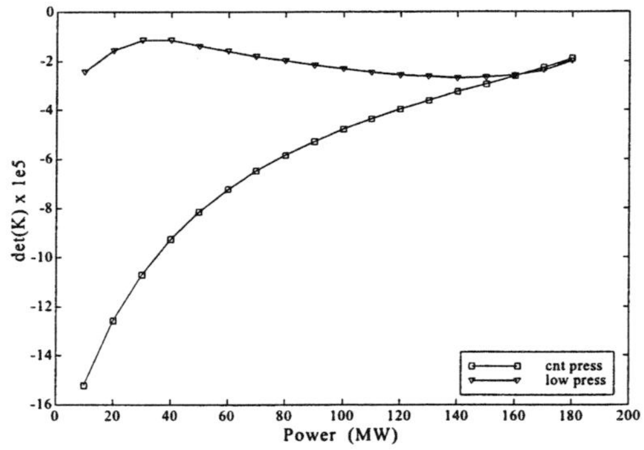

Fig. 7. Determinant of equivalent gain matrix.

4.4 Interaction Compensation Simulations

Simulation experiments were carried out to evaluate the response of the power unit. Fig. 8 shows the unit response to a wide-range unit load demand ramp for both compensated and uncompensated cases. The ramp goes from 50% (80 MW) through 100% (160 MW) base load, with a 5%/min (8 MW/min) load rate, which is a rather fast loading according to American standards. Power and pressure responses are both good. However, a meaningful improvement is obtained in the compensated level response throughout the ramp transition, which means the compensator is effectively reducing the interaction among the control loops, thus helping the controllers to achieve their proposed goals.

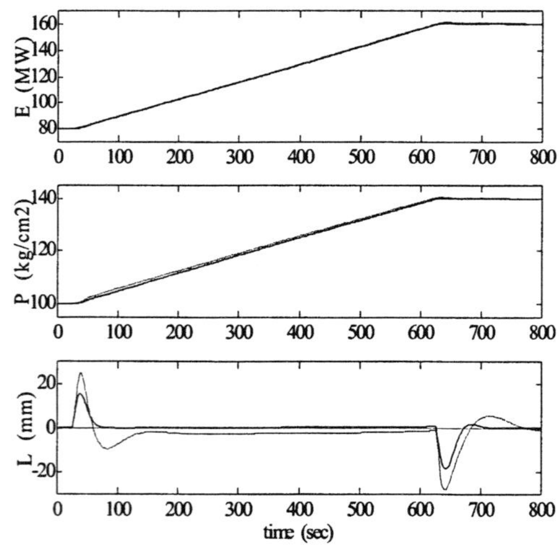

Fig. 8. Response to load ramp. Reference (dotted), uncompensated (dashed), compensated (solid).

Although not shown, it should be mentioned that both the control signals calculated by the controllers (v_1, v_2, and v_3) and those coming out from the interaction compensator (u_1, u_2 and u_3) take allowable values, confirming the feasibility of the approach.

5. SUMMARY AND CONCLUSIONS

This paper presented the feasibility analysis of a compensator to diminish control loop interaction in decentralized control systems to ease wide-range power generation control throughout the power plant operating space. The structure of the proposed interaction compensator was derived from the multivariable case of a feedforward control scheme for disturbance rejection, obtaining a simple inverted decoupling scheme.

The interaction compensator was found to be numerically well-conditioned. The proposed design procedure is a systematic design approach that does not require the solution of an optimization problem, nor years of design experience. Furthermore, its structure provides the compensator with the versatility and necessary characteristics for practical application. Also, the fact that control loops with integrative dynamics can be included in the design without any further complication adds for the generality of the approach. It should be emphasized that this kind of process dynamics are normally avoided by other control strategies such as predictive control. Simulation results demonstrated that inclusion of the control-loop interaction compensator effectively reduced interaction among the feedback control loops, easing their job and improving their effectiveness throughout the power plant operating space.

6. ACKNOWLEDGMENTS

This work was supported in part by NSF under grants INT-9605028 and ECS-9705105, the Pennsylvania State University, the Electrical Research Institute (IIE-Mexico), and Conacyt.

7. REFERENCES

Bell, R.D. and K.J. Astrom (1987). *Dynamic Models for Boiler-Turbine-Alternator Units: Data Logs and Parameter Estimation for a 160 MW Unit, TFRT-3192.* Lund Institute of Technology, Sweden.

Dimeo, R. and K.Y. Lee (1995). Boiler-turbine control system design using a genetic algorithm. *IEEE Trans. on Energy Conversion.* **10(4)**, 752-759.

Garduno, R. and K.Y. Lee (2000). Feedforward compensated multiloop control of a power plant. *Proc. 2000 IEEE PES Summer Meeting.*

Garduno, R. and K.Y. Lee (2002). Power plant coordinated-control with wide-range control-loop interaction compensation. *Proc. IFAC 15th World Congress.*

Luyben, W.L. (1970). Distillation decoupling. *AIChE Journal.* **16(2)**, 198-203.

McAvoy, T.J. (1983). *Interaction Analysis.* Instrument Society of America. USA.

Ray, K.S. and D.D. Majumder (1985). Fuzzy logic control of a nonlinear multivariable steam generating unit using decoupling theory. *IEEE Trans. on Systems, Man, and Cybernetics.* **SMC-15(4)**, 539-558.

Taft, C.W. (1987). Performance evaluation of boiler control system strategies. *Proc. 1987 Conference on Control Systems for Fossil Fuel Power Plants.* **EPRI CS-6049**, 4-1/4-15.

ELSEVIER
IFAC
PUBLICATIONS
www.elsevier.com/locate/ifac

CONDENSER MAINTENANCE COST OPTIMISATION USING GENETIC ALGORITHMS

M. Cregan and **D. Flynn**

The Queen's University of Belfast, N.Ireland

Abstract: Power plant optimisation is one of the most dynamic issues in process and system engineering. This paper discusses the exploitation of genetic algorithm optimisation for improving condenser maintenance scheduling in a thermal power plant. In collaboration with a local utility, an innovative software application has been developed to optimise the cost of scheduling regular condenser maintenance. Typically maintenance is based on a simplistic linear costing formula. However, by utilising the information available from the resident distributed control system (DCS) a complex cost function was developed which more accurately embraces the performance related, contractual and direct maintenance issues. A binary genetic algorithm optimisation scheme was then applied to determine the optimum number and scheduling pattern of maintenance activity.

Keywords: genetic algorithms, steam plants, physical models

1. INTRODUCTION

The primary role of the condenser is to convert the exhaust steam from the turbine into water for reuse in the feed heating cycle. Most common are 'heated water discharge' condensers where the cooling water is drawn from a natural resource such as rivers, lakes, seas, etc. The performance and maintenance requirements of the condenser will be affected by seasonal variations in sea temperature, tidal influences and environmental contamination. Frequently, polluted cooling water will give rise to fouling and blocking of condenser cooling tubes, and hence a reduction in unit efficiency. Traditionally, cooling water inlet and outlet temperatures are monitored to warn of tube fouling and hence signal the need for maintenance. However, given the relative importance of condenser operation in power plant performance, monitoring systems are particularly simplistic and rely upon the user understanding and interpreting the presented data.

The work presented in this paper illustrates how a traditional power station is able to reduce operating costs by improving condenser maintenance using an innovative scheduling application. The industrial partner in this work is Premier Power who operate a 960 MW gas fired power station, which is situated on the north east coast of Ireland. The six units in the station are all fitted with dual pass, single pressure, under slung condensers. By combining a complex cost function which represents condenser operating costs, and a genetic algorithm (GA) optimisation routine, it was possible to search for optimum, least cost, maintenance schedules. The software system implemented on the plant dynamically identifies a range of least cost maintenance schedules, allowing the user to graphically display and compare different options.

2. TRADITIONAL METHODS

For condenser maintenance, the degradation of performance, as a result of a diminishing heat transfer between steam and cooling water, is ultimately of concern. The most common problems affecting condensers are tube fouling and air ingress, both of which limit the heat transfer. These faults are difficult to accurately detect due to the size and shape of the condenser and the fact that the tubes are hidden inside a metal shell, only visible when inspected off line. Unfortunately, as yet, there are no practical on-line non-destructive testing (NDT), or other methods, for determining the condition of cooling tubes.

The most widely accepted methods are based on terminal temperature measurements for both the cooling water and steam flow, giving rise to terminal temperature difference (TTD) and log mean temperature difference (LMTD) parameters. A secondary measurement used in diagnostics is that of vapour temperature from the air extraction pump, indicative of possible air ingress.

Unfortunately, when these calculations are applied to on-line measurements the resulting plots exhibit a high degree of noise and recognisable trends can be difficult to identify. The root cause of this variability stems from the estimation of steam saturation temperature, t_s, which can not be measured directly due to the hazardous environment and the multi-dimensional steam flow inside the condenser hood. Instead, t_s is derived from an internal condenser pressure, referred to as condenser back pressure (ASME, 2000). This becomes especially problematic on generating units subject to significant load changes during normal operation. To reduce noise levels an exponentially weighted moving average (EWMA) algorithm can be employed to filter the measured data (Thomson *et al.*, 2000). However, the sensitivity of the calculations can be adversely affected.

Using the above statistics past performance of the condenser can be analysed. However, such information is insufficient to predict future operation. Instead, it is essential to understand the physical principles behind the condenser. By modelling these processes it then becomes possible to extrapolate and predict future behaviour. The difficulty of this task is evident by the dearth of model-based forecasting tools available in commercial software. Typically the strength of these systems lies in their ability to interface with existing plant instrumentation, archive plant data and provide a configurable graphical user interface. In selected cases the software may also offer pseudo-intelligence, in the form of an expert system, to assist in plant maintenance decisions. While such systems are efficient at what they do they are limited analytically and rely entirely on the user interpreting the presented data.

One of the reasons why many maintenance software systems have only limited functionality is due to the difficulty of accessing plant data and making it available in a usable form. Often this task has proven to be more difficult than anticipated due to the wide variety of communication protocols that exist. The Machine Information Management Open System Alliance (MIMOSA) is a non-profit making organisation whose aspiration is for a commonly accepted open protocol standard. Acceptance of such a standard would resolve the communications issues quickly and cheaply, so that a greater proportion of development funding could be shifted to high value areas such as intelligent diagnostics, prognostics and forecasting.

Advances on the above outlined techniques, or, indeed research publications within the domain of condenser maintenance are very limited. Much of the recent work has been concerned with modifications to the existing ASME and Heat Exchange Institute (HEI) standards (ASME, 1998); (HEI, 1995). The purpose of the ASME standard "is to provide rules for performing condenser acceptance tests", normally included in commercial agreements. Although similar, the HEI standard discusses a methodology for estimating performance based on physical design parameters, such as size and water temperature, and is commonly quoted in design specifications. While both standards can be applied to estimate performance for perfectly functioning, clean condensers neither are suited for continuous on-line monitoring, nor assisting in maintenance related decisions.

3. MAINTENANCE COSTS

A typical approach to the maintenance scheduling problem is illustrated by Putman (2000). By developing a simple linear relationship between condenser losses and time, based on gathered plant data, it is possible to chart the expected losses resulting from an inefficient condenser. Unfortunately, this simple cost model does not provide a true representation. Rarely is condenser fouling linear, instead it varies with time of year, unit loading, etc., and the significant effect of unit contractual costs is not included. This simple approach may have been sufficient when the economics of operating a generating unit were limited to unit efficiency, but in today's open energy markets the autonomy of individual generating companies requires a much more complex decision making process to determine suitable maintenance schedules. Unit efficiency is only one of a myriad of influencing factors.

In order to effectively represent the fiscal expense of operating a condenser, a cost function was developed comprising three components which are listed in order of importance:-

(1) performance related costs, C_η

(2) contractual costs, C_c
(3) direct maintenance costs, C_m

so that the total condenser operational cost, C_{cond}, is represented by:-

$$C_{cond} = f(C_\eta, C_c, C_m) \qquad (1)$$

3.1 Performance Related Costs

Performance related costs embrace the additional expense of operating a unit at reduced efficiency. The deterioration in efficiency resulting from condenser fouling or air ingress is predicted from a composite model which represents the different fouling processes at work within the condenser. By varying the weighting attributed to each of the sub-models within the composite model, variations for time of year and cooling water can be easily accommodated. Potentially, this allows the proposed approach to function with many different types of condensers. The following section reviews different means of estimating condenser efficiency, followed by a discussion of the composite fouling model.

3.1.1. Efficiency As a condenser ages, cooling tubes become fouled, back pressure rises, and efficiency is reduced. Unfortunately, condenser efficiency, η, is not easily quantified due to the large volumes and non-laminar flow of water and steam. In addition, the lack of adequate instrumentation outside of the boiler in most power stations further compounds the problem. ASME and HEI differ in their methods for determining efficiency, with neither ideally suited for on-line measurement.

However, it is possible to estimate efficiency based on condenser back pressure, p_s, and unit heat rate, hr, where the following assumptions are made:-

$$\eta \propto 1/p_s \quad \text{and} \quad \eta \propto 1/hr$$

$$\text{therefore: } hr \propto p_s$$

The relationship between hr and p_s is specific to each boiler - usually this information is available as a standard performance chart from the boiler manufacturer. Consequently, the additional fuel cost required to operate a unit at lower efficiency can be calculated.

3.1.2. Composite fouling model The primary aim of condenser maintenance is to minimize the influence of fouling and air ingress on unit efficiency. Therefore, by modelling the physical processes at work within the condenser it is possible to forecast likely deterioration. Most approaches apply finite element analysis (FEA) and computational fluid dynamics (CFD) techniques to model the influence of geometry on flow characteristics within the condenser (Zhang *et al.*, 1993). Others, such as Putman (1999) and Tsou (1994) have analysed performance from an operational perspective, but neither have directly addressed the problem of identifying the fouling processes at work.

Table 1 summarises the principal fouling processes identified, and a mathematical representation is proposed for each (Bansal and Muller-Steinhagaen, 1993); (Muller-Steinhagen *et al.*, 1988); (Khan *et al.*, 1996) and (Zubair *et al.*, 2000).

Table 1. Condenser fouling processes

	Name	Description	Sub-model
r_1	Power law	high temp. scaling	$At^{1/B}$
r_2	Asymptotic	low temp. scaling	$R_s(1-e^{-t/A})$
r_3	Falling rate	particulate fouling	$A\log(Bt)$
r_4	Linear	plant ageing	$At+B$
r_5	Exponential	air ingress	At^B

For each model t is time, A and B are constants and R_s is saturation resistance. Identifying valid process constants is difficult. However, by combining process data trends, e.g. TTD and LMTD, observations made during routine maintenance and available data from the identified publications it is possible to estimate parameter values.

In practice, fouling will be the cumulative effect of multiple simultaneous processes, dependent on condenser type, cooling water source, etc. The composite model shown in Equation 2 represents this effect by combining the individual process sub-models from Table 1.

$$R_c = f(w_i, r_i, t), \quad s.t. \sum_{i=1}^{5} w_i = 1 \qquad (2)$$

where R_c is the total composite fouling resistance and w_1 to w_5 are weights, which may vary with seasonal conditions, and are attributed to each individual sub-model r_1 to r_5. At Premier Power, the effect of marine shells, represented by r_3, is much more significant during the months when cooling water sea temperatures are at their highest. While each composite model will be unique to a specific unit, the general form of Equation 2 may be modified by adjusting the sub-model weightings w_1 to w_5, and individual A and B coefficients. This will permit its use on other condenser systems.

Now, having obtained a representation for the time varying fouling resistance, R_c, and knowing that $R_c \propto 1/\eta$, the increase in heat rate, hr, and hence the cumulative cost of additional fuel required to operate at reduced unit efficiency can be estimated. The performance related cost component, C_η, from Equation 1, is calculated as the cost of additional fuel over time.

$$C_\eta = f(R_c, t) \tag{3}$$

3.2 Contractual Costs

With the introduction of independent power producers (IPP) to electricity markets, many power stations are now governed by a power purchase agreement (PPA). These contracts specify the details of acceptable performance and the fiscal payments structure between the energy supplier and purchaser. The details of the PPA are confidential. However, a limited amount of information is available in the public domain.

The following list identifies only those costs in the Premier Power PPA which are directly affected by condenser performance and maintenance schedules:-

(1) declared capacity of generating unit
(2) time of year weighting
(3) inflexibility of generating unit

The payment procedure to the power station is scheduled around a thirty minute settlement period for each generating unit. The loss of availability payment to the generator, ap_x, for any settlement period, x, during a maintenance outage is defined as:-

$$ap_x = f(dc_x, w_x, k_x) \tag{4}$$

where the availability payment, ap_x, is a function of dc_x, declared generating capacity, w_x, time of year weighting, and k_x, inflexibility factor of generating unit. The time of year weighting, w_x, can have a significant influence on maintenance costs. It is structured so as to reward the generator most during periods of high energy demand, typically cold winter evenings between 4 - 7 pm. Figure 1 illustrates typical hourly variations in w_x for a complete calendar year.

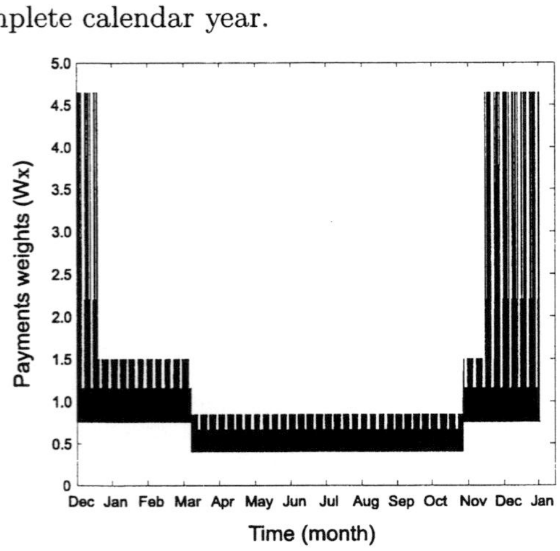

Fig. 1. Hourly payment weights

The consequence of a generator failing to comply with unit performance characteristics, e.g. power output, reserve response, etc., as stipulated in the PPA, results in fines, rp_x, that pass from the generator to the purchaser. Common condenser faults that affect rp_x are:-

(1) A reduction in condenser back pressure may result in a loss of generation capacity.
(2) An unexpected tube blockage or leak might require an unplanned maintenance outage.
(3) Failure to carry out maintenance repair work within the specified time will result in a delay before returning to service.

Hence, the contractual cost associated with rebate payments, rp_x, for any settlement period, x, is described in Equation 5:-

$$rp_x = f(dc_x, w_x, k_x, t_n, t_d) \tag{5}$$

where the combined effect of w_x and k_x are the same as for the availability payment, but the declared capacity, dc_x, may be redeclared downwards if the unit is unable to meet the capacity specified in the PPA. The additional time components are t_n, time between notification of expected outage and maintenance outage, and t_d, time delay in returning to service after outage. With unplanned outages, the longer the notification period given in advance of the outage the less severe the financial rebate rp_x.

Estimating the total contractual maintenance cost C_c, is determined by calculating the combined effect of ap_x and rp_x for the duration of the maintenance outage such that:-

$$C_c = f(ap_x, rp_x) \tag{6}$$

3.3 Direct Maintenance Costs

The direct maintenance cost component, C_m, of Equation 1 represents the expense incurred while carrying out maintenance work. Usually this is composed of labour and material costs, and will vary depending on the nature of the work and regularity of outages. Routine maintenance might involve cleaning the condenser tubes with a power water lance, removing marine shells blocking the cooling water inlet, etc. Costing routine maintenance such as cleaning is straightforward, but estimating the time required to find a tube leak is much more difficult. For atypical tasks a conservative estimate is usually best. Equation 7 demonstrates how the direct maintenance cost is a function of c_{mat}, material costs, c_{lab}, labour costs and, N, frequency of outages.

$$C_m = f(c_{mat}, c_{lab}, N) \tag{7}$$

4. MAINTENANCE SCHEDULING

While the previous section described an integrated approach to representing the true cost of condenser maintenance, the task remains to determine the most effective, least cost maintenance schedule over a specific time period. Typically a twelve month ahead forecast is considered. However, at other times the problem may be determining the least cost solution to keep a unit operational until the next major outage. In either case, an optimisation procedure is only required periodically in response to operational problems, or as part of regular housekeeping, e.g. once every quarter.

In order to estimate the actual costs incurred, the three component costs C_η, C_c and C_m are combined into a single operational cost function, Equation 1. An examination of the cost curve shows that it is discontinuous. This arises from step changes in contract costs, multiple maintenance outages and on/off unit switching. Consequently, traditional gradient based optimisation approaches are not suitable. However, recent developments in evolutionary computing have demonstrated successful application of genetic algorithms (GAs) in similar optimisation problems (Dahal *et al.*, 2000).

4.1 Genetic Algorithms

Genetic algorithms approach problem solving by modelling themselves on evolutionary and natural selection processes. The mechanics of GAs are well documented, but the selection of GA parameters are not as well understood. Despite their success there is no formal methodology for choosing parameters such as population size, generation gap, mutation rate, etc. Instead heuristics and personal preference assist in their selection.

Here the GA configuration was determined following a series of performance tests to evaluate the result of different settings when optimising the operational cost function. The final binary GA configuration implemented employed an elitist selection strategy, stochastic universal sampling (SUS), two point crossover and a run length of fifty generations.

The parameters passed to the operational cost function by the GA for evaluation represent the times at which maintenance is to be carried out, t_1 to t_N. The number of outages, N, is fixed for each GA run but varied between runs. Alternatively, N could be included directly in the cost function as a parameter to be optimised. The value returned by the cost function, C_{cond}, is an estimate of the operating cost over a specific time period, such that:-

$$C_{cond} = cost\_function(t_1. t_2, \ldots, t_N) \quad (8)$$

The GA searches the solution space for global minima by randomly selecting different values for t_1 to t_N. A number of the best or fittest combinations are retained, and these are evolved through a fixed number of generations. At the end of which a list of the best performing individuals is generated. In the example schedule, Table 2, a twelve month forecast estimates possible options for a coastal power station in the period 01 June 2002 to 01 May 2003, after just returning to service from a previous outage.

Table 2. Optimised schedules

Schedule	t_1	t_2	t_3	t_4
#1	15Dec03	-	-	-
#2	01Oct03	09Feb03	-	-
#3	23Aug03	12Nov03	13Apr03	-
#4	15Aug03	10Nov03	05Feb03	04May03

By graphically illustrating how the operational cost varies with the number of outages, Figure 2, the relative cost of the different schedules can be highlighted. In this example there is little difference in cost between 3, 4 and 5 outages. Ignoring any additional external factors, 3 outages would probably be the most effective strategy.

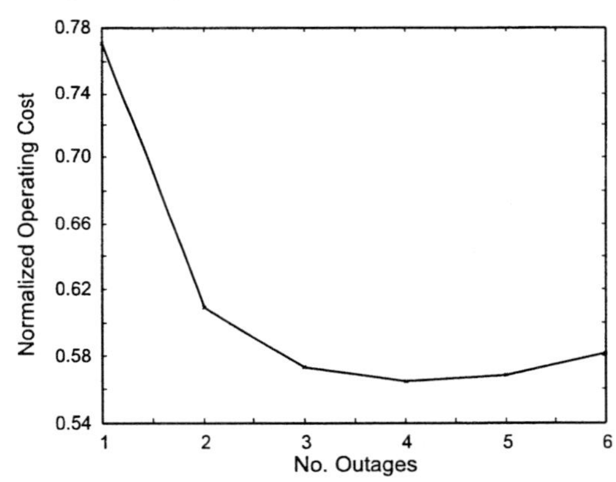

Fig. 2. Outage Costs

The details listed in Table 2 refer to the best or least cost schedules found by the GA. While it is possible to list the second, third, forth, etc. least cost schedules, it was discovered that generally the GA converged on similar solutions. By selecting one or more of the schedules from Table 2 the distribution of costs can be compared graphically, Figure 3.

The reliability of the results generated by the GA and listed in Table 2 is dependent upon the accuracy of the modelled costs. Within the performance related costs it is essential that the condenser fouling model is customized for the application. In Figure 4 three separate fouling sub-models are identified and weighted according to their influence. For example, w_2, is fouling influenced by air temperature, w_3, sea water temperature and, w_5,

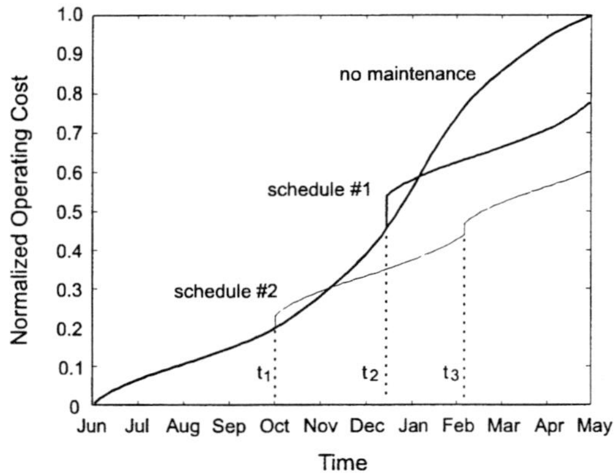

Fig. 3. Maintenance schedules

is constant. These weights relate to the fouling sub-models listed in Table 1 and are identified heuristically. While incremental changes to the weights do not clearly, affect the performance of the model it is, important that the processes are correctly identified.

By combining the modelled costs and GA derived maintenance schedules it is possible to estimate operating costs as illustrated in Figure 3. In this example two maintenance schedules are plotted against time with schedule #2 representing the least cost solution with outage times at t_1 and t_3.

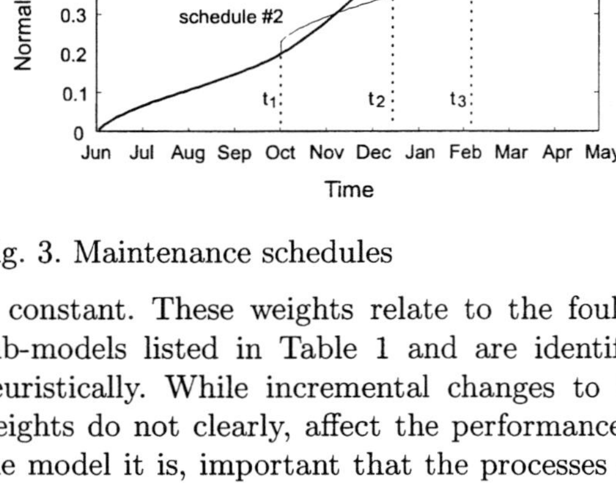

Fig. 4. Sub-model weights

5. CONCLUSION

A method for minimizing condenser operating costs through intelligent maintenance scheduling has been presented. The technique estimates operating costs by combining individual components for performance related, contractual and direct maintenance costs into a single operational cost function. This returns the cost over a specific time period. The contractual costs reflect fiscal payments received, while the direct maintenance costs represent the actual expense of carrying out maintenance work. Finally, the performance related cost estimates the additional fuel required to operate at reduced efficiency as a result of time dependent condenser fouling.

It has been shown that employing a genetic algorithm optimisation routine to search for least cost solutions will successfully identify a range of possible maintenance schedules, and by analysing the results graphically, the user can determine their suitability. These tools have been developed in collaboration with a local utility company and have been applied to improve unit efficiency through intelligent maintenance scheduling on a 960 MW thermal power plant.

REFERENCES

ASME (1998). *Performance Test Code On Steam Surface Condensers (12.2-1998)*. ASME.

ASME (2000). *International Steam Tables For Industrial Use*. ASME.

Bansal, B. and H. Muller-Steinhagaen (1993). Crystallization fouling in plate heat exchangers. *ASME J. Heat Transfer,* **115**, pp. 584–591.

Dahal, K.P., J.R. McDonald and G.M. Burt (2000). Modern heauristic techniques for scheduling generator maintenance in power systems. *Trans. Inst. Measurement and Control,* **22**, pp. 179–194.

HEI (1995). *Standards For Steam Surface Condensers - 9th Edition.* Heat Exchange Institute, Inc.

Khan, M.S., S.M. Zubair, M.O. Budair, A.K. Sheikh and A. Quddus (1996). Fouling resistance model for prediction of $CaCO_3$ scaling in AIAI 316 tubes. *Heat Mass Transfer,* **32**, pp. 73–80.

Muller-Steinhagen, H., F. Reif, N. Epstein and A.P. Watkinson (1988). Influence of operating conditions on particulate fouling. *Can. Jour. Chem. Eng.,* **66**, pp. 42–50.

Putman, R (2000). *Steam Surface Condensers.* ASME.

Putman, R. and D. Karg (1999). Monitoring condenser cleanliness factor in cycling plants,. *ASME Power Div. Int. Joint Power Generation Conf.,* **34**, pp. 133–142.

Thomson, M., P.M. Twigg, B.A. Majeed and N. Ruck (2000). Statistical process control based fault detection of CHP units. *Control Engineering Practice,* **8**, pp. 13–20.

Tsou, J. L. (1994). New approaches to condenser performance analysis. *American Power Conf.,* **56**, pp. 1102–1107.

Zhang, C., A.C.M. Sousa and J.E.S. Venart (1993). The numerical and experimental study of a power plant condenser. *ASME J. Heat Transfer,* **115**, pp. 435–445.

Zubair, S. M., A. K. Sheikh, M. Younas and M.O. Budair (2000). A risk based heat exchanger analysis subject to fouling - part 1 - performance evaluation. *Energy,* **25**, pp. 427–443.

ELSEVIER
IFAC
PUBLICATIONS
www.elsevier.com/locate/ifac

DESIGN APPROACH OF CONTROL SYSTEM FOR ADVANCED NUCLEAR POWER PLANT

**Jin-Koo Kim, Jin-Woong Lee, Ki-Soo Park,
Moon-Jae Choi and Il-Nam Choe**

Korea Power Engineering Company, Inc.
360-9 Mabuk-Ri, Gusong-Eup, Yongin-Si, Kyonggi-Do, 449-713, KOREA

Abstract: This paper describes design concept of control system for Advanced Power Reactor 1400 (APR1400) plant that is designed as advanced nuclear power plant. The control system for APR1400 plant is designed with data communication network-based digital control technologies to enhance the system functionality, reliability, operability, maintainability and economy.
Commercially available PLC (Programmable Logic Controller) and DCS (Distributed Control System) are utilized for Component Control System (CCS) for safety system and non-safety system of the APR1400. Computer-based redundant compact workstations are also adopted for the Main Control Room (MCR) to provide convenient working environment.

This paper describes the approaches and methodologies taken in designing the control system for the APR1400 plant. Also, major design features of the advanced control system to better cope with regulatory requirements are introduced.

Keywords: Distributed Control System, Programmable Logic Controller, Component Control System, Man-Machine Interface System, Compact Workstation.

1. INTRODUCTION

The APR1400 is an Advanced Light Water Reactor (ALWR). It is based on the experience and technology of the Korean Standard Nuclear Power Plant (KSNP) – a 1000Mwe PWR. Shin-Kori 3&4 Nuclear Power Plant (NPP) is first NPP which is designed based on APR1400.

In conventional KSNP, the control system has one dedicated control card per one control component or one analog loop control. For safety system the signals are directly wired from sensors and plant equipment to the control system cabinets. For non-safety system the signal multiplexing is utilized for the all non-safety control components. The multiplexing cabinets are located close by sensor and plant equipment, and these multiplexing cabinets are connected through the network to the controller cabinets centrally located near the main control room (MCR). The microprocessor based single loop controller is also wired to their corresponding indicators and controls in the MCR. The MCR contains large bench boards with many control switches, indicators, recorders, annunciator windows, M/A stations and other monitoring instruments which are separated hard-wired from the I&C cabinets and field instruments.

The conventional design changes dramatically for the design of the control system for an advanced nuclear power plant. Distributed digital I&C systems improve data access and data processing. Multi-loop controllers reduce the number of control cards and cabinets. Signal multiplexing and data highways are used to transmit almost all monitoring and control signals. The compact workstation type Man-Machine Interface (MMI) provides convenient working environment to the control room operators providing plant operating status information to the operators such that the operability and reliability can be enhanced and the human error can be minimized to a great extent.

The control system for APR1400 plant uses advanced designs and as such is, in like manner, designed with data communication, network-based distributed digital control system, and compact workstation-based MMI in the control room. The digitalized common control system platform based on the Distributed Control System (DCS) and Programmable Logic Controller (PLC) is used for non-safety control system as well as safety control system. This design approach will improve functionality, operability and cost effective maintainability of the control system and MMI system.

2. DESIGN CHARACTERISTIC OF COMPONENT CONTROL SYSTEM

The Component Control System (CCS) is a distributed digital multi-loop controller based system that performs component control and data acquisition functions. The CCS provides discrete and continuous control of normally used plant process systems and automatic or manual actuation of Engineered Safety Features (ESFs) system for a nuclear power plant.
The CCS is devided into two major parts: Engineered Safety Feature-Component Control System (ESF-CCS) and Process-Component Control System (Process-CCS). The ESF-CCS accommodates safety-related plant components and systems which include ESF systems. The ESF-CCS consists of four independent divisions to control four independent of mechanical systems. The Process-CCS accommodates non-safety-related plant components and systems related to normal plant process control. These systems are divided into two Process-CCS divisions. The Process-CCS is not mechanically or electrically independent. Figure-1 illustrates overview of I&C system architecture for APR1400.

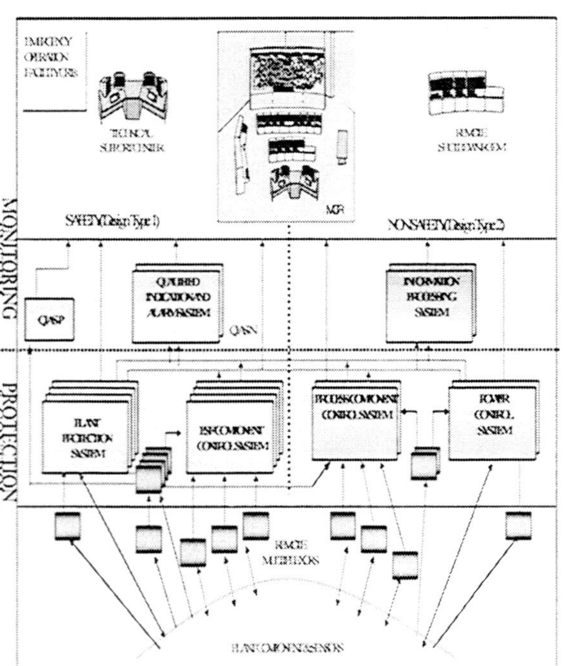

Fig. 1. Overview of I&C System Architecture for APR1400.

The CCS division is segmented to include the Group Controller, Loop Controller, Gateway, Maintenance and Test Panel and data communication networks for the internal and external data transfer. Each segment contains processors and communications interfaces. Each Group Controller supports plant component and data acquisition interface, system I/O data links and performs high-level control functions. Each Loop Controller provides component control level logic and signal multiplexing. The gateway segment supports inter-division communication datalinks and intersystem communication datalinks. The gateway also facilitates the data communication between the control system and the plant protection system. Fiber optic cable is utilized to provide electrical isolation for discrete, datalink and network interfaces where required for inter-channel or geographical location independence. The maintenance and test panel provides the indication for CCS equipment status and this operator interface is used to system maintenance, test and diagnostics.

Major design features of the CCS are summarized as follows:

2.1 Four Quadrant I&C System Architecture

The APR1400 I&C system architecture maintains the four quadrant architecture for both monitoring and control function. This architecture provides distinction between 1) safety and non-safety control systems and 2) monitoring and control systems, making the four quadrants of the design. The distinction between safety and non-safety systems is particularly important because it is the dividing line for design diversity, which provides a means to address the concern of common mode failure of digital I&C systems. The distinction between control and monitoring is also important to separate the process intensive monitoring functions from the less process intensive, yet more time critical, control functions. The four quadrant architecture also applies to the MMI features for control and monitoring function.

Figure-2 illustrates the four quadrant architecture for control and monitoring function, respectively. Safety control systems use diverse hardware and software from non-safety related control system. Safety MMI consist of Class 1E soft control Flat Panel Display (FPD) and fixed position switches for credited control capability of safety equipment. Non-safety control MMI consist of non-safety soft controls on the workstations, which use diverse technology from the safety control MMI.

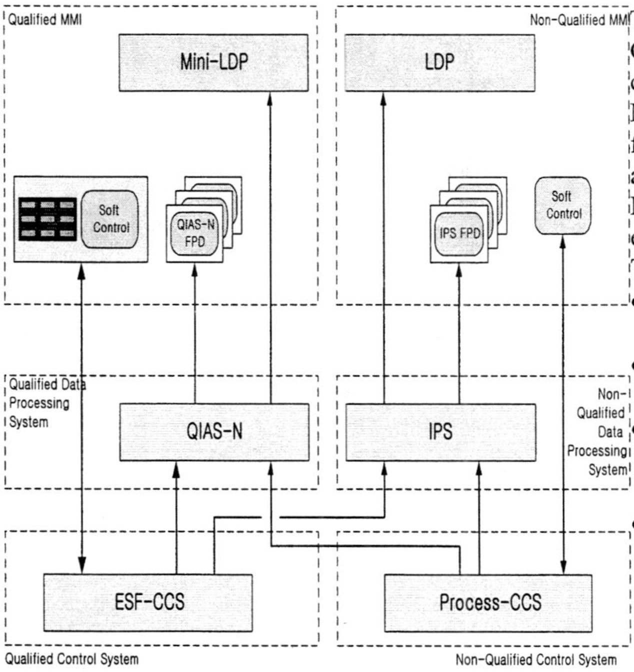

Fig. 2. Four Quadrant Architecture of APR1400.

2.2 Hierarchical and distributed architecture of CCS

The CCS is structured in loop controller level, group controller level and operator interface level.

Loop Controller Level

The distributed loop controller level performs process interface, data acquisition and component control logic. The loop controllers provide hardwired output signals to motor control centers and electrical distribution panels that control plant components. Under normal plant operating conditions, the loop controller provides output signals to the plant components in response to digitized input signals received through a communication network interface.

Group Controller Level

The high-level control function such as system level control logic is performed within group controllers, since the group controller must coordinate control among multiple plant components. Group controllers support a set of loop controllers for plant components and data acquisition interfaces. Appropriate control commands for the components are issued to the loop controllers and various checkback or feedback signals for the process and component conditions are received from loop controllers.

Each group controller executes control programs of a set of inter-related components along with system interlocks associated with that loop controller as well as basic interlocks and protection related to individual component falling under that group controller from plant protection system or control MMI.

Operator Interface Level

The main task of the operator interface level of the CCS is to serve MMI functions to monitor and control the plant under all plant conditions. The MMI devices of the CCS provide operator interface functions for access to all information and control associated with plant operation and safety. These MMI devices are designed to support human factors engineering (HFE) requirements for consistency. The MMI devices of the CCS consist of:

- Workstation including non-safety soft control for non-safety-related components
- Safety soft control to control safety-related components
- Fixed position switches to provide the manual control function with a licensing credit taken for the component level or system level controls
- Maintenance & test panel to provide additional system test capability and to provide an alternate location from the MCR to improve controllability

2.3 Defense-in-Depth & Diversity Design

The defense-in-depth concept is the fundamental principle of the I&C design. A defense-in-depth approach is employed to eliminate common mode software failures as a concern for the digital I&C systems. Common mode failures could defeat the redundancy achieved by the hardware architecture, and could result in the loss of more than one echelon of defense-in-depth provided by the monitoring, control, protection, and engineered safety functions performed by the digital I&C system. Diversity offers the final defense against common mode failures. The safety CCS and non-safety CCS have diverse design architecture by implementing two different digital control systems which have non-identical hardware and software for safety and non-safety control respectively to prohibit the common mode failure.

The safety-related digital equipment are required for a software quality and for a defense-in-depth approach to ensure the integrity of I&C functions including requirements for a backup hardwired manual actuation capability for the system-level actuation of safety functions. To implement adequate diversity for safety-related CCS application, manual ESF actuation switches, which are not based on software, are provided for the control of critical safety functions as well as to accomplish hot standby of the plant condition. These switches are used for protection against common mode failure of the safety-related CCS in the plant in accordance with the requirements of NRC SECY Report 93-087 Enclosure 1, Position Q, II, 4. Thus, diverse manual ESF actuation switches are hardwired in the safety-related CCS architecture to the lowest practical level, and are designed independent and diverse from the safety-related CCS.

Depending on the different task of the I&C function contradictory control commands can be given concurrently by the different control functions to a particular actuator. Therefore, general priority rules must be established, so that any potentially existing command will be assigned to a defined priority level. The following general rules are applied for all safety-related actuator in the plant:

- Higher class functions have priority against commands from lower class functions.
- The principle order of priority is:
  - Control command signal by manual ESF actuation switches against common mode failure of the CCS has the highest priority.
  - Control command signal by ESFAS initiated from Plant Protection System (PPS) has next high priority.
  - Manual control commands of component level has lowest priority.

Additionally, APR1400 design bases are applied as defense-in-depth measure for control diversity in normal and emergency success path. For an example, the ESF-CCS provides auxiliary feedwater system control while the Process-CCS provides main feedwater system control. The ESF-CCS and Process-CCS implementations are diverse and independent to support this design basis.

2.4 Multi-Loop Controller Design

Plant components which are controlled through the conventional single-loop control system have traditionally been controlled using dedicated circuits for each component. If the components of nuclear power plant are grouped relevantly considering the functions and configurations of process system, the reliability, availability and economy can be enhanced by using multi-loop control capability of a digital control system. For another example, a process system with eight inter-dependent components is controlled in the multi-loop control system by one microprocessor. In the conventional single-loop control system design this same process system would be controlled by eight microprocessors. Therefore, the multi-loop control system would demonstrate an 8 to 1 reliability improvement over the conventional single-loop control system. Another benefit of the multi-loop control system can be seen in failure modes and effects analysis. Since components are assigned to multi-loop functional groups in a manner consistent with their process relationship, the effects of failures are predictable and manageable.

Reduction of the quantity of control cards and electronic components and low heat generating electronics due to using the segmented multi-loop controller design can minimize the source of malfunctions, hence the control system maintenance necessity can be reduced.

The digital control system based on a multi-loop controller has automatic self-diagnostic capability. Automatic self-diagnostics performed by software achieves the decreased Mean Time to Repair (MTTR) because the operator can easily access the location and details of the malfunction. The increased Mean Time Between Failure (MTBF) and decreased MTTR result in the enhancement of plant availability.

2.5 Segmentation Design

The functional and physical designs of the CCS are segmented. The designs also explicitly incorporate other functional defensive measures to inhibit the propagation of failures across major dissimilar functional groups. In a multi-loop controller design, each controller independently controls a functionally related group of plant components. Each controller is truly independent of other controllers such that failures, maintenance or testing on one controller does not effect multiple plant functions. Plant components are assigned to controllers in a manner that preserves the redundancy and independence that exists within the mechanical process system being controlled. Components that are mechanically redundant or independent have their associated control functions physically separated into separate controllers.

The CCS is functionally distributed and inter-dependency between subsystems is limited to minimize the impact to the interfaced subsystems by the failure of one subsystem.

The I/O points and control function are grouped functionally and physically on the basis of the division concept. The functional segmented design is to prevent the failure of one controller from affecting the group of equipment performing the same function. However, in case of splitting any of the group is required due to the limitations of controller capability with respect to number of inputs/outputs, controller reaction time, etc., as specified elsewhere, the same can be done by providing more than one set of controllers for such groups. After segmentation is performed, components are rearranged as necessary to form one more compact and functionally cohesive groups as other systems and components are grouped. At the same time, segmentation design is made based on a review of the design as evaluated with respect to a system Failure Mode and Effects Analysis (FMEA).

3. DESIGN CONCEPT OF DATA PROCESSING SYSTEM

The APR1400 has two data processing systems, and the diversity design is applied for these data processing systems. The data processing system consists of Information Processing System (IPS) and Qualified Indication and Alarm System-N (QIAS-N). Qualified monitoring is provided by the QIAS-N and non-qualified monitoring is provided by the diverse IPS.

The IPS is a computer based system that serves to provide operational means for monitoring and control of the plant. The information is derived from other I&C systems and self-contained algorithms called application programs. The IPS makes the information available to the plant operation staff both on a real-time and historical basis. The IPS is additionally designed to enhance overall power plant operation, availability and efficiency. This is accomplished through the use of integrated plant information displays and early warning of conditions that could lead to shutdown. Through analysis of data it assists the plant staff in operating the plant within specified limits while evaluating the performance of the reactor core, primary and secondary plant systems and components. The IPS performs in a supervisory monitoring and control capacity for the NSSS and BOP steam and electrical production processes. It allows the plant operations staff to obtain detailed plant data via its man-machine interfaces. These man-machine interface devices are integrated into the main control room in a manner that meets the HFE requirements. The major functions performed by the IPS include plant wide data acquisition, validation of sensed parameters, the execution of NSSS application programs and BOP performance calculations, the monitoring of plant safety and general status, the presentation of status and calculation results for the workstation displays, the provision of logs, and the determination of alarm conditions.

The QIAS-N is one of two primary data processing systems which make up the monitoring and data handling system within the APR1400 architecture. The main purpose of QIAS-N is to provide qualified display of key information and to allow continued operation upon IPS failure. The QIAS-N monitors the safety-related parameters and key operating parameters in non-safety-related systems to be used by the operators during accidents. The parameters required for Emergency Operation Procedure (EOP) execution, safe shutdown, Critical Operator Action and Tech. Spec. monitoring shall also provided to the operator to be used during and after plant accident condition. In particular, the QIAS-N provides parameter values and alarms using signal validation and alarm processing of the signals to the operator. The operator can access alarms on the FPDs.

In addition, the hardware diversity design concept is also carried through the design of these data processing systems such that both hardware and software diversity is employed between IPS and QIAS-N to eliminate the potential for common mode failures that affect both the IPS and QIAS-N functions. This diversity exists in all software based aspects of these systems, including processor, communication networks and MMI devices. This design approach will ensure availability of main control room information.

4. DESIGN CONCEPT OF MAN-MACHINE INTERFACE

The APR1400 MMI is designed such that the plant can be operated from the MCR under all anticipated operating conditions including design basis accidents. This advanced MMI is designed to meet all regulatory requirements for a nuclear power plant, which include separation, independence, defense-in-depth and diversity requirements for the protection system.

One of the most distinguishing design features of APR1400 MMI is the adoption of compact workstation-based advanced control room compared to the conventional nuclear power plants in the I&C area.

The compact workstation-type operator consoles that can provide convenient working environment to the control room operators facilitate display of plant operating status information for the operators such that the operability and reliability can be enhanced and the human error can be minimized to a great extent.

The compact workstation-based MCR of APR1400 is comprised of the following operating facilities to support the operating staffs for efficient and safe plant operation.

1) Three identical compact workstations

2) A Large Display Panel (LDP) for display of overall plant operational and safety assessment data

3) A safety console which provides controls and information for all class 1E, safety-related components independent of the compact workstations.

Each workstation allows access to all information and controls necessary for one operator to monitor and control all processes associated with nuclear plant operation and safety. Each workstation is comprised of four information FPDs and pointing devices. The information FPDs provide plant operating status information to operator via graphical flow diagrams, alarm displays, plant summary displays, computerized procedures and soft control templates.

The LDP which is located in front of the compact workstation provides convenient and prompt means for the MCR staffs to assess plant operation and safety status. It provides key plant status information including important safety parameters and operating status of essential components, high hierarchy alarms such as critical safety function and success path monitoring alarms. It also provides dynamic display of time dependent variable parameters for the parameter trend monitoring.

A safety console is located on the left of the workstation. Hardwired safety-related fixed position controls are provided on the safety console for manual actuation of the safety systems. Hence the MCR operators can still mitigate the accident and maintain the plant in a safe condition using the safety console even when all three compact workstations fail. Additional information FPDs, operator module FPDs for Plant Protection System (PPS) and PAMI display FPDs are also installed on the safety console.

All of the information display, controls and operator aid functions are centralized into compact workstations in the APR1400 MCR design, and each workstation has access to all plant data and control components. Therefore safety-related system as well as non-safety related system could be operated by software-driven control MMI on the workstation. The APR1400 MMI also provides emergency operation capability on the workstation so that the operator uses the same control MMI on the workstation in a normal as well as in an emergency condition. This design approach reduces the operator error and improves the operability because the operator can use familiar manual controls and operator aid functions such as computerized operating procedure displays in emergency conditions. The APR1400 MMI also provides manual control that can take a licensing credit for safe operation of the plant when a total failure of workstation occurs.

5. DESIGN CONCEPT OF BACKUP CONTROL

In the unlikely event that the MCR should become inaccessible, the operators supervise and control the plant from the remote shutdown room (RSR) to:

- Achieve prompt hot standby of the reactor,
- Maintain the unit in a safe condition during hot shutdown, and
- Achieve cold shutdown of the reactor through the use of suitable procedures.

For safe shutdown, the RSR is equipped with:

- a compact workstation that is the same type as those in the MCR
- shutdown overview display panel that provides the information that the operator need for quick assessment of overall plant status.

Therefore, the operator can access components and variables that are necessary for cold shutdown using the compact workstation.

The MCR workstations and the RSR workstation are placed in separate physical locations, on separate elevations, with separate ventilation systems, and with multiple communication systems between the two. The design includes signal isolation and disabling of all main controls and the transfer of all cold standby controls to the RSR workstation. Therefore, no single credible event that will cause evacuation of the MCR will cause the RSR workstation to be inoperable.

6. HARDWARE DESIGN OF I&C SYSTEM

The APR1400 control system design is based on the use of commercially available programmable equipment. This equipment is industrial-hardened field proven, digital software-based control equipment whose design overcomes the difficulties attributed to traditional component control implementations. Remote multiplexing is incorporated to reduce overall plant wiring by distributing hardwired signal interfaces throughout the plant.

The commercialized and field-proven DCS and PLC are adopted for the control and monitoring system of the APR1400.

The DCS is used for the Process-CCS, IPS and the compact workstation. The control system within the DCS performs the data acquisition and control function of a discrete and continuous loop for non-safety plant process systems (for Process-CCS). The data processing server within the DCS also provides IPS function to perform the plant wide data acquisition, validation of sensed parameters, the execution of NSSS application programs and BOP performance calculations, the monitoring of plant safety and general status, the presentation of status and calculation results on the MMI device, the provision of logs, and the determination of alarm conditions. The DCS workstations are provided as operator interface to control and monitoring components in the MCR and RSR.

The qualified PLC is used for ESF-CCS within the safety boundary. The PLC for the ESF-CCS provides discrete and continuous control of safety-related systems and automatic or manual actuation of Engineered Safety Features Systems for a nuclear power generating station. The PLC for ESF-CCS provides data acquisition of discrete signals indicative of safety-related system and component status and analog signals which represent various monitored and controlled plant parameters to IPS and QIAS-N for data management such as monitoring, display and storage.

The DCS is electrically isolated from the qualified PLC so that no electrical faults in the DCS will propagate to such equipment. Signal interfaces between the DCS and other safety related equipment/systems is implemented via fiber optic data links to assure signal isolation.

7. CONCLUSION

In line with worldwide industry trend and the conceptual design goal and criteria established for APR1400, the control system of APR1400 incorporates advanced digital technologies based on DCS and PLC to enhance plant safety and economy. Major advanced digital technologies implemented in APR1400 are the distributed multi-loop controller, advanced control room design based on compact workstation, redundant and diversity data processing system, and defense-in-depth design to meet the SECY 93-087 requirement. The control systems of the APR1400 are also designed under the following design principle:

- Licensable design of I&C architecture to meet the latest licensing requirements.
- Provision of an effective means of controlling and monitoring safety as well as non-safety systems.

- Reliable and proven design for defense-in-depth against any credible common mode failures of the control system affecting the entire safety system functional performance.
- Simplicity in the design of the system architecture and minimization of the number of interface.
- Using digital equipment and 'off-the-shelf' products.
- Minimization of wide variety of different equipment/products to reduce cost and to simplify the maintenance.

These design approaches taken for the advanced nuclear power plant will enhance safety, operability, reliability, availability, maintainability, constructability, and economy compared to conventional nuclear power plant.

REFERENCES

APR1400 Project Dept. KOPEC (1997). *MMIS Design Basis*

IEEE Std. 603 (1991). *Standard Criteria for Safety Systems for Nuclear Power Generating Stations*

NUREG/CR-6635 (2000) *Soft Control : Technical Basis and Human Factors Review Guidance*

Il-Nam Choe (1998). *Technical and Economic Effect Analysis of Advanced MMIS Implementation in KNGR*

Y.C.Shin (1998) *KNGR MMI Design Approach for Licensing Issue*

Jong-Jae Choi, Moon-Jae Choi, Jae-Youb Byun (2000) *KNGR MMI Development Strategy and Optimized Features, ANS.*

ELSEVIER
IFAC
PUBLICATIONS
www.elsevier.com/locate/ifac

ROBUST DIGITAL CONTROL FOR MULTI-MACHINE POWER SYSTEMS WITH DATA TRANSMISSION DELAYS

Alireza Sanai Sabzevary

Mashhad Institute of Technology
Mashhad P.O. Box. 91735-171, Iran
sanai@pwrs.elec.waseda.ac.jp

Shinichi Iwamoto

Waseda University 3-4-1 Ohkubo,
Shinjuku-Ku Tokyo 169-8555 Japan
iwamoto@mse.waseda.ac.jp

Abstract: In this paper for a multi-machine power system equipped with Unified Power Flow Controller a robust digital control is designed. Data transmission delays are considered during the design. The main source of uncertainty considered is one line outage of double circuit transmission line. Numerical simulations show that the proposed control can stabilize the system under sever fault conditions. *Copyright© 2003 IFAC*

Keywords: power systems, robust control, digital control, data transmission delay

1. INTRODUCTION

Power systems are large interconnected systems in which operating conditions are always varying to satisfy different load conditions. The new deregulated environment for the power industry has triggered new research efforts to model, analyze, simulate and control complex dynamical processes that take place during the electrical power generation, transmission and distribution. The possible interplays among independent power producers, large consumers, co-generators and utility owned generation, require that control systems have the ability of better assessment of all possible control actions through devices already in place in the power system or by the design and installation of new and more flexible controls. However, as power systems are large-scale nonlinear systems, the applications of conventional linear control approaches are limited because they can only deal with small disturbances around an operating point and cannot guarantee the system stability when faults or significant changes of operating conditions occur. Since the introduction of robust control theory various robust control strategies have been proposed to enhance power system stability. Power electronic devices not only improve power system performance by controlling voltage, power flows over the transmission lines and other system variables, but also provide supplementary controls aiming at the improving dynamic and transient stability of power systems. UPFC (Unified Power Flow Controller), is mainly used to control the power transmission line reactance and phase angles, and such power electronic devices have the capability of quickly responding to power system changes and can damp power system oscillations effectively.

On the other hand realization of control schemes in reality encounters delays both in states and control inputs. Data transmission delays are one of the sources of instability and poor performance of dynamical systems (Malek-Zavareh, 1987). Sources of delays in power systems come from measurements and transformation of information between different parts of the system. Therefore it is necessary to consider time delays during the design otherwise the control may be degraded or the expected performance of the overall system may not be obtained.

The problem of robust stabilization of power systems has been studied by many researchers. But, there has been few works on the robust digital control design with delay consideration which can guarantee some performance level. In this paper the authors consider the robust digital control of multi-machine power systems with delays, for a 4-machine power system that is equipped with UPFC. The main source of uncertainty considered is one line outage of double circuit transmission lines. Through numerical simulations it is shown that the proposed control scheme can stabilize the power system under different operating conditions and for different time delay values.

2. POWER SYSTEM MODEL

A four machine 11-bus system model shown in Fig. 1. is considered for numerical simulations (Kunitomo, 1998). The system is equipped with UPFC, which is installed at bus 4 and is influenced by the terminal voltage of generator G1.

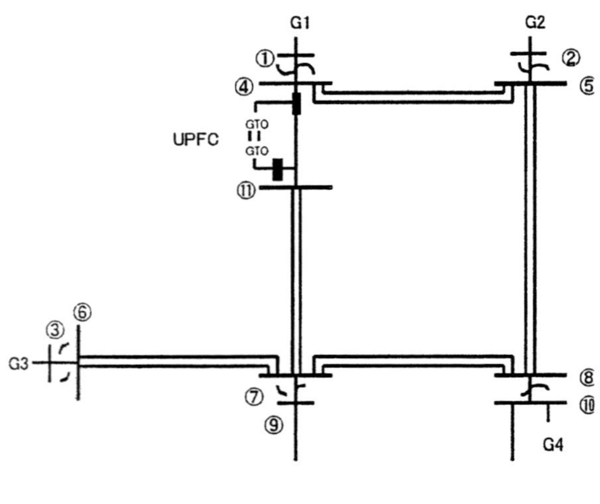

Fig.1. Four machine power system with UPFC

Fig. 2 shows the one line diagram of the UPFC. The main role of UPFC is controlling the reactance and power flow of the system, which in turn increases the damping of system oscillations and may increase the

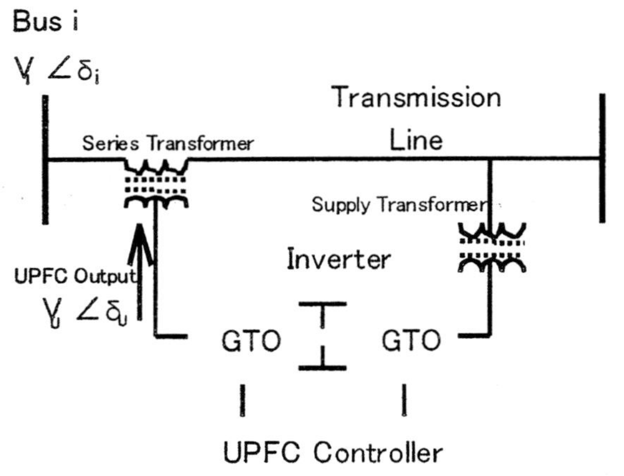

Fig.2. One line diagram of UPFC

Transmission line capabilities and overall system stability margins.

The place where UPFC is located is shown in Fig. 1. Each generator has one excitation control input. The nonlinear differential equations describing the dynamics of the ith. machine are given as follows (see Appendix for nomanclatures)

$$\dot{\delta}_{(i)} = \omega_0 (\omega_{(i)} - 1) \tag{1}$$

$$\dot{\omega}_{(i)} = \frac{P_{m(i)} - P_{e(i)} - D_{(i)}(\omega_{(i)} - 1)}{M_{(i)}} \tag{2}$$

$$\dot{E}'_{q(i)} = \frac{E_{fd(i)} - E'_{q(i)} - (X_{d(i)} - X'_{d(i)})I_{d(i)}}{\tau'_{do(i)}} \tag{3}$$

$$\dot{E}_{fd(i)} = -\frac{E_{fd(i)} - E_{fd0(i)} + K_{a(i)}(V_{t(i)} - V_{t0(i)})}{\tau_{a(i)}} \tag{4}$$

$$V_{d(i)} = X_{q(i)} I_{q(i)} \tag{5}$$

$$V_{q(i)} = E'_{q(i)} - X'_{d(i)} I_{d(i)} \tag{6}$$

$$V_{t(i)} = \sqrt{V_{d(i)}^2 + V_{q(i)}^2} \tag{7}$$

$$P_{e(i)} = V_{d(i)} I_{d(i)} + V_{q(i)} I_{q(i)} \tag{8}$$

They are linearized at a given operating point. Then linearized power system states and their compact form are written as below

$$\dot{x}(t) = A_0\, x(t) + B_0\, u(t)$$

$$x(t) = [\Delta\delta_{(1)} \ \Delta\omega_{(1)} \ \Delta E'_{q(1)} \ \Delta E_{fd(1)} \ \ldots\ldots \ \Delta\delta_{(n)} \ \Delta\omega_{(n)} \ \Delta E'_{q(n)} \ \Delta E_{fd(n)}]^T$$

$$u(t) = [\Delta I_{dU} \ \ \Delta I_{qU} \ \ \Delta\delta_U]^T$$

$$A_0 = \begin{bmatrix} 0 & \omega_0 & 0 & 0 & \cdots & 0 & 0 & 0 & 0 \\ -\frac{K_{1(1,1)}}{M_{(1)}} & -\frac{D_{(1)}}{M_{(1)}} & -\frac{K_{2(1,1)}}{M_{(1)}} & 0 & \cdots & -\frac{K_{1(1,n)}}{M_{(1)}} & -\frac{D_{(1)}}{M_{(1)}} & -\frac{K_{2(1,n)}}{M_{(1)}} & 0 \\ -\frac{K_{3(1,1)}}{\tau'_{d0(1)}} & 0 & -\frac{1}{\tau'_{d0(1)}K_{4(1,1)}} & \frac{1}{\tau'_{d0(1)}} & \cdots & -\frac{K_{3(1,n)}}{\tau'_{d0(1)}} & 0 & -\frac{1}{\tau'_{d0(1)}K_{4(1,n)}} & 0 \\ -\frac{K_{a(1)}K_{5(1,1)}}{\tau_{a(1)}} & 0 & -\frac{K_{a(1)}K_{6(1,1)}}{\tau_{a(1)}} & -\frac{1}{\tau_{a(1)}} & \cdots & -\frac{K_{a(1)}K_{5(1,n)}}{\tau_{a(1)}} & 0 & -\frac{K_{a(1)}K_{6(1,n)}}{\tau_{a(1)}} & 0 \\ \vdots & \vdots & \vdots & \vdots & \ddots & \vdots & \vdots & \vdots & \vdots \\ 0 & 0 & 0 & 0 & \cdots & 0 & \omega_0 & 0 & 0 \\ -\frac{K_{1(n,1)}}{M_{(n)}} & -\frac{D_{(n)}}{M_{(n)}} & -\frac{K_{2(n,1)}}{M_{(n)}} & 0 & \cdots & -\frac{K_{1(n,n)}}{M_{(n)}} & -\frac{D_{(n)}}{M_{(n)}} & -\frac{K_{2(n,n)}}{M_{(n)}} & 0 \\ -\frac{K_{3(n,1)}}{\tau'_{d0(n)}} & 0 & -\frac{1}{\tau'_{d0(n)}K_{4(n,1)}} & 0 & \cdots & -\frac{K_{3(n,n)}}{\tau'_{d0(n)}} & 0 & -\frac{1}{\tau'_{d0(n)}K_{4(n,n)}} & \frac{1}{\tau'_{d0(n)}} \\ -\frac{K_{a(n)}K_{5(n,1)}}{\tau_{a(n)}} & 0 & -\frac{K_{a(n)}K_{6(n,1)}}{\tau_{a(n)}} & 0 & \cdots & -\frac{K_{a(n)}K_{5(n,n)}}{\tau_{a(n)}} & 0 & -\frac{K_{a(n)}K_{6(n,n)}}{\tau_{a(n)}} & -\frac{1}{\tau_{a(n)}} \end{bmatrix}$$

$$B_0 = \begin{bmatrix} 0 & -\frac{K_{7(1)}}{M_{(1)}} & -\frac{K_{8(1)}}{\tau'_{do(1)}} & -\frac{K_{a(1)}K_{9(1)}}{\tau_{a(1)}} \cdots\cdots & 0 & -\frac{K_{7(n)}}{M_{(n)}} & -\frac{K_{8(n)}}{\tau'_{do(n)}} & -\frac{K_{a(n)}K_{9(n)}}{\tau_{a(n)}} \\ 0 & -\frac{K_{10(1)}}{M_{(1)}} & -\frac{K_{11(1)}}{\tau'_{do(1)}} & -\frac{K_{a(1)}K_{12(1)}}{\tau_{a(1)}} \cdots\cdots & 0 & -\frac{K_{10(n)}}{M_{(n)}} & -\frac{K_{11(n)}}{\tau'_{do(n)}} & -\frac{K_{a(n)}K_{12(n)}}{\tau_{a(n)}} \\ 0 & -\frac{K_{13(1)}}{M_{(1)}} & 0 & -\frac{K_{a(1)}K_{14(1)}}{\tau_{a(1)}} \cdots\cdots & 0 & -\frac{K_{13(n)}}{M_{(n)}} & 0 & -\frac{K_{a(n)}K_{14(n)}}{\tau_{a(n)}} \end{bmatrix}^T$$

Initial values of generators are given in Table 1.

Table 1.: Generators initial values

	G1	G2	G3	G4
δ [deg]	31.11	18.18	-10.21	0
ω [pu]	1	1	1	1
E'_q [pu]	1.691	1.153	1.286	1.105
E_{fd} [pu]	6.678	3.587	3.940	3.834
P_e [pu]	4.000	2.000	2.000	2.130
V_t [pu]	1.050	1.050	1.050	1.025
I_t [pu]	3.825	1.927	2.030	2.097

3. ROBUST DIGITAL CONTROL WITH DELAY

In this paper we apply the results of the theory developed by M. Mahmoud, (Mahmoud, 2000). The uncertain system considered with time delays is represented as follows

$$x(k+1) = [A_0 + \Delta A(k)]x(k) + [B_0 + \Delta B]u(k) + D_0 x(k-\tau) = A_\Delta(k)x(k) + B_\Delta(k)u(k) + D_0 x(k-\tau) \quad \text{......... ......(9)}$$

Where $x(t) \in R^n$ is the state, $u(t) \in R^m$ is the control input. A_0, B_0, and D_0 are the system nominal matrices. τ is an unknown constant integer representing the number of delay steps in the state. It is assumed that $0 \le \tau \le \tau^*$ where τ^* being known. $\Delta A(k)$ and $\Delta B(k)$ are time varying parametric uncertainty matrices for A and B respectively. D_0 is the delay matrix which can be determined by measurement devices. The matrices $\Delta A(k)$ and $\Delta B(k)$ have the following form.

$$[\Delta A(k) \quad \Delta B(k)] = L\Delta(k)[M_1 \quad M_2] \quad \text{............(10)}$$

Where L, M_1 and M_2 are known constant matrices and $\Delta(k)$ is an unknown matrix satisfying

$$\Delta^T(k)\Delta(k) \le I \qquad \forall k \qquad \text{.................(11)}$$

the cost function for the uncertain system (1) is

$$J = \sum_{k=0}^{\infty} [x^T k Q x(k) + u^T(k) R u(k)] \quad \text{............(12)}$$

where $Q = Q^T > 0$ and $R = R^T > 0$ are state and control weighting matrices.

To design the robust digital power system controller the parameter dependent Algebraic Riccati Equation (13) should be solved for a positive definite solution $X = X^T > 0$

$$A_0^T X A_0 - X + \eta^{-1} M_1^T M_1 + Q + W - (A_0^T \bar{X} B_0 + \eta^{-1} M_1^T M_2) \times (\hat{R} + B_0^T \bar{X} B_0)^{-1} \times (A_0^T \bar{X} B_0 + \eta^{-1} M_1^T M_2)^T = 0 \quad \text{.....(13)}$$

Where

$$\Lambda\Lambda^T = (D_0 W^{-1} D_0^T + \mu LL^T)$$
$$\bar{X} = (X^{-1} - \Lambda\Lambda^T)^{-1} \qquad \text{.........(14)}$$
$$\hat{R} = R + \mu^{-1} M_2^T M_2$$

η is a constant and W is a weighting matrix. Then the state feedback controller

$$u(k) = K_s x(k)$$

where

$$K_s = -(\hat{R} + B_0^T X B_0)^{-1} (A_0^T \bar{X} B_0 + \eta^{-1} M_1^T M_2)^T \qquad \text{........(15)}$$

will stabilize the power system with a quadratic guaranteed cost control. To design the proposed power system control first we rewrite the eq. (5) in the form

$$X = f(X, Q, W, R)$$

$$f(X_j, Q, W, R) = A_0^T X A_0 - X + \eta^{-1} M_1^T M_1 + Q + W - (A_0^T X B_0 + \eta^{-1} M_1^T M_2) \times (\hat{R} + B_0^T X B_0)^{-1} \times (A_0^T X B_0 + \eta^{-1} M_1^T M_2)^T \qquad \text{........(16)}$$

then, the following algorithm is followed

Algorithm:

Step1: Read in the nominal matrices A_0, B_0, D_0, L, M_1, and M_2.

Step2: Select the weighting matrices Q, R, W and choose a value for delay and let ε be a tolerance level constant.

Step3: Select initial value for X_1 for the Riccati matrix X and set the iteration index j=1

Step4: Compute $X_j^+ = f(X_j, Q, W, R)$

Step5: Update the Riccati matrix X using the following rule

$$X_{j+1} = (1-\alpha) X_j^+ + \alpha X_j \quad 0 \le \alpha \le 1$$

Step6: If $\left\| X_{j+1} - X_j \right\| \le \varepsilon$, then record X_{j+1} and go to step 7 otherwise change α, set $X_j = X_{j+1}$ and j=j+1 then go to step 4.

Step7: Compute the upper bound of cost function

$$J_0 = x_0^T P x_0 + \sum_{\alpha=-\tau}^{0} x^T(\alpha) W x(\alpha) \qquad \text{..................(17)}$$

Where P and W are symmetric matrices satisfying the following Linear Matrix Inequality (LMI)

$$\begin{bmatrix} A_\Delta^T(k) P A_\Delta(k) - P + W & A_\Delta^T(k) P D_0 \\ D_0^T(k) P A_\Delta(k) & -(W - D_0^T P D_0) \end{bmatrix} < 0 \qquad \text{.........(18)}$$

or equivalently satisfy the following Algebraic Riccati Inequality (ARI):

$$A_\Delta^T(k) P A_\Delta(k) - P + A_\Delta^T(k) P D_0 (W - D_0^T P D_0)^{-1} \times D_0^T(k) P A_\Delta(k) + W < 0$$
$$\forall \Delta : \Delta^T(k)\Delta(k) \le I \quad \forall k \qquad \text{......(19)}$$

and $W - D_0^T P D_0 > 0$

if necessary increase the delay bound and W and go to step 3. For more details see [3].

4. SIMULATIONS

For the power system under study a controller is designed which is robust against parametric variations of the system with delay. The main source of uncertainty considered is one line outage of double circuit transmission line between buses No.7 and No. 11. The nominal system matrix A_0 is calculated before and after fault. Then ΔA is put equal to the difference between these two matrices. It is considered that there is no uncertainty in input matrix, i.e. $\Delta B = 0$. Our purpose is to design a robust power system stabilizer that can guarantee the stability of the power system for different possible operating conditions and different values of delays. The calculations can be carried out off line and practically the best possible controller parameters with delay considerations are selected. Since the controller is designed for most sever fault conditions numerical simulations show that it is robust against uncertainties such as one circuit line opening. The iteration procedure was tested for different values of initial guess and it was found that the best initial guess for X is first solving the equation (16) without considering delay and uncertainty. The design procedure was followed, with the following weighting matrices and with a starting value such as $\alpha = 0.5$. After three iterations the parametric ARE, equation (16) converged to a desired solution. For the simulation it is assumed that the data transmission delay is 30 ms for all the δ's. Then the matrix D_0, has zero elements except ones related to δ, i.e.

D0(1,2)=0.1 D0(5,6)=0.1

D0(9,10)=0.1 D0(13,14)=0.1

Q=diag [2 2 2 2 1 1 1 1 1 1 1 2 2 2 2 2 2]

W=100diag[1 1 1 1 1 1 1 1 1 1 2 2 2 2 2 2]

R=diag([2 1 1])

$\eta = 57$

$\alpha = 0.86$

$\tau = 3$ (number of delay steps)

with these values the performance index was calculated as J=50.

The system behavior with the proposed control when a 100ms fault occurs at the (7-11) transmission line plus one circuit line opening of line (4-5) is shown in Fig.3. It is clear from simulation results that the proposed method can enhance the performance and power system stability.

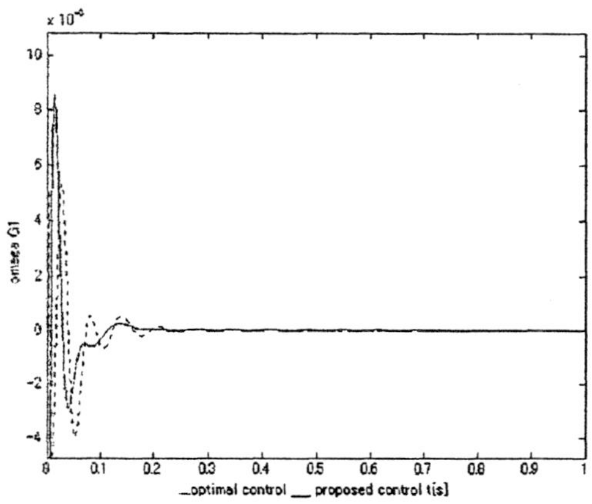

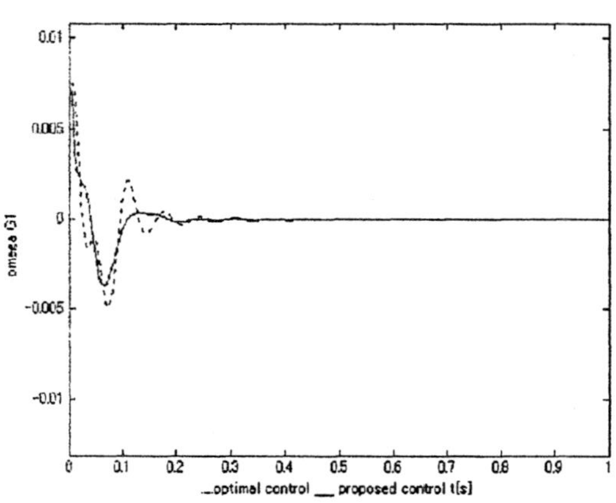

Fig. 3: Comparison of power system with proposed robust digital control and optimal control

5. CONCLUSIONS

Delays in states and control inputs occur because of data measuring, transmission, etc., and thus the effect of these delays should be considered during the design. In this paper a new robust digital power system control which guarantees a performance index with delay considerations in states and input is applied to a power system equipped with UPFC. Simulation results show that the control design method is straight- forward and have shown that the proposed control, which is digital and robust to parametric variation of the system, can stabilize a multi machine power system.

6. REFERENCES

Malek-Zavrei, M. and Jamshidi, M., (1987). *Time Delay Systems: Analysis, Optimization and Applications*, Amsterdam, North Holland.

Kunitomo, O., H. Satomura, S. Iwamoto, "Robust stabilization of power systems with FACTS", Proceedings of 9th. *Annual Conference of Power & Energy Society IEE of Japan August 1998.*

Mahmoud, M., S., Lihua Xie, "Guaraned cost control of uncertain discrete systems with delays", *Int. J. Control, 2000,* **Vol. 73, No. 2**, 105-114.

ELSEVIER
IFAC
PUBLICATIONS
www.elsevier.com/locate/ifac

A NASH SOLUTION TO PREDICTIVE CONTROL PROBLEM FOR POWER SYSTEMS

Choon Ki Ahn and Wook Hyun Kwon

School of Electrical Engineering and Computer Science,
Seoul National University, Seoul, Korea, 151-742
Email: {hironaka, whkwon}@cisl.snu.ac.kr

Abstract: In this paper, we provide a Nash solution to predictive control problem for power systems consisting of a hydraulic turbine and a synchronous generator. In contrast to the existing control method for power systems, our design scheme is based on Nash predictive control idea in the feedback linearization framework. The proposed control law will be shown to optimally and exponentially regulate the terminal voltage, the rotor angle, and the rotor speed to the reference values. This control scheme can be extended to a generally formulated class of nonlinear systems. *Copyright © 2003 IFAC*

Keywords: power system, voltage regulation, Nash solution, predictive control, nonlinear control

1. INTRODUCTION

The past couple of decades has witnessed a tremendous amount of research in development of efficient control algorithm for power systems. Most of the existing control design in power systems are based on linearized model such as IEEE type I control scheme (Kundur, P., 1994). Linear controllers based on classical control algorithms can be used effectively to provide necessary damping and insure asymptotic stability of the equilibrium under a small perturbation. Nonlinear control theory based on differential geometric concept (Isidori, 1989) has provided powerful tools for systematic synthesis of control for power systems. The feedback linearization techniques were fist introduced in the power system applications in (Marino, 1984) and later (Champman and Kaufman, 1993; Mielczarski and Zajaczkowski, 1994). With the introduction of this design tool, the stability and the damping performance of the power system was enhanced. These works considered the reduced 3rd model of the synchronous machine without internal dynamics. In [4], the switching scheme based on the feedback linearization technique was used to enhance the stability and to achieve a good voltage regulation result. But, this scheme assumes that the mechanical power is constant. Also, it is unclear how the switching times can be determined. In (Mielczarski and Zajaczkowski, 1994), the authors considered the power systems which consist of the 3rd order model of the machine and the already controlled steam turbine. This work did not address the regulation problem of the terminal voltage.

Model predictive control is now regarded as a widely used control method. In model predictive control, an open-loop finite horizon optimization is solved. The resulting open-loop trajectory is applied to the system for a fraction of the horizon length. This process is then repeated, resulting in a feedback law. The interest of this control scheme partly due to the availability of faster and cheaper computers as well as efficient numerical algorithms for solving optimization problems. However, there are two disadvantages in current nonlinear predictive control algorithms.

First, since the optimization problem to be solved online is a nonconvex optimization problem, the global optimum can not be achieved. Secondly, it is difficult to apply the existing nonlinear model predictive control algorithm to the real-time applications such as mechanical systems and electrical systems due to an online computational burden. In order to avoid an online computational burden, analytic approaches for predictive control problem have been proposed in various articles. In (Chen W. H., 2000), the authors proposed a closed form control law without online optimization. In (P., 1998), a closed form approximate receding horizon controller was constructed using a linear multi-step-ahead prediction. More recently, Chen's recent result (Chen W. H., Ballanche D. J.,and Gawthrop P. J., 2001) considered the nonlinear systems with ill-defined relative degree and presented an analytic output predictive control law.

In this paper, we provide a Nash solution to predictive control problem for a single machine infinite bus power system. In contrast to the existing result on power system control, our novel design scheme introduces a output predictive Nash control idea in the feedback linearization framework. Our design technique guarantees exponential and optimal regulation result of the terminal voltage, the rotor angle, and the rotor speed to the predefined values. The proposed control concept can be applied to a generally formulated class of nonlinear systems which includes the singularity and the nonminimum-phase.

The class of power systems and the control problem considered in this paper are described in Section 2. In Section 3, A main result on Nash predictive control problem for power systems is presented. The conclusion is given in Section 4.

2. PROBLEM FORMULATION

The class of power systems controlled in this paper consists of the synchronous machine, the hydraulic turbine, and the infinite bus. We consider a two-axis model for the synchronous machine. It is driven by an hydraulic turbine which provides the mechanical power.

The synchronous machine considered in this paper is a reduced form of the well-known two-axis model presented in (M. Anderson and A. Foud, 1993; Peter W. Sauer and M. A. Pai, 1998) as follow:

$$\begin{aligned}
T'_{q0}\dot{E}'_d &= -E_d + (X_q - X'_q)I_q \\
T'_{d0}\dot{E}'_q &= -E_q + (X_d - X'_d)I_d + E_{fd} \\
\frac{2H}{w_s}\dot{w} &= P_m - [(X_q - X'_d)I_dI_q + E'_qI_q + E'_dI_d] \\
\dot{\delta} &= w - w_s \\
V_d &= E'_d - R_sI_d + X'_qI_q \\
V_q &= E'_q - R_sI_q + X'_dI_d
\end{aligned} \quad (1)$$

where I_d, I_q, V_d, and V_q are respectively the current and the voltage of the generator in the d-q reference frame. Also, E'_d, E'_q, δ, and w represent respectively the damper windings dynamics and the field-flux voltage dynamics, the relative rotor angle and the speed of the rotor.

In order to provide the mechanical power, we use an hydraulic turbine (O. Akhrif, F.A. Okou, L.A. Dessaint, 1998) represented as

$$\begin{aligned}
\dot{q} &= \frac{1}{T_w}(1 - \frac{q^2}{A_t^2 g^2}) \\
T_g\dot{g} &= -g + u_g \\
P_m &= \frac{q^3}{A_t^2 g^2}
\end{aligned} \quad (2)$$

where P_m, q, and g are respectively the mechanical power, the turbine flow rate and the gate opening.

The infinite bus is given by the following equations (Peter W. Sauer and M. A. Pai, 1998) as

$$\begin{aligned}
V_d &= V_s\sin(\delta - \phi) + R_eI_d - X_eI_q \\
V_q &= V_s\cos(\delta - \phi) + R_eI_q - X'_dI_d
\end{aligned} \quad (3)$$

where R_e, X_e, V_s, and ϕ are respectively the network resistor and admittance, the RMS value and the phase of the infinite bus voltage.

After some manipuations, we can obtain the combined model as

$$\begin{aligned}
\dot{V}_d &= a_1V_d + a_2V_q + a_3V_s\sin(\delta - \phi) \\
&\quad + a_4V_s\cos(\delta - \phi) + a_5V_s(w - w_s)\cos(\delta - \phi) \\
&\quad - a_6V_s(w - w_s)\sin(\delta - \phi) + g_1E_{fd} \\
\dot{V}_q &= b_1V_d + b_2V_q + b_3V_s\sin(\delta - \phi) \\
&\quad + b_4V_s\cos(\delta - \phi) + b_5V_s(w - w_s)\cos(\delta - \phi) \\
&\quad - b_6V_s(w - w_s)\sin(\delta - \phi) + g_bE_{fd} \\
\dot{\delta} &= w - w_s \\
\dot{w} &= \frac{w_s}{2H}[c_2\cos(\gamma + \delta - \frac{\pi}{2})V_d + c_2\sin(\gamma + \delta - \frac{\pi}{2})V_q] \\
&\quad + \frac{w_s}{2H}[\frac{q^3}{A_t^2g^2} - c_1(V_d^2 + V_q^2)] \\
\dot{q} &= \frac{1}{T_w}(1 - \frac{q^2}{A_t^2g^2}) \\
\dot{g} &= \frac{1}{T_g}(-g + u_g)
\end{aligned} \quad (4)$$

where a_i, b_i, c_i, and g_i are constant parameters.

In (4), we use the d-q reference frame representation of the terminal voltage V_d and V_q as state variables since they are more easily accessible than E'_q and E'_d. In order to achieve the regulation result of the voltage and the speed for power systems, we define the output equations as

$$y_1 = V_t^2 - V_{t_r}^2 = V_d^2 + V_q^2 - 1$$

$$y_2 = \delta - \delta_r \quad (5)$$

where δ_r and V_{t_r} are the steady state values of the power angle and the terminal voltage of the generator.

The control objective in this paper is to construct a controller for power systems to optimally regulate the output to the origin with minimization of the following performance indices:

$$J_1(E_{fd}, u_g) = \int_0^{T_1} [y_1^2(t+\tau) + y_2^2(t+\tau) + r_{11}E_{fd}^2(t) + r_{12}u_g^2(t)]d\tau \quad (6)$$

$$J_2(E_{fd}, u_g) = \int_0^{T_2} [y_1^2(t+\tau) + y_2^2(t+\tau) + r_{21}E_{fd}^2(t) + r_{22}u_g^2(t)]d\tau \quad (7)$$

where T_1, T_2 are the predictive times for costs J_1, J_2 and $y(t+\tau)$ denotes the output prediction in the predictive period. Also, r_{11}, r_{12}, r_{21}, r_{22} are weighting constants. It is note that each input has its own cost. In order to obtain a Nash solution for (4) with regard to J_1 and J_2, we have to consider the following minimization problem:

$$J_1(E_{fd}^*, u_g^*) \leq J_1(E_{fd}, u_g^*) \quad (8)$$

$$J_2(E_{fd}^*, u_g^*) \leq J_2(E_{fd}^*, u_g) \quad (9)$$

where E_{fd}^*, u_g^* are the optimal inputs to minimize the costs J_1, J_2.

3. NASH PREDICTIVE CONTROL FOR POWER SYSTEMS

In order to obtain a Nash solution to predictive control problem for power systems in the feedback linearization framework, we have to check the relative degree of (4). Differentiating each output yields

$$\dot{y}_1 = F_1 + G_{11}E_{fd} \quad (10)$$

$$\ddot{y}_2 = F_2 + G_{21}E_{fd} + G_{22}u_g \quad (11)$$

$$(12)$$

where

$$F_1 = 2V_d[a_1V_d + a_2V_q + a_3V_s\sin(\delta-\phi) + a_4V_s\cos(\delta-\phi) + a_5V_s(w-w_s)\cos(\delta-\phi) - a_6V_s(w-w_s)\sin(\delta-\phi)] + 2V_q[b_1V_d + b_2V_q + b_3V_s\sin(\delta-\phi) + b_4V_s\cos(\delta-\phi) + b_5V_s(w-w_s)\cos(\delta-\phi) - b_6V_s(w-w_s)\sin(\delta-\phi)]$$

$$G_{11} = 2(g_2V_q + g_1V_d)$$

$$F_2 = \frac{w_s}{2H}[\frac{3q^2}{A_t^2g^2T_w}(1-\frac{q^2}{A_t^2g^2}) + \frac{2q^2}{A_t^2g^2T_g} - \frac{R_e}{X_e^2+R_e^2}]F_1 + \frac{V_s}{\cos(\gamma)}[\cos(\gamma+\delta-\frac{\pi}{2})\dot{V}_dv - V_d(w-w_s)\sin(\gamma+\delta-\frac{\pi}{2}) - \cos(\gamma+\delta-\frac{\pi}{2})\dot{V}_d - V_d(w-w_s)\sin(\gamma+\delta-\frac{\pi}{2})]$$

$$G_{21} = \frac{w_sR_e}{2H(X_e^2+R_e^2)}[\frac{V_s}{\cos(\gamma)}g_1\cos(\gamma+\delta-\frac{\pi}{2}) + \frac{V_s}{\cos(\gamma)}g_2\sin(\gamma+\delta-\frac{\pi}{2}) - G_{11}]$$

$$G_{22} = -\frac{w_sq^3}{HA_t^2g^3T_g} \quad (13)$$

Therefore, the relative degree of (4) is 4.

With the introduction of the following diffeomorphism:

$$z = \begin{bmatrix} z_1 \\ z_2 \\ z_3 \\ z_4 \\ z_5 \\ z_6 \end{bmatrix} = \begin{bmatrix} V_d^2 + V_q^2 - 1 \\ \delta - \delta_r \\ w - w_s \\ \dot{w} \\ g_2V_d - g_1V_q \\ q_2 \end{bmatrix} \quad (14)$$

where we let $z_5 = \eta_1$ and $z_6 = \eta_2$, we can obtain the transformed system represented by

$$\begin{aligned} \dot{z}_1 &= F_1 + G_{11}E_{fd} \\ \dot{z}_2 &= z_3 \\ \dot{z}_3 &= z_4 \\ \dot{z}_4 &= F_2 + G_{21}E_{fd} + G_{22}u_g \\ \dot{z}_5 &= \dot{\eta}_1 = \beta_1(z) \\ \dot{z}_6 &= \dot{\eta}_2 = \beta_2(z) \end{aligned} \quad (15)$$

where

$$\beta_1 = (a_1g_2 - b_1g_1)V_d + (a_2g_2 - b_2g_1)V_q + (a_3g_2 - b_3g_1)V_s\sin(\delta-\phi) + (a_4g_2 - b_4g_1)V_s\sin(\delta-\phi) + (a_5g_2 - b_5g_1)V_s(w-w_s)\cos(\delta-\phi) - (a_6g_2 - b_6g_1)V_s(w-w_s)\cos(\delta-\phi)$$

$$\beta_2 = \frac{2}{T_w}[\sqrt{\eta_2} - \frac{2H}{w_s}z_4 - c_1(z_1+1) + c_2\cos(\gamma+\delta-\frac{\pi}{2})V_d + c_2\sin(\gamma+\delta-\frac{\pi}{2})V_q] \quad (16)$$

If we let the feedback linearizing inputs as

$$E_{fd} = \frac{v_1 - F_1}{G_{11}} \quad (17)$$

$$u_g = \frac{v_2 - (F_2 + G_{21}E_{fd})}{G_{22}} \quad (18)$$

then we have

$$\begin{aligned} \dot{z}_1 &= v_1 \\ \dot{z}_2 &= z_3 \\ \dot{z}_3 &= z_4 \\ \dot{z}_4 &= v_2 \\ \dot{z}_5 &= \dot{\eta}_1 = \beta_1(z) \\ \dot{z}_6 &= \dot{\eta}_2 = \beta_2(z) \end{aligned} \quad (19)$$

In this case, the internal dynamics can be expanded as follows:

$$\dot{\eta} = \beta(z) = A_z\bar{z} + A_\eta\underline{z} + \begin{bmatrix} \alpha_1(z) \\ \alpha_2(z) \end{bmatrix} \quad (20)$$

where

$$A_z = \frac{\partial \beta(z)}{\partial \overline{z}}|_{\overline{z}=\underline{z}=0}, \qquad A_\eta = \frac{\partial \beta(z)}{\partial \underline{z}}|_{\overline{z}=\underline{z}=0} \quad (21)$$

$$\beta(z) = \begin{bmatrix} \beta_1 \\ \beta_2 \end{bmatrix}, \qquad \eta = \begin{bmatrix} \eta_1 \\ \eta_2 \end{bmatrix} \quad (22)$$

$$\overline{z} = \begin{bmatrix} z_1 \\ z_2 \\ z_3 \\ z_4 \end{bmatrix}, \qquad \underline{z} = \begin{bmatrix} z_5 \\ z_6 \end{bmatrix} \quad (23)$$

and α_1 and α_2 are the remainder terms. Then, we have

$$\dot{z} = Az + Bv + \alpha \quad (24)$$

where

$$A = \left[\begin{array}{cccc|cc} 0 & 0 & 0 & 0 & 0 & 0 \\ 0 & 0 & 1 & 0 & 0 & 0 \\ 0 & 0 & 0 & 1 & 0 & 0 \\ 0 & 0 & 0 & 0 & 0 & 0 \\ \hline \multicolumn{4}{c|}{A_z} & \multicolumn{2}{c}{A_\eta} \end{array}\right], \qquad B = \begin{bmatrix} 1 & 0 \\ 0 & 0 \\ 0 & 0 \\ 0 & 1 \\ 0 & 0 \\ 0 & 0 \end{bmatrix} \quad (25)$$

$$\alpha = \begin{bmatrix} 0 \cdots 0 \ \alpha_1 \ \alpha_2 \end{bmatrix}^T, \qquad v = \begin{bmatrix} v_1 \\ v_2 \end{bmatrix} \quad (26)$$

Let the new control inputs v_1 and v_2 as

$$\begin{bmatrix} v_1 \\ v_2 \end{bmatrix} = -K \begin{bmatrix} z_1 \\ \vdots \\ z_6 \end{bmatrix} + \begin{bmatrix} v_{11} \\ v_{21} \end{bmatrix}$$

$$\triangleq -\left[\begin{array}{c|c} K_1 & K_2 \\ \hline K_3 & K_4 \end{array}\right] \begin{bmatrix} z_1 \\ \vdots \\ z_6 \end{bmatrix} + \begin{bmatrix} v_{11} \\ v_{21} \end{bmatrix} \quad (27)$$

$$= \begin{bmatrix} -K_1\overline{z} - K_2\underline{z} + v_{11} \\ -K_3\overline{z} - K_4\underline{z} + v_{12} \end{bmatrix} \quad (28)$$

then

$$\dot{z} = A_c z + B \begin{bmatrix} v_{11} \\ v_{21} \end{bmatrix} + \alpha \quad (29)$$

where $A_c = A - BK$ is Hurwitz matrix. In this case, A_c is a stability matrix and there exists a positive definite symmetric matrix P such that

$$PA_c + A_c^T P = -Q_c \quad (30)$$

where Q_c is a positive symmetric matrix.

In receding short horizon time frame, the future output is predicted by Taylor series expansion. If the predictive time is selected to be sufficiently small, the very accurate output prediction can be obtained by the approximation of Taylor series. For short time horizon T_1 and T_2, the first predicted output is given by

$$y_1(t+\tau) = y_1(t) + \tau[F_1 + G_{11}E_{fd}] \quad (31)$$

where $\tau \in [0, T_1]$ or $[0, T_2]$. Also, the second predicted output can be represented by

$$y_2(t+\tau) = y_2(t) + \tau[w - w_s] + \frac{\tau^2}{2} \cdot y_{21} + \frac{\tau^3}{6}[F_2 + G_{21}E_{fd} + G_{22}u_g] \quad (32)$$

where

$$y_{21} = \frac{w_s}{2H}[c_2 \cos(\gamma + \delta - \frac{\pi}{2})V_d + c_2 \sin(\gamma + \delta - \frac{\pi}{2})V_q] + \frac{w_s}{2H}\left[\frac{q^3}{A_t^2 g^2} - c_1(V_d^2 + V_q^2)\right] \quad (33)$$

For the first cost (6), we consider the minimization problem of J_1 with the fixed control input u_g^*. By the predicted outputs, (6) is given by

$$J_1(E_{fd}, u_g^*) = J_{11}E_{fd}^2 + J_{12}E_{fd} + \ldots \quad (34)$$

where

$$J_{11} = r_{11}T_1 + \frac{1}{3}T_1^3 G_{11}^2 + \frac{1}{252}T_1^7 G_{21}^2 \quad (35)$$

$$J_{12} = T_1^2 y_1 G_{11} + \frac{2}{3}T_1^3 F_1 G_{11} + \frac{1}{12}T_1^4 y_2 G_{21} + \frac{1}{15}T_1^5(w - w_s)G_{21} + \frac{1}{36}T_1^6 y_{21} G_{21} + \frac{1}{126}T_1^7(G_{21}F_2 + G_{21}G_{22}u_g^*) \quad (36)$$

From (10), if we define

$$E_{fd} = \frac{v_{11} - F_1 - K_1\overline{z} - K_2\underline{z}}{G_{11}} \triangleq \frac{v_{11} - F_{1d}}{G_{11}} \quad (37)$$

then we have

$$J_1(E_{fd}, u_g^*) = \frac{J_{11}}{G_{11}^2}v_{11}^2 + \left(\frac{J_{12}}{G_{11}} - \frac{2J_{11}F_{1d}}{G_{11}^2}\right)v_{11} + \ldots \quad (38)$$

This is a quadratic minimization problem where there exists an explicit solution. Therefore, the first Nash equilibrium can be constructed by solving the minimization problem of the cost J_1 as

$$v_{11}^* = \frac{2J_{11}F_{1d} - J_{12}G_{11}}{2J_{11}} \quad (39)$$

For the second cost (7), the minimization problem of J_2 with the fixed control input E_{fd}^* is also considered. From the previous argument, we can obtain the following representation of cost J_2 as

$$J_2(E_{fd}^*, u_g) = J_{21}u_g^2 + J_{22}u_g + \ldots \quad (40)$$

where

$$J_{21} = r_{12}T_2 + \frac{1}{252}T_2^7 G_{22}^2 \quad (41)$$

$$J_{22} = \frac{1}{12}T_2^4 y_2 G_{22} + \frac{1}{15}T_2^5(w - w_s)G_{22} + \frac{1}{36}T_2^6 y_{21}G_{22} + \frac{1}{126}T_2^7(G_{22}F_2 + G_{22}G_{21}F_{fd}^*) \quad (42)$$

From (11), if we define

$$u_g = \frac{v_{21} - (F_2 + G_{21}E_{fd}) - K_3\overline{z} - K_4\underline{z}}{G_{22}} \triangleq \frac{v_{21} - (F_{2d} + G_{21}E_{fd})}{G_{22}} \quad (43)$$

and introduce the new control input as

$$v_{21} = v_{22} + l \quad (44)$$

then we have

$$J_2(E_{fd}^*, u_g) = \frac{J_{21}}{G_{22}^2}v_{22}^2 + \left[\frac{J_{22}}{G_{22}} + \frac{2J_{21}[l - (F_{2d} + G_{21}E_{fd}^*)]}{G_{22}^2}\right]v_{22} + ... \tag{45}$$

Also, the minimization problem of this cost can be considered as the quadratic minimization problem of the cost J_2. Therefore, we can obtain the second Nash equilibrium by solving the minimization problem of the cost J_2 as

$$v_{22}^* = -[2J_{21}\{lG_{11} - F_{2d}G_{11} - G_{21}(v_{11}^* - F_{1d})\} + J_{22}G_{22}G_{11}]/[2J_{21}G_{11}] \tag{46}$$

In order to have the notational simplification, we define

$$J_{12} = J_{12d} + \frac{T_1^7}{126}G_{21}G_{22}u_g^* \tag{47}$$

$$J_{22} = J_{22d} + \frac{T_2^7}{126}G_{21}G_{22}E_{fd}^* \tag{48}$$

$$v_{1d1} = 2J_{11}F_{1d} - J_{12}G_{11} + \frac{T_1^7}{126}G_{21}G_{11}F_{2d} - \frac{T_1^7}{126}G_{21}^2F_{1d} \tag{49}$$

$$v_{1d2} = 2J_{11} - \frac{T_1^7}{126}G_{21}^2 \tag{50}$$

$$v_{2d1} = 2J_{21}(-F_{2d}G_{11} + G_{21}F_{1d}) + G_{22}(J_{22d}G_{11} - \frac{T_2^7}{126}G_{21}G_{22}F_{1d}) \tag{51}$$

$$v_{2d2} = 2J_{21}G_{21} - G_{22}^2G_{21}\frac{T_2^7}{126} \tag{52}$$

With these notations, a Nash solution is given by

$$v_{11}^* = \frac{252J_{21}v_{1d1} - T_1^7G_{21}(v_{2d1} + 2J_{21}lG_{11})}{252J_{21}v_{1d2} + T_1^7G_{21}v_{2d2}} \tag{53}$$

$$v_{22}^* = -[126v_{1d2}v_{2d1} + v_{2d2}(126v_{1d1} + T_1^7G_{21}^2l) + 252v_{1d2}J_{21}lG_{11}]/[252J_{21}v_{1d2}G_{11} - T_1^7G_{11}G_{21}v_{2d2}] \tag{54}$$

If we define

$$v_{11d} = \frac{252J_{21}v_{1d1} - T_1^7G_{21}v_{2d1}}{252J_{21}v_{1d2} + T_1^7G_{21}v_{2d2}} \tag{55}$$

$$v_{11d2} = \frac{-2T_1^7G_{21}J_{21}G_{11}}{252J_{21}v_{1d2} + T_1^7G_{21}v_{2d2}} \tag{56}$$

$$v_{22d} = \frac{-126v_{1d2}v_{2d1} - 126v_{2d2}v_{1d1}}{252J_{21}v_{1d2}G_{11} - T_1^7G_{11}G_{21}v_{2d2}} \tag{57}$$

$$v_{22d2} = \frac{-v_{2d2}T_1^7G_{21}^2 - 252v_{1d2}J_{21}G_{11}}{252J_{21}v_{1d2}G_{11} - T_1^7G_{11}G_{21}v_{2d2}} \tag{58}$$

(59)

then we have a Nash solution with simplified notations as

$$v_{11}^* = v_{11d} + v_{11d2} \cdot l \tag{60}$$

$$v_{22}^* = v_{22d} + v_{22d2} \cdot l \tag{61}$$

After the proposed Nash solution is applied, the system is given by

$$\dot{z} = A_c z + v_{d2}l + v_d \tag{62}$$

where

$$v_{d2} = \begin{bmatrix} v_{11d2} \\ 0 \\ 0 \\ v_{22d2} \\ 0 \\ 0 \end{bmatrix}, \quad v_d = \begin{bmatrix} v_{11d} \\ 0 \\ 0 \\ v_{22d} \\ \alpha_1 \\ \alpha_1 \end{bmatrix} \tag{63}$$

With the previous argument, we are ready to state the main theorem of this paper.

Theorem 1. *If we use the proposed Nash solution such as*

$$E_{fd}^* = \frac{v_1^* - F_1}{G_{11}} \tag{64}$$

$$u_g^* = \frac{v_2^* - (F_2 + G_{21}E_{fd}^*)}{G_{22}} \tag{65}$$

$$\begin{bmatrix} v_1^* \\ v_2^* \end{bmatrix} = -K\begin{bmatrix} z_1 \\ \vdots \\ z_6 \end{bmatrix} + \begin{bmatrix} v_{11}^* \\ v_{21}^* \end{bmatrix} \tag{66}$$

$$v_{11}^* = v_{11d} + v_{11d2} \cdot l \tag{67}$$

$$v_{22}^* = v_{22d} + v_{22d2} \cdot l \tag{68}$$

$$l = \left[\sum_{i=1}^{6}(P_{i1}v_{11d} + P_{i4}v_{22d}) + \alpha_1\sum_{i=1}^{6}z_iP_{i5} + \alpha_2\sum_{i=1}^{6}z_iP_{i6}\right] \Big/ \left[\sum_{i=1}^{6}z_i(P_{i1}v_{11d2} + P_{i4}v_{22d2})\right] \tag{69}$$

then, we can obtain the following results at once:
(1) obtain an analytic solution to nonlinear predictive control problem
(2) obtain the exact output regulation result

The outline of proof. Consider the following Lyapunov function candidate

$$V = z^{*T}Pz^* \tag{70}$$

then the time derivative of this function is given by

$$\dot{V} = -z^{*T}Q_cz^* + 2z^{*T}Pv_{d2}l + 2z^{*T}Pv_d \tag{71}$$

If we select the design function l as

$$l = \left[\sum_{i=1}^{6}(P_{i1}v_{11d} + P_{i4}v_{22d}) + \alpha_1\sum_{i=1}^{6}z_iP_{i5} + \alpha_2\sum_{i=1}^{6}z_iP_{i6}\right] \Big/ \left[\sum_{i=1}^{6}z_i(P_{i1}v_{11d2} + P_{i4}v_{22d2})\right] \tag{72}$$

then we have

$$\dot{V} = -z^{*T}Q_cz^* \tag{73}$$

Since the following relations are available

$$\begin{aligned}\dot{V} &= -z^{*T}Q_c z^* \leq -\lambda_m(Q_c)||z^*||^2 \\ &= -\frac{\lambda_m(Q_c)}{\lambda_M(P)}||z^*||^2\lambda_M(P) \leq -\frac{\lambda_m(Q_c)}{\lambda_M(P)}z^{*T}Pz^* \\ &= -\frac{\lambda_m(Q_c)}{\lambda_M(P)}V \triangleq -kV \end{aligned} \quad (74)$$

we can obtain

$$V(t) \leq V(t_0)e^{-k(t-t_0)} \quad (75)$$

From the fact of

$$\lambda_m(P)||z^*||^2 \leq V(t) \leq V(t_0)e^{-k(t-t_0)} \quad (76)$$

the exponential convergence of the state to the origin can be obtained as follows:

$$||z^*|| \leq \sqrt{\frac{V(t_0)}{\lambda_m(P)}}e^{-\frac{k}{2}(t-t_0)} \quad (77)$$

By (14), it is clear that

$$V_t^2 \longrightarrow V_{t_r}^2 \quad (78)$$

$$\delta \longrightarrow \delta_r \quad (79)$$

$$w \longrightarrow w_s \quad (80)$$

Therefore, it is guaranteed that the exact regulation of the terminal voltage, the rotor angle, and the rotor speed for power systems to the reference values can be achieved. □

4. CONCLUSION

In this paper, we propose a feedback linearizing MIMO (Multi-Input Multi-Output) nonlinear controller with output predictive Nash control idea for power systems. The class of considered power systems consists of the two axis synchronous machine, the hydraulic turbine, and the infinite bus. Our novel design scheme for power systems has the optimal and exponential regulation property of the terminal voltage, the rotor angle, and the rotor speed to pre-defined values. The proposed control concept can be applied to a generally formulated class of nonlinear systems.

REFERENCES

Champman, J.W., Ilic M.D. King C.A. Eng L. and H. Kaufman (1993). Stablizing a multimachine power system via decentralized feedback linearizing excitation control. *IEEE Trans. on Power Systems.*

Chen W. H., Ballanche D. J., and Gawthrop P. J. (2000). Optimal predictive control of siso nonlinear systems: a predictive control approach. *Proceedings of International conference on Control'00.*

Chen W. H., Ballanche D. J.,and Gawthrop P. J. (2001). Analytic predictive controllers for nonlinear systems with ill-defined relative degree. *IEE Proc.-Control Theory Appl.*

Isidori, A. (1989). *Nonlinear Control Systems.* Berlin: Springer-Verlag.

Kundur, P. (1994). *Power System Stability and Control.* McGraw-Hill Inc.

M. Anderson and A. Foud (1993). *Power System Control and Stability.* IEEE Press.

Marino, R. (1984). An example of a nonlinear regulator. *IEEE Trans. on Autom. Cont.*

Mielczarski, W. and A.M. Zajaczkowski (1994). Multivariable nonlinear controller for a syncronous generator. *Optimal Conrol Applications and Methods* **15**, 49–65.

O. Akhrif, F.A. Okou, L.A. Dessaint (1998). Application of a multivariable feedback linearization scheme for rotor angle stability and voltage regulation of power systems. *IEEE Trans. on Power Systems* **14**, 620–628.

P., Lu (1998). Approximate nonlinear receding horizon control laws in closed form. *International Journal of Control* **71**, 19–34.

Peter W. Sauer and M. A. Pai (1998). *Power System Dynamics and Stability.* Prentice Hall.

ELSEVIER
IFAC
PUBLICATIONS
www.elsevier.com/locate/ifac

A NONCONVEX QUADRATIC SDP FOR FIXED ORDER H_∞ CONTROL AND ITS APPLICATION TO GENERATOR EXCITATION SYSTEM DESIGN

**Seogjoo Kim * Jong-Moon Kim * Soonman Kwon *
Kook-Hun Kim ** Young-Hyun Moon *****

* *Korea Electrotechnology Research Institute(KERI)
Changwon-city, Kyongnam, 641-120 Korea.
{sjkim,jmkim,smkwon} @keri.re.kr*
** *Korea Nuclear I&C System R&D Center
(KNICS R&D Center)
Changwon-city, Kyongnam, 641-120 Korea.
kimkook@keri.re.kr*
*** *Department of Electrical Engineering,
Yonsei University, Seoul, 120-749 Korea.
moon@yonsei.ac.kr*

Abstract: This paper proposes a computation method for fixed order controller synthesis which often arises in H_∞ or robust control. With the motivation that the fixed order control problem can be casted to a non-convex quadratic objective function minimization subject to linear matrix inequality (LMI) constraints, a quadratic semidefinite programming (SDP) problem is solved by primal dual interior point method (IPM). The advantage of the proposed method is two-fold. First, the existing SDP solving method can be used with a slight modification. Second, since the method directly computes the solution of the problem, much faster results can be obtained than the existing successive LMI relaxation method such as cone complementarity linearization algorithm.
In order to demonstrate the algorithm presented, an excitation system for a synchronous generator connected to an infinite bus is designed and simulated. A proportional-integral type automatic voltage regulator and a power system stabilizer considering H_∞ performance are co-designed to provide appropriate damping for the system. The results show the practical applicability of the method.

Keywords: SDP, LMI, interior point method, fixed order H_∞ control, power system.

1. INTRODUCTION

Since the fact was known that SDP could be solved in polynomial time by IPMs (Nesterov and Nemirovsky, 1994), LMIs have been an active research area in the control engineering. Various kinds of control problems in H_∞ or robust control fields (Boyd *et al.*, 1994) can now be numerically and efficiently solved by IPMs (Gahinet *et al.*, 1995), some of which have no known analytic solutions. Though H_∞ control methodology has many theoretically distinguished advantages, the high order of the resulting controller seems to be a big obstacle to widely spread its application in the industry. Two methods can be thought to tackle this high order problem. One is the traditional *design-then-reduce* method and the other is to directly synthesize a required order controller, which is the subject of this paper. In fact, to control engineers, designing the desired order of modern controllers has been a challenging topic since long and is still a typical open problem.

It is well known that the reduced order H_∞ controller synthesis can be described as a rank minimization problem over LMI constraints (Ghaoui *et al.*, 1997), or an LMI feasiblity problem with some algebraic constraints (Apkarian and Tuan, 1999). These LMI problems are non-convex and there is no polynomial time algorithm up to date. Many efforts have been devoted to solving this rank constrained LMI problems including global and local approaches (Iwasaki, 1999; Iwasaki and Skelton, 1995; Ghaoui *et al.*, 1997; Apkarian and Tuan, 1998; Goh *et al.*, 1994). This paper presents another computational method based on IPM. With the motivation that the rank minimization can be reduced to a quadratic objective function optimization subject to LMI constraints, general primal dual path following interior point approach is employed. The so-called Schur complement matrix and line search method in typical IPMs (Wright, 1997) are slightly modified to solve the problem. In order to demonstrate the algorithm presented, an excitation system for a synchronous generator connected to an infinite bus is designed and simulated. A proportional-integral type automatic voltage regulator and a power system stabilizer considering H_∞ performance are co-designed to provide appropriate damping for the system. Some results with MATLAB implementation show the practical applicability of the method.

The paper is organized as follows. In Section 2, we formulate a class of problems that can be represented as a quadratic SDP and then Section 3 presents a solving method of the quadratic SDP. Section 4 deals with the application of the proposed method to a synchronous generator excitation system design and conclusions are made in the last.

The notation here is quite standard. A^T means transpose of a matrix A and $A \succ 0$ (or $\succeq 0$) denotes the matrix A is symmetric and positive definite (positive semidefinite). $A \prec 0$ (or $\preceq 0$) means the matrix $-A$ is positvie definite (positve semidefintie). $A \succeq B$ means $A - B \succeq 0$.

2. PROBLEM FORMULATION

We consider the following standard H_∞ control in the usual state space notation

$$\begin{bmatrix} \dot{x} \\ z \\ y \end{bmatrix} = \left[\begin{array}{c|cc} A & B_1 & B_2 \\ C_1 & D_{11} & D_{12} \\ C_2 & D_{21} & D_{22} \end{array}\right] \begin{bmatrix} x \\ w \\ u \end{bmatrix}, \tag{1}$$

where $x \in R^{n_x}$ is the state vector and $y \in R^{n_y}$, $z \in R^{n_z}$, $w \in R^{n_w}$, $u \in R^{n_u}$ are the measured output, regulated output, exogenous input and control input vectors, respectively. Without loss of generality, $D_{22} = 0$ is assumed. Further, (A, B_2) and (A, C_2) are supposed to be stabilizable and detectible.

Since the reduced order control problem can be easily casted to a static output control problem by system augmentation technique (Gahinet and Apkarian, 1994), we, hereby, only deals with the static output control problems. Later in Section 4, a system augmentation method is illustrated. The existence condition of a static output control $u = Ky$, which stabilizes the system (1) and makes the closed system H_∞ norm $\| T_{zw} \|_\infty < \gamma$, is equivalent to the existence of symmetric matrices X, Y satisfying the following LMIs (Gahinet and Apkarian, 1994)

$$(\bullet)^T \begin{bmatrix} AX + XA^T & XC_1^T & B_1 \\ C_1X & -\gamma I & D_{11} \\ B_1^T & D_{11}^T & -\gamma I \end{bmatrix} (\bullet) \prec 0 \tag{2}$$

$$(\star)^T \begin{bmatrix} A^TY + YA & YB_1 & C_1^T \\ B_1^TY & -\gamma I & D_{11}^T \\ C_1 & D_{11} & -\gamma I \end{bmatrix} (\star) \prec 0 \tag{3}$$

$$\begin{bmatrix} X & I \\ I & Y \end{bmatrix} \succeq 0 \tag{4}$$

$$\text{rank}\left(\begin{bmatrix} X & I \\ I & Y \end{bmatrix}\right) = n_x \quad \Leftrightarrow \quad XY = I, \tag{5}$$

where

$$(\bullet) = \begin{bmatrix} \mathcal{N}_X & 0 \\ 0 & I \end{bmatrix}, \quad (\star) = \begin{bmatrix} \mathcal{N}_Y & 0 \\ 0 & I \end{bmatrix}$$

and $\mathcal{N}_X$ and $\mathcal{N}_Y$ mean bases of the null spaces of (B_2^T, D_{12}^T) and (C_2, D_{21}), respectively. Rank constraint (5) makes the problem non-convex. Instead of solving this problem directly, the following cone complementarity problem

$$\begin{array}{cc} \min\limits_{X,Y} & \text{Trace}(XY) \\ \text{subject to} & \text{LMIs (2), (3) (4)} \end{array} \tag{6}$$

was solved (Ghaoui *et al.*, 1997). It is proved that a static controller exists if and only if the solution to the problem (6) is n_x. Once the optimal symmetric matrices X and Y satisfying (6) are found, the static output feedback gain K for system (1) can be obtained by solving the following LMI feasibility problem

$$\begin{array}{cc} \text{find} & K \\ \text{subject to} & M + PKQ + (PKQ)^T \prec 0, \end{array} \tag{7}$$

where

$$M = \begin{bmatrix} A^TY + YA & YB_1 & C_1^T \\ B_1^TY & -\gamma I & D_{11}^T \\ C_1 & D_{11} & -\gamma I \end{bmatrix}, P = \begin{bmatrix} YB_2 \\ 0 \\ D_{12} \end{bmatrix}$$

$$Q = \begin{bmatrix} C_2 & D_{21} & 0 \end{bmatrix}$$

Another class of problems that can be described by trace minimization over LMIs is LMI feasibility problems subject to algebraic constraints, which arises in robust control synthesis (Apkarian and Tuan, 1999), has the following form

$$\begin{array}{rl} \text{find} & X \succeq 0, Y \succeq 0, F, G, H, R \\ \text{subject to} & \mathcal{L}(X, Y, F, G, H, R) \succeq 0 \\ & F = GR^{-1}H. \end{array} \quad (8)$$

After introducing slack variables Z and T, the condition that the optimal solution to the following problem

$$\begin{array}{rl} \min\limits_{(Z,T,X,Y,F,G,H,R)} & \text{Trace}(ZT) \\ \text{subject to} & \mathcal{L}(X, Y, F, G, H, R) \succeq 0 \\ & \left[\begin{array}{cc|cc} Z_1 & Z_2 & F & G \\ Z_2 & Z_3 & H & R \\ \hline F^T & H^T & I & 0 \\ G^T & R^T & 0 & I \end{array}\right] \succeq 0 \\ & X \succeq 0, Y \succeq 0, T \succeq 0, \\ & F, G, H, R, T_1 \succeq I, \end{array} \quad (9)$$

where

$$Z = \begin{bmatrix} Z_1 & Z_2 \\ Z_2 & Z_3 \end{bmatrix}, \quad T = \begin{bmatrix} T_1 & T_2 \\ T_2 & T_3 \end{bmatrix},$$

is zero is the existence condition of the solution to (8) in the same manner (Henrion *et al.*, 2000). These two kinds of problem can be generally written as,

$$\min_{X,Y,\cdots} \text{Trace}(XY) \quad \text{subject to LMIs} \cdots. \quad (10)$$

3. QUADRATIC SEMIDEFINTE PROGRAMMING

The LMI based trace minimization problem (10) can be casted as a general quadratic semidefinite program, i.e.,

$$\begin{array}{rl} \min\limits_{x} & \frac{1}{2}x^T Qx + c^T x \\ \text{subject to} & \mathcal{A}^*(x) - A_0 \succeq 0 \end{array} \quad (11)$$

where

$$\mathcal{A}^*(x) = \sum_{i=1}^{m} x_i A_i$$

and $x \in R^m$ is the decision variable, symmetric matrices $A_i \in R^{n \times n}, i = 1, ..m$, $Q \in R^{m \times m}$, $c \in R^m$ are the given data variables. If the matrix Q is positive semidefinite, the problem above becomes convex and can be easily solved. However, the matrix Q here is indefinite since the diagonal terms are all zero and the vector c is zero. Introducing the logarithmic barrier function and slack variable $X \succeq 0$ leads to

$$\begin{array}{rl} \min\limits_{x,X} & \frac{1}{2}x^T Qx + c^T x - \mu \log\det X \\ \text{subject to} & \mathcal{A}^*(x) - A_0 - X = 0. \end{array} \quad (12)$$

If we denote the Lagrangian dual variable as $Z \in R^{n \times n}$, the Lagrangin dual function for (12) becomes

$$\begin{aligned} \mathcal{L}_\mu(x, X, Z) = & \frac{1}{2}x^T Qx + c^T x - \mu \log\det X \\ & - \text{Trace}\{Z(\mathcal{A}^*(x) - A_0 - X)\}. \end{aligned} \quad (13)$$

Karush-Khun-Tucker(KKT) first order necessary condition for the Lagrangian (13) is

$$\mathcal{A}^*(x) - A_0 - X = 0 \quad (14)$$

$$Qx + c - \mathcal{A}(Z) = 0 \quad (15)$$

$$-\mu X^{-1} + Z = 0 \quad \Leftrightarrow \quad XZ = \mu I \quad (16)$$

where

$$\mathcal{A}(Z) = \begin{bmatrix} \text{Trace}(A_1 Z) \\ \text{Trace}(A_2 Z) \\ \vdots \\ \text{Trace}(A_m Z) \end{bmatrix}$$

As the parameter μ is sequentially decreased to zero, the solution to KKT condition converges to an optimal point of (11). Linearization of the KKT condition for the small search steps $(\Delta x, \Delta X, \Delta Z)$ yields the Newton direction

$$\mathcal{A}^*(\Delta x) - \Delta X = R_p \quad (17)$$

$$Q\Delta x - \mathcal{A}(\Delta Z) = r_d \quad (18)$$

$$Z\Delta X + \Delta Z X = R_c \quad (19)$$

where R_p, r_d, R_c are residuals, i.e.,

$$\begin{aligned} R_p &= X + A_0 - \mathcal{A}^*(x) \\ r_d &= \mathcal{A}(Z) - Qx - c \\ R_c &= \mu I - XZ. \end{aligned} \quad (20)$$

For nonlinear complementarity slackness (19), symmetrization strategy is commonly adopted. Employing the symmetrization scheme in (19) and arranging for Δx lead to the following linear system (Details are omitted here)

$$(M + Q)\Delta x = b. \quad (21)$$

M is symmetric positive definite Schur complement matrix. In convex SDP, linear equation (21) is computed by Cholesky factorization method. Since $M + Q$ is indefinite in quadratic SDP, LU or QR factorization must be used. If we find Δx in (21), other search direction $\Delta X, \Delta Z$ can be readily solved. Then primal dual variables are updated by line search technique incorporating merit function. Merit function may be very important because the problem to solve is non-convex. The overall algorithm is similar to other SDP solvers such as SDPA (Yamashita *et al.*, 2002) and SDPT3 (Toh *et al.*, 1999).

Now let's give a simple example from the VTOL helicopter. The system matrices are given as follows

$$A = \begin{bmatrix} -0.0366 & 0.0271 & 0.0188 & -0.4555 \\ 0.0482 & -1.0100 & 0.0024 & -4.0208 \\ 0.1002 & 0.3681 & -0.7070 & 1.4200 \\ 0 & 0 & 1 & 0 \end{bmatrix}$$

$$B_1 = \begin{bmatrix} 0.0468 & 0 \\ 0.0457 & 0.0099 \\ 0.0437 & 0.0011 \\ -0.0218 & 0 \end{bmatrix}, C_1 = \begin{bmatrix} 2 & 0 \\ 0 & 1 \\ 0 & 0 \\ 0 & 0 \end{bmatrix}^T$$

$$B_2 = \begin{bmatrix} 0.4422 & 0.1761 \\ 3.5446 & -7.5922 \\ -5.52 & 4.49 \\ 0 & 0 \end{bmatrix}, C_2 = \begin{bmatrix} 0 \\ 1 \\ 0 \\ 0 \end{bmatrix}^T$$

$$D_{11} = 0_{2\times 2},\ D_{12} = I_{2\times 2}/\sqrt{2},\ D_{21} = 0_{1\times 2}.$$

For this problem, (6) is solved by cone complementarity linearization method (Ghaoui *et al.*, 1997) and the results are compared with those of the proposed method implemented in MATLAB. Popular Mehrotra's predictor-corrector algorithm is adopted and for simplicity, norms of the residuals (20) are used as the merit function.
Implemented algorithm can be outlined as follows;

(Step 0) initialization
$x = 0, X = I, Z = I$
(Step 1) Mehrotra's predictor-corrector step
find Δx by LU factorization then $\Delta X, \Delta Z$
(Step 2) line search using merit function
find $\min_{0<\alpha\leq 1} \parallel (20) \parallel$ such that $X + \alpha\Delta X \succeq 0, Z + \alpha\Delta Z \succeq 0$
(Step 3) update variables
$x = x + \alpha\Delta x, X = X + \alpha\Delta X, Z = Z + \alpha\Delta Z$
goto step 1.

The results are shown in Table 1 and 2. As shown in the tables, the proposed method can get the solution with a fewer iteration.

Table 1. Results of cone complementarity linearization method

outer iter	inner iter	Trace(XY)
1	16	11.05097025983685
2	16	4.41273723389565
3	16	4.13326400053911
4	18	4.00481414088979
⋮	⋮	
13	21	4.00000000581846
13	187 (total inner iter)	

Table 2. Results of the proposed algorithm

iteration	Trace(XY)	barrier parameter μ
1	0.40185843009663	8.740651e-001
2	-0.23490802840503	9.381771e-001
3	1.74381868131373	1.518287e+000
4	-1.81804050533717	1.363926e+000
⋮	⋮	⋮
33	4.00000000002735	7.286327e-012

4. GENERATOR EXCITATION SYSTEM DESIGN

Today's excitation system performs almost all protection and control functions in software programs. It seems, therefore, convincible that automatic voltage regulator (AVR) and power system stabilizer (PSS) should be designed in the same framework. Though there are many advanced methods for excitation systems (Hoang and Tomsovic, 1996; Shu and Chen, 1998), it will be very helpful to field engineers if a simple tuning and testing method for control functions is available. In this section we illustrate such an example from the practical point of view.
A synchronous generator connected to an infinite bus is dealt and a linearized $E_q^{'}$ model (Kundur, 1994) is exploited for the example. As shown in Figure 1, a proportional-integral (PI) regulator type AVR and a PI gain block connected to the output of the phase compensation block of PSS are supposed to be tuned to provide appropriate damping for the system. Operating parameters are as follows;

$$\begin{array}{lll} K_1 = 0.7643 & K_2 = 0.8649 & K_3 = 0.3230 \\ K_4 = 1.4187 & K_5 = -0.1463 & K_6 = 0.4168 \\ T_1 = 0.154 & T_2 = 0.033 & T_3 = 2.365 \\ T_w = 1.4. & & \end{array}$$

System vectors for (1) are

$$z = \begin{pmatrix} \Delta V_{ref} - \Delta E_t \\ \Delta\delta \end{pmatrix}, w = \begin{pmatrix} \Delta V_{ref} \\ \Delta T_m \end{pmatrix}$$

$$y = \begin{pmatrix} y_1 \\ y_2 \end{pmatrix} = \begin{pmatrix} \Delta V_{ref} - \Delta E_t \\ \Delta\omega_{fil} \end{pmatrix}$$

$$u = \Delta E_{fd} = y_1(k_{p1} + \frac{k_{i1}}{s+\tau_{i1}}) + y_2(k_{p2} + \frac{k_{i2}}{s+\tau_{i2}}). \quad (22)$$

This case is chosen to show that a fixed structure control can also be solved in the similar manner (Ibaraki and Tomizuka, 2001). The state space representation of the controller is

$$\begin{bmatrix} \dot{x}_{c1} \\ \dot{x}_{c2} \\ u \end{bmatrix} = \begin{bmatrix} -\tau_{i1} & 0 \\ 0 & -\tau_{i2} \\ k_{i1} & k_{i2} \end{bmatrix} \begin{bmatrix} x_{c1} \\ x_{c2} \end{bmatrix} + \begin{bmatrix} 1 & 0 \\ 0 & 1 \\ k_{p1} & k_{p2} \end{bmatrix} \begin{bmatrix} y_1 \\ y_2 \end{bmatrix}. \quad (23)$$

Augmenting this control equation into system (1) leads to

$$\begin{bmatrix} \dot{x} \\ \dot{x}_{c1} \\ \dot{x}_{c2} \end{bmatrix} = \overbrace{\begin{bmatrix} A & 0 & 0 \\ C_{2,y_1} & -\tau_{i1} & 0 \\ C_{2,y_2} & 0 & -\tau_{i2} \end{bmatrix}}^{\hat{A}} \begin{bmatrix} x \\ x_{c1} \\ x_{c2} \end{bmatrix} + \overbrace{\begin{bmatrix} B_1 \\ D_{21,y_1} \\ D_{21,y_2} \end{bmatrix}}^{\hat{B}_1} w + \overbrace{\begin{bmatrix} B_2 \\ 0 \\ 0 \end{bmatrix}}^{\hat{B}_2} u \quad (24)$$

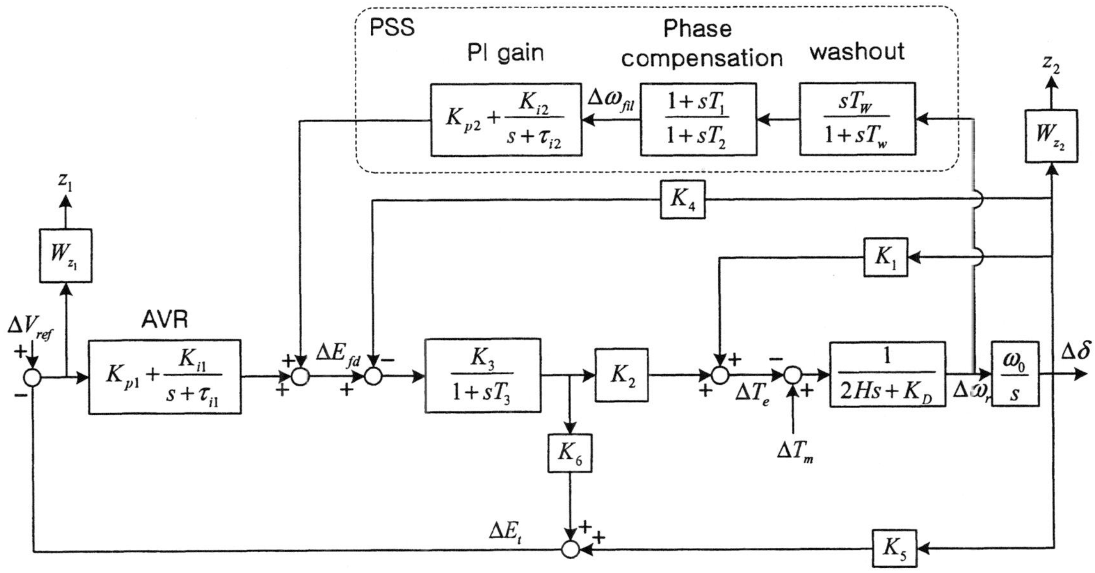

Fig. 1. Block diagram of a synchronous generator connected to an infinite bus (with AVR and PSS)

$$z = \overbrace{\begin{bmatrix} C_1 & 0 & 0 \end{bmatrix}}^{\hat{C}_1} \begin{bmatrix} x \\ x_{c1} \\ x_{c2} \end{bmatrix} + \overbrace{D_{11}}^{\hat{D}_{11}} w + \overbrace{D_{12}}^{\hat{D}_{12}} u \qquad (25)$$

$$\hat{y} = \begin{bmatrix} y \\ y_{a1} \\ y_{a2} \end{bmatrix} = \overbrace{\begin{bmatrix} C_2 & 0 & 0 \\ 0 & 1 & 0 \\ 0 & 0 & 1 \end{bmatrix}}^{\hat{C}_2} \begin{bmatrix} x \\ x_{c1} \\ x_{c2} \end{bmatrix} + \overbrace{\begin{bmatrix} D_{21} \\ 0 \\ 0 \end{bmatrix}}^{\hat{D}_{21}} w. \qquad (26)$$

Now it is straightforward that the control input u of the system can be represented by the static output feedback of the augmented system, i.e.,

$$u = \begin{bmatrix} k_{p1} & k_{p2} & k_{i1} & k_{i2} \end{bmatrix} \begin{bmatrix} y_1 \\ y_2 \\ y_{a1} \\ y_{a2} \end{bmatrix} \qquad (27)$$

The weighting functions are chosen

$$W_{z_1} = 80, \quad W_{z_2} = \frac{95s^2 + 1600s + 2}{s^2 + 110s + 1500}. \qquad (28)$$

Optimal H_∞ performance(when full order controller is adopted) is 31.41. Obtained H_∞ norm is 80.3634 and feedback gains satisfying LMIs (7) are

$$\begin{pmatrix} k_{p1} \\ k_{i1} \end{pmatrix} = \begin{pmatrix} 9.6476 \\ 0.0029 \end{pmatrix}, \begin{pmatrix} k_{p2} \\ k_{i2} \end{pmatrix} = \begin{pmatrix} 168.9671 \\ 0.0272 \end{pmatrix}.$$

Figure 2 shows the result of a conventional design when a 10% step increase in mechanical input T_m occurs in one second. We can see that sufficient damping is not provided in this case. The results by the proposed method in Figure 3, 4, 5, 6 proves that the proposed design can give more damping to the system. One thing to mention is that if more weights are placed on damping, we get a smaller integral gain of AVR. This means that sufficient damping for the system can be obtained at the expense of some degradation in voltage regulation

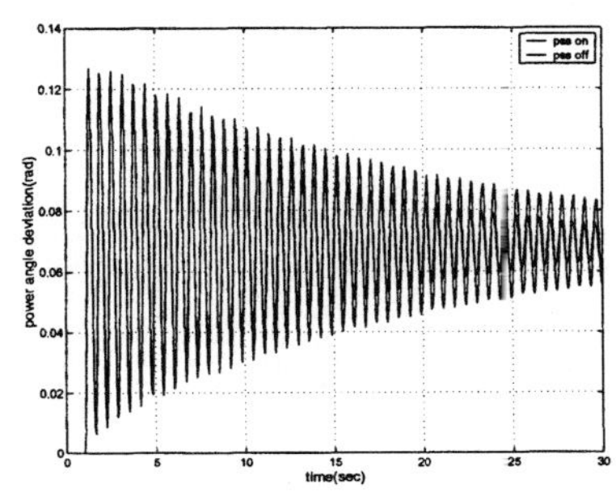

Fig. 2. Power angle response to 10% step increase in T_m (conventional design)

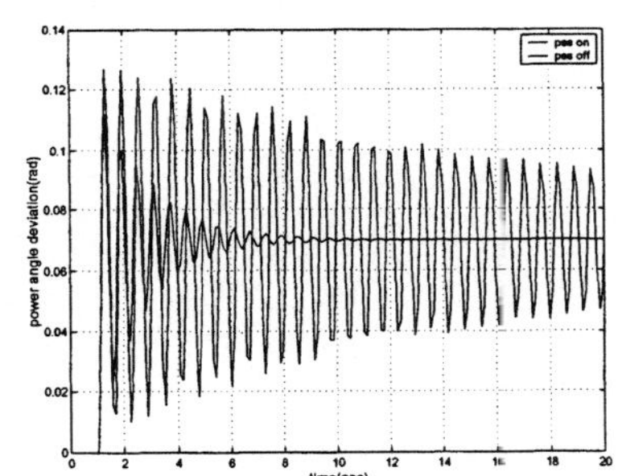

Fig. 3. Power angle response to 10% step increase in T_m (proposed method)

performance.
Besides PI controller, an arbitrary order of stabilizing controllers can be designed easily.

5. CONCLUSION

In this paper we have concentrated on a computational method for a fixed order controller design from the practical point of view. The algorithm has been developed in the conventional SDP framework with a slight modification. Some results with MATLAB implementation show that

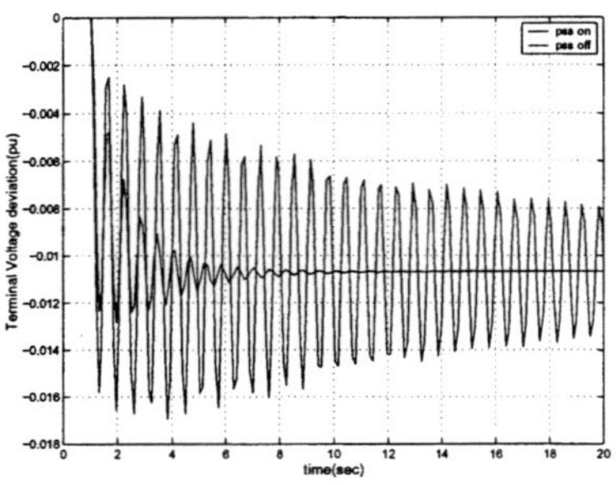

Fig. 4. Terminal voltage response to 10% step increase in T_m (proposed method)

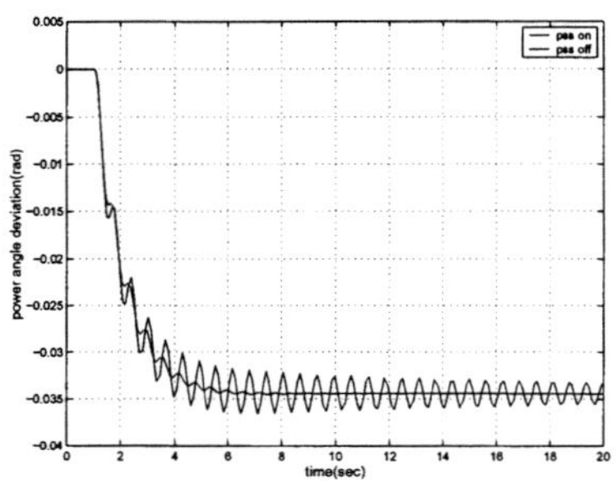

Fig. 5. Power angle response to 2% step increase in V_{ref} (proposed method)

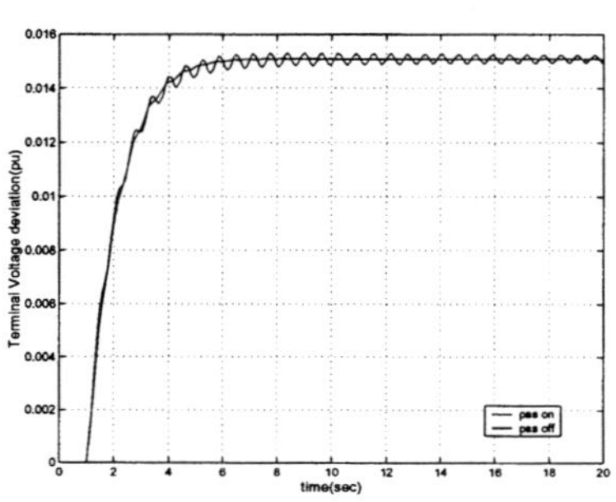

Fig. 6. Terminal voltage response to 2% step increase in V_{ref}(proposed method)

the presented algorithm is efficient and promising. Co-design example of an AVR and PSS for synchronous generator excitation systems shows its practical applicability. Presently, analysis of the complexity and convergence of the algorithm is being investigated.

REFERENCES

Apkarian, P. and H. D. Tuan (1998). Robust control via concave minimization — local and global algorithms. In: *Proc. IEEE Conf. on Decision and Control.* pp. 3855–3860.

Apkarian, P. and H. D. Tuan (1999). Concave programming in control theory. *Journal of Global Optimization* **15**(4), 343–370.

Boyd, S., L. El Ghaoui, E. Feron and V. Balakrishnan (1994). *Linear Matrix Inequalities in System and Control Theory.* Vol. 15 of Studies in Applied Mathematics. SIAM. Philadelphia.

Gahinet, P., A. Nemirovski, A.J. Laub and M. Chilali (1995). *LMI Control toolbox.* The MathWorks Inc.

Gahinet, P. and P. Apkarian (1994). A linear matrix inequality approach to H_∞ control. *International Journal of Robust and Nonlinear Control* pp. 421–448.

Ghaoui, L., F. Oustry and M. Rami (1997). A cone complementarity linearization algorithm for static output feedback and related problems. *IEEE Trans. on Automatic Control* **42**(8), 1171–1176.

Goh, K. C., M. G. Safonov and G. P. Papavassilopoulos (1994). A global optimization approach for the BMI problem. In: *Proc. IEEE Conf. on Decision and Control.* pp. 2009–2014.

Henrion, D., D. Arzelier and D. Peaucelle (2000). Robust stabilization of matrix polytopes with the cone complementarity linearization algorithm: Numerical issues. Technical report. LAAS-CNRS.

Hoang, P. and K. Tomsovic (1996). Design and evaluation of an adaptive fuzzy power system stabilizer. *IEEE Trans. on Energy Conversion* **11**(2), 23–31.

Ibaraki, Soichi and Masayoshi Tomizuka (2001). H_∞ optimization of fixed structure controllers. In: *Proc. of European Control Conference.*

Iwasaki, T. (1999). The dual iteration for fixed order control. *IEEE Trans. on Automatic Control* **44**(4), 783–788.

Iwasaki, T. and R. E. Skelton (1995). The XY-centering algorithm for the dual LMI problem: a new approach to fixed order control design. *International Journal of Control* **62**(6), 1257–1272.

Kundur, Prabha S. (1994). *Power System Stability and Control.* McGraw-Hill.

Nesterov, Yu. and A. Nemirovsky (1994). *Interior Point Polynomial Methods in Convex Programming.* SIAM.

Shu, H. and T. Chen (1998). Robust digital design of power system stabilizers. *International Journal of Control* **70**(2), 227–248.

Toh, K.C., M.J. Todd and R.H. Tütüncü (1999). SDPT3 — a Matlab software package for semidefinite programming. *Optimization Methods and Software* **11**, 545–581.

Wright, Stephen J. (1997). *Primal-Dual Interior-Point Methods.* SIAM.

Yamashita, M., K. Fujisawa and M. Kojima (2002). Implementation and evaluation of SDPA 6.0 (SemiDefinite Programming Algorithm 6.0). Technical report. Tokyo Institute of Technology.

ELSEVIER
IFAC
PUBLICATIONS
www.elsevier.com/locate/ifac

NETWORK REQUIREMENTS OF FUTURE ENERGY CONVERSION TECHNOLOGIES

H. Weber* T. Hamacher T. Haase***

**University of Rostock*
***Max-Planck-Institut für Plasmaphysik Garching*

Abstract: Especially the introduction of intermittent electricity sources and the diffusion of distributed generation will make it necessary to improve the electricity network capacities and control mechanisms and will make the introduction of completely new dispatching philosophies necessary. The paper discusses the expected problems of the UCTE net introduced by the possible installation of 25 GW off-shore wind power in the North and Baltic sea. An up-grade of the transport capacities will become necessary to make the power available all over Germany. Control problems might especially arise, if a rapid shut down of the wind turbines in a storm situation would occur. The dispatching philosophy has to be changed from a demand driven to a supply driven philosophy, which distinguishes between renewable – primary – supplies and back-up – secondary - supplies. *Copyright © 2003 IFAC*

Keywords: Dynamic modelling, Dynamic behaviour, Simulation, Wind speeds, Load Flows

1. INTRODUCTION

There is a lot of evidence that the future energy and especially electricity system will look completely different than the system looks today. Changes will be induced by the increased depletion of fossil fuels, the increased concern about environmental damages – especially climate change – and increasing political tension around the remaining attractive oil and gas reserves. A number of possible solutions are presented: more rational use of energy, energy savings and the introduction of completely new energy carriers like renewable and new nuclear energies.

New electricity sources will come increasingly into operation in the next fifty years. The characteristics of these source: unit size, spatial distribution, availability and so on might be completely different than of the power plants operating these days. The following paper will focus on the impacts of new generation technologies on the electricity network. This work is done in the context of a comprehensive study that will try to elaborate completely new supply systems and their interplay with the demand side. An intermediate step of this study is presented, namely a drastic increase of wind turbines in the German off-shore regions in the North and the Baltic sea. Plans exist to install up to 25 GW of wind power in the region. This installations could cover roughly 20% of the German electricity demand in the year 2030.

2. INSTALLATIONS AND REGULATIONS IN THE UCTE NET

Development and benefits of large electricity networks are discussed elsewhere [1, 2]. With an yearly electricity production of 2.205,6 TWh and a installed capacity of 515 GW (UCTE, 2002) the UCTE net covering Western and Central Europe is one of the biggest networks which is frequency synchronised

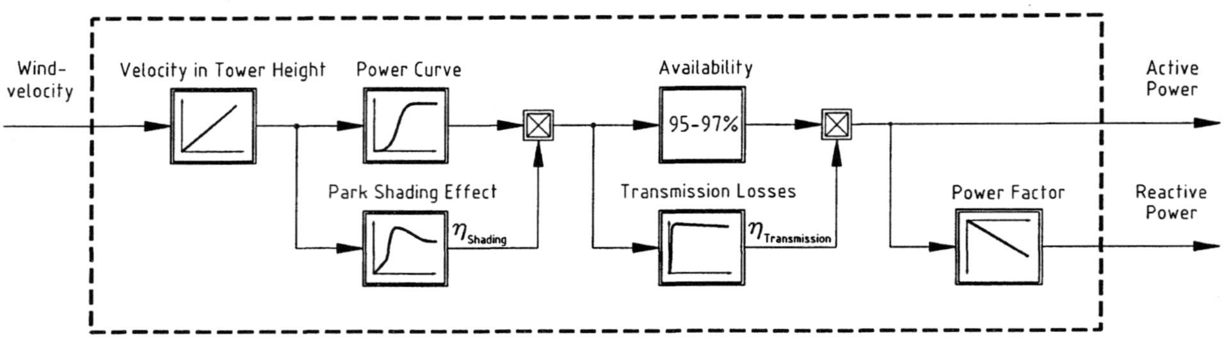

Figure 1: Simple wind park model

and voltage stabilised over the complete network. To guarantee voltage and frequency stability even with changing loads and power plant outages a general control procedure in three steps is agreed on. In case of an rapid increase of the load or an accidental outage the so called primary control is activated. A number of plants distributed over the complete network take part in this control measure. In a very short time up to 3.000 MW of extra power can be activated. This is mainly done by automatic changes in the steam cycle of fossil power plants which lead to an power increase. But this measure can only be kept active for a few minutes. Before this power gets lost again, the secondary control comes online. These are mainly very fast pump-storage hydro plants or specially designated thermal plants which operated not at full-load. In a third step power plants not yet online are activated and go on-line and replace the power plants activated by the secondary control. The UCTE network is divided in various control circles. While all partners in the network contribute to the primary control, the network provider in the control circle, where an outage happened, is responsible to supply the secondary and tertiary control. In the following study it is done as if the whole of Germany would be covered by one control circle.

3. SIMULATION OF FUTURE ELECTRICITY NETWORKS

3.1 Technical simulation with DIgSILENT

Basis for the following analysis is the software tool DIgSILENT PowerFactory [3], which is a computer aided engineering tool for the analysis of industrial, utility, and commercial electrical power systems. It has been designed as an advanced integrated and interactive software package dedicated to electrical power system and control analysis in order to achieve the main objectives of planning and operation optimisation. Modelling of the dynamic behaviour of all power system components down to very short time scales (milliseconds) is possible. The dispatching and control philosophy of various networks can also be adequately simulated with the help of a special block oriented programming language.

Within this tool the UCTE net is modelled using

- 1488 nodes
- 2401 transmission lines in the 380/220-kV-level
- 533 power plants
- 92 transformers
- 948 consumer loads.

Information on the

- production and consumption in the UCTE states
- distribution of produced power on the individual power plants and the individual plant types
- load-curves for high and low demand cases
- special conditions of the net
- international power flows
- dynamic of power plants and consumer loads

are available [4].

A description of the individual plant components for the conventional fossil, hydro and nuclear plants can be found elsewhere [5]. The simulation of wind power plants is done as depicted in Figure 1. Most important input parameter is the wind velocity in 10 m above the sea level. With the logarithmic elevator profile the velocity in tower height (100 m) is calculated. The wind-velocity-power characteristic is assumed to be like the characteristic of a Nordex N80 off-shore-turbine [6]. Impacts like shadings in the wind park and the power transmission losses are considered in efficiency curves. The availability of all plants in the wind park is expected with 95–97%.

3.2 Economic consideration: the optimal power plant scheduling

Operation of power plants is scheduled within the physics and engineering limits by economic consideration. A simple optimisation procedure was installed to model this decision process. Only fuel and variable costs were considered for the optimisation. Figure 5 gives a case for Germany, if 25 GW offshore wind and 20 GW of solar PV power were installed in addition to the existing power plants.

4. IMPACTS OF LARGE OFF-SHORE WINDPARKS IN GERMANY ON THE UCTE GRID

4.1 The German off-shore wind plans

Wind energy increased globally and a considerable fraction of this growth rate was achieved in Germany. The success of wind energy was promoted by two measure: first the so called 250 MW wind programme and than the "Stromeinspeisegesetz" which is now followed by "Erneuerbare Energien Gesetz". Wind energy is strongly promoted in Germany. Network providers are forced to buy the electricity from the wind power plants for a fixed price, namely for 8.5 Eurocent/kWh. The limited number of favourable on-shore sites makes it necessary to go to off-shore locations in the long term if the growth rates stay high. A plan was elaborated to install in 4 phases up to 25 GW wind capacity off-shore till 2030 [7].

On the 11th of June 2001 the German government and the German utilities operating nuclear power plants signed an agreement for the phase out of nuclear power in Germany in the next decades. The agreement states a maximum of electricity that can still be delivered by nuclear power. Nonetheless it is reasonable to assume that in the next two decades most of the German nuclear generation capacity will be switched of an dismantled. One possible solution could be the installation of off-shore wind power, although it should be stated right from the beginning that off-shore wind power will be not able alone to replace base-load nuclear power plants. This will be discussed more in detail in 4.4.

4.2 The missing transport capacities

In a first step the static electricity transport capacities are analysed. Basis for the analysis are the existing capacities of the 380-kV-transmission lines, see Table 1. Just for illustrative purposes Germany is divided in five supply areas (North, West, East, Centre, South).

Transport capacities from the region "North" to the rest are limited to roughly 13,32 GW. Just from this number it becomes obvious, that a production capacity of 25 GW will lead to problems if not a good fraction of the wind power is already consumed in the region "North". The case is if course more delicate since the transmission capacities are not solely reserved for the wind power. Two detailed investigations for a high load winter day and low load summer

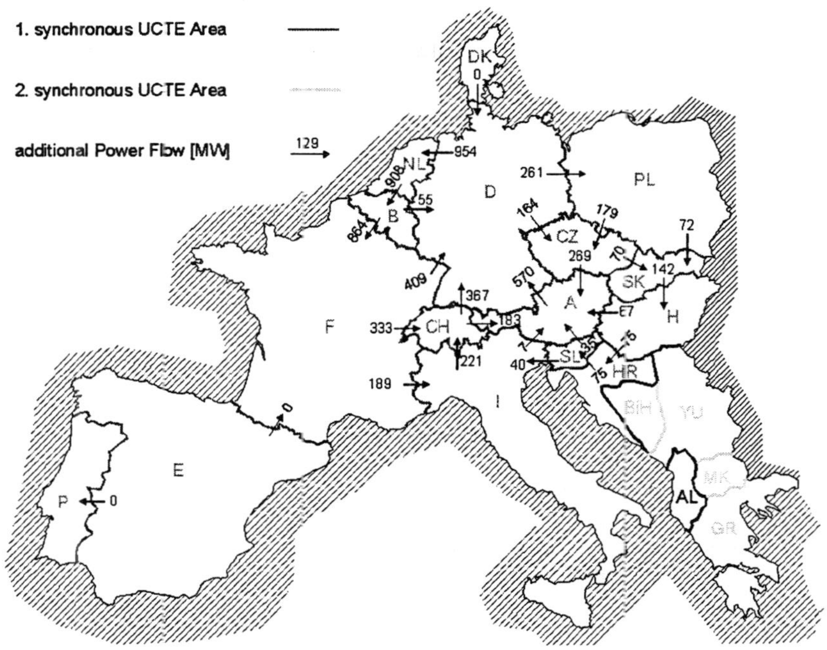

Figure 2: Additional power flow in Europe after feeding 7 GW wind power in North Germany

night very made. Wind velocities are assumed to be the average value for the respective month. It is now necessary to assume the power flows in the complete UCTE net. Power plants are on-line following the usual dispatching philosophy applied today. The investigations show that the winter day case poses the bigger problem, especially if it is assumed that electricity from nuclear power plants in the region "South" are replaced by the wind power. This is of course done on purpose, to understand the possible limitations of the network power needs to be delivered from the off-shore regions even to South-Germany. While in a base case without off-shore wind power plants the transmission capacities are only used up to 20 %, this increases considerable with 7 GW of wind power capacity in North and Baltic sea. Especially the line from the North to the West region would be used by more than 100 %, see Tables 2 and 3.

The wind power production does not only changes power flows in Germany but also in the other European UCTE partners. Figure 2 shows the impact of the wind power production in the north of Germany on the rest of the European UCTE net. Considerable power flows are delivered to the Benelux-countries.

Another negative side effect of the intense power flows are higher losses. The extra losses account for 470 MW, nearly 7 % of the wind power production.

The simulations show that already an installed capacity of 7 GW will lead to shortages of transport capacity in the Germany, if power needs to be delivered to South Germany. If all planed up-grades would be done this number would increase to nearly 20-25% of installed capacity. The at first intended 25 GW capacity needs completely new power lines.

Transmission Capacities [MVA]	North	East	West	Centre	South
North	-	5.580	1.380	6.450	-
East	5.580	-	-	8.890	2.760
West	1.380	-	-	15.340	1.390
Centre	6.450	8.890	15.340	-	13.630
South	-	2.760	1.390	13.630	-

Table 1 Transport Capacity in the German Grid

Line Loading [%]	North	East	West	Centre	South
North	-	8.2	14.8	8.7	-
East	8.2	-	-	10.5	6.6
West	14.8	-	-	18.4	6.3
Centre	8.7	10.5	18.4	-	6.6
South	-	6.6	6.3	6.6	-

Table 2 Line Loading without wind power

Line Loading [%]	North	East	West	Centre	South
North	-	15.2	106.8	53.4	-
East	15.2	-	-	2.8	24.3
West	106.8	-	-	12.0	29.4
Centre	53.4	2.8	12.0	-	19.3
South	-	24.3	29.4	19.3	-

Table 3 Line Loading with 7 GW wind power

4.3 The challenge for the net-stability

In chapter 2 the overall philosophy to guarantee the net stability in the UCTE net was presented. The question is, will the existing power plants, which take part in the net stabilisation, be sufficient to counterbalance possible changes in the output of the wind power or not. If the capacities will not be sufficient the question will be raised what kind of extra power plants need to be installed. Regular measurements of installed wind power plants show that in the year 2000 power changes stayed below 20 % in one hour [8]. Changes are expected to more pronounced in the off-shore case, since the geographical distance between wind turbines will be shorter.

The primary control in the UCTE net is able to counterbalance the loss of 3.000 MW on a second basis. If the changes in the hour range would not be larger that 20 % this should not pose a considerable problem, if 25 GW of installed capacity are assumed. The picture looks different if a rapid shut down becomes necessary. The question is, at which pace would a storm front pass over the North sea and would such an event not be predictable well in advance. For a quantitative examination a typical situation in winter and in summer was studied. In the winter case roughly 80 % of the installed capacity in the UCTE net was on-line. The wind velocity was assumed to be average velocity in January for the North and the Baltic sea. A drop of the wind velocity of 15 % was assumed to happen in a 10 min time period. This corresponded to a power drop from 14.770 MW down to 10.420 MW. This power drop let to a maximum change of frequency df = -170 mHz, depicted in Figure 3. The rapid shut-down of 1.000 MW nuclear power plant in Switzerland would lead to a frequency drop of 50 mHz. At a frequency drop of more than 1 Hz some loads are no longer supplied by the net. Similar assumptions were made for a low load summer night, but the problem here turned out to be less sever. In a second quantitative analysis it was assumed that the whole 14.770 MW capacity had to be switched of because of a storm. Obviously this needs not to be done at once, but at the time, when the storm front reaches the turbines. The time lack in between depends of the velocity of the storm front and the geographical distance of the wind turbines. In the simulation the wind capacity was switched of in four steps with a time delay of each 3 minutes. Figure 4 shows the effect of the net frequency and the Area Control Error (ACE).

4.4 New philosophies of dispatching

The variation in the power output goes from nearly zero during storms or large scale calms up to nearly full load. The rest of the power system needs to balance these fluctuations. Dispatching philosophies do

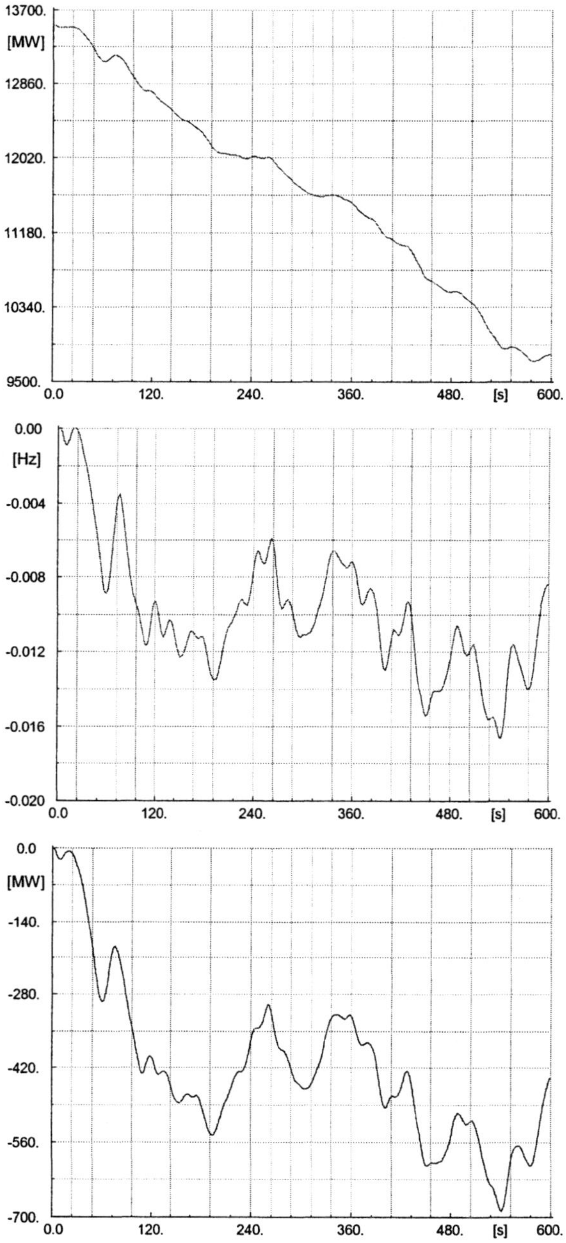

Figure 3: Power, Frequency and ACE during a drop of the wind velocity of 15 %

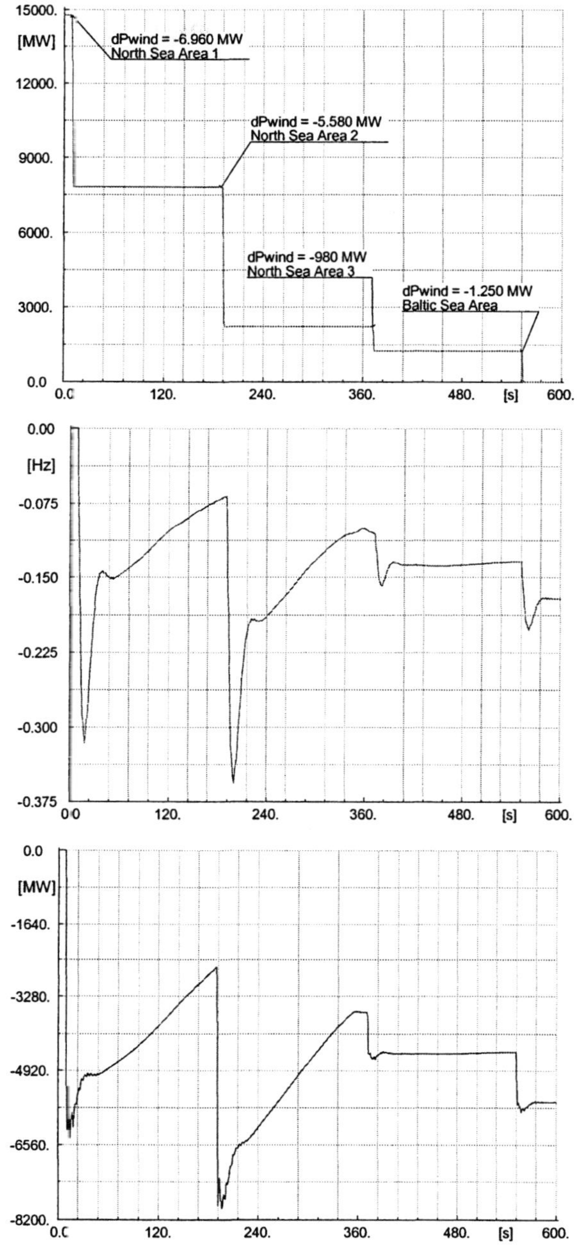

Figure 4: Power, Frequency and ACE during the wind capacity was switched of in four steps

depend on technological as well as on economic circumstances. In principle the load pattern is divided in a base, a middle and a peak load and different technologies are used to supply the three load categories. Liberalisation of the electricity markets might lead to changes in the way power plants are operated. Another change is expected by the strong linkage of electricity and gas market, volatility in gas prices might lead to very different competition situations of power plants using gas. The intensive introduction of intermittent renewable electricity sources and the increase of distributed generation might again lead to a change in the dispatching philosophy. An example is presented in Figure 5. Electricity from wind and solar PV is always used if available, while thermal power plants have to fill the remaining gap. Priority might be given to renewable technologies. The nuclear power plants have today still another large portion of the German energy production [9]. The concept of dispatching would then change from being demand driven as today to being more supply driven. The concepts of base, middle and peak load might be softened and be changes to primary supply and back-up supply. This would of course have a very strong impact on the supply technologies. To study the various impacts a simplified DIgSILENT model was created. This made it possible to study various dispatching philosophies under various conditions over periods of weeks, under very different weather situations.

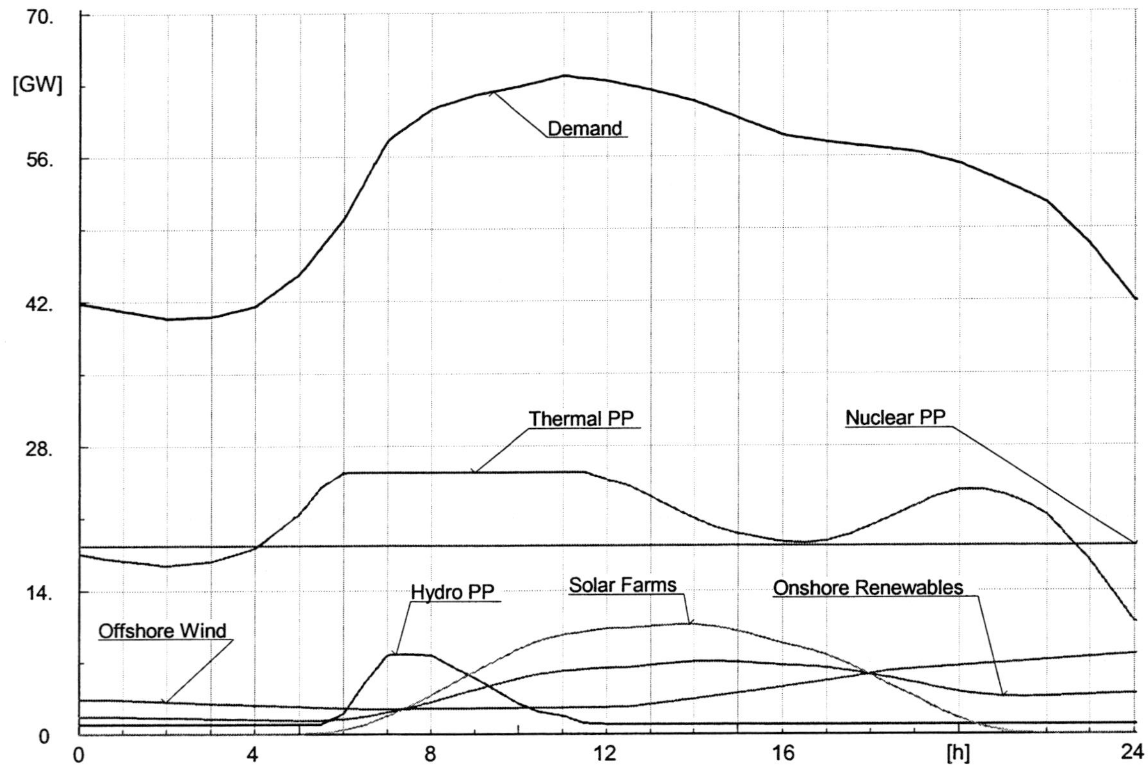

Figure 5: Power plant plan of application

5. CONCLUSION

The introduction of new intermittent electricity sources in the power market requires a revision of the complete power system. In case of the introduction of new off-shore wind capacities in the range of 25 GW in Germany, it will become necessary to strengthen the transmission capacity, to make sure that especially in case of storms no rapid shut-down of the wind-turbines becomes necessary and that a complete new dispatching philosophy is developed, which might make some of the base-load power plants obsolete.

REFERENCES

[1] VDI-Gesellschaft Energietechnik, Global link, - Interkontinentaler Energieverbund, VDI-Verlag GmbH, Düsseldorf 1994

[2] Bedeutung und Aufgaben der Hoch- und Höchstspannungs-Freileitungsnetze (1988), Zeitschrift der Vereinigung Deutscher Elektrizitätswerke – VDEW, Jg. 87 (1988), H. 22, S. 1083-1087

[3] DIgSILENT GmbH, Gomaringen, Germany

[4] UCTE-Jahresbericht 2000

[5] Spanner, M. (1999), Netzpendelungen im erweiterten Verbundnetz bei größeren Leistungstransiten, Dissertation, Universität Stuttgart

[6] Bundesministerium für Umwelt, Naturschutz und Reaktorsicherheit, Weiterer Ausbau der Windenergienutzung im Hinblick auf den Klimaschutz, Berlin 4/2001

[7] Nordex Energy GmbH Rostock, Nordex N-80, Technical Description

[8] Durstewitz, M., B. Hahn and M. Hoppe-Kilpper (2001), Offshore-Windenergienutzung in der AWZ – potenziale, Netzintegration, Stromgestehungskosten -, Im Auftrag des Bundesministerium für Wirtschaft und Technologie Gesch.-Z.: I A 2 -02 08 15 – 48/01, Dezember 2001

[9] Verband der Netzbetreiber – VDN, Leistungsbilanz der allgemeinen Stromversorgung: Vorschau 2003 – 2005

ELSEVIER
IFAC
PUBLICATIONS
www.elsevier.com/locate/ifac

CONTROL SCHEMES OF UTILITY INTERACTIVE MULTIFUNCTIONAL DISPERSED GENERATION

Il-Yop Chung, Sung-Woo Park, Joon-Ho Choi, Seung-Il Moon, and Jae-Eon Kim

School of Electrical Engineering & Computer Science #013
Seoul National University
Shinlim-dong, Kwanak-gu, Seoul, 151-742, Korea
iryop@powerlab.snu.ac.kr

Abstract: In this paper, control schemes of utility interactive *Multifunctional Dispersed Generation* (MDG) are proposed to improve the power quality and the continuity of the electric power supply. The proposed MDG is designed to compensate the power quality event, for example, voltage sag and harmonic distortion, and to supply electric power in parallel operation with power gird. It is also designed to serve a local load for improving the power quality and the continuity of the electric power supply when it is isolated from the power gird. To cope well with this, the instantaneous power theory is used to control the series and shunt DC/AC inverter of the MDG. A numerical formulation of the proposed control scheme is presented. Case study shows the effectiveness and flexibility of the proposed control schemes of the proposed MDG configuration for the multifunctional purpose. *Copyright © 2003 IFAC*

Keywords: Custom Power, Dispersed Generation (DG), Distribution System, Instantaneous Power Theory, Power Quality, and PWM Inverter.

1. INTRODUCTION

Recently, there are two important issues in the power distribution system: power quality and power system deregulation.

Power quality means the ability to keep voltage and current near to constant sinusoidal waveform lest customer equipments should fail or operate incorrectly. Sag, interruption and harmonics are the representative power quality problems. There are many researches about solutions of power quality. Specially, custom power devices including power electronic interface can be the effective solution because they can provide fast response and flexible compensation required by customers.

Custom power devices have various types of structure depending on their inverter connection to the grid, that is, series-connected device, shunt-connected device, and series-and-shunt-connected device. The last device is usually called *Unified Power Quality Conditioner* (UPQC) and it has almost all remedies about power quality problem including voltage quality and current quality. Although there have been several control schemes of UPQC, the instantaneous power theory is the most effective scheme (Akagi, *et al.*, 1984; Aredes, *et al.*, 1998; Li, *et al.*, 2000). However, UPQC cannot supply large active power to customers steadily due to the limitation of power storage.

On the other hand, the opening of the energy market under deregulation brings about the interest in *Dispersed Generation* (DG) because it can provide independence and flexibility to the customer in planning and developing the installation and can give economic benefits in many cases (Joos, *et al.*, 2000). The most fundamental function of DG is to maintain the reliability of electricity supply even though utility cannot supply sufficient power due to faults or other conditions. This function has several sub-operating rules about sharing active power with grid during paralleling mode, maintaining load voltage constantly and controlling frequency during the islanding operating mode (Barsali, *et al.*, 2002).

However, DG has the negative effects and the positive effects to the existing power system. The employment of DG in the existing power system can cause several potential operation problems, i.e. negative effects of DG, such as voltage variation (Choi, *et al.*, 2001a), protection, and harmonics (Choi, *et al.*, 2001b).

The recent trend of DG is to use power electronic inverter in order to interface various types of DGs like micro-gas turbines, fuel cells, and wind power stations (Lasseter, *et al.*, 2001). Shunt connected inverter interface is mostly used because it is insensitive to the grid parameter variation and easy to share active power with the utilities (Marei, *et al.*, 2002). In addition, the recent DG needs to have multiple functions to solve various problems such as reliability and power quality.

This paper proposes *Multifunctional Dispersed Generation* (MDG) that has not only shunt inverter but also series inverter. Therefore MDG can inject both voltage and current to the grid more freely than other dispersed generations that have only one inverter connection. Furthermore, MDG can control both the power quality and quantity during paralleling mode and islanding mode.

Moreover, in this paper, the novel control scheme of MDG is represented for the purpose of paralleling mode and islanding mode. The performance of this system and control scheme is simulated using EMTDC/PSCAD package.

2. MULTIFUNCTIONAL DG (MDG) CONFIGURATION

Recently, the several dispersed generator configurations including power quality solutions have been studied. There are two types of dispersed generator; one uses series inverter only and the other uses shunt inverter only (Marei, *et al.*, 2002). However, the former has difficulties about compensating reactive power and harmonic currents. On the other hand, the latter has hardships about mitigating voltage sag or swell because a shunt inverter requires a significant amount of reactive current and the voltage support capacity at the Point of Common Coupling (PCC). Thus the rating of the inverter may have to be significantly increased (Joos, *et al.,* 2000).

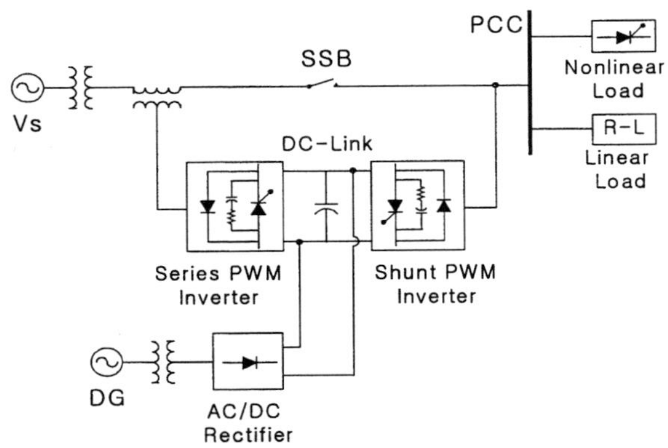

Fig.1 The configuration of the MDG

The configuration of the proposed MDG is shown in Fig.1. The MDG system consists of synchronous generator, rectifier with DC-Link, *Solid State Breaker* (SSB) and two Pulse-Width-Modulation (PWM) inverters. Because proposed MDG has two inverters, series and shunt inverters, it can compensate both voltage disturbance and current distortion. Thus MDG can perform UPQC function to maintain the power quality of the system and power supplying function to exchange real and reactive power with the grid during paralleling mode.

When the severe disturbance is occurred in the utility side, MDG should open its SSB and operate in islanding mode and then it bears the responsibility on supplying power required by loads.

Each component of the proposed MDG has the following characteristics.

2.1 Synchronous generator

The main function of synchronous generator is supplying power in order to charge DC link capacitor within the range of voltage that results in appropriate inverter operations. Because the dynamic response of the inverter is faster than that of synchronous generator used for the source of the MDG, the inverters conduct all controls of the MDG like power quality control and real and reactive power control. Therefore, not only in normal state but also in case of occurring of disturbances such as sags or swells, the synchronous generator performs only a role of maintaining the dc-link voltage in constant set point.

2.2 DC-link

The DC-link connected in common between series and shunt inverter are charged by the synchronous generator through the AC/DC rectifier and discharged by the inverters. This dc-link can ensure a reliable compensation of the disturbances such as sags or swells within about few seconds. If the disturbances are continued beyond few seconds, it is exceeded the compensating capacity and causes the SSB to be opened and then enters the islanding mode. Therefore, the larger the size of capacitor in DC link is, the better the capacity of MDG is.

2.3 Series and shunt inverter

Two inverters in the MDG perform roles of suitable supplying capability of power to the grid from the DC link, and of maintaining the power quality. Especially, as the maintenance of power quality not only from the main source side but also to the load side is possible, this multi-functional DG can achieve outstanding UPQC functions. Fig.1 shows a detail configuration of the series and shunt inverter. The series inverter can supply real power to the grid and compensate harmonics and sag or swell of the voltage. It can also act as a voltage regulator by assigning the reference of the load terminal voltage to a constant sinusoidal waveform. The shunt can supply the reactive power to the load and cancel the harmonics contained in the load current by injecting the compensating current. Furthermore, in islanding mode, by following the load power, it is possible to supply the real and reactive power to the load through the shunt inverter without discontinuity. Table 1 shows the principle functions of these inverters. More detail descriptions for the inverter controls will be following.

Table 1. Main functions of the series and shunt inverters of MDG

	Series Inverter	Shunt Inverter
Paralleling mode	Regulating the load voltage	Supplying power to load Suppressing harmonic current
Islanding mode	-	Maintaining load voltage Supplying power to load

2.4 Solid-state breaker (SSB)

A SSB installed between two inverter connections usually uses GTO thyristor. As the GTO thyristor could be operated regardless of the magnitude of the current, a detachment of the load side from the grid is possible within 1/2 up to 1/4 cycle. Estimating the entering voltage and current from the both ends as the input signals, if the serious disturbances are continued above few second that is previously set up compared to DC-link capacity, then the SSB is opened and changed into the islanding mode. Hereafter if the disturbances are passed, the SSB is closed and changed into the paralleling mode.

3. CONTROL SCHEMES OF MDG

The control of the proposed MDG has two operating modes: paralleling mode and islanding mode. Each mode is selected by the operation of SSB. In paralleling mode, MDG can supply the power to the customer's network in cooperation with power utilities. Besides this function, MDG can also maintain power quality required by customers through series and shunt-connected inverters like UPQC operations.

When the grid voltage supplied by utility can be maintained normally, MDG can be in the paralleling mode. However, if the grid voltage drops below the certain threshold voltage level beyond the certain fixed duration, SSB should detach customer's network from the utility lines. Then MDG has to change its operating mode from paralleling mode to islanding mode. In islanding mode, customers may receive electricity only from MDG, so it should follow the load power and maintain line voltage. If there are several DG in the islanded network, all of the DG units should have droop control function in order to follow a local load demand correctly.

The important functions of MDG are listed in Table 2.

Table 2 The main functions of MDG

Operating mode	Main function
Paralleling mode	Power control Power quality control
Islanding mode	Voltage control Droop control

3.1 Control Scheme of Paralleling Mode

The paralleling mode has two representative functions. One is constant power control function and the other is power quality control function. MDG can accomplish easily these functions by controlling the injection voltage and current because it has series and shunt inverters that have fast response time.

Specifically, series inverter injects the voltage difference between the sinusoidal load voltage reference and the distorted grid voltage in order to compensate load voltage regularly. Therefore, series inverter can maintain voltage quality, for instance, voltage unbalance, sag, and swell. The voltage reference of the series inverter (V_{serREF}) is given by Eq. (1).

$$v_{serREF} = v_{LREF} - v_G \tag{1}$$

In Eq (1), v_{LREF} is the load reference and v_G is the utility supplying voltage.

On the other hand, shunt inverter compensates distorted current so that harmonic current cannot flow from customer's network to grid line. Furthermore, shunt inverter is able to not only improve power quality but also supply active and reactive power to customer's load. The power supplied by MDG is generally fixed value depending on its power rating.

The control of shunt inverter can be easily realized by the instantaneous power theory (Akagi, *et al.*, 1984). This theory can generate inverter voltage or current references from the instantaneous power and the instantaneous imaginary power.

If zero-sequence components can be neglected, the conventional three-phase instantaneous power (p) and the instantaneous imaginary power (q) on the α-β coordinates can be defined as Eq.(2).

$$\begin{aligned} p &= v_\alpha \cdot i_\alpha + v_\beta \cdot i_\beta \\ q &= v_\alpha \cdot i_\beta - v_\beta \cdot i_\alpha \end{aligned} \tag{2}$$

Where v_α, v_β and i_α, i_β mean the voltage and current of each α-β axis, respectively.

The transformation from a-b-c coordinates to α-β coordinates becomes Eq.(3) and the current transformation is the same way.

$$\begin{bmatrix} v_\alpha \\ v_\beta \end{bmatrix} = \sqrt{\frac{2}{3}} \cdot \begin{bmatrix} 1 & -1/2 & -1/2 \\ 0 & \sqrt{3}/2 & -\sqrt{3}/2 \end{bmatrix} \cdot \begin{bmatrix} v_a \\ v_b \\ v_c \end{bmatrix} \tag{3}$$

Generally, in the single-phase circuit, the instantaneous power has two oscillatory terms whose amplitude are the conventional active power ($P_1 = V \cdot I \cdot cos\Phi$) and the conventional reactive power ($Q_1 = V \cdot I \cdot sin\Phi$), respectively. However, in the balanced three-phase circuit, the instantaneous power becomes constant that is the three-phase conventional active power ($3P_1$). Therefore, the instantaneous power cannot give the information of the reactive power.

However, the instantaneous imaginary power is numerically equal to three times the conventional reactive power ($3Q_1$). In other words, injecting the instantaneous imaginary power can compensate the reactive power on the load side.

If three-phase circuits have unbalance or harmonic components, oscillatory AC components happen to the instantaneous power and the instantaneous imaginary power like Eq.(4).

$$\begin{aligned} p &= \bar{p} + \tilde{p} \\ q &= \bar{q} + \tilde{q} \end{aligned} \tag{4}$$

Where, $\bar{p}$ and $\tilde{p}$ are the DC and AC components of the instantaneous power and $\bar{q}$ and $\tilde{q}$ are the DC and AC components of the instantaneous imaginary power.

The shunt inverter should compensate $\tilde{p}$ and q in order to improve power factor and power quality, especially harmonic current problems. Moreover, for the purpose of sharing power supply with the grid, the shunt inverter should provide the set point active power P_{sp}. Therefore, from Eq.(2), the reference injection current of shunt inverter can be obtained like Eq.(5).

$$\begin{bmatrix} i_{\alpha REF} \\ i_{\beta REF} \end{bmatrix} = \begin{bmatrix} v_\alpha & v_\beta \\ -v_\beta & v_\alpha \end{bmatrix}^{-1} \cdot \begin{bmatrix} P_{sp} + \tilde{p} \\ q \end{bmatrix} \quad (5)$$

Where, $i_{\alpha REF}$, $i_{\beta REF}$ are the reference currents on the α-β coordinates and v_α, v_β are the fundamental voltage of the point where the shunt inverter is connected to the grid line.

The detailed control block diagram is illustrated in Fig.2.

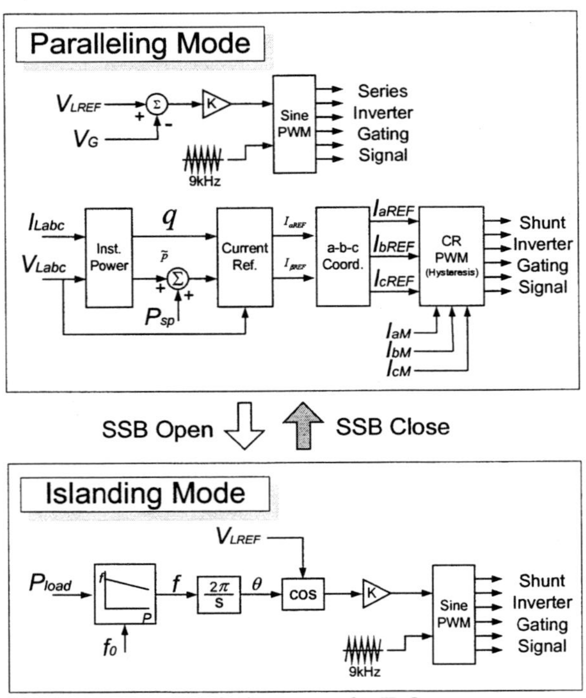

Fig. 2. Control block diagram of MDG

3.2 Control Scheme of Islanding Mode

When the SSB opens, MDG is operated in islanding mode. In this mode, MDG should keep the load voltage constant and follow up the load power variation. Therefore, the active and reactive power providing is defined automatically by the load condition.

If there is only one DG in the islanding customer's network, an isochronous control that provides load power with constant frequency is mostly used. However, when two or more DG units are located in the islanding network, the droop control should be adopted in order to assign the supplying power of each unit automatically. MDG can control droop characteristic more easily by regulate the frequency of the voltage reference.

Since SSB detaches the series inverter from customer's network, only shunt inverter can supply electricity to the load. The reference voltage of the shunt inverter output is given by Eq.(6).

$$v_{shunt} = v_m \sin[\theta(t)] \quad (6)$$

Where, v_m is set point load voltage magnitude and $\theta(t) = 2\pi f t + \theta_0$ is the phase angle. v_m is given as the constant voltage and $\theta(t)$ can be given by the droop control.

Fig. 2 shows the overall control schemes of MDG.

4. SIMULATIONS

4.1 Simulation Model

The test distribution system for simulation is illustrated in Fig.3. MDG is located in one of the two feeder lines of which line voltage is 22.9kV. In addition, there are linear and nonlinear sensitive loads under MDG. Nonlinear load emits harmonic currents into the grid line. Therefore, MDG should supply certain portion of load power and eliminate load harmonic currents.

On the other hand, line to ground fault happens in the other feeder line that has no MDG. Therefore, the sensitive loads under MDG will suffer voltage sag or interruption. If voltage sag occurs, series inverter should inject missing voltage to the line in order to maintain load voltage constantly. However, when voltage drops so deeply that interruption occurs, SSB isolates customers network from the grid line because the grid line cannot supply power to the sensitive load. In this case, MDG can maintain load voltage and supply load power through shunt inverter only.

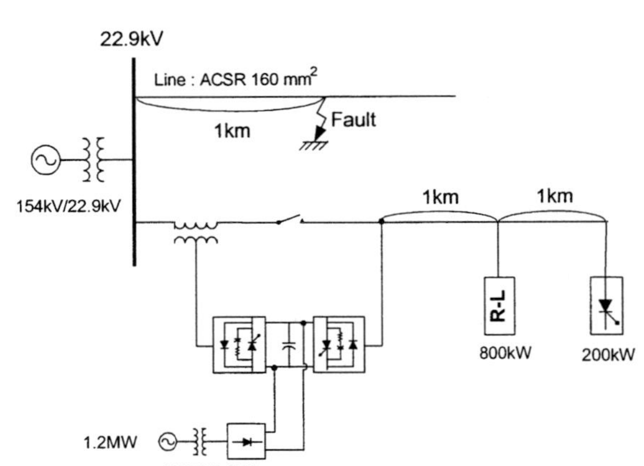

Fig. 3 Test distribution system

The synchronous generator of small-scale DG has some different characteristics, in inertia and transient behaviour, and distinguished from a large generating unit. The rating of the synchronous generator used for MDG is 1.2[MVA] and 480[V]. The model of the exciter, which regulates the output voltage, is IEEE AC1 type exciter. Furthermore, the model of the governor, which controls the frequency of the generator, is hydro-governor. The parameters of the synchronous generator, exciter, and governor are taken from (Gish, 1986; Kundur, 1993).

Table 3 shows the distribution parameters of the test system.

Table 3 The parameters of distribution system

Categories	Parameters
Transformer	Winding Ratio: 154kV/22.9kV Rating power: 45MVA Impedance: 15%
Line (ACSR160mm^2)	Positive seq.: 3.86+j7.42 %Z/km Zero seq.: 9.87+j22.68 %Z/km (Base: 100MVA)
Load	Nonlinear: 200kW Linear: 800kW
DC Link	Energy Capacity: 2MJ Capacitance: 8000uF
Synchronous Generator	Rated power: 1.2MVA Rated voltage: 480V

4.2 Simulation Results

Fig.4 shows the simulation results. The simulation sequences are as follows.

□ 0.2 sec: Beginning MDG operation
(Paralleling mode)

□ 0.3 sec: Constant power control
(P_{sp} = 0.5MW)

□ 0.4 sec: Occurrence of voltage sag

□ 0.5 sec: Occurrence of interruption
(Islanding mode)

MDG operates in paralleling mode from 0.2 sec to 0.5 sec. In the duration, MDG has to improve power quality and supply electric power to the load. Fig.4 (b) shows the grid line current of which low frequency harmonic components are nearly eliminated after 0.2 sec and Fig.4 (e) shows that MDG compensates reactive power. Then, after 0.3 sec, MDG starts to supply active power following active power set point value 0.5MW. Furthermore, after 0.4 sec, MDG compensates voltage sag as illustrated in Fig.4 (c).

After 0.5 sec, the grid line cannot supply power to the load because of interruption. Then, MDG should be operated in islanding mode. In this mode, MDG supplies load voltage and power required by loads. It can be seen that MDG provides full load power well in Fig.4 (d) and load voltage is maintained constantly in Fig.4 (c).

Therefore, it can be thought that all the functions of paralleling mode and islanding mode executes well.

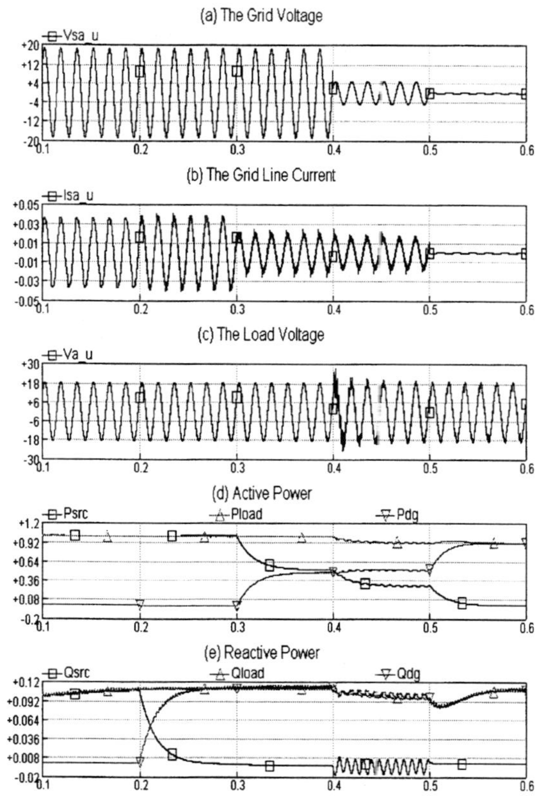

Fig 4. The simulation results

(a) The voltage supplied by grid
(b) The grid line current
(c) The load voltage
(d) Active power of grid supplying, MDG supplying, and load consuming
(e) Reactive power of grid supplying, MDG supplying, and load consuming

CONCLUSION

Since the proposed MDG consists of synchronous generator, rectifier with DC link, SSB, and series and shunt inverter, it can supply electrical power like a general DG and compensate various power quality events like an UPQC. The advantages and flexibility of the proposed MDG configuration and the role of each component are presented. In addition, the control schemes of the MDG are presented and also its numerical formulations are presented.

In the case study, the power quality improvement and reliable power supply in the paralleling operation and islanding operation is evaluated. From the results, proposed MDG can improve the power reliability and quality under the severe disturbances such as voltage sag, harmonic distortion, and unexpected isolation.

The proposed MDG can be an attractive power source meets the various functions in the power distribution systems.

REFERENCES

Akagi. H, Y. Kanazawa, and A. Nabae (1984). Instantaneous Reactive Power Compensators Comprising Switching Devices without Energy

Storage Components. *IEEE Trans. on Industry Application*, **Vol. 1A-20, No.3, May/June**, pp625-630.
Aredes. M, K. Heumann, and E.H. Watanabe (1998). An Universal Active Power Line Conditioner. *IEEE Trans. on Power Delivery*, **Vol.13, No.2, April**, pp545-551.
Barsali. S, M. Ceraolo, P. Pelacchi and D. Poli (2002). Control techniques of Dispersed Generators to improve the continuity of electricity supply, *PES Winter Meeting, IEEE*, **Vol.2**, pp789-794.
Choi. J and J. Kim (2001a). Advanced Voltage Regulation Method of Power Distribution Systems Interconnected with Dispersed Storage and Generation Systems (Revised), *IEEE Trans. on Power Delivery*, **Vol.16, No.2, April**, pp329-334.
Choi. J (2001b). Operation and Integration Strategies of Dispersed Generations in Power Distribution System, Ph.D Dissertation, Soongsil University.
Gish. W.B (1986). Small Induction Generator and Synchronous Generator Constants for DSG Isolation Studies, IEEE Trans. on Power Delivery, **Vol.1**, April, pp.231-239.
Kundur. P (1993). *Power System Stability and Control,* Chapter 4, McGraw-Hill Inc.
Lasseter. R (2001). Dynamic Models for Micro-Turbines and Fuel Cells, *PES Summer Meeting, IEEE*, **Vol.2**, pp761-766.
Li. R, A.T. Johns, and M.M Elkateb (2000). Control Concept of Unified Power Line Conditioner, *PESWinter Meeting, IEEE*, **Vol. 4**, pp2594–2599.
Marei. M.I, E.F. El-Saadany, and M.M.A. Salama (2002). Flexible Distributed Generation: (FDG), *PES Summer Meeting, IEEE*, **Vol.1**, pp49-53.
Joos. G, B.T. Ooi, D. McGillis, F.D. Galiana and R, Marceau (2000). The Potential of Distributed Generation to Provide Ancillary Sevices, *PES Summer Meeting, IEEE*, Vol.3, pp1762-1767.

ACKNOWLEDGEMENT

This work has been supported in part by Electrical Engineering & Science Research Institute under grant 00-JEON-01 which is funded by Korea Electric Power Co.

BIOGRAPHIES

Il-Yop Chung received his B.S. and M.S. degrees in School of Electrical and Computer Engineering at Seoul National University, Seoul, Korea in 1999 and 2001 respectively. He is currently Ph. D candidate at Seoul National University. His research interests include power quality analysis and diagnosis, custom power devices and dispersed generation.
E-mail: iryop@powerlab.snu.ac.kr

Sung-Woo Park received his BS degrees in Electrical Engineering at Korea University, Seoul, Korea in 2001. He is currently MS student at Seoul National University. His research interests include power quality analysis, custom power devices and dispersed generation.
E-mail: swp0706@powerlab.snu.ac.kr

Joon-Ho Choi received his B.S., M.S., and Ph. D. degrees in electrical engineering from Soongsil University, Korea, in 1996, 1998, and 2002, respectively. He was a Post-Doctoral Fellow of School of Electrical and Computer Engineering at Seoul National University, Korea in 2003. Currently, he is a full-time lecturer in Chonnam National University. He is a member of IEEE and KIEE. From 2001, he has been the reviewer of IEEE PES Transactions Paper. His areas of interest are operation and integration strategies of dispersed generation, distribution automation system, and intelligent application of power system control.
E-mail: joono@korea.com

Seung-Il Moon received his B.S. degree in electrical engineering from Seoul National University, Korea in 1985, and M.S. and Ph. D. degrees from Ohio State University in 1989 and 1993, respectively. Currently he is an associate professor of School of Electrical and Computer Engineering at Seoul National University, Korea. He is a member of IEEE and KIEE. His research interests include analysis, control, and modelling of the power system, power quality and FACTS.
E-mail: moonsi@plaza.snu.ac.kr

Jae-Eon Kim received his B.S. and M.S. degrees from the University of Hanyang in 1982 and 1984, respectively. He was with KERI (Korea Electrotechnology Research Institute) as a researcher from 1984 to 1992. He received his Ph. D. degree from Kyoto University, Japan, in 1996. He has since been KERI as a senior researcher. He has been an associate professor at the Chungbuk National University since 1998. He is a member of IEEE and member of KIEE. His current interest is optimal operation and design of power distribution systems with non-utility generation systems.
E-mail: jekim@trut.chungbuk.ac.kr

ELSEVIER
IFAC
PUBLICATIONS
www.elsevier.com/locate/ifac

ANALYSIS OF SYSTEM IMPACT OF THE DISTRIBUTED GENERATION USING EMTP WITH PARTICULAR REFERENCE TO PROTECTION STRATEGIES

S. M. Yeo, C. H. Kim

Information and Computer Engineering, Sungkyunkwan University
300, Cheoncheon-dong, Jangan-gu, Suwon-city, Gyeonggi-do, Korea

Abstract: With the advent of distributed generation, the overall power system are impacted on stability and power quality. The distributed generation has a positive impact on system restoration following a fault, higher reliability, and mitigation of effect due to voltage sag. However, the distributed generation also has a negative impact on decrease of reliability such as changes of protective device setting and mal-operation.
Because bulk power systems consist of various sources and loads, it is complicated to analyze power system with distributed generation. The types of distributed generation usually are classified both rotating machinery and inverter-based system.
In this paper, the distributed generation is designed by synchronous generator, and the distribution system with model of the distributed generation is simulated using EMTP. In addition, this paper shows the simulation results according to the types of distributed generation. *Copyright © 2003 IFAC*

Keywords: Distribution system, Electric power systems, Fault, Generator, Power-system voltages, Protection

1. INTRODUCTION

Since the complexity of modern power systems is increasing (longer lines, increased power transfer over existing lines due to the limitations imposed by environmental pressures, etc.), traditional analogue relaying is no longer able to cope with the performance requirements, and hence the advantage of employing digital protection relays which are much better suited to cope with the modern-day protection problems, particularly in terms of speed and accuracy (Kim, *et al.*, 2000). Recently there has been a considerable issue connecting new generation to the distribution network and this is known as embedded, distributed or dispersed generation (Nick, *et al.*, 2000; Schweer, 1997).

Such distributed generation has some advantages and disadvantages. When fault occurs in power system, protective devices have mal-operation and power system reliability is decrease. Distributed generation in distribution system has various effects such as network voltage changes, increase fault levels.

Many researchers have studied distribution network including distributed generation (Milanovic and Gnativ, 2001; Girgis and Brahma, 2000; Miller, *et al.*, 2001). Particularly, J. V. Milanovic and R. Gnativ (2001) studied characteristics of voltage sags in radial network with dynamic loads and embedded generator, and A. Girgis and S. Brahma (2000) studied effect of distributed generation of protective device coordination.

In this paper, the distribution system with distributed generator is simulated for various system conditions in order to analyze and understand system impact of the distributed generation with particular reference to protection strategies. So, a model of the distribution system and distributed generation is designed as synchronous generator, and simulated using EMTP. In this paper, simulation results is represented and impacts of distributed generation are analyzed.

2. DISTRIBUTED GENERATION

Distributed generation was investigated be working groups, CIGRE (The International Conference on Large High Voltage Electric Systems) and CIRED (The International Conference on electricity Distribution Networks). Distributed generation consists of synchronous or induction generator, or power electronic device. And they have some common attributes as follows generation (Nick, *et al.*, 2000; Schweer, 1997).

(1) Not centrally planned
(2) Not centrally dispatched
(3) Normally smaller than 50-100MW
(4) Usually connected to the distribution system.

The CIRED survey and the CIGRE report represented reasons of encouraging distributed generation. Some reasons are followed (Nick, *et al.*, 2000).

(1) Reduction in gaseous emissions
(2) Energy efficiency or rational use of energy
(3) Deregulation or competition policy
(4) Diversification of energy sources
(5) Ease of finding sites for smaller generators

Because environmental impact is a major factor in the consideration of electrical power scheme, distributed generation is classified into combined heat and power generation and renewable energy resources such as wind power, micro-hydro, solar photovoltaic, landfill gas, etc The different classification of distributed generation is for types of generation, such as synchronous, induction generator or inverter-based system.

A distributed generator is connected to an electrical distribution network. Distributed generator is used to supply electric power for consumer typically but, sometimes electric power generated by distributed generator flows reversely distributed network. So, it is important to check that they will not degrade the quality of supply for the consumer. Fig. 1 shows power flow at distribution network with distributed generation.

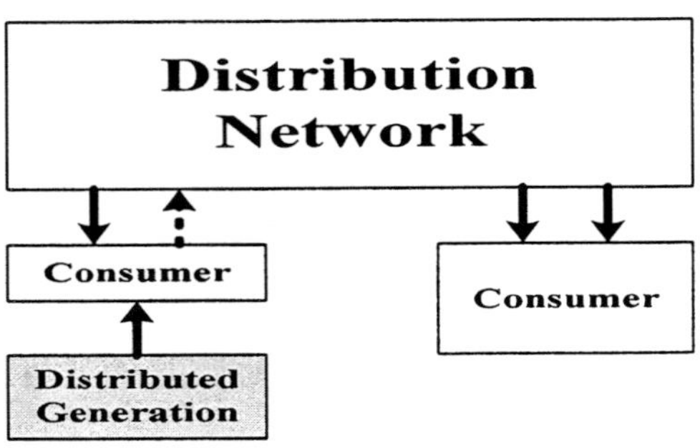

Fig. 1. Power flow at distribution network

In this paper, we design for a model of synchronous generator using EMTP for simulation.

3. SIMULATION AND RESULTS

3.1. System model studied

Model system of distribution network for simulation is shown in Fig. 2. Model system has 5 buses, 3 transformers, and 3 loads. Total simulation time is 1.5[sec], and sampling frequency is 3840[Hz].

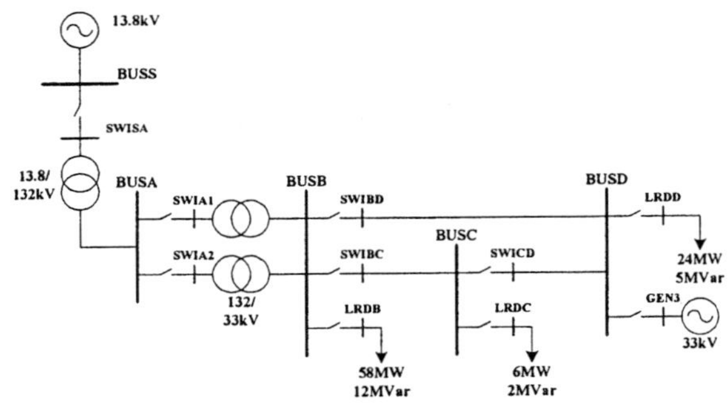

Fig. 2. System diagram for maximum load

Model system is introduced by reference (Nick, *et al.*, 2000) and is implemented by EMTP. BUSS has been chosen as the slack bus, so this generator will supply the active power required to balance the system. So, BUSS is taken as ideal source. Table 1 is system parameters. Simulation was performed in 3 cases of system conditions.

(1) Maximum load, no distributed generation (DG)
(2) Maximum load, synchronous generator at BUSD
(3) Minimum load, synchronous generator at BUSD

And, a distributed generation under fault condition is simulated to analyze the system impacts. Table 2 shows fault types and fault locations. Fault starting time is 0.25[sec] and duration is 0.25[sec].

Table 1 System parameters for networks

From bus	To bus	Type	Resistance	Reactance
BUSS	BUSA	transformer	0.0	0.06670
BUSA	BUSB	transformer	0.00994	0.20880
BUSA	BUSB	transformer	0.00921	0.21700
BUSB	BUSC	line	0.04460	0.19170
BUSB	BUSD	line	0.21460	0.34290
BUSC	BUSD	line	0.23900	0.41630

Table 2 Fault types and locations

Fault type	Single line-to-ground (SLGF) 3-phase fault
Fault location	BUSA, BUSC, LineBD (line between BUSB and BUSD)

Load as shown in Fig. 2 is maximum load. Fig. 3 is shown for minimum load, where the loads have been set at 10% of the maximum loads.

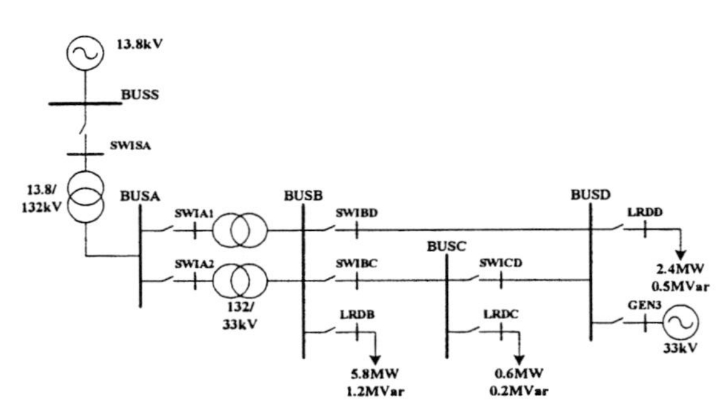

Fig. 3. System diagram for minimum load

3.2 No distributed generation system

At each bus and line, voltages and currents can be obtained, so RMS values are calculated. Per unit quantities are introduced, and baseMVA is 50MVA. Table 3 lists voltages at each bus under normal condition. As shown in figure 4, because voltage was dropped by loads and line impedances, a bus which was far from source has lower voltage than 1[pu]. For example, voltage at bus BUSD is about 0.98[pu].

Table 3 Voltages at each bus under normal condition

Bus	BUSS	BUSA	BUSB	BUSC	BUSD
Voltage [pu]	1.0	1.0	0.994	0.991	0.982

Table 4 lists currents through each switch connected with line, bus and transformer. Currents flow from bus BUSS to each load at buses, and each current has correlation as shown in the following equation (1).

$$i(SWISA) = i(SWIA1) + i(SWIA2) = i(SWIBC) + i(SWIBD) + i(SWICD) \quad (1)$$

Table 4 Currents through each switch

Switch	SWISA	SWIA1	SWIA2	SWIBC	SWIBD	SWICD
Current [pu]	2.6	1.4	1.2	1.08	0.79	0.73

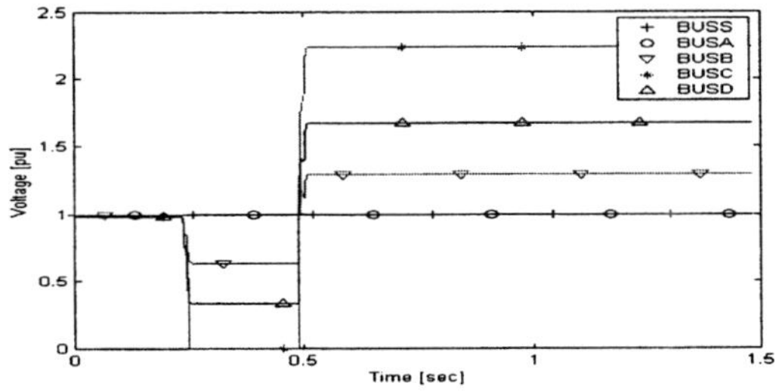

Fig. 6. Voltages at each bus under SLGF (BUSC)

Fig. 6 shows waveforms of voltage at each bus under single line-to-ground fault at bus BUSC with duration of 0.25[sec]. At this, post-fault voltage is not restored to pre-fault voltage, moreover it is increased larger than pre-fault voltage. Fig. 7 presents currents through each switch.

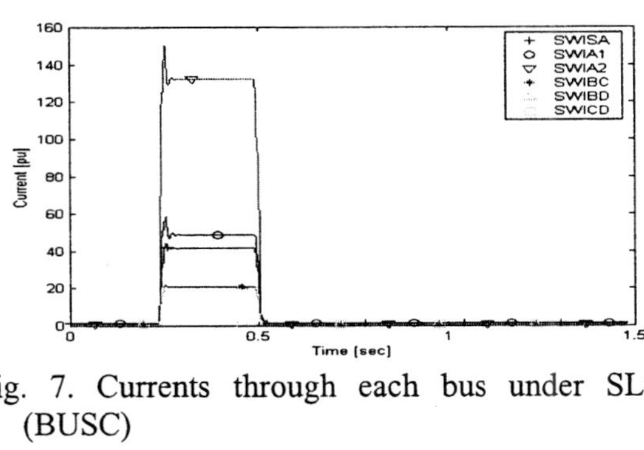

Fig. 7. Currents through each bus under SLGF (BUSC)

Figures 8-9 show waveforms of voltage and current under 3-phase fault at bus BUSC, respectively.

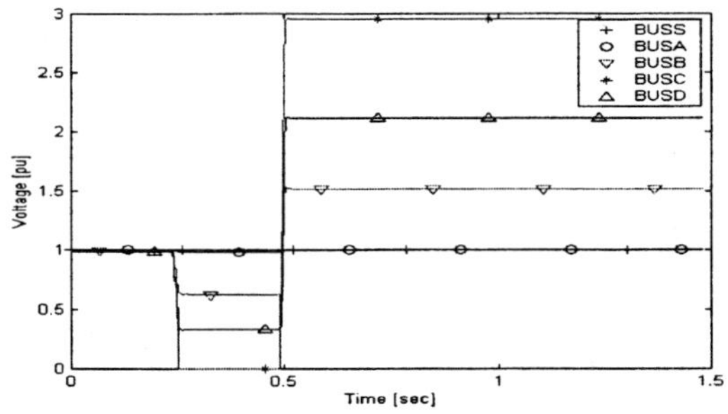

Fig. 8. Voltages at each bus under 3-phase fault

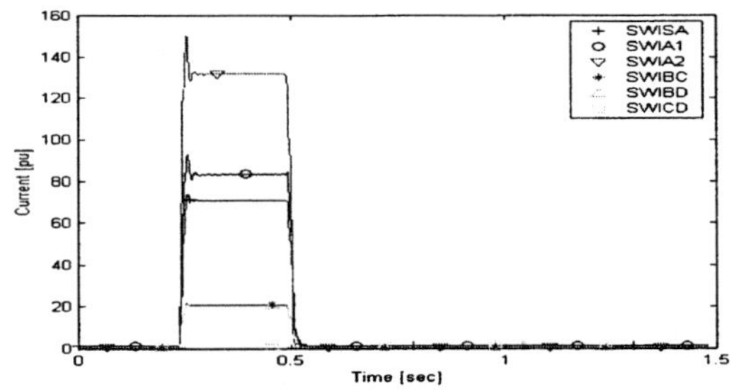

Fig. 9. Currents through each switch under 3-phase fault

As shown in figures, currents are magnified during fault and magnitude of voltage after fault is inverse proportion to distance from source. Figure 10 presents waveforms of voltage at bus BUSD under 3-phase fault. At this, we know that voltages of each phase are not balanced.

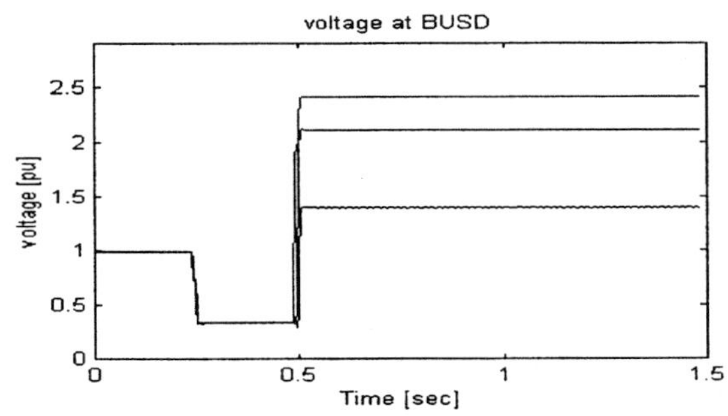

Fig. 10. voltages at BUSD under 3-phase fault

3.3 Maximum load, synchronous generator

Figures 11-12 show waveforms of voltages and currents under transient single line-to-ground fault at bus BUSC, respectively.

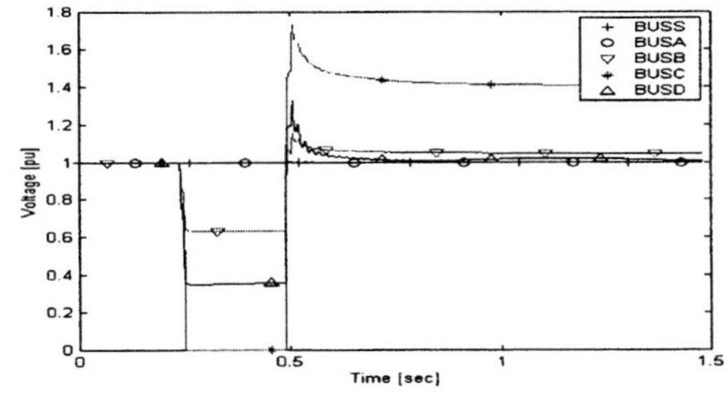

Fig. 11. Voltages at each bus under SLGF (BUSC) with DG

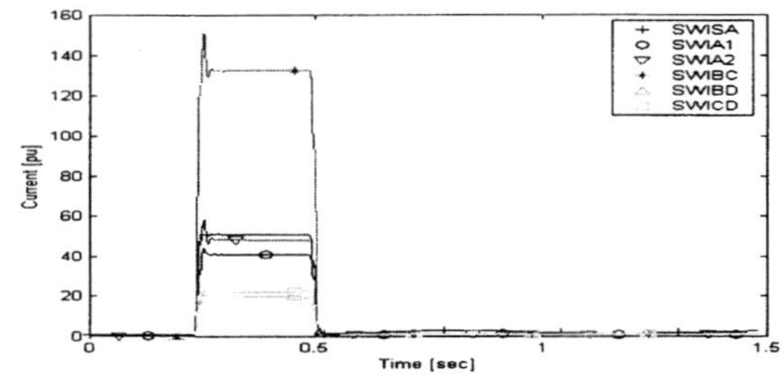

Fig. 12. Currents through each switch under SLGF (BUSC)

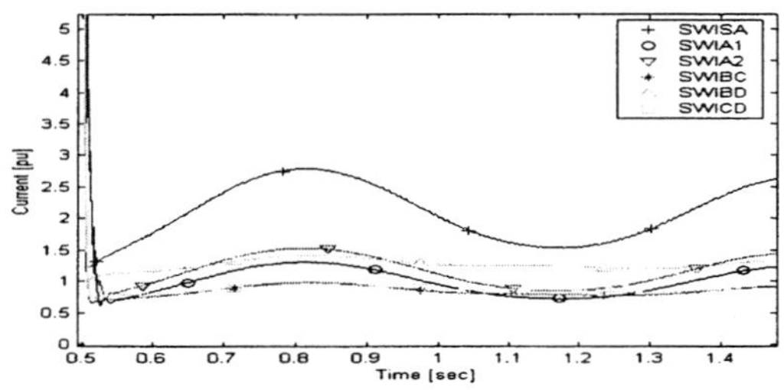

Fig. 13. Enlarge of figure 12

In this case, post-fault voltage is restored to below 1.1[pu] against cases of no distributed generation. Also, currents are similar magnitude during fault, but currents are larger than ones in the case of no distributed generation after fault. Figure 13 present enlarge of figure 12.

Figure 14 shows waveforms of voltage under 3-phase fault. Voltages of all buses except faulted bus are restored to pre-fault voltage like case of single line-to-ground fault. But, phenomenon of power swing at bus BUSD connected with distributed generation can be observed, and voltages at other buses are larger than ones in case of single line-to-ground fault.

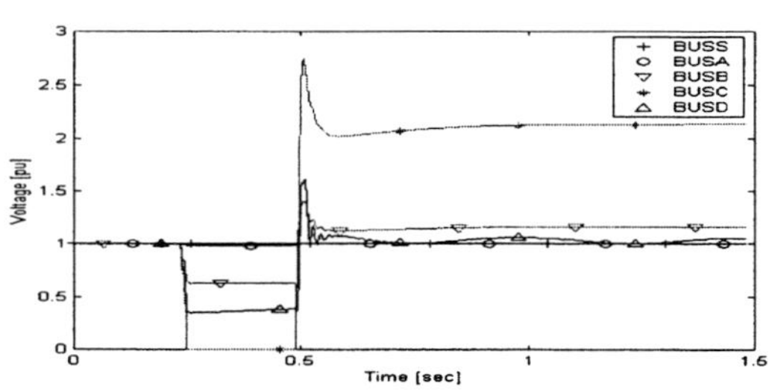

Fig. 14. Voltages at each bus under 3-phase fault with DG

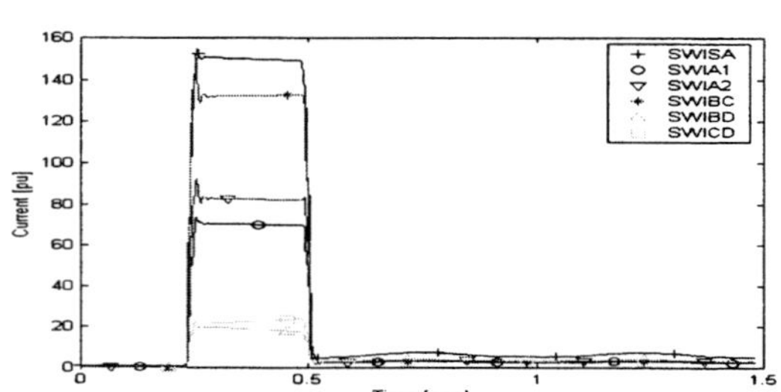

Fig. 15. Currents through each switch under 3-phase fault

Figure 15 shows waveforms of currents through each switch. Because of the power generated by synchronous generator, trajectory of currents is different from the case of no distributed generation.

Figure 16 shows the effect of single line-to ground fault at bus BUSC with distributed generation at bus BUSD in comparison with no distributed generation. As shown in figure 15, if distributed generation exists, voltages at each bus are restored to 1.0[pu] approximately.

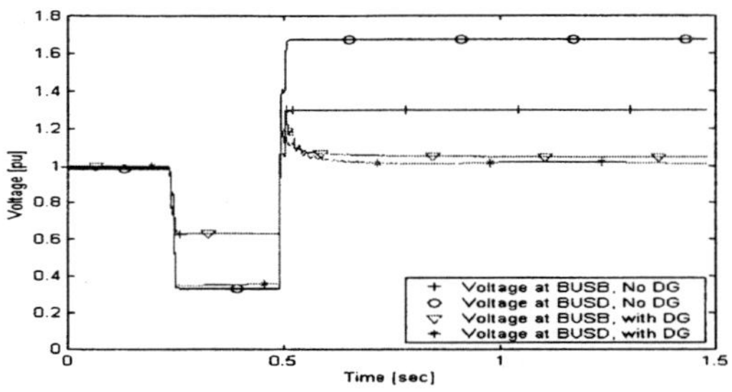

Fig. 16. Comparison of "No DG" case with "with DG" case

3.4 Minimum load, synchronous generator

In order to show the effects of distributed generation in case of minimum load, the case of minimum loads in figure 4 is simulated. Figures 17-18 present waveforms of voltage and current under single line-to-ground fault, respectively.

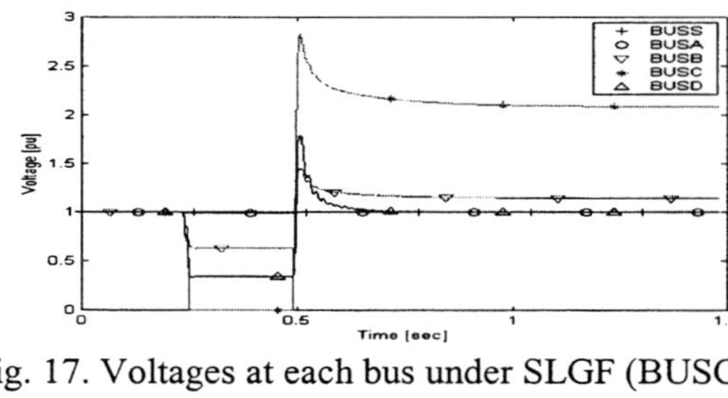

Fig. 17. Voltages at each bus under SLGF (BUSC)

As shown in figure 18, in before-fault state, the magnitude of currents is very small values about 0 [pu], but, during fault, it's magnitude is same values as shown of figures 7, 9.

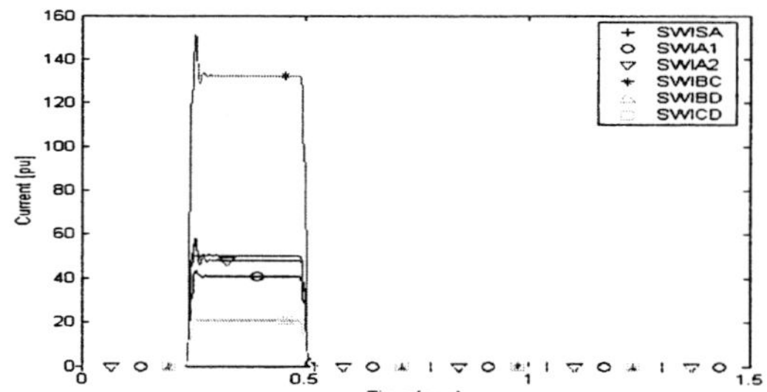

Fig. 18. Currents through each switch under SLGF (BUSC)

Figure 19 show waveforms of voltage at bus BUSD under single line-to-ground fault in three types of system condition. Voltage is restored to 1.0[pu] by synchronous generator after fault, and voltage in case of minimum load quickly converges on pre-fault voltage against the case of maximum load.

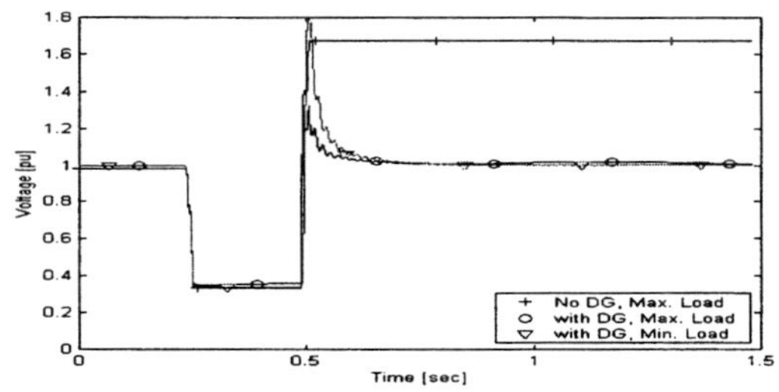

Fig. 19. Comparison of various cases of simulation

3.5 Summary

Faults are shown similar results at other location such as BUSA, LineBD between BUSB and BUSD. Figure 20 presents waveforms of voltages at bus BUSB and BUSD under three fault locations and two system condition.

Table 5 shows RMS values of voltages and currents under other fault location. Each value is measured in maximum load and DG case after fault clear.

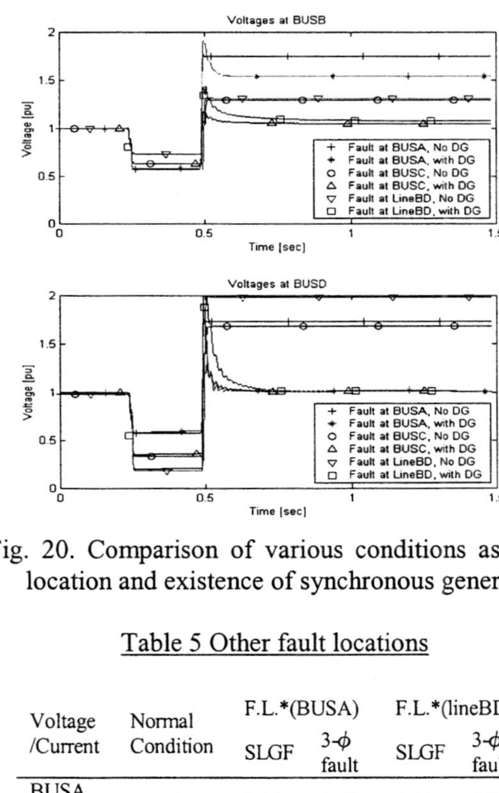

Fig. 20. Comparison of various conditions as fault location and existence of synchronous generator

Table 5 Other fault locations

Voltage /Current	Normal Condition	F.L.*(BUSA) SLGF	F.L.*(BUSA) 3-ϕ fault	F.L.*(lineBD) SLGF	F.L.*(lineBD) 3-ϕ fault
BUSA [pu, V]	1.00	3.20	3.57	1.00	1.00
BUSB [pu, V]	1.00	1.54	1.65	1.08	1.10
BUSC [pu, V]	1.00	1.29	1.36	1.04	1.08
BUSD [pu, V]	1.00	1.01	1.1	1.01	1.07
lineBC [pu, I]	0.77	0.94~ 1.21	3.92~ 4.85	0.76~ 1.00	2.60~ 3.25
lineBD [pu, I]	1.04	1.43~ 1.74	5.89~ 7.01	1.16~ 1.43	3.90~ 4.68
lineCD [pu, I]	0.71	0.99~ 1.12	3.93~ 4.77	0.80~ 0.97	2.77~ 3.20

* F.L.: Fault Location
** Voltage: maximum value among phase A, B, C
*** Current: from minimum value to maximum value

Voltage at BUSD with DG is shown about 1.01~1.10 pu. But, currents have large magnitude, and fluctuation. Particularly, currents are very large magnitude during fault. So, this situation is related to the capacity of breaker, and lead on breaker failure. Figure 21 shows waveform of voltage of phase A at bus BUSD under different fault duration. Simulated fault also is single line-to-ground fault. As shown in figure 21, the shorter clearing time is, the better system stability will be. If fault continues above several duration, then synchronous generator losses synchronism and system is unstable.

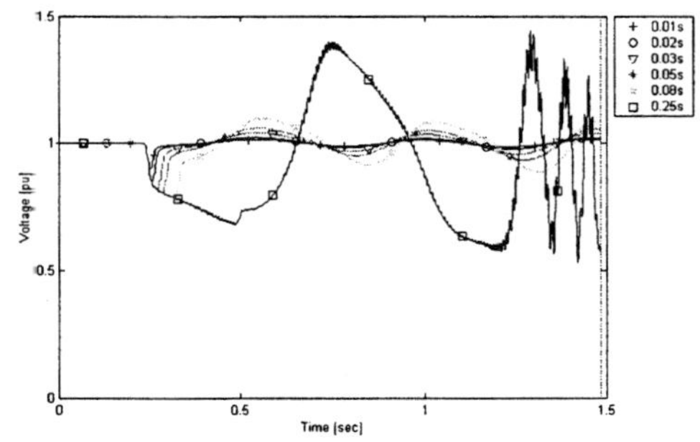

Fig. 21. Voltages at BUSD for different duration of fault

As shown in figure 21, if fault duration is below 0.02~0.03 [sec] (about 1~2 cycles), post-fault voltage restores to pre-fault voltage approximately. But, if fault has longer clearing time, fluctuation of voltage will be severe, what is worse, synchronous generator will loss synchronism.

4. CONCLUSION

Typically, when fault occur, current is several times as large as normal current. So, power system engineer choose breaker that suitable to clear fault. But, if power system is connected DG, fault current will be larger than no DG case and currents may fluctuate after fault clear. Sometimes, synchronous generator may loss synchronous.

So, distribution network with distributed generation is simulated, and the system impact is analyzed by using EMTP. This simulation results are shown serious impact of distributed generation under fault. Therefore power system engineer can choose new breakers, or establish new protection strategies. For example, when fault occur, the fault will be more quickly cleared and islanding distributed generator.

5. ACKNOWLEDGEMENT

The authors would like to thank Korea Ministry of Commerce, Industry and Energy and Korea Electric Power Research Institute for their support.

REFERENCES

Girgis A. and Brahma S. (2000), Effect of Distributed Generation on Protective Device

Coordination in Distribution System, *Power Engineering, LESCOPE '01*, pp. 115-119.

Kim C.H., Lee M.H., Aggarwal R.K. and Johns A.T. (2000), Educational Use of EMTP MODELS for the Study of a Distance Relaying Algorithm for Protecting Transmission Lines, *IEEE Trans Power Systems*, **Vol. 15, No. 1**, pp. 9–15.

Milanovic J. V. and Gnativ R. (2001), Characteristics of Voltage Sags in Radial Networks with Dynamic Loads and Embedded Generators, *2001 IEEE Porto Power Tech Conference, Porto, Portugal*, **Vol. 1**.

Miller N.W., Walling R.A. and Achilles A.S. (2001), Impact of Distributed Resources on System Dynamic Performance, *Transmission and Distribution Conference and Exposition, 2001 IEEE/PES*, pp. 951-952.

Nick Jenkins, Ron Allan, Peter Crossley, David Kirschen and Goran Strbac (2000), *Embedded Generation*, IEE Power and Energy Series, London, United Kingdom.

Schweer, A. (1997), Special report for Session 3, *CIGRE Symposium on Impact of demand side management, integrated resource planning and distributed generation, Neptun, Romania*, Paper no. 300-00.

ELSEVIER
IFAC
PUBLICATIONS
www.elsevier.com/locate/ifac

PROTECTION OF A NEW 500 KVA STATIC VAR GENERATOR

Chen, Xianming Xu, Heping Wang, Xiaohong Wang, Tong
Nanjing Automation Research Institute
P.O.Box 323, Nanjing, 210003, PRC Email: ecti@nari-china.com

Abstract: The paper briefs the main configuration of a new 500 kvar static var generator (SVG) designed and manufactured by the affiliated workshop of our Institute (NARI) and its features, then describes many different protection measures adopted for safety and smooth operation of SVG, specially for protection of gate turned-off thyristors(GTOs) in voltage source inverter (VSI) of SVG. Some interesting test results are also given. *Copyright © 2003 IFAC*

Keywords--Protection, Static Var Generator (SVG), Three-level Inverter (TLI), Fast fuse, GTO

I INTRODUCTION

For the time being SVG (statcom) is one of the best controllable equipments for power network.. Of course, the operation reliability of SVG is rather important. In addition, the cost of SVG is still quite high nowadays. Specially, the price of powerful switch device such as GTO or IGCT is much higher than that of power capacitor, so the protection of SVG from damage is one of major factor in the design consideration of SVG.

Generally speaking, the reliability of GTO is much less than that of general use thyristor because of the structure reason. The main failure for GTO is happened when GTO turns off the current bigger than its rated value.

Fig .1 shows the single–line diagram of 500 kvar SVG connected to a power network. The configuration of the SVG is rather compact [1].

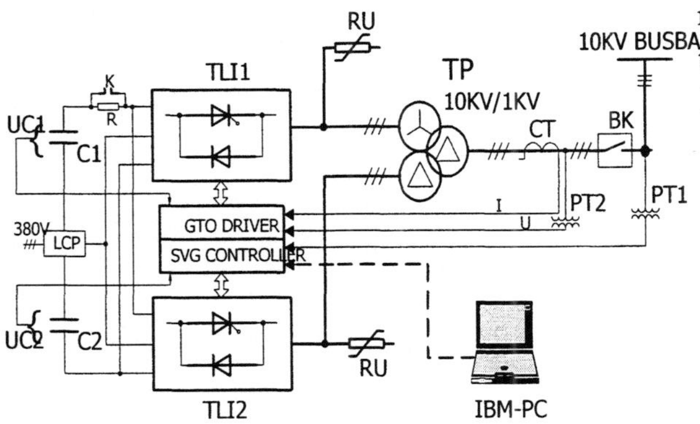

Fig.1 Single-line diagram of 500 kvar SVG GTOs, and two neutral point clamped diodes Du.

It mainly consists of a three-phase three-winding split transformer (TP) and two GTO voltage source inverters, (VSI) formed by a neutral point-clamped three-level inverter (TLI). In addition, several other components such as: capacitor banks C1,C2 for supporting dc voltage of VSI, equipments for SVG starting(resistor R, contactor K), loop current protection (LCP) for limiting rise speed of faulty current flowing through GTOs, control and regulation part for SVG, some PTs and CTs for detecting voltage, current concerned and synchronization, a note-book computer for man-machine interact, etc.

Now 500kvar SVG is connected on a 10.5 kv power network in the enterprise which is producing calcium carbide with an electric arc furnace (EAF) .The capacity of the transformer for EAF is 5 MVA.

II. PROTECTION OF 500 Kvar SVG

The main cost of 500 kvar SVG is contributed by expensive GTO thyristors, so that protection them from damage is rather a big task There several measures are adopted for protection of SVG so as to guarantee them to work normally.

II.1 *Proper gate drive for GTOs of TLI*

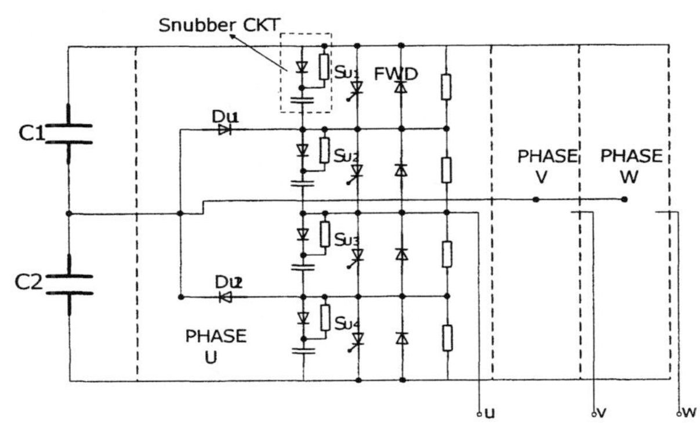

Fig.2 A phase of TLI with capacitors C1,C2 in dc link

Fig.2 shows one phase circuit of TLI, it has four

GTOs, Su , four free wheel diodes(FWD), snubber circuit for GTOs, and two neutral point clamped diodes Du.

In normal operation the phase output of TLI has three electric level such as: 1,0,-1 when Su1 Su2,or Su2 Su3, or Su3 Su4 switch on respectively.

If accidentally Su1,Su2,Su3 or Su2,Su3,Su4, namely Su1 (or Su2) and its complementary Su3 (or Su4) or all switch devices in one phase of TLI turn on simultaneously ,the serious short-circuit will happen via capacitors C1,C2. Thus C1 or/and C2 will discharge immediately with huge current (hereafter call it "straight through"),because there is not any significantly impedance presented on that circuit. This will often cause the damage of switch device.

In order to avoid such trouble, the PWM output pulses from SVG controller to switch device GTOs and its complementary was designed in accordance with following principle.

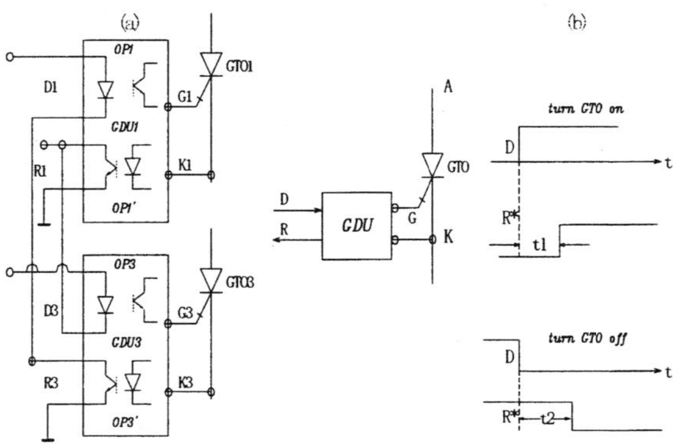

Fig.3 GTO drive unit(b) and mutual lock device(a)

Fig.3b shows GTO drive unit(GDU) adopted in 500 kvar SVG. There are two optical isolator used in a GDU of GTO.

One(op1)is for input PWM pulses D1 from the controller to gate of GTO1, the other (op1') is for output a return signal R1 from GDU1 to inform the controller that GTO1 has just turned on or turned off. Then the controller is enable to send PWM pulses to its complementary, GTO 3.

Fig.3b also shows the electric level of the drive signal D and return signal R in both turning on and turning off time. Apparently, both D and R are high electric level when GTO is turning on.

It can be seen that there is some time delay(t1) after signal D arrival to truly conduction of GTO. On the other hand, the same is for the turning off GTO, but time delay (t2) depends on the current at the instant and much large than t1.

II.2 *"Straight through" detector (STD)*

In order to protect GTO of TLI from "straight through" caused by error pulse emerged by whatsoever reason, a STD was designed shown in Fig.4

The main idea for STD is first to detect situation about conduction of GTOs in TLI, namely, to detect drive signal D and return signal R of GDU of switch devices Sx1, Sx2, Sx3 and Sx2, Sx3, Sx4 in three phase of TLI (where x=u,v,w)

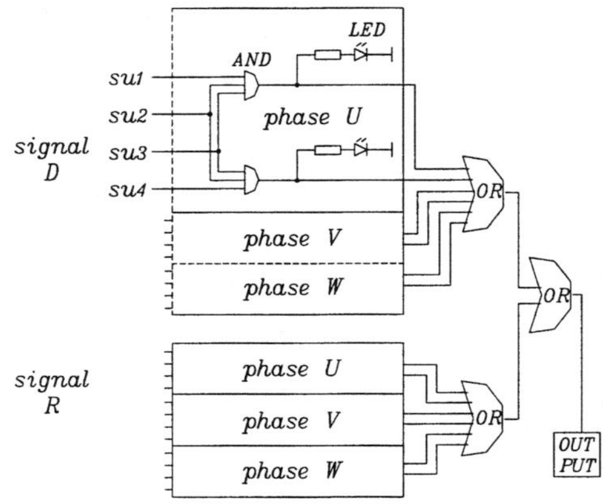

Fig.4 Circuit diagram of STD

If the result of two 'logic AND' gates of these three D and R signals in one phase inputs 'logic OR' with that of other two phase, the final output is high level, that means "straight through" has happened. And the controller is going to lock all PWM pulses straight away and to indicate the location of faulty GTOs by signal LEDs as well.

II.3 *Loop current protection (LCP) device*

The overload capability of GTO is much less than that of general use thyristor because of structure reason.

The most serious failure is three or all GTOs in one phase are turn on simultaneously, namely, "straight through".The faulty current will increase rapidly so that not enough time for the controller to locking all pulses.

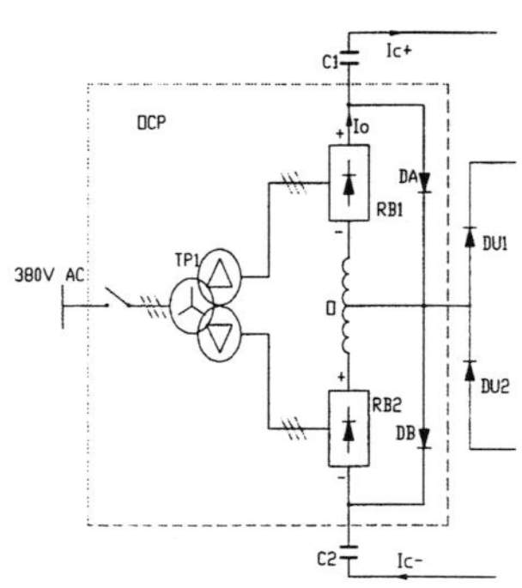

Fig.5 LCP circuit of 500 kvar SVG

That is why loop current protection circuit used. Fig.5 shows the circuit diagram of LCP used. It is only for protect GTOs in TLIs. It consists of one three-phase three-winding transformer TP1, two three-phase rectifier bridges RB1,RB2, two diodes Da, Db and a inductor L with a center tap. As a matter of fact, the upper part of LCP is used only to protect upper GTOs Sx1, Sx2 of three phase in TLIs, and the lower part to GTOs Sx3, Sx4.

For example, the loop current Io in upper part of LCP is predetermined in accordance with the requirement of GTO protection

At first, assuming that the discharge current Ic+ of C1, which runs through Sx1,Sx2 is zero, Io goes through from RB1+,DA and upper half of inductor L, then returns to RB1-. When Ic+<=Io, the dc current flowing on the upper half of the inductor L is a constant Io, and the inductance of the inductor L has no effect at all.

If the discharge current Ic+ of C1 larger than Io, then the electromotive force induced in inductor L will reluct against the further rise of Ic+, that gives enough time for SVG controller to shut down the drive units.

II.4 *Over current and over voltage limitation by software*

For 500 kvar SVG it is permitted to operate with 1.1 rated load continuously, 1.5 for 20 seconds. That is very helpful for power network to maintain voltage level while troubles happens. The major control principle of 500 kvar SVG is to maintain voltage level at the point of power network to which SVG is connected. If voltage of the network is lower than the specific value SVG outputs reactive power to network..

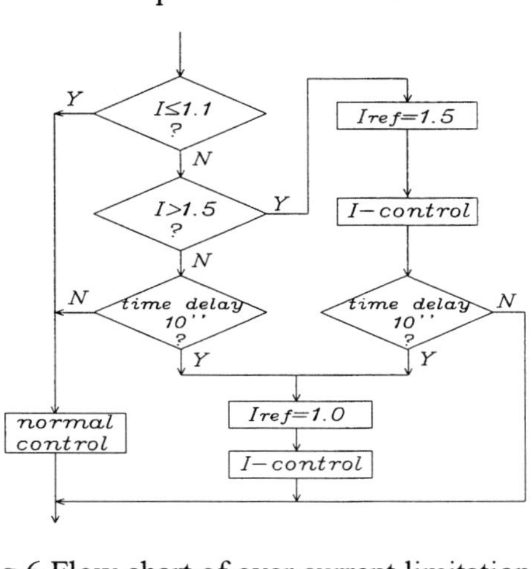

Fig.6 Flow chart of over current limitation

If the former higher than the specific value SVG absorbs reactive power from the network.

When voltage of the network drops largely due to whatsoever happened, SVG may output the current over its allow value. That is why the over current and over voltage limitation routines were designed for SVG.

In 500 kvar SVG reactive power regulation is implemented by controlling phase angle δ between output voltage of SVG and voltage of network.

In order to prevent undue output current of SVG, angle δ is limited in both direction (positive and negative). Fig.6 shows the outline of flow chart of over current limitation program.

It starts to compare SVG current I with 1.1 rated value, if it is not over, nothing will happen. Otherwise and if the current not larger than 1.5 rated, it is going to sustain 10 sec. Then the control of SVG transfers to constant current control mode and keeps the current as rated current, in spite of the network voltage. If the current is larger than 1.5 rated value, the control of SVG transfers to constant current control with 1.5 rated current immediately and sustain for 10 second. Then the current returns to its rated value in constant current control mode. A check program for checking if the network voltage returns to normal is being executed all the time during the limitation process. If it does, SVG waits for a couple of seconds, then the control returns to constant voltage control automatically.

Generally speaking, if SVG has enough capacity, the voltage of network at point where SVG connected can maintain as desired value. Otherwise, voltage of network at the point SVG connected may be over and under its normal value. In 500 kvar SVG if the deviation of network voltage with its normal value $\triangle U$: $\pm 5\% < \triangle U < \pm 10\%$ the control of SVG transfers to constant current control with current no more than rated value. SVG will exit from the network if $\triangle U > 10\%$. The over voltage limitation program and its automatic return program is analogous to that for over current limitation.

As an alternative SVG may be controlled to output arbitrary reactive power as required without consideration of voltage of the network, if necessary. The over current and over voltage limitation program are still in effect here.

II.5 *Hardware over-current protecting (HOCP) device for TLI*

In 500 kvar SVG HOCP is designed to protect GTOs of each TLI if over current limitation is failed. The current for HOCP acting adopted is larger than the current for over current limitation by software. HOCP acquires currents of upper arm, lower arm of all three phase from hall-effect current transducers, for instance, HL1, HL4 for phase U shown in Fig.8a.

All outputs of these six transducers are added to a 'logic OR' gate via diode gates. The direction of diodes should guarantee that the detected currents are flowing through GTOs, not FWDs of TLI.

Output of 'logic OR' gate gets to a comparator via a high frequency filter for isolating the current spike caused by charge or discharge current of snubber capacitors of GTOs during the commutation process between pairs of complementary GTOs, (for instance, Sx1,Sx3 or Sx2,Sx4 in phase x, x=u,v,w)

In comparator the current signal of TLI compare with limit value, if the former bigger than the latter, then a command issues from HOCP for lock all PWM pulses for GTOs, trip off main breaker of SVG. In the meantime, signal lamps concerned are lighting to indicate which arm's current caused the actuation of HOCP.

It is difficult to avoid malfunction of HOCP caused by current spikes. So the proper selection of high frequency filter should be dealt carefully. Specially, the time delay of the filter is unfavorable for fast protecting of HOCP.

II.6 *Protecting GTOs with the help of fast fuses*

One of the conventional method to protect thyristor is using fast fuse. For protecting GTOs this method

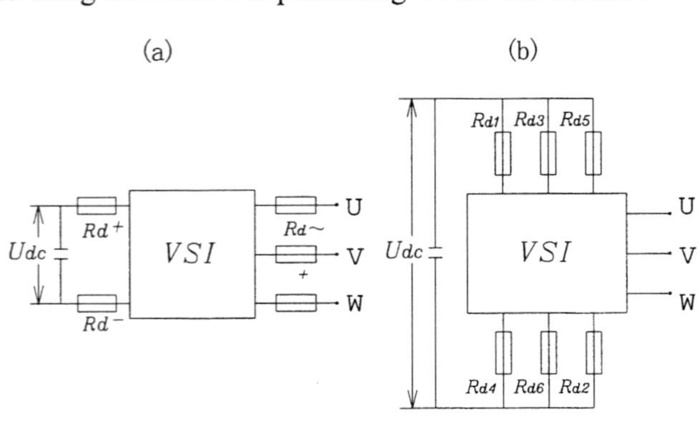

Fig.7 Two ways of fast fuse placing in VSI

is not very effective. Anyway, they are still used in 500 kvar SVG. At least they can help to isolate the failure of GTOs and prevent proliferation of the failure. But the location for fast fuses in TLI should be chosen reasonably.

Furthermore it should estimate the rms value of the maximum permissible current flowing through respective GTOs for matching the rating of the fuses, in order to protect GTOs properly.

Usually, there are two ways to place fast fuses in a VSI as shown in Fig.7. In Fig.7a fuses Rd+, Rd- are used for protection of 'straight through', and fuses Rd~ for protection of over current caused by external circuit. On the other hand, the fuses arrangement of Fig.7b can protect GTOs of inverter from the fault caused by both 'straight through' and external circuit.

In 500 kvar SVG the latter method was adopted, but in TLI there are lots of location for fuses placed, such as point A,B,C,D of Fig.8a

(a)

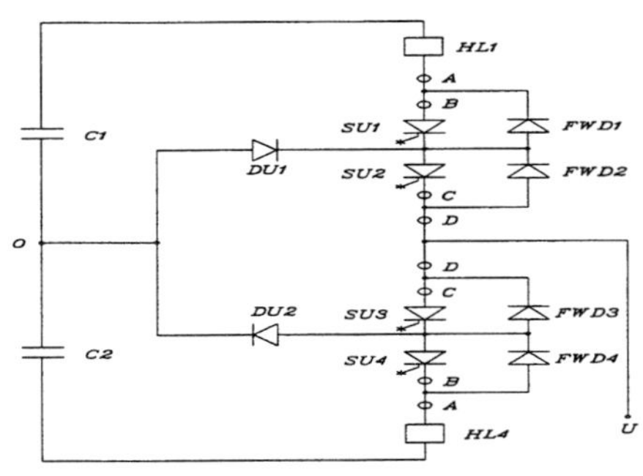

(b)

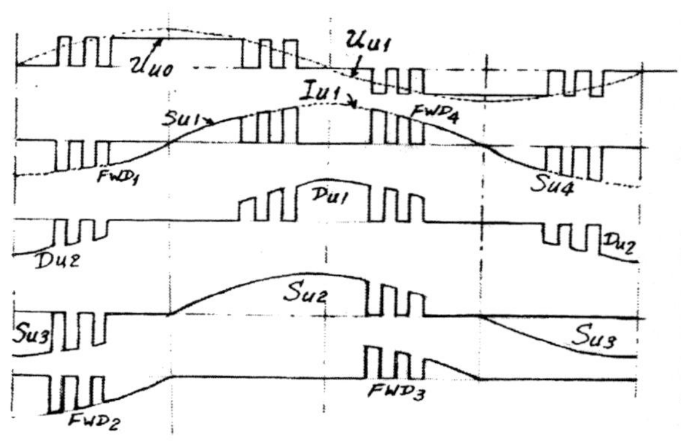

Fig.8 One phase of TLI(a),current in different devices(b)

Fig.8b shows the TLI output voltage Uuo of phaseU to neutral point in dc link of TLI and its fundamental component Uu1, also the output fundamental current Iu1,which is 90 electric

degree lagging Uu1,while SVG outputs reactive power to network. In addition, Fig.8b also illustrates the currents flowing through different components such as Su, Fwd, Du, etc. in phase U of TLI.
From Fig.8b it can be seen that current flowing through Su1 or Su4 is much less than that of Su2 or,Su3 ,because the current in latter are the sum of current flowing through both Su1(or Su4) and Du1(or Du2).

It is easy to know from Fig.8b that in ideal circumstance currents flow through points A,B,C,D equal to ≈25%,<25%, >25% and <50%, 50% of output current of TLI respectively.

But in practice currents flowing through point A,B,C,D are much more due to the presence of charge or discharge current of snubber circuit and some harmonic current of TLI. The rms value of current flowing through point A measured in field was almost 50% of output current of TLI.
In considering that the FWD diode has strong overload ability, the ideal place for fast fuse may be point B,C which are suitable for protecting GTO only. Apparently fuse location of point C is better than that of point B. Considering that the convenience of GTO module structure, in 500 kvar SVG the fast fuses are placed at point D of TLI. The rating of fast fuses are selected from 60% of phase current TLI outputs.

Ⅱ.7 *Improvement of snubber circuit for GTOs in TLI*

In 500 kvar SVG improvement of snubber circuit adopted results in reducing voltage spike happened while GTO carrying heavy current turns off.[2]

Ⅱ.8 *Over voltage suppression by zinc-oxide nonlinear resistor (ZONR)*

This is the traditional method for suppression over voltage. There is a set delta-connection ZONR connected to ac output of each TLI for suppression over voltage caused by switching transient. Also ZONRs are used in dc link of TLI for over voltage suppression on capacitors.

Ⅲ. TEST RESULTS

Fig.9 shows the current at point A in Fig.8a of TLI, which was taken in workshop and obviously agreed with what shown in Fig.8b. Fig.10 shows the test about over current limitation. In practice, for easy implementation, instead of voltage drop

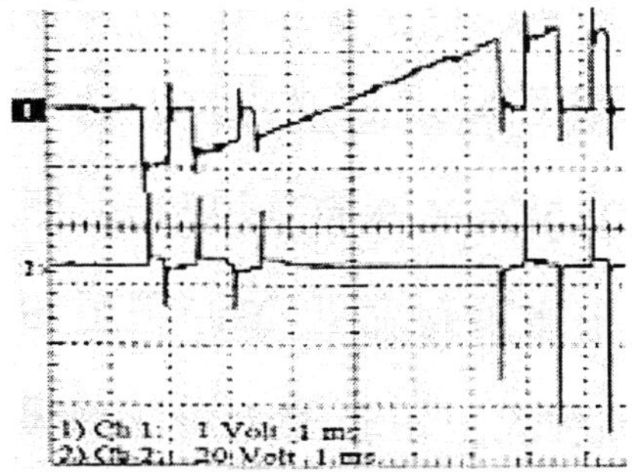

Fig.9 Current at point A of Fig.8a taken in workshop (1) current (2) drive pulse for gate of GTO

of the network, the setting value of voltage of SVG, Uref was raised abruptly, and sustained duration t1. That caused SVG output current I1 rising to the value determined by threshold of δ, and soon decreased to I2 under the constant current control. Current I2 sustained for duration t2, then dropped to I3 permissible for long time operation. If time t1 lapsed and the network voltage recovered normal, SVG output current returned I1 automatically after waiting a couple of seconds.

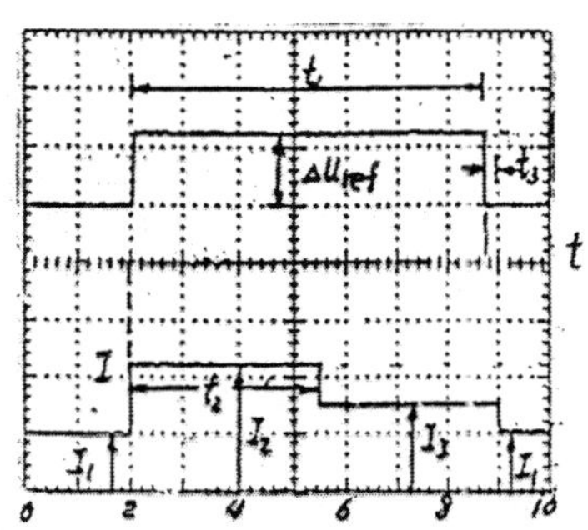

Fig.10 Over current limitation test of 500 kvar SVG

The maximum time for HOCP acting from the over current detection to lock PWM pulse on gates of GTO in TLI here is no more than 65 μs, and the current rise speed is 1.8a per μs. The current tolerance of GTOs of TLI is more than 200a while SVG output 1.5 rated current. So LCP is able to satisfy over current protection requirements
In the commissioning tests of 500 kvar SVG on site, it was found SVG can't normally operate caused by electromagnetic interference (EMI) when the vacuum switch (VS) for the 5 MVA transformer of EAF turns off and turns on. The more serious situation happened when the VS turns on if 5 MVA transformer with full load. Specially, once a error pulse shown in Fig.11 caused by EMI was caught, which was traveling along the 5-meter

control wires, connecting the controller to the gate of GTOs in VSI . That may cause damage of GTOs.

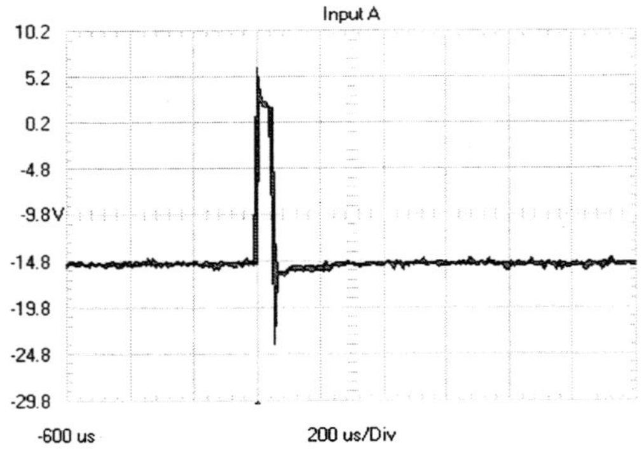

Fig.11 An error (Interference) pulse caught

In consequence, a hardware mutual lock (HML) device shown in Fig.3a is provided for gate control units of GTO with its complementary on same phase. It can be seen that the drive signal D1 for GDU1 of GTO1 is serially connected to return signal R3 of GDU3 for its complementary, GTO3. So optical isolator op1 turns on while drive signal D1 is high, then optical isolator op1'for return signal R1 is open after t1 delay. Because input optical isolator op3 of GDU3 is serially connected to optical isolator op1' of GDU1,not any pulse including interference pulse is able to enter the gate of GTO3..

When drive signal D1 becomes low,op1 is open,GTO1 turns off under the negative pulse produced by GDU1. And after time delay t2 op1' turns on, which is enable the condition for GTO3's conduction,

If GTO3 is conducted, op3' of GDU3 turns off .In this case not any pulse including interference pulse has chance to enter GDU1. That is why the interference pulse is locked outside the GDUs.

In fact, the concrete HML is a bit more complex than that shown in Fig.3a. However SVG is able to operate safely after HMLs installed during VS turns on.

Another problem met in commissioning test was frequent actuation of HOCP while VS turned on, which caused SVG tripping off from the network.

In 500 kvar SVG sampling rate is 300 time a second, namely, six times a period. The fastest regulation rate of software is at least 3.3ms, and the slowest 6.6ms.

The voltage drop of the network is as big as about 10% at the instant VS turns on, but SVG output can not change abrupt ,because of the capacitors in dc link of TLIs. That results fast increase of output current so that over current limitation by software failed to act properly, but HOCP was really acted.

In order to solve this problem, when HOCP acts, only let lock all pulses for GTOs , but not trip off SVG . Then SVG wait a couple of seconds to see if voltage of the network recovers and to detect if all GTOs are right, if it does, then all pulses for GTOs will issue again and return to the original situation step by step.

All these process is implemented automatically. It just like the manipulation of auto-reclosure in power network..

Ⅳ CONCLUSION

Thanks to the all measures mentioned above 500 kvar SVG can operate safely, smoothly and satisfactorily for long time. The protection measures adopted for 500 kvar SVG is basically successful. The main problem was caused by under estimation of importance of electric-magnetic compatibility in SVG design.

In addition, if the capacity of SVG is big enough relative to 5Mva transformer for the EAF, there is no need to use 'auto-reclosure'. Furthermore, it seems that the sampling rate used (300 per second) in 500 kvar SVG ought to be increased further.

V. REFERENCES

1). Chen Xianming et al " Design considerations and test results of a new 500 kvar Static Var Generator" *Proceedings of International Conference on Power System , CIGRE* pp200-205 Sept.3-5 2001, Wuhan,China

2).Chen Xianming et al " Experimental Studies of Three- level Inverter with PWM Control for Static Var Generator" *Proceedings of ICEE'2001,* **Vol.1 of 3** pp513-518 July22-26,2001 Xi'an , China

ELSEVIER
IFAC
PUBLICATIONS
www.elsevier.com/locate/ifac

A STUDY FOR AN AUTONOMOUS OPTIMAL PROTECTION SYSTEM OF A POWER DISTRIBUTION NETWORK

Byung-Woon Min, Kwang-Ho Jung, Seung-Jae Lee, Myeon-Song Choi, ,Seung-Ho Hyun

Next-Generation Power System Technology Center, Myongji University,
Yongin, 449-728, Kyunggido, Korea

Abstract: In this paper, Every digital over current relay(OCR) is developed as an agent by adding its own intelligence, self-tuning and communication ability. The ability of digital over current relay is one of autonomous optimal protection system. A protection system using a Multi-Agent concept for power distribution networks is proposed. The main advantage of the Multi-Agent concept is that a group of agents work together to achieve a global goal which is beyond the ability of each individual agent. In order to cope with frequent changes in the network operation condition and faults, an OCR agent, suggested in this paper, is able to detect a fault or a change in the network and find its optimal parameters for protection in an autonomous manner considering information of the whole network obtained by communication between other agents. Through this kind of coordination and information exchanges, not only a local but also a global protective scheme is completed. Simulations in a simple distribution network show the effectiveness of the suggested protection system.

Keywords Multi-Agent System, OCR(Over Current Relay), Communication, Protection of distribution network.

1. INTRODUCTION

A multi-agent system is, basically, a kind of decentralized system. An agent performs its own job with given information and functions. Sometimes a group of agents work together to achieve a global goal, in which case each agent does the given job only.

Recently, the rapid development in digital technology and communication has drawn considerable attention of power system engineers to the digital protection systems. Especially, under the circumstances where the size of power systems grow rapidly, and the requirements of the customers for the quality of electricity is also increased, it is quite natural for the engineers to strengthen the ability of each digital protection device.

Authors would like to express their sincere thanks to Korea Ministry of Science and Technology and Korea Science and Engineering Foundation for their support through ERC program.
Manuscript received : June 4, 2002 accepted : August 2, 2002
B.W. Min, K.H. Jung, S.J. Lee, M.S. Choi, S.H. Kang and S.H. Hyun are with Next Generation Power Technology Center in Department of Electrical Engineering, Myongji Univeersuty, Yongin, Kyongdo, Korea.

A distribution network inevitably meets many kinds of faults and sudden changes in its loads or structure. In each case, if we can provide sufficient information to all the digital protection devices and every device has an adequate level of intelligence, then a very reliable and efficient automatic protection of a large scale distribution network can be realized. Especially, if the concept of a multi-agent system is considered from the design stage of a distribution network, the efficiency of design, construction and operation of a distribution network will be enhanced to a considerable degree. That is the ultimate goal of the application of a multi-agent concept to a distribution protection system.

As a first step of the multi-agent protection system, correction methodology of changing pick-up values in an over current relay(OCR) is suggested in this paper. An OCR trips a circuit breaker(CB) to cut the line when the current is beyond the pick-up value. The pick-up value is selected between the maximum load current through the line and minimum fault current at that point, both of which are dependent upon the structure and operation condition of the distribution network concerned. Furthermore, the structure and operation condition change from time to time. Therefore, the auto parameter setting of an OCR with network information considered is the first area to which the multi-agent concept is applied with efficiency.

In this paper, a multi-agent protection system is suggested for a distribution network. Every OCR is defined as an agent. An agent calculates and corrects its parameter by itself through communication with neighboring agents and its own intelligence. No human effort for calculation and re-setting of parameters are needed at all. The proposed concept is applied to a simple network to show its soundness and effectiveness.

2. Multi-Agent System

Several years ago, the concept of an agent system was introduced to software engineering with communication. There are many kinds of definitions and taxonomies on the agent system[3]. A common characteristic is that an agent has 'autonomy, i.e. self starting ability' and 'intelligence'. It has its own knowledge base, reasoning and judging ability. With these tools, an agent performs some functions in an autonomous manner[3-5]. Recently, great efforts are being concentrated on the multi-agent system[2,6-7]. Several agents group themselves automatically and they exchange information with each other and, lastly, they work together and achieve a global goal beyond the ability of an individual agent.

Since a power network is spread over a wide range of areas with many OCR's, the decentralized concept seems to be very useful in control and operation. Especially, recent remarkable developments in digital protection devices and in communication technology enable us to adopt the multi agent concept. It would be very efficient if every OCR corrected its parameters by itself every time a change occurs. Furthermore, the optimal operation can be realized without centralized control action if each agent has an adequate level of intelligence and information. That is why the multi agent concept is focused on as a promising research area in the protection of power systems in which sudden changes or faults may occur frequently and in which there are many devices able to work as agents and in which coordination between devices is absolutely required. A multi-agent system is realized most efficiently if it is considered from the start of the design stage. And this concept should be applied not to an individual device but to the whole system.

In the next section, the application of a multi-agent system to the protection of a distribution network is suggested.

3. Protection system of a Distribution Network using MultI-Agent System

3.1 Basic concept

In a distribution network, an over current relay(OCR) is frequently used as a protection device. The pick up value of an OCR is selected among the values between maximum load current and minimum fault current. Therefore, every time the operation condition changes, the pick up value should be also changed according to it. In this paper, each OCR is defined as an agent so that the frequent changes in a distribution network may be considered automatically.

Another type of agent is the feeder agent which represents the source of electrical power in the distribution network concerned. It works as a starting point of a branch and also gives the OCR agents information on the source, e.g. the source capacity, source impedance and so on.

The last agent is the data agent, or equivalently display agent. This is not a prerequisite for real field application. It gathers data and displays it for an operator or a developer to watch the current state.

Each agent gives and takes information to and from neighboring agents not only to perform its own job but also for other agents to do so. The schematic diagram is given in Fig. 1.

3.2 Design of agents

3.2.1 OCR Agent

The main function of an OCR is to protect the network from a fault by measuring the current and comparing this with a pick up value which is calculated in advance. The main functions of an OCR agent are summarized in the following. Since a distribution network is, in general, operated in the radial structure, we use the notations 'upper' and 'lower' to represent the direction to source and to load respectively. In other words, the end of the 'upper' agent is the feeder agent and the end of opposite side, the OCR agent connected to the load directly.

♦ Self correction

When it receives either the signal of network change or requirement of correction, it sends this signal to neighboring agents and inquires of the upper agent about source impedance.

When it receives the source impedance from the upper agent, the line impedance between the receiver and the sender of source impedance is added to the source impedance received. This updated source impedance is sent to the lower agents. In addition, it calculates the minimum fault current.

It calculates the maximum load current with the load both in lower agents and in itself considered. It sends this maximum load current to the upper agents and decides the pick up current value.

It corrects itself when required

It trips the circuit breaker when required.

It sends data changed to the data agent.

♦ Protection

It measures the current and trips the circuit breaker concerned when the current exceeds the pick up value.

It sends all the protective action data to the data agent.

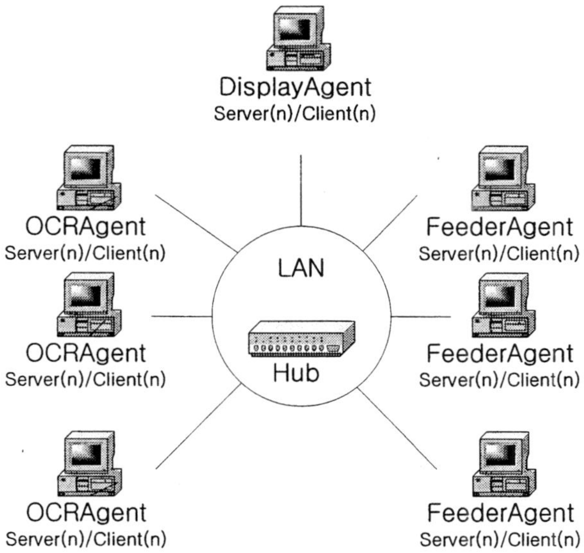

Fig. 1. Schematic diagram of a distribution network protection system with multi agent concept.

3.2.2 Feeder Agent

The feeder agent is located at the secondary winding sides of a transformer. It checks if any change in source impedance happens. And when a lower agent requires the source impedance, this agent responds to it.

3.2.3 Display Agent

The display agent is designed to show the current operation state of the power network to the operator. Furthermore, it displays the behavior of every agent and data exchanged between agents to confirm the soundness of the suggested algorithm. This agent has no function of control so that it may have no influence on the activities of other agents at all. In real field application, the display agent can be omitted.

Fig. 2 is an example of a display agent showing the current status of each agent.

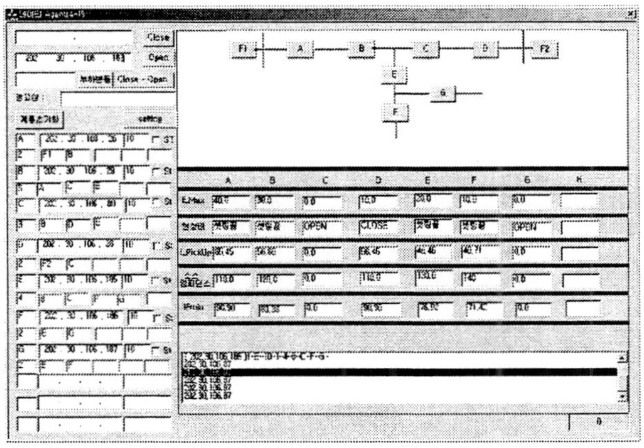

Fig.2. An example of a display agent

3.3 Communication between Agents

One of the main advantages in the agent-system application is that a global goal can be achieved by grouping several agents and their co-working. An OCR agent protects its responsible area such as generally used OCRs. The optimal or sub-optimal protection of the whole network, which may be regarded as the global goal, is realized by the coordination of agents. For example, if a change in the structure of the network concerned occurs, the nearest agent from the changed point recognizes it and sends other agents this fact. Some OCR agents must correct their pick up values in this case. The information needed for this correction is obtained through communication with other agents: source impedance from upper agents, load current from lower agents, etc. Fig. 3 shows a data flow when a change in structure of the studied network occurs.

The key words used in the communication between OCR agents are tabulated in table 1. If any functions of OCR agents are to be added, the numbers or kinds of keywords can be extended easily.

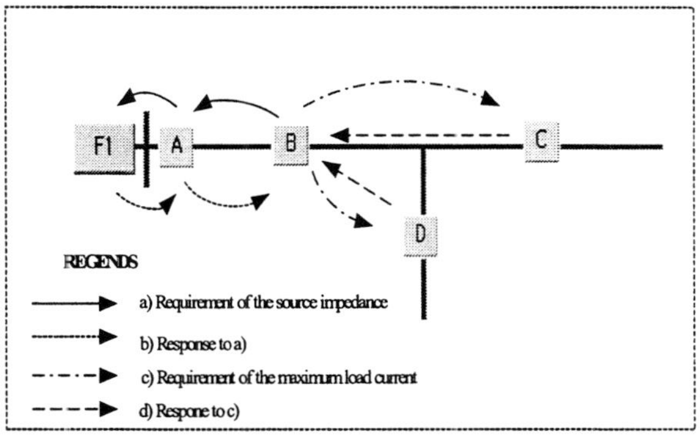

Fig. 3. Data flow in deciding the pick up value

Table 1. Key words for communication

KEYWORD	CONTENT
OPEN	To open CB
CLOSE	To close CB
RESET	To Reset the data of an agent and to require the source impedance of upper agents
SOURCE_IMP RECEIVE	To receive source impedance from upper agents
IL_MAX RECEIVE	To receive maximum load current from lower agents
IL_SELF CHANGE	To change self load value
CH_OPEN	To open the changes in network
CH_ CLOSE	To close the changes in network

4. SIMULATION RESULTS

4.1 Test network

A simple distribution network is selected with 2 transformer buses and 7 OCR's as shown in Fig. 4. There are a total of 10 agents, 7 for OCR agents, 2 for feeder agents and one for a display agent. Every agent is realized in an individual PC and 10 PC's are connected through Local Area Network(LAN) for communication.

In Fig. 4, F1 and F2 are feeder agents linked to a transformer and from A to G represent the OCR agents with a circuit breaker. A regular black square means that the circuit breaker concerned is closed, and a white one vice versa. For example, the circuit breaker in agent C is open: consequently, feeder F2 supplies electric power only to the load at agent D. For a simple illustration, the operation condition is chosen as simple values. It is supposed that the load at each agent is 10kVA, line impedance is 1p.u./km and the distance between agents is equally 10km. The set value of each agent is given in table 2.

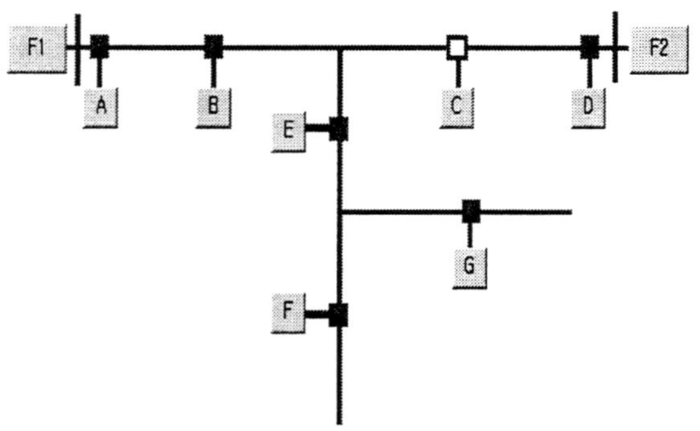

Fig. 4. Simple diagram of the studied network

Table 2. Initial setting status of 2 OCR Agents

[A]

	A OCR	B OCR	C OCR	D OCR	E OCR	F OCR	G OCR
PU	70.45	61.66		56.45	53.46	40.71	40.71
Max. I_load	50.0	40.0		10.0	30.0	10.0	10.0
Min. I_fault	90.90	83.33	Open	90.90	76.92	71.42	71.42
I_self	10.0	10.0		10.0	10.0	10.0	10.0
Source feeder	F1	F1		F2	F1	F1	F1

Max. I_load : Maximum Load Current
Min. I_fault : Minimum Fault current
I_self : Self Load Current

4.2 Fault case and Re-closing case

It is assumed that a line to ground fault occurred at area ① which is the protection area of OCR G as shown in Fig. 5. Then, We suppose the fault was fixed and the circuit breaker at G closes the line.

Agent G tripped the circuit breaker successfully. Then, in the whole network, every OCR have resetting for adaptation on changed network by communication among Agent.

Fig. 5 shows the signal flow in the above procedure and the resultant correction values are given in table 3.

The fault was fixed and the circuit breaker at G closes the line again. Every OCR have resetting for restoration network by communication among Agent. Fig. 6 is the signal flow of this procedure and the results is corrected original value in table 2. As it is respected, the resultant corrected values are the same as in the pre-fault stage, since no change occurred during the term from fault to re-closing.

Table 3. Correction results after the fault

[A]

	A OCR	B OCR	C OCR	D OCR	E OCR	F OCR	G OCR
P.U.	65.45	56.66		56.45	48.46	40.71	
Max. I_load	40.0	30.0		10.0	20.0	10.0	
Min. I_fault	90.90	83.33	Open	90.90	76.92	71.42	Open
I_self	10.0	10.0		10.0	10.0	10.0	
Source feeder	F1	F1		F2	F1	F1	

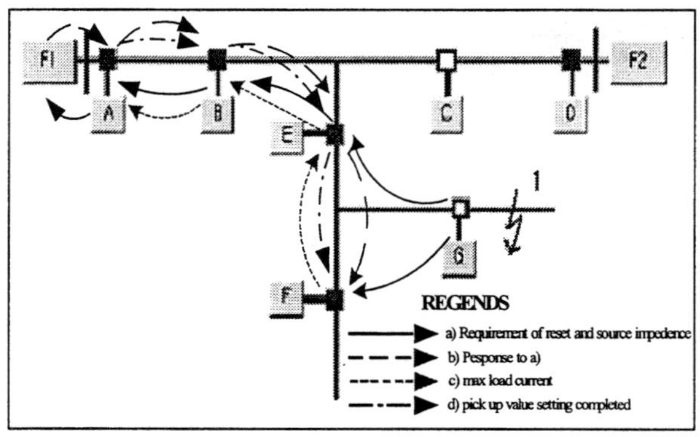

Fig. 5. Signal flow after the fault

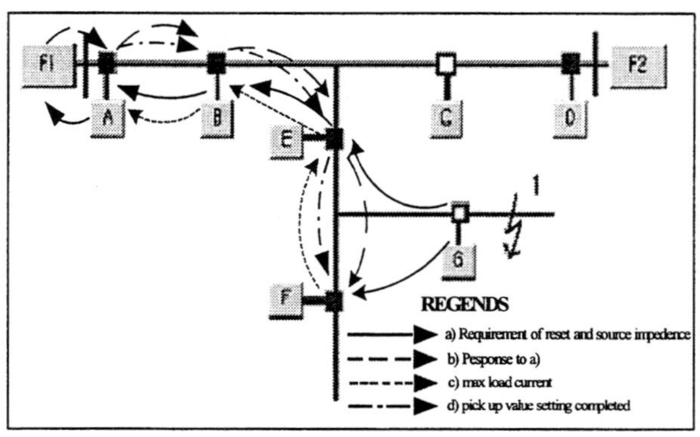

Fig. 6. Signal flow when the CB G is re-closed.

4.3 Structure change case

In this case, it is assumed that the source of load at E, F and G is transferred from feeder 1 to feeder 2. The circuit breaker at C should be closed and the circuit breaker at B should be open. The procedure is very similar to those of previous subsections except that the open operation of circuit breaker B is performed after it is confirmed that CB C is closed. It is necessary to avoid a black out of area E, F, and G in case CB B opens the circuit ahead of the closing of CB C.

In this situation, the actions of OCR agent can be described as follows;

- Agent C sends a reset message to whole agents as shown in Fig. 7
- After CB C is closed, Each Agent send a network-change message to whole agents to confirm network as shown in Fig. 8

• Agent B sends a reset message for Opening CB B to whole agents as shown in Fig. 9. Then next step is same previous simulation(fault-case).

The results are tabulated in table 4.

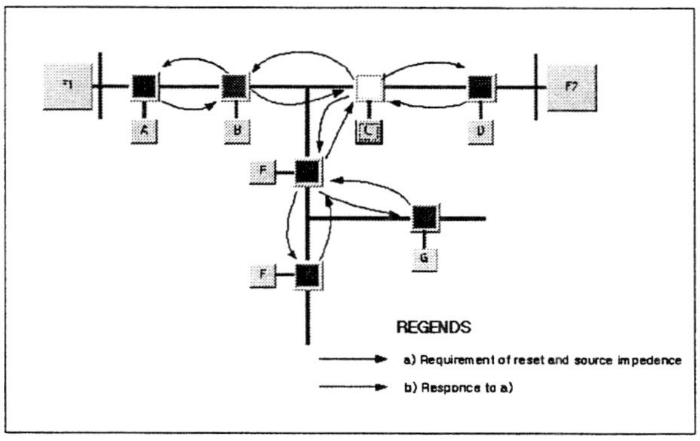

Fig. 7. Network Change Signal flow Step 1

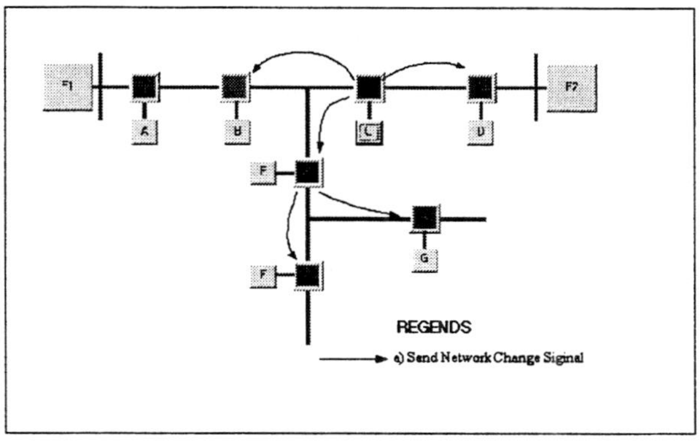

Fig. 8. Network Change Signal flow Step 2

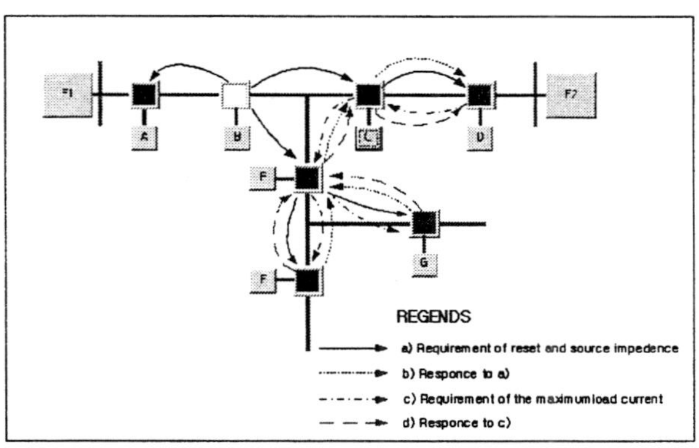

Fig. 9. Network Change Signal flow Step 3

Table 4. Correction results after a change in structure

[A]

	A OCR	B OCR	C OCR	D OCR	E OCR	F OCR	G OCR
P.U.	50.45		61.66	70.45	53.46	40.71	40.71
Max. I_load	10.0		40.0	50.0	30.0	10.0	10.0
Min. I_fault	90.90	Open	83.33	90.90	76.92	71.42	71.42
I_self	10.0		10.0	10.0	10.0	10.0	10.0
Source feeder	F1		F2	F2	F2	F2	F2

5. CONCLUSION

In this paper, the 'Multi-Agent' scheme is applied to a protection system of a distribution network.

A digital over current relay(OCR) is developed as an agent by adding abilities of self-tuning, intelligence and communication.

The detection of changes in the network, and tuning of pick up value in an OCR according to the changes are performed autonomously by the decision ability in every agent and exchange of information between agents. This type of protection system is mostly adequate for a distribution network protection system where the structure is complicated and sudden changes take place frequently.

It can be concluded that the agent concept should be extended to other protection devices such as distance relay, and that the coordination schemes in the operation of every protective device should be developed in order for the total protection system to be realized.

Acknowledgement

The authors would like to express their sincere thanks to the Korea Ministry of Science and Technology and Korea Science and Engineering Foundation for their support through the ERC program.

REFERENCES

[1]Joong Goo Jang, "Multi-Agent System having Autonomous and Adaptive Optimal Protection Function for Power Distribution Functions for Power Distribution Systems", Ph.D. Thesis, 2001

[2]Juhwan Jung, "Multi-Agent Technology for Vuluera-bility Assessment and Control", IEEE Power Engineering Society Summer Meeting, 2001 , Vol. 2 , pp. 1287 –1292

[3]A Taxonomy for Autonomous Agents, [http://www. Msci.memphis.edu/~franklin/AgentProg.html]

[4]Intelligent Software Agent [http:// Retriever.cs.umbc.edu:80]

[5]Joongmin Choi, " Concept of Agent and Research Direction" , Trans of Korean Information Science Society 1997. Vol. 15

[6]Oren Etzioni and Danniel S. Weld, " Intelligent agents on the internet : Fact, Fiction and Forecast" , 1995

[7]Nagata, T.; Sasaki, H., "A multi-agent system for power system restoration, Power Engineering Society Winter Meeting, 2001 IEEE, Volume: 3, 2001 pp. 1359 –1364.

[8]Azevedo, G.P.; Feijo, B.; Costa, M., Control centers evolve with agent technology, IEEE Computer Applications in Power, Vol. 13 Issue: 3 , pp. 48 –53.

ELSEVIER
IFAC
PUBLICATIONS
www.elsevier.com/locate/ifac

ANALYSIS OF PROTECTIVE RELAYING OPERATION AND RELATED POWER SYSTEM INTERACTION

Mladen Kezunovic, Slavko Vasilic

Texas A&M University
Department of Electrical Engineering
College Station, TX 77843-3128, U.S.A.

Abstract: Advanced techniques for analysing operation of protective relays and related interactions with the power system are discussed. Three major issues are addressed: evaluation of existing physical relays, study of new and/or existing relays using relay models, and assessment of relay operation and its interaction with the power system using models of both relays and power system. The underlying approach is based on an advanced modelling and simulation environment as well as the use of digital power system simulators capable of interfacing with physical relays. Such advanced analysis techniques are needed to provide improved protective relaying reliability and system performance. *Copyright © 2003 IFAC*

Keywords: Power systems, Protection, Modelling, Simulation, Simulators, Transient analysis, Man-machine interfaces.

1. INTRODUCTION

This paper is related to the issue of advanced analysis of operation of protective relays and their interactions with the power system where they are installed. The history of this subject is rather interesting: the analysis of the relaying operation went from "the art" in the early times to a rather precise engineering analysis in the most recent times.

The driving factors for the need to have better and more precise analysis methodology and tools are multiple. They range from the change in the relaying requirements that were introduced by the deregulation to the new relaying performance characteristics that are now available with the microprocessor-based relays. In both cases, evaluating, studying their performance, and designing relays is becoming more demanding and requires different engineering approach where the transient behaviour of relays, instrument transformers and power systems needs to be observed (Kezunović and Kasztenny, 2000).

As a result of the new demand for more precise analysis of relay characteristics as well as relay interactions with the power system and other relays, the use of new tools such as digital simulators and relay models is required. This paper provides a justification for the new needs and presents examples of the new tools and their uses.

Three scenarios for analysis of protective relay operation are identified: evaluation of physical relays, study of the relay performance using relay models, and assessment of an interaction between relays and power systems using modelling and simulation. Evaluation of physical relays requires the use of digital simulators (Kezunović, *et al.*, 1998). Study of the behaviour of the existing or new relay designs may be performed using software models of the device (McLaren, *et al.*, 2001). The interactions between relays and power systems can be presented using advanced software tools that represent both relays and power system in one simulation environment (Kasztenny and Kezunović, 2000).

The paper starts with a background section that introduces the reasons for needing the new analysis tools. After that, each type of the tool is discussed in detail: digitals simulators, relay models, and combined solutions for modelling and simulating operation of multiple relays and related interactions in a given power system. Some conclusions and references are given at the end.

2. BACKGROUND

This section discusses three important background issues associated with an increased need for better analysis of protective relaying operation. The issues have especially new meaning when an increased relaying performance is sought.

2.1 Deregulation, Restructuring and Increased Competition

The mentioned developments have brought forward a need to have more secure and dependable protective relay operation. As a result, standard analysis tools for evaluation, study and assessment of relay operation are insufficient. The tools that allow transient analysis, that resembles the actual power operation, are needed. With this capability it will be possible to investigate protective relay operation for the cases where the system is loaded very close to the limits. Other examples are the cases when new relaying schemes are needed to accommodate for more stringent requirements for protecting the system under cascading failures and/or specific interfaces among independently owned power system parts. All the cases are caused by the competition in the industry creating a requirement that the protection operates more reliably enabling the power system's down time to be reduced as much as possible.

2.2 New Protective Relay Designs and Features

In striving for better performance of the relaying operation, it is important to consider new relay designs and features. This translates into an ability to analyse both new relay products and new protective relay concepts. The new products are introduced to overcome the shortcomings of the previous products and to offer some new capabilities. Typically, to confirm the performance of the new designs, one needs a set of tools that can facilitate the evaluation using transients. Some difficult applications may require introduction of new concepts for relaying. Studying the new concepts requires extensive modelling a simulation capability since the concepts may not yet be implemented in a particular hardware.

2.3 Availability of Advanced Modelling and Simulation Tools

This requirement is brand new. Earlier analysis methodologies for protective relaying did not include extensive use of relay models. As the technology for implementing the analysis tools has improved, the modelling and simulation tools have slowly emerged and evolved. The state of the art on the market is to offer a short circuit programs that will have the relay coordination capabilities based on the simplified (phasor-based) relay models. The next step is to have similar tools except tied to the electromagnetic transient programs and aimed at transient evaluation of relay operation. Such programs are ideal for investigating an interaction among large number of relays protecting a given power system.

3. EVALUATION OF PHYSICAL RELAYS

This section discusses the tools for evaluation of physical relays using "real-world" fault transients (application testing). This is achieved through utilization of digital simulators that allow one to assess both the trip/no trip performance as well as the speed of operation (in the case the relay is expected to operate). The application testing is not a common practice today but may have to be used much more in the future if one wants to evaluate new products and troubleshoot the existing ones.

3.1 Requirements for Digital Simulators

The requirements for the simulator system performing application testing with transients are summarized bellow:

- Possibility to import waveforms with native file formats for different types of recording devices.
- Availability of several signal-editing features to enable customizing replaying of pre-recorded transients.
- Flexibility of utilizing customized interface for simulation tools such as ATP (CanAm UG, 2002).
- Possibility to generate the tests automatically by configuring simulation tools for such a task.
- Capability to perform automated transient testing.
- Ability to create test reports automatically.
- Capability for high-performance waveform reconstruction with respect to D-to-A resolution and sampling rate.
- Possibility of using high power voltage and current amplifiers.

3.2 Portable Digital Simulator Solutions

The simulator technology has matured to the pint where low-cost high-performance solutions may be found on the market (TLI, Inc, 2002). The next step is to provide the diversity needed to meet a variety of testing needs. The hardware interface options to accommodate for different output power levels of the simulators are shown in Figure I.

The software features are expected to provide more flexibility in running transient tests aimed at evaluating application properties of a relay. The methodology of application evaluation will need to be better understood so that an agreeable set of tests may be standardized. At present there is very little guidance on how to perform transients test to assess the application performance of a relay (Kezunovic, *et al.*, 1997). A software configuration that represents a good start for further enhancements in this area is shown in Figure II.

The real-time simulator configurations will become more compact and eventually will be implemented using portable PCs. This will bring the ultimate in flexibility to the users and they will be able to perform the transient tests rather easily both in the field and labs.

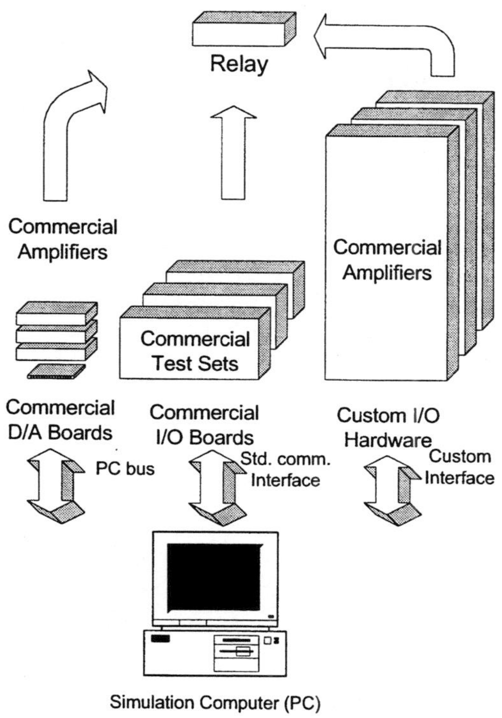

Fig. I. An advanced hardware configuration of an open-loop simulator

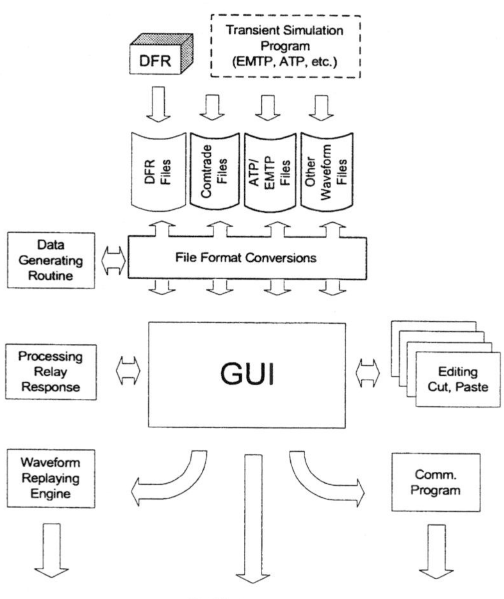

Fig. II. An advanced software configuration for an open-loop simulator

4. STUDY OF EXISTING AND NEW RELAYS USING MODELING AND SIMULATION

This section discusses the application of relay models for study of existing or new relaying concepts. The ability to use the models allows the overall capability to tune performance of the selected relaying solution even better. The next sections give examples of modelling needs and approaches when implementing a modelling and simulation environment for representing distance relays using either an existing commercial or new custom designed software environment for modelling and simulation.

4.1 Use Of a Commercial, General Purpose Software

MATLAB, a high-performance language for technical computation, integrates calculation, visualization and programming in an easy-to-use environment (Mathworks, Inc., 2002). Particular factors that support the choice of MATLAB in modelling the relays are:

(a) The high-volume low-price distribution, which makes the package readily available for all the targeted audiences.

(b) Flexible software structure of MATLAB comprising libraries, models and programs enables one to integrate different relay components in one package conveniently.

(c) MATLAB and its time domain solver, SIMULINK, create a friendly and open system. New models and libraries may be just added to the package without a deep knowledge or modification of the existing parts. This is very useful if the user is involved in further development of the software (modifications, extensions or customisation).

(d) Fast-development with MATLAB using powerful calculation and visualization means of the package enables one to expand the software quickly and efficiently without developing any extra programming tools.

(e) A wide selection of Toolboxes, including comprehensive collections of pre-defined functions for solving application-specific problems, is already available with MATLAB and seems it will be growing even faster in the future. The examples of toolboxes relevant for the protective relaying application include the signal processing, optimisation, neural network and wavelet toolboxes.

(f) Power System Block set, one of the latest extensions of MATLAB enabling one to model the basic components of power systems, provides a vital alternative to EMTP/ATP particularly by permitting modelling of both the power system and its controls in the same environment facilitating closed-loop simulation

Fig. III introduces, as an example, a pre-defined SIMULINK model of impedance (Distance) relaying principle. The model consists of:

- Single-phase model of a two-machine three-line power system with models of the fault and instrument transformers.
- Measuring blocks estimating the voltage and current phasors as well as impedance components.
- Zone comparators checking the impedance location against three impedance zones.
- Displaying elements.

4.2 Use Of Customized Software For Modelling New Relay Designs

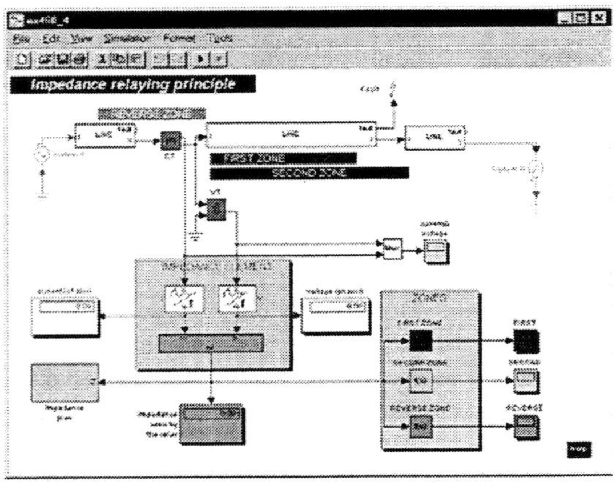

Fig. III. Matlab and Simulink implementation of a Distance Relay model

An example of a neural network based protective relay has been recently widely discussed (Vasilić, *et al.*, 2002). The relay is based on self-organized neural network that establishes the prototypes of typical input patterns, and afterwards employs the prototypes for classification of new inputs (Fig. IV). The design of distance relay using this concept exceptionally depends on the availability of detailed and accurate software model of power network as well as definition and feasibility of a set of scenarios that imitate a realistic network operation.

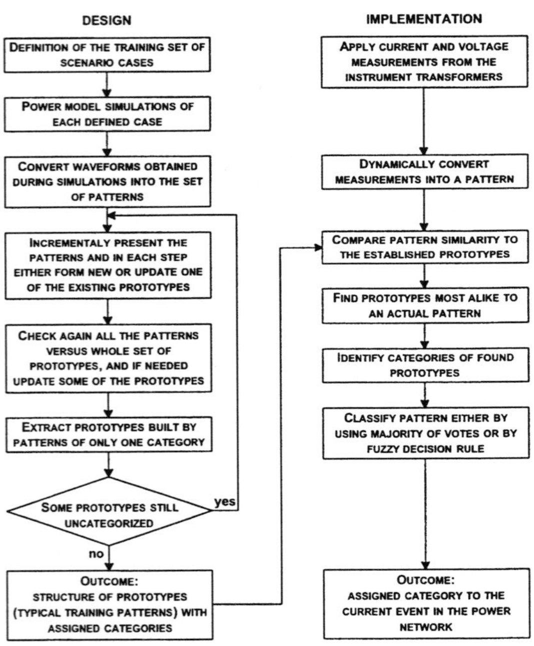

Fig. IV. Design and implementation steps of neural network based relay

Various classification tasks can be inserted into the relay. The most common function is to detect the fault appearance and fault type in selected area in both the forward and reverse direction. Another very important feature is to recognize the zone of protection where fault occurred. For each input pattern that represents an event in the power network the relay assigns a category to that pattern (event) and the number of categories depends on the assigned classification task. If the relay has to detect a type of the fault, there are 12 categories (11 for faults and one for no-fault situation). If it is supposed to simultaneously detect a type of the fault and, for instance, its occurrence in either first or second zone of protection there are 25 categories (11x2 for fault and one for no-fault). The relay model can be trained for both purposes.

The design starts with simulating various fault and disturbance events and operating states, by changing power network topology and parameters. The result of each generated case, voltage and current measurements at the relay location, is being converted into a training pattern by filtering, sampling and using a desired data window. The obtained samples build a pattern with inherited characteristics of simulated event. In the training process, the patterns are presented incrementally and pattern prototypes are constructed. The training is complex incremental procedure of adding new and updating existing prototypes. The outcome of design process is structure of prototypes where each belongs to one of the categories.

In the implementation step the prototypes established during training are dynamically compared with measurements taken from the instrument transformers. The measurements are converted into the patterns before the training process. Then a pattern similarity to all the prototypes is numerically calculated. A subset of the most resembling prototypes is retrieved and their categories identified. The pattern is being classified according to the majority of their categories, or by using fuzzy decision rule. Depending on the recognized event and selected designer's logic, the relay either has to operate or to stay unaffected.

5. ASSESSMENT OF RELAY OPERATION AND RELATED POWER SYSTEM INTERACTION

The problem of proper representation of relay interactions with the power system becomes very important due to the complex topology of the power network. This has created a need for efficient and easy way for modelling and simulating a variety of scenarios by changing the network topology, selecting single or multiple network elements, and choosing the corresponding network parameters. The set of relevant interaction that may appear in the network must be widely studied in advance to properly select relay actions for ambiguous and not easily predicted cases. The techniques and tools available today are not capable of handling a variety of changeable network conditions in an easy way.

5.1 Simulation Environment for representing relay interactions

The solution has to be capable of interconnecting a program for data processing, optimisation procedures, file management, and user-friendly interface. The solution could become powerful tool for fault analysis by overcoming usually tedious tasks of manually creating each single scenario and plugging the input data into the relays.

An example of such a tool, already utilized to enhance relay design and evaluation, is presented in Fig. V. Alternate Transient Program (ATP) for power system transient simulation and MATLAB computational tool have been interfaced for efficient simulation of relay interaction. One of MATLAB's utilities is able to manage ATP input file for all defined cases. It interactively updates ATP input file and invokes ATP program to simulate each scenario from a defined set. Utility for database management organize sets of network outputs to be readily available for later use. The user can pull out outputs of specific simulated scenarios and show its waveforms, use them for the relay design and evaluation, and for showing estimated phasors and impedance trajectories. The libraries of scenario cases, network responses, design parameters, designed relays and relay responses, can be established and used for particular needs.

The utility for relay input signal pre-processing extracts the data obtained through power system simulations, selects desired measurements and converts them into the format needed by the relays. Selecting an analogue filter (type, order and cut-off frequency), sampling frequency, and data window length is also possible.

In the example shown in Fig. V, a software model of a classical distance relay performs voltage and current phasor estimation (using one cycle Fourier algorithm) and impedance estimation. Furthermore, it does overcurrent and undervoltage condition detection, grounded or ungrounded fault detection, R-X impedance trajectory identification and its comparison with selected operating characteristics. The relay final output is information about detected fault type and fault zone. Design parameters for distance relay are distance limits for zones of protection, impedance characteristic selection, delay between zones, overcurrent, undervoltage and zero-sequence thresholds, voltage and current instrument transformers ratios, and positive and zero sequence line impedance.

User interface facilitates creation of scenarios including selecting the combinations of events and setting their parameters. It manages database of simulated cases, relay models and their responses obtained during evaluation. Furthermore it is used for relay input data pre-processing, and to set up relay design parameters. It also shows simulated results (like V/I time waveforms, estimated V/I phasor values, and locus of measured impedance in R-X plain) at the relay location.

5.2 Selecting and implementing complex simulation scenarios

Scenarios are artificially divided into a group of static and dynamic scenarios. Static scenarios include the

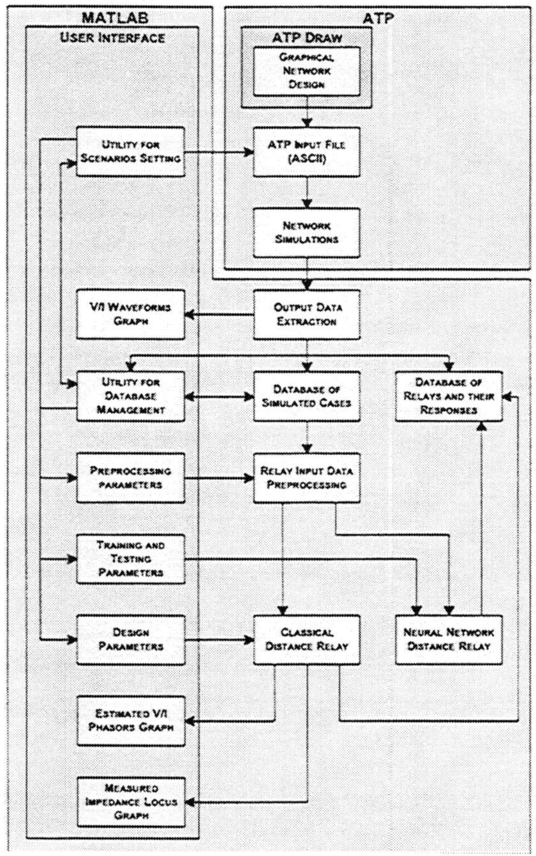

Fig. V. A new software for power system modelling and simulation as well as protective relay design and evaluation

changes in the network topology as well as parameters that stay unchanged in all consecutive simulations for various dynamic events. Examples of such scenarios are week in-feed (by disconnecting some of the equivalent sources), line out of service, line with one end open, system frequency variation, source/load variation, and arbitrary combinations among them (Ristanović, *et al.*, 2001). Line switching is controlled through circuit breaker opening and closing times, corresponding to various fault inception times. Changes of equivalent network sources are established through varying voltage amplitude, phase shift and source impedance. Changes of power system frequency affect all the equivalent sources and frequency dependent lines. If set of values for source parameters and system frequency is given, the running of dynamic scenario cases is repeated for each value. In addition, the general simulation parameters, like simulation end-time and integration step-time are set at this level, providing the user capability to select the simulation time frame and desired accuracy.

Dynamic scenarios are characterized by changing at least one parameter between two consecutive simulation cases. The dynamic scenarios are line fault, line switching, switching on-to fault, cross-country fault, evolving fault, etc. External control of transmission line fault simulations is based on changing the line section lengths to emulate varying fault distance, changing the opening and closing times of phase-to-phase and phase-to-ground switches to imitate fault type and inception time, and varying fault distance. In addition, for evolving faults, varying the delay between primary and secondary fault is included. The possible fault locations cover first, second, third and reverse zone of relay protection, as well as faults on certain number of lines out of these zones. For cross-country faults, the fault is placed on corresponding positions of two parallel lines. As an example, Table 1 gives a listing of the target scenario parameters and corresponding action for updating power network model.

Table 1 Target scenario parameters and corresponding action for updating power network model

Target	Action
Fault type	Adjusting opening/closing times of the phase-to-phase and phase-to-ground switches covering all fault types and various fault angles inside one cycle.
Fault inception time	
Fault location	Adjusting line section lengths from both line ends to the faulted point and covering the faults along entire line.
Fault impedance	Adjusting fault impedance.
Delay between evolving faults	Adjusting various angles between primary and secondary fault inside one cycle.
Line out of service	Hold line breakers permanently open.
Line switching	Adjusting opening/closing times for the circuit breakers and covering various switching angles inside one cycle.
Source/load variation	Adjusting voltage amplitude, phase shift and impedance.
System frequency variation	Adjusting system frequency and frequency of equivalent sources.
Relay location	Activating specific measurement points in the network.
Integration step time	Ajdusting general simulation parameters.
Simulation end time	

All simulation cases can be performed for deterministic or random values of event parameters. Deterministic generation of scenarios is based on specifying several values for each of the event parameters and combining these values to achieve the whole set of scenarios. The other way to create scenarios is only to specify the boundaries (usually imposed by the nature of the event) and then generate randomly the values of parameters for each event to evaluate designed solution in heuristic, not foreseen situations.

6. CONCLUSIONS

The following conclusions can be drawn based on the discussion given in the paper:

- Transient analysis of relay operation and interaction with the power system is essential in improving relaying performance
- The tools to perform the transient analysis are not readily available and used today
- Future development of the required techniques and tools for such an analysis needs to support evaluation of physical relays as well as modelling of relays, power systems and related interactions.

REFERENCES

CanAm EMTP UG (2002). *Alternative Transients Program*. Online: http://www.emtp.org/

Kasztenny, B. and M. Kezunović (2000). "A Method for Linking Different Modeling Techniques for Accurate and Efficient Simulation". *IEEE Trans. on Power Systems*, Vol. **15**, No. **1**, pp. 65-72.

Kezunović M., Y.Q. Xia, Y. Guo, C.W. Fromen and D.R. Sevcik (1997). "Distance Relay Application Testing Using a Digital Simulator". *IEEE Trans. on Power Delivery*, Vol. **12**, No. **1**, pp. 72-82.

Kezunović, M., B.A. Pickett, M.G. Adamiak, G.E. Alexander, K.R. Carr, G. Chirco, J. Esztergalyos, M. Jerosolimski, W.C. Kotheimer, R. Marttila, J.L. McElray, S.M. McKenna, P.G. McLaren, R.J. Murphy, J. Nordstrom, R. Ryan, V. Skendzic, S.I. Thompson, D.A. Tziouvaras, F.M. Phillips, T. Sakaguchi and R.B. Sollero (1998). "Digital Simulator Performance Requirements for Relay Testing". *IEEE Trans. on Power Delivery*, Vol. **13**, No. **1**, pp. 78-84.

Kezunović, M. and B. Kasztenny (2000). "Design Optimization and Performance Evaluation of the Relay Algorithms, Relays and Protective Systems Using Advanced Testing Tools". *IEEE Trans. on Power Delivery*, Vol. **15**, No. **4**, pp. 1129-1135.

MathWorks, Inc., (1997). MATLAB – Reference manual.

McLaren, P.G., C. Henville, V. Skendzic, A. Girgis, M. Sachdev, G. Benmouyal, K. Mustaphi, M. Kezunović, Lj. Kojovic, M. Meisinger, C. Simon, T. Sidhu, R. Marttila and D. Tziouvaras (2001). "Software Models for Relays". *IEEE Trans. on Power Delivery*, Vol. **16**, No. **2**, pp. 238-246.

Ristanović, D., S. Vasilić and M. Kezunović (2001). "Design and Implementation of Scenarios for Evaluating and Testing Distance Relays". *North American Power Symp.*, College Station, U.S.A.

TLI, Inc. (2002) Relay Assistant product brochures, http://www.tli-inc.com/RelayAssistant.htm

Vasilić, S., M. Kezunović and D. Šobajić (2002). "Optimising the Performance of a Transmission Line Relaying Algorithm Implemented Using an Adaptive Self-Organized Neural Network". *14th Power Systems Computation Conf*, Seville, Spain.

ELSEVIER
IFAC
PUBLICATIONS
www.elsevier.com/locate/ifac

A DIRECTIONAL RELAY ALGORITHM USING POSITIVE-SEQUENCE SUPERIMPOSED QUANTITY FOR TRANSMISSION LINE PROTECTION

Myoung Soo Lee*, Seok Ku You

Department of Electrical Engineering, Hanyang University, Korea,

Abstract: Directional elements are fundamental to protection scheme security and selectivity, performing such critical tasks as supervising distance elements and controlling overcurrent elements. But, conventional operating principles for directional detection, based on negative or zero sequence voltages or currents polarization in some cases do not satisfy the requirements for improved sensitivity and fast operation under vary system conditions.

In this paper, new algorithm for directional elements is proposed. The proposed algorithm use the positive-sequence superimposed voltages and currents in order to be used in all fault conditions. Also, because this algorithm uses a voltage compensation method, it can be well operated under strong source conditions.

Keywords: Superimposed Positive-sequence Quantity, Directional Element, Transmission Line Protection

1. INTRODUCTION

Directional elements are fundamental to protection scheme security and selectivity, performing such critical tasks as supervising distance elements and controlling overcurrent elements. Conventional directional detection methods are based on the changes in the phase relationship between a measured phasor and a reference (polarizing) phasor (Apostolv, 1999). Conventional negative-sequence directional elements base their measurements on the phase angle between the negative-sequence voltage and current. This directional element can be shown that the negative-sequence voltage developed during a fault is inversely proportional to the strength of the source either behind relay for a forward fault or in front of the relay for a reverse fault. That is, the stronger the source (the lower the source impedance), the less the negative-sequence voltage is produced. This fact, coupled with the effect of fault resistance, which tends to lower the amount of fault current (and negative-sequence current) available during a fault, produces a minimum sensitivity for the conventional negative sequence-directional element (Fleming, 1998).

Conventional zero-sequence directional elements use their measurements on the phase angle between the zero-sequence voltage and current. The zero-sequence polarized directional elements are reliable in the applications which parallel lines with common buses at both line ends do not cause zero-sequence polarized directional elements to make incorrect fault direction declarations. But, zero-sequence polarized directional elements can misoperate in parallel lines with high zero-sequence mutual coupling (Z_{0M}) and isolated zero-sequence sources (Roberts, 1994).

In this paper, new algorithm for directional elements is proposed. The proposed algorithm use the positive-sequence superimposed voltages and currents in order to be used in all fault conditions. Also, because this algorithm uses a voltage compensation method, it can be well operated under strong source conditions.

2. CONVENTIONAL DIRECTIONAL ELEMENT

A. Negative-Sequence Directional Element

The conventional negative-sequence element is one that looks at the negative-sequence voltage on, and negative-sequence current through, a transmission line and compares the relative phase angles of the two quantities. Shown in Fig. 1 is negative-sequence element characteristic.

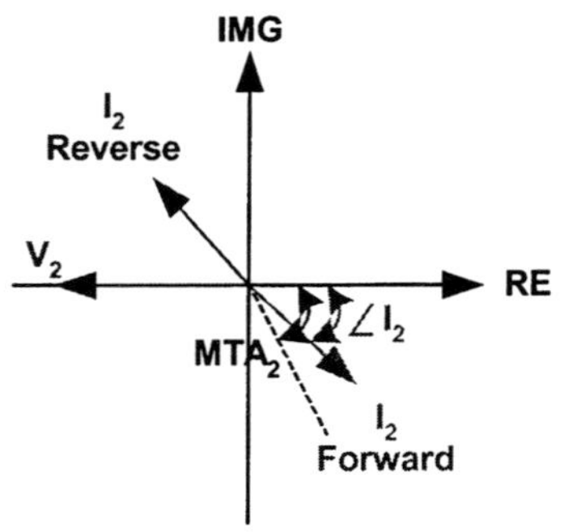

Fig. 1 Negative-Sequence Directional Element Characteristics

This characteristic is described by the equation (1).

$$|V_2|\cdot|I_2|\cdot\cos(\angle -V_2-(MTA_2-\angle I_2))\ge|V_{2SET}||I_{2SET}| \qquad (1)$$

Where MTA_2 is the characteristic angle of the transmission line. This directional element can be shown that the negative-sequence voltage developed during a fault is inversely proportional to the strength of the source either behind relay for a forward fault or in front of the relay for a reverse fault. That is, the stronger the source (the lower the source impedance), the less the negative-sequence voltage is produced. This fact, coupled with the effect of fault resistance, which tends to lower the amount of fault current (and negative-sequence current) available during a fault, produces a minimum sensitivity for the conventional negative sequence-directional element (Fleming, 1998).

B. Zero-Sequence Directional Element

The conventional zero-sequence element is one that looks at the zero-sequence voltage on, and zero-sequence current through, a transmission line and compares the relative phase angles of the two quantities. Shown in Fig. 2 is zero-sequence element characteristic.

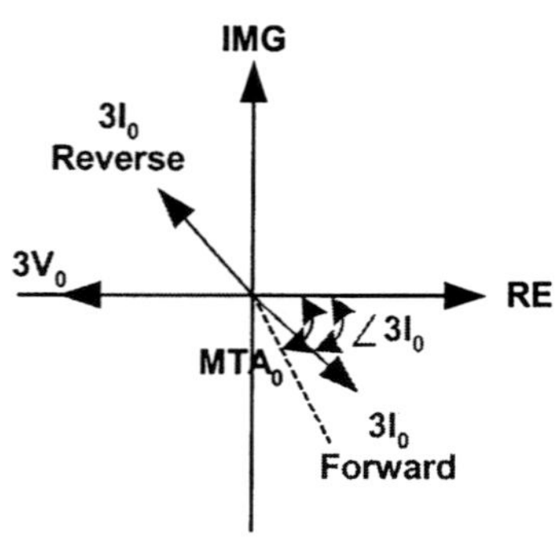

Fig. 2 Zero-Sequence Directional Element Characteristics

This characteristic is described by the equation (2).

$$|3V_0|\cdot|3I_0|\cdot\cos(\angle -3V_0-(MTA_0-\angle 3I_0))\ge|3V_{0SET}||3I_{0SET}| \qquad (2)$$

Where MTA_0 is the characteristic angle of the transmission line. The zero-sequence polarized directional elements are reliable in the applications which parallel lines with common buses at both line ends do not cause zero-sequence polarized directional elements to make incorrect fault direction declarations. But, zero-sequence polarized directional elements can misoperate in parallel lines with high zero-sequence mutual coupling (Z_{0M}) and isolated zero-sequence sources (Roberts, 1994).

New Directional Algorithm based on Positive-Sequence Superimposed Quantity

A. Selection of Sequence Component for New Directional Algorithm

The directional algorithm can be control operation of distance relay element or overcurrent relay element under all fault conditions. Table 1 represents available sequence component quantities for all fault conditions.

TABLE 1 AVAILABLE SEQUENCE COMPONENTS

Sequence Quantity	Positive	Negative	Zero
Single-Line-to-Ground	O	O	O
Phase-to-Phase	O	O	X
Double-Line-to-Ground	O	O	O
3 Phase	O	X	X

From the Table 1, the available sequence quantities under all conditions are only positive-sequence quantity. Therefore, directional algorithm must use positive-sequence current and voltage to do so that direction distinction may be applicable under all fault conditions.

But, positive-sequence current measured by relay can be misoperate by effect of load current. Also, conventional directional algorithm cannot determine correct direction if it does not use polarity voltage or memory voltage because all 3-phase voltage become 0 when 3-phase solid fault happened at relay location.

Therefore, in this paper, superimposed positive-sequence current and superimposed positive-sequence voltage is used.

B. New Directional Algorithm based on Superimposed Positive-Sequence Quantity

Shown in Fig. 3 is the fault component system of a forward fault (F_1) and a reverse fault (F_2).

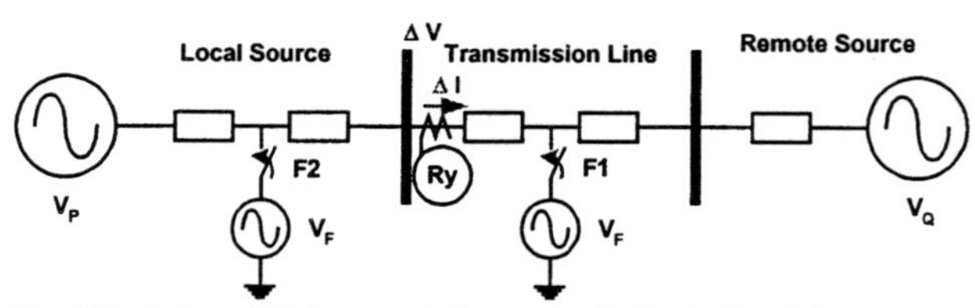

Fig. 3 Variation of Voltage and Current under Fault Condition

The superimposed positive-sequence current and voltage as seen at relay location in Fig. 3 are the changes ΔV_1 and ΔI_1 caused by the application of the fault at location F_1 or F_2. This superimposed positive-sequence voltage and current measured by subtracting pre-fault signals at the relay location from the corresponding fault signals (McLaren, 1995)(He, 1998).

$$\Delta V_1 = V_1 - V_{1P} \quad (3)$$

$$\Delta I_1 = I_1 - I_{1P} \quad (4)$$

Where,

ΔV_1	;	Superimposed positive-sequence voltage
V_1	;	Positive-sequence voltage
V_{1P}	;	Pre-fault positive-sequence voltage
ΔI_1	;	Superimposed positive-sequence current
I_1	;	Positive-sequence current
I_{1P}	;	Pre-fault positive-sequence current

From the equation (3) and (4), the time-delay between the signals subtracted must be an integral number of cycles and it is this time delay which determines the duration of the superimposed signals (McLaren, 1995). In this paper, this time delay uses 5 cycles for considering the operating time of instantaneous relay element.

From the Fig. 3, when fault occurs in fault location F_1, the direction of superimposed current (ΔI_1) passes from source to line and the magnitude of superimposed voltage (V_1) decreases, therefore become opposite direction almost with direction of positive-sequence voltage V_1. when fault occurs in fault location F_2, the direction of superimposed current (ΔI_1) passes from line to source and the magnitude of superimposed voltage (V_1) is equal with voltage change amount direction of positive-sequence voltage (ΔV_1) in fault location F_1 .

Fig. 4 represents the phase angle relationship between superimposed voltage and current.

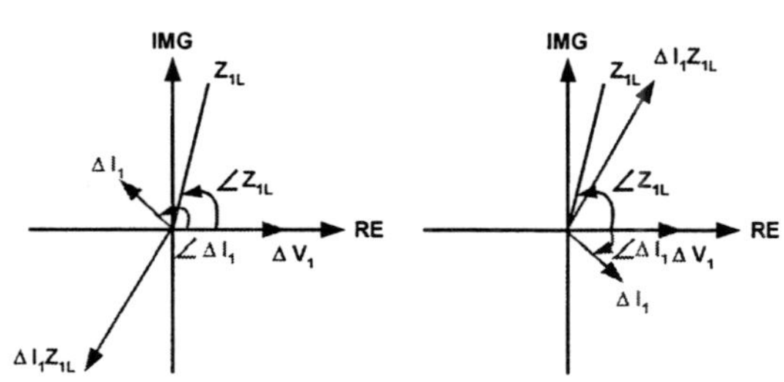

(a) Forward Fault (b) Reverse Fault

Fig 4 Phase Relationship between ΔV_1 and ΔI_1

In this paper, proposed algorithm is implemented using an amplitude comparator. An amplitude comparator is a generic type of relay function measurement technique that compares the magnitude of an "Operate" quantity versus "Restraint" quantity to determine if the relay function should be operate or not (GE). "Operate" quantity and "Restraint" quantity can be obtained by equation (7), (8).

$$Operate = |\Delta V_1 - \Delta I_1 Z_{1L}| \quad (7)$$

$$Restraint = |\Delta V_1 - \Delta I_1 Z_{1L}| \quad (8)$$

Fig. 5 represents the relationship between "Operate" quantity and "Restraint" quantity.

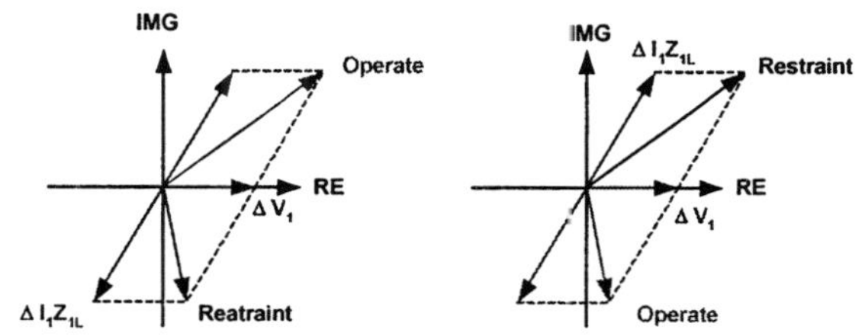

(a) Forward Fault (b) Reverse Fault

Fig. 5 Relationship between Operate and Restraint

Because back source impedance is small, in case of strong source that voltage measured by relay in high impedance fault occurrence in remote is small, superimposed positive-sequence voltage (ΔV_1) measured by relay can close to 0, therefore conventional directional algorithm may not act reliably.

Directional element can be shown that the sequence voltage developed during a fault is inversely proportional to the strength of the source either behind relay for a forward fault or in front of the relay for a reverse fault. That is, the stronger the source (the lower the source impedance), the less the sequence voltage is produced. This fact, coupled with the effect of fault resistance, which tends to lower the amount of fault current (and negative-sequence current) available during a fault, produces a minimum sensitivity for the conventional negative sequence-directional element.

Therefore, the proposed algorithm use voltage compensation techniques based on polarity voltage in relay location. So, even if magnitude of ΔV_1 closes to 0 in relay location, it makes ΔV.

This compensation technique is implemented by modifying "Operate" quantity and "Restraint" quantity as equation (9), (10).

$$Operate = |\Delta V_1 - (1+K) \cdot \Delta I_1 Z_{1L}| \quad (9)$$

$$Restraint = |\Delta V_1 - (1-K) \cdot \Delta I_1 Z_{1L}| \quad (10)$$

Where K is a "Compensation Factor" (Andrichak). In this paper, this factor sets 0.5. Direction criterion use "Operate" and "Restraint" calculated by equation (9), (10). This direction criterion is represented by equation (11).

$$Direction = Operate - Restraint \quad (11)$$

From equation (11), if Direction > 0, forward fault is declared. Otherwise, reverse fault is declared.

3. PERFORMANCE EVALUATION

The performance of the proposed directional algorithm was evaluated for different faults on a 154 kV 25 km overhead transmission line as shown in

Figure 6. The system parameters used in the simulations are given in Table 2. The EMTDC was used to generate fault current and voltage signals under different faults.

The sampling frequency was set to 1440-Hz : 24 samples per cycle in a 60-Hz system and fault point is 95% point of total line length from A bus. And fault resistance was set to 30 Ω.

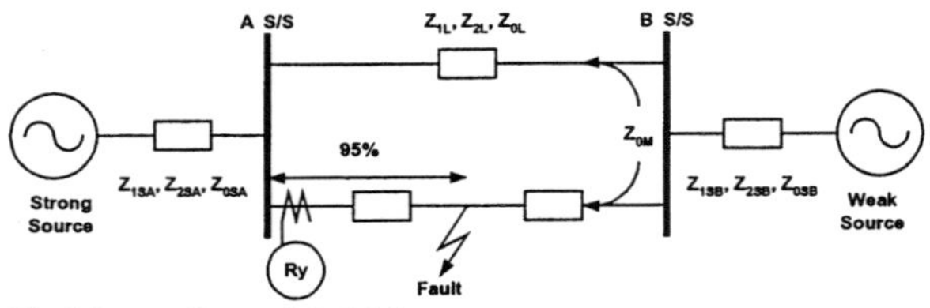

Fig 6 Strong Source Model System

TABLE. 2 MODEL SYSTEM PARAMETER

		Positive	Zero	
			Self	Mutual
Source (Ω)	Strong	0.7+j4.95	2.1+j14.85	
	Weak	0.8+j3.89	3.7+j18.85	
Line(Ω/km)		0.1006+j0.4667	0.309+j1.301	0.208+j0.696

To simulate the strong source which characteristic is small voltage drop, the back source impedance must be set to low. Therefore, in this paper, the back source impedance decrease from 1 to 0.01 times of strong source impedance in Table 2.

Table 3 shows comparison of proposed algorithm with conventional directional algorithm when changed back source impedance.

TABLE 3. VARIATION OF BACK SOURCE IMPEDANCE

Source Impedance	Fault Type	Directional Algorithm Negative	Zero	New
1	AG	Operate	Operate	Operate
	BCG	Operate	Operate	Operate
	BCS	Operate	Nonoperate	Operate
	ABCS	Nonoperate	Nonoperate	Operate
0.5	AG	Nonoperate	Operate	Operate
	BCG	Operate	Operate	Operate
	BCS	Operate	Nonoperate	Operate
	ABCS	Nonoperate	Nonoperate	Operate
0.1	AG	Nonoperate	Operate	Operate
	BCG	Operate	Operate	Operate
	BCS	Operate	Nonoperate	Operate
	ABCS	Nonoperate	Nonoperate	Operate
0.05	AG	Nonoperate	Operate	Operate
	BCG	Operate	Operate	Operate
	BCS	Operate	Nonoperate	Operate
	ABCS	Nonoperate	Nonoperate	Operate
0.01	AG	Nonoperate	Nonoperate	Operate
	BCG	Nonoperate	Nonoperate	Operate
	BCS	Nonoperate	Nonoperate	Operate
	ABCS	Nonoperate	Nonoperate	Operate

From the table 3, the lower the back source impedance, the less the sequence voltage is produced. That is, it becomes near to strong source, negative-sequence directional relay algorithm and zero-sequence directional relay algorithm misoperate and the proposed algorithm can well operate regardless of change of equivalence back source impedance.

Fig 7 and 8 represent the single-line-to-ground fault and 3-phase fault in 1 times back source impedance, respectively.

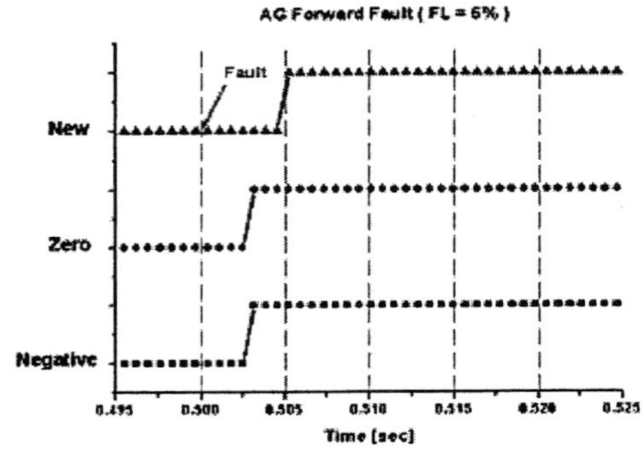

Fig. 7 AG Fault(5% Forward Fault)

From the Fig. 7 Zero-sequence directional algorithm and negative-sequence directional algorithm operated at 2.78 ms and proposed algorithm operated at 4.86ms

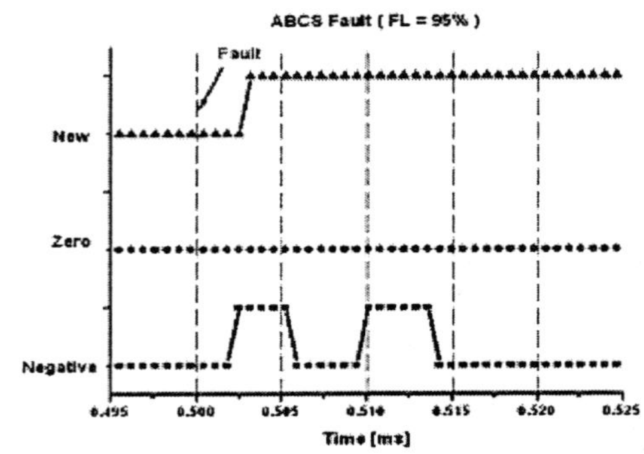

Fig. 8 3-Phase Fault(95% Forwrad Fault)

From the Fig. 8, conventional algorithm not operated and proposed algorithm operated at 2.78ms

4. CONCLUSION

In this paper, conventional directional algorithm cannot operate balanced fault such as 3-phase fault but proposed algorithm could distinguish fault direction exactly in balanced fault as well as unbalanced fault because took advantage of superimposed positive-sequence voltage and current regardless of fault type. Also, the lower the back source impedance, the more conventional algorithm is misoperated, but proposed algorithm is well operated regardless of change of back source impedance.

REFERENCES

Andrichak J. G. and Patel S. C., *Polarizing Source for Directional Ground Relay*

Apostolov A.(1999), Implementation of a Transient Energy Method for Directional Detection in Numerical Distance Relays, *IEEE Transmission and Distribution Conference*

Fleming Bill, *Negative-Sequence Impedance Directional Element, 1998*

GE, *DLP-Digital Line Protection,* GET8037

He Beneteng, Jin Huafeng, Li Ju(1998), The Energy Based Directional Protection of Transmission Lines, *International Conference of POWERCON98.* Vol. **2**.

McLaren SM P.G., etc(1995), A New Directional Element for Numerical Distance Relays, *IEEE Trans. on Power Delivery*, Vol. **10**, No. 2, April.

Roberts J. and Guzman A.(1994), "Directional Element Design and Evaluation", *WPRC*, October